Communications in Computer and Information Science   1602

More information about this series at https://link.springer.com/bookseries/7899

Davide Ciucci · Inés Couso · Jesús Medina ·
Dominik Ślęzak · Davide Petturiti ·
Bernadette Bouchon-Meunier ·
Ronald R. Yager (Eds.)

Information Processing and Management of Uncertainty in Knowledge-Based Systems

19th International Conference, IPMU 2022
Milan, Italy, July 11–15, 2022
Proceedings, Part II

Springer

Editors
Davide Ciucci ⓘ
University of Milano-Bicocca
Milan, Italy

Jesús Medina ⓘ
University of Cádiz
Cádiz, Spain

Davide Petturiti ⓘ
University of Perugia
Perugia, Italy

Ronald R. Yager ⓘ
Iona College
New Rochelle, NY, USA

Inés Couso ⓘ
University of Oviedo
Oviedo, Spain

Dominik Ślęzak ⓘ
University of Warsaw
Warsaw, Poland

Bernadette Bouchon-Meunier ⓘ
Sorbonne Université
Paris, France

ISSN 1865-0929   ISSN 1865-0937 (electronic)
Communications in Computer and Information Science
ISBN 978-3-031-08973-2   ISBN 978-3-031-08974-9 (eBook)
https://doi.org/10.1007/978-3-031-08974-9

This Springer imprint is published by the registered company Springer Nature Switzerland AG
The registered company address is: Gewerbestrasse 11, 6330 Cham, Switzerland

Preface

We are very pleased to present you with the proceedings of the 19th International Conference on Information Processing and Management of Uncertainty in Knowledge-Based Systems (IPMU 2022). The conference was held during July 11–15, 2022, in Milan, Italy. The IPMU conference is organized every two years with the aim of bringing together scientists working on methods for the management of uncertainty and aggregation of information in intelligent systems. Since 1986, the IPMU conference has been providing a forum for the exchange of ideas between theoreticians and practitioners working in these areas and related fields.

Following the IPMU tradition, the Kampé de Fériet Award for outstanding contributions to the field of uncertainty and management of uncertainty was presented. Past winners of this prestigious award are Lotfi A. Zadeh (1992), Ilya Prigogine (1994), Toshiro Terano (1996), Kenneth Arrow (1998), Richard Jeffrey (2000), Arthur Dempster (2002), Janos Aczel (2004), Daniel Kahneman (2006), Enric Trillas (2008), James Bezdek (2010), Michio Sugeno (2012), Vladimir N. Vapnik (2014), Joseph Y. Halpern (2016), Glenn Shafer (2018) and Barbara Tversky (2020). In this 2022 edition, the award was given to Tomaso Poggio for his interdiciplinary work on human and machine intelligence and his fundamental research in computational neuroscience, in particular concerning the computational analysis of vision and learning.

The program included the keynote talk of Tomaso Poggio, as recipient of the Kampé de Feriet Award, and keynote talks by Cesar Hidalgo (Artificial and Natural Intelligence Toulouse Institute, France), Marianne Huchard (Laboratory of Informatics, Robotics, and Microelectronics, France) and Andrzej Skowron (University of Warsaw, Poland).

To celebrate the 40th anniversary of the seminal paper "Rough Sets" by Z. Pawlak, a panel session on rough sets was organized. The panel session witnessed the participation and discussion of renowned researchers on rough sets, including Salvatore Greco, Ernestina Menasalvas, and Andrzej Skowron, to whom we are grateful for their contribution. The participants shared their memories and experiences related to rough set-based decision making, applications of rough set approximations, and rough set contributions to machine learning, emphasizing the strong points of rough sets and the ways of using them in hybrid solutions.

The IPMU 2022 program consisted of 14 special sessions and 124 papers authored by researchers from 38 different countries. The conference followed a single-blind review process, respecting the usual conflict-of-interest standards. All submitted papers were judged by at least two reviewers and in most cases by three or more – even up to five – referees. Furthermore, all the papers were examined by the program chairs. As a result of the reviewing process, 124 submissions were accepted as full papers, which are included in the two volumes of the proceedings.

The organization of the IPMU 2022 conference was possible with the assistance, dedication, and support of many people and institutions. We are particularly thankful to the organizers of special sessions. Such sessions, dedicated to a variety of topics and organized by experts, have always been a characteristic feature of IPMU conferences. We would like to pass on our special thanks to Célia Da Costa Pereira, who helped evaluate all special session proposals. We would like to acknowledge all members of the IPMU 2022 Program Committee, as well as the additional reviewers who played an essential role in the reviewing process, ensuring a high-quality conference. Thank you very much for all your work and efforts. We gratefully acknowledge the technical co-sponsorship of the IEEE Computational Intelligence Society and the European Society for Fuzzy Logic and Technology (EUSFLAT).

We also acknowledge the support received from the University of Milano-Bicocca, and in particular from the Department of Information, Systems and Communications; from the Springer team who managed the publication of these proceedings; and from the EasyChair platform used to handle submissions and the review process. Our very special and greatest gratitude goes to the authors who submitted the results of their work and presented them at the conference. Without you this conference would not take place. Thank you!

We hope that these proceedings provide the readers with multiple ideas leading to numerous research activities, significant publications, and intriguing presentations at future IPMU conferences.

July 2022

Davide Ciucci
Inés Couso
Jesús Medina
Dominik Ślęzak
Davide Petturiti
Bernadette Bouchon-Meunier
Ronald R. Yager

Organization

General Chair

Davide Ciucci — University of Milano-Bicocca, Italy

Program Chairs

Inés Couso — University of Oviedo, Spain
Jesús Medina — University of Cádiz, Spain
Dominik Ślęzak — University of Warsaw, Poland

Executive Directors

Bernadette Bouchon-Meunier — LIP6, CNRS, France
Ronald R. Yager — Iona College, USA

Special Session Chair

Célia da Costa Pereira — Université Côte d'Azur, France

Publication Chair

Davide Petturiti — University of Perugia, Italy

Virtual Conference Chair

Rafael Peñaloza — University of Milano-Bicocca, Italy

Web Chair

Marco Viviani — University of Milano-Bicocca, Italy

International Advisory Board

Joao Paulo Carvalho — Instituto Superior Tecnico/INESC-ID, Portugal
Giulianella Coletti — University of Perugia, Italy
Miguel Delgado — University of Granada, Spain
Mario Fedrizzi — University of Trento, Italy
Laurent Foulloy — Université de Savoie, France

Salvatore Greco	University of Catania, Italy
Julio Gutiérrez-Ríos	Universidad Politécnica de Madrid, Spain
Eyke Hüllermeier	Paderborn University, Germany
Uzay Kaymak	Eindhoven University of Technology, The Netherlands
Anne Laurent	University of Montpellier, France
Marie-Jeanne Lesot	Universite Pierre et Marie Curie - Paris 6, France
Luis Magdalena	Universidad Politécnica de Madrid, Spain
Christophe Marsala	Universite Pierre et Marie Curie - Paris 6, France
Benedetto Matarazzo	University of Catania, Italy
Jesús Medina	University of Cádiz, Spain
Manuel Ojeda-Aciego	University of Malaga, Spain
Maria Rifqi	LEMMA, Université Panthéon-Assas, France
Lorenzo Saitta	Università del Piemonte Orientale, Italy
Olivier Strauss	Université de Montpellier, France
Enric Trillas	Universidad Politécnica de Madrid, Spain
Llorenç Valverde	Universitat de les Illes Balears, Spain
José Luis Verdegay	University of Granada, Spain
María Amparo Vila	University of Granada, Spain

Program Committee

Michał Baczyński	University of Silesia in Katowice, Poland
Gleb Beliakov	Deakin University, Australia
Vaishak Belle	University of Edinburgh, UK
Rafael Bello	Universidad Central "Marta Abreu" de las Villas, Cuba
Radim Bělohlávek	Palacky University Olomouc, Czech Republic
Salem Benferhat	CNRS, Université d'Artois, France
Isabelle Bloch	LTCI, Télécom Paris, France
Ulrich Bodenhofer	University of Applied Sciences Upper Austria, Austria
Humberto Bustince	UPNA, Spain
Joao Paulo Carvalho	Instituto Superior Tecnico/INESC-ID, Portugal
Giulianella Coletti	University of Perugia, Italy
Ana Colubi	University of Oviedo, Spain
María Eugenia Cornejo Piñero	Universidad de Cádiz, Spain
Chris Cornelis	Ghent University, Belgium
Keeley Crockett	Manchester Metropolitan University, UK
Bernard De Baets	Ghent University, Belgium
Guy De Tre	Ghent University, Belgium
Sébastien Destercke	CNRS, Université de Technologie de Compiègne, France

Antonio Di Nola	University of Salerno, Italy
Didier Dubois	IRIT-CNRS, France
Sylvie Galichet	LISTIC, Université de Savoie, France
Lluis Godo	Artificial Intelligence Research Institute, IIIA-CSIC, Spain
Fernando Gomide	University of Campinas, Brazil
Gil González-Rodríguez	University of Oviedo, Spain
Przemysław Grzegorzewski	Polish Academy of Sciences, Poland
Janusz Kacprzyk	Systems Research Institute, Polish Academy of Sciences, Poland
Uzay Kaymak	Eindhoven University of Technology, The Netherlands
Jim Keller	University of Missouri, USA
Frank Klawonn	Ostfalia University of Applied Sciences, Germany
Erich Peter Klement	Johannes Kepler University Linz, Austria
Lászlo T. Kóczy	Budapest University of Technology and Economics, Hungary
Vladik Kreinovich	University of Texas at El Paso, USA
Tomas Kroupa	CTU in Prague, Czech Republic
Rudolf Kruse	OVGU Magdeburg, Germany
Christophe Labreuche	Thales R&T, France
Jérôme Lang	CNRS, Université Paris-Dauphine, France
Anne Laurent	University of Montpellier, France
Marie-Jeanne Lesot	Universite Pierre et Marie Curie - Paris 6, France
Weldon Lodwick	University of Colorado at Denver, USA
Luis Magdalena	Universidad Politécnica de Madrid, Spain
Christophe Marsala	Universite Pierre et Marie Curie - Paris 6, France
Trevor Martin	University of Bristol, UK
Sebastià Massanet	University of the Balearic Islands, Spain
Gilles Mauris	Université de Savoie, France
Jerry Mendel	University of Southern California, USA
Radko Mesiar	STU Bratislava, Slovakia
Enrique Miranda	University of Oviedo, Spain
Javier Montero	Universidad Complutense de Madrid, Spain
Susana Montes	University of Oviedo, Spain
Jacky Montmain	École des Mines d'Alès, France
Serafín Moral	University of Granada, Spain
Zbigniew Nahorski	Polish Academy of Sciences, Poland
Vilém Novák	University of Ostrava, Czech Republic
Manuel Ojeda-Aciego	University of Malaga, Spain
Endre Pap	Singidunum University, Serbia
Gabriella Pasi	Università degli Studi di Milano-Bicocca, Italy

Irina Perfilieva	University of Ostrava, Czech Republic
Fred Petry	Naval Research Lab, USA
Vincenzo Piuri	University of Milan, Italy
Olivier Pivert	IRISA-ENSSAT, France
Henri Prade	IRIT-CNRS, France
Anca Ralescu	University of Cincinnati, USA
Mohammed Ramdani	Hassan II University of Casablanca, Morocco
Eloísa Ramírez-Poussa	Universidad de Cádiz, Spain
Marek Reformat	University of Alberta, Canada
Adrien Revault d'Allonnes	LIASD, France
Maria Rifqi	LEMMA, Université Panthéon-Assas, France
Thomas A. Runkler	Siemens Corporate Technology, Germany
Daniel Sánchez	University of Granada, Spain
Mika Sato-Ilic	University of Tsukuba, Japan
Roman Słowiński	Poznań University of Technology, Poland
Grégory Smits	IRISA/University of Rennes 1, France
Joao Sousa	IST, University of Lisbon, Portugal
Martin Štěpnička	University of Ostrava, Czech Republic
Umberto Straccia	ISTI-CNR, Italy
Eulalia Szmidt	Systems Research Institute, Polish Academy of Sciences, Poland
Marco Elio Tabacchi	University of Palermo, Italy
Andreja Tepavčević	University of Novi Sad, Serbia
Settimo Termini	University of Palermo, Italy
Vicenç Torra	University of Skövde, Sweden
Barbara Vantaggi	Sapienza Università di Roma, Italy
Marley Vellasco	Pontifical Catholic University of Rio de Janeiro, Brazil
José Luis Verdegay	University of Granada, Spain
Thomas Vetterlein	Johannes Kepler University Linz, Austria
Susana Vieira	Universidade de Lisboa, Portugal
Qiang Wei	State Key Laboratory of Mathematical Engineering and Advanced Computing, China
Sławomir Zadrożny	Systems Research Institute, Polish Academy of Sciences, Poland

Special Session Organizers

Stefano Aguzzoli	University of Milan, Italy
Michał Baczyński	University of Silesia in Katowice, Poland
Valerio Basile	University of Turin, Italy
Salem Benferhat	CNRS, Université d'Artois, France
Matteo Bianchi	University of Milan, Italy

Irina Perfilieva	University of Ostrava, Czech Republic
Silvia Prieto Herráez	Universidad de Salamanca, Spain
Raúl Pérez-Fernández	Universidad de Oviedo, Spain
Barbara Pekała	University of Rzeszów, Poland
Rosana Rodríguez-López	Universidad de Santiago de Compostela, Spain
Luciano Sánchez	Universidad de Oviedo, Spain
Teresa Scantamburlo	Ca' Foscari University of Venice, Italy
Olivier Strauss	Université de Montpellier, France
Karim Tabia	Artois University, France
Sara Ugolini	Artificial Intelligence Research Institute, IIIA-CSIC, Spain
Amanda Vidal	Artificial Intelligence Research Institute, IIIA-CSIC, Spain
Anna Wilbik	Maastricht University, The Netherlands

Additional Reviewers

Angulo Castillo, Vladimir
Antonucci, Alessandro
Badia, Guillermo
Baldi, Paolo
Behounek, Libor
Ben Amor, Nahla
Benavoli, Alessio
Bianchi, Matteo
Cabañas, Rafael
Cao, Nhung
Carvalho, Thiago
Casalino, Gabriella
Cornejo Piñero, Maria Eugenia
Diaz-Garcia, J. Angel
Doria, Serena
Elouedi, Zied
Erreygers, Alexander
Fernandez-Peralta, Raquel
Figueroa-García, Juan Carlos
Flaminio, Tommaso
Garcia Calvés, Pere
Godo, Lluis
González-Arteaga, Teresa
Gupta, Megha
Gupta, Vikash Kumar
Guyot, Patrice
Helbin, Piotr

Hoffmann, Frank
Holcapek, Michal
Hryniewicz, Olgierd
Jabbour, Said
Kmita, Kamil
Kreinovich, Vladik
Król, Anna
Lapenta, Serafina
Leray, Philippe
Llamazares, Bonifacio
Mir, Arnau
Miś, Katarzyna
Murinová, Petra
Muñoz-Velasco, Emilio
Nanavati, Kavit
Ojeda-Aciego, Manuel
Ojeda-Hernandez, Manuel
Peelman, Milan
Pelessoni, Renato
Petturiti, Davide
Pekala, Barbara
Quaeghebeur, Erik
Riera, Juan Vicente
Rodriguez, Ricardo Oscar
Rodríguez, Domingo
Romaniuk, Maciej
Ruiz, M. Dolores

Runkler, Thomas A.
Rutkowska, Aleksandra
Seliga, Adam
Stupnanova, Andrea
Toulemonde, Gwladys
Troffaes, Matthias
Truong, Phuong

Vannucci, Sara
Vemuri, Nageswara Rao
Vicig, Paolo
Yang, Xiang
Yepmo, Véronne
Yoon, Jin Hee

Contents – Part II

E-Health

Fuzzy Methods in Data Mining and Knowledge Discovery

Soft Computing and Artificial Intelligence Techniques in Image Processing

Soft Methods in Statistics and Data Analysis

Uncertainty, Heterogeneity, Reliability and Explainability in AI

Weak and Cautious Supervised Learning

Contents – Part I

Generalized Sets and Operators

**Information Fusion Techniques based on Aggregation Functions,
Pre-aggregation Functions, and Their Generalizations**

Interval Uncertainty

Theoretical and Applied Aspects of Imprecise Probabilities

Data Science and Machine Learning

Nonlinear Weighted Independent Component Analysis

Andrzej Bedychaj[✉], Przemysław Spurek, Aleksandra Nowak, and Jacek Tabor

Jagiellonian University, Kraków, Poland
andrzej.bedychaj@gmail.com

Abstract. Independent Component Analysis (ICA) aims to find a coordinate system in which the components of the data are independent. In this paper we construct a new nonlinear ICA model, called WICA, which obtains better and more stable results than other algorithms. A crucial tool is given by a new efficient method of verifying nonlinear dependence with the use of computation of correlation coefficients for normally weighted data. In addition, we propose a new nonlinear mixing to perform baseline comparison experiments and a reliable measure that allows for a fair evaluation of nonlinear models. Code for methods presented in the paper is available on our GitHub.

Keywords: Independent Component Analysis · Signal processing · Signal retrieving

1 Introduction

The goal of Independent Component Analysis (ICA) is to find such an unmixing function of the given data that the resulting representation has statistically independent components. Common tools solving this problem are based on maximizing some measure of nongaussianity, e.g. kurtosis [4,10] or skewness [21]. An obvious limitation of those approaches is the assumption of linearity, as the real-world data usually contains complicated and nonlinear dependencies. Examples of such problems can be found for instance in [17,24].

In this paper, we present a competitive approach to nonlinear independent components analysis – Nonlinear Weighted ICA (WICA). A crucial role in our method is played by the conclusion from [3], which proves that to verify nonlinear independence it is sufficient to check the linear independence of the normally weighted dataset (see Fig. 1 for a visualization of this claim). Based on this result we introduce *weighted independence index (*wii*)* which relies on computing weighted covariance and can be applied to the verification of the nonlinear independence. Consequently, the constructed WICA algorithm is based on simple operations on matrices, and therefore is ideal for GPU calculation and parallel processing. We construct it by incorporating the introduced cost function in a commonly used in ICA problems auto-encoder framework [6,18]. The role of the decoder is to limit the unmixing function so that the learned by the encoder independent components contained the information needed to reconstruct the inputs.

D. Ciucci et al. (Eds.): IPMU 2022, CCIS 1602, pp. 3–16, 2022.
https://doi.org/10.1007/978-3-031-08974-9_1

We verify our algorithm by applying it to the source signal separation problem, which is a common setting for benchmarking ICA algorithms [6,21]. In Sect. 7, we present the results obtained by WICA for nonlinear mixes of images and for the decomposition of electroencephalogram signals. It occurs that WICA outperforms other methods of nonlinear ICA, both with respect to unmixing quality and the stability of the results over an increasing number of dimensions, see Fig. 7.

To fairly evaluate various nonlinear ICA methods in the case of higher dimensional datasets, we introduce a new measure index called Optimal Transport Spearman (OTS), which is based on Spearman's rank correlation coefficient and optimal transport. In consequence, the obtained measure is sensitive to higher-order dependencies and returns a one-to-one mapping between the original sources and the retrieved signal. The second property is especially important in a higher-dimensional space, where the correspondence between components is non-trivial.

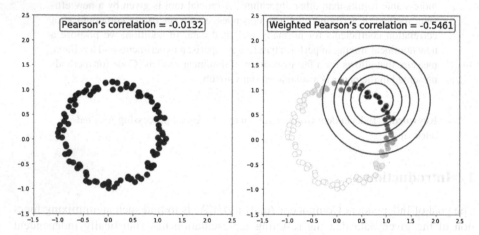

Fig. 1. Sample from a random vector in which Pearson's correlation is equal to zero (left), but the components are not independent. Since the components are not independent, one can choose Gaussian weights so that Pearson's correlation of weighted dataset is not zero (right).

2   Related Work

Designing an efficient and easily implementable nonlinear analogue of ICA is a much more complex problem than its linear counterpart. A crucial complication is that without any limitations imposed on the space of the mixing functions or on the space of observations the problem of nonlinear-ICA is ill-posed, as there are infinitely many valid solutions [13].

As an alternative to the fully unsupervised setting of the nonlinear ICA one can assume some prior knowledge about the distribution of the sources, which allows obtaining identifiability [11,14]. Several algorithms exploiting this property have been recently proposed, either assuming access to segment labels of the sources [11], a temporal dependency of the sources [12] or, generally, that the sources are conditionally independent, and the conditional variable is observed along with the mixes [14, 15]. However, such models are hard to generalize and inapplicable to a fully unsupervised setting, where prior knowledge or the quality of the data itself remains unknown for the researcher.

An additional complication in devising nonlinear-ICA algorithms lies in proposing an efficient measure of independence, which optimization would encourage the model to disentangle the components. One of the most common nonlinear methods is MISEP [1] which, similar to the popular INFOMAX algorithm [4], uses the mutual information criterion. In consequence, the procedure involves the calculation of the Jacobian of the modelled nonlinear transformation, which often causes a computational overhead when both the input and output dimensions are large.

Another approach is applied in Nonlinear Independent Component Estimation (NICE) [7]. Authors propose a fully invertible neural network architecture where the Jacobian is trivially obtained. The independent components are then estimated using the maximum likelihood criterion. The drawback of both MISEP and NICE is that they require choosing the prior distribution family of the unknown independent components. An alternative approach is given by Adversarial Nonlinear ICA (ANICA) [6], where the independence measure is directly learned in each task with the use of GAN-like adversarial method combined with an autoencoder architecture. However, the introduction of a GAN-based independence measure results in an often unstable adversarial training.

It is worth mentioning that nonlinear-ICA is closely related to disentanglement learning. In recent research [9] authors find that a spectral method that explicitly optimizes local isometry and non-Gaussianity consistently finds the correct latent factors, while baseline deep autoencoders do not. Nonetheless, autoencoders offer a built-in decoder and also enable out-of-sample extensions. These are two valuable advantages that their method lacks.

Finally, one can also find some connections between ICA models and data perspectivism discussed in [2]. An application of such algorithms would aim to find maximally independent perspectives in some data with multiple mixed sources. It would result in initial ground-truthing of the dataset in question for annotators. Thus simplifying the collected set of perceptions, opinions, and judgments which could be maximally representative.

3   Weighted Independence Index

Let us consider a random vector $\mathbf{X}$ in $\mathbb{R}^d$ with density f. Then $\mathbf{X}$ has independent components iff f factorizes as: $f(x_1, x_2, \ldots, x_d) = f_1(x_1) \cdot f_2(x_2) \cdot \ldots \cdot f_d(x_d)$, for some marginal densities f_i, where $i \in \{1, 2, \ldots, d\}$.

A related, but the much weaker notion, is the uncorrelatedness. We say that $\mathbf{X}$ has uncorrelated components if the covariance of $\mathbf{X}$ is diagonal. Contrary to independence, the correlation has fast and easy to compute estimators. The independence of the components implies uncorrelatedness, but the opposite is not valid, see Fig. 1.

Let us mention that there exist several measures which verify the independence. One of the most well-known measures of independence of random vectors is the *distance correlation* (dCor) [22], which is applied in [19] to solve the linear ICA problem. Unfortunately, to verify the independence of components of the samples, dCor needs $2^d N^4$ comparisons, where d is the dimension of the sample and N is the sample size. Moreover, even a simplified version of dCor which checks only pairwise independence has high complexity and does not obtain very good results (which can be seen in experiments from Sect. 7). This motivates the research into fast, stable and efficient measures of independence, which are adapted to GPU processing.

3.1   Introducing wii Index

In this subsection, we fill this gap and introduce a method of verifying independence which is based on the covariance of the weighted data. The covariance scales well for the sample size and data dimension, therefore the proposed covariance-based index inherits similar properties.

To proceed further, let us introduce weighted random vectors.

Definition 1. *Let* $w : \mathbb{R}^d \to \mathbb{R}_+$ *be a bounded weighting function. By* $\mathbf{X}_w$ *we denote a weighted random vector with a density*[3] $f_w(x) = \frac{w(x)f(x)}{\int w(z)f(z)dz}$.

Observation 3.1. *Let* $\mathbf{X}$ *be a random vector that has independent components, and let* w *be an arbitrary weighting function. Then* $\mathbf{X}_w$ *has independent components as well.*

One of the main results of [3] is that the strong version of the inverse of the above theorem holds. Given $m \in \mathbb{R}^d$, we consider the weighting of $\mathbf{X}$ by the standard normal gaussian with centre at m: $\mathbf{X}_{[m]} = \mathbf{X}_{\mathrm{N}(m,\mathbb{I})}$. We quote the following result which follows directly from the proof of Theorem 2 from the paper mentioned above:

Theorem 1. *Let* $\mathbf{X}$ *be a random vector, let* $p \in \mathbb{R}^d$ *and* $r > 0$ *be arbitrary. If* $\mathbf{X}_{[q]}$ *has linearly independent components for every* $q \in B(p,r)$, *where* $B(p,r)$ *is a ball with centre in* p *and radius* r, *then* $\mathbf{X}$ *has independent components.*

Given sample $X = (x_i) \subset \mathbb{R}^d$, vector $p \in \mathbb{R}^d$, and weights $w_i = \mathrm{N}(p,\mathbb{I})(x_i)$, we define the weighted sample as: $X_{[p]} = (x_i, w_i)$. Then the mean and covariance for the weighted sample $X_{[p]} = (x_i, w_i)$ is given by: $\mathrm{mean}X_w = \frac{1}{\sum_i w_i} \sum_i w_i x_i$ and $\mathrm{cov}X_w = \frac{1}{\sum_i w_i} \sum_i w_i (x_i - \mathrm{mean}X_w)^T (x_i - \mathrm{mean}X_w)$.

The informal conclusion from the above theorem can be stated as follows: *if* $\mathrm{cov}X_{[p]}$ *is (approximately) diagonal for a sufficiently large set of* p, *then the sample* X *was generated from a distribution with independent components.*

Let us now define an index that will measure the distance from being independent. We define the *weighted independence index* $(\mathrm{wii}(X, p))$ as

$$\mathrm{wii}(X, p) = \frac{2}{d(d-1)} \sum_{i<j} c_{ij}, \text{ for } c_{ij} = \frac{2z_{ij}^2}{z_{ii}^2 + z_{jj}^2}$$

where d is the dimension of X and $Z = [z_{ij}] = \mathrm{cov}X_{[p]}$.

---

[3] This is just the normalization of $w(x)f(x)$.

Observation 3.2. *Let us first observe that c_{ij} is a close measure to the correlation ρ_{ij}, namely: $c_{ij} \leq \rho_{ij}^2$, where the equality holds iff the i-th and j-th components in $X_{[p]}$ have equal standard deviations.*

Proof. Obviously $\rho_{ij}^2 = \frac{z_{ij}^2}{z_{ii} \cdot z_{jj}}$. Since $ab \leq \frac{1}{2}(a^2 + b^2)$ (where the equality holds iff $a = b$), we obtain the assertion of the observation.

Consequently, $\text{wii}(X, p) = 1$ iff all components of $X_{[p]}$ are linearly dependent and have equal standard deviations. Thus, the minimization of wii simultaneously aims at maximizing the independence and increasing the difference between the standard deviations.

We extend the index for a sequence of points $\{p_1, p_2, \ldots, p_n\}$, as the mean of the indexes for each p_i to $\text{wii}(X; \{p_1, p_2, \ldots, p_n\}) = \frac{1}{n} \sum_{i=1}^{n} \text{wii}(X, p_i)$.

To implement the weighted independence index in practice, we need to find the optimal choice of weighting centers (p_i). We argue that an effective choice of those points is to take the mean of d randomly selected vectors from $\mathbf{X}$.

This leads to the following definition:

Definition 2. *For the dataset $X \subset \mathbb{R}^d$, we define*

$$\text{wii}(X) = \mathbb{E}\{\text{wii}(Y, p) : p \text{ a mean of random } d \text{ elements of } Y\},$$

where Y is a componentwise normalization of X and $\mathbb{E}$ stands for expected value.

4　WICA Algorithm

In this section, we propose the WICA algorithm for nonlinear ICA decomposition which exploits the $\text{wii}(X)$ index in practice.

Following [6], we use an auto-encoder (AE) architecture, which consists of an encoder function $\mathcal{E} : \mathbb{R}^d \to \mathcal{Z}$ and a complementary decoder function $\mathcal{D} : \mathcal{Z} \to \mathbb{R}^d$. The role of the encoder is to learn a transformation of the data that unmixes the latent components, utilizing some measure of independence (we use the $\text{wii}(X)$ index). The decoder is responsible for limiting the encoder so that the learned representation does not

---

Algorithm 1 WICA

1. Take mini-batch X' from the dataset X.
2. Normalize componentwise $\mathcal{E}X'$, to obtain Y
3. Compute $p_1, \ldots, p_d$, where p_i is the mean of randomly chosen d elements from Y,
4. Minimize:

   $$\text{rec\_error}(X'; \mathcal{E}, \mathcal{D}) + \beta \text{wii}(Y; p_1, \ldots, p_d).$$

---

lose any information about the input. In practice, this is implemented by simultaneously minimizing the reconstruction error: $\text{rec\_error}(X; \mathcal{E}, \mathcal{D}) = \sum_{i=1}^{d} \|x_i - \mathcal{D}(\mathcal{E}x_i)\|^2$.

Reducing the difference between the input and the output is crucial to recover unmixing mapping close to the inverse of the mixing one. Thus our final cost function is given by $\text{cost}(X; \mathcal{E}, \mathcal{D}) = \text{rec\_error}(X; \mathcal{E}, \mathcal{D}) + \beta \text{wii}(\mathcal{E}X)$, where β is a hyperparameter that aims to weigh the role of independence for the reconstruction (analogous to β-VAE [8]).

The training procedure follows the steps described in Algorithm 1.

5 Nonlinear Mixing

Another important ingredient of this paper is the introduction of a new and fully invertible nonlinear mixing function.

In the case of linear ICA, one can easily construct many experiment settings that can be used to evaluate and compare different methods. Such standards are unfortunately not present in the case of nonlinear ICA. Therefore it is not clear what kind of nonlinear mixing should be used in the benchmark experiments.

In most papers, authors use mixing functions that correspond to the architecture of the model in question and are typically implemented by repeatedly applying linear transformations interlaced with nonlinear activations [1,6]. The main drawback of such an approach is that the obtained observations may not be complex enough, or even degenerate (see Figs. 2 and 3).

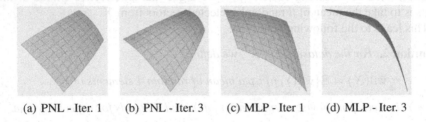

(a) PNL - Iter. 1   (b) PNL - Iter. 3   (c) MLP - Iter 1   (d) MLP - Iter. 3

Fig. 2. Results of the nonlinear mixing techniques proposed in [6] on a normalized synthetic lattice data. Post nonlinear mixing model (PNL) introduced only slight nonlinearities, which are not hard enough to solve even for the linear algorithms. On the other hand, the multi-layer perceptron mixing (MLP) technique collapses after just a couple of iterations.

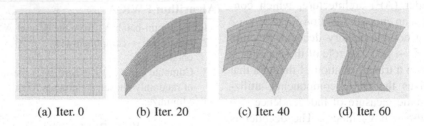

(a) Iter. 0   (b) Iter. 20   (c) Iter. 40   (d) Iter. 60

Fig. 3. Results of our proposition of mixing over normalized synthetic lattice data. One may observe that after multiple iterations of the proposed mixing, results become highly nonlinear but do not degenerate into any obscure solutions.

In contrast to such methodology, we propose a new iterative nonlinear mixing function inspired by the flow models [7,16]. This method does not relate to the internal design of our network architecture, is invertible, and allows for chaining the task complexity by varying the number of iterations, making it a useful tool in the verification of the nonlinear ICA models.

Let S be a sample of vectors with independent components. We apply a random isometry on S, by taking $X = (UV^T) S$, where UV^T comes from the Singular Value Decomposition on a random matrix $A_{ij} \sim N(0, 1)$. Next, we split $X \in \mathbb{R}^d$ into half:

$$(x_i, x_j) \rightarrow (x_i, x_j + \phi(x_i)),$$

similarly as it was done in [16]. Function ϕ is a randomly initialized neural network with two hidden layers and tanh activations after each of them. This approach can be repeated over multiple times to achieve the desired level of nonlinearity of the mixing by taking $(x_i, x_j) \rightarrow (x_i + \phi(x_j), x_j)$ for even and $(x_i, x_j) \rightarrow (x_i, x_j + \phi(x_i))$ for odd iterate.

The proposed mixing procedure scales well in higher dimensions by iteration over the splits in $\mathbb{R}^d$. Additionally, it is easily invertible, therefore there is a guarantee that the source components may be retrieved.

6 Measuring the Quality of Obtained Results

In this section, we define Optimal Transport Spearman (OTS) measure and discuss its properties.

6.1 Optimal Transport Spearman Measure

For the benchmark experiments, we want to be able to measure the similarity between the obtained results Z and the original sources S. In the case of linear mixing the common choice is the maximum absolute correlation over all possible permutations of the signals [5, 10–12, 14, 20, 23, 23] denoted hereafter as max_corr.

However, this measure is based on Pearson's correlation coefficient and therefore is not able to catch any high order dependencies. To address this problem we introduce a new measure based on the nonlinear Spearman's rank correlation coefficient and optimal transport.

Let Z denote the signal retrieved by an ICA algorithm and let the $r_s\left(z^j, s^k\right)$ be the Spearman's rank correlation coefficient between the j-th component of Z and k-th component of S. We define the Spearman's distance matrix $M(Z, S)$ as $M(Z, S) = \left[1 - \left|r_s(z^j, s^k)\right|\right]_{j,k=1,2,\ldots d}$, where the zero entries indicate a monotonic relationship between the corresponding features. This matrix is then used as the transportation cost of the components. Formally, we compute the value of the optimal transport problem formulated in terms of integer linear programming:

$$\text{OTS} = 1 - I_s(Z, S),$$

$$I_s(Z, S) = \min_{\gamma} \frac{1}{D} \sum_{j,k} \gamma_{j,k} M(Z, S)_{j,k},$$

subject to:

$$\sum_k^d \gamma_{j,k} = A_j \text{ for all } j \in \{1, 2, \ldots, d\},$$

$$\sum_{j}^{d} \gamma_{j,k} = A_k \text{ for all } k \in \{1, 2, \ldots, d\},$$

$$\gamma_{j,k} \in \{0, 1\} \text{ for all } j, k \in \{1, 2, \ldots, d\},$$

where $A_j = A_k = 1$.

As a result of the last constraint, the obtained transport plan γ defines a one-to-one map from the retrieved signals to the original sources. In addition, the proposed Spearman-based measure (OTS) is sensitive to monotonic nonlinear dependencies and also relatively easy to compute with the use of existing tools for integer programming.

6.2   Discussion on OTS Properties

A popular measure used to compare the signals obtained by an ICA method with the original sources is the maximum mean correlation (max_corr). In this subsection we elaborate more about the differences between max_corr and the proposed by us Optimal Transport Spearman (OTS) measure.

One of the characteristics distinguishing OTS from max_corr is that the latter favours stronger disentanglement of few components, while the first one gives lower results for outcomes that decompose the observation more equally.

In other words, consider an experiment in which n signals were mixed. However, assume that some (nonlinear) ICA algorithms failed to unmix all but one component (i.e. only one unmixed component matches exactly one source signal, while the rest is still highly unrecognizable). In such a situation, the max_corr value will be significantly higher than OTS, although only a small portion of the base dataset was recovered.

To empirically demonstrate this property, we artificially mixed a multidimensional grid using the mixing function from Sect. 5 of the main article. Next, we randomly swapped one of the mixed signals with the original signal from the base dataset. We compared this *mixed-and-swapped* data to the source signals using max_corr and OTS. The results over different mixing iterations are presented in Fig. 4. One may observe that max_corr values are always above the OTS ones, suggesting that max_corr measure prefers such a recovery more than OTS. Naturally, in the case when all signals are far different from the true sources, values for max_corr and OTS are almost the same (see Fig. 5).

In consequence, the max_corr measure can help to assess the maximum of informativeness from the retrieved signal. This can be desired in situations that favour the good decomposition of few components at the cost of lower correlatedness of the remaining ones (which may happen, for instance, in denoising problems). In the case where approximately equal recovery of all the signals is requested, the OTS measure would be a better choice.

2 dimensions   4 dimensions   6 dimensions   8 dimensions

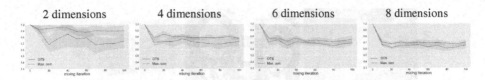

Fig. 4. Results of the experiment wherein n–dimensional mixed observation one component was swapped with a randomly chosen source signal. One may observe that max_corr almost always prefers such a situation, while OTS seems to be more rigorous.

2 dimensions   4 dimensions   6 dimensions   8 dimensions

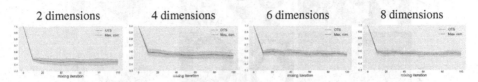

Fig. 5. Results for the OTS and max_corr values for fully mixed dataset. One may observe that both measures in this case give similar outcomes.

7 Experiments

In this section, we present several simulated experiments to validate the WICA algorithm empirically.

7.1 Qualitative Results

We start from the simulated example of the ICA application in the case of the images separation problem. We use this regime because the results can be understood with the naked eye of a reader.

To construct this experiment one needs to apply some artificial mixing function (i.e. linear transformation or mixing function from Sect. 5) on the independent source signals. Such mixture is then passed to the ICA model in question to perform the unmixing task.

To compare the WICA algorithm to other nonlinear ICA approaches we evaluated the models' performance in the case of separation of artificially mixed images. Results of this toy example are presented in Fig. 6. Besides the retrieved images and their scatter plots, we also demonstrated the projection of marginal densities. The desired goal is to achieve similar images and marginal densities as in the source (original) pictures.

One can easily spot that FastICA and dCor seem to only rotate the mixed signals. The ANICA method, on the other hand, transformed the observations to a high extent, but the recovered signals are visually worse than the original pictures. PNLMISEP and WICA algorithms also performed some nontrivial shift on the marginal densities, but in this case the retrieved densities resemble the original ones more naturally.

This experiment was fully qualitative and the outcome is subject to one's perception. We demonstrated the images as a visualization of the different ICA models performance in simple nonlinear setup. We report quantitative results in the next subsection.

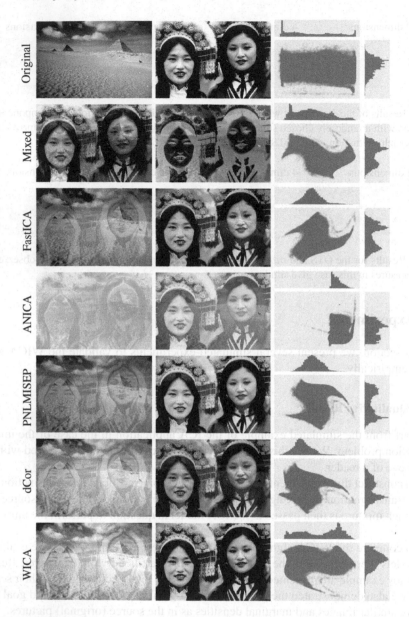

Fig. 6. Two-dimensional example of the problem of unmixing natural images. One can easily spot that WICA has the smallest amount of artefacts remained after retrieving the signals. All of the scatter plots were normalized and are presented on the same scale. It is valuable to look at the attached marginal histograms also, where some of the similarities between the original signal and its retrieved counterpart may be observed.

7.2   Quantitative Results

From the preliminary results reported in the previous subsection, we moved to a more complex scenario in which we quantitatively evaluated ICA methods in a higher dimensional setup – where d signals were synthetically mixed i number of times.

Table 1. Comparison between nonlinear ICA methods (PNLMISEP, dCor, ANICA, WICA) and the classical linear ICA approach (FastICA) on images separation problem (with different dimensions) by using max correlation and OTS measures. In the experiment, we tuned and trained four models (excluding FastICA, which is a linear model) and present mean and standard deviation in the tabular form.

Measure	Dim	Mixes	WICA	FastICA	ANICA	dCor	PNLMISEP
OTS	2	10	0.798 ± 0.048	0.652 ± 0.001	0.938 ± 0.088	0.899 ± 0.076	**0.948 ± 00.041**
	4	10	**0.890 ± 00.065**	0.582 ± 0.001	0.784 ± 0.062	0.554 ± 0.264	0.652 ± 0.360
	6	10	**0.807 ± 00.043**	0.419 ± 0.001	0.571 ± 0.056	0.666 ± 0.064	0.779 ± 0.046
	8	10	**0.784 ± 00.025**	0.457 ± 0.058	0.431 ± 0.054	0.594 ± 0.051	0.769 ± 0.097
	10	10	0.742 ± 0.030	0.405 ± 0.077	0.405 ± 0.032	0.556 ± 0.041	**0.758 ± 00.103**
	2	50	0.759 ± 0.097	**0.872 ± 00.001**	0.866 ± 0.128	0.771 ± 0.090	0.811 ± 0.122
	4	50	0.774 ± 0.050	0.692 ± 0.001	0.573 ± 0.067	0.756 ± 0.086	**0.825 ± 00.108**
	6	50	**0.769 ± 00.030**	0.562 ± 0.001	0.465 ± 0.055	0.695 ± 0.059	0.721 ± 0.102
	8	50	**0.831 ± 00.061**	0.773 ± 0.021	0.798 ± 0.087	0.668 ± 0.053	0.711 ± 0.022
	10	50	**0.819 ± 00.052**	0.756 ± 0.026	0.796 ± 0.087	0.644 ± 0.032	0.738 ± 0.084
Max correlation	2	10	0.771 ± 0.013	0.965 ± 0.001	0.631 ± 0.112	0.901 ± 0.058	0.942 ± 0.045
	4	10	**0.910 ± 00.065**	0.710 ± 0.001	0.588 ± 0.063	0.552 ± 0.278	0.645 ± 0.363
	6	10	**0.821 ± 00.041**	0.578 ± 0.001	0.505 ± 0.062	0.696 ± 0.059	0.808 ± 0.063
	8	10	**0.814 ± 00.058**	0.759 ± 0.046	0.769 ± 0.065	0.658 ± 0.044	0.812 + 0.085
	10	10	0.812 ± 0.049	0.770 ± 0.058	**0.837 ± 00.042**	0.658 ± 0.041	0.820 ± 0.077
	2	50	0.759 ± 0.097	**0.872 ± 00.001**	0.866 ± 0.128	0.771 ± 0.090	0.811 ± 0.122
	4	50	0.774 ± 0.050	0.692 ± 0.001	0.573 ± 0.067	0.756 ± 0.086	**0.825 ± 00.108**
	6	50	**0.769 ± 00.030**	0.562 ± 0.001	0.465 + 0.055	0.695 ± 0.059	0.721 ± 0.102
	8	50	**0.831 ± 00.061**	0.773 ± 0.021	0.798 ± 0.087	0.668 ± 0.053	0.711 ± 0.022
	10	50	**0.819 ± 00.052**	0.756 ± 0.026	0.796 ± 0.087	0.644 + 0.032	0.738 ± 0.084

Performance Across Different Dimensions. We plotted the results of this experiment for $i = 50$ mixes[1] with respect to the data dimension d in Fig. 7.

The outcomes demonstrated that the WICA method outperformed any other nonlinear algorithm in the proposed task by achieving high and stable results regardless of the considered data dimension. In the case of the stability of the results, WICA losses only to the linear method – FastICA – which, unfortunately, cannot satisfactorily factorize nonlinear data.

This experiment demonstrated that WICA is a strong competitor to other models in a fully unsupervised environment for nonlinear ICA.

It is also worth mentioning the difference between the results measured by OTS and max_corr for the ANICA and FastICA models applied in high dimensions. We hypothesize it may indicate that those algorithms were able to retrieve very well only a small subset of true components, while the remaining variables were still highly mixed.

---

[1] We considered the setting with 50 mixing iterations as the hardest one.

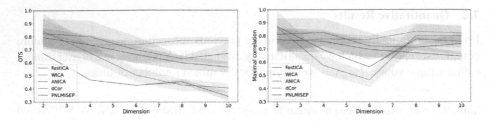

Fig. 7. Comparison between standard ICA methods (PNLMISEP, dCor, ANICA, FastICA) and our approach by using OTS (left) and max_corr (right) measures in the setup where 50 mixing iterations were performed. In the experiment we train five models and present the mean and standard deviation of each of the used measures (the higher the better). One can observe that WICA consistently obtains good results for all of the dimensions and outperforms the other methods in higher dimensions. Moreover, it has the lowest standard deviation across all the nonlinear algorithms. More numerical results of the experiment are presented in Table 1.

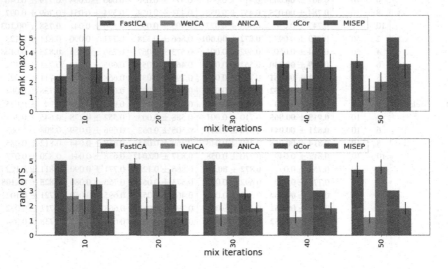

Fig. 8. The mean rank results for different mixes measured by max_corr (top) and OTS (bottom). The lower the better.

Performance Across Different Number of Mixes. For every model we evaluated the mean OTS and max_corr score on a given dimension d and number of mixing iterations i. Then, for each pair (d, i) we ranked the tested models based on their performance. We report the mean rank of models for each mixing iteration i in Fig. 8 (the lower the better).

One may be observed that for tasks relatively similar to the linear case (where the number of mixes is equal to 10) the PNLMISEP method performs the best both on max_corr and OTS. However, as the number of mixes increases, the WICA algorithm usually outperforms all the other methods in both measures, achieving the lowest mean rank. The numerical results measured in all experiments with 10 and 50 iterations are presented in Table 1.

8 Conclusions

In this paper, we presented a new approach to the nonlinear ICA task, which proves to be matching the results of all other tested nonlinear algorithms.

In addition to the investigation of WICA method, we proposed a new mixing function for validating nonlinear tasks in a structured manner. Our mixing scales to higher dimensions and is easily invertible.

Lastly, we defined OTS, a measure that can catch nonlinear dependence and is easy to compute. The OTS measure and the proposed mixing have the potential to become benchmark tools for all future work in this field.

References

1. Almeida, L.B.: Misep-linear and nonlinear ICA based on mutual information. J. Mach. Learn. Res. **4**(Dec), 1297–1318 (2003)
2. Basile, V., Cabitza, F., Campagner, A., Fell, M.: Toward a perspectivist turn in ground truthing for predictive computing. CoRR abs/2109.04270 (2021). https://arxiv.org/abs/2109.04270
3. Bedychaj, A., Spurek, P., Struski, Ł., Tabor, J.: Independent component analysis based on multiple data-weighting. arXiv preprint arXiv:1906.00028 (2019)
4. Bell, A.J., Sejnowski, T.J.: An information-maximization approach to blind separation and blind deconvolution. Neural Comput. **7**(6), 1129–1159 (1995)
5. Bengio, Y., Courville, A., Vincent, P.: Representation learning: a review and new perspectives. IEEE Trans. Pattern Anal. Mach. Intell. **35**(8), 1798–1828 (2013)
6. Brakel, P., Bengio, Y.: Learning independent features with adversarial nets for non-linear ICA. arXiv preprint arXiv:1710.05050 (2017)
7. Dinh, L., Krueger, D., Bengio, Y.: Nice: Non-linear independent components estimation. arXiv preprint arXiv:1410.8516 (2014)
8. Higgins, I., et al.: beta-VAE: learning basic visual concepts with a constrained variational framework. In: ICLR (2017)
9. Horan, D., Richardson, E., Weiss, Y.: When is unsupervised disentanglement possible? In: Advances in Neural Information Processing Systems 34 Pre-proceedings (NeurIPS 2021) (2021)
10. Hyvärinen, A.: Fast and robust fixed-point algorithms for independent component analysis. IEEE Trans. Neural Netw. **10**(3), 626–634 (1999)
11. Hyvarinen, A., Morioka, H.: Unsupervised feature extraction by time-contrastive learning and nonlinear ICA. In: Advances in Neural Information Processing Systems, pp. 3765–3773 (2016)
12. Hyvärinen, A., Morioka, H.: Nonlinear ICA of temporally dependent stationary sources. In: International Conference on Artificial Intelligence and Statistics, pp. 460–469. Microtome Publishing (2017)
13. Hyvärinen, A., Pajunen, P.: Nonlinear independent component analysis: existence and uniqueness results. Neural Netw. **12**(3), 429–439 (1999)
14. Hyvärinen, A., Sasaki, H., Turner, R.E.: Nonlinear ICA using auxiliary variables and generalized contrastive learning. In: The 22nd International Conference on Artificial Intelligence and Statistics, pp. 859–868 (2019). (J. Mach. Learn. Res.)
15. Khemakhem, I., Kingma, D.P., Hyvärinen, A.: Variational autoencoders and nonlinear ICA: a unifying framework. arXiv preprint arXiv:1907.04809 (2019)

16. Kingma, D.P., Dhariwal, P.: Glow: generative flow with invertible 1x1 convolutions. In: Advances in Neural Information Processing Systems, pp. 10215–10224 (2018)
17. Larson, L.E.: Radio frequency integrated circuit technology for low-power wireless communications. IEEE Pers. Commun. **5**(3), 11–19 (1998)
18. Le, Q.V., Karpenko, A., Ngiam, J., Ng, A.Y.: Ica with reconstruction cost for efficient overcomplete feature learning. In: Advances in Neural Information Processing Systems, pp. 1017–1025 (2011)
19. Matteson, D.S., Tsay, R.S.: Independent component analysis via distance covariance. J. Am. Stat. Assoc. 1–16 (2017)
20. Spurek, P., Nowak, A., Tabor, J., Maziarka, L., Jastrzbski, S.: Non-linear ICA based on Cramer-Wold metric. CoRR abs/1903.00201 (2019). http://arxiv.org/abs/1903.00201
21. Spurek, P., Tabor, J., Rola, P., Ociepka, M.: ICA based on asymmetry. Pattern Recogn. **67**, 230–244 (2017)
22. Székely, G.J., Rizzo, M.L., Bakirov, N.K., et al.: Measuring and testing dependence by correlation of distances. Ann. Stat. **35**(6), 2769–2794 (2007)
23. Zheng, C.H., Huang, D.S., Li, K., Irwin, G., Sun, Z.L.: MISEP method for postnonlinear blind source separation. Neural Comput. **19**, 2557–2578 (2007). https://doi.org/10.1162/neco.2007.19.9.2557
24. Ziehe, A., Muller, K.R., Nolte, G., Mackert, B.M., Curio, G.: Artifact reduction in magnetoneurography based on time-delayed second-order correlations. IEEE Trans. Biomed. Eng. **47**(1), 75–87 (2000)

Statistical Models for Partial Orders Based on Data Depth and Formal Concept Analysis

Hannah Blocher[✉], Georg Schollmeyer, and Christoph Jansen

Department of Statistics, Ludwig-Maximilians-Universität Munich, Munich, Germany
{hannah.blocher,georg.schollmeyer,christoph.jansen}@stat.uni-muenchen.de

Abstract. In this paper, we develop statistical models for partial orders where the partially ordered character cannot be interpreted as stemming from the non-observation of data. After discussing some shortcomings of distance based models in this context, we introduce statistical models for partial orders based on the notion of data depth. Here we use the rich vocabulary of formal concept analysis to utilize the notion of data depth for the case of partial orders data. After giving a concise definition of unimodal distributions and unimodal statistical models of partial orders, we present an algorithm for efficiently sampling from unimodal models as well as from arbitrary models based on data depth.

Keywords: Partial orders · Partial rankings · Data depth · Formal concept analysis · Unimodality · Quasiconcavity

1 Introduction

Orders play a role in a broad range of scientific disciplines. In many of these disciplines like revealed preference theory, social choice theory, decision making under uncertainty, social-economics (Human Development Index, costumer preference rankings etc.) or statistics and machine learning, studying *partial* orders has attracted more and more researchers (see [5,10,18,19,21,28,33] and [13] for recent works in the respective discipline). As an example, one can consider pair comparison data sets as in [9] and [12]. Consequently, there are many approaches that can deal with partial orders. However, in most approaches known to the authors, the incompleteness of the involved orders is interpreted as stemming from missing data, see, e.g., [25,30]. In other words, an explicit missing mechanism is modeled or at least assumed. In contrast, in this paper we explicitly assume that the incompleteness of the order is not due to missing of data. Instead, we understand incomparabilities within observed partial orders as precise observation of a factual incomparability that actually exists.

The aim of this paper is to define statistical models on the set of partial orders that do not assume a missing mechanism and that can be easily specified by defining a location and a scale parameter. Beyond this methodological contribution we propose an efficient algorithm for sampling from such models. Our

© Springer Nature Switzerland AG 2022
D. Ciucci et al. (Eds.): IPMU 2022, CCIS 1602, pp. 17–30, 2022.
https://doi.org/10.1007/978-3-031-08974-9_2

statistical models are based on formal concept analysis (**FCA**) and the concept of data depth. Via the chosen representation of the data set using a formal context (which is a formalization of a cross table defined in FCA), the information that a pair of elements does not exist in a partial order is explicitly included. Furthermore, we embed the notion of depth function in the theory of FCA. Data depth functions are commonly used in robust and nonparametric statistics and can be viewed as a generalization of univariate quantiles. Applications range from outlier detection over nonparametric statistical tests and confidence intervals to robust regression, for a short overview see, e.g., [26] and the references therein. Generally, a depth function measures the outlyingness and centrality of an observation w.r.t. a data cloud or an underlying probability measure. While common depth functions are defined for data in $\mathbb{R}^d$, accompanied by a geometric intuition, within this paper, we aim at generalizing the concept of data depth to the abstract setting of data points that are objects of a formal context. Unlike many of the existing distance based models for partial order data (cf., e.g., [14,25,27,30]), we are therefore able to use data depth functions instead of a distance measure on the set of partial orders. While depth functions compute the depth value with respect to all other data points, distance measures compute only the distance between the data points of a partial order with respect to a predefined partial order representing the center, see e.g. [6]. Note further that many distance measures are based on the linear extensions of the partial orders involved and thus do not take into account the incomparable character, but mimic an underlying true linear order. Since we want to define a simple location-scale type statistical model, we also further define unimodality in the context of FCA and we consider in particular depth functions that satisfy this property. As far as the authors are aware, there is no other work that directly links FCA and data depth or uses FCA to define a model for the set of partial orders.

In Sect. 2 we begin with an overview of currently used distance measures and point out their explicit and implicit assumptions. In Sect. 3 we give a short introduction to FCA and define unimodality in this framework. Then, in Sect. 4 and 5 we define and discuss the formal context representing the set of partial orders, and, afterwards, propose some concrete depth functions. In Sect. 6, we introduce an algorithm for sampling from the proposed statistical models and finally we give a brief conclusion.

2   Motivation and Related Work

To illustrate how the currently used distance measures implicitly mimic the missing mechanism and other counter-intuitive structures, let us start by discussing the current approaches that use distance measures for (partial) orders. There are several proposals for adequately defining a meaningful distance concept between (partial) orders in the literature (cf, e.g., [6,11]) which can be used to establish distance based statistical models for partial orders. Throughout the paper let $\mathcal{X}$ be a finite ground space with $n \geq 3$ elements and let $\mathscr{P}$ denote the set of all partial orders[1] (i.e., all reflexive, transitive and anti-symmetric binary relations) on $\mathcal{X}$. Two prominent distance measures for partial orders are discussed for example in [6]: The *nearest neighbour* and the *Hausdorff distance*. Both of these distances rely on the idea of first computing the set of all linear extensions of the considered partial orders and then, each in its own manner, generalizing the well-known *Kendall's τ-distance* (see [22]) for linear orders (i.e. counting pairs that are ranked oppositely by the considered orders). However, such an approach has the following counter-intuitive property: The nearest neighbour distance systematically assigns lower distance values if sparse partial orders are involved. The nearest neighbour distance is defined as

$$d_{NN}(P_1, P_2) := \min_{L_1 \in lext(P_1)} \min_{L_2 \in lext(P_2)} \tau(L_1, L_2)$$

for two orders P_1, P_2 where $lext(P)$ denotes the set of all linear extensions of a partial order P and τ denotes the Kendall's τ-distance for linear orders mentioned before. Then it is immediate from the definition that $d_{NN}(\tilde{P}_1, P_2) \leq d_{NN}(P_1, P_2)$ for arbitrary partial orders $\tilde{P}_1 \subseteq P_1$, since this implies $lext(P_1) \subseteq lext(\tilde{P}_1)$ and therefore the minimum is taken over a super-set of the original one. Most extremely, the minimal distance is attained whenever one of the considered partial orders is the trivial one consisting solely of the diagonal $D_{\mathcal{X}} := \{(x, x) : x \in \mathcal{X}\}$, whereas two partial orders differing only in few pairs receive non-minimal distance value. This seems to be a very counter-intuitive property of this generalized distance measure. An analogous line of argumentation applies when the nearest neighbour distance is replaced by the directed Hausdorff hemi-metric

$$m_H(P_1, P_2) := \max_{L_1 \in lext(P_1)} \min_{L_2 \in lext(P_2)} \tau(L_1, L_2).$$

Then, in a dual manner, $D_{\mathcal{X}}$ (if seen as the first argument in the Hausdorff hemi-metric) has always the maximal distance to other orders whereas a linear order L has always a smaller distance to other orders compared to any other partial order $P \subseteq L$. Similar arguments can be given for the usual symmetrized non-directed Hausdorff distance defined by

$$d_H(P_1, P_2) := max\{m_H(P_1, P_2), m_H(P_2, P_1)\}.$$

Alternatively, one could directly generalize Kendall's τ to partial orders without looking at linear extensions. This would result in one of the two expressions

$$\tau_s(P_1, P_2) := |\Delta(P_1, P_2)| = |(P_1 \cup P_2) \setminus (P_1 \cap P_2)| \quad \text{or}$$
$$\tau_a(P_1, P_2) := |\{(x, y) \mid x \neq y, (x, y) \in P_1, (y, x) \in P_2\}|,$$

both, in a way, generalizing the idea of counting pairs that are ranked oppositely by the considered *partial* orders. However, whereas τ_a has the same problem like the nearest neighbour distance, the expression τ_s would lead, as will be shown in

---

[1] In the sequel, we will also shortly say order instead of partial order.

Sect. 3, Example 1, to statistical models that are not completely quasiconcave, which means that it seems to be impossible to build a simple unimodal model with such a distance (cf., Definition 1). Furthermore, τ_s treats pairs which are in the relation and pairs being not in the relation in exact the same way, and one can ask if this is natural. As we will see later, our approach that uses a depth function treats pairs being in the relation or not seemingly differently. (Note that a partial order is transitive but not necessarily negatively transitive, so there is in fact some asymmetry between a pair being in the relation or not.) With these problems in mind, we propose statistical modelling of partial orders based on a depth function. The model idea is analogous to a distance based version

$$P(X = x) = C_\lambda \cdot \Gamma\left(\lambda \cdot d\left(\mu, x\right)\right),$$

where, C_λ is a normalizing constant, $d : \mathscr{P} \times \mathscr{P} \longrightarrow \mathbb{R}_{\geq 0}$ is a distance function, $\Gamma : \mathbb{R}_{\geq 0} \longrightarrow \mathbb{R}_{\geq 0}$ is a (weakly decreasing) decay function, $\mu \in \mathscr{P}$ is a location parameter and $\lambda \in \mathbb{R}_{>0}$ is a scale parameter. Now, instead of a distance function, in this paper we work with a depth function and a statistical model given by

$$P(X = x) = C_\lambda \cdot \Gamma\left(\lambda \cdot (1 - D^\mu\left(x\right))\right) \tag{1}$$

where now D^μ is a depth function that is maximal at partial order μ. Since depth functions are usually only used for data in $\mathbb{R}^d$ we have to adapt the notion of data depth to partial order data, for which we use FCA.

3   Formal Concept Analysis, Data Depth and Unimodality

In this section we only touch a few aspects about the theory of formal concept analysis (FCA) and we refer the reader to [17] for more details. The basis of FCA is the definition of a **formal context** $\mathbb{K} = (G, M, I)$ which is a generalization and formalization of a cross table. Here, G is a set of objects, M a set of binary attributes and $I \subseteq G \times M$ a relation. We say that an object g has an attribute m if $(g, m) \in I$ is true. For example Cross Table 1 describes a formal context with $G = \{\mu, g, h, i\}$, $M = \{m_1, \ldots, m_6\}$ and the relation I is given by the crosses. By the use of the following **derivation operators**, we obtain a description of the relation between the object and attribute set:

$$\Psi : 2^G \to 2^M : A \mapsto \{m \in M \mid \forall g \in A : (g, m) \in I\}$$

$$\Phi : 2^M \to 2^G : B \mapsto \{g \in G \mid \forall m \in B : (g, m) \in I\}.$$

Here $\Psi(A)$ contains all the attributes that each object in A has, and $\Phi \circ \Psi(A) \subseteq G$ are all objects that have all attributes in $\Psi(A)$. The tuple $(\Phi \circ \Psi(A), \Psi(A))$ for $A \subseteq G$ is called a **formal concept**, $\Phi \circ \Psi(A)$ its **extent**, and $\Psi(A)$ its **intent**. The construction of the two derivation operators allows to determine the relation I when the set of all formal concepts is known. Note that $\Psi(A) = \Psi \circ \Phi \circ \Psi(A)$ holds, and thus each formal concept is uniquely described by its extent or intent. Moreover, the set of extents and the set of intents yield a closure system with

$\Phi \circ \Psi$ and $\Psi \circ \Phi$, respectively, the corresponding closure operator. Note that if $A \subseteq G$ lies in an extent E, then the closure operator $\Phi \circ \Psi$ ensures that every object having all attributes of $\Psi(A)$ is also an element of E. Thus, $A \subseteq E$ implies that $\Phi \circ \Psi(A) \subseteq E$. With this, we say that the pair $A, B \subseteq G$ is an **(object) implication** (we denote this by $A \to B$) if $\Phi \circ \Psi(A) \supseteq \Phi \circ \Psi(B)$ holds. Moreover, one can show that the set of all implications that follow from the extent set completely describes the extent set itself, see, e.g., [17, p. 80, Proposition 20]. Within this paper, we use formal implications between objects to model a notion of betweenness. For example $\{g, h\} \longrightarrow \{i\}$ can be interpreted as "object i lies between object g and object h" (or "object i lies in the space that is spanned by the objects g and h"), because object i has all attributes that are shared by both g and h. (Note that we do not restrict the premise of a formal implication to have exactly two objects.) For further discussion of a family of implications, see [2] and [17]. If non-binary attributes are considered, then they are converted into a set of binary attributes by using a so-called conceptual scaling method (see Sect. 4).

Our approach is to represent the set of partial orders by a formal context and, using the properties of a formal context, to define the notion of unimodality and depth function. By using the following properties that a function $f \colon G \to \mathbb{R}$ can satisfy on a formal context $\mathbb{K}$, we define the notion of unimodality.

Definition 1. *Let* $\mathbb{K} = (G, M, I)$ *be a formal context and let* $f : H \longrightarrow \mathbb{R}$ *with* $H \subseteq G$ *be a function. Then* f *is called*

   i)   *isotone if for all* $g, h \in H$ *we have* $\{g\} \longrightarrow \{h\} \implies f(g) \leq f(h)$;

  ii)   **2-quasiconcave** *if for arbitrary objects* $g, h, i \in H$ *we have* $\{g, i\} \longrightarrow \{h\} \implies f(h) \geq \min\{f(g), f(i)\}$;

 iii)   **completely quasiconcave** *if for every finite set of objects* $\{g_1, \ldots, g_n\} \subseteq H$ *we have* $\{g_1, \ldots, g_{n-1}\} \longrightarrow \{g_n\} \implies f(g_n) \geq \min\{f(g_1), \ldots f(g_{n-1})\}$;

 iv)   **strongly quasiconcave** *if for every finite set* $\{g_1, \ldots, g_n\} \subseteq H$ *of size* $n \geq 2$ *we have* $\{g_1, \ldots, g_{n-1}\} \longrightarrow \{g_n\} \implies f(g_n) > \min\{f(g_1), \ldots f(g_{n-1})\}$;

  v)   **star-shaped** *if there exists a center* $c \in H$ *such that for all* $g \in H$ *we have* $\{c, g\} \longrightarrow \{h\} \implies f(h) \geq \min(f(c), f(g))$.

Additionally, a probability measure P *on a finite* G *is called* **unimodal (strictly unimodal)** *if its probability function, restricted to its support* $\{g \in G \mid P(\{g\}) > 0\}$*, is completely quasiconcave (strongly quasiconcave).*

In general, depth functions measure outlyingness and centrality of an observation w.r.t. a data cloud or an underlying probability measure. We apply the concept of data depth to partial order data represented by a formal context and we denote it by $D \colon G \to \mathbb{R}_{\geq 0}$. Note that it depends on the formal context. Moreover, if we ensure that the depth function is completely quasiconcave (strongly quasiconcave), then the statistical model given in (1) is unimodal (strictly unimodal).

Our notion of quasiconcavity is an adaption of classical quasiconcavity which was already used (e.g., in [29]) for classical data depth for $\mathbb{R}^d$. In particular, here we emphasize (complete) quasiconcavity because it most adequately renders the idea of an unimodal distribution of partial orders that would be induced by a statistical model that uses a quasiconcave depth function: Quasiconcavity would ensure that we have no point that is a local minimum of the probability function w.r.t. the notion of betweenness that is appropriate for a FCA view on partial orders. Another nice feature of complete quasiconcavity is the fact that this property is equivalent to the property that the upper level sets $D_\alpha := \{g \in G \mid D(g) \geq \alpha\}$ of the depth function D are extents. Thus, every upper level set can be nicely described by a formal concept which makes them descriptively accessible, especially the fact that they cannot only be exactly described by objects, but also by attributes, is very convincing.

Example 1. Let $\mathbb{K} = (G, M, I)$ be given by Cross Table 1. Then, the depth function D^μ with mode μ given by $D^\mu(g) := |\Psi(\mu) \cap \Psi(g)|$, together with the conceptual scaling of Sect. 4 can be shown to be exactly the depth-based formulation of a distance based approach with τ_s. It is 2-quasiconcave but in general not completely quasiconcave and therefore is not appropriate to define a unimodal distribution. Note that for arbitrary contexts, D^μ is generally not 2-quasiconcave. Note further that D^μ is at least star-shaped for arbitrary contexts. Furthermore, a generalization of Tukey's depth $\mathcal{T}$ (cf., [31]) and a localized version of Tukey's depth $\mathcal{T}^\mu$ with mode μ can be defined via

$$\mathcal{T}(g) := 1 - \max_{m \in M \setminus \Psi(\{g\})} \frac{|\Phi(\{m\})|}{|G|}; \quad \mathcal{T}^\mu(g) := 1 - \frac{\max\limits_{\substack{m \in M \setminus \Psi(\{g\}), \\ \mu I m}} |\Phi(\{m\})|}{|G|}, \quad (2)$$

respectively. (Here the empty maximum is defined as 0.) Both $\mathcal{T}$ and $\mathcal{T}^\mu$ are completely quasiconcave functions.

Table 1. Illustration of the difference between complete and 2-quasiconcavity.

	m_1	m_2	m_3	m_4	m_5	m_6
μ		x	x	x	x	x
g		x				
h	x				x	x
i	x		x	x		

4 Formal Context Defined by All Partial Orders

In our case the set G is exactly the set $\mathscr{P}$ of all partial orders on $\mathcal{X}$. Note that we regard a partial order not necessarily as a linear order together with a missing mechanism. Therefore, as attributes we also include the property of being incomparable pairs and get

$$M := \underbrace{\left\{ \text{``}x_i \leq x_j\text{''} \mid i, j = 1, \ldots, n, \, i \neq j \right\}}_{=:M_{\leq}} \cup \underbrace{\left\{ \text{``}x_i \nleq x_j\text{''} \mid i, j = 1, \ldots, n, \, i \neq j \right\}}_{:=M_{\nleq}}.$$

Since we consider only reflexive relations the attributes "$x_i \leq x_i$" and "$x_i \nleq x_i$" are redundant and therefore not included here. Note that each order g has $n(n-1)$ many attributes $B = \Psi(\{g\})$ which can be divided into the set $B_{\leq} \subseteq M_{\leq}$ and $B_{\nleq} \subseteq M_{\nleq}$. In particular, we have that either (x_i, x_j) lies in g or not and thus we can conclude $(g, \text{``}x_i \leq x_j\text{''}) \in I \Leftrightarrow (g, \text{``}x_i \nleq x_j\text{''}) \notin I \,\&\, (g, \text{``}x_j \nleq x_i\text{''}) \in I$. This means if a pair (x_i, x_j) exists then the attribute "$x_i \nleq x_j$" cannot hold, but "$x_j \nleq x_i$" must be true. The same is true for the reverse. Indeed, ensuring that a pair $(x_i, x_j), i \neq j$ is in an order g or not has a different strength of restriction, i.e., if we assume that $(x_i, x_j) \in g$, then $g^{-1} := \{(x_j, x_i) \mid (x_i, x_j) \in g\}$ satisfies the condition $(x_i, x_j) \notin g^{-1}$. Thus, the number of orders $\tilde{g}$ fulfilling the condition $(x_i, x_j) \notin \tilde{g}$ is larger than the number of orders g fulfilling $(x_i, x_j) \in g$. Additionally, because of symmetry these numbers are independent of the concrete pair (x_i, x_j).

First let us go one step back and consider the formal context given only by the attribute set $M_{\leq}$. In this case, for an isotone depth function D and two orders g, h with $g \subseteq h$ we have $D(g) \leq D(h)$. Thus, we would obtain again a depth concentration on linear orders. Furthermore, if the depth function is additional 2-quasiconcave and we consider the space of all partial orders, then at least half of all partial orders must have equal depth. More precisely, the depth must be minimal. To see this, let g be an order and let g^{-1} be the inverse order. We obtain that $\{g, g^{-1}\} \longrightarrow G$ and therefore either the depth of g or the depth of g^{-1} is minimal. Thus, because the map $g \mapsto g^{-1}$ is a bijection, half of all orders have minimal depth. Note that the stronger property of strong quasiconcavity cannot be fulfilled by any depth function: Assume we have four orders $g_1, \ldots, g_4$ where all pairs of orders have no attribute $m \in M_{\leq}$ in common. Then $\{g_1, g_2\} \longrightarrow G$ and therefore $\min\{D(g_1), D(g_2)\} < D(g_i), i = 3, 4$. But since $\{g_3, g_4\} \longrightarrow G$ this is a contradiction to $\min\{D(g_3), D(g_4)\} < D(g_i), i = 1, 2$. Note that for $|\mathcal{X}| \geq 3$ there exist four linear orders fulfilling this property. Let us now return to the formal context given by the entire attribute set M.

Then, the same argument for the non-existence of a strongly quasiconcave depth function from above would still apply for the extended attribute set M. Beyond this, now the context defined here contains no two different orders g and h such that $\Psi(g) \subseteq \Psi(h)$. Thus, for an isotone depth, isotonicity alone does not imply that the depth value of one order is constrained by the depth value of any other order. In contrast, 2-quasiconcavity would still lead to some restriction on the depth function: Let g be an order such that the complement order (i.e. $g^c := (\mathcal{X} \times \mathcal{X} \setminus g) \cup D_{\mathcal{X}})$ is also an order. Then one of the orders must have minimal depth, since $\{g, g^c\} \longrightarrow G$. If we take G as the set of all partial orders, then examples of such orders are exactly the linear orders. If, in contrast, one had chosen G as the set of all quasiorders, then exactly all negatively transitive orders g would have the property that also g^c is in G and therefore one of g or g^c would have minimal depth.

5 Specifying Unimodal Distributions of Partial Orders

In this section we want to analyze how one can specify a generic non-null model (i.e. a model that is different from a uniform distribution, see [27]) with the help of a depth function and Eq. (1) by specifying only two parameters, namely location and scale. More specifically, we discuss methods for generating unimodal distributions of partial orders based on three concrete depth functions. Firstly, we will discuss Tukey's depth defined by Eq. (2). Secondly, we define a generalization of the convex hull peeling depth. The classical convex hull peeling depth for $\mathbb{R}^d$-valued data was introduced in [4] and we will generalize it here for the case of data represented by a formal context. We will call this depth function peeling depth. It is sometimes said that the convex hull peeling depth has the disadvantage that it can only order the data points from outwards to inwards. In contrast, in our situation, we are able to directly specify a mode of the distribution and therefore we know beforehand, where 'the inwards', i.e., the mode, is exactly located. With this, we can in fact order the data points from inwards to outwards by starting from the mode and successively enclosing further layers. Thus, thirdly, we can define a new depth function that we call enclosing depth, here. The generalization of Tukey's depth for data values or probability distributions on arbitrary complete lattices or formal contexts was introduced in [31] and applied to the case of ranking data in [32] and in the context of social choice theory in [20]. The definition is given in Eq. (2). Before discussing all three data depths, we firstly define the remaining two:

Definition 2. *Define the **peeling depth** $\mathcal{P}$ by $\mathcal{P}(extr(G)) := \frac{1}{|G|}$ and*

$$\mathcal{P}\left(extr\left(G\backslash\mathcal{P}^{-1}\left(\left[0,\frac{i}{|G|}\right]\right)\right)\right) = \frac{i+1}{|G|}, \quad i = 1, 2, \ldots$$

*Additionally, define the **localized peeling depth** $\mathcal{P}^{\mu}$ w.r.t. mode $\mu \in G$ simply by adding a high enough amount of objects which have exactly the same attributes as μ to the original context G. The operator extr is here the extreme point operator which maps a set A to the set of all its extreme points.[2] Note that this definition is only well defined if the underlying context is meet-distributive.[3] Furthermore,*

---

[2] A point $g \in A$ is an extreme point of A if $A\backslash\{h \in G \mid \Psi(\{h\}) = \Psi(\{g\})\} \twoheadrightarrow \{g\}$.

[3] A context is called meet-distributive, if every extent is generated by all extreme points of the extent. In our situation, the underlying context is not meet-distributive, but it is possible to replace the extreme point operator by another appropriate operator that maps a set A to a set $B \subseteq A$ that implies A and is minimal w.r.t. this property. Note that for such an operator the obtained depth function is generally not quasiconcave anymore. Another possibility would be the operator $\widetilde{extr}(A) := extr(A) \cup A\backslash(\Phi \circ \Psi)(extr(A))$. This operator would lead to a completely quasiconcave depth function. Note further that this operator is generally not minimal which means that the number of depth layers is usually lower compared to a minimal operator.

*we define the **enclosing depth** $\mathcal{E}^\mu$ w.r.t. mode μ by $\mathcal{E}^\mu\left((\Phi \circ \Psi)(\{\mu\})\right) = 1$ and*

$$\mathcal{E}^\mu\left(encl\left((\mathcal{E}^\mu)^{-1}\left(\left[\frac{i}{|G|}, 1\right]\right)\right)\right) = \frac{i-1}{|G|}; \quad i = |G|, |G|-1, \ldots$$

Here, encl denotes an operator which we would like to call an enclosing operator. Concretely, we have in mind an operator $encl : H \longrightarrow 2^G$ with $H \subseteq 2^G$ that for all $A \in H$ satisfies the three properties i): $encl(A) \cap A = \emptyset$, ii): $encl(A) \longrightarrow A$ and iii): $(\Phi \circ \Psi)(encl(A))$ is minimal w.r.t. properties i) and ii).

Now we discuss, how one can specify with the above depth functions a unimodal distribution of orders with a given mode and one scale parameter. The simplest distribution, which can be always defined in a finite setting, is the uniform distribution. To specify a distribution that is in some certain sense distributed around a given mode, one simple approach would be to first generate every partial order exactly one time (this would correspond to a uniform distribution) and then to simply add a big amount of partial orders that are identical to the mode. Then, based on the corresponding data depth that is obtained for this data set, one can define a distribution according to Eq. (1). (Note that generally the obtained distribution is different from a mixture of a uniform distribution and a distribution that equals the mode with probability one.) However, for Tukey's depth, due to reasons of symmetry one can show that the obtained distribution of orders would assign the mode one probability p and every other order that differs from the mode exactly one of two probability values q or r. More concretely, the localized Tukey's depth could then be written as

$$\mathcal{T}^\mu(g) = 1 - \max\left\{ \max_{(p,q)\in\mu\backslash g} \alpha_{p,q}, \max_{(p,q)\in g\backslash\mu} \beta_{p,q} \right\} \text{ with } \alpha_{p,q}, \beta_{p,q} \in [0,1],$$

[4]where actually $\alpha_{p,q}$ and $\beta_{p,q}$ do not depend on p or q. This seems to be somehow unsatisfying. Of course, one can use Tukey's depth based on another (empirically or analytically) given distribution, but then, in the first place one is back at the "... major outstanding problem in ranking theory ..." and has to specify a "... suitable population of ranks in non-null cases..." ([23]). Alternatively, one can replace $\alpha_{p,q}$ and $\beta_{p,q}$ by other weights that depend on the pairs (p, q), actually fortunately without losing the quasiconcavity. For this weighted Tukey-type depth function one would have to specify only n^2 values instead of $2^{n^2/4}$ or more values (cf., [24]), which would be needed for a completely nonparametric approach. Because this can still be very demanding, we will later use an analysis of the enclosing depth to get a rough guidance for specifying the weights. For the peeling depth there seems to be not so much ties compared to Tukey's depth. However, it seems a little bit counter-intuitive to specify a distribution of orders that are distributed around a mode by not locally looking at a neighbourhood of the mode but instead by globally ordering the data points from outwards to inwards. Compared to other applications of data depth where one does not know

---

[4] This also shows an asymmetry between pairs that are in relation and pairs that are not in relation.

the location beforehand but where the problem is actually the estimation of the mode of the distribution, here we are in the comfortable situation that we can simply specify the mode. Therefore, in the sequel, we will focus on the enclosing depth (applied for the case $G := \mathscr{P}$). Also here, because of the high amount of symmetries there are many ways of defining an enclosing operator and corresponding depth layers. One way out of this would be to compute in a first step for every partial order the expected depth value under a stochastic choice of the layer that is built in every step. This is of course possible and also a simulation from such a model can be exactly done. However, the obtained depth function is not completely quasiconcave. Therefore, one can in a second step build the closure of every depth contour to obtain a completely quasiconcave depth function. For this, one has to analyze in detail, how the expected depth values of the first step exactly look like, which seems to be a very difficult problem. Therefore, we only analyze the situation for total orders and a totally ordered mode μ under a conceptual scaling of the partial orders that uses only $M_\leq$. With this analysis we are able to roughly oversee the situation for the enclosing depth and we will use the results to guide the specification of the weights within the modified Tukey's depth (see above) under a conceptual scaling that uses both $M_\leq$ and $M_\nleq$: Let $(p, q) \in \mu$. Define $\Delta_\mu(p, q)$ simply as the "distance" between p and q w.r.t. the mode μ measured by the number of pairs between p and q w.r.t. the covering relation of μ. Furthermore, for $x \in \mathscr{P}$ define

$$s_\mu(x) := \max_{(p,q) \in \mu \backslash x} \Delta_\mu(p, q).$$

Then one can show that total orders x with a higher $s_\mu(x)$ have a lower depth value w.r.t. the enclosing depth $\mathscr{E}^\mu$. Thus, for a weighted version of Tukey's depth function one can weight pairs (p, q) with a higher $\Delta_\mu(p, q)$ correspondingly with a higher weight, e.g., via $\alpha_{p,q} \propto \Delta_\mu(p, q)$. (For pairs with the same value it would be natural to choose the same weight.) Now, the problem is to specify the corresponding weights for pairs $(p, q) \in x \backslash \mu$. Because we would like to think from the direction of the mode μ and not from the perspective of x, we do not want to simply change the roles of μ and x. The problem here is that it seems to be somehow difficult to order pairs (p, q) w.r.t. the mode μ that are not in relation w.r.t. μ. However, there are some possibilities to rank such pairs. The following definition is somehow inspired by the work in [15]: For $(p, q) \notin \mu$ one could define[5]

$$\Delta_\mu(p, q) := |\{r \in \mathcal{X} \mid p \wedge_\mu q \leq_\mu r \leq_\mu p \vee_\mu q\}| - 1.$$

This definition extends the original definition of Δ_μ and it can be used to specify the weights (e.g., via $\alpha_{p,q} \propto \Delta_\mu(p, q); \beta_{p,q} \propto \Delta_\mu(p, q)$) for the modified Tukey's depth that uses the whole attribute set M for the conceptual scaling.

---

[5] If the considered partial order does not build a complete lattice one could simply compute the Dedekind-MacNeille completion beforehand.

6   Simulation

By representing the set of partial orders as defined in Sect. 4 and applying one of
the data depth functions of Sect. 5, we obtain a statistical model on the set of partial orders on $\mathcal{X}$ by Eq. (1). In this section we derive an algorithm to sample from
such a statistical model. The algorithms is based on the acceptance-rejection
method and the idea of the algorithm is based on [16]. For a small number of
elements, we can directly compute all reflexive, transitive, and anti-symmetric
orders. Thus, we can easily draw a sample from one of the above distributions.
Since the runtime of the computation of all partial orders grows with the number of elements faster or equal to $2^{n^2/4}$ (see [24]), the direct computation is not
feasible for larger n. Therefore, we provide an algorithm based on the following
structure: First, we systematically draw a partial order and calculate the number
of possible paths to obtain this partial order. Finally, we compute the acceptance
probability such that we sample with probability of interest f. The algorithm
uses that each partial order is a subset of at least one linear order. A linear order
has $\frac{1}{2}(n-1)n$ many pairs of the form (x_i, x_j) with $i \neq j$ and, in particular, if
we randomly delete some of these pairs, then, by computing the transitive hull,
we obtain a partial order. To obtain step 1, we first take a uniform sample of
a linear order and then randomly delete some pairs by a uniform variable on
all subsets. Note that drawing a linear order is only a tool to obtain a partial
order, and due to the definition of the acceptance probability, does not affect
the probability of selecting a partial order. By computing all linear extensions,
we can compute the probability that this partial order was sampled. Finally, we
adjust the acceptance probability so that the sample ends up consisting of the
probability function f we are interested in. More precisely, the probability that
a given order g is computed in step 1 is:

$$P_{algo\_select}(g) = |lext(g)| \cdot 2^{|g|-|reduc(g)|} \cdot \left(n! 2^{n(n-1)/2} \right)^{-1}$$

where $reduc(g)$ is the transitive reduction[6] of g and $n! \cdot 2^{n(n-1)/2}$ is the number
of all paths to obtain a partial order by the procedure above. Since the number
of pairs of each linear order is the same, the probability that the partial order g
is sampled is identical for each linear order from the linear extension of g. Let
f be the probability function from which we want to draw a sample, then the
acceptance function is given by

$$acc(g) = f(g) \cdot \left(P_{algo\_select}(g) \cdot n! 2^{n(n-1)/2} \right)^{-1}. \tag{3}$$

Lemma 1. *Algorithm 1 samples a partial order with probability function f on
all partial orders of G.*

---

[6] The transitive closure of a relation is the smallest transitive relation containing it,
and the transitive reduction is a minimal relation having the same transitive closure.

Algorithm 1: Sampling a partial order

Input: n: number of items;
f: probability function with the set of all partial orders as domain;
Result: partial order sampled according to the probability given by f.
repeat

    # sampling the order

    LIN_ORDER ← sample uniformly a linear order;

    DEL_PAIRS ← sample uniformly a subset of $\{1, \ldots, (n-1)n/2\}$;

    PARTIAL_ORDER ← uniformly delete DEL_PAIRS many pairs and compute the transitive closure;

    # compute the acceptance probability (thereby we have to compute the transitive reduction)

    ACCEPT_PROB ← computation of (3);

until *random]0,1]* ≤ ACCEPT_PROB;

The proof is analogous to the one given in [16]. Note we could use also a modified version of the acceptance function: $\tilde{acc} = c \cdot acc$ with constant $c \geq \max_g f(g)/P_{algo\_select}(g)$. This modified version must assure that for all partial orders g, $f(g) \leq c \cdot P_{algo\_select}(g)$ is true. Unfortunately, the computation of all linear extensions is # P-complete (see [7]). Note that for some subsets of all partial orders the running time of the computation of the linear extension is smaller, i.e., if we consider only the set of trees (see [3]). Additionally, to improve the runtime of the algorithm we generally could also use an approximation for the number of all linear extensions $|lext(g)|$, for which e.g. [8] gives approximation approaches.

7   Conclusion

In this paper, we developed statistical models for partial orders based on data depth and FCA. Therefore we embedded the terms data depth and unimodality into FCA. We think that with this approach, opposed to statistical models based on distances, we are in fact able to appropriately incorporate the incomparability of two items as well as the notion of unimodality of a statistical model for partial orders. In particular, we think that a notion of unimodality based on concepts of lattice theory is more appropriate compared to notions based on metrics or based on the embedding of partial orders into a linear space. In particular, the simulation approach allows to perform simulation studies in order to analyze the statistical behavior of certain statistical procedures for partial order data. In addition, this enables the construction of parametric bootstrap procedures to quantify the uncertainty inherent in any statistical procedure for partial order data. What is left open for further research is the question how to exactly specify the decay function and the weights within the approach that uses Tukey's depth. A further analysis of the newly developed enclosing depth, especially w.r.t. the question whether this depth function can be also applied if one does not know the mode beforehand, is also of high interest. Additionally, the application of our approach to concrete data situations is another important line of further research.

Acknowledgments. Hannah Blocher and Georg Schollmeyer gratefully acknowledge the financial and general support of the LMU Mentoring program. Further, we thank the three anonymous reviewers for their constructive and insightful comments that helped to improve the paper.

References

1. Alvo, M., Cabilio, P.: Rank correlation methods for missing data. Canad. J. Stat. **23**(4), 345–358 (1995)
2. Armstrong, W.: Dependency structures of data base relationships. Int. Fed. Inf. Process. Congress **74**, 580–583 (1974)
3. Atkinson, M.: On computing the number of linear extensions of a tree. Order **7**, 23–25 (1990)
4. Barnett, V.: The ordering of multivariate data (with discussion) J. Roy. Stat. Soc. Ser. A **139**(3), 318–352 (1976)
5. Boutilier, C., Rosenschein, J.: Incomplete information and communication in voting. In: Moulin, H., Brandt, F., Conitzer, V., Endriss, U., Lang, Procaccia, A. (eds.) Handbook of Computational Social Choice, pp. 223–258, Cambridge University Press (2016)
6. Brandenburg, F.J., Gleißner, A., Hofmeier, A.: Comparing and aggregating partial orders with kendall tau distances. In: Rahman, M.S., Nakano, S. (eds.) WALCOM 2012. LNCS, vol. 7157, pp. 88–99. Springer, Heidelberg (2012). https://doi.org/10.1007/978-3-642-28076-4_11
7. Brightwell, G., Winkler, P.: Counting linear extensions is # P-complete. In: Proceedings 23rd ACM Symposium on the Theory of Computing, pp. 175–181 (1991)
8. Bubley, R., Dyer, M.: Faster random generation of linear extensions. Discret. Math. **201**(1–3), 81–88 (1999)
9. Collins-Thompson, K., Callan, J.: A language modeling approach to predicting reading difficulty. In: Proceedings of the Human Language Technology Conference of the North American Chapter of the Association for Computational Linguistics: HLT-NAACL, pp. 193–200 (2004)
10. Comim, F.: Beyond the HDI? Assessing alternative measures of human development from a capability perspective. In: Background paper of the Human Development Report. UNDP Human Development Report (2016)
11. Critchlow, D.: Metric methods for analyzing partially ranked data. Lecture Notes in Statistics, 34. Springer (1985)
12. Dittrich, R., Hatzinger, R., Katzenbeisser, W.: Modelling the effect of subject-specific covariates in paired comparison studies with an application to university rankings. J. R. Stat. Soc. Ser. C **47**(4), 511–525 (1998)
13. Fahandar, M., Hüllermeier, E., Couso, I.: Statistical inference for incomplete ranking data: The case of rank-dependent coarsening. In: Proceedings of the 34th International Conference on Machine Learning, vol. 70, pp. 1078–1087 (2017)
14. Fligner, M., Verducci, J.: Distance based ranking models. J. Roy. Stat. Soc. B **48**(3), 359–369 (1986)

15. Gäbel-Hökenschnieder, T., Schmidt, S.: Generalized metrics and their relevance for FCA and closure operators. Concept Lattices and their Applications, pp. 175–186 (2016)
16. Ganter, B.: Random extents and random closure systems. Concept Lattices and their Applications, pp. 309–318 (2011)
17. Ganter, B., Wille, R.: Formal Concept Analysis: Mathematical Foundations. Springer Science & Business Media (2012)
18. Jansen, C., Blocher, H., Augustin, T., Schollmeyer, G.: Information efficient learning of complexly structured preferences: elicitation procedures and their application to decision making under uncertainty. Int. J. Approximate Reasoning **144**, 69–91 (2022)
19. Jansen, C., Schollmeyer, G., Augustin, T.: Concepts for decision making under severe uncertainty with partial ordinal and partial cardinal preferences. Int. J. Approximate Reasoning **98**, 112–131 (2018)
20. Jansen, C., Schollmeyer, G., Augustin, T.: A probabilistic evaluation framework for preference aggregation reflecting group homogeneity. Math. Soc. Sci. **96**, 49–62 (2018)
21. Jena, S., Lodi, A., Palmer, H., Sole, C.: A partially ranked choice model for large-scale data-driven assortment optimization. Informs J. Optim. **2**(4), 297–319 (2020)
22. Kendall, M.: A new measure of rank correlation. Biometrika **30**(1/2), 81–93 (1938)
23. Kendall, M.: Discussion on symposium on ranking methods. J. Roy. Stat. Soc. B **12**, 153–162 (1950)
24. Kleitman, D., Rothschild, B.: The number of finite topologies. Proc. Am. Math. Soc. **25**(2), 276–282 (1970)
25. Lebanon, G., Mao, Y.: Non-parametric modeling of partially ranked data. J. Mach. Learn. Res. **9**(10), 2401–2429 (2008)
26. Liu, R., Parelius, J., Singh, K.: Multivariate analysis by data depth: descriptive statistics, graphics and inference (with discussion and a rejoinder by Liu and Singh). Ann. Stat. **27**(3), 783–858 (1999)
27. Mallows, C.: Non-null ranking models. I. Biometrika **44**(1/2), 114–130 (1957)
28. Mangaraj, B., Aparajita, U.: Constructing a generalized model of the human development index. Socio-Econ. Plann. Sci. **70**, 100778 (2020)
29. Mosler, K.: Depth statistics. In: Becker, C., Fried, R., Kuhnt, S. (eds.) Robustness and Complex Data Structures, pp. 17–34. Springer, Heidelberg (2013)
30. Nakamura, K., Yano, K., Fumiyasu, K.: Learning partially ranked data based on graph regularization. arXiv:1902.10963 (2019)
31. Schollmeyer, G.: Lower quantiles for complete lattices. Technical Report 207. Department of Statistics. LMU Munich (2017)
32. Schollmeyer, G.: Application of lower quantiles for complete lattices to ranking data: Analyzing outlyingness of preference orderings. Technical Report 208. Department of Statistics. LMU Munich (2017)
33. Stewart, R.: Weak pseudo-rationalizability. Math. Soc. Sci. **104**, 23–28 (2020)

BEUD: Bifold-Encoder Uni-Decoder Based Network for Anomaly Detection

Mohith Rajesh[✉][iD], Chinmay Kulkarni[iD], and S. S. Shylaja

PES University, Bengaluru, India

mohithraj9301@gmail.com, camkoolkarni@gmail.com, shylaja.sharath@pes.edu

Abstract. Anomaly detection is a very critical and significant data analysis mission given the raft of cyber-attacks these days. Used to identify thought-provoking and emerging patterns, predispositions, and irregularities in the data, it is an important tool to perceive abnormalities in many different domains, including security, finance, power automation, health, computer network intrusion detection, etc. Deep learning-based AutoEncoders have shown great potential in identifying anomalies. However, state-of-the-art anomaly scores are still based on reconstruction errors, which do not take advantage of the available anomalous data samples during the training phase. Towards this direction, we demonstrate a novel extension to the AutoEncoder that not only maintains the AutoEncoder's ability to discover non-linear features of non-anomalies but also uses the existing anomalous samples to assist in learning the features of the data better. Since the model architecture is designed to have two encoders and one decoder network, we name our model as Bifold-Encoder Uni-Decoder (BEUD) network. In this paper, we discuss two different ways of using the BEUD model to predict the anomalies in the data. BEUD is conceptually analogous to AutoEncoder but empirically more powerful. The experimental results of this architecture demonstrated a fairly good performance compared to the standard AutoEncoder architecture evaluated for the anomaly detection task.

Keywords: Anomaly detection · Neural network · AutoEncoder · Contrastive learning · Imbalanced dataset · Representation learning

1 Introduction

Bearing in mind the swift spike in E-business and digital payment systems, there is an uptick in frauds pertaining to financial and banking transactions allied to credit cards. Analysts and researchers have been vigorously trying to discover solutions to overpower them and one of the solutions is to use data mining approaches. However, the collected credit card data can be a reasonable challenge for researchers because of the data characteristics that contain an imbalanced class distribution.

We need to recognise that the magnitude of the significant classes in the data, which are mostly deceitful transactions, is usually marginal. This is the

© Springer Nature Switzerland AG 2022
D. Ciucci et al. (Eds.): IPMU 2022, CCIS 1602, pp. 31–43, 2022.
https://doi.org/10.1007/978-3-031-08974-9_3

main challenge in the problem of anomaly detection. Generally, it is easy for learning algorithms to unearth their regularities if they have adequate records. But when their number is insignificant, as is typically the case for anomalies, unearthing their regularities and generalising their definite conclusive regions using learning algorithms becomes difficult. Many learning algorithms are envisioned to maximise accuracy, which leans more towards majority classes and against minorities. Generally, for classification, models like Deep Neural Networks demand that all the classes have an adequate number of data samples. But it is strenuous to find data samples for anomalies.

In this paper, we discuss the AutoEncoder methodology for the anomaly detection problem. An AutoEncoder is used to perceive fraudulent transactions (anomalies) based on reconstruction error; only data with normal instances is used to train the AutoEncoder, which is why this method stands apart from other deep learning methods. We present an analysis showing that exposing the existing anomalous samples to the network is crucial. We also present the **BEUD** architecture, a model architecture that instead of relying entirely on an encoder representation of non-anomalies to draw global dependencies between input and output flags (anomaly or non-anomaly), allows for significantly better encoder representations by making use of existing anomaly examples.

2   Related Work and Background

Anomaly detection is a significant endeavour that has been extensively studied within various research areas and application domains. A long line of literature has thus been proposed, including reconstruction-based [9–14], density-based [1–8], one-class classifier [15,16], and self-supervised [17–19] approaches. Overall, a majority of recent literature is concerned with (a) modeling the representation to better encode normality [18,20], and (b) defining a new detection score [16,19]. Recently, significant progress has been made in the self-supervised domain, with the use of contrastive learning [22].

Contrastive learning extracts a strong inductive bias from multiple (similar) views of a sample by letting them attract each other, yet repelling them to other samples. It is a machine learning technique used to learn the general features of a dataset by teaching the model which data points are analogous or different.

We first found that the existing AutoEncoder method is already reasonably effective for detecting anomalous samples with a proper detection score. We further observe that one can improve its performance by utilising available anomalous samples while training. In particular, the existing AutoEncoder method predicts anomalies by learning the reconstructions of the non-anomalous samples and using the appropriate detection score; whereas the BEUD model additionally pushes the encoding of anomalies and normal data apart while pulling the encoding of normal data closer. The model now learns a new task of discriminating between anomalies and non-anomalies, in addition to the original task of learning the reconstructions of non-anomalous data.

3   Dataset

The AutoEncoder methodology demands non-anomalous data for training and predicts whether the given data is anomalous or not. On the other hand, our proposed BEUD model requires both anomalies and non-anomalies for training. So we require data which represents both the anomalous and non-anomalous proceedings.

The dataset (https://www.kaggle.com/mlg-ulb/creditcardfraud) prepared by Worldline and the Machine Learning Group (http://mlg.ulb.ac.be) of ULB (Université Libre de Bruxelles) and provided by Kaggle perfectly matched our requirements. The dataset contains transactions made by credit cards in September 2013 by European cardholders. This dataset contains transactions of two days, where there are 492 frauds out of 284,807 transactions. It is highly imbalanced, the positive class (frauds) account for 0.172% of all transactions.

The attributes are amount of money involved in transactions and time with other 28 columns labelled as "V1", "V2", "V3", till "V28". The results of the transactions (whether anomaly or non-anomaly) are given in the "Class" column, where 1 represents anomalous transactions and 0 represents normal transactions. This dataset contains only numeric input variables which are the result of a PCA (Principal Component Analysis) transformation.

Additionally, we perform a Min/Max scaling before training, which is a normalization method that consists of re-scaling the range of features and the data is divided into 80% Train set and 20% Test set.

4   BEUD: Bifold-Encoder Uni-Decoder Network

AutoEncoders are trained by unsupervised learning to learn reconstructions that are close to their original input; we leverage neural networks for the task of **representation learning**. Explicitly, we design a neural network architecture such that a bottleneck is enforced in the network which compels a **compressed knowledge representation** of the original input.

The encoder and decoder with a single hidden layer can be shown as Eq. 1 and Eq. 2, respectively. W and b are the weight and bias of the network respectively. T is the nonlinear transformation function, and the subscript xe represents that x is the input with which W is multiplied, and e is the output of the equation.

$$e = T(W_{xe}x + b_{xe}) \tag{1}$$

The encoder in Eq. 1 maps an input vector x to a hidden representation performing an affine mapping followed by a non-linear transformation.

$$d = T(W_{ex}e + b_{ex}) \tag{2}$$

The decoder in Eq. 2 maps the hidden representations back to the original input space as a Reconstruction.

$$e_t = |x - d| \tag{3}$$

The deviation of Reconstruction from the original input vector is called the Reconstruction error represented as e_t in Eq. 3.

An AutoEncoder learns to minimize the mean of Reconstruction error given by Eq. 4.

$$MAE = \frac{1}{n}\sum_{t=1}^{n} |e_t| \tag{4}$$

The data representing normal instances is used to train the AutoEncoder; When testing, the AutoEncoder will reconstruct normal data precisely, while failing to do so with anomalous data, as it has never run into the latter before. Reconstruction error is used as an anomaly score. If it is above the threshold, the data is flagged as anomalous.

The AutoEncoder method illustrated above can be effectively applied for anomaly detection in a dataset with few or no anomalous data samples accessible for training, as it requires only normal data for training. The anomalous samples existing in the training dataset are not taken advantage of while training the AutoEncoder. We present the BEUD model architecture, which is designed analogous to the AutoEncoder with an additional encoder to utilise anomalous data samples accessible for training.

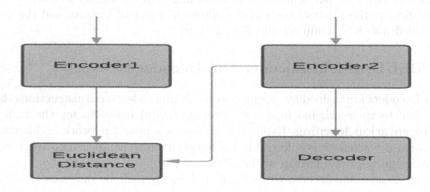

Fig. 1. BEUD model

BEUD model architecture is depicted in Fig. 1. The encoder and decoder on the right represent the AutoEncoder. We will be referring to this encoder as encoder2, whereas the encoder on the left is an additional component to the AutoEncoder in the BEUD architecture. We will be referring to this encoder as encoder1. Encoders are named distinctly just to ease the explanation of different phases of training the BEUD model; otherwise, they are homogeneous, i.e., they share weights.

The BEUD model goes through two stages of training. The first phase is similar to the AutoEncoder approach. The encoder2 and decoder have been trained to minimise the Mean Average Error (MAE) of the normal data reconstruction.

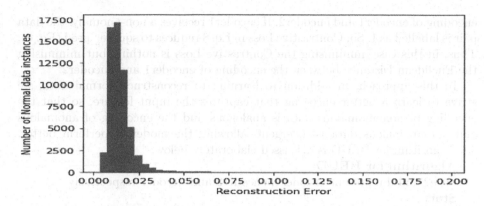

Fig. 2. Number of non-anomalous test examples v/s reconstruction error

Phase-1 has a single objective, so it may be completed with greater accuracy. The second phase of BEUD model training involves simultaneously minimising MAE and the Contrastive loss as indicated in Eq. 8. In essence, weights learned in phase-1 are readjusted in phase-2 to maintain and reduce the MAE to some extent while also achieving the additional goal of effectively reducing the Contrastive loss given by Eq. 8. The training phase-2 is made easier because the weights were taught to minimise MAE in phase-1.

The encoder2 receives and encodes only non-anomalous data during the training phase-1, thus the decoder learns to decode only non-anomalous data encoding. Even in the second phase of training, encoder2 only receives and encodes non-anomalous data, while encoder1 can receive either anomalous or non-anomalous input samples. If encoder1 gets a non-anomaly, encoder1's output encoding and encoder2's output encoding must be analogous. If encoder1 gets an anomaly, the encoding of encoder1 and encoder2 should be divergent. This is accomplished by minimising the Contrastive Loss, as shown in Eq. 8.

The Euclidean Distance Node in Fig. 1 calculates the Euclidean Distance between the encoding of encoder1 and encoder2. Encoder1 is used to push the encoding of anomalies and non-anomalies distinctively and bring closer the encoding of two non-anomalies, consequently utilising the existing data of anomalies. So, pairs of data instances is passed to BEUD: if encoder1 (the first element of the pair) receives an anomaly, the data pair is labelled as 0, and if encoder1 receives a non-anomaly, the data pair is labelled as 1. The second element of the pair is always a non-anomaly, which is passed to encoder2. Thus, these pairs of data, along with labels, are used in training phase-2. Equation 6 calculates the square of Euclidean Distance between the encoding of encoder1 and encoder2. Margin in Eq. 7 is a hyper-parameter usually set to 1.

If encoder1 (the first element of the data pair) receives an anomaly since the label for such a pair is 0, Contrastive Loss in Eq. 8 reduces to margin_square (Eq. 7). Thus, in the event that encoder1 receives an anomaly, minimising the Contrastive Loss is nothing but maximising the Euclidean Distance between the encoding of encoder1 and encoder2. If encoder1 receives a non-anomaly, the data pair is labelled as 1. So, Contrastive Loss in Eq. 8 reduces to square_pred (Eq. 6). Thus, in this case, minimising the Contrastive Loss is nothing but minimising the Euclidean Distance between the encoding of encoder1 and encoder2.

In this approach, in addition to learning to reconstruct normal data, we strive to learn a better encoding that captures the input feature, so that the encoding of non-anomalous data is analogous and the encoding of anomalous and non-anomalous data is divergent, allowing the model to perform better. The algorithm for BEUD is discussed elaborately below.

Algorithm for BEUD:

e1, e2 and d represent encoder1, encoder2 and decoder respectively

Start

$\phi \rightarrow$ parameters of e1 and e2

$\theta \rightarrow$ parameters of d

$\alpha \rightarrow$ threshold reconstruction error

$x\_train, x\_test \rightarrow$ input features of train data and test data.

$y\_train \rightarrow$ labels(0 for non-anomaly 1 for anomaly) of train data

$N \rightarrow$ size of test data

$\phi, \theta \leftarrow$ train the network using the second element in the pair of input features, minimising MAE for ϕ(encoder2) and θ for n1 number of epochs.

$\phi, \theta \leftarrow$ train the network using the pair of input features along with their appropriate labels, minimising Contrastive loss for ϕ(encoder1 and encoder2) and MAE for ϕ(encoder2) and θ for n2 number of epochs.

if method1: //Reconstruction error as anomaly score

　　Testing:

　　　　for i = 1 to N:

　　　　　　//error in reconstructing the original input by encoder2
　　　　　　and decoder sub-networks of BEUD

　　　　　　Reconstruction error(i)=

$$\|x\_test(i) - d_\theta(e2_\phi(x\_test(i)))\| \tag{5}$$

　　　　　　if Reconstruction error(i) $> \alpha$:

　　　　　　　　x_test(i) is an Anomaly

　　　　　　else:

　　　　　　　　x_test(i) is not an Anomaly

　　if method2: Training classifier for the output encoding returned by the encoder of BEUD

　　　　$newTrainX = e2_\phi(x\_train)$

　　　　new Train Y = y_train

　　　　classifier.fit(new Train X, new Train Y)

　　　　Testing:

　　　　　　for i = 1 to N:

　　　　　　　　if classifier.predict(x_test(i)) > 0.5:

　　　　　　　　　　x_test(i) is an Anomaly

else:

  x_test(i) is not an Anomaly

Stop

Contrastive Loss:

$\hat{y}$ is the predicted Euclidean Distance between the encoding of encoder1 and encoder2 and y is the label for the data pair.

$$square\_pred = (\hat{y})^2 \qquad (6)$$

$$margin\_square = (max(margin - \hat{y}, 0))^2 \qquad (7)$$

$$Loss = mean(y * square\_pred + (1 - y) * margin\_square) \qquad (8)$$

In this paper we discuss 2 different ways of using BEUD model to predict the anomalies in the data.

4.1 Reconstruction Error as the Anomaly Score

The encoder2 and decoder of the BEUD model were initially trained to minimise the Mean Average Error (MAE) of the reconstructions of normal data, and later, the BEUD model was additionally trained to minimise the Contrastive loss along with MAE, reconstructions of normal data using encoder2 and decoder would be precise in comparison to the reconstructions of anomalous data. Thus, how well the data is constructed by the encoder2 and decoder could be used as the score to classify the data as an anomaly.

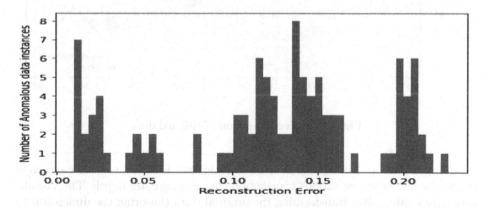

Fig. 3. Number of anomalous test examples v/s reconstruction error

The BEUD model has three layers in both the encoder and decoder sub-networks and was trained on 76,600 data pairs, with the first element being either an anomaly or non-anomaly, but the second element being always non-anomaly. The data pairs are generated from the training dataset, and each pair is labelled appropriately, as stated previously. The labels 0 and 1 have an equal class distribution, which implies that there are 38,300 data pairs with the labels 0 and 1. However, since the availability of anomalies throughout the training data is minimal, 383 anomalies are duplicated to generate 38,300 data pairs for label 0, but the data pairs are distinct since the second element of each pair (normal data) is not duplicated. In phase 1, BEUD is exclusively trained on the second element in 76,600 data pairs to minimize MAE. It is done for 100 epochs, and then 76,600 data pairs with labels are utilised to train(phase-2) the BEUD for 500 epochs to minimise both MAE and Contrastive loss.

The threshold for data reconstruction error is determined; if the error in data reconstruction exceeds the threshold, the data sample is flagged as anomalous. Figure 2 shows that reconstruction errors for most normal examples are less than 0.05, while Fig. 3 shows that reconstruction errors for many anomalous occurrences are greater than 0.1. A recall value of 75.23% and a precision of 75.93% were observed when the threshold reconstruction error was set to 0.1.

4.2   Training Classifier for the Output Encoding Returned by the Encoder of BEUD

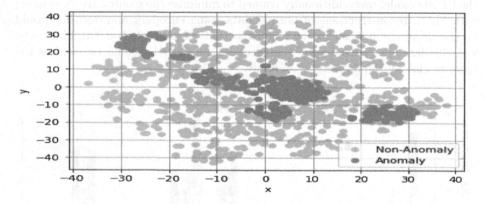

Fig. 4. Reduced dimension of original data

Using T-SNE (t-Distributed Stochastic Neighbor Embedding), we can investigate the nature of our various transaction classes in greater depth. The visualisation generated after transforming the original data (lowering the dimension to 2) is displayed in Fig. 4. There is no satisfiable decision boundary that separates an anomaly from a non-anomaly. In this section, we'll look at how the BEUD model may be used to develop improved representations of the input so that anomalous and non-anomalous data can be distinguished.

The BEUD model is built using four layers in the encoder and three layers in the decoder sub-network, and it is trained on 2,298 data pairs (the first element being an anomaly or non-anomaly, but the second being non-anomaly always) to get improved representations of the data. As mentioned in the preceding

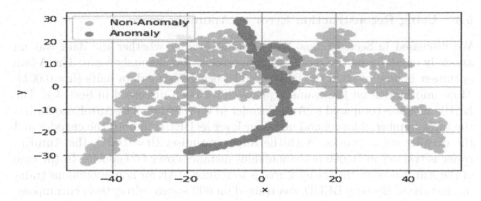

Fig. 5. Reduced dimension of the encoding returned by encoder of BEUD

section, data pairs are constructed from the train dataset, with each pair being suitably labelled and each label having an equal class distribution (4.1). To minimise MAE, the BEUD model is first trained just on the second element in the 2,298 data pairs. This initial phase (phase-1) lasts 10 epochs, after which 2,298 data pairs with labels are utilised to train the BEUD for 30 epochs (phase-2) in order to reduce MAE and Contrastive loss. We can tell the difference between anomalous and non-anomalous cases by their inner nature. BEUD's encoder receives the original data, output of encoder is then translated into a lower dimension via the T-SNE transformation. We can see in Fig. 5 that both classes are now visible, and the anomalies exist in a specific region, i.e., they are not distributed across the normal data. It's worth noting that the T-SNE transformation is only utilised to obtain data visualisation insights; it's not employed in either the BEUD training or the prediction stage.

After the BEUD has been fully trained, we want to find the latent space representation of the input that the network has learned (encoder). The hidden representations of both non-anomalous and anomalous classes are then obtained. With the encoding of 50,000 non-anomalies and the encoding of 383 anomalies returned by the encoder, a new training dataset with labels is produced where label 1 represents the encoding of anomalous data and label 0 represents encoding of non-anomalous data. This information is utilised to train Classifiers like SVM.

5   Experimental Results

In this section, we cover the outcomes of the BEUD and AutoEncoder anomaly detection methods, as well as the importance of leveraging existing anomalous data with additional encoder. We examine the performance of AutoEncoder and BEUD for anomaly detection on the credit card dataset.

5.1   Using Reconstruction Error as Anomaly Score

We discussed in Sect. 4.1 how BEUD could predict whether the data was an anomaly or not using the reconstruction error as the anomaly score. The Adam optimiser is used to train the BEUD model using 76,600 data pairs (lr = 0.001). More information on the training phases have been discussed in Sect. 4.1. The BEUD model is compared to AutoEncoder in this section. The AutoEncoder has the same number of layers and units per layer as BEUD in both the encoder and the decoder sub-networks. With the Adam optimiser (lr = 0.001), the AutoEncoder is trained on 76,600 non-anomalous instances over 600 epochs. In the case of the AutoEncoder, the training goal is to reduce MAE for non-anomalous training instances. Because BEUD was trained on 600 epochs altogether, encompassing both phases, the AutoEncoder was also trained on 600 epochs to ensure a fair comparison. On the CreditCard test dataset, Fig. 6 shows the plot of precision vs recall for BEUD and AutoEncoder for various threshold reconstruction errors ranging from 0.01 to 0.2 with an increment of 0.0001. The BEUD model performed substantially better than the AutoEncoder. The precision of the BEUD model is much higher than the AutoEncoder for the same recall.

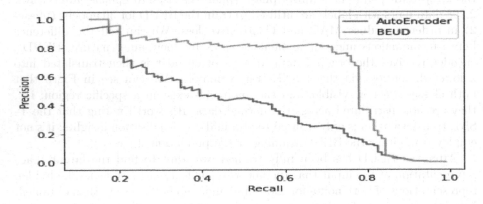

Fig. 6. Precision v/s recall for BEUD and AutoEncoder on the CreditCard test dataset

5.2   Training Classifier for the Output Encoding Returned by the Encoder of BEUD

In this technique, the BEUD model is trained on 2,298 data pairs using the Adam optimiser (lr = 0.001). More information on the training phases can be found in Sect. 4.2. In this section, we compare the BEUD model against the AutoEncoder (with same number of layers and units per layer as BEUD in both the encoder and the decoder sub-networks), which was trained with Adam optimiser on 2,298 non-anomalous examples over 40 epochs (lr = 0.001). Two distinct training datasets are constructed, each with labels(label 1 represents the encoding of anomalous data and label 0 represents encoding of non-anomalous data): one is made up of the encoding returned by the BEUD encoder, called new dataset1, and the other is built up of the encoding returned by the AutoEncoder encoder, called new dataset2. Both datasets encode 50,000 non-anomalies and 383 anomalies, and they are used to train two separate SVM classifiers, SVM1 on new dataset1 and SVM2 on new dataset2. On test data, SVM1, which uses the encoding of test data supplied by the encoder of BEUD, performs significantly better than SVM2, which uses the encoding of test data returned by the encoder of AutoEncoder (Table 1). The Confusion matrix for the predictions of SVM1 using the input as the encoding of test data supplied by the encoder of BEUD is shown in Fig. 7.

Table 1. Performance comparison of AutoEncoder and BEUD.

	Recall	Precision
AutoEncoder	70.64%	85.55%
BEUD	80.73%	88.88%

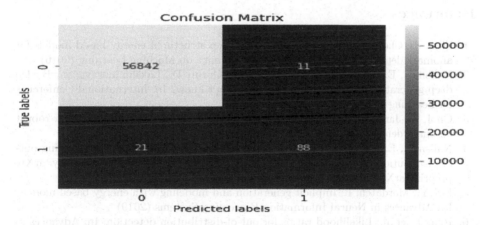

Fig. 7. Confusion matrix for the predictions of SVM using the input as the encoding of test data supplied by the encoder of BEUD

6 Discussions and Conclusion

In terms of anomaly detection on a credit card dataset, the BEUD model outperformed the AutoEncoder model, as shown in the preceding section; the BEUD model's results are fairly better compared to the AutoEncoder-based method discussed in "Applications of Machine Learning in Fin-tech Credit Card Fraud Detection" [21].

By altering the threshold for reconstruction error of BEUD, method 1 (Sect. 4.1) can be used to control false-positives and false-negatives. As a result, precision and recall can be managed. BEUD is adaptable to a variety of applications. When it comes to bank transactions, recall is crucial. Declaring non-spam material as spam in spam mail detection could have disastrous consequences, as critical mail would be missed. Precision is of the utmost significance in this application.

Despite the fact that the BEUD model performed better than the AutoEncoder, the performance of the BEUD model necessitates being compared with other anomaly detection methods. The performance of BEUD needs to be evaluated on other anomaly scores and on image datasets. This model poses fresh prospects for research since the results of BEUD are quite good and enhancing the performance marginally will assist many anomaly detection applications.

References

1. Zhai, S., Cheng, Y., Lu, W., Zhang, Z.: Deep structured energy based models for anomaly detection. In: International Conference on Machine Learning (2016)
2. Nalisnick, E., Matsukawa, A., Teh, Y.W., Gorur, D., Lakshminarayanan, B.: Do deep generative models know what they don't know? In: International Conference on Learning Representations (2019)
3. Choi, H., Jang, E., Alemi, A.A.: WAIC, but why? Generative ensembles for robust anomaly detection. arXiv preprint arXiv:1810.01392 (2018)
4. Nalisnick, E., Matsukawa, A., Teh, Y.W., Lakshminarayanan, B.: Detecting out-of-distribution inputs to deep generative models using a test for typicality. arXiv preprint arXiv:1906.02994 (2019)
5. Du, Y., Mordatch, I.: Implicit generation and modeling with energy based models. In: Advances in Neural Information Processing Systems (2019)
6. Ren, J., et al.: Likelihood ratios for out-of-distribution detection. In: Advances in Neural Information Processing Systems (2019)
7. Serrà, J., Álvarez, D., Gómez, V., Slizovskaia, O., Núñez, J.F., Luque, J.: Input complexity and out-of-distribution detection with likelihood-based generative models. In: International Conference on Learning Representations (2020)
8. Grathwohl, W., Wang, K.-C., Jacobsen, J.-H., Duvenaud, D., Norouzi, M., Swersky, K.: Your classifier is secretly an energy based model and you should treat it like one. In: International Conference on Learning Representations (2020)
9. Schlegl, T., Seeböck, P., Waldstein, S.M., Schmidt-Erfurth, U., Langs, G.: Unsupervised anomaly detection with generative adversarial networks to guide marker discovery. In: International Conference on Information Processing in Medical Imaging (2017)

10. Zong, B., et al.: Deep autoencoding gaussian mixture model for unsupervised anomaly detection. In: International Conference on Learning Representations (2018)

11. Deecke, L., Vandermeulen, R., Ruff, L., Mandt, S., Kloft, M.: Image anomaly detection with generative adversarial networks. In: Joint European Conference on Machine Learning and Knowledge Discovery in Databases (2018)

12. Pidhorskyi, S., Almohsen, R., Doretto, G.: Generative probabilistic novelty detection with adversarial autoencoders. In: Advances in Neural Information Processing Systems (2018)

13. Perera, P., Nallapati, R., Xiang, B.: OCGAN: one-class novelty detection using GANs with constrained latent representations. In: IEEE Conference on Computer Vision and Pattern Recognition (2019)

14. Choi, S., Chung, S.-Y.: Novelty detection via blurring. In: International Conference on Learning Representations (2020)

15. Schölkopf, B., Williamson, R.C., Smola, A.J., et al.: Support vector method for novelty detection. In: Advances in Neural Information Processing Systems (2000)

16. Ruff, L., Vandermeulen, R., Goernitz, N., et al.: Deep one-class classification. In: International Conference on Machine Learning (2018)

17. Golan, I., El-Yaniv, R.: Deep anomaly detection using geometric transformations. In: Advances in Neural Information Processing Systems (2018)

18. Hendrycks, D., Mazeika, M., Kadavath, S., Song, D.: Using self-supervised learning can improve model robustness and uncertainty. In: Advances in Neural Information Processing Systems (2019)

19. Bergman, L., Hoshen, Y.: Classification-based anomaly detection for general data. In: International Conference on Learning Representations (2020)

20. Hendrycks, D., Lee, K., Mazeika, M.: Using pre-training can improve model robustness and uncertainty. In: International Conference on Machine Learning (2019)

21. Lacruz, F., Saniie, J.: Applications of machine learning in fintech credit card fraud detection. In: 2021 IEEE International Conference on Electro Information Technology (EIT). IEEE (2021)

22. He, K., Fan, H., Wu, Y., Xie, S., Girshick, R.: Momentum contrast for unsupervised visual representation learning. In: Proceedings of the IEEE/CVF Conference on Computer Vision and Pattern Recognition, pp. 9729–9738 (2020)

Prescriptive Analytics for Optimization of FMCG Delivery Plans

Marek Grzegorowski[1]($\boxtimes$) (iD), Andrzej Janusz[1,2] (iD), Stanisław Łażewski[2,4],
Maciej Świechowski[2] (iD), and Monika Jankowska[3]

[1] Institute of Informatics, University of Warsaw, Banacha 2, 02-097 Warsaw, Poland
M.Grzegorowski@mimuw.edu.pl
[2] QED Software Sp. z o.o., Warsaw, Poland
{andrzej.janusz,stanislaw.lazewski,maciej.swiechowski}@qed.pl
[3] FitBoxY.com, Kraków, Poland
m.jankowska@fitfoodpoland.pl
[4] Institute of Computer Science, AGH University of Science and Technology,
Kraków, Poland

Abstract. Where the predictive analysis answers the question of what is going to happen, the prescriptive analytics aims at indicating how we can make something happen. In this paper, we propose a prescriptive framework for optimizing sales and profits of companies operating in the fast-moving consumer goods (FMCG) market. We discuss the key aspects of the framework and demonstrate its effectiveness in a series of experiments on real-world data. We compare results obtained for two different optimization heuristics (i.e., hill climbing and simulated annealing) and investigate a possibility to support the product delivery planning process by providing experts with several diverse recommendations. We argue that our solution is in line with the idea of Industry 5.0 and can provide significant value to the FMCG industry.

Keywords: Prescriptive analytics · FMCG · Heuristic search · Sales optimization

1 Introduction

Modern businesses base their decisions on data and facts instead of experts' intuitions, feelings, or hunches. The general availability of cloud solutions with flexible cost models has significantly facilitated access to more sophisticated data analysis solutions [4]. This allowed for the adoption of advanced analytics in small and medium-sized enterprises. Reporting tools that enable descriptive analytics, as well as advanced machine learning (ML) predictive solutions, became commonly available and today are broadly applicable in various domains of the industry [6,11]. Descriptive and predictive analytics provide a comprehensive

Supported by Polish National Centre for Research and Development (NCBiR) grant no. POIR.01.01.01-00-0963/19-00 and Polish National Science Centre (NCN) grant no. 2018/31/N/ST6/00610.

picture of the current and (expected) future state of the business. Therefore, those techniques allow to monitor the situation in a company and make the essential decisions in advance.

Prescriptive analytics describes, explains, and predicts how to suggest the course of action needed in the future [7,9]. Based on reliable predictive models and their interpretable visualization, prescriptive analytics aims at guiding the decision process to take advantage of the predicted future [5]. This way, it can bring great value to businesses. One of the industries where automated decision-making is essential is the fast-moving consumer goods (FMCG) market, and the food industry in particular. Especially when the goods are distributed by vending machines, typically characterized by a relatively small capacity, and products have a short expiry date, such as lunch sets or fruits [8].

The proposed approach to prescriptive analytics combines the predictive models with heuristic search algorithms exploring the space of models' inputs, constrained to the variables which values can be influenced by human decisions. This way, by exploring a vast number of potential settings, we may provide the optimal recommendation. The developed method prepares K best recommendations of sufficiently high quality, which is defined as a percentage of the best solution score, and are significantly different from each other (according to the distance metric of choice and diversification strategy). The prepared framework supports various prediction models and optimization heuristics. It provides multi-criteria assessment of recommendations by determining the expected quality and diversification of results. Therefore, it is in line with the idea of Industry 5.0, supporting work effectiveness and creativity of experts by diversified recommendations, which may lead to better exploration of the market potential.

In the case study, we present the application of the developed framework in the FitBoxY.com delivering lunch meals to the dispersed vending machines in hundreds of office buildings (cf. Fig. 1). The short shelf life of meals and the limited capacity of vending machines enforce a relatively high frequency of deliveries that ought to be planned. With a considerable number of locations, this implies a severe complexity and workload.

Fig. 1. A sales process implemented with a mobile app and tablet.

In this study, we present an approach to automating the process of delivery planning by the supply prescription based on multivariate time series, demand predictions with a trained XGBoost model, and heuristic optimisation. We aim to prepare delivery plans that optimize sales profitability, that is the essential financial metrics. The particular contributions of this paper are as follows:

1. We propose a prescriptive analytics framework that recommends K optimal yet diversified solutions.
2. We discuss the key characteristics of our solution and demonstrate its performance in a series of experiments on real-life data in the FMCG market.
3. We compare results obtained for two different optimization heuristics, i.e., hill climbing and simulated annealing, and two diversification strategies.

The rest of the paper is organized as follows. In Sect. 2, we review the related literature and recall the essential preliminary knowledge. Section 3 outlines the proposed prescriptive analytics framework and the high-level architecture of the developed solution. Section 4 presents the conducted experiments. Wherein, the analyzed data and the experimental setup are presented in Sects. 4.2 and 4.1, respectively. In Sect. 4.3, we discuss the results. Finally, in Sect. 5, we summarize the paper.

2   Related Works

Demand forecasting for fast moving consumer goods (FMCG) is a complex task, with a significant impact on a company's profitability and sustainability, especially in the food sector [12]. Differences between the planned supply and actual demand cause additional costs for the company, related to unnecessary food waste due to surplus production. The application of predictive analytics with machine learning methods to address this challenge has been broadly studied, reporting some promising results [3,8]. However, when the business operates a big network of unmanned shops or dispersed vending machines with a limited storage capacity, it might be not enough. Even though predictive analytics is insightful, it still requires a lot of human effort in the delivery planning [15]. From a business perspective, it is particularly interesting to embrace optimizing and automating the operational processes related to production and delivery planning. The main challenge is the formation of such an assortment for each vending machine, the realization of which will bring maximum profit and minimize the food waste at the same time [10].

ML algorithms commonly applied in this context are often categorized into simple and intrinsically interpretable univariate methods, such as ARIMA or BATS [4], and multivariate models, like deep neural networks or eXtreme Gradient Boosting trees (XGBoost) [14]. In practice, the latter approach often yields better results when dealing with missing data and various disruptions in the ingestion process. Yet other popular methods, allowing decision-makers to obtain near-optimal results within a relatively shorter period of time, are optimization algorithms like genetic algorithm, hill climbing (HC), or simulated annealing (SA) [1]. Metaheuristics were successfully applied to food processing, including transportation, storage, warehousing, and production planning [13].

The state-of-the-art predictive models for sale forcasting like GRU-Prophet predict upon future values of the time series target [6], whereas prescriptive analytics uses optimization algorithms to attain possible outcome on what has to be done to attain better output at the future and provide advice on the possibilities

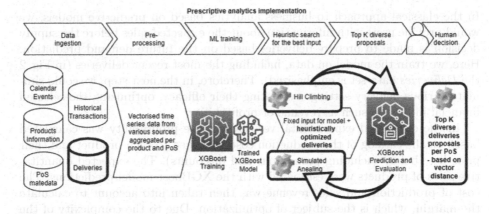

Fig. 2. High-level process and solution overview.

for optimal outcome [9]. Both machine learning and optimization techniques are already a well established and largely adopted, however it is the combination of predictions and optimization that enables the most promising possibilities for decision support [2]. This way, prescriptive analytics, based on reliable predictive models, can simulate possible decisions by searching the space of opportunities to provide the best recommendations together with their quantitative evaluation. Still, compared to the predictive and descriptive, prescriptive analytics is less mature and established, with a limited number of methods and applications reported in literature [5].

3   Solution Overview

The developed solution (cf. Fig. 2) is based on the classic approach to predictive analytics. First of all, data acquisition from several sources and initial preprocessing occurs, including data cleansing and encoding. Afterward, the data is represented as a number of multivariate time series. In the presented case study, the multivariate time series correspond to the historical sales of *FitBoxY* products at each point of sale (i.e., at each vending machine in a particular location), which are additionally enriched with product metadata (such as ingredients list or nutrition table), location metadata, features extracted from the calendar (such as the number of working days in the next week), and information about the availability of products.

In general, the purpose of prescription is to select the optimal input parameters for the model. Of course, not all values are worth analyzing. The ones that are subject to a decision are essential. In particular, we can not influence the historical sales, but we can manage the type and quantity of products delivered to each point of sale (PoS). In the proposed approach, we represent time series as a feature vector containing sales of each product in a given PoS aggregated daily, enriched with the metadata indicated above (i.e., PoS and product information, calendar events, and deliveries). Next, the predictive model is trained.

In the classical approach to business analytics based on predictive models, we would provide users with information about the expected sales before the supply decision is made to prepare deliveries based on the future demand prediction. Here, we train the model on data, including the most recent deliveries (in Fig. 2 the *Deliveries* data-set is emphasized). Therefore, in the next step, we could simulate various delivery settings evaluating their efficacy, optimizing the selected financial indicator, such as revenue or profitability.

In the conducted experimental verification, the profitability was calculated based on the catalog of the production costs assigned to each product and their price list (before including promotions and discounts). The expected quantitative sales of products were estimated with the XGBoost model predictions. The cost of production and sales revenue was then taken into account to calculate the margin, which is the subject of optimization. Due to the complexity of the problem, which is exponential with respect to the number of products, it was necessary to use heuristics for the search process, like hill climbing.

The developed framework follows the idea of Industry 5.0 by performing complex machine learning studies yielding recommendations that are subjected to experts' decisions. This way, we can optimize human work efficacy, not hindering creativity and the possibility to introduce changes. For that reason, we assume that users, apart from the best configuration, should be presented with a few more options, which are as diverse as possible and do not decrease quality compared to the best one (e.g., with a sales forecast at most 20% lower than the optimal one). The proposed methodology is very flexible, allowing to use of various prediction models, optimization heuristics, and distance metrics. The high-level overview of the whole process and its specific implementation are presented in Fig. 2.

4   Experiments

4.1   Experimental Setup

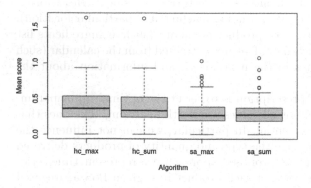

Fig. 3. Profitability gain ratio.

In the experimental evaluation of the developed prescriptive analytics framework, we consider the costs of production and delivery of goods. Therefore, we can analyze the opportunity of increasing profitability of FitBoxY company due to the appropriate selection of products and better planning their deliveries. For this purpose, we train a decision tree-based gradient boosting model on available

---

Algorithm 1: The pseudocode of the experiment.

Data:
TS: Time series data;
$ProdCostsAndPrices$: production costs and product price lists;
$OptH$: Optimisation algorithm, e.g., Simulated Annealing or Hill Climbing;
ML: ML algorithm with parameters, in our study: XGBoost;
$DistMetric$: Distance metric, in our study: Euclidean distance;
K: number of recommendations;
$DistH$: Distance diversification heuristic, e.g., "sum" heuristic
Result:
Top K recommendations with their evaluation scores

1   $\mathcal{R} \leftarrow \emptyset$;
2   TS data initial preprocessing;
3   $oracleML \leftarrow$ Train ML on TS data;
4   **for** *each PoS-Prod time series ts in TS* **do**
5   **for** *each delivery.day in ts* **do**
6   Calculate profit from original stocking with $ProdCostsAndPrices$
       while $OptH.should.continue$ **do**
7   Generate a delivery candidate with $OptH$;
8   Predict products sales with $oracleML$;
9   Calculate candidate profitability with $ProdCostsAndPrices$;
10   Establish top K diversified candidates with $DistMetric$ and $DistH$;
11       $\mathcal{R} \leftarrow \mathcal{R} \cup$
        $\{(\text{PoS, Prod, ts.delivery.day, top K candidates with profit gain scores})\}$

12   *return $\mathcal{R}$*

---

data to treat it as an *Oracle*. Such an approach allows us to carry out the entire analysis based on historical data, avoiding conducting experiments with clients before confirming the quality of the solution. In the performed evaluation, we decided to use the XGBoost model since, for our data, this technique turned out to be relatively insensitive to the parameters tuning. The chosen hyperparameter values were obtained by tuning the model using a small validation set (10% of the data). Eventually, the max depth of trees was set to 6, the number of trees was 1000, and the learning rate was set to 0.01. We also added some regularization to the model to avoid overfitting (i.e., L1 regularization parameter $\alpha = 0.01$, L2 regularization $\lambda = 1$, and the fraction of attributes used for constructing a single tree was set to 0.5). The final model was trained with the squared error objective function. Its validation RMSE, R^2, and MAE was 1.1942545, 0.7196729, and 0.6045681, respectively.

We verify two optimization heuristics, namely: Hill Climbing (HC) and Simulated Annealing (SA). Both, with the search space limited in a way that the overall quantity of goods in each delivery equals the historical values prepared by experts. Such assumptions allow us to fairly compare the profitability generated by the developed method with that obtained by experts and prevent violating physical constraints, such as the available capacity of a vending machine.

We search for $K = 3$ optimal yet significantly different solutions in the experiment. For this purpose, we rely on the Euclidean distance. We analyze two heuristic approaches to providing various product vectors, which we call "sum" and "max" methods. In both, the final set of solutions contains the best product vector found by the developed framework. However, in the "sum" heuristic, the best solution is supplemented with two additional settings to maximize the sum of the distances between all 3 product vectors. In the "max" heuristic approach, we iteratively add a new supply vector that maximizes the distance from the selected element. The second approach is computationally more effective (essential for large values of K). However, it provides less diversified results. The HC and SA parameters were adjusted empirically so that the calculation didn't exceed the assumed execution time of six hours. For SA, the initial temperature of 500 was decreasing towards 0 with a random step of 1 to 3. All experiments were implemented in R-language and executed in the Intel(R) Core(TM) i7-6900K CPU @ 3.20 GHz (8 cores × 2 HT) with the same pre-trained *Oracle* (i.e., the same XGBoost model).

4.2   Data

The data used in the experiment were collected by *FitBoxY* company from the selected 49 vending machines (located in office buildings in four major cities in Poland) operating between December 1, 2020, and July 2, 2021. In this period, there were 101 days when the deliveries were planned (for each deliveries days, several locations are typically stocked). In total, there were 1054 supplies investigated in the conducted study. Apart from information from each PoS transaction, the constructed time series data representation contained product metadata such as a nutrient table, ingredients lists, category, and product availability. We also used information about production costs and selling prices, and VAT taxes of all 249 lunch meals in production. This allowed us to compare the actual revenues with those that could be achieved by stocking PoS with recommendations and estimating expected profits. Due to the limited capacity of the vending machine, the number of different products considered for a delivery is typically far smaller and averages about 23.

4.3   Experimental Results of Profitability Optimization

The analysis was carried out only on the days of actual deliveries of FitBoxY vending machines to compare the historically observed sales with the prescritption results. The quality expressed as the ratio of the profit obtained from the best prescription (estimated with the Oracle) to the actual one. The results revealed that the greedy HC optimization achieved on average 35% higher profit compared to the actual sales. SA allowed obtaining on average 28% higher profit. The box-plot in Fig. 3 summarise the profitability increase ration achieved by the delivery prescriptions generated with each of the combinations of optimisation algorithm and diversity heuristic.

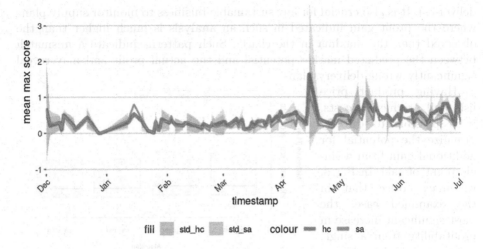

Fig. 4. The ratio of top recommendations' profit to historical.

In Fig. 4, we can see the ratio of the profitability gain obtained through the top HC and SA prescriptions during the whole test period. Most of the time, the HC heuristic performs slightly better. Apart from some periods of low sales, such as the turn of the year, when we do not observe any potential for better profit (i.e., the plot in Fig. 4 is around zero), there is a great promise of financial gain while maintaining the same production size (recall that the optimization algorithms were constrained to keep the same number of goods as the historical

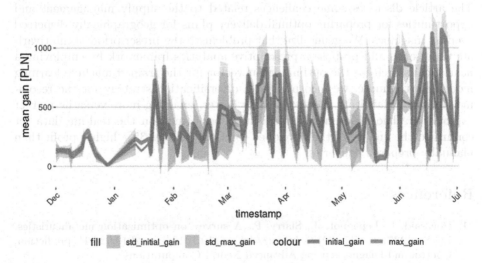

Fig. 5. The profit gain in local currency [PLN] averaged per each PoS. (Color figure online)

deliveries). It is also crucial for any sustainable business to monitor supply plans where the profit gain indicated in such an analysis is much higher than the observed (i.e., the maxima in the chart). Such patterns indicate a mismatch between the predicted market potential and the actual profit and may reveal significantly wrong delivery plans.

Having product price lists and production costs, it is possible to directly visualize the potential for additional gain from a single point of sale. In Fig. 6, we may notice that in the examined case, the most significant increase in profitability from a single point of sale is achieved thanks to the hill-climbing greedy optimization (HC).

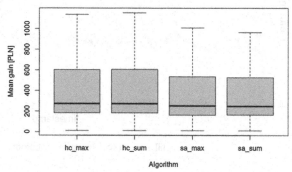

Fig. 6. The profit gain in PLN.

In Fig. 5, we may observe the impact of diversification of the recommendations (with the Euclidean metric and the "sum" heuristics) on their profitability over the evaluated test period. We may notice that the average gain from the top $K = 3$ solutions (the red line) is just slightly lower than the extra profit achieved with the optimal prescription (blue line).

5   Summary

The article discusses some challenges related to the supply management and opportunities for preparing optimal delivery plans for geographically dispersed vending machines. We generalize that problem to the prescription of multivariate time series and propose a prescriptive analytics framework by employing a heuristical search for the optimal input vector for the trained machine learning model. Furthermore, we incorporated a diversification strategy for the recommendations to facilitate users' creativity better, enabling more versatile market exploration. The experimental verification performed on the real-life data set confirmed the quality of the developed solution yielding 35% higher profit than manually prepared deliveries.

References

1. Boussaid, I., Lepagnot, J., Siarry, P.: A survey on optimization metaheuristics. Inf. Sci. **237**, 82–117 (2013). https://doi.org/10.1016/j.ins.2013.02.041. prediction, Control and Diagnosis using Advanced Neural Computations
2. Frazzetto, D., Nielsen, T.D., Pedersen, T.B., Šikšnys, L.: Prescriptive analytics: a survey of emerging trends and technologies. VLDB J. **28**(4), 575–595 (2019). https://doi.org/10.1007/s00778-019-00539-y

3. Garre, A., Ruiz, M.C., Hontoria, E.: Application of machine learning to support production planning of a food industry in the context of waste generation under uncertainty. Oper. Res. Perspect. **7**, 100147 (2020). https://doi.org/10.1016/j.orp.2020.100147

4. Grzegorowski, M., Zdravevski, E., Janusz, A., Lameski, P., Apanowicz, C., Ślęzak, D.: Cost optimization for big data workloads based on dynamic scheduling and cluster-size tuning. Big Data Res. **25**, 100203 (2021). https://doi.org/10.1016/j.bdr.2021.100203

5. Lepenioti, K., Bousdekis, A., Apostolou, D., Mentzas, G.: Prescriptive analytics: literature review and research challenges. Int. J. Inf. Manage. **50**, 57–70 (2020). https://doi.org/10.1016/j.ijinfomgt.2019.04.003

6. Li, Y., Yang, Y., Zhu, K., Zhang, J.: Clothing sale forecasting by a composite GRU-prophet model with an attention mechanism. IEEE Trans. Industr. Inf. **17**(12), 8335–8344 (2021). https://doi.org/10.1109/TII.2021.3057922

7. L'Héritier, C., Imoussaten, A., Harispe, S., Dusserre, G., Roig, B.: Identifying criteria most influencing strategy performance: application to humanitarian logistical strategy planning. In: Medina, J., Ojeda-Aciego, M., Verdegay, J.L., Perfilieva, I., Bouchon-Meunier, B., Yager, R.R. (eds.) IPMU 2018. CCIS, vol. 855, pp. 111–123. Springer, Cham (2018). https://doi.org/10.1007/978-3-319-91479-4_10

8. Malefors, C., Secondi, L., Marchetti, S., Eriksson, M.: Food waste reduction and economic savings in times of crisis: the potential of machine learning methods to plan guest attendance in Swedish public catering during the Covid-19 pandemic. Socio-Econ. Plan. Sci. 101041 (2021). https://doi.org/10.1016/j.seps.2021.101041

9. Poornima, S., Pushpalatha, M.: A survey on various applications of prescriptive analytics. Int. J. Intell. Netw. **1**, 76–84 (2020). https://doi.org/10.1016/j.ijin.2020.07.001

10. Semenov, V.P., Chernokulsky, V.V., Razmochaeva, N.V.: Research of artificial intelligence in the retail management problems. In: 2017 IEEE II International Conference on Control in Technical Systems (CTS), pp. 333–336 (2017). https://doi.org/10.1109/CTSYS.2017.8109560

11. Ślęzak, D., et al.: A framework for learning and embedding multi-sensor forecasting models into a decision support system: a case study of methane concentration in coal mines. Inf. Sci. **451–452**, 112–133 (2018)

12. Tarallo, E., Akabane, G.K., Shimabukuro, C.I., Mello, J., Amancio, D.: Machine learning in predicting demand for fast-moving consumer goods: an exploratory research. IFAC-PapersOnLine **52**(13), 737–742 (2019). https://doi.org/10.1016/j.ifacol.2019.11.203. 9th IFAC Conference on Manufacturing Modelling, Management and Control MIM 2019

13. Wari, E., Zhu, W.: A survey on metaheuristics for optimization in food manufacturing industry. Appl. Soft Comput. **46**, 328–343 (2016). https://doi.org/10.1016/j.asoc.2016.04.034

14. Zhai, N., Yao, P., Zhou, X.: Multivariate time series forecast in industrial process based on XGBoost and GRU. In: 2020 IEEE 9th Joint International Information Technology and Artificial Intelligence Conference (ITAIC), vol. 9, pp. 1397–1400 (2020). https://doi.org/10.1109/ITAIC49862.2020.9338878

15. Zhang, H., Li, D., Ji, Y., Zhou, H., Wu, W., Liu, K.: Toward new retail: a benchmark dataset for smart unmanned vending machines. IEEE Trans. Industr. Inf. **16**(12), 7722–7731 (2020). https://doi.org/10.1109/TII.2019.2954956

A Multilevel Clustering Method for Risky Areas in the Context of Avalanche Danger Management

Fanny Pagnier[✉], Frédéric Pourraz, Didier Coquin, Hervé Verjus, and Gilles Mauris

LISTIC - Université Savoie Mont Blanc, 5 chemin de Bellevue, Annecy-le-Vieux, 74940 Annecy, France
{fanny.pagnier,frederic.pourraz,didier.coquin,herve.verjus, gilles.mauris}@univ-smb.fr

Abstract. In the context of avalanche risk management, we study the spatial variability of rainfall conditions, which is one of the main parameters that induce natural avalanches. This paper focuses on the geographical variability of the snow overload due to recent precipitations. We propose a generic approach applicable at a larger scale and, for this reason, without relying on any expert knowledge. Our proposal is a multilevel clustering process based on classical methods processed in sequence to take advantage of each one. As a result, the multilevel clustering process outputs four main detected weather trends that affect the French Alps. The developed process is generic enough to be used in other areas. Our work is intended to positively impact and improve the current and future decision support methods and tools for mountain practitioners.

Keywords: Clustering · Unsupervised methods · Decision making

1 Introduction

In avalanche risk management, avalanche observations are a particularly effective indicator of the current danger level and a direct evidence of snow instability [7]. Avalanche observations are indeed one of the input factors of several decision support methods. Until now, this factor remains considered as a warning sign and is never quantified or measured in these methods. The reader can refer to the most recent survey considering the factors and methods used by experts [6]. For example, the Obvious Clues Method [10] remains vague by solely considering *"Avalanches in the area in the last 48h"* but not formalizing the boundaries of the area under consideration. However, to take correct decisions, it is beneficial to estimate where similar conditions as the one leading to the observed avalanche are likely to be encountered. This paper proposes a first approach to quantify the size of the area to take into consideration. Although several parameters can be taken into account to consider this notion of similar conditions, the paper focuses on the amount of accumulated snow during the last 24 h.

© Springer Nature Switzerland AG 2022
D. Ciucci et al. (Eds.): IPMU 2022, CCIS 1602, pp. 54–68, 2022.
https://doi.org/10.1007/978-3-031-08974-9_5

In this general context, the main objective is to identify which French Alps areas are likely to receive similar amounts of fresh snow, according to different weather trends. It is equivalent to studying the variability in rainfall conditions at lower altitudes, where the density of automatic measuring stations is higher. Such studies and proposed work in this paper are intended to positively impact and to improve the mountain practitioners' decision support methods and tools. Scientific developments in avalanche science adopt a multidisciplinary approach covering field observations and experiments as well as mathematical and physical analysis modeling [9]. Most researchers are interested in the internal snowpack variability at various scales [16] but experts consider the snowpack's overload as one of the main criteria causing natural avalanches. For this reason, this paper focuses on the geographical variability of the snow overload due to recent precipitations. This work is part of a French-Swiss Interreg project (see Acknowledgements) and is currently being evaluated at the French Alps scale. We aim to propose a generic approach applicable at a larger scale and, for this reason, without relying on any expert knowledge. Several works [1,5,8] have already addressed the study of precipitations in different countries and regions but there is no general method that could be applied on a wide scale. While two locations may encounter the same rainfall conditions under a given weather trend, they may behave differently under a different one. Moreover, some areas may be affected by several weather trends. Thus, when observing a new natural avalanche, it is necessary to estimate the current weather trend of this specific day to know which area makes sense for the given observation.

The specific objective of this study is thus to classify days that are similar in terms of the location of the main rainfall totals. This way, we bring out the major weather trends affecting a given mountain area and we can determine the influencing zone of an avalanche observation.

We work on a dataset based on 12 years of rainfalls measurements collected during the winter season. These measurements are recorded hourly from 90 stations spread on the French Alps. Data comes from EDF-DTG's[1] automatic measuring stations.

Early, we understood that applying clustering methods on the whole dataset does not give the expected result. It classifies days according to the total amount of new rainfalls instead of depending on the location of the main rainfall totals. Figure 1 illustrates this assessment, showing three clusters with the same rainfalls location but various intensities. Then, to address this problem, we decided to work on subsets, which considerably improve the results: when reducing the size of the dataset (i.e., the number of individuals), we noticed that the clustering better captured the location of the precipitations rather than focusing on the total amount of precipitations. Applying clustering methods on 12 separated data subsets implies merging the results obtained on each separated subset into a final global one. That is why we then develop a two-level clustering process.

The paper is structured as follows: Sect. 2 presents step-by-step the developed process and the methods used. Then, Sect. 3 shows the obtained results

---

[1] *Électricité de France*, the French electricity production and supply company.

and validation cases. We discuss in Sect. 4 the developed process and give some perspectives relative to this work. Finally, Sect. 5 conclude the paper.

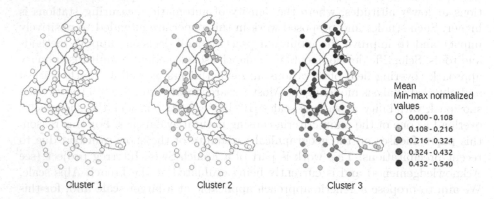

Fig. 1. Clustering (PCA → HC → K-means) on the entire dataset (i.e., 12 winter seasons). For details about the use of the sequence PCA → HC before K-means, see Sect. 2.1.

2   Multilevel Clustering Method

Machine learning approaches can be classified into supervised learning, unsupervised learning, and reinforcement learning. Since we do not have any labeled data and we can not do trial-and-error experiments in the context of risk management, we focus on unsupervised methods. We present a classification approach that 1) meets our previously stated objectives, 2) is an automatic process that does not require any parameter refinement nor specification, and 3) is a transferable process working end-to-end without expert assistance. For this purpose, we first use classical methods of statistical analysis and machine learning: Principal Component Analysis (PCA) [15], Hierarchical Classification (HC) [14], and K-means [4,19] successively, and to refine the result, we use Affinity Propagation (AP) [3].

The whole multilevel clustering process is presented in Fig. 2. In addition, Fig. 4, which illustrates the whole process once applied to our data (see Sect. 3), gives also a general overview of the multilevel clustering process.

For our problem, we consider a dataset structured as follows: individuals are days, and variables are rainfall totals recorded over 24 h in a given meteorological station and for a given day. We thus manage both temporal and spatial components. The temporal component, first, makes it possible to split the initial dataset into 12 data subsets (to work on each winter season separately). Then, as mentioned here, the temporal component permits us to identify the studied individuals : we classify days. For each individual, the variables correspond to measures of rainfall totals received in 90 stations spread over the French Alps, which induces an indirect spatial component as every measuring station corresponds to a specific location on the territory. But, the spatial component relative to the geographical location of the measuring stations is never taken into account (we do not consider information as latitude or longitude). However, this emerges due to PCA whose main components reflect the geographical location of the stations (see Sect. 2.1 and [13] for more details). In addition, the spatial component is visible when visualizing the results on maps. We will see that results are consistent with the spatial location of the measuring stations. To summarize, the clustering process aims to group individuals (i.e., days) that are similar in terms of the location of the main rainfall totals (i.e., days receiving the main rainfall totals on the same measuring stations). At the end of the first-level clustering process, each cluster (i.e., group of days) corresponds to a meteorological trend (all days received similar amounts of precipitations at the same location). That finally permits the identification of geographical extents that are impacted in the same way according to a given trend.

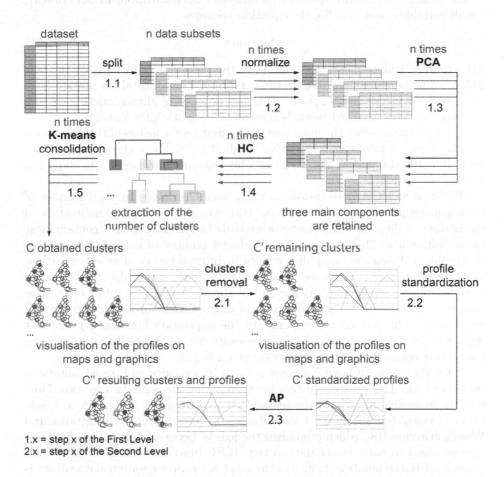

Fig. 2. Schematic representation of the multilevel clustering method

2.1   First Level (PCA → HC → K-means)

The first level contains 5 successive steps. First, for the reason mentioned in the introduction, the initial dataset is split into subsets (see step 1.1 on Fig. 2). Here, we decide to process each winter season separately. Thus we obtain n data subsets. Each data subset contains x individuals and y variables (x and y can be slightly different depending on the processed year).

Data subsets have a feature that requires pre-processing (see step 1.2 on Fig. 2): the variables have different orders of magnitude whereas all of them need to have the same importance for the following treatments. As pre-processing, we normalize the range of the variables to values between 0 and 1 and give them all the same importance for the subsequent treatments. All values are divided by the maximum value recorded on the station over the 12 seasons (i.e. the whole available data). That corresponds to a min-max normalization. In our context, for all variables: $min = 0$. So, the equation becomes:

$$value/max \tag{1}$$

All variables will translate the idea that the recorded rainfalls over the last 24 h are small or high according to the maximum rainfall total the station have ever received in the last 12 years. This pre-processing allows working on the importance of the received rainfalls relative to each station's features.

After pre-processing, the main part of the first level consists of three methods used in sequence to obtain a clustering result [12,17]. These three methods are: PCA [15], HC [14] and K-means [4,19]. This sequence corresponds to steps 1.3, 1.4 and 1.5 on Fig. 2.

The goal of **PCA** is to transform the y variables into a reduced number of principal components. Each component corresponds to a linear combination of the initial variables, and only the firsts are truly informative as they contain most of the variability. Then, to work on a reduced number of components reduces the problem dimension, simplifies results in interpretation after classification, and removes a part of the noise contained in the data subset due to the nature of the studied phenomenon. At the end of the PCA stop, we select only three principal components as they capture most of the initial variability and are well explainable. The first one corresponds to the opposition between dry and wet days, whereas the second and third components correspond to North/South and East/West oppositions. More details are given in [13].

After the PCA, HC and K-means are run in sequence to take advantage of both methods. **HC** creates clusters by aggregating elements two by two. During the successive iterations, the method creates from $x - 1$ clusters to a sole cluster (which contains all the x individuals). We use the euclidean distance and Ward's criterion [18], which minimizes the loss of between-clusters inertia when aggregating two clusters. Thanks to the HCPC function from the FactoMineR package of R, the number of clusters the most appropriate for our data subset is automatically obtained. The partition is the one with the biggest relative loss of inertia. We call this number of clusters c_i (where i represents each data subset, i.e., $i \in [1, n]$). This method has, however, a drawback: when it misclassifies an individual, it remains misclassified until the end. Using the **K-means** algorithm improves the classification. It offers each individual the possibility to move from one cluster to another during successive iterations. The K-means algorithm has to be initialized, which is possible thanks to the result given by the HC and avoids a random initialization. As an objective was to develop a generic process suitable for other data (for example, in the Swiss Alps) without any expert knowledge, we looked for a method suggesting a first estimation of the k value. This way, K-means is not set randomly but corresponds to a consolidation of the HC's result.

This sequence (PCA → HC → K-means) is applied to each data subset. At the end of the first level clustering, it gives in total C clusters, where $C = \sum_{i=1}^{n} c_i$. The number of clusters (c_i), automatically estimated by the process, may differ on each data subset. Hence, here are some drawbacks:

– The process does not always detect every weather trend, even if they are effectively present in the data subset;
– The process sometimes returns several clusters corresponding to the same weather trend (i.e., same location of the precipitations) but with different intensities.

These drawbacks explain why we use a multilevel clustering process:

– Repeating the first level clustering on several data subsets (so on several winter seasons) increases the possible detection of all trends;
– A second level of clustering is required to group similar clusters (i.e., clusters that correspond to the same area impacted by the rainfalls but with various intensity), whether they are obtained on one data subset or among the different winter seasons.

Of course, at the end of the whole process, the aim is to detect all the main trends, and to obtain only one final cluster per trend, i.e., to classify together all the clusters that have similar rainfalls locations but different intensities. To explain that, we introduce the notion of *rainfalls profile*.

The rainfalls profile of a day is the graphical representation of the received rainfalls on each of the 90 measuring stations (i.e., on the 90 variables). This profile can be represented on a graphic or a map (according to the location of the stations), and permits the identification of the location the most affected by the rainfalls. The profile can represent the raw data or the normalized ones (given after Eq. 1). Figure 3 shows, for the 08th January 2018, the normalized profile on a graphic (Fig. 3-a, values between 0 and 1) and the raw profile on a map (Fig. 3-b, values corresponding to rainfalls in mm). The most affected stations (and by extension areas) correspond to peaks on the graphic or to dark points on the map. The profile of a cluster (which is a group of days) corresponds to the mean of the days' profiles.

Profiles that have similar shape (on the graphic) correspond to the same trend as the same area is impacted. For a given trend, profiles may vary in values (that corresponds to more or less intense rainfalls). As the need is to classify together all the clusters that have similar rainfalls locations even if they have different intensities, the process needs to group profiles that are similar in shape even if they are different in values. Indeed, a given weather trend impacts the same areas, whatever the rainfall total.

In this context, for the second level clustering described below, we now consider the clusters profiles as new individuals. As first level gives C clusters, C different profiles are considered. Clusters' centroids are not taken as new individuals for the second level because the values on the 3 components given by PCA are in a specific referential given by the calculation of the PCA on each data subset independently. Coming back to all the initial variables makes it possible to have a common referential for all clusters.

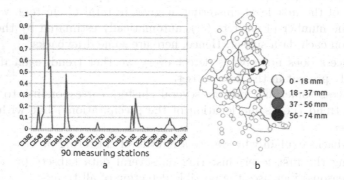

Fig. 3. Illustration of the notion of *rainfalls profile* (08th January 2018)

2.2 Second Level (AP)

At this stage, the clusters composed of days with no or few precipitations can be removed. Indeed, as we are working in the context of avalanche risk, there is no interest for these clusters as they do not correspond to an important enough overload to impact the snowpack's stability and generate avalanche activity. We decide to remove all the clusters for which the profile is, on each variable, lower than 0.2 (see step 2.1 on Fig. 2). Note that, when removing days with no or few precipitations (i.e., values lower than 0.2 on each variable) before starting the multilevel clustering method, the first level process outputs anyway clusters corresponding to few rainfalls (i.e., profile lower than 0.2 on each variable, because values upper than 0.2, which induce to keep these days, are sparse, and not on the same stations). Thus, the sole deletion of clusters, at step 2.1, allows a sole data deletion. In addition, it makes it possible to keep the most available information of the initial data for the PCA step in order to better highlight the spatial components described above. As variables have been normalized (see above step 1.2), 0.2 means that precipitations were less than 20% of the maximum rainfall total observed in 24 h on the station over the 12 seasons. This threshold is arbitrary, but some experimentations with 0.1 and 0.3 thresholds were carried out and variations had no impact on the final result. At this stage, p profiles are removed. It remains then C' profiles considered as individuals for the second level, where $C' = C - p$.

Then, as the need is to group profiles similar in shape but different in intensity, we have to standardize (i.e., center and reduce) them before starting a new clustering method. This way, profiles similar in shape become also more similar in values (see step 2.2 on Fig. 2). For each profile, values are replaced by:

$$(value - \mu)/\sigma, \tag{2}$$

where μ is the mean and σ the standard deviation of the 90 values that composed the profile.

The C' standardized profiles can then be classified. For this clustering level (see step 2.3 on Fig. 2), we decide to use the **AP** method [3] as it does not require setting the number of clusters nor choosing an initial set of points. This algorithm considers each individual as a potential exemplar and uses a similarity matrix between all individuals two by two. This method fits our objectives as it aims to classify together similar profiles.

As presented in the literature from other fields [11], the AP's unsupervised version (which consists in using the preference value by default, i.e., the median value of the similarity matrix) usually gives too big of a number of clusters [3]. It was our case when we tried to use AP directly after PCA at the first level clustering (instead of HC and K-means). But here, as we are working on pre-processed data and mean individuals obtained through the first level clustering, the default preference value ($pref$) gives an acceptable number of clusters in the output. The ratio between $pref$ and the maximum value of the similarity matrix is approximately 40%. This ratio was approximately 1% when using AP at the first level clustering. This difference explains the former observation. The higher $pref$ is, the lower is the number of final clusters. So, at the second level clustering, the value of $pref$ set by default is adapted to give a satisfying number of clusters in the output.

Thus, thanks to AP at this second level, the process finally gives a single cluster per weather trend. It correctly returns all the detected weather trends spotted during the previous level. The process finally returns C'' final clusters. Each resulting cluster corresponds to a group of profiles similar in shape (i.e., profiles that correspond to similar rainfalls locations). Then, at this stage, we can visualize on maps the resulting profiles (which are means of standardized profiles processed at this second level). This gives a first idea of the rough areas mainly affected by rainfalls according to each detected weather trend.

3 Results and Validation

3.1 Multilevel Clustering Applied to Our Data: Results

This section gives the result of the whole process PCA → HC → K-means + AP applied to our dataset (Fig. 4).

Our dataset contains 12 winter seasons (i.e., days from 1st of December to 31th of March) from 2009–2010 to 2020–2021. That corresponds to a sum of 1455 individuals. Once split, there are 12 data subsets (i.e., $n = 12$), each corresponding to a winter season. According to the winter season, data subsets contain 121 or 122 individuals and 90 variables.

We pre-process each data subset and apply the first level clustering. When keeping the three first principal components of the PCA, we keep between 84.5% and 90.5% of the initial variability of the data subsets. HC suggests generating between three and seven clusters (i.e., $c_i \in [3; 7]$) per the data subset. We finally apply K-means to consolidate results. This way, on the 12 data subsets, we obtain 48 clusters in total (i.e., $C = 48$). We then calculate the 48 corresponding profiles.

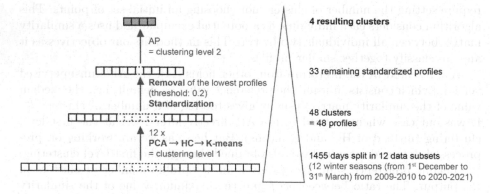

Fig. 4. Multilevel clustering applied to our data

Before the second level clustering, we pre-process the data by removing the profiles corresponding to a few rainfalls (with a 0.2 threshold), that is to say here 15 profiles (i.e., $p = 15$). We then calculate the reduced profiles on the 33 remaining ones (i.e., $C' = 33$). Then, we apply the second level clustering, i.e., AP. The process gives at the end four clusters (i.e., $C'' = 4$) as result. Figure 5 shows their profiles on maps. The four resulting clusters correspond to four fictitious days, which are typical of the four main weather trends that the process detects (Fig. 5). These trends are:

1. Most impacted area located in the southern part of the French Alps that corresponds to a flux mostly coming from South or South-West directions;
2. Most impacted area located in the northern part of the French Alps that corresponds to a flux coming from North-West;
3. Most impacted area located in the eastern side of the French Alps that corresponds to a flux coming back from the East;
4. Most impacted area located in the western part of the French Alps that corresponds to a flux coming mostly from the West, affecting, the pre-alpine massifs.

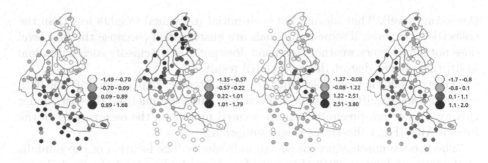

Fig. 5. Visualization of the reduced mean profiles of the resulting clusters

We presently know that at least four different meteorological trends may be detected in the 12 winter seasons dataset. Let us see which result gives K-means algorithm directly leads on the three main components extracting by PCA on the whole dataset (i.e., 12 winter seasons), with random initialization and $k = 4$. In this case, we detect two different trends (see Fig. 6, clusters 1 and 4 correspond to two different trends, with the same range of intensity) and one of these trends appears with three different intensities (see clusters 2, 3, and 4 on Fig. 6). Thus, applying K-means on the entire dataset does not give an optimum result (as solely two out of the four possible trends are detected, and as a sole trend is given several times): even by imposing $k = 4$, it does not output the four possible trends. It emphasizes that the multilevel clustering process (working on a split dataset in the first level and merging the results in the second one) produces better outputs.

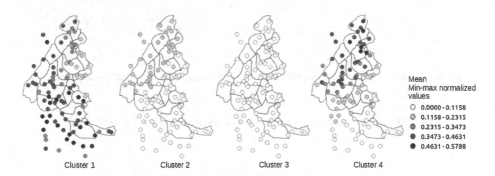

Fig. 6. Clustering (PCA → K-means) on the entire dataset (i.e., 12 winter seasons), with k = 4

3.2   Two Validation Cases

The process' strength is its global result (obtained through two levels clustering and the calculation of means to generate the new individuals processed in the second level). That means that each initial individual weights less than the collective does: even if some individuals are misclassified (because the first level does not detect every weather trend and does not strictly classify each individual well), they do not distort the final global result.

The two following examples illustrate this outcome. Experts identify the 08th and 09th January 2018 as two days affected by a flux coming from the east. Under this weather trend, precipitations are located mostly on the eastern part of the French Alps (Fig. 7 illustrates this assumption).

The process misclassified these two individuals. The location of the rainfalls of the 09th of January 2018 (Fig. 7) differs from the profile of the final cluster to which it has contributed (see cluster 2 on Fig. 5). The rainfalls of the 08th of January 2018 were locally considerable (more than 60 mm, see Fig. 7-a) whereas the process classifies this day in a finally removed cluster due to too few rainfalls on average.

We now present whether the final result of the process is valid in correctly classifying these days and decide which weather trends affect them. We determine that by comparing the profiles of these days (the one visible on Fig. 7-b, obtained after applying Eqs. 1 and 2 on their raw data) with those of the final clusters (FC), which are typical of each detected weather trend. We calculate the euclidean distances between all profiles. The day to analyze is then associated with the closest final cluster (i.e., to the most similar fictive day given by the multilevel clustering process) and corresponds to the respective trend.

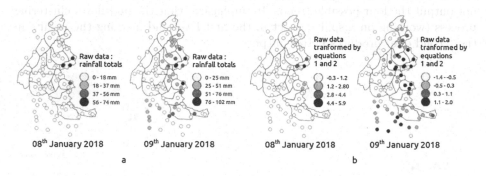

Fig. 7. Rainfall totals and profiles of the 08th and 09th January 2018

	FC 1	FC 2	FC 3	FC 4
08th Jan. 2018	13.71	13.94	9.97	15.07
09th Jan. 2018	12.82	13.94	8.30	15.50

Fig. 8. Distances between days to analyze and final clusters

According to the result (Fig. 8), both days are affected to the resulting cluster number 3, i.e., the one corresponding to a flux coming back from the east. This conclusion agrees with our first assumption and the expert knowledge we have for these days. Thus, the result of the process is valid as it permits correctly classifying the days. It corrects their cluster, even if they contribute wrongly, during the process. That shows the weight of the global and the validity of the final result given by the multilevel clustering process.

The four detected trends output by the multilevel clustering process have been validated based on expert knowledge. Indeed, we cannot establish a correct classification rate since the expected cluster for each day is not known in advance (no complete ground truth on the dataset). In addition, the cluster given to each day individually does not matter so much as the clustering process is useful to finally output and detect the main trends affecting the studied area. Thus, the multilevel clustering process highlights the different areas impacted by precipitation according to different trends. This result is particularly useful in avalanche danger management as these areas fit the ones impacted by spontaneous avalanches that occurred on the corresponding days. Finally, we will massively test the framework and do some further validation on future data (recorded for the following winters).

4 Discussion

First, the multilevel clustering process can be adapted and used with other clustering methods than those used here. Depending on the data, some clustering methods are more suitable than others. Depending on the expected shape of the clusters, it will be, for example, more appropriate to use K-means (circular) or GMM (ellipsoidal) [20], or for any other reason, another clustering method.

In our work, we tried with AP or GMM instead of K-means in the first level clustering. But, these two methods are not only a consolidation of IIC results, as was the case for K-means. Then, the sequence of methods was less fluent with AP: the parameter of preference has to be adapted to obtain the number of clusters given by HC, which needs several tries. Nevertheless, the result obtained at the end was very close. With GMM instead of K-means, results were only slightly different. But as we did not dispose of adequate ground truth (only a few days among 1455 available), we did not have any possibility to be sure that one result was better than another. In addition, GMM requires adapting some parameters to obtain the best possible result (most relevant with the data) and, without any ground truth, it was impossible to estimate what changes were needed. We can not derive conclusions on any improvement or degradation of the result. With the constraints to work without any ground truth and any expert intervention, we prefer keep using the K-means method which was a consolidation of HC and does not need any other parameter to be set and refined.

A possible follow up to this work is the refinement of the profiles of the four obtained fictitious days, corresponding to typical representants of the four detected weather trends. The validation step showed that days, which the process misclassifies, were finally well estimated based on their distances to the four final fictitious days (see Sect. 3.2). This way, by recalculating to which final cluster each day should belong to, each cluster's profile can be updated. Thus, the updated result will be based on well-classified individuals, which should improve the correctness and precision. Thus it will be a better base to estimate, for each weather trend, which area encounters similar rainfall conditions.

Then, what should follow this work is to study, for each weather trend, which stations have the same features. That means studying which stations receive similar amounts of new rainfalls according to different weather trends.

Even if we already get a rough idea of these zoning through the map representing the profiles (see Fig. 5), some additional work is required to precise which areas are similar in terms of rainfall conditions, for each trend. It will take place in the general objective firstly mentioned in the introduction and the specific context of our work, to finally link areas impacted by rainfalls and avalanche observations for given days.

5   Conclusion

Our multilevel clustering process is based on two clustering levels that use classical methods in sequence: 1) PCA → HC → K-means, and 2) AP. Clustering in two levels masks the misclassification of some individuals by giving a correct global result. At the end, the process gives four typical profiles, which are equivalent to fictitious days presenting the features of the detected weather trends.

The first level clustering does not always detect all the trends present in the data subset. That is why, to increase the capacity to detect all the possible main trends at least once, we increase the number of studied days in the input of the process. We divide these days into several data subsets. Otherwise, the obtained result does not correspond to what we are looking for: it classifies days according to intensity instead of the location of the rainfalls.

On the other hand, the first level clustering sometimes gives several clusters that correspond to only one trend (i.e., profiles with the same shape but differences in values), whereas we want to detect a single final cluster for each weather trend. We do not stop only after this first level, to aggregate the similar clusters. At the end of the second level clustering, we obtain only one final cluster for each detected weather trend. The role of this second level is also to merge the results obtained due to the n iterations of the first level clustering.

Moreover, if some days could be misclassified, the global result is in adequacy with what is intended. It is possible to extract the associated area affected by precipitations. That is why the principle of our process is not to check precisely in which cluster each day is affected but to consider only the global result obtained thanks to means and two-level clustering.

The strength of our process is that despite some misclassifications on initial individuals (i.e., days), the weight of the collective gives a satisfying final result. Indeed, when we use the result to estimate which weather trend a day should be associated with, the estimation is correct. By associating a day to the closest fictitious day among those given in result by our multilevel clustering process, we can affect the right trend to the analyzed day. That means the output of our process is good enough to correctly decide: which weather trend affects a day among the main detected trends.

In the future, the developed process will be integrated into the decision support method CRISTAL [2]. As the process automatically assesses the extent of the area which makes sense for a given avalanche observation, it will fill one of the six parameters on which CRISTAL relies. Indeed, as the multilevel clustering result gives a better understanding of the areas affected by precipitation under different meteorological trends, it makes it possible to specify a criterion that is vague when it is taken into account in the existing decision-making frameworks (as mentioned, for example, in the Obvious Clues Method). Thanks to this result, we can better formalize the boundaries of the area subject to similar danger as the one that led to avalanching.

Acknowledgements. The CIME project is supported by the European cross-border cooperation program Interreg France-Switzerland 2014–2020 and has been awarded a European grant (European Regional Development Fund) covering 60% of the total French cost.

We thank the snow expert Alain Duclos, who provided us the information on some typical days representative of the trends, especially concerning the third detected one.

References

1. Casado, M.J., Pastor, M.A., Doblas-Reyes, F.J.: Links between circulation types and precipitation over Spain. Phys. Chem. Earth **35**, 437–447 (2010)
2. Duclos, A.: Nivologie pratique: les 4 modes de vigilance encadrée. Neige et avalanches **160**, 7–9 (2018)
3. Frey, B.J., Dueck, D.: Clustering by passing messages between data points. Science **315**(5814), 972–976 (2007)
4. Huan, M., Lin, R., Uang, S., Xing, T.: A novel approach for precipitation forecast via improved K-nearest neighbor algorithm. Adv. Eng. Inform. **33**, 89–95 (2017)
5. Irannezhad, M., Ronkanen, A.K., Kiani, S., Chen, D., Klove, B.: Long-term variability and trends in annual snowfall/total precipitation ratio in Finland and the role of atmospheric circulation patterns. Cold Reg. Sci. Technol. **143**, 23–31 (2017)
6. Landrø, M., Hetland, A., Engeset, R., Pfuhl, G.: Avalanche decision-making frameworks: factors and methods used by experts. Cold Reg. Sci. Technol. **170**, 102897 (2020)
7. Landrø, M., Pfuhl, G., Engeset, R., Jackson, M., Hetland, A.: Avalanche decision-making frameworks: classification and description of underlying factors. Cold Reg. Sci. Technol. **169**, 102903 (2020)
8. Lemus-Canovas, M., Lopez-Bustins, J.A., Trapero, L., Martin-Vide, J.: Combining circulation weather types and daily precipitation modelling to derive climatic precipitation regions in the Pyrenees. Atmos. Res. **220**, 181–193 (2019)
9. Louchet, F.: Snow Avalanches: Beliefs, Facts, and Science (2021). 112 p. ISBN: 9780198866930
10. McCammon, I.: Obvious clues method: a user's guide. Avalanche Rev. **25**(2), 8–9 (2006)
11. Meng, J., Hao, H., Luan, Y.: Classifier ensemble selection based on affinity propagation clustering. J. Biomed. Inform. **60**, 234–242 (2016)

12. Murtagh, F., Legendre, P.: Ward's hierarchical agglomerative clustering method: which algorithms implement ward's criterion? J. Classif. **31**(3), 274–295 (2014). https://doi.org/10.1007/s00357-014-9161-z
13. Pagnier, F., Coquin, D., Pourraz, F., Verjus, H., Mauris, G.: Classification des précipitations sur les massifs alpins français. ORASIS 2021, Centre National de la Recherche Scientifique [CNRS], Saint Ferréol, France, September 2021. (hal-03339626)
14. Praene, J.P., Malet-Damour, B., Radanielina, M.H., Fontaine, L., Rivière, G.: GIS-based approach to identify climatic zoning: a hierarchical clustering on principal component analysis. Build Environ. **164**, 106330 (2019)
15. Richman, M.B., Adrianto, I.: Classification and regionalization through kernel principal component analysis. Phys. Chem. Earth **35**, 316–328 (2010)
16. Schweizer, J., Kronholm, K., Jamieson, J.B., Birkeland, K.W.: Review of spatial variability of snowpack properties and its importance for avalanche formation. Cold Reg. Sci. Technol. **51**, 253–272 (2008)
17. Tufféry, S.: Data mining et statistique décisionnelle: L'intelligence des données, 4ème édition. Editions Technip, Paris (2012)
18. Ward, J.H.: Hierarchical grouping to optimize an objective function. J. Am. Stat. Assoc. (1963). https://doi.org/10.1080/01621459.1963.10500845
19. Zahraie, B., Rooszbahani, A.: SST clustering for winter precipitation prediction in southeast of Iran: comparison between modified K-means and genetic algorithm-based clustering methods. Expert Syst. Appl. **38**, 5919–5929 (2011)
20. Zhao, L., et al.: Adaptive parameter estimation of GMM and its application in clustering. Future Gener. Comput. Syst. **106**, 250–259 (2020)

Fast Text Based Classification of News Snippets for Telecom Assurance

Artur Simões[1,2] and Joao Paulo Carvalho[1,2](✉)(iD)

[1] Instituto Superior Técnico, Universidade de Lisboa, Lisbon, Portugal
artur.f.f.simoes@tecnico.ulisboa.pt
[2] INESC-ID Lisboa, Lisbon, Portugal
joao.carvalho@inesc-id.pt

Abstract. The quality of Telecom companies' mobile service can be seriously compromised by the occurrence of different types of events, whether they are expected or not. The goal of this work is to automatically identify online news that report such events. Three possible topics are searched for: "fire"; "meteorologic" and "public gatherings". Remaining news' topics should be ignored. Each category was specifically chosen by its relevance towards the most known network providers' problems. The data is highly unbalanced.

We compare different lightweight models for text classification using information collected from several Portuguese online newspapers: Support-Vector Machines (SVM), Fuzzy Fingerprints and K-Nearest Neighbours (KNN). More complex deep models, such as Bert or RoBerta, are dismissed due to the requirement of a fast response. The proposed models predict the categories based entirely on the title and the short news' snippets that are freely available. Preliminary results indicate F1-scores above 0.78 for each of the three topics.

Keywords: Event classification · Short texts classification · Support vector machines · Fuzzy fingerprints · kNN

1 Introduction

Customer Experience Management (CEM) transforms traditional network-centric telecom monitoring into customer-centric telecom Service Quality Management (SQM). CEM demands insight into customers' perceptions of service quality and experience. With this in place, it is important to monitor network performance based on analysis and forecasts of the influence of events on customer experience. Thus deliver high-level customer service assurance. The SQM enables you to become both proactive and active when dealing with telecom

This work was supported by Fundação para a Ciência e a Tecnologia (FCT), through Portuguese national funds Ref. UIDB/50021/2020, Agência Nacional de Inovação (ANI), through the project CMU-PT MAIA Ref. 045909 and Inova-Ria associated with Altice Labs, S.A.

D. Ciucci et al. (Eds.): IPMU 2022, CCIS 1602, pp. 69–81, 2022.
https://doi.org/10.1007/978-3-031-08974-9_6

service quality by preventing customers from experiencing service problems and finding the nature of those problems, thanks to processing a large amount of information like the latest news, weather conditions, major events, public gatherings, etc. The digital era has raised customer expectations about the range and quality of services delivered by any company. Traditional network and CEM is no longer enough to keep your customers satisfied. Today, you need to manage service quality proactively, improve troubleshooting, so that you can prevent any customer issues before they even arise.

Access to abundant information has become the key advantage of online news media. In addition to news supplied by mainstream news organizations, people also have online access to alternative news and views from social media. Amongst those considered unique to the Internet are Rich Site Summary (RSS) feeds, news aggregators, news blogs and news wikis [1]. This work will focus on online newspapers as telecom related rich source of information. Studies suggest that the shift towards online news sources stems from the attraction of the Internet as news' medium [2]. Online news media has some important attributes considering the goal of this project, such as the 24/7 delivery of breaking news (news immediacy), the free access to news snippets, and the capacity to find information faster than using offline media. In addition, online news stories can be "linked" to more in-depth information, giving news and information a detailed and richer historical background. In this work, online news are used as a tool to prevent possible telecom network problems: by analysing and classifying the most recent news feeds in close to real-time, one has the opportunity to improve CEM assurance by anticipating service quality troubles.

The goal of this project is, based on the nature of some relevant news types, to identify events that could compromise the quality of service of a network provider. Each piece of news should be labeled with one category from a set of possible four: forest fires ("fires"), "public gatherings", meteorologic disruptions ("meteorologic") and "others". The category "public gatherings" combines both social and cultural events (e.g.: political rallies, musical festivals, etc.). The used data set is highly unbalanced. Each category was specifically chosen by its relevance towards the most known network providers' problems.

The proposed models predict the categories based entirely on the title and short news description. In this work SVM, KNN and Fuzzy Fingerprints (FFP) classifiers were used. Our experiments reveal that the text available of each news contain features that allow to predict its nature, with a minimum of 0.78 F1-score. The resulting models can also be used with other short news sources, such as web pages, text blogs, social media posts, etc.

2   Related Work

For years, a wide range of methods has been applied to Text Classification problems (TC), ranging from hand-coded rules to supervised and unsupervised machine learning. Some of the most well-known and commonly applied methods for text classification tasks [3] include: Naïve Bayes variants, k-Nearest Neighbor (kNN) [4], Logistic Regression or Maximum Entropy [4–6], Decision Trees [4], Neural Networks, and Support Vector Machines (SVM) [4,6]. Although many approaches have been proposed, automatic text classification is still an active area of research, mostly because existing text classifiers are still far from perfect in these tasks.

The goal of this work is essentially to automatically classify text from news into a set of topics. This process is broadly known in Natural Language Processing (NLP) as Text Categorization, and consists of finding the correct topic for each document, given a set of categories (topics) and a collection of text documents [7]. Topic classification problems define a short and generic set of categories, ranging from politics to sports and the documents will often belong to at least one of those categories. It is very rare that a news does not fit into any topic. The difficulty of a classification task varies across tasks and becomes substantially high as the number of categories/classes increase. Moreover, in mul ticlass text classification tasks, an increased number of classes demand larger sets of training data, and some of those classes will always be more difficult than others to classify. Reasons for that may be: i) few positive training examples for the class, and/or ii) lack of good predictive features for that class [3]. It is possible to find several works regarding topic classification. For example, in [4], an attempt is made to classify Twitter Trending Topics into 18 broad categories, such as: sports, politics, technology, etc., and their experiments on a database of randomly selected 768 trending topics (over 18 classes) show that, using text-based and network-based classification modeling, a classification accuracy up to 65% and 70% can be achieved, respectively. Another interesting article [8], have built a news processing system that identifies the tweets corresponding to late breaking news. Issues addressed in their work include removing the noise, determining tweet cluster of interest using online methods, and identifying relevant locations associated with the tweets.

In what concerns text classification, K-Nearest Neighbors (kNN) [9] and the Support Vector Machine (SVM) [10] are amongst the most widely used and best performing classifiers when lightweight models are a must. In [11], Yang and Liu, performed several tests in a controlled study and reported that SVM and kNN are at least comparable to other well-known classification methods, including Neural Networks and Naive Bayes, and that significantly outperform the other methods when the number of positive training instances per category are small. In this paper we also use an adaptation of the Fuzzy Fingerprints introduced in [12] and applied for topic classification in [13,14], since they proved to be simple, fast and efficient text classifiers when in the presence of a large number of classes.

The text contained in each each news is the most relevant source of information for classification. However, text is an unstructured form of data that classifiers and learning algorithms cannot directly process [7]. For that reason, our news must be converted into a more manageable form, during a preprocessing step.

In what concerns the features, text classification tasks where computational efficiency is a must, usually rely on unigrams, also known as bag-of-words representation [4–6]. The bag-of-words representation can be combined with different weighting schemes, including: i) binary or Bernouli weights that indicate whether a given word occurs in the document; ii) the words frequency that provide more information to the model; and iii) TF-IDF that is a scoring method that can measures the importance of a word or term in a collection of documents. TF-IDF is a powerful weighing schema, commonly reported in the literature to text classification tasks [6,14,17,18].

As succinctly explained in [19], TF-IDF assigns a weight to a term in document that is: 1) highest when the term occurs many times within a small number of documents; 2) lower when the term occurs fewer times in a document, or occurs in many documents; 3) lowest when the term occurs in virtually all documents;

Recent works in text classification are mostly based in very large models such as Bidirectional Encoder Representations from Transformers (BERT) or its evolutions such as RoBERTa. BERT is a transformer-based machine learning technique for natural language processing (NLP) pre-training developed by Google in 2018. Even though such models outperform most other simpler techniques in most NLP tasks, they are too computationally demanding for the application addressed in this paper.

2.1 SVM

Support Vector Machines (SVM) are commonly used for text classification tasks [4,6,20]. This classification is based on set of hyperplanes in a high-dimensional space. A good separation is achieved by the hyperplane that has the largest functional margin, the distance to the nearest training data point of any class. In general, a larger margin means a lower classifier generalization error. SVMs can efficiently perform linear and non-linear classifications using what is called the kernel trick, implicitly mapping their inputs into high-dimensional feature spaces. SVM are also very fast and effective text classifiers when used as binary relevance classifiers. According to [21] "every category has a separate classifier and documents are individually matched against each category".

An interesting property of SVM is that the decision surface is determined only by support vectors, which are the only effective elements in the training set; if all other points were removed, the algorithm will learn the same decision function. This characteristic makes SVM theoretically unique and different from other methods where all the data points in the training set are used to optimize the decision function [11].

2.2 kNN

kNN is an example-based classifier, commonly referred in the literature [3,20, 22]. The representations of training data are simply stored together with their category labels. In order to decide whether a document d belongs to a category c, kNN checks if the k training documents most similar to d belong to c. If the answer is positive for a sufficiently large proportion of documents then the result is positive. An appropriate value of k is of the utmost importance. While k=1 can be too simplistic, as the decision is made according only to the nearest neighbor, a high value of k can lead to too much noise and favor dominant categories. kNN is known to be affected by noisy data, still it is considered one of the simplest and best performing text classifiers, whose main drawback is the relatively high computational cost of classification, because for each test document it must compute its similarity to all of the training documents. The training is fast, but the classification is slow because it implies computing all the similarities between a document that has not been categorized and the existing a collection of training documents. However, given that in our work we are using short text snippets, kNN should be adequate at least as a baseline while the dataset is in its earliest stages.

2.3 Fuzzy Fingerprints

We use the Fuzzy Fingerprints classification method described in [14] to tackle the problem of news snippets classification. The idea behind the method is to have a fingerprint library (containing the fingerprint of each topic), and to check the similarity of each news snippet with each of the fingerprints in the library.

Building the Fingerprint Library. In order to build the fingerprint library, the proposed method goes over the training set, which, in this situation, are news from different four topics. For each news it adds each word in the news to a topic table alongside with its counter of occurrences. Only the top-k most frequent words are considered.

Due to the small size of each news snippet, its words should be as unique as possible in order to make the fingerprints distinguishable amongst the various topics. Therefore, in addition to counting each word occurrence, we also account for of its Inverse Topic Frequency (ITF), an adaptation of the Inverse Document Frequency, where we use the total number of topics and the number of topics where the word is present.

After obtaining the top-k list for a given topic, we take the same approach as the original method, and use the membership function (1) to build the fingerprint, where k is the size of the top-k fingerprint and i represents the membership index.

$$\mu_{ab}(i) = \begin{cases} 1 - (1-b)\frac{i}{kb}, i < a \\ \frac{a(1-\frac{i-a}{k-a})}{k}, i \geq a \end{cases} \tag{1}$$

The fingerprint is a k sized bi-dimensional array containing in the first column the list of the top-k words, and in the second column its membership value $\mu_{ab}(i)$ obtained by the application of (1).

Snippet-Topic Similarity Score. Since news snippets do not contain many words, it does not make sense to count the number of individual word occurrences in each news snippet to build a snippet fingerprint. Instead, a Snippet-topic Similarity Score (T2S2) is used to test how much a news snippet fits to a given topic. The T2S2 function, Eq. (2), provides a normalized value ranging between 0 and 1, that takes into account the size of the (preprocessed) news snippet (i.e., its number of features).

$$T2S2(\phi, T) = \frac{\sum_v \mu_\phi(v) : v \subset (\phi \cap T)}{\sum_{i=0}^{j} \mu_\phi(w_i)} \qquad (2)$$

In Eq. (2) ϕ is the topic fingerprint, T is the set of words of the (preprocessed) news snippet, $\mu_\phi(v)$ is the membership degree of word v in the topic fingerprint, and j is the number of features of the news snippet. Essentially, T2S2 divides the sum of the membership values $\mu_\phi(v)$ of every word v that is common between the news and the topic fingerprint, by the sum of the top j membership values in $\mu_\phi(w_i)$. Equation 2 will tend to 1.0 when most to all features of the news snippet belong to the top words of the fingerprint, and tends to 0.0 when none or very few features of the news snippet belongs to the bottom words of the fingerprint.

3 Experimental Setup

This section details the corpus of the experiment and preprocessing stage, methods of classification and finally the evaluation metrics used to evaluate the models.

3.1 Corpus

In this work, all the news were retrieved on a daily basis from several Portuguese online newspapers from the 29th December 2021 until the 4th of February 2022, containing a total of 3136 up-to-date news. The data was retrieved using Really Simple Syndication (RSS) technology available on the newspapers websites and stored in different Excel files. RSS is a web feed that allows users and applications to have access to updates to websites in a standardized computer-readable format. This method of data extraction was used to have access to approximately the latest 15 news from each online newspaper in Extensible Markup Language (XML) format. Figure 1 presents the data extraction and filtering methods. The extracted list of news in XML from RSS contains a lot of heterogeneous information but only a portion of that information is needed to our goal. File "Raw News" contains a standardized and filtered list of news. Each element of the list has the following attributes: Title of the news, small description of the news and link to the website. Unfortunately is not possible to retrieve the body of the news using RSS, so the title and the description are the only used text in our approach. The link is a unique identifier to prevent duplicated entries. Finally, file "Labeled news" contains the data used in our experiments, where each news was manually labeled with just one of four possible categories. The 3136 records are randomly shuffled and splitted into two different sets: *training set* and *validation set*, respectively with 70% and 30% of the data set.

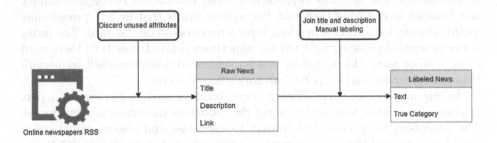

Fig. 1. Collecting and filtering architecture

Each record contains text and a topic. Commonly, each text consists of only a couple of sentences and has on average 37.2 words, including stopwords, and has on average 22.7 words, excluding stopwords. The reduced number of words makes the text based classification task a difficult problem. Figure 2 shows an histogram of the frequency of news per topic. Revealing that the data set is highly unbalanced. There are 3136 records with the following distribution *fires*: 220, *meteorologic*: 222, *public gatherings*: 134 and *other*: 2560. The minor category *public gatherings* is assigned to only 4.27% of the records and the largest category *others* is assigned to 81.% of the records. It is important to note that it was decided to keep the the real world dataset unbalance in order not to bias the models. This allows us to expect a similar model performance when under real use cases.

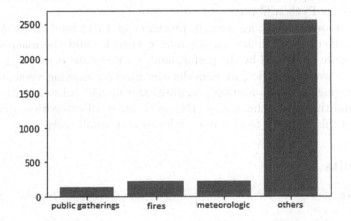

Fig. 2. Number of news labeled with a given category.

3.2 Preprocessing

For each scenario, we tested the influence of removal of *stopwords* and *stemming* on the results. The first was implemented using *nltk.corpus* Portuguese library and consists in removing from text Portuguese words that do not provide any useful information to decide in which topic a text should be classified. The latter is the process of reducing inflected (or sometimes derived) words to their word stem, base or root. This technique was implemented using *Snowball* framework Portuguese version and simplifies category's *bag-of-words*.

In our models, the preprocessing stage, implemented in Python, is a part of reading the news' text and creating the model as the execution progresses. The processing stage consist of convert to lowercase and remove punctuation and characters and then strip, removing stopwords and stemming. In KNN and SVM we use TF-IDF. In FPP and adaptation of this technique called TF-ICF is used [14].

3.3 Methods

SVM and KNN were modeled using a Python library *scikit-learn*. The FFP method was adapted from [16] and [15]. The exact same training data sets and test data sets were used for all methods. Several test scenarios were built to find each algorithm optimal performance setting. SVM model was applied as a binary classifier and as a multi-class classifier. KNN e FFP were applied as multi-class classifiers.

3.4 Evaluation Metrics

Three metrics were used to evaluate our approaches: Precision, Recall and F1-Measure, computed based on the calculation of a confusion matrix using *sklearn.metrics* Python library.

The micro-averaged scores (recall, precision and F1) tend to be dominated by the classifier's performance on common categories and the macro-averaged scores are more influenced by the performance on rare categories [11].In order to compute the overall metrics, we consider the macro-averaging version over the micro-averaging. A macro-average computes the metric independently for each category and then takes the average (hence treating all categories equally), the objective of this choice is to give more relevance to small categories.

4 Results

4.1 SVM

The SVM models were trained using bag-of-words with TF-IDF weighting and C value as soft-margin. Word embeddings were tested but produced worse results, probably due to the lack of data to create a good model.

For the multi-class model, a hyper-parameter C needs to be tuned to get the best results. The algorithm was trained with a C value in the range of 1 to 20. At the end, the best results were obtained using a C = 2. Table 1 resumes the algorithm's performance. We also tested the linear and non-linear models and obtained better results with the linear model.

Table 1. Classification results for multi-class SVM.

Category	Precision	Recall	F1-Score	Support
Fires	0.97	0.95	0.96	66
Meteorologic	0.96	0.82	0.89	67
Public gatherings	0.91	0.53	0.67	40
Others	0.96	0.99	0.97	768
Macro avg	0.95	0.82	0.87	941

For the binary SVM, one model was created for each of the four categories. In order to train each model we selected news labeled with the corresponding category as positive samples and all the other news as negative samples. At the end of the classification, each news is assigned to the category with the highest score. Table 2 resumes the binary SVM performance.

Table 2. Classification results for binary SVM.

Category	Precision	Recall	F1-Score	Support
Fires	0.97	0.95	0.96	66
Meteorologic	0.95	0.87	0.91	67
Public gatherings	0.84	0.65	0.73	40
Others	0.97	0.99	0.98	768
Macro avg	0.93	0.86	0.89	941

The results obtained for SVM binary classifier were slightly better than SVM multi-class and better than the other classifiers. The category "Public gatherings" is the one with the worst scores, which is understandable since it groups events of very different nature, and hence using a larger range of vocabulary.

4.2   K-Nearest Neighbors

The KNN model used bag-of-words with TF-IDF weighting. K, the number of nearest neighbours, was tuned to get the best results. The algorithm was trained with a K value in the range of 1 to 20. The best results were obtained using a K = 6. Table 3 resumes the obtained results.

Table 3. Classification results for multi-class KNN.

Category	Precision	Recall	F1-Score	Support
Fires	0.88	0.88	0.88	66
Meteorologic	0.89	0.88	0.89	67
Public gatherings	0.95	0.47	0.63	40
Others	0.96	0.98	0.97	768
Macro avg	0.92	0.80	0.84	941

4.3   Fuzzy Fingerprints

The fuzzy fingerprint parameters were optimized using the training set. We ended up with a fingerprint size $K = 150$ and a threshold value T2S2 = 0.17.

Table 4 shows the influence of stemming and the inclusion of stopwords. It shows the macro averaging F1-score of all categories excluding "Others".

Table 4. FFP score using different techniques (T2S2 = 0.17).

Techniques	F1-score
Stopwords, stemming	0.705
Stopwords, no stemming	0.683
No stopwords, stemming	0.737
Stopwords, no stemming	0.684

Overall results with T2S2 = 0.17, no stopwords and stemming are shown in Table 5. The performance is acceptable, but the results are below the SVM models and are heavily penalized by the performance in the minor "Public gatherings" category. The results are not totally unexpected, since FFP usually demand longer texts to perform well and FFP tend to outperform better other classifiers when the number of categories is much larger [14].

Table 5. FFP multi-class with no stopwords and stemming topics scores.

Category	Precision	Recall	F1-Score	Support
Fires	0.84	0.80	0.82	66
Meteorologic	0.92	0.88	0.90	67
Public gatherings	0.79	0.38	0.51	40
Others	0.94	0.98	0.96	768
Macro avg	0.89	0.76	0.81	941

Table 6. Comparison of all models.

Model	Precision	Recall	F1-Score
SVM binary	0.93	**0.86**	**0.89**
SVM multi-class	**0.95**	0.82	0.87
Fingerprints multi-class	0.89	0.76	0.81
KNN multi-class	0.88	0.80	0.83

Table 7. Category F1-Score comparison.

Category	SVM binary	SVM multi-class	FFP	KNN
Fires	**0.96**	**0.96**	0.82	0.88
Meteorologic	**0.91**	0.89	0.90	0.85
Public gatherings	**0.73**	0.67	0.51	0.61
Others	**0.98**	0.97	0.96	0.97

5 Conclusion and Future Work

This paper describes multi-label text classification experiments over a dataset containing around 3000 records of Portuguese short text news. We have performed experiments using four classification approaches, multi-class and binary SVM, KNN and Fuzzy Fingerprints. Our real world dataset is highly unbalanced (the frequency of the largest category is 4 times greater then the other 3 categories combined). Our results reveal that despite the short text size (title and description of a piece of news) it is possible to predict its category with an overall performance of about 89% F1-Score when using a binary SVM approach (Table 6). When comparing all methods, it is evident that the SVM binary outperforms the others. The proposed method achieves the best values for recall and f-score value. The Fingerprints strategy, while achieving a similar precision achieves almost 10% lower recall. One reason could be the small amount of words per piece of news, which jeopardizes heavily fingerprints' performance. And the KNN classifier, although perform better than both FFP, it has 5% less score on both precision and recall than SVM binary. FPP performance is superior to other when dealing with large number of classes which is not the situation. One reason than can explain why SVM performs better than kNN is because it can be better fine tuned than the latter. Additional testing showed that non-linear kernels displayed the same kind of results as kNN, exactly for the same reason of being unable to handle an unbalanced data set.

The category "Public gatherings" is not only the one with less training examples, but also the most heterogeneous one, which justifies why it has the worst score in every model (Table 7). Possible solutions to improve its score could be to split the category into more specific ones or increase data training "Public gatherings" records.

With the future inclusion of different news sources such as web pages, text blogs or social media posts, it is likely that longer texts are available, which might change the relative performance of the models.

Future work includes adding other news sources in order to build a more extensive and comprehensive dataset, improving the performance in the "Public Gatherings" category by dividing it into homogeneous subcategories, considering other text representation models, and experimenting with other features such as extracting locations in the news by searching for a city, region, district, etc.

References

1. Tewksbury, D., Rittenberg, J.: News on the Internet: Information and Citizenship in the 21st Century. Oxford University Press, New York (2012)
2. Gonçalves, T.: Públicos e consumos de média : o consumo de notícias e as plataformas digitais em portugal e em mais dez países. Entidade Reguladora para a comunicação social, 36 (2015)
3. Ikonomakis, M., Kotsiantis, S., Tampakas, V.: Text classification using machine learning techniques. WSEAS Trans. Comput. 4(8), 966–974 (2005)
4. Lee, K., et al.: Twitter trending topic classification. In: 2011 IEEE 11th International Conference on Data Mining Workshops (ICDMW), pp. 251–258. IEEE (2011)
5. Batista, F., Ribeiro, R.: Sentiment analysis and topic classification based on binary maximum entropy classifiers. Procesamiento de Lenguaje Natural 50, 77–84 (2013)
6. Wang, S., Manning, C.D.: Baselines and bigrams: simple, good sentiment and topic classification. In: Proceedings of the 50th Annual Meeting of the Association for Computational Linguistics: Short Papers-Volume 2. Association for Computational Linguistics (2012)
7. Feldman, R., Sanger, J.: The Text Mining Handbook. Cambridge University Press (2007)
8. Sankaranarayanan, J., Samet, H., Teitler, B.E., Lieberman, M.D., Sperling, J.: TwitterStand: news in tweets. In: Proceedings of the 17th ACM SIGSPATIAL International Conference on Advances in Geographic Information Systems, pp. 42–51 (2009)
9. Aha, D., Kibler, D., Albert, M.: Instance-based learning algorithms. Mach. Learn. 6(1), 37–66 (1991)
10. Cristianini, N., Shawe-Taylor, Y.: An introduction to support Vector Machines: and other kernel-based learning methods. Cambridge Univ Press (2000)
11. Yang, Y., Liu, X.: A re-examination of text categorization methods; Carnegie Mellon University; Proceedings of the 22nd Annual International ACM SIGIR Conference on Research and Development in Information Retrieval, pp 42–49 (1999)
12. Homem, N., Carvalho, J.: Authorship Identification and Author Fuzzy Fingerprints. In: Proceedings of the NAFIPS2011 - 30th Annual Conference of the North American Fuzzy Information Processing Society, 2011, IEEE Xplorer (2011)
13. Marujo, L., Carvalho, J.P., Gershman, A., Carbonell, J., Neto, J.P., de Matos, D.M.: Textual Event Detection Using Fuzzy Fingerprints. In: Angelov, P., et al. (eds.) Intelligent Systems'2014. AISC, vol. 322, pp. 825–836. Springer, Cham (2015). https://doi.org/10.1007/978-3-319-11313-5_72

14. Rosa, H., Batista, F., Carvalho, J.P.: Twitter topic fuzzy fingerprints. In: WCCI2014, FUZZ-IEEE,: IEEE World Congress on Computational Intelligence. International Conference on Fuzzy Systems, 2014, pp. 776–783 (2014)

15. Felgueiras, M., Batista, F., Carvalho, J.P.: Creating classification models from textual descriptions of companies using crunchbase. In: Lesot, M.-J., Vieira, S., Reformat, M.Z., Carvalho, J.P., Wilbik, A., Bouchon-Meunier, B., Yager, R.R. (eds.) IPMU 2020. CCIS, vol. 1237, pp. 695–707. Springer, Cham (2020). https://doi.org/10.1007/978-3-030-50146-4_51

16. Batista, F., Carvalho, J.P.: Text based classification of companies in Crunch-Base. In: FUZZ-IEEE2015, 2015 IEEE International Conference on Fuzzy Systems, IEEE, August 2015

17. Salton, G., Buckley, C.: Term-weighting approaches in automatic text retrieval. Inf. Process. Manage. **24**(5), 513–523 (1988)

18. Sood, S., et al.: TagAssist: automatic tag suggestion for blog posts. In: ICWSM (2007)

19. Rajaraman, A., Ullman, J.D.: Mining of Massive Datasets. Cambridge University Press (2011)

20. Cardoso-Cachopo, A., Oliveira, A.L.: An empirical comparison of text categorization methods. In: Nascimento, M.A., de Moura, E.S., Oliveira, A.L. (eds.) SPIRE 2003. LNCS, vol. 2857, pp. 183–196. Springer, Heidelberg (2003). https://doi.org/10.1007/978-3-540-39984-1_14

21. Kondachy, M.: Text Mining Application Programming; Charles River Media (2006)

22. Lim, H.S.: Improving kNN based text classification with well estimated parameters. In: Pal, N.R., Kasabov, N., Mudi, R.K., Pal, S., Parui, S.K. (eds.) ICONIP 2004. LNCS, vol. 3316, pp. 516–523. Springer, Heidelberg (2004). https://doi.org/10.1007/978-3-540-30499-9_79

23. Yang, Y.: An evaluation of statistical approaches to test categorization. J. Inf. Retrieval **1**(1/2), 67–88 (1999)

Decision Making Modeling and Applications

Reciprocal Preference-Aversion Structures

J. Tinguaro Rodríguez[1]([⊠])(iD), Camilo Franco[2](iD), and Javier Montero[1](iD)

[1] Faculty of Mathematics, Complutense University of Madrid, Madrid, Spain
{jtrodrig,monty}@mat.ucm.es
[2] European Commission - Joint Research Centre (JRC), Ispra, Italy
camilo.franco-de-los-rios@ec.europa.eu

Abstract. This paper presents an interesting particular instance of the Preference-Aversion (P-A) model, which combines the rich expressive capability allowed by the bipolar approach of that model with the simplicity resulting from imposing additive reciprocity constraints on the basic, positive/negative standard fuzzy preference structures from which the P-A model is built. Thus, the proposed reciprocal P-A model is instead built from a pair of compatible reciprocal preference structures, respectively allowing to express preference/aversion as well as lack of preference/aversion. The conjunctive combination of those basic reciprocal structures and its subsequent refinement then leads to the complete reciprocal P-A structure, which offers a different and somehow wider spectrum of representable decisional notions than that of the standard fuzzy preference structures, and in which ambivalence, a conflicting equilibrium between alternatives, provides the most notable difference with respect to standard and reciprocal fuzzy preference structures.

Keywords: Fuzzy preference structures · Preference-aversion model · Compatible reciprocal preference structures

1 Introduction

Preference relations have been a central tool for representing decision-makers attitudes and decompose the complexity of real decision-making problems for already almost a century. A main feature of preference relations is the possibility they enable to be organized/decomposed into preference structures, in which several complementary relations coexist – each one representing a distinguishable decisional notion, such as strict preference or indifference. This allows for a more rigorous and clear specification of a decision model's semantics and expressive capability, i.e. the amount and nature of the different decisional notions it can represent. However, although more expressive preference structures enable a more detailed and realistic representation of a decision problem, it is also true

This research has been partially supported by the Government of Spain (grant PGC2018-096509-B-100) and Complutense University of Madrid (group 910149).

© Springer Nature Switzerland AG 2022
D. Ciucci et al. (Eds.): IPMU 2022, CCIS 1602, pp. 85–98, 2022.
https://doi.org/10.1007/978-3-031-08974-9_7

that simpler models, with a reduced expressive capability, usually require less initial information to build the model for each pair of alternatives to be compared. Therefore, it is important to develop models that achieve an adequate balance regarding this trade-off between simplicity and expressive capability.

In this sense, fuzzy preference models (see e.g. [11]) have the ability to extend the expressive power of the classical (crisp) preference models by allowing gradable preference intensities, thus requiring basically the same amount of information as classical models while offering a more flexible framework to express the classical decisional notions of strict preference P, indifference I and incomparability J. In turn, fuzzy reciprocal preference models can be understood as a simplification of the standard fuzzy preference model [4], requiring half the initial information needed by the latter, but being able to represent just strict preference and indifference (see [8]) or, in a more general way, just strict preference and lack of preference (see [2]). At the other extreme, the Preference-Aversion (P-A) model (see [5]) requires twice the information needed by the standard model, but is able to express a rich spectrum of up to 10 different decisional notions. Let us recall that the P-A model can be understood as a bipolar [10] extension of the standard fuzzy model, built from the combination of a couple of fuzzy standard preference structures (the basic P-A structures), one devoted to capture positive aspects of the comparison between alternatives and the other devoted to represent negative aspects.

The objective of this work is to propose a new intermediate fuzzy preference model, balancing simplicity and expressive power while offering a different and somehow wider spectrum of representable decisional notions than that of the standard fuzzy model. To this aim, our proposal is to simplify the bipolar approach of the P-A model by imposing the reciprocity condition on both P-A basic structures. Therefore, the proposed reciprocal P-A model is built from a pair of compatible reciprocal structures (see again [2]), with respectively a positive and negative character. The complete reciprocal P-A structure then emerges from the conjunctive combination of these basic reciprocal structures and its subsequent refinement by grouping equivalent relations.

Then, although requiring a similar amount of initial information as the standard model, the proposed complete reciprocal P-A structure allows representing a diversity of 5 decisional notions, ranging from strong preference to strong aversion through semi-preference and semi-aversion, all of which allow to choose between a given pair of alternatives, as well as global lack of preference and ambivalence, which constitute indecisive situations. Particularly, ambivalence expresses a conflicting situation in which the positive and negative aspects gathered for the comparison between alternatives do oppose each other. As such, this ambivalence constitutes a novel decisional notion in relation with the set of notions representable by the basic reciprocal structures and the standard fuzzy model.

This paper is structured as follows: Sect. 2 recalls some basic about fuzzy standard preference structures, the P-A model and compatible reciprocal preference structures. Section 3 is devoted to present the proposed reciprocal P-A model. Finally, some conclusions are shed in Sect. 4.

2 Preliminaries

Given a set of alternatives $\mathbb{A}$, and for any pair $a, b \in \mathbb{A}$, classical (crisp) preference modelling understands the preference predicate $R(a, b)$ as "*a is at least as good as b*" or "*a is at least as wanted as b*" [4,11,14]. This predicate is represented by the relation $R : \mathbb{A}^2 \to \{0, 1\}$, usually referred to as the weak preference relation. Notice that R has to be a *reflexive* relation, such that $R(a, a) = 1$, for all $a \in \mathbb{A}$. The classical preference model assumes that the weak relation R can be decomposed in terms of three binary relations, namely *strict preference P*, *indifference I* and *incomparability J*, leading to the notion of preference structure $R = \langle P, I, J \rangle$, which is somehow characterized by the following properties:

$$P \cup I = R, \tag{1}$$

$$P \cup J = R^d, \tag{2}$$

where $R^d = \neg R^{-1}$, and $R^{-1}(a, b) = R(b, a) \; \forall a, b \in \mathbb{A}$.

2.1 Standard Fuzzy Preference Structures

Fuzzy preferences [16] were introduced to capture the *approximate* degree in which any alternative is considered to be preferred to another one. Here, the preference relation over any pair of alternatives $a, b \in \mathbb{A}$ is taken as a *valued* preference relation, such that now the weak preference predicate associated to R corresponds with the mapping $R : \mathbb{A}^2 \to [0, 1]$. Such a valued weak preference relation is also assumed to be reflexive, i.e., $R(a, a) = 1, \forall a \in \mathbb{A}$.

The standard fuzzy preference model [4] is built on the Independence of Irrelevant Alternatives axiom, assuming the existence of functions $p, i, j : [0, 1]^2 \to [0, 1]$, such that the values of $P(a, b)$, $I(a, b)$ and $J(a, b)$ depend only on the pair (a, b) through the weak preference functions $x = R(a, b)$ and $y = R(b, a)$. In this way, it is $P(a, b) = p(x, y), P^{-1}(a, b) = p(y, x), I(a, b) = i(x, y)$ and $J(a, b) = j(x, y)$. Furthermore, the model assumes that p, i, j are monotonic functions, such that $p(x, n(y))$, $i(x, y)$ and $j(n(x), n(y))$ are non-decreasing over both arguments, where n is a strict negation, and functions $i(x, y)$ and $j(x, y)$ are assumed to be symmetric (see again [4]).

The standard model of fuzzy preferences is developed from the fuzzy translation of properties (1) and (2), which is carried out through strong *De Morgan triples* (T_L, S_L, N), where T_L denotes the Lukasiewicz t-norm, S_L is its dual t-conorm and N is a strong negation. This translation results in the standard system of functional equations [4], given by:

$$x = S_L(p(x, y), i(x, y)), \tag{3}$$

$$N(y) = S_L(p(x, y), j(x, y)). \tag{4}$$

Different alternative solutions to this system provide different choices for the standard fuzzy structure (p, i, j) and its semantics, being the most relevant for the purpose of this study the solution given by

$$p(x, y) = \max\{x - y, 0\}, \tag{5}$$

$$i(x, y) = \min\{x, y\}, \tag{6}$$

$$j(x, y) = \min\{1 - x, 1 - y\}, \tag{7}$$

which, besides fulfilling the system of Eqs. (3)–(4), defines strict preference p as an asymmetrical relation, and also allows i and j to be simultaneously verified (up to a degree).

2.2 The Preference-Aversion Model

In [5], a bipolar extension of the standard fuzzy preference model was proposed in order to allow distinguishing between arguments in favor and against possible decisions (a postulate supported by relevant results in distinct scientific fields, see e.g. [1,3,15]), thus introducing a gains-losses rationality (using the terms of Cumulative Prospect Theory, see [12]) in the context of preference representation. This extension was named the Preference-Aversion (P-A) model, since its main feature is the consideration of a couple of preference structures (as that just exposed of the standard model) to separately represent such positive (preference) and negative (aversion) preference dimensions (see also [6]).

Therefore, the P-A model considers two valued weak relations for preference and aversion, $R_+, R_- : \mathbb{A}^2 \to [0, 1]$, associating with each of them an standard fuzzy structure, respectively $R_+ = \langle P, I, J \rangle$ and $R_- = \langle Z, G, H \rangle$, where the relations P, I, J retain their above-exposed meaning, while Z, G, H are just their negative counterparts –i.e. Z denotes *strict aversion*, G *indifference on aversion* and H *incomparability on aversion*.

Similarly to the standard fuzzy model, the aversion structure $R_- = \langle Z, G, H \rangle$ is built by assuming the existence of functions $z, g, h : [0, 1]^2 \to [0, 1]$ such that for any $a, b \in \mathbb{A}$ it is $Z(a, b) = z(u, v), Z^{-1}(a, b) = z(v, u), G(a, b) = g(u, v)$ and $H(a, b) = h(u, v)$, where $u = R_-(a, b)$ and $v = R_-(b, a)$. In turn, the semantics of this structure is developed through the solutions z, g, h of a system of functional equations obtained by respectively substituting p, i, j by z, g, h, and x, y by u, v in Eqs. (3) and (4). Particularly, the semantic solution given in Eqs. (5)–(7) can also be associated to this aversion structure (again substituting p, i, j by z, g, h, and x, y by u, v).

The complete P-A structure is then obtained by combining the preference and aversion basic structures through a conjunctive operator T, producing up to 16 different situations, which in turn can be characterized by means of 10 distinct relations. For instance, for any $a, b \in \mathbb{A}$ an *ambivalence* relation PZ is obtained as $PZ(a, b) = T(P(a, b), Z(a, b))$, and similarly strong preference is obtained as $PZ^{-1}(a, b) = T(P(a, b), Z^{-1}(a, b))$ (see [5] for more details).

Example 1. Let's consider the following example to illustrate the complete P-A structure, which combines the preference and aversion basic structures. Under a decision problem for selecting the type of energy fuel for municipal buildings, the local authority considers using electricity (E), gas (G), or solar photo-voltaics (S). For each fuel type, there are positive and negative reasons to choose one or the other.

Comparing the alternatives on both positive and negative arguments, we have the following preference matrices, where E is compared with all the other alternatives in the first row, and in the same way, G in the second row and S in the third row. To the left, the separate pairwise comparison values for R_+, and to the right, the corresponding values for R_-.

$$R_+ = \begin{bmatrix} 1 & 0.9 & 0.9 \\ 0.6 & 1 & 0.7 \\ 0.8 & 0.8 & 1 \end{bmatrix} \qquad R_- = \begin{bmatrix} 1 & 0.1 & 0.2 \\ 0.7 & 1 & 0.8 \\ 0.6 & 0.3 & 1 \end{bmatrix}$$

Following the complete P-A structure, the computation of the 10 distinct P-A relations allow describing the complete decision problem, given the four separate valuations for $R_+(a, b), R_+(b, a)$ and $R_-(a, b), R_-(b, a)$.

Notice that expressing the same decision problem by the standard model requires to somehow resolve the aggregation of positive and negative reasons. It entails that the decision maker or analyst somehow reveals a net preference value, incurring in an evident simplification with its associated loss of information. For example, if the net preference takes the average, minimum or maximum of both positive and negative reasons, the resulting standard preference matrices are the following, respectively.

$$R_{mean} = \begin{bmatrix} 1 & 0.5 & 0.6 \\ 0.7 & 1 & 0.8 \\ 0.7 & 0.6 & 1 \end{bmatrix} \qquad R_{min} = \begin{bmatrix} 1 & 0.1 & 0.2 \\ 0.6 & 1 & 0.7 \\ 0.6 & 0.3 & 1 \end{bmatrix} \qquad R_{max} = \begin{bmatrix} 1 & 0.9 & 0.9 \\ 0.7 & 1 & 0.8 \\ 0.8 & 0.8 & 1 \end{bmatrix}$$

Evidently, depending on the type of aggregation for both positive and negative reasons, the resulting assessment of the decision problem can be expected to differ.

2.3 Compatible Reciprocal Preference Structures

A reciprocal fuzzy preference relation is a fuzzy relation $Q : \mathbb{A}^2 \to [0, 1]$ such that the reciprocity condition $Q(a, b) + Q(b, a) = 1$ holds $\forall a, b \in \mathbb{A}$. Because of this constraint, the reciprocal model requires half the initial information needed by the standard model. However, this condition also entails that $Q(a, a) = 0.5 \ \forall a \in \mathbb{A}$, and thus reciprocal relations can not be reflexive, and neither interpreted as weak nor strict preference relations [7]. Although this fact poses an apparent incompatibility between the standard and reciprocal fuzzy preference models, various authors have tried to formally relate both models in order to acquire a better understanding of the reciprocal model in terms of the well-established semantics of the standard fuzzy preference model.

In this way, in [8] it is shown that, under the assumption that all alternatives in $\mathbb{A}$ are comparable (thus ruling out incomparability), reciprocal relations are in fact isomorphic to strongly complete weak fuzzy preference relations (i.e., such that either $R(a,b) = 1$ or $R(b,a) = 1$ for any $a, b \in \mathbb{A}$). This correspondence thus understands reciprocal relations as only being able to express strict preference P and indifference I.

From a broader perspective, in [2] it is shown that reciprocal relations can be seen as a particular aggregation of the standard preference structure of weak fuzzy relations. Particularly, let the function $f : [0,1]^2 \rightarrow [0,1]$ be given by $f(x,y) = \frac{1+x-y}{2}$. Then the following result holds:

Theorem 1. *[2] Let Q be a reciprocal relation, and let R be a weak fuzzy relation such that $Q(a,b) = f(x,y)$ for all $a, b \in \mathbb{A}$, where $x = R(a,b)$ and $y = R(b,a)$. If $\langle P, I, J \rangle$ is the preference structure associated to R through any solution of the system of functional equations of the standard model given by Eqs. (3) and (4), then $Q(a,b) = P(a,b) + 0.5(I(a,b) + J(a,b))$ for all $a, b \in \mathbb{A}$.*

This result and some of its consequences led the authors of [2] to state that reciprocal relations can be decomposed in a general way in terms of a pair of relations, strict preference P and *lack of preference* L, where $L(a,b) = S_L(I(a,b), J(a,b))$ for all $a, b \in \mathbb{A}$. Thus, this relation of lack of preference comprises in an indistinguishable manner both indifference and incomparability. Therefore, the approach in [2] does not rule incomparability out, but just asserts that in the framework of reciprocal relations both indifference and incomparability collapse in a single epistemic situation of lack of preference. Similarly to indifference and incomparability, this lack of preference relation can be shown to be symmetric, i.e., it is $L(a,b) = L(b,a)$ for any $a, b, \in \mathbb{A}$.

Furthermore, in [2] it is also shown that reciprocal relations can be endowed with a particular kind of preference structure $Q = \langle P, L \rangle$. This means that the relations P and L in the previous structure can be obtained directly from any given reciprocal relation Q, and *vice versa*, the relation Q can be also recovered from the structure $\langle P, L \rangle$. Moreover, the semantics of this reciprocal preference structure can be established in a compatible way with that of the standard fuzzy preference model through the verification of a system of functional equations extending the one already described in Eqs. (3) and (4) (see [2] for more details). A particular solution of this extended system, providing an specific semantic choice for reciprocal preference structures, can be expressed for any $a, b \in \mathbb{A}$ as follows:

$$P(a,b) = \max\{2Q(a,b) - 1, 0\} \tag{8}$$

$$L(a,b) = 1 - |2Q(a,b) - 1| \tag{9}$$

Notice that, following Theorem 1 above, the reciprocal relation Q can be recovered through the expression $Q(a,b) = S_L(P(a,b), 0.5L(a,b))$ from these relations P and L. Furthermore, it can be shown that for each pair of alternatives $(a,b) \in \mathbb{A}^2$ the three relations P, P^{-1} and L define a fuzzy partition, in the sense of verifying $P(a,b) + P^{-1}(a,b) + L(a,b) = 1$. It is also important to point out that the above solution for reciprocal preference structures in Eqs. (8) and (9) is connected to the solution of standard fuzzy preference structures given in Eqs. (5)–(7). This means that it is necessary to assume that particular solution of the standard structure for the solution of the reciprocal structure to acquire its declared semantics of strict preference and lack of preference in a compatible way with the standard semantics (see [2]).

3 Reciprocal P-A Structures

As just described, the P-A model is built from a pair of independent preference structures, one associated with a positive character of preference or desire, and the other with a negative character of aversion. On the other hand, reciprocal relations have been shown to admit a decomposition in terms of its own preference structure, whose semantics can be established rigorously in a compatible way with the well-established semantics of the standard model. Thus, our proposal here is to combine the bipolar approach of the P-A model with the simplicity of the reciprocal model by imposing the reciprocity condition on both dimensions of the P-A model. That is, our aim is to develop a reciprocal P-A model, in which the basic preference and aversion structures are built from reciprocal relations.

3.1 Reciprocal Basic Preference and Aversion Structures

Let us start by considering the positive dimension of preference/desire, which is now built from a fuzzy reciprocal relation Q_+ instead of from a weak fuzzy preference relation R_+. Following the exposition above, this reciprocal relation Q_+ can be associated with a reciprocal preference structure $Q_+ = \langle P, L_+ \rangle$, where P stands for strict preference and L_+ denotes lack of (positive) preference. The relations P and L_+ are obtained from Q_+ following Eqs. (8) and (9). Therefore, attending only to this positive dimension, three relevant situations can hold up to a degree for any pair of alternatives $(a, b) \in \mathbb{A}^2$: strict preference P, its inverse P^{-1} and lack of preference L_+. Let us remark that, due to their definition, L_+ is a symmetric relation, and strict preference P and its inverse P^{-1} are mutually exclusive, i.e. if $P(a, b) > 0$ then $P^{-1}(a, b) = 0$, and reciprocally if $P^{-1}(a, b) > 0$ then $P(a, b) = 0$.

This setting is replicated for the negative dimension of aversion, which is thus now built from a fuzzy reciprocal relation Q_- instead of from a weak fuzzy relation R_-. Again, Q_- is associated through Eqs. (8) and (9) with a reciprocal preference structure $Q_- = \langle Z, L_- \rangle$, where Z denotes strict aversion and L_- denotes *lack of aversion*. Let us remark that Q_-, and hence also Z and L_-, are assessed independently from the positive structure $Q_+ = \langle P, L_+ \rangle$, and are devoted to represent negative aspects in the comparison between alternatives in $\mathbb{A}$. In this sense, $Z(a, b) \in [0, 1]$ evaluates the degree up to which alternative a is strictly more rejected (i.e., worse) than b. And $L_-(a, b) \in [0, 1]$ represents the degree up to which neither a nor b are more rejected than the other, either because of them being equally worse (negative indifference G) or because their negative aspects can not be compared (negative incomparability H). In other words, in the same way as (positive) lack of preference L_+ agglutinates indifference I and incomparability J, the lack of aversion relation L_- can be understood as the disjunction of G and H, comprising in an indistinguishable manner both negative indifference and negative incomparability. Of course, due to their definition through Eqs. (8) and (9) from Q_-, L_- is also a symmetric relation, and Z and its inverse Z^{-1} are mutually exclusive, rendering again three relevant situations, Z, Z^{-1} and L_-, regarding the comparison of the negative aspects of a pair of alternatives in $\mathbb{A}$.

Let us stress that for a given pair of alternatives $(a, b) \in \mathbb{A}^2$ the reciprocal basic P-A structures $Q_+ = \langle P, L_+ \rangle$ and $Q_- = \langle Z, L_- \rangle$ for this pair are built from just the comparisons $Q_+(a, b)$ and $Q_-(a, b)$, since the reciprocity condition is imposed at both positive and negative dimensions of the P-A model. This sensibly reduces the amount of initial information needed to develop this reciprocal P-A model when compared with the general P-A model.

3.2   The Complete Reciprocal P-A Structure

The basic reciprocal P-A structures $Q_+ = \langle P, L_+ \rangle$ and $Q_- = \langle Z, L_- \rangle$ allow capturing separately, in an independent way, both positive and negative aspects relevant for the comparison between pair of alternatives in $\mathbb{A}$. However, for decision-making purposes both separate dimensions have to be combined in order to provide an overall preferential status regarding each pair of alternatives.

Following the P-A methodology proposed in [5], such a combination is accomplished by *crossing* both basic reciprocal P-A structures $Q_+ = \langle P, L_+ \rangle$ and $Q_- = \langle Z, L_- \rangle$ in a conjunctive way, attending to the relevant situations allowed at each dimension. That is, the relations that compose the complete reciprocal P-A structure, to be denoted by R_{P-A}, are built from the conjunctive combination of elements in $\{P, P^{-1}, L_+\} \times \{Z, Z^{-1}, L_-\}$, for each $(a, b) \in \mathbb{A}^2$. Conjunction can be represented by different symmetric binary operators T, such as t-norms or general overlap functions [9], although non-symmetric operators may be instead convenient whenever it is needed to consider different contributions of the positive and negative dimensions in the aggregation process.

Finally, although in principle 9 (3×3) different relations will arise from this combination step, it is important to take into account that some pairs of these 9 relations are related through the inversion operation (similarly to how P is related to P^{-1}), and thus the effective number of relevant, informative relations that actually compose the complete reciprocal P-A structure R_{P-A} will be smaller. In this sense, R_{P-A} will account for the different preference-related notions that can be expressed for any given pair of alternatives $a, b \in \mathbb{A}$.

Therefore, let us first describe the 9 relations arising from the combination of the positive and negative basic reciprocal P-A structures. Let $(a, b) \in \mathbb{A}^2$. We denote the resulting relations by the concatenation of the symbols for the combined basic relations:

- $PZ(a,b) = T(P(a,b), Z(a,b))$: This relation corresponds to *ambivalence*, as there is simultaneously an strict preference of a over b and an strict aversion of a over b.
- $P^{-1}Z^{-1}(a,b) = T(P^{-1}(a,b), Z^{-1}(a,b))$: This corresponds to *inverse ambivalence*, as there is both strict preference of b over a and strict aversion of b over a.
- $PZ^{-1}(a,b) = T(P(a,b), Z^{-1}(a,b))$: Represents *strong preference* of a over b, as a is strictly preferred to b while b is strictly worse than a.
- $P^{-1}Z(a,b) = T(P^{-1}(a,b), Z(a,b))$: Corresponds to *inverse strong preference*, or equivalently to *strong aversion* of a over b, since b is strictly preferred to a while also a is strictly worse than b.
- $PL_-(a,b) = T(P(a,b), L_-(a,b))$: This relation can be understood as *semi-strong preference* of a over b, as it combines an strict preference of a over b with a lack of aversion between both alternatives.
- $P^{-1}L_-(a,b) = T(P^{-1}(a,b), L_-(a,b))$: It can be understood as *inverse semi-strong preference*, due to b being strictly preferred to a while both alternatives are equally averted.
- $L_+Z(a,b) = T(L_+(a,b), Z(a,b))$: This relation can be interpreted as *semi-strong aversion* of a over b, since there is a lack of preference between the alternatives while also a being considered strictly worse than b.
- $L_+Z^{-1}(a,b) = T(L_+(a,b), Z^{-1}(a,b))$: Corresponds to *inverse semi-strong aversion*, as there is both lack of preference between alternatives and strict aversion of b over a.
- $L_+L_-(a,b) = T(L_+(a,b), L_-(a,b))$: This last relation has to be understood as *global lack of preference*, as there is a simultaneous lack of (positive) preference and lack of aversion between the alternatives.

Notice that the above interpretation of the previous 9 relations is guided by the combination of the positive and negative preferential statuses of the pair of alternatives, as expressed by the basic reciprocal P-A structures. Such interpretation already hints to some couples of these 9 relations being related through the relation-inversion operation, in the sense of one relation of the couple affirming the same about the pair (a, b) as the other relation about the inverse pair (b, a). These so-related couples of relations should therefore be interpreted as expressing a single preferential notion, and thus should be grouped into a single element in the complete reciprocal P-A structure R_{P-A}.

To analyze the relationships between these 9 resulting relations, let us first introduce the notation $\mathcal{R}$ for the set containing all those relations, that is

$$\mathcal{R} = \{PZ, P^{-1}Z^{-1}, PZ^{-1}, P^{-1}Z, PL_-, P^{-1}L_-, L_+Z, L_+Z^{-1}, L_+L_-\}, \quad (10)$$

and consider the binary relation $\sim$ on $\mathcal{R}$ given for all $A, B \in \mathcal{R}$ by

$$A \sim B \iff (A = B) \vee (A = B^{-1}) \quad (11)$$

Elements of $\mathcal{R}$ related through $\sim$ should be associated to a single object in R_{P-A}. This is the same as to say that the complete reciprocal P-A structure has to be identified with the quotient set of $\mathcal{R}$ by $\sim$, that is, we reach the following definition of the complete reciprocal P-A structure:

$$R_{P-A} = \mathcal{R}/\sim \tag{12}$$

The following proposition guarantees the soundness of this definition.

Proposition 2. *The binary relation $\sim$ defined in Eq. (11) is an equivalence relation on $\mathcal{R}$. Furthermore, it holds that*

$$\mathcal{R}/\sim = \{[PZ], [PZ^{-1}], [PL_-], [L_+Z], [L_+L_-]\}$$

where the equivalence classes in $\mathcal{R}/\sim$ are composed as follows:

$$[PZ] = \{PZ, P^{-1}Z^{-1}\}$$

$$[PZ^{-1}] = \{PZ^{-1}, P^{-1}Z\}$$

$$[PL_-] = \{PL_-, P^{-1}L_-\}$$

$$[L_+Z] = \{L_+Z, L_+Z^{-1}\}$$

$$[L_+L_-] = \{L_+L_-\}$$

Proof. The relation $\sim$ is clearly reflexive and symmetric. And it is transitive on $\mathcal{R}$ since the elements in $\mathcal{R}$ are only related to themselves and to their inverse, thus forming two-elements equivalence classes, except for L_+L_-, which is itself symmetric. Let see this last claim first, that follows from the symmetry of L_+ and L_-: $(L_+L_-)^{-1}(a,b) = L_+L_-(b,a) = T(L_+(b,a), L_-(b,a)) = T(L_+(a,b), L_-(a,b)) = L_+L_-(a,b)$ for all $a,b \in \mathbb{A}$, and hence L_+L_- is a symmetric relation. It is also straightforward to see that the elements in the remaining 4 equivalence classes above are each other inverse. For instance, $(PZ)^{-1}(a,b) = PZ(b,a) = T(P(b,a), Z(b,a)) = T(P^{-1}(a,b), Z^{-1}(a,b)) = P^{-1}Z^{-1}(a,b)$, and so $(PZ)^{-1} = P^{-1}Z^{-1}$, hence $PZ \sim P^{-1}Z^{-1}$. The identities $(PZ^{-1})^{-1} = P^{-1}Z$, $(PL_-)^{-1} = P^{-1}L_-$ and $(L_+Z)^{-1} = L_+Z^{-1}$ can be proved by a similar reasoning.

Then, only 5 of the 9 relations initially obtained by crossing the basic reciprocal P-A structures are actually informative, since relations belonging to the same equivalence class in $\mathcal{R}/\sim$ convey the same information. The representatives we have chosen for these 5 equivalence classes are ambivalence PZ, strong preference PZ^{-1}, semi-strong preference PL_-, semi-strong aversion L_+Z and global lack of preference L_+L_-. With a kind abuse of notation, we can therefore write

$$R_{P-A} = \langle PZ, PZ^{-1}, PL_-, L_+Z, L_+L_- \rangle \tag{13}$$

Let us discuss with more detail how to understand these 5 representative relations that compose the complete reciprocal P-A structure R_{P-A}, as well as some basic guidelines regarding their exploitation for decision-making purposes.

Thus, in first place, notice that strong preference of a over b is equivalent to strong aversion of b over a, that is $PZ^{-1}(a,b) = P^{-1}Z(b,a)$. In other words, an alternative a is only strongly preferred to another alternative b if b is at the same time strongly rejected regarding a. This amounts to a situation in which the positive and negative aspects of the alternatives being compared align so to produce a conclusive, clear preference of one alternative over the other.

However, semi-strong preference PL_- and semi-strong aversion L_+Z are not each other inverse, and thus an alternative a can be semi-strongly preferred (resp. averted) to another alternative b without the latter being semi-strongly averted (resp. preferred) regarding the former. In this sense, both PL_- and L_+Z may be regarded as intermediate preferential statuses, in between strong preference and strong aversion. In these intermediate situations, some conclusive evidence appears in one of the dimensions, for instance (positive) strict preference P in the case of semi-strong preference (resp. strict aversion Z in the case of semi-strong aversion), with the evidence at the other dimension neither aligning with it to provide a strong conclusion, nor opposing it in a conflicting way, for instance due to lack of aversion L_- (resp. due to lack of preference L_+). Thus, semi-strong preference $PL_-(a,b)$ (resp. semi-strong aversion $L_+Z(a,b)$) can be understood as allowing to choose a over b (resp. b over a), although with a weaker confidence than that allowed by strong preference (resp. strong aversion) of a over b.

Otherwise, when the evidence at the basic reciprocal P-A dimensions oppose each other, a preferential status of ambivalence is reached. The capability of expressing such ambivalence is perhaps the more interesting feature enabled by P-A models like the reciprocal P-A model here proposed, as it allows representing a epistemic state of conflict, in the sense of contradicting evidence regarding the decision to be made for a pair of alternatives, that does not appear at the level of the basic reciprocal P-A structures. What to chose when there is ambivalence between a pair of alternatives? Clearly, ambivalence prevents from making a choice between the pair of considered alternatives, since the positive aspects of one alternative outperforms those of the other, while the negative aspects of the former are also more detrimental than those of the latter. Therefore, ambivalence between a pair of alternatives necessarily points to the existence of a complex, underlying multicriteria decision context, in which the relative prevalence of the different criteria, or the presence (or absence) of a *compelling argument* that leans the balance towards one of the alternatives, has to be reexamined in order to help understanding and solve the conflict (see e.g. [13] for a discussion on the relevance of conflict for understanding decisions, where delaying the choice or seeking new alternatives or criteria are likely outcomes in the presence of conflict). Anyway, it is necessary to remark that the capability of identifying this conflicting complexity is associated exclusively with the complete reciprocal P-A structure, and only available once it is built.

Finally, it is also possible that neither dimension provides conclusive evidence regarding the preference or aversion of one alternative over the other. That is the case when lack of preference L_+ and lack of aversion L_- simultaneously hold. Somehow both alternatives are considered equally indifferent or incomparable regarding both their positive and negative aspects, leading to a status of global lack of preference L_+L_-. It is important to remark that, contrary to ambivalence, this lack of global preference does not have to be understood as a conflicting situation –it just expresses a tie between the alternatives. This tie may be due to different causes: either both alternatives are too similar (or too different) to choose one over the other, or there is a general lack of information regarding both positive and negative aspects of the alternatives, or even because of the relative irrelevance of the positive and negative aspects at which the alternatives differ. In either way, this situation of global lack of preference seems to request searching for more information regarding the alternatives –in case such a non-conflicting tie needs to be broken.

Example 2. Continuing the previous Example 1, the representation of the decision problem according to the reciprocal P-A structure depends on how the basic fuzzy reciprocal relations Q_+, Q_- are built. In this way, if the upper triangular matrix is taken from R_+ and R_-, the resulting relations are:

$$Q_+ = \begin{bmatrix} \text{NA} & 0.9 & 0.9 \\ & \text{NA} & 0.7 \\ & & \text{NA} \end{bmatrix} \quad Q_- = \begin{bmatrix} \text{NA} & 0.1 & 0.2 \\ & \text{NA} & 0.8 \\ & & \text{NA} \end{bmatrix}$$

On the other hand, if the lower triangular matrix is taken from R_+ and R_-, the resulting relations are:

$$Q_+ = \begin{bmatrix} \text{NA} & & \\ 0.6 & \text{NA} & \\ 0.8 & 0.8 & \text{NA} \end{bmatrix} \quad Q_- = \begin{bmatrix} \text{NA} & & \\ 0.7 & \text{NA} & \\ 0.6 & 0.3 & \text{NA} \end{bmatrix}$$

Then, according to how the initial reciprocal fuzzy values are stated, the corresponding representation of the decision problem can be developed under the complete reciprocal P-A structure. As it is evident, the P-A model is simplified at the expense of losing information when reciprocity is forced, but such a simplification still allows maintaining the separate valuations for the positive reasons acting in favor of the alternatives, together with the negative reasons acting against them.

4   Conclusions

A novel fuzzy preference representation formalism, the reciprocal P-A model, has been proposed in this work, which constitutes a particular instance of the general P-A model [5] offering a more balanced equilibrium between simplicity and expressive capability. This particularization is attained by imposing the additive reciprocity condition on the two basic standard preference structures of the P-A model, thus reducing the amount of initial information needed to compute the model. Therefore, the proposed model preserve the bipolar approach to preference representation of the general P-A model, based on establishing a two-dimensional framework where positive and negative aspects of the comparison between pairs of alternatives are modelled in a separated, independent manner. However, instead of being based on a couple of standard fuzzy preference structures, the basic structures of the proposed reciprocal P-A model are given by two compatible reciprocal preference structures [2], that respectively enable to represent preference plus lack of preference, and aversion plus lack of aversion. The conjunctive combination of these basic reciprocal structures and its subsequent refinement lead then to the complete reciprocal P-A structure. This structure offers a different and somehow wider expressive capability than that of the standard fuzzy preference model, in which the most notable novelty is the capability of representing the decisional notion of ambivalence, a conflicting equilibrium between alternatives due to contradictory positive and negative assessments.

Future work in this line will include the detailed analysis of particular instances of the complete P-A structure obtained by applying specific conjunctive operators for the combination of the basic reciprocal structures, as well as possible extensions by means of more complex instances of fuzzy sets, such as interval-valued fuzzy sets or type-2 fuzzy sets, allowing to represent higher-level imprecision at the estimation of the preference degrees. Some work will be also devoted to assess the validity of the proposed reciprocal P-A model to harbor a need-desire decision rationality, in the line of [5], as well as to investigate the viability of a mixed P-A model combining both reciprocal and standard basic preference structures.

References

1. Cacioppo, J., Gardner, W., Berntson, G.: Beyond bipolar conceptualizations and measures: the case of attitudes and evaluative space. Pers. Soc. Psychol. Rev. **1**, 3 25 (1997)
2. Castiblanco, F., Franco, C., Rodríguez, J.T., Montero, J.: A characterization of reciprocal fuzzy preference structures and its compatibility with standard fuzzy preference structures. Fuzzy Sets Syst. **422**, 48–67 (2021)
3. O'Doherty, J., Kringelback, M., Rolls, E., Hornak, J., Andrews, C.: Abstract reward and punishment representations in the human orbitofrontal cortex. Nat. Neurosci. **4**, 95–102 (2001)
4. Fodor, J., Roubens, M.: Fuzzy Preference Modelling and Multicriteria Decision Support. Kluwer Academic Publishers, Dordrecht (1994)
5. Franco, C., Montero, J., Rodríguez, J.T.: A fuzzy and bipolar approach to preference modeling with application to need and desire. Fuzzy Sets Syst. **214**, 20–34 (2013)
6. Franco, C., Rodríguez, J.T., Montero, J.: Building the meaning of preference from logical paired structures. Knowl.-Based Syst. **83**, 32–41 (2017)
7. Herrera-Viedma, E., Herrera, F., Chiclana, F., Luque, M.: Some issues on consistency of fuzzy preference relations. Eur. J. Oper. Res. **154**, 98–109 (2004)
8. Martinetti, D., Montes, S., Díaz, S., de Baets, B.: On the correspondence between reciprocal relations and strongly complete fuzzy relations. Fuzzy Sets Syst. **322**, 19–34 (2017)

9. de Miguel, L., Gómez, D., Rodríguez, J.T., Montero, J., Bustince, H., Dimuro, G.P., Sanz, J.A.: General overlap functions. Fuzzy Sets Syst. **372**, 81–96 (2019)
10. Montero, J., Bustince, H., Franco, C., Rodríguez, J.T., Gómez, D., Pagola, M., Fernández, J., Barrenechea, E.: Paired structures and bipolar knowledge representation. Knowl.-Based Syst. **100**, 50–58 (2016)
11. Montero, J., Tejada, J., Cutello, C.: A general model for deriving preference structures from data. Eur. J. Oper. Res. **98**, 98–110 (1997)
12. Tversky, A., Kahneman, D.: Advances in prospect theory: cumulative representation of uncertainty. J. Risk Uncertain. **5**, 297–323 (1992)
13. Tversky, A., Shafir, E.: Choice under conflict: the dynamics of deferred decision. Psychol. Sci. **3**, 358–361 (1992)
14. van der Walle, B., de Baets, B., Kerre, E.: Characterizable fuzzy preference structures. Ann. Oper. Res. **80**, 105–136 (1998)
15. Yacubian, J., Gläscher, J., Schroeder, K., Sommer, T., Braus, D., Büchel, Ch.: Dissociable systems for gain-and loss-related value predictions and errors of prediction in the human brain. J. Neurosci. **26**, 9530–9537 (2006)
16. Zadeh, L.A.: Similarity relations and fuzzy orderings. Inf. Sci. **3**, 177–200 (1971)

Calibration of Radiation-Induced Cancer Risk Models According to Random Data

Luis G. Crespo[1]([⊠]), Tony C. Slaba[1], Floriane A. Poignant[2],
and Sean P. Kenny[1]

[1] NASA Langley Research Center, Hampton, VA 23681, USA
Luis.G.Crespo@nasa.gov
[2] National Institute of Aerospace, Hampton, VA 23681, USA

Abstract. This paper presents formulations for the calibration of computational models according to random data. Uncertainty in the data might be caused by a poor metrology system, measurement noise, missing or uncontrollable input variables, or by the inability to directly measure the inputs and/or outputs of interest. The forward approach performs the calibration in the space of the model's output thereby requiring repeated model simulations. Conversely, the inverse approach leverages an ensemble of solutions to an inverse problem in order to perform the calibration in the space of the model's parameters. This approach not only has a lower computational cost but also allows for the identification of more suitable parameter model classes, which in turn yield better calibrated models. The proposed formulations are used to calibrate a radiation model that informs cancer risk projections for future outer space missions according to distinct datasets.

Keywords: Model calibration · Space radiation · Uncertainty quantification

1 Introduction

Model calibration entails seeking the value of the parameters that minimize the offset between the prediction and the observations. Measurement- and model-form uncertainty often inhibit confidently prescribing constant values for such parameters. When the parameters are prescribed as a path-connected set we obtain an *Interval Predictor Model* (IPM), whereas their characterization as a random vector yields a *Random Predictor Model* (RPM) [5]. This paper proposes strategies for calibrating parametric RPMs.

The most common approach to model calibration is Bayesian inference [9]. This is a method of statistical inference in which Bayes' Theorem is used to update the probability for a hypothesis according to evidence in order to produce a RPM. The resulting random vector that characterizes the parameters, called the posterior, depends on an assumed prior random vector, and the likelihood function which in turn depends on the data, and on the model structure.

© Springer Nature Switzerland AG 2022
D. Ciucci et al. (Eds.): IPMU 2022, CCIS 1602, pp. 99–114, 2022.
https://doi.org/10.1007/978-3-031-08974-9_8

From the point of view of Bayesian inference, the *Maximum Likelihood* (ML) is a special case of maximum a posteriori estimation that assumes a uniform prior distribution. In frequentist inference, ML is a special case of an extremum estimator, with the objective function being the likelihood. Bayesian inference does not make any limiting assumptions on the manner in which the output depends on the model's parameters, nor on the structure of the resulting posterior. In spite of its high computational demands, and the potentially high sensitivity of the posterior to the assumed prior, this method is regarded as a benchmark for model calibration.

In many science and engineering applications the metrology system is accurate and the data is plentiful thereby making standard calibration techniques applicable. Unfortunately, many biological and environmental systems lack these attributes thereby making such techniques inadequate. This paper proposes model calibration techniques in which each input-output datum is characterized as a random vector of possibly dependent variables. Next we explain the reason this feature requires using a new framework. Bayesian inference and ML approaches require evaluating the likelihood function at the data points. When a datum is prescribed as a point its likelihood is a scalar. However, when a datum is random, its likelihood becomes a random variable. Hence, the likelihood of a random data is a random variable free to take on any value within a possibly wide range. Worst-case approaches, for which the lower limit of this interval is maximized, often lead to overly conservative predictors in which rare observations drive the calibration process.

This paper proposes forward and inverse approaches that maximize the expected log-likelihood or one of its quantiles. These approaches are used to calibrate a radiation quality model used by NASA to perform cancer risk assessments [7]. Of particular interest are the uncertainties associated with radiation quality effects, as they remain the dominant source of uncertainty for cancer risk projections of astronauts participating in deep space missions. In this application, the predicted variable of interest, which quantifies the increased biological effectiveness of energetic ions compared to gamma rays, cannot be directly measured. Instead, raw biological measurements e.g., tumor incidence, chromosome aberrations, are analyzed with well-documented methods to infer surrogate data. The uncertainty introduced by this process, along with the underlying aleatory uncertainty in the raw data, originate the random data.

2   Problem Statement

A *Data Generating Mechanism* (DGM) is postulated to act on a vector of input variables, $x \in \mathbb{R}^{n_x}$ in X, to produce the output variable, $y \in \mathbb{R}$ in Y. Assume that n_m input-output pairs are drawn from a DGM. In the standard case, these pairs constitute the deterministic data sequence

$$\mathcal{D} = \left\{ \left(x^{(i)}, y^{(i)} \right) \right\}_{i=1}^{n_m}, \tag{1}$$

where $x^{(i)}$ is a vector and $y^{(i)}$ is a scalar. Conversely, aleatory uncertainty yields the random data sequence

$$\mathcal{D} = \left\{ f_{x^{(i)} y^{(i)}} \right\}_{i=1}^{n_m}, \tag{2}$$

where $f_{x^{(i)} y^{(i)}}$ is the joint *Probability Density Function* (PDF) of the i-th input-output pair. Assume that a continuous *Input-output Model* of the DGM, given by

$$y(x, p), \tag{3}$$

where $p \in \mathbb{R}^{n_p}$ is a parameter, is available. The parameter p will be characterized as a vector of continuous and possibly dependent random variables, i.e.,

$$p(\theta) \sim f_p(p; \theta), \tag{4}$$

where $\theta \in \mathbb{R}^{n_\theta}$ is the hyper-parameter of the distribution. Equation (4) will be called the *Parameter Model*. Our goal is to calibrate (3) according to (2) by seeking a suitable value for θ in (4). The resulting model is called a RPM because the predicted output at any given input value is a random variable. Strategies for calibrating RPMs based upon the developments in [6] are presented next.

3   Forward Calibration Approach

The strategies below are forward formulations because they require propagating the uncertainty from the parameter space to the input-output space using (3).

For clarity in the presentation we first consider the case in which the data is deterministic. The ML estimate of (4) corresponding to (1) is

$$\theta_{\mathrm{ML}} = \underset{\theta \in \Theta}{\operatorname{argmax}} L(\theta), \tag{5}$$

where

$$L(\theta) = \sum_{i=1}^{n_m} \log z_i(\theta), \tag{6}$$

is the log-likelihood of the data,

$$z_i(\theta) = f_y\left(y^{(i)}; x^{(i)}, p(\theta)\right), \tag{7}$$

and $f_y(y; x, p(\theta))$ is the likelihood function. Having deterministic input-output pairs make the z_i's and therefore $L(\theta)$ be scalar quantities.

The extension to random data is considered next. In contrast to the deterministic case, random data makes z_i a random variable. This variable results from propagating the i-th random input-output pair through the likelihood function. Changes in θ yield changes in this function, and therefore in the distribution of all z_i's. The sum of these n_m independent random variables, $L(\theta)$, will change accordingly. In this setting the analyst could maximize any statistic of the random likelihood including its expected value,

$$\theta_{\text{FML}}^{\text{mean}} = \underset{\theta \in \Theta}{\text{argmax}} \, \mathbb{E}[L(\theta)], \tag{8}$$

or its α-quantile[1],

$$\theta_{\text{FML}}^{\text{chance}}(\alpha) = \underset{\theta \in \Theta, \, \gamma > 0}{\text{argmax}} \left\{ \gamma : F_{L(\theta)}^{-1}(\alpha) \geq \gamma \right\}, \tag{9}$$

where $\alpha \in [0,1]$ is the *Risk*, and F_L is the *Cumulative Distribution Function* (CDF) of L. These approaches will be called *Forward Maximum Likelihood* (FML) formulations.

FML approaches are computationally expensive because evaluating each candidate θ entails (a) generating $n \gg 1$ samples of p from (4), (b) propagating these samples through (3) to obtain n output samples at many input values for each of the n_m elements of $\mathcal{D}$, (c) estimating the distribution of $z_i(\theta)$ for all $i = 1, \ldots n_m$ from the output samples at such inputs, (d) estimating the expected value or the α-quantile of L, and (e) computing its gradient and providing a new candidate θ.

4   Inverse Calibration Approach

An approach to model calibration requiring several solutions to an inverse problem is presented next. These solutions allow representing each input-output pair in the data sequence $\mathcal{D}$ as an object in the parameter space. The inverse approach follows the same rationale of the FML approach with the key difference being that it measures the offset between the data and the prediction in parameter space, thereby eliminating the need to use (3) during the search for θ. However, the performance of the resulting RPM is often suboptimal because only a finite number of inverse solutions are used. The mapping from p to y for a fixed x defined by $y(x,p)$ is one-to-one, but the inverse mapping from y and x to p is generally one-to-many. The calibration approach proposed, based on one selection of such mappings, is described in the following two-step procedure. The fist step maps each input-output datum to a finite collection of points in parameter space whereas the second step maximizes a statistic of their likelihood.

4.1   Data Mapping

As before, calibration according to the deterministic data sequence in (1) is considered first. In this setting each input-output pair is represented by the parameter point

$$p^{(i)} = \underset{p \in P}{\text{argmin}} \left\{ C(p, \mathcal{D}) : y^{(i)} = y\left(x^{(i)}, p\right) \right\}, \tag{10}$$

---

[1] Chance-constrained programs such as (9) trade-off a reduction in the performance of a $100\alpha\%$ of all possible outcomes for an improved performance for the remaining $100(1 - \alpha)\%$.

where C is a cost function. Hence, $p^{(i)}$ is the parameter point that minimizes C while attaining a zero prediction error for the i-th datum. The solution to (10) for $i = 1, \ldots n_m$ yields the parameter sequence

$$\mathcal{P} = \left\{ p^{(i)} \right\}_{i=1}^{n_m}. \tag{11}$$

Hence, each input-output pair in (1) is mapped to a parameter point in (11). As such, the unobservable elements of (11) might be regarded as "surrogate data."

Random data is considered next. To this end, we first draw n_r samples from $f_{x^{(i)}y^{(i)}}$ for each $i = 1, \ldots n_m$ in (2), and denote them as $(x^{(i,k)}, y^{(i,k)})$ with $k = 1, \ldots n_r$. Each input-output pair in (2) will be represented by the collection of n_r parameter points, each of them given by

$$p^{(i,k)} = \underset{p \in P}{\operatorname{argmin}} \left\{ C(p, \mathcal{D}) : y^{(i,k)} = y\left(x^{(i,k)}, p\right) \right\}. \tag{12}$$

Hence, $p^{(i,k)}$ is the parameter point that minimizes C while attaining a zero-prediction error for the k-th sample point of the i-th input-output pair. The points $p^{(i,k)}$ can be interpreted as samples of a random vector with joint density $f_{p^{(i)}}$, whose sequence is

$$\mathcal{P} = \left\{ f_{p^{(i)}} \right\}_{i=1}^{n_m}. \tag{13}$$

Hence, each element of (2) is mapped to a random vector having the joint PDF in (13). The unobservable elements of (13) might be regarded as probabilistic "surrogate data." The sequence in (11) is a particular case of (13). An approach to compute $\mathcal{P}$ is proposed next. This approach requires the calculation of an IPM, whose mathematical structure and algorithmic implementation are presented next.

Interval Predictor Models (IPM): The IPM corresponding to (3), the input $x \in X$, and the set $P \subset \mathbb{R}^{n_p}$ is defined as

$$I_y(y; x, P) \triangleq \{ y(x, p) \text{ for all } p \in P \}. \tag{14}$$

I_y is an interval when $y(x, p)$ is continuous in p for a fixed x and P is closed, bounded, and path-connected. The IPM can be written as

$$I_y(y; x, P) = \left[\underline{y}(x, P), \, \overline{y}(x, P) \right] = \left[\min_{p \in P} y(x, p), \, \max_{p \in P} y(x, p) \right], \tag{15}$$

where $\underline{y}$ and $\overline{y}$ are the lower and upper IPM boundaries respectively. Therefore, an IPM is an interval-valued function of x given by the mapping of all possible values of p in P through the input-output model.

The multi-point IPM corresponding to the parameter sequence $\mathcal{Q} = \left\{ q^{(j)} \right\}_{j=1}^r$, where $q^{(j)} \in \mathbb{R}^{n_p}$, is given by

$$I_y(y; x, \mathcal{Q}) = \left[\underline{y}(x, \mathcal{Q}), \, \overline{y}(x, \mathcal{Q}) \right] = \left[\min_{j=1\ldots r} y\left(x, q^{(j)}\right), \, \max_{j=1\ldots r} y\left(x, q^{(j)}\right) \right]. \tag{16}$$

The spread of this IPM is defined as

$$S(\mathcal{Q}) \triangleq \mathbb{E}_x \left[\overline{y}(x, \mathcal{Q}) - \underline{y}(x, \mathcal{Q}) \right], \tag{17}$$

where $\mathbb{E}_x[\cdot]$ is the expectation with respect to the input x in X.

The following algorithms yield parameter points that minimize the IPM's spread[2].

Algorithm 1 (Deterministic Data): Set $\mathcal{Q} \leftarrow \{p_{\mathrm{LS}}\}$, where $p_{\mathrm{LS}} = \arg\min_p \sum_i^{n_m} \|y^{(i)} - y(x^{(i)}, p)\|^2$, and $i \leftarrow 1$.

1. Solve (10) with $C = S(\{\mathcal{Q}, p\})$ for $(x^{(i)}, y^{(i)})$ to obtain $p^{(i)}$.
2. If $i = n_m$ stop. If $k = n_r$ then make $i \leftarrow i + 1$, $\mathcal{Q} \leftarrow \{\mathcal{Q}, p^{(i)}\}$, and go to Step 1.

Algorithm 2 (Random Data): Set $\mathcal{Q} \leftarrow \{p_{\mathrm{LS}}\}$, where $p_{\mathrm{LS}} = \arg\min_p \sum_i^{n_m} \sum_k^{n_r} \|y^{(i,k)} - y(x^{(i,k)}, p)\|^2$, $i \leftarrow 1$ and $k \leftarrow 1$.

1. Solve (12) with $C = S(\{\mathcal{Q}, p\})$ for $(x^{(i,k)}, y^{(i,k)})$ to obtain $p^{(s)}$.
2. If $s = n_m n_r$ stop. Otherwise, $\mathcal{Q} \leftarrow \{\mathcal{Q}, p^{(s)}\}$.
3. If $k = n_r$ then $i \leftarrow i + 1$ and $k = 1$. Otherwise $k \leftarrow k + 1$. Go to Step 1.

The desired parameter sequence $\mathcal{P}$ is given by the resulting $\mathcal{Q} \setminus p_{\mathrm{LS}}$.

4.2 Optimal Calibrated Model

Next we propose strategies that maximize a statistic of the likelihood of the surrogate data in $\mathcal{P}$. This approach, called the *Inverse Maximum Likelihood* (IML), follows the same rationale of the FML approach so details are omitted. The IML formulation corresponding to the expected log-likelihood in (8) is given by

$$\theta_{\mathrm{IML}}^{\mathrm{mean}} = \underset{\theta \in \Theta}{\mathrm{argmax}} \, \mathbb{E}[L_p(\theta)], \tag{18}$$

where

$$L_p(\theta) = \sum_{i=1}^{n_m} \log \nu_i(\theta), \tag{19}$$

is the log-likelihood of the surrogate data, and ν_i is the random variable that results from propagating the random vector with density $f_{p^{(i)}}$ through the likelihood function $f_p(p; \theta)$. Likewise, the analogous formulation to maximal α-quantile in (9) is[3]

$$\theta_{\mathrm{IML}}^{\mathrm{chance}}(\alpha) = \underset{\theta \in \Theta, \, \gamma > 0}{\mathrm{argmax}} \left\{ \gamma : F_{L_p(\theta)}^{-1}(\alpha) \geq \gamma \right\}. \tag{20}$$

---

[2] The *Least-Squares* (LS) parameter point will be denoted as p_{LS} hereafter.

[3] If n_r samples of each ν_i are available, the corresponding samples of $L_p(\theta)$, $\{\sum_{i=1}^{n_m} \log \nu_i^{(j)}(\theta)\}_{j=1}^{n_r}$, can be readily used to estimate the cost functions in (18) and (20). Using a piece-wise linear CDF in (9) and (20) makes gradient-based algorithms applicable.

The effectiveness of the inverse approach depends on two factors. The first factor is how well the predicted outputs corresponding to the surrogate parameter points represent the data. The second factor is the ability of the parameter model to attain a high likelihood at such points. Hence, choosing a parameter model capable of characterizing the possibly strong parameter dependencies exhibited by $\mathcal{P}$ is critical. Sliced distributions [3,4] have served this goal well. A key benefit of the inverse approach is that the elements of (11) and (13) enable the analyst to infer a suitable distribution class in (4). This is in sharp contrast to forward approaches for which this class must be chosen upfront often without any justification other than mathematical convenience. This choice might lead to under-performing calibrated models, e.g., a RPM having a Gaussian parameter model whose θ is the ML estimate might still attain an inadequately low likelihood of the data, from which a better distribution class can not be identified.

5   Performance Analysis

5.1   Risk Analysis

A metric for quantifying the loss incurred by taking the risk α when seeking either $\theta_{\mathrm{FML}}^{\mathrm{chance}}(\alpha)$ or $\theta_{\mathrm{IML}}^{\mathrm{chance}}(\alpha)$ is

$$\ell(\alpha) = \alpha \mathbb{E}\left[L(\theta)|L(\theta) \leq F_{L(\theta)}^{-1}(\alpha)\right], \tag{21}$$

whereas the corresponding gain is

$$g(\alpha) = (1-\alpha)\mathbb{E}\left[L(\theta)|L(\theta) > F_{L(\theta)}^{-1}(\alpha)\right]. \tag{22}$$

An optimal trade-off can be sought by studying the dependency of these metrics on α.

5.2   Comparative Analysis of Competing Models

The optimality condition sought by the FML and the IML approaches differ in general. A metric for comparing the performance of an RPM with parameter θ relative to a reference RPM with parameter θ_{REF} in the probability range $[a, b]$ is

$$\rho\left(\theta, \theta_{\mathrm{REF}}, a, b, \mathcal{D}\right) = \int_a^b F_{L(\theta, \mathcal{D})} - F_{L(\theta_{\mathrm{REF}}, \mathcal{D})}\, dF, \tag{23}$$

where $0 \leq a < b \leq 1$, and $\mathcal{D}$ is the dataset the two RPMs are compared against. Positive ρ values denote an improvement whereas negative values denote a performance reduction.

6   Example: Calibration of a Radiation Dose-Effect Model

In this example we consider a radiation dose-effect model used to estimate *Relative Biological Effectiveness* (RBE) values. RBEs spanning multiple biological systems and radiation types are required to characterize the increased carcinogenic risk of the space radiation environment compared to terrestrial radiation. In this setting, the input-output model is [2]:

$$R(D, p) = p_1 D + p_2 D^2, \tag{24}$$

where D is the input (radiation dose), $p = [p_1, p_2]$ is the parameter, and R is the biological response. This model may be applied to background-subtracted experimental data. Models such as (24) must be fitted to data obtained from distinct experimental systems and endpoints such as tumor prevalence in irradiated mice and chromosome aberrations in irradiated cells. This example investigates a scaling procedure that enables these seemingly unrelated experimental data sets to be combined and therefore modeled in aggregate. Figure 1 shows the data[4] corresponding to Harderian gland tumor prevalence [2], $\mathcal{D}_1$, scaled mutation frequency [1], $\mathcal{D}_2$, and scaled chromosome aberrations [8], $\mathcal{D}_3$. In this setting, the dose $D > 0$ in Gy is the input and the non-dimensional biologic response R is the output. Each random input-output pair in (2) is given by an uncorrelated Normal distribution.

The parameter model will be given by either a Normal or a *Sliced-Normal* (SN) distribution. In the former case, p is modeled as a correlated Gaussian, so $\theta \in \mathbb{R}^5$ prescribes its mean and covariance. The linear parameter dependency in (24) along with a Normal parameter model enables the calculation of $f_R(R; D, \theta)$ in closed-form. In the latter case, p is given by a SN distribution of degree $d = 2$, so $\theta \in \mathbb{R}^{20}$. In this setting, which is the most commonly found in practice, $f_R(R; D, \theta)$ must be estimated by simulation. This not only makes the FML approach computationally expensive but also renders gradient-based algorithms inapplicable.

---

[4] Each mutational frequency data point of [1], denoted as $y_{i,M}$ for irradiation dose $D_{i,M}$, is scaled according to $y_{i,\text{M-scaled}} = y_{i,M} R(D_{i,M}, p_{HG})/R(D_{i,M}, p_M)$, where p_{HG} and p_M are the LS regression estimates of the parameters associated with the Harderian gland tumor prevalence data of [2] and mutational frequency data of [1], respectively. Similarly, each chromosome aberration data point of [8], denoted as $y_{i,CA}$ for irradiation dose $D_{i,CA}$, is scaled according to $y_{i,\text{CA-scaled}} = y_{i,CA} R(D_{i,CA}, p_{HG})/R(D_{i,CA}, p_{CA})$, where p_{CA} is the LS regression estimate of the parameters associated with the chromosome aberration data of [8].

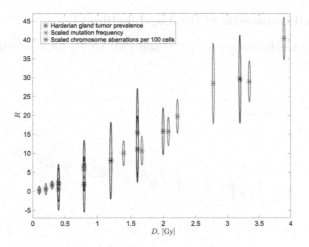

Fig. 1. Level sets of the joint density containing 95% of the data and corresponding means.

6.1 Maximal Expected Log-Likelihood

Next we apply the formulations in (8) and (18) to $\mathcal{D}_1$ by assuming a Normally distributed parameter model. Figure 2 shows the resulting RPMs. Note that the FML RPM encloses the data more tightly than the IML RPM. For both RPMs, however, the probability of the lower and upper tails of several data points in the R direction are overly small. This deficiency is addressed by the chance-constrained formulations in Sect. 6.2. Further notice that ellipsoidal level sets of the joint densities not only have a different center but also have principal axes that are not aligned.

Recall that the IML approach requires finding the solution to many inverse problems. These solutions, which result from applying Algorithm 1, are shown in Fig. 3. Each solution, shown as a colored asterisk on the bottom subplot, falls onto a linear zero-error manifold, shown as a correspondingly colored line. The distribution of the points in $\mathcal{P}$ depends strongly on the input-output pair that originates them. Other cost functions $C(p, \mathcal{D}_1)$'s led to different clouds of surrogate data (not shown) for which the corresponding RPM underperformed.

The performance[5] of the RPMs is $\mathbb{E}[L(\theta_{\mathrm{FML,N}}^{\mathrm{mean}})] = -21.76$ and $\mathbb{E}[L(\theta_{\mathrm{IML,N}}^{\mathrm{mean}})] = -22.13$. As such, the performance of the IML RPM is comparable to that of the FML RPM for a significant lower cost (see details below).

---

[5] Specifics on the setting used to compute the optimal θ will be given in the notation used hereafter. In particular, the first subscript will refer to either the FML or the IML formulation, the second subscript will refer to either the Normal or the SN parameter model, and the superscript will refer to either the mean or the quantile cost function.

6.2 Maximal α-quantile of the Log-Likelihood

Next we apply the formulations in (9) and (20) to $\mathcal{D}_1$ for $\alpha = 0.01$, $\alpha = 0.05$, and $\alpha = 0.1$.

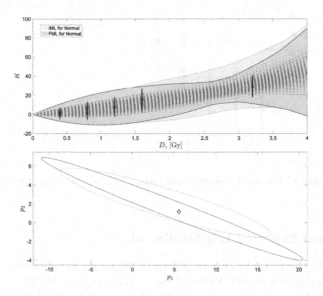

Fig. 2. Top: 2.5 quantiles of two RPMs. Data points are shown in black. Bottom: 95% prediction set for the Normal parameter model. The LS parameter point is shown as a diamond.

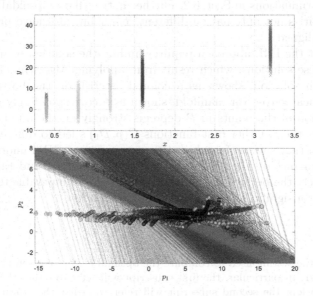

Fig. 3. Top: color-coded input-output data. Bottom: parameter points on the zero-error manifolds.

Normal Parameter Model: Figure 4 shows the RPMs and the parameter models resulting from the FML formulation. Note that the greater the risk α, the smaller the overall variance of R. As expected, the lowest-likelihood portion of the data being ignored falls onto the lower and upper tails of the data points in the R direction. The bottom figure shows level sets of the corresponding parameter models. The optimal value of the correlation coefficient is almost minus one in all three cases, thereby indicating that the parameter model accounts for parameter dependencies. The size of these sets is inversely proportional to the value of α and all three sets are centered about the LS prediction. More importantly, the principal axes of the ellipsoids are practically aligned, thereby meaning that the goal of the chance-constrained formulation can be simply attained by scaling the covariance matrix of the Normal parameter model.

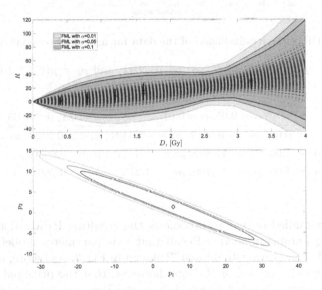

Fig. 4. Top: 2.5 quantiles of the RPMs resulting from the FML formulation for a Normal parameter model. Previous conventions apply.

The CDFs of the log-likelihood of the data L corresponding to these RPMs are shown in Fig. 5. Recall that the greater L, the better the performance. As intended, the α quantile of the CDFs for the RPMs with parameter $\theta_{\mathrm{FML}}^{\mathrm{chance}}(\alpha)$ takes on the largest values. This figure illustrates the trade-off carried out by the chance-constrained formulation: the range of the greatest $100(1-\alpha)\%$ of the log-likelihood values corresponding to $\theta_{\mathrm{FML}}^{\mathrm{chance}}(\alpha)$ takes on larger values than that of the other RPMs, whereas the range of the lowest $100\alpha\%$ of the log-likelihood values often takes smaller values. Table 1 lists the value of ρ in (23) for several RPMs. The negative entries in the first row illustrate the extent of the loss in performance, whereas the second row shows the improvement. The metric in the third row quantifies the overall effect. The application of the

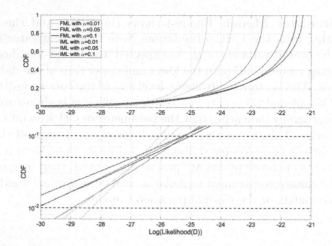

Fig. 5. CDF of the log-likelihood of the data for a Normal parameter model.

Table 1. $\rho_1 = \rho\left(\theta_{\mathrm{FML,N}}^{\mathrm{chance}}(\hat{\alpha}), \theta_{\mathrm{FML,N}}^{\mathrm{chance}}(\tilde{\alpha}), \alpha_1, \alpha_2, \mathcal{D}_1\right)$, and $\rho_2 = \rho\left(\theta_{\mathrm{IML,N}}^{\mathrm{chance}}(\hat{\alpha}), \theta_{\mathrm{IML,N}}^{\mathrm{chance}}(\tilde{\alpha})\right.$, $\left.\alpha_1, \alpha_2, \mathcal{D}_1\right)$.

	$\hat{\alpha} = 0.05,\ \tilde{\alpha} = 0.01$	$\hat{\alpha} = 0.1,\ \tilde{\alpha} = 0.05$
$\alpha_1 = 0.00,\ \alpha_2 = 0.05$	$\rho_1 = -0.0112,\ \rho_2 = -0.0421$	$\rho_1 = -0.0429,\ \rho_2 = -0.0411$
$\alpha_1 = 0.05,\ \alpha_2 = 1.00$	$\rho_1 = 1.4892,\ \rho_2 = 1.3177$	$\rho_1 = 0.8030,\ \rho_2 = 0.4098$
$\alpha_1 = 0.00,\ \alpha_2 = 1.00$	$\rho_1 = 1.4780,\ \rho_2 = 1.2757$	$\rho_1 = 0.7601,\ \rho_2 = 0.3687$

IML to $\mathcal{D}_1$ is studied next. Figure 6 shows the resulting RPMs along with the corresponding parameter model. Recall that this parameter model maximizes the likelihood of the surrogate dataset $\mathcal{P}$ shown in Fig. 3. As before, increasing α reduces the overall variance of R. Note, however, that the principal axes of the PDF are no longer aligned, whereas their center does not coincide with the LS solution. The performance of the IML RPMs is illustrated in Fig. 5 and Table 1. Note that the IML RPMs underperform the FML RPMs at the value of α used for the calibration. This is caused by the IML RPMs being optimal in p-space rather than in the desired input-output space. Table 1 shows that the trade-off enforced by the chance-constrained formulation holds for both the FML and IML formulations. This outcome is not true in general. Notice that the loss in performance in the IML RPM relative to the FML RML is small at the chosen value of α, i.e., the α quantile of the likelihood for $\theta_{\mathrm{FML}}^{\mathrm{chance}}(\alpha)$ is greater than that for $\theta_{\mathrm{IML}}^{\mathrm{chance}}(\alpha)$. However, Fig. 5 and Table 2 indicate that the RPMs based on IML perform better than the FML overall, i.e., the CDF for $\theta_{\mathrm{IML}}^{\mathrm{chance}}(\alpha)$ is to the right of the CDF for $\theta_{\mathrm{FML}}^{\mathrm{chance}}(\alpha)$ for most α values.

Table 2. Performance metric $\rho\left(\theta_{\text{FML,N}}^{\text{chance}}(\hat{\alpha}), \theta_{\text{IML,N}}^{\text{chance}}(\hat{\alpha}), \alpha_1, \alpha_2, \mathcal{D}_1\right)$.

	$\hat{\alpha} = 0.05$	$\hat{\alpha} = 0.1$
$\alpha_1 = 0.00, \ \alpha_2 = 0.05$	0.0438	0.0520
$\alpha_1 = 0.05, \ \alpha_2 = 1.00$	−0.4781	−0.0948
$\alpha_1 = 0.00, \ \alpha_2 = 1.00$	−0.4342	−0.0428

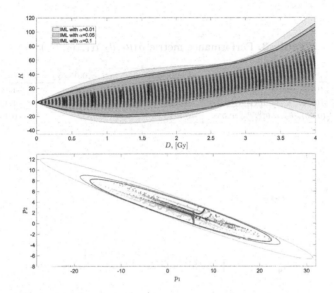

Fig. 6. IML-based RPM for a Normal model. The surrogate data is superimposed at the bottom.

Sliced-Normal Parameter Model: The IML approach was used to compute $\theta_{\text{IML,SN}}^{\text{chance}}(0.05)$. The corresponding RPM is in Fig. 7. In contrast to the mean-based RPMs above, this RPM attains a much higher likelihood over the full support sets of the random data. Besides, and in contrast to the SN-based RPM above, this RPM is more informative because its overall predicted variance is smaller. Note that the resulting parameter model is bimodal over a support set that aligns well to the optimal parameter models found above. The corresponding CDF of the log-likelihood of the data is shown in Fig. 8.

Using $\theta_{\text{IML}}^{\text{chance}}(0.05)$ as the starting point in the search for $\theta_{\text{FML,SN}}^{\text{chance}}(0.05)$ yields an RPM (not shown) that closely resembles the one in Fig. 7. The CPU time required to calculate the FML RPM is about 10 times greater than that for the IML RPM. This ratio includes the time needed to compute 500 solutions to the inverse problem for each random input-output pair. These calculations, which can be used to fit any distribution class, took about 90% of the time. This ratio decreases as the chosen number of such solutions increases. However, a moderate number of solutions is often sufficient. The CDFs corresponding to these RPMs are shown in Fig. 8. The IML RPM for the SN outperforms the IML RPM for the Normal at $\alpha = 0.05$. That is also the case for the FML RPMs. These trends, however, do not hold for other values of α. The metrics in Table 3 show that the IML RPMs outperform the FML RPMs overall, whereas the SN parameter model generally yields better RPMs than those based on a Normal at the chosen α. The enhanced flexibility of the SN yields a discontinuity in the slope of the CDF for the FML solution. However, the overall performance of $\theta_{\text{IML,N}}$ is better than that of $\theta_{\text{IML,SN}}$.

Table 3. Performance metric $\rho\left(\theta_1, \theta_2, \alpha_1, \alpha_2, \mathcal{D}_1\right)$.

	$\alpha_1 = 0,\ \alpha_2 = 0.05$	$\alpha_1 = 0.05,\ \alpha_2 = 1$	$\alpha_1 = 0,\ \alpha_2 = 1$
$\theta_1 = \theta_{\text{FML,SN}}^{\text{chance}}(0.05),\ \theta_2 = \theta_{\text{IML,SN}}^{\text{chance}}(0.05)$	0.2110	-1.3675	-1.1566
$\theta_1 = \theta_{\text{FML,SN}}^{\text{chance}}(0.05),\ \theta_2 = \theta_{\text{FML,N}}^{\text{chance}}(0.05)$	0.0606	-1.5267	-1.4662
$\theta_1 = \theta_{\text{IML,SN}}^{\text{chance}}(0.05),\ \theta_2 = \theta_{\text{IML,N}}^{\text{chance}}(0.05)$	-0.1213	-0.1332	-0.2546

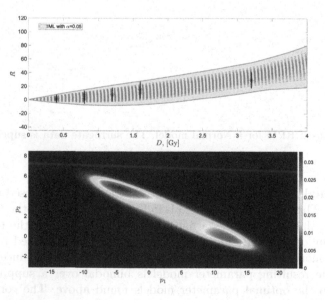

Fig. 7. IML RPM for a SN parameter model. Previous conventions apply. The joint PDF is shown at the bottom.

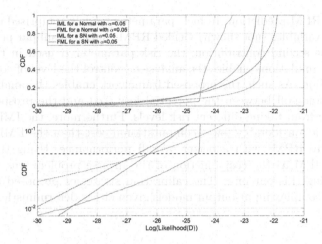

Fig. 8. CDFs of the log-likelihood of the data for several RPMs. Previous conventions apply.

6.3 Generalization Properties

The ability of RPMs with parameters $\theta_1 = \theta_{\text{FML,N}}^{\text{chance}}(0.05, \mathcal{D}_1)$, $\theta_2 = \theta_{\text{FML,N}}^{\text{chance}}(0.05, \mathcal{D}_2)$, and $\theta_3 = \theta_{\text{FML,N}}^{\text{chance}}(0.05, \mathcal{D}_3)$ to accurately characterize all three datasets is evaluated next using the metric in Table 4. The better the RPM the closer ρ is to zero. Note that $\theta_{\text{FML,N}}^{\text{chance}}(\alpha, \mathcal{D}_1)$ provides the best characterization but the performance of all three RPMs is significantly lower than that of $\theta_{\text{FML,N}}^{\text{chance}}(\alpha, \mathcal{D}_1 \cup \mathcal{D}_2 \cup \mathcal{D}_3)$, which is the preferred RPM.

Table 4. Performance metric $\rho\left(\theta, \theta_{\text{FML,N}}^{\text{chance}}(\alpha, \mathcal{D}_1 \cup \mathcal{D}_2 \cup \mathcal{D}_3), \alpha_1, \alpha_2, \mathcal{D}_1 \cup \mathcal{D}_2 \cup \mathcal{D}_3\right)$.

	$\alpha_1 = 0, \alpha_2 = 0.05$	$\alpha_1 = 0.05, \alpha_2 = 1$	$\alpha_1 = 0, \alpha_2 = 1$
$\theta_{\text{FML,N}}^{\text{chance}}(\alpha, \mathcal{D}_1)$	0.0131	-4.2705	-4.2574
$\theta_{\text{FML,N}}^{\text{chance}}(\alpha, \mathcal{D}_2)$	-4.4118	-1.8828	-6.2947
$\theta_{\text{FML,N}}^{\text{chance}}(\alpha, \mathcal{D}_3)$	-11.2133	-25.7719	-36.9852

7   Concluding Remarks

The level of conservatism in the prediction depends on both the structure of the assumed parameter model and on the manner by which the likelihood of the random data drives the optimization-based search for the calibrated model. In regard to the first contribution, using a parameter model that accounted for dependencies was instrumental in eliminating unnecessary conservatism in the resulting RPM prediction. In fact, parameter models comprised of independent random variables (not shown) yielded RPMs having a subpar performance. Regarding the second contribution, the risk parameter α used in the chance-constrained formulations enables the analyst to control the level of conservatism in the prediction. As such, the proposed framework enables the analyst to seek the optimal trade-off between the potential loss $\ell(\alpha)$ and the prospective gain $g(\alpha)$ resulting from taking different risk levels. Furthermore, the IML approach not only had a considerably lower computational cost than the FML approach but also led to RPMs with a better overall performance. Using the minimal spread of an IPM as the cost function of the inverse problem was instrumental in obtaining this outcome. The calibration strategies proposed are readily applicable to complex input-output models given either deterministic or random data.

References

1. Belli, M., Cera, F., Cherubini, R., Haque, A.: Inactivation and mutation induction in V79 cells by low energy protons: re-evaluation of the results at the LNL facility. Int. J. Radiat. Biol. **63**, 331–337 (1993)
2. Chang, P., Cuccinotta, F., Bjornstad, K.: Harderian gland tumorigenesis: low-dose and LET response. Radiat. Res. **185**(5), 449–460 (2016)
3. Crespo, L., Colbert, B., Kenny, S.: On the quantification of aleatory and epistemic uncertainty using sliced-normal distributions. Syst. Control Lett. **2019**, 134 (2019)
4. Crespo, L., Colbert, B., Slagel, S., Kenny, S.: Strategies for the robust estimation of sliced-distributions. In: IEEE CDC Conference (2021)
5. Crespo, L., Kenny, S., Giesy, D.: Staircase predictor models for reliability and risk analysis. Struct. Saf. **75**, 35–44 (2018)
6. Crespo, L., Slaba, T., Kenny, S., Swinney, M., Giesy, D.: Calibration of a radiation quality model for sparse and uncertain data. Appl. Math. Model. **95**, 734–759 (2021)
7. Cucinotta, F., Kim, M., Chappell, L.: Space radiation cancer risk projections and uncertainties. NASA Technical Paper 2013–217375 1(1) (2013)
8. George, K., Cuccinotta, F.: The influence of shielding on the biological effectiveness of accelerated particles for chromosome damage. Adv. Space Res. **39**(1), 1076–1081 (2007)
9. Kennedy, M., O'Hagan, A.: Bayesian calibration of computer models. J. R. Stat. Soc. B **63**(3), 425–464 (2001)

A Novel Variable Selection Approach Based on Multi-criteria Decision Analysis

Shengkun Xie$^{(\boxtimes)}$ (iD) and Jin Zhang

Global Management Studies, Ted Rogers School of Management, Ryerson University,
350 Victoria Street, Toronto, ON M5B 2K3, Canada
shengkun.xie@ryerson.ca
https://sites.google.com/view/shengkunxie/

Abstract. Real-world complex systems, such as transportation and insurance systems, have constantly produced massive data, and the variables used to capture their variability may be overwhelming. Therefore, it is important to balance the model's interpretability and prediction accuracy when building a predictive model. Keeping a balance on these two aspects may significantly improve prediction reliability and maintain key knowledge or information from complex systems so that the overall control and management of the complex system are statistically optimal. The paper proposes a novel approach for variable selection based on the importance measures from different data sources or different types of measures of importance obtained from machine learning models. The method formulates the variable selection problem in terms of multi-criteria decision analysis. It aims to bring a systematic way for decision-making in terms of variable selection to build more interpretable predictive models.

Keywords: Variable selection · Variable importance measure · TOPSIS · Machine learning · Complex systems

1 Introduction

In modelling massive data sets involving many different variables, selecting important variables that significantly contribute to model building is critical in real-world predictive modelling applications. Traditional approaches to variable selection include all possible subset regression, stepwise regression or partial least squares regression, to name a few [7,12,19]. These methods are more suitable for dimension reduction where a small or medium-sized predictive model is desired, and the variables are primarily numerical. The application of variable selection for dimension reduction is often linked to a model, where the model type has been proven to be a reasonable choice for a given problem. For example, generalized linear models have been widely used for predicting the financial or insurance risk [17,18]. The idea of traditional variable selection problems tends to be conditional on the selected model. However, traditional

© Springer Nature Switzerland AG 2022
D. Ciucci et al. (Eds.): IPMU 2022, CCIS 1602, pp. 115–127, 2022.
https://doi.org/10.1007/978-3-031-08974-9_9

approaches are based on measuring the data variation associated with each variable. A decision is made by running statistical tests (e.g. F-test), which may depend on whether the assumption of the underlying sampling distribution is valid. In the case that we have to deal with many variables, but we aim to create an interpretable statistical model, which can capture the major data pattern, the noise contained in the data may affect the performance on selecting importance variables [6,13]. Because of this, variable importance measures via different approaches have been developed, including interpretable model-based methods and model-agnostic methods. By ranking the variable importance measures, one can determine variables included in the interpretable statistical model.

Variable importance measures or feature selections have been an essential aspect of machine learning, and much of current research has been focusing on this type of study, particularly for random forest or decision tree models [1,3,14, 15]. The current research has been focusing on improving the performance of the selected features based on their measured importance, including the prediction performance and the model stability. In [9] ensemble-based methods were used for radar data to select essential features to reduce the dimension of input data. The study in [9] shows that the reduced features make better performances in predicting convective storms. In [8], a sensitivity-based feature importance measure using failure probability was proposed. The method effectively describes the influence of input variables on the likelihood of failure in the predictive model. Finally, in [11], to avoid an over-fitting problem in a random forest, a method that computes the loss reduction on the out-of-bag instead of the in-bag training samples was used. The idea is to improve the stability of the calculated feature values.

The main reason for studying variable importance is that their measures enable better interpretability of machine learning techniques; therefore, one can better understand obtained results and the corresponding analysis. However, the current study on variable importance techniques mainly measures the variables used for a particular machine learning model or a single dataset. Data sets from various observation years are often available in forecasting or model prediction using statistical machine learning. Due to the variability of data sets, different data sets may produce different results on the variable importance measures for the model predictors. Also, different techniques may be introduced to measure the importance of predictors. For example, in Generalized Linear Models (GLM), an Analysis of Variance (ANOVA) table is created. Variation associated with each predictor is then used to measure the importance of the given variables. When using random forest, many decision trees are created for model prediction and feature importance values are obtained in terms of mean decease accuracy or mean decrease using the Gini coefficient. However, there is no guarantee that these approaches applied to the given problem will produce consistent results. This may cause difficulty in selecting critical features for building a model.

This work proposes a variable selection approach based on the Technique for Order of Preference by Similarity to Ideal Solution (TOPSIS). TOPSIS is a multi-criteria decision analysis method, which was initially proposed by [16]. It is based on the concept that chosen alternatives should have the shortest Euclidean distance from the positive ideal solution (PIS) [2]. These chosen alternatives correspond to our model variables or variables for selection for our proposed method. The positive ideal solution corresponds to the highest feature values for each criterion or dimension that we consider. The criteria we consider in this work are the feature importance measures from random forest models obtained using different years' data sets. In addition, various measures of feature value from the random forest model are also considered. We aim to combine the variable importance measures of the variables based on the results obtained from a single data set to further decide variable selection. In this work, we also consider combining feature values from different models. For example, we combine the feature importance from the random forest model and ANOVA of GLM. To our best knowledge, it is the first time to use TOPSIS as a decision-making approach to evaluate the importance of variables in machine learning problems.

2   Methods

As we mentioned in the introduction section, we aim to rank variables based on their feature values, obtained from either different models or different measures associated with the same model or different data sets. Therefore, different ways of producing feature values are considered different criteria in decision analysis using TOPSIS.

2.1   Variable Importance Ranking by TOPSIS

Let us assume that we have m variables, $X_1, X_2, \cdots, X_m$, needed for ranking in terms of their importance for predictive modelling, and n criteria, denoted by $C_j, j = 1, 2, \ldots, n$, that we can use to evaluate these variables. The values in each C_j are variable importance measures obtained from some repeated measures dataset (i.e., data sets corresponding to different years of observation) or different methods applied to measure variable importance. These data can be then formed into a data matrix $(x_{ij})_{m \times n}$. This data matrix is an evaluation matrix used as an input data matrix for our TOPSIS approach. In the original TOPSIS, the criteria are used to determine the preference for selecting alternatives. In our case, we will determine the ranking for each variable using the criteria we consider. The process used to carry out the TOPSIS ranking for our variables is given as follows:

1. Normalize the matrix $(x_{ij})_{m \times n}$ to form the new data matrix $R = (r_{ij})_{m \times n}$ by rescaling each element using the method $r_{ij} = \frac{x_{ij}}{\sqrt{\sum_{k=1}^{m} x_{kj}^2}}, i = 1, 2, \ldots, m, j = 1, 2, \ldots, n$. Since different variable importance measures may appear to have different scales, it is important to rescale the data corresponding to different criteria.

2. Reweighing the normalized evaluation matrix, $(r_{ij})_{m \times n}$, by multiplying the column weight values w_j for each row, that is $t_{ij} = r_{ij} \cdot w_j$, for $i = 1, 2, \ldots, m$,

$j = 1, 2, \ldots, n$. Here, the weight values associated with each column, w_j, are pre-determined and will be discussed later.

3. Determine the least importance (V_l) and the most importance (V_b) as follows: $V_l = \{\langle \max(t_{ij} \mid i = 1, 2, \ldots, m) \mid j \in J_- \rangle, \langle \min(t_{ij} \mid i = 1, 2, \ldots, m) \mid j \in J_+ \rangle\}$ and $V_b = \{\langle \min(t_{ij} \mid i = 1, 2, \ldots, m) \mid j \in J_- \rangle, \langle \max(t_{ij} \mid i = 1, 2, \ldots, m) \mid j \in J_+ \rangle\}$, where, $J_+ = \{j = 1, 2, \ldots, n \mid j\}$ associated with the criteria having a positive impact to variable importance, and $J_- = \{j = 1, 2, \ldots, n \mid j\}$ associated with the criteria having a negative impact to variable importance.

4. Re-define $V_l \equiv \{t_{lj} \mid j = 1, 2, \ldots, n\}$, and $V_b \equiv \{t_{bj} \mid j = 1, 2, \ldots, n\}$. Calculate the squared Euclidean-distance between the target variable X_i and the least importance condition V_l, denoted by $d_{il} = \sqrt{\sum_{j=1}^{n}(t_{ij} - t_{lj})^2}$, for $i = 1, 2, \ldots, m$, and the squared Euclidean-distance between the target variable X_i and the most importance condition V_b, denoted by $d_{ib} = \sqrt{\sum_{j=1}^{n}(t_{ij} - t_{bj})^2}$, for $i = 1, 2, \ldots, m$.

5. Calculate the similarity to the condition that is considered to be least important, which is given as: $s_{il} = d_{il}/(d_{il} + d_{ib}), 0 \le s_{il} \le 1, i = 1, 2, \ldots, m$. $s_{il} = 1$ if and only if the variable has the condition of being the most important; and $s_{il} = 0$ if and only if the variable has the condition that is least important. s_{il} is referred to performance scores.

6. Rank the variable importance according to performance scores s_{il}, for $i = 1, 2, \ldots, m$. The highest value of s_{il} is ranked the most important variable, and the lowest value is ranked the least important variable.

2.2   Determine Weight Values for Criteria in TOPSIS

The weight values for each criterion used to evaluate the variable importance are needed to be defined before we can run the TOPSIS analysis. Suppose that we assign the weight values W_j to each criterion $C_j, j = 1, 2, \ldots, n$, then the actual weight used for the TOPSIS is a standardized value of W_j, defined as

$$w_j = W_j \Big/ \sum_{k=1}^{n} W_k, \, j = 1, 2, \ldots, n, \tag{1}$$

so that $\sum_{i=1}^{n} w_i = 1$. One approach to assigning the weight value is to let W_j be a constant number. This implies that we take equally the criteria toward ranking the variables. In the literature, the entropy method is the most popular approach used to determine the weight value W_j [4,5,10]. However, TOPSIS was proposed to deal with decision-making problems that mainly involve the count data for each criterion. Because of this, the relative frequencies $F_{ij} = r_{ij}/\sum_{i=1}^{m} r_{ij}$ are then taken as the probability to calculate the entropy for each criterion C_j, which can be computed as follows:

$$E(j) = -\frac{1}{\ln m} \sum_{i=1}^{m} F_{ij} \ln F_{ij}. \tag{2}$$

We have numerical criteria in this work, and the relativity frequency is less applicable. For a continue random variable with density function $f(x)$, entropy of the density function $f(x)$ can be approximated by first approximating $f(x)$ with a histogram of the observations $f(x_{ij})$, and it can be calculated by

$$H(C_j) = -\frac{1}{\ln w} \sum_{i=1}^{m} f(x_{ij}) \ln f(x_{ij}), \tag{3}$$

where w is the optimal bin width used to calculate the histogram $f(x_{ij})$. After obtaining the entropy for each criterion C_j, the weight values are then defined as follows:

$$w_j = \frac{1 - H(C_j)}{\sum_{k=1}^{n}(1 - H(C_j))}, j = 1, 2, \ldots, n. \tag{4}$$

Since the entropy is a measure of uncertainty and when the pattern among variables is more specific, the weight value to be assigned should be higher than when the pattern is less clear, or there is more uncertainty for the pattern among variables. So, the higher the entropy values, the higher the uncertainty; therefore, the lower the weight values.

3   Results

This work analyzes a complex and massive data set coming from Transport Canada to illustrate our proposed method of using the TOPSIS approach to select variables. This database contains all police-reported motor vehicle collisions on public roads in Canada and includes many categorical variables that have many levels. The variables (data elements) under consideration are related to fatal and injury collisions from 2014 to 2017. Therefore, we have four years of data. The variables, the associated number of levels, and the abbreviated description are given in Table 1. These variables shown in Table 1 are potentially valuable for studying the fatality of car accidents. However, these variables' effects on predictive modelling may be different. We will explore which variables are more dominant in contributing to a model for predicting the fatal accident rate.

The main reason for considering the variable selection for modelling these data sets is that some categorical variables contain many levels. For example, the vehicle sequence number takes 98 levels, which means that there are 98 different outcomes associated with this variable. Therefore, when combined with other variables with many variable levels, the predictive models will involve an extremely high volume of parameters estimates. Assuming that the number of ith variable has p_i level, the total number of parameters we have to estimate becomes $\prod_{i=1}^{n}(p_i - 1) + 1$, for a total number of categorical variables n. This number could be huge if each p_i is high. So, variable selection of such a modelling problem involving many variable levels becomes necessary.

Table 1. The table lists the variables that we consider in this work, and their abbreviated description. The dataset is from national collision database of transport Canada.

Variable	Variable level	Definition
Collision level data elements		
C_YEAR	4	Year
C_MNTH	12	Month
C_WDAY	7	Day of week
C_HOUR	24	Collision hour
C_SEV	2	Whether collision has at least one fatality
C_VEHS	Numeric	Number of vehicles involved in collision
C_CONF	18	Collision configuration on how it is occurred
C_RCFG	12	Roadway Types
C_WTHR	7	Weather condition
C_RSUR	9	Road surface
C_RALN	6	Road alignment
C_TRAF	18	Traffic control
Vehicle level data elements		
V_ID	98	Vehicle sequence number
V_TYPE	17	Vehicle types
V_YEAR	Numeric	Vehicle model year
Person level data elements		
P_ID	99	Person sequence number
P_SEX	2	Person sex
P_AGE	Numeric	Person age
P_PSN	13	Person position
P_ISEV	3	Medical treatment required
P_SAFE	7	Safety device used
P_USER	5	Road user class

We first present the feature importance plots obtained from random forest models for the 2014 to 2017 data sets. The results shown in Fig. 1 correspond to the ones using 400 different decision trees for constructing a random forest. The average decrease in the model accuracy and Gini index determines the feature importance. The results show that different data sets produce similar feature importance plots, but they are not the same in terms of feature importance ranking. We also observe that various measures to capture the feature's importance lead to different results. They appear to have some level of similarity, but the results are inconsistent among all. This may cause difficulty in drawing the final decision on dominant variables that we must select for the predictive

Table 2. The TOPSIS scores using the data matrix consisting of four years of feature values obtained from random forest, and ANOVA of GLM, the combination of feature values from random forest, and all method combined, respectively.

Variable	TOPSIS scores				
	Mean decreased accuracy	Mean decreased gini index	ANOVA method	Accuracy & Gini	All combined
P_ISEV	1.000	1.000	0.960	1.000	0.980
C_CONF	0.216	0.100	0.923	0.158	0.406
C_HOUR	0.235	0.150	0.016	0.189	0.159
C_VEHS	0.149	0.068	0.206	0.109	0.140
V_YEAR	0.152	0.139	0.002	0.145	0.122
P_AGE	0.091	0.168	0.042	0.140	0.121
C_WDAY	0.168	0.091	0.015	0.128	0.109
V_TYPE	0.178	0.028	0.071	0.114	0.105
C_RSUR	0.162	0.034	0.036	0.105	0.092
C_MNTH	0.118	0.094	0.012	0.105	0.089
C_RALN	0.150	0.040	0.033	0.100	0.087
C_TRAF	0.152	0.014	0.052	0.096	0.087
C_WTHR	0.148	0.033	0.037	0.097	0.086
P_SAFE	0.079	0.017	0.086	0.052	0.063
C_RCFG	0.106	0.016	0.018	0.068	0.059
P_ID	0.083	0.024	0.039	0.056	0.052
V_ID	0.079	0.036	0.000	0.058	0.049
P_PSN	0.070	0.020	0.018	0.048	0.042
P_USER	0.050	0.000	0.020	0.032	0.029
P_SEX	0.000	0.008	0.030	0.006	0.016

model, particularly when requiring only a small set of variables to achieve model interpretability.

We apply our proposed method to overcome this difficulty in selecting variables based on their importance. We consider both random forest and generalized linear models to produce variable importance measures. Furthermore, both average decrease in accuracy and Gini index are used to measure the variable importance. The amount of variation explained by the variables computed from the ANOVA of GLM is also considered a measure of variable importance. Because of this, we have two different ways of measuring variable importance. In addition, we have four different data sets to be applied. Therefore, each combination of the approach we are taking will be treated as a criterion for decision-making of variable importance. This is why we refer to our decision-making as a multi-criteria decision-making approach. The obtained results of their performance scores and ranks are displayed in Figs. 2 and 3. Figure 2 shows the results for the random forest model, while Fig. 3 includes the results from the ANOVA model. In addition, to give more details of the values, we present the obtained performance scores and ranks in Table 2 and 3, respectively. We see that the TOPSIS based on the feature values obtained from the random forest method performs more consistently than the ANOVA of GLM. In addition, TOPSIS with multiple measures tends to perform better than a single measure approach. This may imply that the machine learning method is more powerful than the traditional statis-

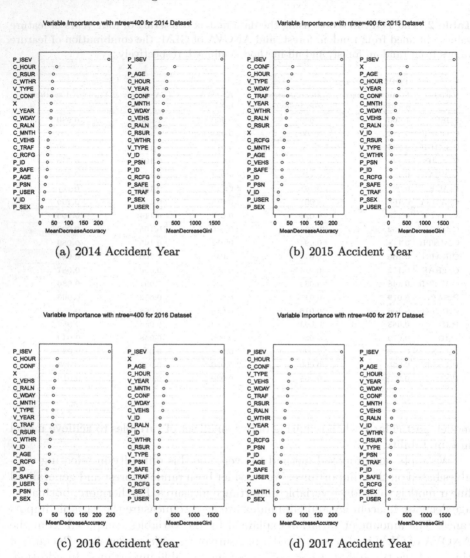

Fig. 1. The feature importance plots obtained from a random forest approach with 400 decision trees for different accident years data, where feature values were measured by mean decrease accuracy and Gini coefficients, respectively.

tical method. Also, the variability of ranking based on the different techniques may be due to the slight difference in performance scores.

More specifically, the results show that the ranks obtained from the TOP-SIS using a single input data matrix fluctuate around the results obtained by combining all information as a complete data matrix. We observe a similar pattern when the plots are based on the performance scores. However, the variation corresponding to these two different output measures (i.e. scores versus ranks)

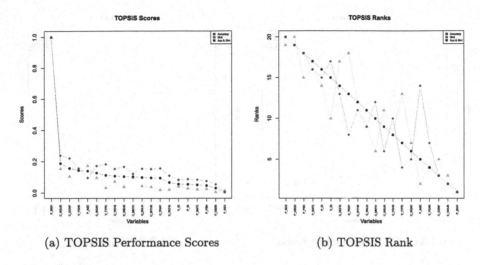

(a) TOPSIS Performance Scores   (b) TOPSIS Rank

Fig. 2. The plots of TOPSIS performance scores and ranking for the variables of interest for all methods and measures we consider, including feature values based on mean decrease accuracy and mean decrease Gini index from random forest models applied to different years of data.

Table 3. The TOPSIS ranks using the data matrix consisting of four years of feature values obtained from random forest, and ANOVA of GLM, the combination of feature values from random forest, and all method combined, respectively.

| Variable | TOPSIS ranks | | | | |
	Mean decreased accuracy	Mean decreased gini index	ANOVA method	Accuracy & Gini	All combined
P_ISEV	1	1	1	1	1
C_CONF	3	5	2	3	2
C_HOUR	2	3	16	2	3
C_VEHS	10	8	3	8	4
V_YEAR	7	4	19	4	5
P_AGE	14	2	7	5	6
C_WDAY	5	7	17	6	7
V_TYPE	4	13	5	7	8
C_RSUR	6	11	10	9	9
C_MNTH	12	6	18	10	10
C_RALN	9	9	11	11	11
C_TRAF	8	18	6	13	12
C_WTHR	11	12	9	12	13
P_SAFE	16	16	4	17	14
C_RCFG	13	17	15	14	15
P_ID	15	14	8	16	16
V_ID	17	10	20	15	17
P_PSN	18	15	14	18	18
P_USER	19	20	13	19	19
P_SEX	20	19	12	20	20

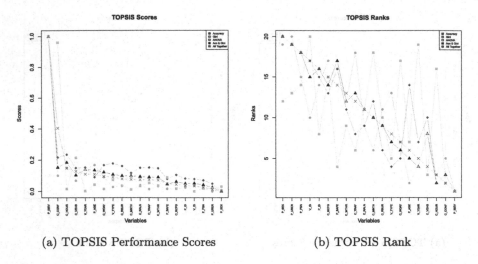

(a) TOPSIS Performance Scores   (b) TOPSIS Rank

Fig. 3. The plots of TOPSIS performance scores and ranking for the variables of interest for all methods and measures we consider, including feature values based on mean decrease accuracy and mean decrease Gini index from random forest models, and ANOVA of generalized linear models applied to different years of data.

is different. The performance scores are more stable than the ranks. The results obtained from these two models (i.e., random forest and GLM) are significantly different. The results obtained by including the ANOVA approach for measuring feature importance are more volatile than those obtained with only the random forest model. This may suggest that the decision-making is more volatile when the TOPSIS involves multiple models. Also, the performance scores lead to less volatile results. This may indicate that rankings strictly based on the performance scores to decide the order may not be ideal. Instead, decision-making based on the smoothing approach may lead to a better solution in terms of variable selection since a slight difference in the values of performance scores may imply the statistical insignificance of the testing difference on two variables. Essentially, combining all available information from different data sets, different models, and different types of variable importance measures will achieve this smoothing effect, which can be easily observed from both Figs. 2 and 3.

In Table 4, we summarize the first ten most important variables and the last ten least important variables in terms of the ranking of TOPSIS performance scores. From the table, we see that the most important variable is whether or not there is an injury or fatal event that occurred after the collision. Since our response variable in the model is the collision severity, it is intuitively making sense that P_ISEV, whether or not medical attention is required, has the most significant impact on the collision severity measures. The second important variable is C_HOUR, which tells us when the collision happened. Fatal events were often occurred due to impaired driving, such as stunt driving, and often, stunt driving occurs during a particular period of a day, such as after midnight. There-

Table 4. The table summarizes the first ten most important variables and the last ten least important variables. The variables for the most important ones are sorted in decreasing order, while the variables for the least important ones are in ascending order instead.

Ten most important variables	Ten least important variables
P_ISEV	P_SEX
C_HOUR	P_USER
C_CONF	P_PSN
V_YEAR	P_SAFE
P_AGE	P_ID
C_WDAY	V_ID
V_TYPE	C_RCFG
C_VEHS	C_TRAF
C_RSUR	C_WTHR
C_MNTH	C_RALN

fore, we believe that this variable is also a key variable in predicting the fatality of car accidents. On the other hand, we obtained P_SEX and P_USER is the least important variable in predicting the fatality of collision, where P_SEX reflects the person's gender P_USER corresponds to the road user class. This implies no particular pattern in the fatal event, whether caused by a motor vehicle driver, bicyclist, motorcyclist, or pedestrian.

4   Conclusions

In this work, a novel variable selection approach was proposed to overcome the difficulty in selecting variables for predictive modelling that involves many different categorical variables with many variable levels. A major challenge in variable selection problems is repeated measures, different uses of a model or different measures of variable importance, which often leads to an inconsistent result of variable ranking. Our entropy-based TOPSIS method was proposed to develop a more systematic measuring and ranking of the variable importance by considering all defined criteria. Because of this effort to improve the results, we can determine the variable importance without losing information from any particular data set or results from the intermediate process. Furthermore, by identifying the more dominant variables from the given data sets, we may improve the model interpretability by including only those dominant variables in the predictive models.

Regarding the variable ranking, we conclude that other methods we considered lead to inconsistent results when a single criterion is considered, such as a single accident year data set, a single model or a single measure. Our proposed method can capture the major variable ranking pattern, among other ways we considered. Because of this, we achieve better variable importance decision-making for the process that involves massive information. This proposed method will be helpful for other applications of decision-making using multiple resource information. In the future, we will investigate how the weight method used in TOPSIS affects the ranking of the variables. We plan to extend our study to more complicated data sets containing more variables.

References

1. Archer, K.J., Kimes, R.V.: Empirical characterization of random forest variable importance measures. Comput. Stat. Data Anal. **52**(4), 2249–2260 (2008)
2. Assari, A., Mahesh, T., Assari, E.: Role of public participation in sustainability of historical city: usage of topsis method. Indian J. Sci. Technol. **5**(3), 2289–2294 (2012)
3. Auret, L., Aldrich, C.: Empirical comparison of tree ensemble variable importance measures. Chemom. Intell. Lab. Syst. **105**(2), 157–170 (2011)
4. Chen, P.: Effects of normalization on the entropy-based topsis method. Expert Syst. Appl. **136**, 33–41 (2019)
5. Chen, P.: Effects of the entropy weight on topsis. Expert Syst. Appl. **168**, 114186 (2021)
6. Chong, I.G., Jun, C.H.: Performance of some variable selection methods when multicollinearity is present. Chemom. Intell. Lab. Syst. **78**(1–2), 103–112 (2005)
7. Chun, H., Keleş, S.: Sparse partial least squares regression for simultaneous dimension reduction and variable selection. J. Roy. Stat. Soc. Ser. B (Stat. Methodol.) **72**(1), 3–25 (2010)
8. Kala, Z.: New importance measures based on failure probability in global sensitivity analysis of reliability. Mathematics **9**(19), 2425 (2021)
9. Lee, H., Kim, J., Jung, S., Kim, M., Kim, B., Kim, S.: Variable importance measures based on ensemble learning methods for convective storm tracking. In: 2020 Joint 11th International Conference on Soft Computing and Intelligent Systems and 21st International Symposium on Advanced Intelligent Systems (SCIS-ISIS), pp. 1–6. IEEE (2020)
10. Li, M., Sun, H., Singh, V.P., Zhou, Y., Ma, M.: Agricultural water resources management using maximum entropy and entropy-weight-based topsis methods. Entropy **21**(4), 364 (2019)
11. Loecher, M.: Unbiased variable importance for random forests. Commun. Stat.-Theory Meth. **51**, 11413–1425 (2020)
12. Ma, Y., Zhu, L.: A review on dimension reduction. Int. Stat. Rev. **81**(1), 134–150 (2013)
13. Ojha, P.K., Roy, K.: Comparative QSARS for antimalarial endochins: importance of descriptor-thinning and noise reduction prior to feature selection. Chemom. Intell. Lab. Syst. **109**(2), 146–161 (2011)
14. Strobl, C., Boulesteix, A.L., Kneib, T., Augustin, T., Zeileis, A.: Conditional variable importance for random forests. BMC Bioinform. **9**(1), 1–11 (2008)
15. Strobl, C., Boulesteix, A.L., Zeileis, A., Hothorn, T.: Bias in random forest variable importance measures: Illustrations, sources and a solution. BMC Bioinform. **8**(1), 1–21 (2007)
16. Tzeng, G.H., Huang, J.J.: Multiple Attribute Decision Making: Methods and Applications. CRC Press, Cambridge (2011)

17. Valecký, J.: GLM analysis applied on claim severity of motor hull insurance portfolio: an empirical study. In: Actuarial Science in Theory and in Practice, p. 161 (2013)
18. Xie, S., Lawniczak, A.T.: Estimating major risk factor relativities in rate filings using generalized linear models. Int. J. Fin. Stud. **6**(4), 84 (2018)
19. Ye, G.B., Xie, X.: Learning sparse gradients for variable selection and dimension reduction. Mach. Learn. **87**(3), 303–355 (2012)

Fintech Lending Decisions: An Interpretable Knowledge-Base System for Retail and Commercial Loans

Swati Sachan[✉]

University of Liverpool Management School, Chatham St., Liverpool L69 7ZH, UK
swati.sachan@liverpool.ac.uk

Abstract. The FinTech lending industry is continuously looking forward to technologies that can revolutionize its customer experience and products. This paper presents an innovative decision-making methodology to speed up the underwriting process by accommodating both the heuristic knowledge of underwriters and supervised learning from data. The paper presents an approach to capture expert knowledge in a rule-base and map these rules with data from multiple sources in a lending firm. It is a transparent system. The reasoning behind the decision for a loan application can be explained by the activation weight of lending rules and the contribution of attributes in each rule. The system's inner dynamics can be explained to end-users and non-technical stockholders. It enables them to comply with existing laws and regulatory requirements related to AI.

Keywords: Fintech · Loan · Lending · Artificial intelligence · Interpretable

1 Introduction

Many FinTech (Financial Technology) startups are focused on offering mainstream lenders retail and commercial-purpose mortgage loans to underserved customer groups. The underwriting process of less than perfect loan applications is detailed and individualised for a given application [1]. An underwriter assesses affordability, credit, repayment, and security to thoroughly understand a given loan application. Individualistic evaluation of an application enables precise estimation of risk or ability of a borrower to repay the loan obligations at a scheduled time. A general decision-making approach adopted by most lenders consists of following several administratively pre-established rules and transformation of data gathered from credit bureau to a composite score by machine learning models.

Most high street banks have standard lending rules that enable them to undertake an automated decision-making approach. However, many lenders prefer a common-sense lending approach to provide personalised decisions to the borrower without any discrimination. The common-sense lending approach needs a robust system to back-test the underwriter's decision-making approach to ensure a thorough loan application assessment by addressing strict compliance with underwriting guidelines [2, 3]. The

© Springer Nature Switzerland AG 2022
D. Ciucci et al. (Eds.): IPMU 2022, CCIS 1602, pp. 128–140, 2022.
https://doi.org/10.1007/978-3-031-08974-9_10

lending firm aims to transform into a FinTech firm by incorporating AI algorithms to augment lending decisions. This transformation would minimise stress and unrecognised biases and errors in complex decision making for thousands of loan applications by co-evolving the relationship between human experts and machines.

FinTech transformation has two main challenges. First, the data captured from multiple sources is uncertain due to uncertain information. Quantifying the reliability of data sources and the quality of attributes to provide trustworthy decisions is very challenging. Second, the regulatory landscape has become more complex, and non-compliance consequences are even more rigorous. The lending regulations change with time, resulting in a change in the decision task environment. This implies that the training data processed by an AI algorithm must reflect the changes in the lending policy. However, the historical data would reflect decision-making under previous policies, making redundant the data and data-driven model [1].

This research presents a hybrid knowledge-based system based on belief-rule-base (BRB). It can utilise uncertain data from heterogeneous sources such as historical credit data from external agencies and lending institutions' internal data. The knowledge-base has inductive and conjunctive IF-THEN belief rules. The belief-degree of rules with zero or significantly less data support in the training set is obtained from domain experts such as underwriters and loan officers. The belief-degree of some most frequent loan applications represented by conjunctive rules having a large number of samples in training data is optimised. Expert knowledge and credit data are independent and complementary. The captured data is utilised for training parameters in a rule-base, such as the outcome of each rule as a degree of belief, the importance of each rule and attributes. A trained BRB establish the association between the default feature and the classified default status space [3]. The proposed methodology can determine the nonlinear relationships between attributes that contribute to the rejection or acceptance of a loan application. The hybrid belief rule-based model is not a black-box simulator. It can explicitly represent the underwriter's domain-specific knowledge and the judgment from historical data. The main purpose of the paper is present a methodology to develop an interpretable knowledge-based system to process loan applications by data-driven approach and by subjective judgment of the experts.

The paper is organised as follows. Section 2 describes the methodology to develop an interpretable BRB decision-making system. Section 3 demonstrates the methodology to prepare data for the BRB system. Section 4 describes the procedure to capture knowledge from the experts to codify it as IF-THEN rules in the knowledge-base. Section 5 presents the result of a dataset obtained from a UK based lending firm. The paper is concluded in Sect. 6.

2   Methodology: Interpretable Belief-Rule-Based System

An expert system has two components, knowledge-base and inference procedures. The BRB rule base rules represent the knowledge of domain experts such as loan underwriters under uncertainty. The inference procedures of BRB consist of input transformation, rule activation weight calculation, belief update and rule aggregation using evidential reasoning [4].

A belief rule extends the traditional IF-THEN rule, where a belief structure is used in the consequent part. The antecedent part of the belief rule consists of one or more antecedent attributes with associated referential values, while the consequent part consists of one consequential attribute. The IF-THEN disjunctive and conjunctive rule is considered an appropriate representation of human knowledge [5]. A disjunctive rule joint set of referential values by "or" ($\vee$) condition. A disjunctive rule activates if one of the referential values of an attribute match with the attribute in input data. A conjunctive rule joint set of referential value by "and" ($\wedge$) condition. The conjunctive rule is activated when all the referential values of the attributes in a rule match with the value in the input data. Expression (1) represents the structure of a disjunctive and conjunctive belief rule in a rule-base [4, 6, 7].

$$R_k \begin{cases} \text{disjunctive} \begin{cases} IF : \left(A_1 \ is \ A_{v,1}^k\right) \vee \cdots \vee \left(A_t \ is \ A_{v,t}^k\right) \vee \cdots \vee \left(A_{T_k} \ is \ A_{v,T}^k\right) \\ THEN : \left\{(D_1, \beta_{1,k}), \ldots, (D_n, \beta_{n,k}), \ldots, (D_N, \beta_{N,k})\right\} \\ sum \ of \ belief \ \beta_{n,k} : \sum_{n=1}^N \beta_{n,k} \leq 1 \\ rule \ weight \ 0 \leq \theta_k \leq 1 \\ attribute \ weight \ \delta_1^k, \ldots, \delta_t^k, \ldots, \delta_T^k \end{cases} \\ \text{conjunctive} \begin{cases} IF : \left(A_1 \ is \ A_{v,1}^k\right) \wedge \cdots \wedge \left(A_t \ is \ A_{v,t}^k\right) \wedge \cdots \wedge \left(A_{T_k} \ is \ A_{v,T}^k\right) \\ THEN : \left\{(D_1, \beta_{1,k}), \ldots, (D_n, \beta_{n,k}), \ldots, (D_N, \beta_{N,k})\right\} \\ sum \ of \ belief \ \beta_{n,k} : \sum_{n=1}^N \beta_{n,k} \leq 1 \\ rule \ weight \ 0 \leq \theta_k \leq 1 \\ attribute \ weight \ \delta_1^k, \ldots, \delta_t^k, \ldots, \delta_T^k \end{cases} \end{cases} \quad (1)$$

where, A_t, $t \in \{1, \ldots, T\}$ is the attributes in k^{th}, $k \in \{1, \ldots, K\}$ rule and $A_{v,t}^k$, $v \in \{1, \ldots, V_T\}$ is a referential value of a numerical or a categorical attribute. The referential value of consequence attributes is represented by D_n and $\beta_{n,k}$, $n \in \{1, \ldots, N\}$ is the belief degree of the consequence. If the sum of belief degrees in a rule is one, then the belief rule is considered complete. If it is less than 1, then is incomplete to accommodate uncertainty in expert knowledge. The belief-degree of each rule is acquired from experts; however, they may not have sufficient knowledge to provide judgment for each rule. They can give incomplete outcomes for a rule. For example, in the case of a fraudulent loan application, the consequence attribute is a fraud indication, and its set of referential values is $\{detected, notdetected\}$. Then experts can provide three types of belief-degrees for the rules; $\{(detected, 1), (notdetected, 0)\}$ or $\{(detected, 0), (notdetected, 1)\}$ to represent the certainty of judgment, $\{(detected, 0.60), (notdetected, 0.40)\}$ to represent the uncertainty, and $\{(detected, 0.70), (notdetected, \_)\}$ to represent both uncertainty and incomplete judgment by an expert. A rule and an attribute's weight in a rule are represented by $\theta_k \in [0, 1]$ and $\delta_t^k \in [0, 1]$, respectively. The inductive rules reduce the number of rules in the rule-base because BRB aims to cover exhaustive combinations to provide a decision by assessing joint evidence. The number of rules can be reduced by fusing the small set of interrelated attributes by maximum likelihood evidential reasoning (MAKER) [1, 8]. The MAKER framework combines multiple pieces of evidence by considering the reliability and weight of evidence for a given proposition in a powerset of frame of discernment.

Input data is transformed before the execution of the inference process. The relationship between input data and each referential value (for numerical or continuous data) or sub-categories (for categorical data) in the attributes of a rule is determined to generate an activation weight for each rule. Input transformation consists of the distribution of the input value of an antecedent attribute over its different referential values. The distributed value with each referential value of the antecedent attribute is called the matching degree or the degree of belief. An input data point is represented by $x_{m,t}$ where $m \in \{1, \ldots, M\}$ are the number of instances in a dataset, and $t \in \{1, \ldots, T\}$ are the number of attributes in a dataset. Expression (2) represents transformed input data point $x_{m,t}$ into a belief distribution.

$$S(x_{m,t}) = \left\{ (A_{v,t}, \alpha_{v,t}(x_{m,t})), v \in \{1, \ldots, V_t\}, \sum_{v=1}^{V} \alpha_{v,t}(x_{m,t}) = 1 \right\} \quad (2)$$

where, $\alpha_{v,t}(x_{m,t})$ is the matching degree of the mth data point to the vth referential value of the tth antecedent attribute.

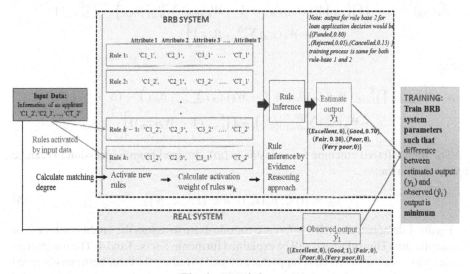

Fig. 1. BRB inference

The Methodology to transform numerical and categorical data is shown in [1]. The matching degree $S(x_{m,t})$ of a transformed data point for each attribute activates multiple rules in a rule-base. The activation weight of the activated rules is w_k. The value of $w_k = 0$ for a kth if it is not activated. The most relevant rule has the highest activation weight and vice-versa. The weight of antecedent attributes within a rule is δ_t^k. Usually, before data-driven training the initial value of θ_k, and δ_t^k is considered equal to one. BRB has main three parameters, $\beta_{n,k}$, θ_k, and δ_t^k. These parameters can be trained if sufficient data is available. The rules with high data support can be trained; however, the rules with less or zero data support could rely on knowledge acquired from domain experts. The activation weight of a kth inductive (Expression 3a) and conjunctive rules (Expression

3b) is obtained by:

$$W_k(x_m) = \frac{\theta_k \sum_{t=1}^{T}(\alpha_{v,t})^{\overline{\delta^k}_t}}{\sum_{k=1}^{K}\left[\theta_k \sum_{t=1}^{T}(\alpha_{v,t})^{\overline{\delta^k}_t}\right]} \quad \text{and} \quad \overline{\delta^k}_t = \frac{\delta_t^k}{max_{t\in\{1,\ldots,T\}}\{\delta_t^k\}} \tag{3a}$$

$$W_k(x_m) = \frac{\theta_k \prod_{t=1}^{T}(\alpha_{v,t})^{\overline{\delta^k}_t}}{\sum_{k=1}^{K}\left[\theta_k \prod_{t=1}^{T}(\alpha_{v,t})^{\overline{\delta^k}_t}\right]} \quad \text{and} \quad \overline{\delta^k}_t = \frac{\delta_t^k}{max_{t\in\{1,\ldots,T\}}\{\delta_t^k\}} \tag{3b}$$

The final inference output is aggregated belief degree of all activated and non-activated rules. The evidential reasoning (ER) approach aggregates the belief-degree of rules. It establishes a non-linear causal relationship between the antecedent attributes and consequent attributes. The following expression calculates the aggregated degree of belief:

$$\beta_n(x_m) = \mu \times \left[\prod_{k=1}^{K}\left(W_k(x_m)\beta_{n,k} + 1 - W_k(x_m)\sum_{n=1}^{N}\beta_{n,k}\right) - \prod_{k=1}^{K}(1 - W_k(x_m)\sum_{n=1}^{N}\beta_{n,k})\right] \tag{4}$$

where

$$\mu = \left[\sum_{n=1}^{N}\prod_{k=1}^{K}\left(W_k(x_m)\beta_{n,k} + 1 - W_k(x_m)\sum_{n=1}^{N}\beta_{n,k}\right) - (N-1)\prod_{k=1}^{K}(1 - W_k(x_m)\sum_{n=1}^{N}\beta_{n,k}) - \prod_{k=1}^{K}(1 - W_k(x_m))\right]^{-1}$$

The final inferred outcome for an instance (m) obtained from Expression (4) can be represented in the:

$$y_m = \{(D_n, \beta_n(x_m)), n \in \{1, \ldots, N\}\} \tag{5}$$

Figure 1 demonstrates BRB inference for rule-base to assess the creditworthiness of loan applicants. This rule-base will be explained further in Sects. 3 and 4. The parameters of rules which has sufficient data support are trained to reduce the error between observed and expected decisions by satisfying as following set of constraints:

$$\begin{cases} 0 \le \beta_{n,k} \le 1 \\ 0 \le \theta_k \le 1 \\ 0 \le \delta_t^k \le 1 \\ \sum_{n=1}^{N}\beta_{n,k} = 1 \end{cases} \tag{6}$$

The inferred output from BRB can be interpreted by the activation weight of the rule, matching degree, and attribute weight. The activation weight of the rule measures the importance of a rule for a data point, shown in Eq. (7). The sum of the importance of all rules in a rule-base is equal to one, shown in Eq. (8).

$$\mathbb{I}_k(x_m) = W_k(x_m) \tag{7}$$

$$\sum_{k=1}^{K} \mathbb{I}_k(x_m) = 1 \tag{8}$$

The contribution of a referential value of an attribute towards the final decision by the system can be measured by matching degree and attribute weight. The attribute weight is the weight of a referential value for the attribute in a given rule. By Eq. (3a & 3b), the contribution of an attribute in a rule has zero importance if the matching degree or attribute weight is zero. The importance of an attribute of an activated rule is given by:

$$\Delta_{v,t}^{k}(x_m) = (\alpha_{v,t})^{\overline{\delta_t^k}} \quad \forall v \in \{1, \dots, V\} \text{ and } \forall k \in \{1, \dots, K\} \tag{9}$$

3   Integration of Credit Data and Lending Institution Data

The data is collected from a UK based retail and commercial lending institution. This institution retrieves credit data from credit reporting agencies such as Equifax and Experian to understand applicants' credit history. Detailed credit data, including payment history, credit limit and percentage of utilization in previous months, high and low balance in the account, and any debt recovery actions about a potential applicant. The historical credit dataset has over 900 attributes (columns), excluding applicant reference and account reference number. The dataset contains two important attributes credit score and application status. The credit agencies summarize the credit profile of applicants by a credit score. Credit profile enhances with the increase of credit score. The referential values of the credit score band considered by the lending institution are shown in Fig. 2.

Credit Score	
Excellent	500+
Good	466
Fair	419
Poor	366
Very Poor	0 -278

Fig. 2. Referential values of credit score

The lending institution initializes the underwriting process by first gathering expenditure, income, employment status and other personal and loan details to assess the affordability. The collected applicant data can be associated with credit data to find the applicant's creditworthiness, which can be integrated with loan details such as loan-to-value ratio, security, and product type to suggest the final decision of funding or rejecting the applications.

Data collection and preparation is an initial and crucial step aspect of developing any data-driven model. It is a challenging and time-consuming task. The entire process involves data collection from multiple sources, integration, cleaning, feature extraction, and transformation, as illustrated in Fig. 3. Exploratory and descriptive data analysis to gain an adequate understanding of independent datasets is conducted before the commencement of the data integration process. The appropriate data preparation process is crucial in achieving a hybrid expert system that can accommodate expert knowledge and utilize data to optimize parameters in independent rules if the rule has sufficient data support.

A complete credit assessment data for retail and commercial loans can be collected from three sources – credit bureau, personal information, and loan details extracted from the electronic loan application. These multiple data sources can be integrated to configure two datasets for two different hierarchical rule-base of a BRB system. A hierarchical structure of BRB is explained in detail in Sect. 4. The data integration process requires the manipulation of relational tables using applicant reference and account reference. The credit data from a credit bureau is largely missing compared to lending institution internal data. The integrated data is pre-processed to identify codes, special characters, and missing values in the credit data. The BRB model can deal with the missing data of consequence attribute (THEN part of the rule); however, it requires complete data for the antecedent attributes (IF part of the rule) to activate all significant rules. Incomplete data will not trigger all significant rules within the rule-base. The missing data were pre-processed by the maximum likelihood evidential reasoning (MAKER) algorithm [8]. The importance of relevant features was validated by Shapley values [9] and ER feature selection approach [10].

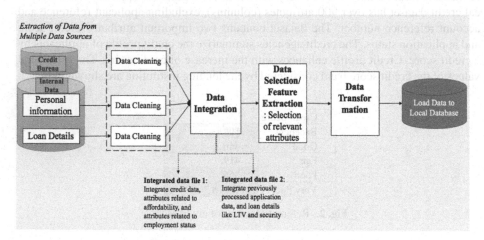

Fig. 3. Data collection and preparation

4   Knowledge Acquisition and Rule-Base Development

An algorithmic decision-making system aggregates information from multiple sources mapped with the rules in a rule-base. A rule-base represents complete knowledge of the experts in a domain to provide trustworthy decisions. A BRB system can have a non-hierarchical or hierarchical structure. A non-hierarchical BRB is a bottom-up approach where higher-level attributes at the subsequent rule-base aggregate multiple pieces of evidence for the lowest level attributes at the preceding rule-base. The rule-base at a higher level aggregates all evidence to provide a final decision.

Figure 4 demonstrates the hierarchical BRB for retail and commercial loan application assessment. It consists of two rule bases: rule-base 1 and rule-base 2. The output from rule-base 1 is utilized as input data in rule-base 2. Rule-base 1 has three different independent sub-rule-base to assess an applicant's creditworthiness by aggregating applicant credit history, affordability, and employment information. The set of referential values in rule base 1 is {*Excellent, Good, Fair, Poor, VeryPoor*}. For example, the creditworthiness belief distribution of an applicant could be {(*Excellent*, 0.1), (*Good*, 0.81), (*Fair*, 0.09), (*Poor*, 0), (*VeryPoor*, 0)}. The rule-base 2 aggregates the creditworthiness of an applicant obtained from rule base 1 and loan details such as loan-to-value (LTV), product type, and security at the subsequent hierarchy level. The aggregated output of rule-base 2 provides the final decision on the application – {*Funded, Rejected*}.

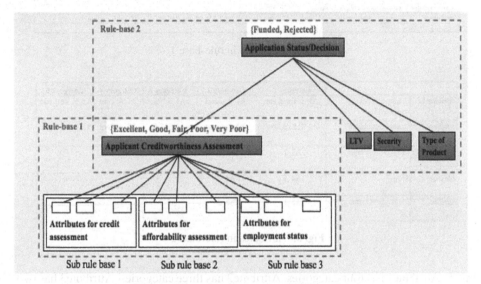

Fig. 4. Hierarchical structure of BRB model

The developers designed the hierarchal structure of BRB and selected attributes in each rule-base and sub-rule-base with the help of underwriters. The rule-base 1 assess the creditworthiness by aggregating the rules assembling referential values of attributes related to lending institution decline rules, credit defaults, applicant personal information such expenditure and employment, and outcome from affordability calculator [1] based monthly expenditure must be selected to assess the creditworthiness, shown in Fig. 5. Similarly, the rules in rule-base 2 has attributes related to loan and creditworthiness status obtained from rule-base 1, shown in Fig. 6. The attributes in rule-base 1 and 2 are denoted by C and A, respectively.

The number of rules in rule-bases increases with the number of attributes and their referential values. In Table 2, rule-base 1 has attributes 1 to 13 for credit assessment, attributes 14 to 16 for employment status assessment, and attributes 17 as the affordability test outcome. Both rule-base1 and rule-base2 have categorical data. For example, in

		Category 1	Category 2	Category 3	Category 4
Characteristics 1	Bankruptcy, County Court Judgments (CCJ), and Individual Voluntary Arrangement (IVA) Characteristics				
Attribute 1	Number of CCJ, IVA and Bankruptcy in last x year (satisfied, unsatisfied)	C1_1: greater than equal to 2	C1_2: equal to 1	C1_3: equal to zero	C1_4: No information/Unknown
⋮	⋮				
Characteristics 2	Unsecured Loan Characteristics (credit card, store cards, and Payday loans)				
Attribute 4	Number of unsecured loans	C4_1	C4_2	C4_3	
Attribute 5	Number of Credit card arrears in last x month	C5_1	C5_2	C5_3	
⋮	⋮				
Characteristics 3	Secured Loans Characteristics (Mortgage)				
Attribute 11	Number of secured loans	C11_1	C11_2	C11_3	
⋮	⋮				
Characteristics 4	Debt and Credit Search Characteristics				
Attribute 14	Number of debt collector searches last 24 month	C14_1	C14_2	C14_3	
⋮	⋮				
Characteristics 5	Affordability Characteristics				
Attribute N	Monthly Affordability	C17_1	C17_2		
⋮	⋮				

Fig. 5. Attributes in rule-base 1

		Category 1	Category 2	Category 3	Category 4	Category 5
Attribute 1	Creditworthiness of Applicant 1	A1_1: Excellent	A1_2: Good	A1_3: Fair	A1_4: Poor	A1_5: Very poor
Attribute 2	Creditworthiness of Applicant 2	A2_1: Excellent	A2_2: Good	A2_3: Fair	A2_4: Poor	A2_5: Very poor
Attribute 3	Loan-to-value ratio	A3_1	A3_2	A3_3		
Attribute 4	Security	A4_1	A4_2			
Attribute 5	Employment status	A5_1	A5_2	A5_3		

Fig. 6. Attributes in rule-base 2

Fig. 5, Attribute1 has four categories, Attribute2 has three categories, Attribute3 has two categories and so on. The number of rules in the rule-base is the sum of disjunctive and conjunctive rules. The number of conjunctive rules is an exhaustive combination of all referential values of attributes, a product of the numbers of referential values in each attribute.

5 Results

In this study, the data was obtained by a FinTech lending firm based in the UK. The data from a credit reporting agency, loan details, and personal information such as the expenditure of the applicants were cleaned and aggregated to map with the rules in the knowledge-based system. The IF-THEN belief rules represent the heuristic knowledge of underwriters. The aggregated dataset had 6030 samples of the UK based retail and commercial loans. Rule-base 1 had eighteen attributes, and rule-base 2 had seven attributes. In this lending firm, an underwriter categorizes a loan application into five levels 'excellent', 'good', 'fair', 'poor', and 'very poor'. These levels correspond with the classes defined by the credit bureau. An underwriter gives a final decision to fund or reject a loan by considering loan details and collateral. The dataset had 31% of 'very poor' and 'poor' loan applications, and the underwriters rejected 81.24% of these applications. The distribution of decisions is shown in Table 1.

Table 1. Percentage of decisions

Rule-base	Decision	Percentage
Rule-base 1	Excellent	11%
	Good	22%
	Fair	36%
	Poor	19%
	Very poor	12%
Rule-base 2	Funded	26%
	Rejected	74%

The BRB system was tested by 10-fold cross-validation. Table 2 demonstrates the test results of each fold. The system was trained by 9/10 of the instances in the dataset and tested on 1/10 of the remaining instances. F-score is a harmonic mean of recall and precision. It does not vary much for test data in 10-folds. Figure 7 illustrates the micro-average and macro-average ROC curves of the 10-fold cross-validation set. An analysis of false-rejection and false-funding decisions by the system gave insight that this system cannot capture inconsistencies in a loan application and unforeseeable changes in customer circumstances.

Table 2. Average metrics in 10-fold cross-validation set

Accuracy	Precision	Recall	F-score
0.957	0.955	0.99	0.975

The BRB system has hierarchical structure. The output from rule-base 1 is used as the input for rule-base 2. Both rule-base are trained separately and independently. The average accuracy of 10-fold cross validation set is 0.957.

A decision for each instance provided by a BRB decision-making system can be interpreted by analyzing the magnitude of parameters such as rule weight (θ_k) and attribute weight (δ_i^k). Both parameters and the matching degree ($\alpha_{v,t}$) of transformed data are accountable for the activation of IF-THEN belief rules. The set of rules activated by transformed data of an instance m $=$ 832 and its importance is shown in Fig. 8. Rule-base 1 has activated four rules, and rule-base 2 has activated seven rules. A rule consists of referential values of attributes in the dataset. The contribution of referential values towards a decision in rule-base is shown in

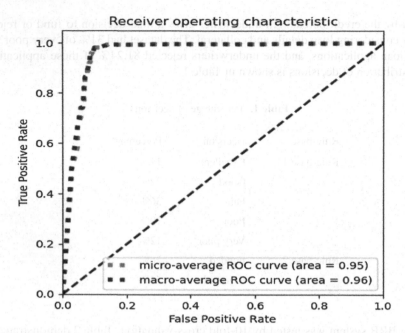

Fig. 7. Micro-average and macro-average ROC curve of 3-fold cross-validation set

Fig. 9. The aggregated belief-degree of activated rules by ER approach provided result {(*Excellent*, 0.0), (*Good*, 0.02), (*Fair*, 0.08), (*Poor*, 0.20), (*VeryPoor*, 0.70)}   and {(*Funded*, 0.37), (*Rejected*, 0.63)} for the rule-base 1 and rule-base 2, respectively.

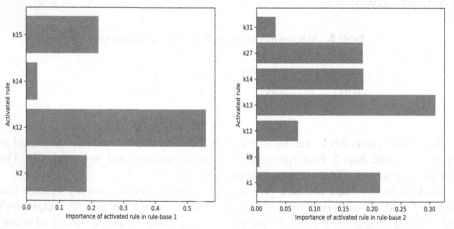

Fig. 8. Importance of rules activated by the transformed data of an instance (m = 832) in the dataset) which led to the strong conclusion of 'very poor' loan application

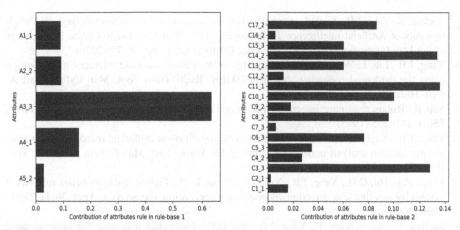

Fig. 9. Normalized contribution of each attribute towards an outcome for an instance ($m = 832$ in the dataset) which led to the strong conclusion of 'very poor' loan application. The attributes in rule-base 1 and 2 are denoted by C and A, respectively.

6 Conclusion

This paper presents a methodology to develop a hybrid knowledge base system based on BRB to obtain interpretable decisions for a loan application. A knowledge base can have inductive, conjunctive, and joint evidence rules by the MAKER framework. Combining multiple pieces of evidence and inductive rules reduces the number of rules in the knowledge base, which reduce space and training time complexity. The inference engine in BRB is based on the evidential reasoning approach that aggregates the degree of belief of all rules activated by data of an instance or a loan application. The system is both interpretable and explainable due to its ability to unravel the inner dynamics of the model and explain the reasoning behind a decision for a given instance, respectively. It can accommodate the subjective knowledge of domain experts as a degree of belief. Credit data captured from multiple sources can train the degree of belief, rule weight, and attribute weight in each rule. The rules which have scarce data support are not trained and rely on the subjective judgment of experts. The development of such a system is challenged by the labor-intensive and time-consuming task of expert knowledge acquisition. The FinTech firms need to allocate human resources to elicit the knowledge of loan underwriters into a set of rules in a knowledge base. In this paper, the hierarchical rule-base is the training and inference independently. Future work concerns developing a hierarchical BRB system that can be trained globally, and its non-linear transformation can be interpreted.

References

1. Sachan, S., Yang, J.B., Xu, D.L., Benavides, D.E., Li, Y.: An explainable AI decision-support-system to automate loan underwriting. Expert Syst. Appl. **144**, 113100 (2020)
2. Krovvidy, S.: Custom DU: a web-based business user-driven automated underwriting system. AI Mag. **29**(1), 41 (2008)

3. Sachan, S., Yang, J.B., Xu, D.L.: Global and local interpretability of belief rule base. In: Developments of Artificial Intelligence Technologies in Computation and Robotics: Proceedings of the 14th International FLINS Conference (FLINS 2020), pp. 68–75 (2020)
4. Yang, J.B., Liu, J., Wang, J., Sii, H.S., Wang, H.W.: Belief rule-base inference methodology using the evidential reasoning approach-RIMER. IEEE Trans. Syst. Man Cybern.-Part A: Syst. Hum. **36**(2), 266–285 (2006)
5. Sun, R.: Robust reasoning: integrating rule-based and similarity-based reasoning. Artif. Intell. **75**(2), 241–295 (1995)
6. Yang, J.B., Xu, D.L.: Nonlinear information aggregation via evidential reasoning in multiattribute decision analysis under uncertainty. IEEE Trans. Syst. Man Cybern. – Part A: Syst. Hum. **32**(3), 376–393 (2002)
7. Zhou, Z.J., Hu, C.H., Yang, J.B., Xu, D.L., Zhou, D.H.: Online updating belief rule-based system for pipeline leak detection under expert intervention. Expert Syst. Appl. **36**(4), 7700–7709 (2009)
8. Sachan, S., Almaghrabi, F., Yang, J.B., Xu, D.L.: Evidential reasoning for preprocessing uncertain categorical data for trustworthy decisions: an application on healthcare and finance. Expert Syst. Appl. **185**, 115597 (2021)
9. Lundberg, S.M., Lee, S.I.: A unified approach to interpreting model predictions. Adv. Neural Inf. Process. Syst. **30**, 4768–4777 (2017)
10. Almaghrabi, F., Xu, D.L., Yang, J.B.: An evidential reasoning rule based feature selection for improving trauma outcome prediction. Appl. Soft Comput. **103**, 107112 (2021)

Intuitionistic Fuzzy Selected Element Reduction Approach (IF-SERA) on Service Quality Evaluation of Digital Suppliers

Esra Çakır[1] (✉) ⓘ, Mehmet Ali Taş[2] ⓘ, and Emre Demircioğlu[1]

[1] Department of Industrial Engineering, Galatasaray University,
34349 Ortakoy, Istanbul, Turkey
{ecakir,edemircioglu}@gsu.edu.tr
[2] Department of Industrial Engineering, Turkish-German University,
34820 Beykoz, Istanbul, Turkey
mehmetali.tas@tau.edu.tr

Abstract. In the application of fuzzy multi-criteria decision-making methods, criteria weights directly affect the evaluation. The methods in the literature are used to calculate the weights of subjective or objective criteria with various techniques. In addition to these methods, the effect of criteria reduction on weighting can be investigated for fuzzy decisions. This study developed a new method for weighting criteria in decision-making problems in intuitionistic fuzzy environment. The Intuitionistic Fuzzy Selected Element Reduction Approach (IF-SERA) is based on the change caused by the reduction of a criterion in an intuitionistic fuzzy decision matrix. The evaluations of decision-makers are used to determine the overall ranking. The influence of a chosen criterion on the findings is then determined by eliminating it from the evaluation. As a result, the criterion that causes the most change is assigned the most weight. The approach yields results that are directly proportional to the criterion weights. With the implementation of the service quality evaluation for digital suppliers, the novel fuzzy weighting approach is introduced to the literature.

Keywords: Selected element reduction approach · Fuzzy multi criteria decision making · Intuitionistic fuzzy set · Service quality evaluation · Digital suppliers

1 Introduction

The disruptive transformation created by the Industry 4.0 revolution changed supply chain structures and processes, like everything else, and led to the emergence of the concept of digital supply chain [1]. The digital supply chain is a transformed supply chain with a technological approach that includes the development and implementation of information systems and the integration of the supply chain [2]. Suppliers that have incorporated advanced technologies into their processes play an important role in increasing the performance of digital supply chains [3]. There are criteria to be considered in the selection problem of the digital supplier with the highest service quality [4].

© Springer Nature Switzerland AG 2022
D. Ciucci et al. (Eds.): IPMU 2022, CCIS 1602, pp. 141–150, 2022.
https://doi.org/10.1007/978-3-031-08974-9_11

Multi-criteria decision-making (MCDM) methods are beneficial in evaluating criteria and alternatives [5]. The subjective evaluations of the decision makers are crucial while determining the criteria weights. With the aid of fuzzy numbers, these evaluations can be reflected numerically and adapted in MCDM methods applications [6]. As a fuzzy criteria weighting method, measuring the effect of excluding the criteria from the process can be adopted as an approach.

In the existing literature of weighting criteria, many researchers have developed different methods to determine the criteria weights of the multi-criteria decision making problem. The weighted sum method (WSM) is known to be the oldest and one of the most used method [7]. This approach was later changed to weighted product method (WPM) to overcome some of the loopholes associated with it [8]. The analytical hierarchy process (AHP) has become a widely used method since it was presented by Saaty [9]. Despite its ease of use, it is difficult to determine the procedure for processing information obtained from decision makers. ELECTRE (Elimination and Selection Conversion Reality) was introduced to deal with ranking relations by ranking alternatives from best to worst using pairwise comparisons between alternatives that treat each criterion separately [10]. Despite the rank relationship, the decision maker may take the risk of evaluating one alternative better than the other. The VIKOR method focuses on ranking and selection from a set of alternatives and determines the closest compromise solution to the ideal solution [11]. According to the technique of ranking preferences by similarity to the ideal solution (TOPSIS), the best alternative is the closest to the positive-ideal solution and the farthest from the negative-ideal solution [12]. The Entropy method is used to evaluate the weight of a particular problem because with this method the decision matrix for a set of candidate materials contains a certain amount of information [13]. Other commonly used methods are PROMETHEE (Preference Ranking Organization METHod for Enrichment of Evaluations) [14], COPRAS (COmplex Proportional ASsessment) [15], EDAS (Evaluation based on Distance from Average Solution) [16], BWM (Best-Worst Method) [17], SECA (Simultaneous Evaluation of Criteria and Alternatives) [18], MEREC (Method Based on the Removal Effects of Criteria) [19], and so on. All the methods used for weighting in the literature have been adapted for fuzzy numbers to make decisions in fuzzy environment.

In this study, existing methods are examined in detail and the SERA [20] is developed to be used in intuitionistic fuzzy environment by making a synthesis. In the proposed method named "Intuitionistic Fuzzy Selected Element Reduction Approach (IF-SERA)" criteria weights are derived from intuitionistic fuzzy (IF) decision matrices. Unlike the classical fuzzy set, Atanassov [21] defined IFS by notations that express the degrees of membership and non-membership. Numerous studies that use intuitionistic fuzzy sets on MCDM methods exist in the literature [22–25].

The application of the IF-SERA is illustrated on service quality evaluation of digital suppliers. In the literature, among the MCDM methods, SWARA and WASPAS were used in the selection of digital suppliers [3]. In another study, digital supplier selection was executed in the textile industry using interval type-2 fuzzy TOPSIS [26]. Similarly, a methodology containing BWM and WASPAS methods was also implemented in digital supplier evaluation for an online retail shopping company [27]. Evaluation of the service quality of the suppliers is an essential type of the problem in which MCDM methods are used [28, 29].

This paper is organized as follows. The preliminaries on intuitionistic fuzzy sets are given in Sect. 2. Following that, Sect. 3 presents the proposed new IF-MCDM procedure step by step with a detailed flowchart. Section 4 illustrate the IF-SERA on evaluation of service quality for digital suppliers. Section 5 concludes the study by discussed the future directions of proposed approach.

2 Intuitionistic Fuzzy Sets

Following the introduction of ordinary fuzzy sets by Zadeh in 1965 [30], Atanassov [21] developed fuzzy concept to present the intuitionistic fuzzy set framework that takes the degree of membership and non-membership into account. This fuzzy set defines uncertainty with the restriction that the total of membership and non-membership degrees cannot be more than one. The definitions of intuitionistic fuzzy sets are given as follows:

Definition 1: Let $X = \{x_1, x_2, \ldots, x_n\}$ be a universe of discourse, an intuitionistic fuzzy set (IFS) A in X is given by

$$A = \{< x, u_A(x),\ v_A(x) > x \in X\} \tag{1}$$

where $u_A : X \rightarrow [0, 1]$ and $v_A : X \rightarrow [0, 1]$ with the conditions $0 \leq u_A(x) + v_A(x) \leq 1, \forall x \in X$. The numbers are the membership degree $u_A(x)$ and $v_A(x)$ non-membership degree of the element x to the set A, respectively.

Given an element x of X, the pair $(u_A(x), v_A(x))$ is called an intuitionistic fuzzy value (IFV) [21]. For convenience, it can be denoted as $\tilde{a} = (u_{\tilde{a}}, v_{\tilde{a}})$ such that $u_{\tilde{a}} \in [0, 1]$, $v_{\tilde{a}} \in [0, 1]$ and $0 \leq u_{\tilde{a}} + v_{\tilde{a}} \leq 1$. The indeterminacy degree is denoted by $\pi_{\tilde{a}}$, with the conditions of $\pi_{\tilde{a}} \in [0, 1]$ and $\pi_{\tilde{a}} = 1 - u_{\tilde{a}} - v_{\tilde{a}}$.

The membership and non-membership degrees in a given universe are interpreted as the points on the intuitionistic fuzzy interpretation triangle (IFIT) for the classical formulation of IFS in Definition 1. Fig. 1 shows the geometrical representation of IFS.

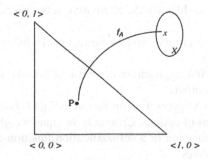

Fig. 1. Illustration of intuitionistic fuzzy set in IFIT [32]

Definition 2: [31] Let $\tilde{a} = (u_{\tilde{a}}, v_{\tilde{a}})$ be an IFV, a score function S and an accuracy function H of the IFV $\tilde{a}$ is defined as the difference of membership and non-membership function, as follows.

$$S(\tilde{a}) = u_{\tilde{a}} - v_{\tilde{a}} \text{ where } S(\tilde{a}) \in [-1, 1] \tag{2}$$

$$H(\tilde{a}) = u_{\tilde{a}} + v_{\tilde{a}} \text{ where } H(\tilde{a}) \in [0, 1] \tag{3}$$

Definition 3: [33] Let $\tilde{a}_j = \left(u_{\tilde{a}_j}, v_{\tilde{a}_j}\right)$ be a set of IF pairs. Then, the IF weighted averaging (IFWA) operator is defined as follows.

$$IFWA_W(\tilde{a}_1, \tilde{a}_2, \ldots) = \; < 1 - \prod_{j=1}^{n} \left(1 - u_{\tilde{a}_j}\right)^{w_j}, \prod_{j=1}^{n} v_{\tilde{a}_j}^{w_j} > \tag{4}$$

where $W = \{w_1, \ldots, w_n\}$ is the weighting vector of IF pairs with $w_j \in [0, 1]$ and $\sum_{j=1}^{n} w_j = 1$.

3 Methodology

The Selected Element Reduction Approach (SERA) is a relatively new method [20] for weighting criteria in a fuzzy environment proposed by the authors of this paper, regardless of whether the criterion weight is acquired directly from a decision maker or an expert. The impact of criteria on the outcomes can also be used to assess their importance. The importance of the criteria is determined by comparing the results obtained by excluding a criterion. In a fuzzy environment, linguistic scales are helpful for the expert to express thir opinions. The philosophy of the novel methodology suggested in this research is based on the relationship of element reduction with weighting in multi-criteria decision making. Furthermore, this procedure is suitable to the use of any fuzzy sets. This paper illustrates the usage of intutionistic fuzzy sets on SERA applications.

The following are the steps of the Intuitionistic Fuzzy Selected Element Reduction Approach (IF-SERA):

Step 1: Define the case. Identify the alternative sets, criteria, and decision-makers (DMs).

Step 2: Based on the criteria, collect intuitionistic fuzzy decision matrices on alternatives from the DMs.

Step 3: Using the IFWA aggregation operator in Eq. (4), compute the combined fuzzy decision matrix of options.

Step 4: Using the IFWA aggregation operation in Eq. (4) (same as in Step 3), obtain aggregated fuzzy decisions of options. Criteria have equal weights.

Step 5: Use a score function or a defuzzification function to calculate the overall score ($S_{i_{OA}}$) of the alternatives.

Step 6: For each alternative, select a criterion to reduce and compute its score ($S_{i_{RC}}$). It signifies that the score of the alternatives are computed without the criteria that were reduced.

Step 7: Calculate the effect of the reductive criterion using the equation:

$$E_{RC_j} = \sum_i \left| S_{i_{OA}} - S_{i_{RC_j}} \right| \tag{5}$$

Step 8: Assign the criterion weights by normalizing the impact of the reduced criterion.

Note that the aggregation operator and score (defuzzification) function differ depending on the fuzzy set type. The IF-SERA is depicted in the flowchart as in Fig. 2.

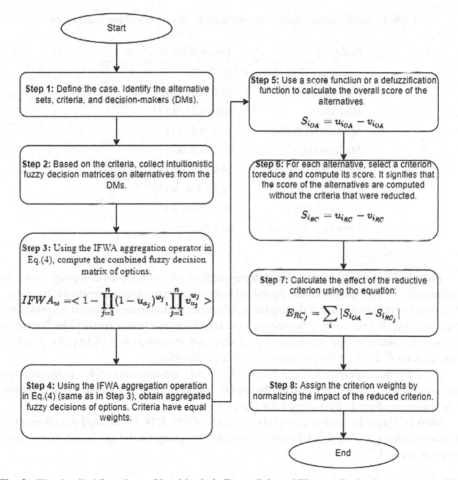

Fig. 2. The detailed flowchart of Intuitionistic Fuzzy Selected Element Reduction Approach (IF-SERA).

4 Application

Intuitionistic Fuzzy Selected Element Reduction Approach (IF-SERA) is applied on evaluation of service quality of digital suppliers to illustrate its usage on MCDM problems. Based on the previous MCDM in the literature, this study investigates four alternatives with four criteria. The new methodology presented in the previous section has been applied step by step. Table 1 lists linguistic terms and their intuitionistic fuzzy analogues.

Table 1. Intuitionistic Fuzzy linguistic scale for decision making evaluations.

Scale	Intuitionistic fuzzy set
Extremely High (EH)	[1; 0; 0]
Very High (VH)	[0.75; 0.1; 0.15]
High (H)	[0.6; 0.25; 0.15]
Medium High (MH)	[0.5; 0.4; 0.1]
Medium (M)	[0.5; 0.5; 0]
Medium Low (ML)	[0.4; 0.5; 0.1]
Low (L)	[0.25; 0.6; 0.15]
Very Low (VL)	[0.1; 0.75; 0.15]
Extremely Low (EL)	[0.1; 0.9; 0]

Step 1: The quality improvement department of an online sales company wants to evaluate the service quality of the digital suppliers that four companies provide for them. As a result of the evaluation of the digital suppliers according to the criteria determined by the company, alternatives will be interviewed. It is discussed how effective which criteria are in this evaluation. The alternative digital suppliers are named as $A = \{A_1, A_2, A_3, A_4\}$. Four criteria $C = \{C_1 : Resposiveness\ level, C_2 : Interface,$

$C_3 : Digital\ experience, C_4 : Technology\ level\}$ are determined for judgement of service quality. Three experts from the quality improvement department of the online sales company have been assigned for this process.

Step 2: Three decision makers $DM = \{DM_1, DM_2, DM_3\}$ evaluated the alternative digital suppliers according to the criteria as follows, using the intuitionistic fuzzy scale given in Table 1.

$$DM_1 = \begin{array}{c} \\ A_1 \\ A_2 \\ A_3 \\ A_4 \end{array} \begin{array}{cccc} C_1 & C_2 & C_3 & C_4 \\ \left[\begin{array}{cccc} MH & L & VL & H \\ L & EL & H & EH \\ VL & MH & M & VL \\ H & EH & L & VH \end{array} \right] \end{array}$$

$$
DM_2 = \begin{array}{c} \\ A_1 \\ A_2 \\ A_3 \\ A_4 \end{array}
\begin{array}{c} C_1 \quad C_2 \quad C_3 \quad C_4 \end{array}
\left[\begin{array}{cccc}
L & M & EL & ML \\
ML & L & VH & H \\
L & H & VH & L \\
ML & VH & M & M
\end{array} \right]
$$

$$
DM_3 = \begin{array}{c} \\ A_1 \\ A_2 \\ A_3 \\ A_4 \end{array}
\begin{array}{c} C_1 \quad C_2 \quad C_3 \quad C_4 \end{array}
\left[\begin{array}{cccc}
M & ML & M & VH \\
VL & L & ML & MH \\
EL & M & H & EL \\
L & MH & MH & ML
\end{array} \right]
$$

Step 3: For the intuitionistic fuzzy sets, the aggregation operator was given in Eq. (4). The aggregated intuitionistic fuzzy decision matrix of alternatives by IFWA is calculated as follows:

$$
DM_{Aggr} = \begin{array}{c} \\ A_1 \\ A_2 \\ A_3 \\ A_4 \end{array}
\left[\begin{array}{cccc}
C_1 & C_2 & C_3 & C_4 \\
(0.40; 0.51, 0.10) & (0.37, 0.54, 0.10) & (0.17, 0.77, 0.06) & (0.56, 0.30, 0.13) \\
(0.22, 0.63, 0.15) & (0.14, 0.78, 0.08) & (0.56, 0.30, 0.13) & (0.67, 0.23, 0.10) \\
(0.14, 0.78, 0.08) & (0.53, 0.39, 0.08) & (0.61, 0.30, 0.09) & (0.14, 0.78, 0.08) \\
(0.39, 0.47, 0.14) & (0.72, 0.19, 0.09) & (0.40, 0.51, 0.10) & (0.53, 0.39, 0.08)
\end{array} \right]
$$

Step 4: Using Eq. (4), the aggregated intuitionistic fuzzy decisions of alternatives is given as follows.

$$
\tilde{D}_{Aggr} = \begin{array}{c} A_1 \\ A_2 \\ A_3 \\ A_4 \end{array}
\left[\begin{array}{c}
(0.34, 0.56, 0.09) \\
(0.32, 0.55, 0.13) \\
(0.28, 0.63, 0.10) \\
(0.49, 0.40, 0.11)
\end{array} \right]
$$

Step 5: With the help of the Eq. (2), the intuitionistic fuzzy numbers are defuzzified. Therefore, the overall score $(S_{i_{OA}})$ of alternatives are calculated as $S_{i_{OA}} = \{ S_{A1_{OA}} = -0.217, S_{A2_{OA}} = -0.222, S_{A3_{OA}} = -0.348, S_{A4_{OA}} = -0.094 \}$

Step 6: Alternative scores are obtained by excluding one element from the evaluation at a time. Return to Step 2 and retrieve the criterion column from the matrices. The criteria used for re-evaluation are now $n-1$. As a result of decreasing the chosen criterion, the following alternative scores are obtained:

$$
Si_{RC_1} = \left\{ S_{A1_{RC_1}} = -0.249, S_{A2_{RC_1}} = -0.142, S_{A3_{RC_1}} = -0.197, S_{A1_{RC_1}} = 0.159 \right\}
$$

$$
Si_{RC_2} = \left\{ S_{A1_{RC_1}} = -0.233, S_{A2_{RC_1}} = 0.012, S_{A3_{RC_1}} = -0.458, S_{A1_{RC_1}} = -0.022 \right\}
$$

$$
Si_{RC_1} = \left\{ S_{A1_{RC_1}} = -0.022, S_{A2_{RC_1}} = -0.337, S_{A3_{RC_1}} = -0.482, S_{A1_{RC_1}} = 0.172 \right\}
$$

$$
Si_{RC_1} = \left\{ S_{A1_{RC_1}} = -0.332, S_{A2_{RC_1}} = -0.364, S_{A3_{RC_1}} = -0.197, S_{A1_{RC_1}} = 0.080 \right\}
$$

Step 7: By Eq. (5), the effects of the reducted criterion are given as follows:

$$E_{RC_j} = \{E_{RC_1} = 0.328, E_{RC_2} = 0.479, E_{RC_3} = 0.521, E_{RC_4} = 0.422\}$$

Step 8: The weightes of criterion are calculated by normalizing the criterion effects as follows:

$$w_j = \{w_{C_1} = 0.187, w_{C_2} = 0.274, w_{C_3} = 0.298, w_{C_4} = 0.241\}$$

According to the results, the most effective criterion on service quality of digital suppliers for online sales company is C_3 : *Digital experience*.

5 Conclusion

This study investigated the SERA method, which is an approach that can be applied with all fuzzy sets, in an intuitionistic fuzzy environment. In SERA, it is aimed to find out how effective the criteria are in the selection according to the alternative-criteria evaluation matrix. The effect of sequentially reducing the selected criteria from the process is examined and this effect ratio is evaluated as the weight of the criterion. The first intuitionistic fuzzy application of the method, which has just been proposed by the authors, has been applied in the evaluation of service quality of digital suppliers. The findings are consistent with previous research. As a future direction, it is suggested that SERA be developed using different fuzzy sets from the literature and that the method be compared to existing criterion weighting methods. In addition, hybrid approaches can be developed with MCDM methodologies such as AHP, TOPSIS, VIKOR, COPRAS, BWM etc.

Acknowledgement. "This work has been supported by the Scientific Research Projects Commission of Galatasaray University under grant number # FBA-2022-1085".

References

1. Garay-Rondero, C.L., Martinez-Flores, J.L., Smith, N.R., Morales, S.O.C., Aldrette-Malacara, A.: Digital supply chain model in Industry 4.0. J. Manuf. Technol. Manag. **31**(5), 887–933 (2020)
2. Ageron, B., Bentahar, O., Gunasekaran, A.: Digital supply chain: challenges and future directions. Supply Chain Forum Int. J. **21**(3), 133–138 (2020)
3. Sharma, M., Joshi, S.: Digital supplier selection reinforcing supply chain quality management systems to enhance firm's performance. The TQM Journal (2020)
4. Resende, C., Geraldes, C., Lima, F.: Decision models for supplier selection in industry 4.0 era: a systematic literature review. Procedia Manuf. **55**, 492–499 (2021). https://doi.org/10.1016/j.promfg.2021.10.067
5. Aruldoss, M., Lakshmi, T.M., Venkatesan, V.P.: A survey on multi criteria decision making methods and its applications. Am. J. Inf. Syst. **1**(1), 31–43 (2013)
6. Kuo, M.S., Liang, G.S.: A soft computing method of performance evaluation with MCDM based on interval-valued fuzzy numbers. Appl. Soft Comput. **12**(1), 476485 (2012)

7. Triantaphyllou, E., Mann, S.H.: An examination of the effectiveness of multi-dimensional decision-making methods: a decision-making paradox. Decision support systems **5**(3), 303–312 (1989)
8. Senapati, T., Yager, R.R.: Some new operations over Fermatean fuzzy numbers and application of Fermatean fuzzy WPM in multiple criteria decision making. Informatica **30**(2), 391–412 (2019)
9. Saaty, T.L.: A scaling method for priorities in hierarchical structures. J. Math. Psychol. **15**(3), 234–281 (1977)
10. Benayoun, R., Roy, B., Sussman, B.: ELECTRE: Une méthode pour guider le choix en présence de points de vue multiples. Note de travail **49**, 2–120 (1966)
11. Opricovic, S.: Multicriteria optimization of civil engineering systems. Fac. Civil Eng. Belgrade **2**(1), 5–21 (1998)
12. Yoon, K.P., Hwang, C.L.: Multiple Attribute Decision Making: An Introduction. Sage publications, Thousand Oaks (1995)
13. Zou, Z.H., Yi, Y., Sun, J.N.: Entropy method for determination of weight of evaluating indicators in fuzzy synthetic evaluation for water quality assessment. J. Environ. Sci. **18**(5), 1020–1023 (2006)
14. Brans, J.P.: L'Ingéniérie de la Décision. Elaboration d'Instruments d'Aide à la Décision. Méthode PROMETHEE. Université LAVAL, Colloque d'Aide la Décision, Québec, Canada, pp. 183–213 (1982)
15. Zavadskas, E.K., Kaklauskas, A., Turskis, Z., Tamošaitiene, J.: Selection of the effective dwelling house walls by applying attributes values determined at intervals. J. Civil Eng. Manag. **14**(2), 85–93 (2008)
16. Keshavarz Ghorabaee, M., Zavadskas, E.K., Olfat, L., Turskis, Z.: Multi-criteria inventory classification using a new method of evaluation based on distance from average solution (EDAS). Informatica **26**(3), 435–451 (2015)
17. Rezaei, J.: Best-worst multi-criteria decision-making method. Omega **53**, 49–57 (2015)
18. Keshavarz-Ghorabaee, M., Amiri, M., Zavadskas, E.K., Turskis, Z., Antucheviciene, J.: Simultaneous evaluation of criteria and alternatives (SECA) for multi-criteria decision-making. Informatica **29**(2), 265–280 (2018)
19. Keshavarz-Ghorabaee, M., Amiri, M., Zavadskas, E.K., Turskis, Z., Antucheviciene, J.: Determination of objective weights using a new method based on the removal effects of criteria (MEREC). Symmetry **13**(4), 525 (2021)
20. Çakır, E., Taş, M.A., Demircioğlu, E.: A new weighting method in fuzzy multi-criteria decision making: selected element reduction approach (SERA). In: Applications of Fuzzy Techniques: Proceedings of the 2022 Annual Conference of the North American Fuzzy Information Processing Society NAFIPS 2022 (2022)
21. Atanassov, K.: Intuitionistic Fuzzy Sets. Fuzzy Sets Syst. **20**, 87–96 (1986)
22. Sadiq, R., Tesfamariam, S.: Environmental decision-making under uncertainty using intuitionistic fuzzy analytic hierarchy process (IF-AHP). Stochastic Environ. Res. Risk Assess. **23**(1), 75–91 (2009)
23. Wang, J.Q., Nie, R.R., Zhang, H.Y., Chen, X.H.: Intuitionistic fuzzy multi-criteria decision-making method based on evidential reasoning. Appl. Soft Comput. **13**(4), 1823–1831 (2013)
24. Shen, F., Ma, X., Li, Z., Xu, Z., Cai, D.: An extended intuitionistic fuzzy TOPSIS method based on a new distance measure with an application to credit risk evaluation. Inf. Sci. **428**, 105–119 (2018)
25. Mishra, A.R., et al.: Novel multi-criteria intuitionistic fuzzy SWARA–COPRAS approach for sustainability evaluation of the bioenergy production process. Sustainability **12**(10), 4155 (2020)
26. Özbek, A., Yildiz, A.: Digital supplier selection for a garment business using interval type-2 fuzzy topsis. Text. Apparel **30**(1), 61–72 (2020)

27. Torkayesh, S.E., Iranizad, A., Torkayesh, A.E., Basit, M.N.: Application of BWM-WASPAS model for digital supplier selection problem: a case study in online retail shopping. J. Ind. Eng. Decis. Mak. **1**(1), 12–23 (2020)
28. Ghorbani, M., Arabzad, S.M., Tavakkoli-Moghaddam, R.: Service quality-based distributor selection problem: a hybrid approach using fuzzy ART and AHP-FTOPSIS. Int. J. Prod. Qual. Manag. **13**(2), 157–177 (2014)
29. Ma, P., Yao, N., Yang, X.: Service quality evaluation of terminal express delivery based on an integrated SERVQUAL-AHP-TOPSIS approach. Math, Prob. Eng. **2021**, 1–10 (2021). https://doi.org/10.1155/2021/8883370
30. Zadeh, L.A.: Fuzzy sets. Inf. Control **8**, 338–356 (1965)
31. Chen, S., Tan, J.: Handling Multicriteria fuzzy decisionmaking problems based on vague set theory. Fuzzy Sets Syst. **67**(2), 163–172 (1994)
32. Hatzimichailidis, A.G., Papadopoulos, B.K.: A new geometrical interpretation of some concepts in the intuitionistic fuzzy logics. NIFS **11**(2), 38–46 (2005)
33. Xu, Z.: Intuitionistic fuzzy aggregation operators. IEEE Trans. Fuzzy Syst. **15**(6), 1179–1187 (2007)

A New Approach to Polarization Modeling Using Markov Chains

Juan Antonio Guevara[1]([⊠])(iD), Daniel Gómez[1](iD), Javier Castro[1](iD),
Inmaculada Gutiérrez[1](iD), and José Manuel Robles[2](iD)

[1] Faculty of Statistics, Complutense University of Madrid, Madrid, Spain
{juanguev,inmaguti}@ucm.es, {dagomez,jcastro}@estad.ucm.es
[2] Faculty of Economics, Complutense University of Madrid, Madrid, Spain
jmrobles@ucm.es

Abstract. In this study, we approach the problem of polarization modeling with Markov Chains (PMMC). We propose a probabilistic model that provides an interesting approach to knowing what the probability for a specific attitudinal distribution is to get to an i.e. social, political, or affective Polarization. It also quantifies how many steps are needed to reach Polarization for that distribution. In this way, we can know how risky an attitudinal distribution is for reaching polarization in the near future. To do so, we establish some premises over which our model fits reality. Furthermore, we compare this probability with the polarization measure proposed by Esteban and Ray and the fuzzy polarization measure by Guevara et al. In this way, PMMC provides the opportunity to study in deep what is the performance of these polarization measures in specific conditions. We find that our model presents evidence that in fact, some distributions will presumably show higher risk than others even when the entire population holds the same attitude. In this sense, according to our model, we find that moderate/indecisive attitudes present a higher risk for polarization than extreme attitudes and should not be considered the same scenario despite the fact that the entire population maintains the same attitude.

Keywords: Polarization · Markov Chains · Fuzzy measures

1 Introduction

The phenomenon of polarization has been studied in several fields (e.g.: [1, 6, 7, 9]) and can be understood in general terms as the split of a given society or population into two opposite groups where the more similar in size these groups are the more polarized the population is. The measurement of polarization has been also studied from different perspectives such as economics [6], religion [9]

Supported by national research projects funded by the Spanish Government, with reference R&D&I, PGC2018-096509B-I00, PR108/20-28 and PID2019-106254RB-100 funding: MINECO (Period: 2020-2024).

© Springer Nature Switzerland AG 2022
D. Ciucci et al. (Eds.): IPMU 2022, CCIS 1602, pp. 151–162, 2022.
https://doi.org/10.1007/978-3-031-08974-9_12

or even from the perspective of the fuzzy sets [7]. However, one step forward should be made as long as only measures to identify the levels of polarization in a specific instant of time have been proposed in the literature. Although some studies have led their aims to model the interpersonal influence over attitudes that might end in polarized situations [2], from the best of our knowledge it has not been tackled from a stochastic perspective. In this vein, Markov Chains can provide adequate resources to approach this task. We consider this an interesting approach that will enable researchers to know, for specific conditions and given a specific distribution of attitude in a population, what is the risk to finally reaching the maximum level of polarization from a probabilistic approach. Markov Chains have been also used to model phenomena along with the literature and among diverse fields such as genetics [10], urban network management [12], or customer relationship in marketing [14].

Despite the fact that there are many approaches in the literature that has tackled the modelization of dynamic phenomena such as the use of differential equations in the dynamic market model, Markov Chains represent one of the simplest and more traditional mathematical models for random phenomena evolving in time [11]. One of the main characteristics of Markov Chains is that they allow us to model time-based phenomena according to a probabilistic approach, being one of the most used stochastic processes. Thus, Markov Chains hold a non-deterministic nature, where the evolution of the process depends on both casual and random variables. In the next paragraphs, we explain the basic properties of Markov Chains [11]. So that, we think that the use of Markov Chains as a first step to model polarization as a dynamic process might be interesting. In this way, knowing the performance of the model may bring some ideas to tackle this task from others approaches more complex in nature.

The aim of this study is to model the dynamism of the phenomenon of polarization using Markov Chains. It will provide the opportunity to understand in deep the performance of some polarization measures proposed in the literature [6,7]. To deal with it, we make use of transition probabilities and the number of steps that might take a given state - a specific attitude distribution - to get to the maximum level of polarization. Also, we contrast them with specific levels of polarization provided by polarization measures.

This paper is structured as follows. In Section 1, we introduce the measurement of polarization and Markov Chains models. In Sect. 3, we propose the probabilistic model for modeling polarization. We also provide a toy example of polarization modeling with some hypotheses and assumptions. Finally, we end this paper with some conclusions in Sect. 4.

2   Preliminaries

2.1   Polarization Measures

The measurement of polarization has been widely analyzed in the last thirty years in different fields. In this study, we focus on a traditional measure [6] and on a newer one [7]. Typically one of the most cited and studied [3–5,13] polarization measures was the one proposed by Esteban and Ray back in 1994 [6]. The authors designed their metric not only from a mathematical approach but also from a sociological one. In this vein, some sociological concepts such as *hemophilia* or *heterogeneity* are included into the measure. Based on this idea, Guevara et al. [7] proposed a new polarization measure on the basis of fuzzy sets [16]. The authors apply these *hemophilia* and *heterogeneity* ideas to their measure by quantifying the membership degree of each individual to both extreme poles. We explain in more detail these measures to compare them with the results provided by modeling polarization with Markov Chains.

- **Esteban and Ray.** They proposed their polarization measure (ER) in the framework of economics as a natural evolution of the inequality index [6]. The authors start from three basic hypotheses to measure polarization:
  1. High degree of homogeneity intra-group.
  2. High degree of heterogeneity inter-groups.
  3. Few numbers of groups with significant size.

  Thus, they focus on measuring polarization with the so-called *effective antagonism approach*, T. This approach takes into account how much an individual i feels close to the group to which it belongs, and how far i is from another individual j with which it is being compared. Formally,

$$P(\pi, \mathbf{y}) = \sum_{i=1}^{n} \sum_{j=1}^{n} \pi_i \pi_j \mathbf{T}(\mathbf{I}(\pi_i), a(\delta(y_i, y_j))) \tag{1}$$

  where *identification* (I) is a function that depends on the size of the group π_i for an individual i, and *alienation* (a) is the absolute distance between the groups y_i and y_j. T is the product between I and a assuming symmetrical alienation between individuals. This measure shows its highest value when 50% of the population presents the same position and the other 50% another consensual position with a high distance between values. In contrast, the lowest value is given when all of the population presents the same value.

- **JDJ measure.** Guevara et al. [7] start from the premise that a given individual might feel identified with a specific political party but it may also agree with some proposals of other political parties as well. To accomplish this idea, *JDJ* is characterized in terms of fuzzy sets [16], using the membership degrees of each individual towards the extreme poles of the attitudinal axis by which polarization is being measured. Authors of *JDJ* include the high homogeneity within groups and high heterogeneity between groups approach by comparing the membership degrees of each individual with others.

  Let X denote a one-dimensional variable, and let us assume that X has two extreme pole/positions, X_A and X_B being μ_{X_A}, μ_{X_B} the membership degree functions, so that $\mu_{X_A}, \mu_{X_B} : N \longrightarrow [0, 1]$ are functions $\forall i \in N$. Authors consider the risk of polarization between two individuals like the possibility these two situations occur at the same time to define *JDJ*:
  1. How individual i is close to the pole X_A and j is close to the pole X_B.

2. How individual i is close to the pole X_B and j is close to the pole X_A.

$$JDJ(X) = \sum_{i,j \in N, i \leq j} \varphi\left(\phi(\mu_{X_A}(i), \mu_{X_B}(j)), (\phi(\mu_{X_B}(i), \mu_{X_A}(j)))\right) \qquad (2)$$

where $\phi : [0,1]^2 \longrightarrow [0,1]$ is an overlapping aggregation operator and $\varphi :$ $[0,1]^2 \longrightarrow [0,1]$ is a grouping function. JDJ reaches its highest value when 50% of the population hold $\mu_{X_A} = 1$ and $\mu_{X_B} = 0$ and the other 50% hold $\mu_{X_A} = 0$ and $\mu_{X_B} = 1$. In contrast to ER, JDJ shows its lowest value not only when all the population presents the same value but also when their attitudes are extreme. Thus, it is considered that there is a difference between the scenario in which all population is radicalized and the scenario in which all the population just presents the same value. This last case might represent a position of uncertainty that could mean that people do not hold a strong identification feeling being more susceptible to polarizing themselves in the future.

2.2 Markov Chains

We think of a chain as a process in time in which a random variable X_n changes along the time. These chains can be considered in both cases: *discrete time*: $n \in \mathbb{Z}^+ = \{0, 1, 2, ...\}$ or *continuous time*: $n \in \mathbb{R}^+ = [0, \infty)$.

In this work, we focus on discrete-time processes where $(X_n)_{n \geq 0}$. Thus, let S be the whole set of states, where each state $i \in S$. We denote by p_{ij} the transition probability from a state i to another state j where $i, j \in S$.

$$p_{ij} = P(P_n = j \mid X_{n-1} = i), \quad \text{with } p_{ij} > 0 \text{ and } \sum_{j=1}^{m} p_{ij} = 1$$

When $p_{ij} > 0$, we say that a state i can communicate with another state j. We call *transition matrix* P with $s \times s$ dimensions the following:

$$P = [p_{ij}]_{i,j \in \{1,...,s\}} = \begin{bmatrix} p_{11} & p_{12} & \cdots & p_{1s} \\ p_{21} & p_{22} & \cdots & p_{2s} \\ \cdots & \cdots & \cdots & \cdots \\ p_{s1} & p_{s2} & \cdots & p_{ss} \end{bmatrix}$$

The probability to start the chain from a state i is defined as $p_i^{(0)}$. Then, $p_{ij}^{(t)}$ is defined as the probability of the chain to reach the state j after t steps, when the chain has started in the state i. So that, the transition probability in t steps $p_{ij}^{(t)}$ is defined as: $p_{ij}^{(t)} = P(X_t = j \mid X_0 = i)$.

Taking the Markov property into account - *the probability of a given future value of a random variable only depends on its present value* - we have that, for $t \geq 2$ as the chain should have passed for at least one of the s possible states in the $t - 1$ phase, it holds:

$$p_{ij}^{(t)} = \sum_{k=1}^{s} P(X_t = j, X_{t-1} = k \mid X_0 = i)$$

Applying this logic to the transition matrix P, we can know the transition probabilities between all the states at t-steps by calculating the t power of P.

In addition, we say that a state i is an absorbing state when once it is reached there is zero probability to change to another state, so that $p_{ii} = 1$ and $p_{ij} = 0$, for $i \neq j$ and $j = 1, ..., s$.

3   A New Problem: Modeling Polarization with Markov Chains

In this section, we propose a novelty probabilistic model for Polarization modeling with Markov Chains (PMMC). We set some postulations to assure that the probabilistic model has a minimum degree of concordance about how the polarization should behave according to its theory. This approach is completely new in the literature to the best of our knowledge.

3.1   Hypotheses

We enunciate some hypotheses needed to fix our model with reality.

1. **Number of individuals**. It refers to the number of individuals to be considered in polarization modeling, being $N = \{1, , ..., k, ..., n\}$ the whole set.
2. **Attitude measurement**. We refer to this point as the variable by which the attitudes of individuals are measured to compute polarization. Let $Z = \{1, ..., z\}$ denote a categorical variable to measure the attitude of N with z levels. We assume that Z has two poles, 1 and z. Finally, let Z_k denote the position of the individual k in Z, being $k \in N$.
3. **Independence of the behavior between individuals**. We propose to specify whether the behavior of each individual is independent - or not - with respect to the others. In other words, the researcher has to decide whether $\forall k, l \in N$ with $k \neq l$, Z_k is independent of Z_l or not.
4. **Nature of the poles**. When people reach extreme attitudes, they tend to radicalize themselves which led to a lesser probability of attitude change. This fact shall be taken into consideration in order to decide whether an individual that holds an extreme position has still a probability to change its attitude or not. In this scenario, where people are in extreme positions, the researcher has to decide whether the chain ends - because extreme attitudes will presumably not change - or not.
5. **Immobility degree**. Every individual presents a probability > 0 of remaining in the same position along with the variable by which their attitudes are being measured. Also, it must be determined whether the quantity of this probability is dependent or not on the units of time.
6. **Attitude change**. The range of the attitude change by units of time should be considered. We understand that a parameter to include is the degree in which individuals change their attitudes. Depending on this parameter, different approaches may appear being (a) *gradualness*, considering that the

attitude of individuals does not change abruptly but softly and (b) *sudden-ness* which states that individuals can abruptly skip within the values of the attitudinal variable.

Let Z denote a categorical variable, Z_k the value that individual k holds in Z. Then, we have that $Z_k \pm d$ denotes the possible values in Z that individual k might take in the next unit of time, being d the units to skip in Z.

7. **Symmetry of the change.** This premise refers to the symmetry of the probabilities of change of individuals from their current positions to others. Thus, it is important to indicate if the probabilities of change are the same according to: (a) *change direction* and (b) *closeness to the poles*, i.e. if for a given $k \in N$, Z_k will change depending on $\pm d$ or μ_{X_A}/μ_{X_B}.

3.2   A New Probabilistic Model: PMMC

A Markov Chain in PMMC is a stochastic process in discrete time in which attitudes measured by variable Z change along the time. When modeling polarization, Markov Chains allow us to represent the transition probabilities between attitudinal distributions in a population. We can infer the evolution of the attitudes of a population in order to know when, how, and under what conditions these individuals get polarized. In this sense, the attitudinal distribution of the population moves from a given state to another with a specific probability.

Definition 1 (States in PMMC). *We call a state in terms of PMMC a specific and static distribution for a population $N = \{1, , \ldots, k, \ldots, n\}$ along the attitudinal feature Z where $Z = \{1, \ldots, z\}$. Thus, let S denote the whole set of states with length $z^n = s$ where $i \in S$ if and only if $i = [Z_1, \ldots, Z_k, \ldots, Z_n]$. Also, we define polarized states the distributions in which a significant part of the population is placed by one pole of the variable and another significant part of the population is placed by the other pole.*

Definition 2 (Transition probabilities in PMMC). *Given $i, j \in S$ we define as P_{ij} as the probability to reach a population opinion j from a population opinion i. Additionally, we say that i is an adjacent state to j when it holds a transition probability > 0 from i to j. Values of P_{ij} depend directly on the rules stated in the hypotheses proposed above.*

Remark 1. Note that having previously defined the states in definition 1 and the transition probabilities in definition 2 the transition matrix T with $s \times s$ dimensions can be constructed according to previous sections.

Example 1. Let $\alpha = 0.8$ denote the probability of remaining in the same position for a given individual, $n = 2$, $z = 3$, $d = 1$, we assume symmetry in the position change. Let $S = \{[1,1], [1,2], [1,3], [2,1], [2,2], [2,3], [3,1], [3,2], [3,3]\}$ be the whole set of states. Also, we assume as absorbing positions extreme values of Z. Then, we have that:

- $P_{[1,1][1,1]} = P_{[3,3][3,3]} = P_{[1,3][1,3]} = P_{[3,1][3,1]} = 1$.

- $P_{[1,2][1,2]} = P_{[2,1][2,1]} = P_{[2,3][2,3]} = P_{[3,2][3,2]} = 1 \cdot 0.8 = 0.8$
- $P_{[2,2][2,2]} = 0.8 \cdot 0.8 = 0.64$
- $P_{[1,2][1,1]} = P_{[1,2][1,3]} = P_{[2,1][1,1]} = P_{[2,1][3,1]} = P_{[2,3][1,3]} = P_{[2,3][3,3]} = P_{[3,2][3,1]} = P_{[3,2][3,3]} = (1 - 0.8)/2 = 0.1$
- $P_{[2,2][1,1]} = P_{[2,2][1,2]} = P_{[2,2][1,3]} = P_{[2,2][2,1]} = P_{[2,2][2,3]} = P_{[2,2][3,1]} = P_{[2,2][3,2]} = P_{[2,2][3,3]} = (1 - 0.64)/8 = 0.045$
- Finally, we have that the remaining transition probabilities are equal to 0.

According to this, the transition matrix T is the next (Table 1):

Table 1. Transition matrix for $\alpha = 0.8$ and $z = 3$

	1,1	1,2	1,3	2,1	2,2	2,3	3,1	3,2	3,3
1,1	1.00	0.00	0.00	0.00	0.00	0.00	0.00	0.00	0.00
1,2	0.10	0.80	0.10	0.00	0.00	0.00	0.00	0.00	0.00
1,3	0.00	0.00	1.00	0.00	0.00	0.00	0.00	0.00	0.00
2,1	0.10	0.00	0.00	0.80	0.00	0.00	0.10	0.00	0.00
2,2	0.045	0.045	0.045	0.045	0.64	0.045	0.045	0.045	0.045
2,3	0.00	0.00	0.10	0.00	0.00	0.80	0.00	0.00	0.10
3,1	0.00	0.00	0.00	0.00	0.00	0.00	1.00	0.00	0.00
3,2	0.00	0.00	0.00	0.00	0.00	0.00	0.10	0.80	0.10
3,3	0.00	0.00	0.00	0.00	0.00	0.00	0.00	0.00	1.00

Once the transition matrix is defined we can compute every Markov Chain process, such as transition probabilities in t steps or detecting absorbing states. PMMC provides useful information: (1) probability of a population to finally reach a polarized state, (2) number of necessary steps to reach polarized states, (3) distribution of a population in n epochs given an initial distribution or (4) number of visits to recurrent states given ∞ epochs, among others. In this way, the aggregation of this information is a useful tool for the measurement of polarization.

3.3   Illustrative Example

In this section we provide an illustrative example of PMMC in which we assume specific values for the proposed hypotheses in the last section. To do so, once the rules have been set and the transition matrix characterized, we use the R package *"markovchain"* [15] to compute the different Markov Chain processes.

1. **Number of individuals.** Due that almost all the polarization measures are computed by the comparison between two individuals and on the grounds of simplicity, we apply PMMC for the simplest scenario in which we have two individuals being $n = 2$. This example is equivalent to the situation in which 50% of the population holds the same value, and the other 50% of the population holds another value, equal for all of them.

2. **Attitude measurement.** We consider their attitudes in a 5-likert scale $Z = \{1, 2, 3, 4, 5\}$.
3. **Independence of the behavior between individuals.** We assume that there is no independence between the behavior of the individuals. In this sense, the states in which there are no individuals changing their behavior more than 1 value in Z are equiprobable.
4. **Nature of the poles.** We assume 1 and 5 as absorbing positions. Then, states in which all the individuals have extreme positions are considered absorbing states, being $[1, 1]$, $[1, 5]$, $[5, 1]$, and $[5, 5]$.
5. **Immobility degree.** Every individual presents a probability of remaining in the same position of $\alpha > 0$. In this example, we vary the α parameter to observe different results. We have that $\alpha \in [0.1, 0.9]$ by 0.1. We avoid extreme values for $\alpha = 0$ and $\alpha = 1$ in order to not contradict our premise about the *nature of the poles*.
6. **Attitude change.** In this example we follow the *gradualness* approach, assuming that individuals do not abruptly change their attitudes. We assume that a given individual k can only change its opinion to adjacent values of the variable Z by units of time, where $d = 1$ from Z_k to $Z_k \pm 1$ in the next step. For example, if $Z_k = 3$, Z_k only can take the values 2 and 4 in the next step of the Markov Chain.
7. **Symmetry of the change.** We assume that the probabilities of changing the position are equal in any direction $(\pm d)$, and they are not affected by the pole closeness $(\mu_{X_A}$ or $\mu_{X_B})$.

On this basis, we have that $S = \{[1, 1], [1, 2], [1, 3], ..., [3, 5], [4, 5], [5, 5]\}$ is the whole set of states, with length $s = z^n = 5^2$. Then, we calculate the transition matrix $T = 25 \times 25$ for each value of α where $\alpha \in [0.1, 0.9]$ by 0.1. Once all transition matrices are ready we simulate our Markov Chains in order to know what is the probability for a given non-polarized state to be polarized and how many steps the process takes to reach these specific states according to each value of α. To do so, we run the *rmarkovchain* function from the "*markovchain*" R package [15], which allows us to generate hundreds of future steps according to T underlying stochastic process. First, we consider every state from S but the absorbing ones ($[1, 1]$, $[1, 5]$, $[5, 1]$ and $[5, 5]$) as initial states. Then, we start the simulation until this initial state reaches an absorbing state. So that, we compute the frequentist probability for this state to reach an absorbing one and we count how many steps take for this state to reach them. We compute 10000 iterations for each initial state and value of $\alpha \in [0.1, 0.9]$ by 0.1.

Then, we calculate the mean results of all the 10000 iterations for each α and for each initial state as units of analysis. In the next, we focus on the probabilities and steps that might take a transient state to reach both polarized states $[1, 5]$ and $[5, 1]$. The mean standard deviation of the states for reaching both polarized states in terms of steps along the variation of α is 17.341 while the *sd* for the probability to reach them is 0.004. We can conclude that the variation of α does not affect the probability of a given state for reaching another polarized state but the steps required. To illustrate this, we show in Fig. 1 the mean steps taken for each state to reach both polarized states along with the variation of α.

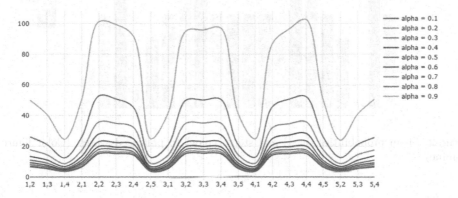

Fig. 1. Mean steps to reach polarized states along with the variation of α parameter.

In Fig. 2 we can see the mean probabilities and mean steps along with all the values of α for each transient state to reach both polarized states, where the higher probability and the lesser steps required by a given state the higher risk of polarization holds. As we can see in these results, Fig. 2 shows coherence between the closeness of the states to the polarized states and their probability to reach it as well as the lesser required steps to reach it. Thanks to this, we can see the evolution of two main groups - or pair of individuals - with the same number of individuals and same attitude intra-group along with the future according to their attitude distribution and their probability to be polarized. From this example we find evidence that some distributions present a higher risk to be polarized in the near future than others even when the entire population holds the same attitude position such as states [2, 2] or [3, 3]. Moreover, we can see that despite the fact that i.e. [3, 1] and [3, 2] present the same probability to reach a polarized state the mean steps required to reach it make [3, 1] a state with a higher polarization risk.

Finally, we compute the polarization scores of two polarization measures ER [6] and JDJ where $\phi = product$ [7] along all the transient states of our example. Also, in Fig. 3 we plot the summed probabilities of reaching both poles for each initial state. Note that probabilities for reaching a polarized state do not change along with the variation of α, showing in Fig. 3 the real probabilities to polarization according to our model. We find that both measures follow the same tendency as the probabilities provided by PMMC. In fact, correlations between polarization measures and probabilities of PMMC are high being $r_{ER} = 0.759$ and $r_{JDJ} = 0.976$. These results are bound to the specific parameters imposed on the model, which is important when analyzing the results. Also, these results bring light to the performance of these measures according to specific conditions, concluding that PMMC can help researchers to know under what conditions a polarization measure might be more adequate than others.

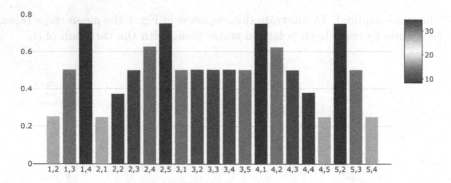

Fig. 2. Mean probabilities (bars) and steps (color) to polarized states. (Color figure online)

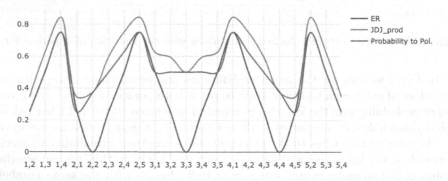

Fig. 3. Polarization scores and polarization probability for each initial state.

4   Conclusion

In this paper, we address the polarization modeling problem using Markov Chains. Thanks to the stochastic processes that underlay Markov Chains we can model social phenomena in terms of future events and probability. In this case, we understand as states specific attitudinal distributions and we start from some premises that allow us to model polarization from a realistic point of view. Thanks to the establishment and variation of these parameters we can simulate different situations that provide us the opportunity to know how a specific scenario might evolve in the future in terms of probability. Thanks to the simulation of PMMC we can know the probabilistic risk for a specific attitude distribution to finally reach a polarized distribution and the steps taken. We have put an example in which attitudes have been measured using a 5 Likert scale that could be political attitude for example in the simplest population with $n = 2$. This scenario can be understood as a pair of individuals or the existence of two groups with a high degree of homogeneity within groups in which all their individuals hold the same position on an attitude scale.

The simulation of this example has brought some interesting results. First, we can know which states present a higher risk to be polarized in the near future and how many units of time might take them to reach the maximum level of polarization. Also, PMMC seems to support the premise that those situations in which all the population holds the same attitude are not equal in terms of risk of being polarized in the future. In this way, we conclude that neutral or central positions present a higher risk for polarization than extreme positions, being significantly different from the scenario in which all of the population is placed by the center of the variable which could be understood as an indecisive position with the scenario in which the entire population has a radicalized attitude. Then, this last scenario supports the fact that radical attitudes are more difficult to change than moderate ones and therefore they should be considered by polarization measures. Finally, we have also compared the probability for a given distribution to be polarized with two polarization measures. We have found a high degree of correlation between this probability and polarization scores, where the fuzzy measure seems to fit better with the PMMC approach than the other traditional measure. Nonetheless, it is important to highlight that this comparison has been made under specific conditions, premises, and hypotheses by which polarization has been modeled and different examples might bring different results.

Thus, this study presents some limitations. First of all, the hypotheses proposed here must be validated and readjusted according to real cases by which we could propose a model that better reflects reality. In addition, as it has been stated in the introduction, there are many approaches in the literature apart from Markov Chains that model dynamic processes, such as differential equations or other time series methods more complex than Markov Chains. In this study, we make a first step to assess the modelization of polarization by stochastic processes. In this sense, as the nature of the model gets more adequate and fits the reality better, contemplating other types of dynamics might improve the performance of this approach.

On the other hand, the study and application of the phenomenon of polarization can be useful for other fields. In general terms, the study of polarization involves the comparison between elements whether they are similar or opposite. Previous research has demonstrated the usefulness of using polarization measures to improve community detection algorithms in the field of Social Network Analysis [8]. In this sense, this approach in which Markov Chains are used could be applied to approaches that model i.e. heterogeneity in data, such as clustering or other unsupervised learning methods.

For future research, the application of this approach to real data will provide key information about the performance of the model as well as for the validation of the hypotheses proposed. Also, we will focus on the study of under what conditions our model premises take specific values. To know the probability of changing the attitudes of a given population as well as knowing if this probability is symmetric in any direction and if it is constant along the attitude axis or whether it changes according to the closeness to the poles, can significantly improve the tuning of the model parameters.

Furthermore, in our example, we work with discrete Markov Chains but it would be also interesting to propose a model using continuous Markov Chains as well as other types of dynamics. Also, we find interesting the development of a new polarization measure based on the probabilities of reaching a polarized distribution with PMMC. Finally, PMMC provides us the opportunity to study in a deeper way what is the real performance of polarization measures allowing us to identify under what conditions a given polarization measure is more adequate than others.

References

1. Apouey, B.: Measuring health polarization with self-assessed health data. Health Econ. **16**, 875–894 (2007)
2. Baldassarri, D., Bearman, P.: Dynamics of political polarization. Am. Sociol. Rev. **72**(5), 784–811 (2007)
3. Bauer, P.C.: Conceptualizing and measuring polarization: A review, September 2019
4. Duclos, J.Y., Esteban, J., Ray, D.: Polarization: concepts, measurement, estimation. Econometrica **72**(6), 1737–1772 (2004)
5. Esteban, J., Ray, D.: Comparing polarization measures. In: Oxford Handbook of Economics of Peace and Conflict, pp. 127–151 (2012)
6. Esteban, J.M., Ray, D.: On the measurement of polarization. Econometrica J. Econom. Soc. **62**, 819–851 (1994)
7. Guevara, J.A., Gómez, D., Robles, J.M., Montero, J.: Measuring polarization: a fuzzy set theoretical approach. In: Lesot, M.-J., et al. (eds.) IPMU 2020. CCIS, vol. 1238, pp. 510–522. Springer, Cham (2020). https://doi.org/10.1007/978-3-030-50143-3_40
8. Gutiérrez, I., Guevara, J.A., Gómez, D., Castro, J., Espínola, R.: Community detection problem based on polarization measures: an application to Twitter: the COVID-19 case in Spain. Mathematics **9**(4), 443 (2021)
9. Montalvo, J.G., Reynal-Querol, M.: Religious polarization and economic development. Econ. Lett. **80**(2), 201–210 (2003)
10. Nix, A.E., Vose, M.D.: Modeling genetic algorithms with Markov chains. Ann. Math. Artif. Intell. **5**(1), 79–88 (1992)
11. Norris, J.R.: Markov Chains. Cambridge Series in Statistical and Probabilistic Mathematics. Cambridge University Press, Cambridge (1998)
12. Osorio-Lird, A., Chamorro, A., Videla, C., Tighe, S., Torres-Machi, C.: Application of Markov chains and Monte Carlo simulations for developing pavement performance models for urban network management. Struct. Infrastruct. Eng. **14**(9), 1169–1181 (2018)
13. Permanyer, I.: The conceptualization and measurement of social polarization. J. Econ. Inequality **10**(1), 45–74 (2012)
14. Pfeifer, P.E., Carraway, R.L.: Modeling customer relationships as Markov chains. J. Interact. Mark. **14**(2), 43–55 (2000)
15. Spedicato, G.A.: Discrete time Markov chains with R. R J. (2017). https://journal.r-project.org/archive/2017/RJ-2017-036/index.html. r package version 0.6.9.7
16. Zadeh, L.A.: Fuzzy sets. Inf. Control **8**(3), 338–353 (1965)

Intervals and Possibility Degree Formulae for Usage Prioritization of Cartagena Coastal Military Batteries

Juan Miguel Sánchez-Lozano[1]([✉])(iD), Manuel Fernández-Martínez[1](iD), Marcelino Cabrera-Cuevas[2], and David A. Pelta[3](iD)

[1] University Centre of Defence at the Spanish Air Force Academy, MDE-UPCT, San Javier, Murcia, Spain
{juanmi.sanchez,manuel.fernandez-martinez}@cud.upct.es
[2] Department of Languages and Informatics Systems, Universidad de Granada, Granada, Spain
mcabrera@ugr.es
[3] Department of Computer Science and AI, Universidad de Granada, Granada, Spain
dpelta@ugr.es

Abstract. Several projects have been proposed in recent years with the goal of restoring some of Cartagena's (a Spanish city) old military batteries so that they can be turned into touristic, scientific, or cultural sites. In previous work, a combination of AHP and TOPSIS (A+T, hereafter) was applied to prioritize such batteries. In this paper, we address the following question: *would it be possible to obtain the same prioritization using a simpler approach?* In this regard, we are focused on an approach that relies on assigning score intervals (rather than single values) to the alternatives and compares them using a possibility degree formulae (we shall denote as I+P in the sequel). That approach's main advantage is that it does not require explicit weights but it only relies on a linear ordering of criteria to reflect the user's preferences, instead.

Our results show that the prioritization obtained by means of I+P, A+T, and a simple additive weighting scheme (SAW) are nearly identical, thus providing a positive response to the question posed above.

Keywords: Multi-criteria decision making · Real world application · Intervals-based calculations · Possibility function

1 Introduction

In 1913, the city of Cartagena (in southern Spain) was designated as the great naval base due to both its orography and strategic position. At the end of the Spanish Civil War, there were 25 active batteries in its coast [3]. Notwithstanding, the push suffered by weapon engineering as a result of WW2 made these batteries obsolete and their majority were left. Figure 1 illustrates the battery of San Isidoro, Santa Florentina and Santa Ana. Recently, several projects have

© Springer Nature Switzerland AG 2022
D. Ciucci et al. (Eds.): IPMU 2022, CCIS 1602, pp. 163–172, 2022.
https://doi.org/10.1007/978-3-031-08974-9_13

been suggested with the aim to restore some Cartagena's batteries so they could be transformed into touristic, scientific, or cultural sites. It is worth mentioning that lately, Cartagena has submitted its application to UNESCO to be designated as a World Heritage Site. In fact, that application is based on the municipality's complex of coastal military fortifications due to its outstanding relevance [2].

The large number of batteries in Cartagena coast makes it necessary prioritizing one over the others for conversion and restoration purposes. Also, to determine which batteries should be reconditioned, a wide variety of factors should be considered such as the number of potential visitors, convenience of access, and closeness to electricity and water supplies, to name some of them. In this way, Sánchez-Lozano et al. contributed in 2020 a study combining geographic information systems (GIS) with multi-criteria decision making (MCDM) (c.f. [7]) which constitutes the starting point of this paper. It is worth mentioning that there were considered criteria consisting of the distance of a battery to a range of locations such as main roads, railway stations, electrical grids, water tanks, beaches, towns, Natura 2000 network, ..., etc. Also, orographic based criteria including slope, altitude, or area of the batteries were also taken into account. The relative importance of such criteria (and hence, an order of preference for them) was calculated by the AHP (Analytic Hierarchic Process, [6]) through the knowledge from an advisory group. Afterwards, a ranking of alternatives was obtained by applying the Technique for Order of Preference by Similarity to Ideal Solution (TOPSIS, [4]).

As it is well known, putting AHP in practice is a time-consuming and hard task. The question we posed here is: *would it be possible to obtain the same prioritization using a simpler approach?*

In this context, the aim of the paper is to explore the application of a methodology, originally developed for selecting solutions of interest in multiobjective optimization problems [8], that relies on assigning score intervals to the alternatives and comparing them using a possibility degree formulae. The key point of the methodology is that it does not require explicit weights and just use a linear ordering of the criteria to reflect the user preferences.

The paper is organized as follows. Section 2 contains the basics on the intervals and possibility based approach. The data this case of study is based on are described in Sect. 3. The experiments and results are discussed in Sect. 4. Finally, Sect. 5 collects our main conclusions.

2   Intervals and Possibility Functions to Rank Solutions According to the User's Preferences

Let's depart from a normalized decision matrix M, where every row i represents an alternative A_i and every column j, a criterion C_j. The position $M_{ij} \in [0,1]$ represents the value attained by the A_i under criterion C_j. Besides, suppose the user provides a set of weights $W = \{w_1, w_2, \ldots, w_n\}$ where if the criterion c_i is more relevant than C_j for the decision maker, then $w_i \geq w_j$. It should hold that

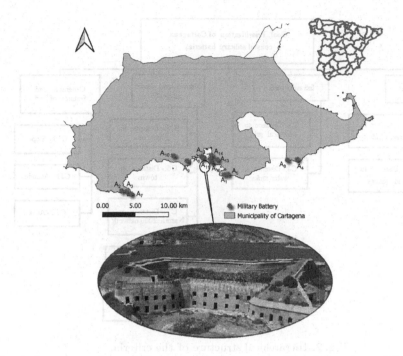

Fig. 1. Locations of coastal batteries in the municipality of Cartagena. Visualization of the San Isidoro, Santa Florentina, and Santa Ana battery (Alternative A_{11}).

$\sum_{j=1}^{m} w_j = 1$. Under these assumptions, a score q_i for the alternative A_i can be calculated as $\sum_{j=1}^{m} w_j \times M_{ij}$.

But this approach has a problem. Let's suppose we have just three criteria and the given preference order is C_2, C_1, C_3. Then, the weights should be defined in such a way that $w_2 \geq w_1 \geq w_3$ with $w_1 + w_2 + w_3 = 1$. As the reader may notice, there are infinite values for w_i that verifies both conditions, and every possible set of values will give a different score for the alternative.

Instead of a single score value, the authors in [8] proposed to calculate *an interval of the potential scores that a solution can achieve* and then sort the solutions in terms of their intervals. The main steps of the proposal are detailed below.

1. Normalize the decision matrix of alternatives and criteria: normalization is needed because different measurements should be combined.
2. Intervals calculation: the user's preferences are given as an ordinal relation among criteria denoted as $C_1 \succeq_p C_2 \succeq_p \ldots \succeq_p C_n$. The symbol $\succeq_p$ is to be read as "at least as preferred to". This implies that the weights are ordered as $w_1 \geq w_2 \geq \ldots \geq w_m$.

   All the potential scores that an alternative A_i can achieve, under the premise of $w_1 \geq w_2 \geq \cdots \geq w_n$, represent an interval. This interval is denoted as $I_i = [L_i, U_i]$ where L_i, U_i are respectively the lowest and highest scores

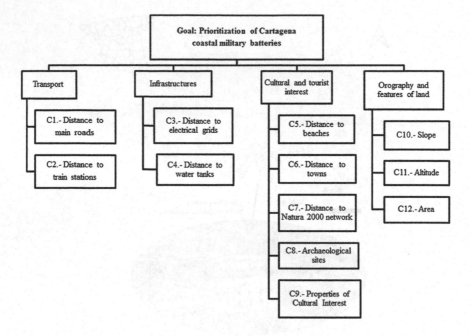

Fig. 2. Hierarchical structure of the criteria.

that an alternative A_j can achieve. I_i is gotten by solving two basic linear programming problems where $w_i, i = 1, \ldots, n$ are the decision variables. (For more information, see [8]).

From a computational standpoint, solving these linear algorithms is now simple and fast. However, as mentioned in [1], an algorithm like Simplex is not required to solve these specific LP problems because "*only the extreme points of the ranked weights need to be considered to effect the desired optimum and they are readily available*". This means that instead of solving the LP problems, just the extreme points should be scored (three sets of weights). The lowest and maximum values are then set to L_i and U_i, respectively, and the operation is completed.

Let n be the number of criteria, thus, the sets of weights to consider are:

  a. $W = (1, 0, \ldots, 0)$: all the weight is assigned to the most preferred criterion.
  b. $W = (1/(n-1), 1/(n-1), \ldots, 1/(n-1))$: the least preferred criterion is assigned $w_n = 0$ while the others get $w_i = 1/(n-1), \forall i \neq n$.
  c. $W = (1/n, 1/n, \ldots, 1/n)$: all the criteria are equally important.

3. Select a reference alternative A^*: the reference alternative A^* is the one having the highest lower interval value. In other words, is the alternative that in the worst case, allows to obtain the highest score.

4. Compare the alternatives against A^*: the corresponding intervals are compared using a possibility function that assesses the degree to which one alternative is superior to another by comparing their corresponding intervals.

Let's look at two alternatives X, Y with the corresponding non-negative intervals $X = [x_l, x_r]$, $Y = [y_l, y_r]$ and $x_l, x_r, y_l, y_r \in R_0^+$. The possibility degree of X being higher than Y is denoted $P(X \geq Y)$ (as in [8]). Considering a user with a neutral attitude, the possibility is calculated as follows [5]:

(a) if $X \cap Y = \varnothing$ (intervals do not overlap),

$$P(X \geq Y) = \begin{cases} 0 & x_r \leq y_l \\ 1 & x_l \geq y_r. \end{cases}$$

(b) if $X \cap Y \neq \varnothing$ (intervals overlap),

$$P(X \geq Y) = \frac{x_r - y_l}{x_r - x_l + y_r - y_l}.$$

5. Ranking of alternatives: The values $P(A_i \geq A^*)$ are determined for each alternative. Then they're ranked based on the degree of possibility they have.

3   Data Description

As stated above, the starting point of our study departs from the information available in [7], which we partially reproduce here in Table 1 for the sake of the completeness.

The table summarizes the information from 15 coastal batteries with respect to the set of 12 criteria related with the following categories: transport (C_1 and C_2), infrastructure availability (C_3 and C_4), cultural and touristic interest (C_5 to C_9), and orography and terrain characteristics (C_{10} to C_{12}). Figure 2 highlights the hierarchical structure of such criteria. The information in the table were obtained after processing thematic layers in a GIS software. Distance criteria such as distance to main roads or railway stations (from C_1 to C_7) coexist with quantity criteria as the number of archaeological sites or properties of cultural interest (C_8 and C_9) or even with orographic criteria (altitude, area or slope) of the terrain (from C_{10} to C_{12}). In this study we deal with the already normalized decision matrix (Table 1).

The geographical locations of the batteries appear in Fig. 1. Also, a view of the battery named San Isidoro, St. Florentina and St. Ana (which corresponds to the alternative A_{11}), is also shown.

After the application of the AHP approach process, the authors in [7] obtained the criteria weights as shown in Table 2. From those weights, the following linear preference order for the 12 criteria is derived:

$$C_1 \succeq_p C_3 \succeq_p C_4 \succeq_p C_6 \succeq_p C_2 \succeq_p C_7$$
$$\succeq_p C_5 \succeq_p C_{10} \succeq_p C_8 \succeq_p C_{11} \succeq_p C_9 \succeq_p C_{12}.$$

As a result, the criterion C_1 (distance to main roads) was found to be the most important, followed by the criteria C_3 (distance to electric grids), C_4 (distance to water tanks), C_6 (distance to towns), and C_2. (distance to railway stations).

Table 1. Normalized decision matrix involving all the 15 coastal batteries by the set of 12 criteria as provided in [7].

Battery	Criterion											
	C_1	C_2	C_3	C_4	C_5	C_6	C_7	C_8	C_9	C_{10}	C_{11}	C_{12}
A_1	0.08	0.17	0.01	0.43	0.30	0.20	0.03	0.21	0.23	0.30	0.19	0.18
A_2	0.36	0.53	0.46	0.24	0.39	0.51	0.00	0.06	0.01	0.31	0.40	0.09
A_3	0.43	0.50	0.50	0.29	0.47	0.49	0.00	0.05	0.01	0.29	0.29	0.41
A_4	0.17	0.18	0.25	0.13	0.20	0.23	0.00	0.45	0.02	0.25	0.36	0.34
A_5	0.07	0.18	0.01	0.45	0.34	0.21	0.00	0.24	0.22	0.35	0.24	0.17
A_6	0.24	0.08	0.16	0.15	0.14	0.07	0.34	0.24	0.36	0.32	0.09	0.34
A_7	0.51	0.49	0.53	0.34	0.55	0.49	0.00	0.04	0.01	0.31	0.25	0.12
A_8	0.11	0.18	0.22	0.22	0.04	0.23	0.00	0.48	0.02	0.12	0.04	0.38
A_9	0.34	0.15	0.19	0.18	0.10	0.13	0.00	0.18	0.30	0.21	0.19	0.27
A_{10}	0.27	0.20	0.18	0.11	0.09	0.12	0.00	0.17	0.25	0.20	0.56	0.14
A_{11}	0.15	0.08	0.10	0.20	0.02	0.07	0.45	0.26	0.34	0.12	0.01	0.42
A_{12}	0.21	0.08	0.15	0.13	0.16	0.07	0.31	0.24	0.37	0.29	0.04	0.03
A_{13}	0.10	0.09	0.05	0.27	0.09	0.08	0.41	0.29	0.33	0.32	0.33	0.29
A_{14}	0.08	0.06	0.10	0.19	0.07	0.04	0.54	0.28	0.39	0.06	0.02	0.07
A_{15}	0.20	0.10	0.08	0.24	0.03	0.09	0.35	0.24	0.32	0.20	0.06	0.10

On the other side, the attributes C_{12} (battery area), C_9 (culturally significant properties), and C_{11} (altitude) appeared to be the least relevant criteria.

It should be noted that the linear order can be obtained by any other means. There is no need to derive it from an AHP process and weights. It is easier for a user to express his/her preferences through a linear ordering than through a set of pairwise comparisons.

4   Experiments and Results

Up to this point, we have a normalized decision matrix, a set of criteria weights, and a linear order of the criteria. In order to prioritize the batteries, three methods are considered:

- **A+T:** the AHP+TOPSIS approach from [7].
- **SAW:** a simple additive weighting scheme using the weights in Table 2.
- **I+P:** the intervals and possibility-based method as described in Sect. 2.

4.1   Analysis of Scores

Table 3 shows the score intervals that every battery can obtain under the given linear ordering of the criteria. Also, the single score given by the SAW approach is provided.

Table 2. Criteria weights as in [7].

Criterion	Weights	Preference order
C_1	0.37	1
C_2	0.07	5
C_3	0.12	2
C_4	0.10	3
C_5	0.05	7
C_6	0.07	4
C_7	0.06	6
C_8	0.04	9
C_9	0.03	11
C_{10}	0.05	8
C_{11}	0.03	10
C_{12}	0.02	12

The alternative achieving the greatest L_b values should be considered as the reference solution. From the table, we observe that $A^* = A_{14}$ is such solution, with $L_b = 0.77$ and $U_b = 0.92$. In this case, this alternative also achieved the highest SAW score.

Finally, the possibility values $P(A_i \geq A_{14})$ for every alternative are calculated. These values are shown also in Table 3, under the $P(A_i \geq A_{14})$ column.

Figure 3 provides a visual representation of the intervals. The X axis corresponds to the alternatives (sorted according with the $P(A_i \geq A_{14})$ values) while the Y axis, to the score attained. The red points corresponds to the SAW score of each alternative. The best alternative is the rightmost one.

The dotted line indicates the greatest L_b value. According with the possibility degree formulae, any alternative whose interval lies completely below this line (as $A_7, A_3, A_2, \ldots$), will never score higher than A_{14} under the provided linear order of the criteria. In the other cases, the possibility degrees varied from 0.21 to 0.41.

It is interesting to note the position of the red points. In the worst alternatives (those in the left side of the plot), the **SAW** approach gives a score closer to the upper bound, while as the quality of the alternatives increased, the SAW scores tend to the middle of the intervals. In other words, it seems that the **SAW** approach tends to overestimate the quality of the worst solutions, while being more conservative for the better ones.

4.2   Analysis of Rankings

Table 4 displays the ranking of batteries according with the three methods considered. As it is observed, three blocks for prioritization are shown.

Table 3. Scores for the batteries. The values are rounded to 2 decimals places. The reference alternative is A_{14}.

Battery	I+P			SAW	A+T
	L_b	U_b	$P(A_i \geq A_{14})$	Score	Ratio
A_1	0.70	0.92	0.41	0.81	0.81
A_2	0.53	0.64	0.00	0.61	0.33
A_3	0.53	0.57	0.00	0.56	0.22
A_4	0.70	0.83	0.21	0.78	0.73
A_5	0.69	0.93	0.40	0.81	0.80
A_6	0.72	0.76	0.00	0.77	0.65
A_7	0.49	0.54	0.00	0.53	0.16
A_8	0.75	0.89	0.40	0.82	0.82
A_9	0.66	0.73	0.00	0.74	0.48
A_{10}	0.70	0.73	0.00	0.76	0.5
A_{11}	0.75	0.85	0.32	0.82	0.79
A_{12}	0.74	0.79	0.07	0.79	0.69
A_{13}	0.72	0.90	0.38	0.81	0.83
A_{14}	**0.77**	**0.92**	**0.50**	**0.85**	0.84
A_{15}	0.75	0.80	0.15	0.80	0.73

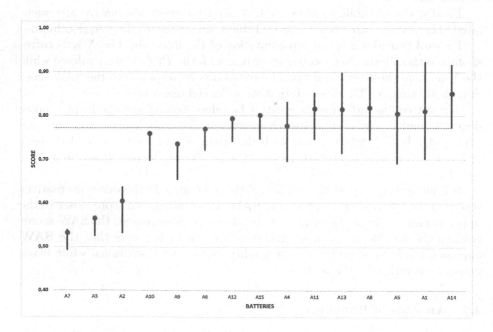

Fig. 3. Intervals of scores for every alternative. Red points corresponds to the **SAW** score. (Color figure online)

Table 4. Ranking of the batteries according with the three methods considered. Three priority blocks are clearly observed.

	Batteries		
Rank	I+P	A+T	SAW
1	A_{14}	A_{14}	A_{14}
2	A_1	A_{13}	A_8
3	A_5	A_8	A_{11}
4	A_8	A_1	A_{13}
5	A_{13}	A_5	A_1
6	A_{11}	A_{11}	A_5
7	A_4	A_4	A_{15}
8	A_{15}	A_{15}	A_{12}
9	A_{12}	A_{12}	A_4
10	A_6	A_6	A_6
11	A_9	A_{10}	A_{10}
12	A_{10}	A_9	A_9
13	A_2	A_2	A_2
14	A_3	A_3	A_3
15	A_7	A_7	A_7

The top ranked alternative is always the same: battery A_{14}. Then, the following 5 alternatives are the same in all of the rankings: $A_1, A_5, A_8, A_{13}, A_{11}$. Although they appear in different order, the small differences in their corresponding scores(ratio) makes hard to assess if one is better than the other.

The alternatives ranked in positions between 10 and 15 are the same in the three rankings with just a minor discordance given by a reversal in positions 11 and 12 in the **I+P** ranking. The alternatives ranked in positions between 7 and 10 are exactly the same for **I+P** and **A+T** methods.

5   Conclusions

In a previous work (c.f. [7]), the **A+T** methodology allowed the authors obtaining a ranking involving 15 coastal batteries that could be restored and reconditioned for touristic, scientific, and cultural purposes. Such a prioritization was obtained on the basis of 12 selected criteria. In this paper, we asked if it would be possible to achieve a similar result by applying a more straightforward approach. As a result, we found that the prioritization of such batteries provided by the **I+P** methodology is in line with the one proposed by **A+T**. We would like also to mention that similar results were obtained by the **SAW** procedure, a simple additive weighting scheme. In fact, all those approaches suggested the battery A_{14} as the top ranked alternative.

One of the main advantages of using **I+P** lies in the fact that it only requires a linear ordering of the criteria which reflects the decision maker's preferences. **I+P** assigns score intervals rather than single values to the alternatives, thus allowing to compare them using a possibility degree formulae. As can be seen in Table 4, **I+P**, **A+T**, and **SAW** methodologies agree that there are prioritization clusters for alternatives, which means that several batteries should be restored with higher priority than the others. Despite **A+T** yields a score for each alternative, it is not always clear why a given alternative should be chosen over another just comparing them by their **A+T** scores. For instance, in this case, the **A+T** ratios of the first ranked alternatives are quite close to each other. This situation makes prioritizing the alternatives by groups a more reasonable strategy than just ranking them. The modelization of the problem leaves out several factors that should be taken into account *a posteriori* to finally make a decision (such as economic aspects).

Another relevant finding is related to the tendency of the **SAW** approach to overestimate the quality of the worst alternatives, while it behaves more conservative when dealing with the best ones. Whether this behaviour is just valid or not for our example will be analyzed in future research.

Acknowledgments. Authors acknowledge support from projects PID2020-112754GB-I0, MCIN/AEI /10.13039/501100011033 and FEDER/Junta de Andalucía-Consejería de Transformación Económica, Industria, Conocimiento y Universidades/Proyecto (B-TIC-640-UGR20).

References

1. Ahn, B.S., Park, K.S.: Comparing methods for multiattribute decision making with ordinal weights. Comput. Oper. Res. **35**(5), 1660–1670 (2008). Part Special Issue: Algorithms and Computational Methods in Feasibility and Infeasibility
2. Cartagena, C.R.: Cartagena basará su candidatura a patrimonio de la humanidad en las fortificaciones y el patrimonio defensivo. Link to article (2020)
3. Celdrán, J.: Los espacios urbanos de la ciudad de cartagena y sus procesos de reforma. Butlletí Del Cercle D'estudis Històrics I Socials Guillem Oliver (KessE) **42**, 9–16 (2008)
4. Hwang, C., Yoon, K.: Multiple Attribute Decision Methods and Applications. Springer, Heidelberg (1981)
5. Liu, F., Pan, L.H., Liu, Z.L., Peng, Y.N.: On possibility-degree formulae for ranking interval numbers. Soft Comput. **22**(8), 2557–2565 (2018). https://doi.org/10.1007/s00500-017-2509-7. https://doi.org/10.1007/s00500-017-2509-7
6. Saaty, T.L.: The Analytic Hierarchy Process. McGraw Hill International, New York (1980)
7. Sánchez-Lozano, J.M., Salmerón-Vera, F.J., Ros-Casajús, C.: Prioritization of cartagena coastal military batteries to transform them into scientific, tourist and cultural places of interest: A gis-mcdm approach. Sustainability **12**, 9908 (2020). https://doi.org/10.3390/su12239908
8. Torres, M., Pelta, D.A., Lamata, M.T., Yager, R.R.: An approach to identify solutions of interest from multi and many-objective optimization problems. Neural Comput. Appl. **33**(7), 2471–2481 (2020). https://doi.org/10.1007/s00521-020-05140-x

Stability of Preferences over Time: Preferences on COVID-19 Vaccine from Spanish and French People

Silvia Prieto-Herráez[1] and Rocio de Andrés Calle[2]

[1] University of Salamanca, E37007 Salamanca, Spain
silvi_ph@usal.es
[2] BORDA and PRESAD Research Groups, Research Excellence Unit GECOS and Multidisciplinary Institute of Enterprise (IME), University of Salamanca, E37007 Salamanca, Spain
rocioac@usal.es

Abstract. Nowadays, there is not an effective COVID-19 disease treatment, then vaccination provides the best hope to restrain the dissemination of the disease. However, a portion of the population doubts the safety of vaccination and many people are unwilling to get vaccinated because of their risk perception. The aim of this study is to understand people's behavior by measuring the stability of their preferences on getting vaccinated against COVID-19 over time. Considering this goal, this contribution suggests a new methodological framework for measuring preference stability over time, and it includes two cases of study. The first one explores the characteristics of Spanish citizens' preferences on COVID-19 vaccination from September 2020 to September 2021 and the second one attempts to evaluate the stability of French citizens' preferences on COVID-19 vaccination from May 2020 to October 2021.

Keywords: Preference stability · Decision stability · Complete pre-orders · Loss memory effect

1 Introduction

The SARS-CoV-2 virus pandemic has caused disastrous damages. In February 2022, Global COVID-19 caseload has surpassed 411 million, including 5,81 million deaths since the end of January 2020 [10]. Nowadays, there is not an effective COVID-19 disease treatment, then the only way to contain the pandemic is the use of preventive measures. One of the most effective actions to restrain the dissemination of any infectious disease is vaccination [4]. The collective vaccination gives rise to individual and also collective immunity (*herd immunity*) [9]. In order to generate an immune response front COVID-19 and slow the pandemic, one of the most essential actuations is a mass vaccination.

From January 2020 so far, several safe and effective COVID-19 vaccines have been developed [12] but their fast getting has generated fear, mistrust and uncertainty in many people [2]. Most of the national opinion polls show that a portion

© Springer Nature Switzerland AG 2022
D. Ciucci et al. (Eds.): IPMU 2022, CCIS 1602, pp. 173–184, 2022.
https://doi.org/10.1007/978-3-031-08974-9_14

of the population is dubious about the safety of vaccination and many people are unwilling to get vaccinated [8], even though all National Public Health institutions over the world have promoted vaccination. To get a favorable outcome of the vaccination programs, it is critical to identify the factors that influence people's intention to get vaccinated and also to develop effective interventions that promote acceptance of COVID-19 vaccines.

Over these months, several theories and approaches have been postulated to explain the factors associated with vaccine uptake [7]. Using concepts relatives to *Behavioral Economics* and *Decision Theory* in the health domain has been increasing in recent years because they allow to identify specific individual decision-making biases [11]. Traditional decision-making approaches from Economic Theory establish perfect rationality of individuals as their main hypothesis. In this way, people establish their behavior according to their expected utility (*Expected Utility Theory*). Following this theory, people's decisions entail a cost-utility analysis and they are usually made under conditions of uncertainty. In the health context, such a cost-utility analysis involves, e.g., a patient must decide about the benefits and costs of taking a treatment or not, being the uncertainty associated worthy of attention. However, a recent literature has emerged that offers contradictory findings about the paradigm of perfect rationality [6].

This paper aims to contribute to this growing area of research by defining a new methodological approach to measure the stability of preferences over time. This contribution provides a classical group decision-making problem where a set of agents shows their preferences for a set of alternatives for distinct moments of time. The solution to this problem involves determining a society's preference for each period and then measuring how much stability includes such society's preferences. To determine the stability of the society's preferences, it is necessary to get an overarching preference of the society for each period. This paper proposes the use of the method included in [1] for computing a social consensus solution for each moment. Considering these starting points, the methodological approach taken in this study is the proposed by Andrés Calle, Cascón and González-Arteaga in [3] that it is useful in measuring how many preferences change over time when preferences are complete preorders. Finally, it is hoped that this research will contribute to a deeper understanding of the proposed method, including two real cases of study.

This work is composed of four sections. The first section of this paper summarizes the notation, the proposed framework, as well as some basic definitions. Section 3 develops two real cases of study, with in-depth analysis of preferences behavior and of measuring their stability. Finally, Sect. 4 presents the conclusions of the research.

2   Background

Prior to commencing our proposal, this section includes notation, it reviews some previous definitions and it establishes the methodological framework used for measuring preferences stability.

2.1 Notation and Starting Hypothesis

Let $\mathbf{N} = \{1, \ldots, n\}$ a society of n individuals and $\mathbf{X} = \{x_1, \ldots, x_k\}$ a finite set of alternatives, $\mid \mathbf{X} \mid \geq 2$. There is no loss of generality in assuming that individuals provide their preferences on alternatives by complete preorders[1] on $\mathbf{X}$. This assumption is introduced in order to capture the reality of decision life-situations. $\mathbf{W}(\mathbf{X})$ denotes the set of all complete preorders on $\mathbf{X}$.

Let $\mathcal{R} \in \mathbf{W}(\mathbf{X})$ be a complete preorder on $\mathbf{X}$, the notation $x_s \succ_R x_k$ means that the alternative x_s is strictly preferred to the alternative x_k, $x_s \sim_R x_k$ means that the alternative x_s and the alternative x_k are equally preferred and $x_s \succcurlyeq_R x_k$ means that the alternative x_s is at least as good as the alternative x_k.

Let $\mathbf{T} = \{t_0, \ldots, t_T\}$ be an ordered time sequence, namely, a *temporal set*.[2] Let $\mathcal{R}_j^i \in \mathbf{W}(\mathbf{X})$ be a *time-related preference* of the individual $j \in \mathbf{N}$ on the set of alternatives $\mathbf{X}$ at the moment of time $t_i \in \mathbf{T}$.

Let $\mathcal{P}^i = (\mathcal{R}_1^i, \mathcal{R}_2^i, \ldots, \mathcal{R}_n^i) \in \mathbf{W}(\mathbf{X})^n$ be a *time-related preference profile* of the society $\mathbf{N}$ on the set of alternatives $\mathbf{X}$ at the moment of time t_i.

Taking into account the overall temporal set, the *society temporal preference profile* on the set the alternatives $\mathbf{X}$ over $T + 1$ moments of time is defined as follows:

$$\mathcal{P} = (\mathcal{R}^0, \mathcal{R}^1, \ldots, \mathcal{R}^T) \in \mathbf{W}(\mathbf{X})^{T+1}$$

The $\mathcal{R}^i \in \mathcal{P}$ profile element is the *society temporal preference* on $\mathbf{X}$ at the moment of time $t_i, i \in \{0, \ldots, T\}$. Let $\mathbf{P}(\mathbf{X})$ denote the set of all temporal preferences profiles, that is, $\mathbf{P}(\mathbf{X}) = \bigcup_{T \geq 1} \mathbf{W}(\mathbf{X})^{T+1}$.

2.2 Codification Procedure of Preferences

Since 1981, the codification of preferences by numerical vectors has been extensively used in theoretical and practical situations. For the purpose of measuring the stability of people's preferences on vaccination, this contribution suggests the use of the codification procedure defined by González-Artega, Alcantud and de Andrés Calle [5] because their approach generates a strong codification of the preferences that allows to compare complete preorders without loss of preference consistency. Let us now introduce it in this specific context.

Given a time-related preference $\mathcal{R}_j^i \in \mathbf{W}(\mathbf{X})$ for $j \in \mathbf{N}$ and $i \in \mathbf{T}$, its corresponding canonical codification is a real vector namely, the *canonical codified time-related preference*: $\mathbf{c}_{\mathcal{R}_j^i} = (c_1^{\mathcal{R}_j^i}, \ldots, c_k^{\mathcal{R}_j^i}) \in (\{1, \ldots, k\})^k$, being $c_g^{\mathcal{R}_j^i}$ the number of alternatives classifying at most as good as the alternative x_g at the

---

[1] Technically speaking, a complete preorder $\mathcal{R}$ on X means a complete and transitive binary relation on $\mathbf{X}$.

[2] It should be noted that the moments of time in the temporal set would not be in need of equidistantly distributed.

moment of time t_i for the individual j. The set of all possible canonical codified time-related preference associated with $\mathbf{W}(\mathbf{X})$ is denoted by $\mathbf{F} = \mathbf{F}(\mathbf{W}(\mathbf{X}))$.

Given a time-related preference profile $\mathcal{P}^i \in \mathbf{W}(\mathbf{X})^n$, its corresponding *canonical codified time-related preference profile* is a $n \times k$ real matrix given by

$$\mathcal{M}_{\mathcal{P}^i} = (\mathbf{c}_{\mathcal{R}_1^i}, \dots, \mathbf{c}_{\mathcal{R}_n^i}) \in \mathbb{M}_{n \times k}$$

where the element j-th indicates the canonical codified time-related preference $\mathbf{c}_{\mathcal{R}_j^i}$ at the moment of time $t_i \in \mathbf{T}$. A *canonical codified society temporal preference profile* associated with the society temporal preference profile $\mathcal{P} = (\mathcal{R}^0, \dots, \mathcal{R}^T) \in \mathbf{W}(\mathbf{X})^{T+1}$ is a $(T+1) \times k$ real matrix, namely $\mathcal{M}_{\mathcal{P}} = (\mathbf{c}_{\mathcal{R}^0}, \dots, \mathbf{c}_{\mathcal{R}^T})$, where element $\mathcal{R}^i$ represents the *canonical codification society temporal preference* at the moment of time t_i. Let $\mathbf{M}$ denote the set of all $(T+1) \times k$ real matrices.

2.3   Methodological Framework and Basic Definitions

Before proceeding to accomplish our proposal, it is important to establish the methodological framework and the basic definitions used for it.

This contribution provides a classical group decision-making problem where a set of agents shows their preferences for a set of alternatives for distinct moments of time (time-related preference profile). The solution to this problem involves determining a society's preference for each period (society temporal profile) and then measuring how much stability includes such society's preferences. Figure 1 presents an overview of the aforementioned decision-making process.

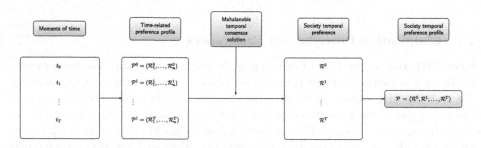

Fig. 1. Scheme of our procedure.

To determine the stability of the society's preferences, it is necessary to get an overarching preference of the society for each period. This paper proposes the use of the method included in [1] for computing a social consensus solution for each moment.

Definition 1. *The* Mahalanobis temporal consensus solution $\hat{R}^i \in \mathbf{W}(\mathbf{X})$ *of a given time-related preference profile* $\mathcal{P}^i = (\mathcal{R}_1^i, \mathcal{R}_2^i, \ldots, \mathcal{R}_n^i) \in \mathbf{W}(\mathbf{X})^n$ *is an ordinal ranking obtained by solving:*

$$\min_{\mathcal{R}^i \in \mathbf{W}(\mathbf{X})} \mathcal{MC}_{\Sigma, \mathcal{P}}(\mathcal{R}^i)$$

where $\mathcal{MC}_{\Sigma, \mathcal{P}}(\mathcal{R}^i)$ *is the Mahalanobis consensus distance function.*[3]

Once a society's preference for each moment of time (society temporal preference) has been obtained by Definition 1, it is possible to measure the stability of the society's preferences over time.

Considering these starting points, the methodological approach taken in this study is the proposed by Andrés Calle, Cascón and González-Arteaga in [3] that it is particularly useful in measuring how much preferences change over time when preferences are considered as complete preorders. This approach includes two particular measures: the *local* and the *global preference stability measure* that verify several significant mathematical properties. Moreover, the model includes a parameter, the λ-*parameter*, that depicts the *memory loss effect* of people on preferences over time[4]. These definitions are formulated hereunder.

Definition 2. *Let* $\mathcal{P} = (\mathcal{R}^0, \ldots, \mathcal{R}^T) \in \mathbf{P}(\mathbf{X})$ *be a society temporal preference profile and its canonical codified temporal preference profile* $\mathcal{M}_{\mathcal{P}} = (\mathbf{c}_{\mathcal{R}^0}, \ldots, \mathbf{c}_{\mathcal{R}^T}) \in \mathbf{M}$. *The* local preference stability measure *between the social temporal preferences at the moments of time* t_{i-1} *and* t_i *is a mapping* $\theta_i : \mathbf{W}(\mathbf{X}) \longrightarrow [0, 1]$ *given by*

$$\theta_i(\mathcal{P}) = \theta_{[i-1,i]}(\mathcal{P}) = 1 - \frac{\| \mathbf{c}_{\mathcal{R}^{i-1}} - \mathbf{c}_{\mathcal{R}^i} \|_1}{r}$$

where $\mathbf{c}_{\mathcal{R}^{i-1}}$ *and* $\mathbf{c}_{\mathcal{R}^i}$ *are the canonical codified vectors associated with the society temporal preferences* $\mathcal{R}^{i-1}$ *and* $\mathcal{R}^i$, *respectively;* $\| \cdot \|_1$ *denotes the* l_1-*norm, and finally,* $r = \max_{\mathbf{c}, \mathbf{c}' \in \mathbf{F}} \| \mathbf{c} - \mathbf{c}' \|_1$. *Therefore:*

$$\theta_i(\mathcal{P}) = \theta_{[i-1,i]}(\mathcal{P}) = 1 - \frac{\sum_{h=1}^{k} | c_h^{\mathcal{R}^{i-1}} - c_h^{\mathcal{R}^i} |}{\max_{\mathbf{c}, \mathbf{c}' \in \mathbf{F}} \| \mathbf{c} - \mathbf{c}' \|_1}$$

---

[3] $\mathcal{C}_{\Sigma, \mathcal{P}}(\mathcal{R}^i) = \sum_{j=1}^n d_{\Sigma}(\mathbf{c}_{\mathcal{R}_j^i}, \mathbf{c}_{\mathcal{R}^i}) = \sum_{j=1}^n (\mathbf{c}_{\mathcal{R}_j^i} - \mathbf{c}_{\mathcal{R}^i}) \Sigma^{-1} (\mathbf{c}_{\mathcal{R}_j^i} - \mathbf{c}_{\mathcal{R}^i})^t$.

[4] In this regard, if individuals express their preferences taking into account more intensively the most recent, $\lambda > 0$; if individuals express their preferences, taking into account equally all, $\lambda = 0$.

Definition 3. *Let* $\mathcal{P} \in \mathbf{P}(\mathbf{X})$ *a society temporal preference profile and* $\mathcal{M}_{\mathcal{P}} \in \mathbf{M}$ *its corresponding canonical codification matrix. The* global preference stability measure *for the society temporal preference profile* $\mathcal{P} \in \mathbf{P}(\mathbf{X})$ *is de mapping* $\Theta : \mathbf{P}(\mathbf{X}) \times \mathbb{R}^+ \longrightarrow [0,1]$ *given by*

$$\Theta(\mathcal{P}, \lambda) = \sum_{i=1}^{T} w_{i,T}(\lambda) \cdot \theta_i(\mathcal{P})$$

where $w_{i,T}(\lambda) = A_T(\lambda) \, e^{-\lambda(T-i)}$ *with* $A_T(\lambda) = \begin{cases} \frac{1-e^{-\lambda}}{1-e^{-\lambda T}} & \text{if} \;\; \lambda > 0, \\ \frac{1}{T} & \text{if} \;\; \lambda = 0 \end{cases}$ *for any* $\lambda \in \mathbb{R}^+$ *and* $i = 1, \ldots, T$. *Note that* $\sum_{i=1}^{T} w_{i,T}(\lambda) = 1$.

3 Cases of Study

In the following pages, the methodology proposed in Sect. 2 is implemented in two real cases of study. The first explores the characteristics of Spanish citizens' preferences on COVID-19 vaccination from September 2020 to September 2021. The second attempts to evaluate the stability of French citizens' preferences on getting vaccinated from May 2020 to October 2021.

3.1 Spanish Case: Stability of Spaniards' Preferences on Getting COVID-19 Vaccine

Data regarding Spaniards' opinions and their perceptions on COVID-19 pandemic were obtained by the survey collection: *"Sanitary barometer"* compiled by the *Spanish Center of Sociological Research* (CIS). This set of studies deals with information related to the COVID-19 pandemic and its effects on Spaniards' life[5]. This contribution is focused on the above surveys and it includes interviews for 3,645 randomly chosen Spanish people for each month from September 2020 to September 2021[6]. Spaniards were asked about their preferences on COVID-19 vaccination and they had to express their preferences over the following set of alternatives:

$\mathbf{X} = \{x_1$: get vaccinated now; x_2: wait to be vaccinated; x_3: reject vaccination$\}$

at twelve moments of time from September 2020 to September 2021. Table 1 presents an overview of the individuals' time-related preference for each moment of time.

---

[5] All data are available through the website: http://www.cis.es/cis/export/sites/default/-Archivos/Marginales/3300_3319/3302/FT3302.pdf.

[6] It is worth mentioning that the vaccination program started in January 2020.

Table 1. Spaniards' time-related preferences on vaccination from September 2020 to September 2021.

Date	Agents	Time-related preference	Date	Agents	Time-related preference
Sep. 2020	1,333	$x_3 \succ x_1 \succ x_2$	Nov. 2020	1,333	$x_3 \succ x_1 \succ x_2$
	2,312	$x_3 \succ x_1 \succ x_2$		2,312	$x_1 \succ x_3 \succ x_2$
Oct. 2020	1,792	$x_1 \succ x_3 \succ x_2$	Dec. 2020	227	$x_1 \sim x_2 \succ x_3$
	1,853	$x_3 \succ x_1 \succ x_2$		436	$x_3 \succ x_1 \succ x_2$
				2,982	$x_1 \succ x_3 \succ x_2$
Jan. 2021	2,720	$x_1 \succ x_3 \succ x_2$	May. 2021	2,720	$x_1 \succ x_2 \succ x_3$
	925	$x_1 \succ x_2 \succ x_3$		925	$x_1 \succ x_3 \succ x_2$
Feb. 2021	2,080	$x_1 \succ x_2 \succ x_3$	Jun. 2021	2,539	$x_1 \succ x_3 \succ x_2$
	640	$x_1 \succ x_3 \succ x_2$		1,106	$x_1 \succ x_2 \succ x_3$
	925	$x_1 \succ x_2 \sim x_3$			
Mar. 2021	227	$x_1 \succ x_2 \sim x_3$	Jul. 2021	663	$x_1 \succ x_2 \succ x_3$
	1,183	$x_1 \succ x_3 \succ x_2$		1,310	$x_1 \succ x_3 \succ x_2$
	2,235	$x_1 \succ x_2 \sim x_3$		1,672	$x_1 \succ x_2 \sim x_3$
Apr. 2021	3,645	$x_1 \succ x_2 \succ x_3$	Sep. 2021	227	$x_1 \succ x_2 \succ x_3$
				2,493	$x_1 \succ x_3 \succ x_2$
				925	$x_1 \succ x_2 \sim x_3$

To measure the stability of Spaniards' preferences taking into account the individuals' time-related preferences from Table 1, it is necessary to obtain a society temporal preference for each month. For this purpose, Definition 1 is applied and a society temporal preference for each month is obtained. These are shown in Table 2.

Table 2. Spaniards' society temporal preference for each moment of time.

Date	$\mathcal{R}^i$	Date	$\mathcal{R}^i$
Sep. 2020	$\mathcal{R}^0 : x_2 \succ x_1 \sim x_3$	Jan. 2021	$\mathcal{R}^4 : x_2 \succ x_3 \succ x_1$
Oct. 2020	$\mathcal{R}^1 : x_2 \succ x_1 \sim x_3$	Feb. 2021	$\mathcal{R}^5 : x_1 \succ x_2 \succ x_3$
Nov. 2020	$\mathcal{R}^2 : x_2 \succ x_1 \succ x_3$	Mar. 2021	$\mathcal{R}^6 : x_2 \succ x_3 \sim x_1$
Dec. 2020	$\mathcal{R}^3 : x_2 \sim x_3 \succ x_1$	Apr. 2021	$\mathcal{R}^7 : x_1 \succ x_2 \succ x_3$
		May. 2021	$\mathcal{R}^8 : x_1 \succ x_2 \succ x_3$
		Jun. 2021	$\mathcal{R}^9 : x_1 \succ x_3 \succ x_2$
		Jul. 2021	$\mathcal{R}^{10} : x_2 \sim x_3 \succ x_1$
		Sep. 2021	$\mathcal{R}^{11} : x_2 \succ x_3 \succ x_1$

By means of the society temporal preference for each month, the local preference stability measures can be calculated by Definition 2. The results obtained are presented in Table 3.

Table 3. Values of the local preference stability measures for Spanish society.

i										
1	2	3	4	5	6	7	8	9	10	11
$\theta_i(\mathcal{P})$ 1	0.8	0.4	0.8	0.6	0.8	0.8	1	0.6	0.8	0.8

Finally, the global preference stability measure is computed by Definition 3. In order to analyze the effect of the λ-parameter, that is, the *loss memory effect* of the Spanish society on getting vaccinated over time, different values of the λ-parameter are considered. Table 4 shows the different weights values of the local preference stability measures, i.e., $w_{i,T}(\lambda)$, and Table 5 presents the global preference stability measures for different values of λ-parameter.

Table 4. Weights values of the local preference stability measures for Spanish society, $w_{i,T}(\lambda)$, for different values of the λ-parameter.

λ	$w_{i,T}(\lambda)$ 1	2	3	4	5	6	7	8	9	10	11
0.2	0.0276	0.0337	0.0412	0.0503	0.0614	0.0750	0.0916	0.1119	0.1366	0.1669	0.2039
0.4	0.0061	0.0091	0.0136	0.0203	0.0303	0.0452	0.0674	0.1005	0.1500	0.2237	0.3338
0.6	0.0011	0.0020	0.0037	0.0068	0.0123	0.0225	0.0410	0.0747	0.1361	0.2480	0.4518
0.8	0.0002	0.0004	0.0009	0.0020	0.0045	0.0101	0.0224	0.0500	0.1112	0.2475	0.5508
1	0.0000	0.0001	0.0002	0.0006	0.0016	0.0043	0.0116	0.0315	0.0855	0.2325	0.6321

Spaniards' preferences on vaccination have been changed over time although they are stables even for different values of the λ-parameter as shown in Table 5 and Fig. 2a.

Table 5. Global preference stability measures for Spanish society attending to several values of λ.

	λ					
	0	0.2	0.4	0.6	0.8	1
$\Theta_i(\mathcal{P}, \lambda)$	0.7636	0.7718	0.7798	0.7840	0.7865	0.7880

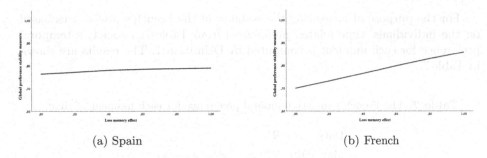

(a) Spain	(b) French

Fig. 2. Graphical representation of the global preference stability measures for different values of the λ-parameter.

3.2 French Case: Stability of the French's Preferences on Getting COVID-19 Vaccine

Data regarding the French's opinions on getting vaccinated against COVID-19 was obtained by the survey collection: *"Campagne santè mentale et coronavirus: En parler, c'est déjà se soigner"*, compiled by the *French Institute of Public Health*. This set of questionnaires deals with information related to COVID-19 pandemic on the French citizens[7]. This contribution is focused on this survey collection and it includes interviews for $1,728$ randomly chosen French people in May 2020, December 2020, March 2021, May 2021 and October 2021[8]. The French were asked about their preferences on COVID-19 vaccination and they had to express their preferences over the following set of alternatives:

$$\mathbf{X} = \{x_1: \text{get vaccinated now}; \ x_2: \text{wait to be vaccinated};$$

$$x_3: \text{only get vaccinated if compulsorily}; \ x_4: \text{reject vaccination}\}$$

at five different moments of time. Table 6 provides the individuals' time-related preference for each studied moment.

Table 6. The French's time-related preferences on vaccination from May 2020 to October 2021.

Date	Agents	Time-related preference	Date	Agents	Time-related preference
May. 2020	1,053	$x_4 \succ x_3 \succ x_2 \succ x_1$	Dec. 2020	720	$x_4 \succ x_3 \succ x_2 \succ x_1$
	420	$x_2 \succ x_3 \succ x_1 \succ x_4$		327	$x_4 \succ x_3 \succ x_1 \succ x_2$
	255	$x_1 \succ x_2 \succ x_3 \succ x_4$		255	$x_3 \succ x_4 \succ x_2 \succ x_1$
Mar. 2021	856	$x_3 \succ x_2 \succ x_1 \succ x_4$	Oct. 2021	675	$x_1 \succ x_2 \succ x_3 \succ x_4$
	617	$x_4 \succ x_3 \succ x_2 \succ x_1$		617	$x_4 \succ x_3 \succ x_2 \succ x_1$
	255	$x_2 \succ x_1 \succ x_3 \succ x_4$		436	$x_2 \succ x_3 \succ x_1 \succ x_4$
May. 2021	675	$x_2 \succ x_1 \succ x_3 \succ x_4$			
	617	$x_4 \succ x_3 \succ x_2 \succ x_1$			
	436	$x_2 \succ x_3 \succ x_1 \succ x_4$			

---

[7] All data are available through the website: https://www.santepubliquefrance.fr/dossiers/coronavirus-covid-19/etudes-et-enquetes-covid-19.

[8] It is worth mentioning that the vaccination program started in January 2020.

For the purpose of measuring the stability of the French's preferences based on the individuals' time-related preferences from Table 6, a society's temporal preference for each moment is computed by Definition 1. The results are shown in Table 7.

Table 7. The French's society temporal preference for each moment of time.

Date	$\mathcal{R}^i$
May. 2020	$\mathcal{R}^0 : x_2 \succ x_1 \succ x_3 \sim x_4$
Dec. 2020	$\mathcal{R}^1 : x_4 \succ x_1 \succ x_2 \sim x_3$
Mar. 2021	$\mathcal{R}^2 : x_2 \succ x_1 \succ x_3 \sim x_4$
May. 2020	$\mathcal{R}^3 : x_1 \succ x_2 \sim x_3 \succ x_4$
Oct. 2021	$\mathcal{R}^4 : x_1 \succ x_2 \sim x_3 \succ x_4$

Once the society temporal preference for each moment has been computed, the local preference stability measures for French society can be calculated by Definition 2. The results are shown in Table 8.

Table 8. Values of the local preference stability measures for French society.

i	1	2	3	4
$\theta_i(\mathcal{P})$	0.6	0.6	0.6	1

To finish, the global preference stability measure is computed by Definition 3 attending to different values of the λ-parameter. Table 9 shows the different weights values of the local preference stability measures, i.e., $w_{i,T}(\lambda)$, and

Table 9. Weights values of the local preference stability measures for French society, $w_{i,T}(\lambda)$, for different values of the λ-parameter.

λ	$w_{i,T}(\lambda)$			
	1	2	3	4
0.0	0.2500	0.2500	0.2500	0.2500
0.2	0.1807	0.2207	0.2695	0.3292
0.4	0.1244	0.1856	0.2769	0.4131
0.6	0.0820	0.1495	0.2723	0.4962
0.8	0.0521	0.1159	0.2579	0.5741
1.0	0.0321	0.0871	0.2369	0.6439

Table 10 presents the global preference stability measures for French society considering different values of the λ-parameter. The French's preferences for vaccination have been changed over time and they are more stable than the Spanish one for high values of the λ-parameter as shown in Fig. 2b.

Table 10. Global preference stability measures for French society attending to several values of λ.

λ					
0	0.2	0.4	0.6	0.8	1
$\Theta_i(\mathcal{P}, \lambda)$ 0.7000	0.7317	0.7652	0.7985	0.8296	0.8576

4 Conclusions

The main goal of the current study has been to determine a new general framework to measure preference stability over time considering complete pre-orders. This research provides useful in expanding our understanding of how preferences change over time from a theoretical and practical perspective. Moreover, the present contribution includes two real cases of study to explore the stability of preferences on COVID-19 vaccination from Spanish and French citizens.

Acknowledgements. De Andrés Calle is grateful to the Junta de Castilla y León and the European Regional Development Fund (Grant CLU-2019-03) for the financial support to the Research Unit of Excellence "Economic Management for Sustainability" (GECOS).

References

1. Cascón, J., González-Arteaga, T., de Andrés Calle, R.: Reaching social consensus family budgets: the Spanish case. Omega **86**, 28–41 (2019)
2. Caserotti, M., Girardi, P., Rubaltelli, E., Tasso, A., Lotto, L., Gavaruzzi, T.: Associations of covid-19 risk perception with vaccine hesitancy over time for Italian residents. Soc. Sci. Med. **272**, 113688 (2021)
3. de Andrés Calle, R., Cascón, J., González-Arteaga, T.: Preferences stability: a measure of preferences changes over time. Decis. Support Syst. **129**, 113169 (2020)
4. Gallagher, K., LaMontagne, D., Watson-Jones, D.: Status of HPV vaccine introduction and barriers to country uptake. Vaccine **36**(32, Part A), 4761–4767 (2018)
5. González-Arteaga, T., Alcantud, J., de Andrés Calle, R.: A new consensus ranking approach for correlated ordinal information based on Mahalanobis distance. Inf. Sci. **372**(Supplement C), 546–564 (2016)
6. Jensen, A.F.: Bounded rational choice behaviour: applications in transport. Transp. Rev. **36**(5), 680–681 (2016)
7. Li, L., Wang, J., Nicholas, S., Maitland, E., Leng, A., Liu, R.: The intention to receive the covid-19 vaccine in china: Insights from protection motivation theory. Vaccines **9**(5) (2021)

8. Motta, M.: Can a covid-19 vaccine live up to Americans' expectations? a conjoint analysis of how vaccine characteristics influence vaccination intentions. Soc. Sci. Med. **272**, 113642 (2021)
9. Randolph, H.E., Barreiro, L.B.: Herd immunity: understanding covid-19. Immunity **52**(5), 737–741 (2020)
10. World Health Organization: Coronavirus disease (covid-19) weekly epidemiological update and weekly operational update. World Health Organization, Tech. rep. (2022)
11. Xue, X., Reed, W.R., Menclova, A.: Social capital and health: a meta-analysis. J. Health Econ. **72**, 102317 (2020)
12. Zimmer, C., Corum, J., Wee, S.L.: Coronavirus vaccine tracker. The New York Times (2021). https://www.nytimes.com/interactive/2020/science/coronavirus-vaccine-tracker.html. Accessed 10 May 2021

On the Notion of Influence in Sensory Analysis

Jacky Montmain$^{(\boxtimes)}$, Abdelhak Imoussaten, Sébastien Harispe,
and Pierre-Antoine Jean

EuroMov Digital Health in Motion, University of Montpellier, IMT Mines Ales, Ales, France
{jacky.montmain,abdelhak.imoussaten,sebastien.harispe,
pierre-antoine.jean}@mines-ales.fr

Abstract. When a group of agents is faced with collective decisional tasks, the agents may have to cooperate to establish which alternative appears to be a convenient consensus. The influence an agent may have upon the other ones may change the collective decision. Modelling of influence in a social network assumes that each player has an initial inclination for one of the alternative that may be different from his final decision. The point of departure of such studies is the concept of the Hoede-Bakker index, which computes the overall decisional power of a player in a social network when the decision is binary. We propose to extend the notions of decision power and influence indices when the decision is not simply binary, but associated with a scale of success: an individual can more or less adhere to the collective decision. Initially, the idea was motivated by the collective evaluation process in sensory analysis. Indeed, sensory analysis is a set of methods for measuring sensory perceptions. When the panel is made up of trained individuals who master the vocabulary of the domain, sensory analysis has been the subject of numerous evaluation protocols. On the other hand, when the evaluators are inexperienced evaluators not mastering the descriptors, the sensory characterization may strongly be impacted by influential panelists. In this paper, we are interested in measuring such a qualitative influence. Framing our work in game theory models, we propose to extend the notion of influence index to the qualitative process of graded decision in sensory analysis, justifying our choices on the basis of our feedback on the process of sensory analysis by a collective of panelists.

Keyword: Decision-making power - Index of influence - Social network - Sensory analysis

1 Introduction

In industry, when a sensory analysis of an odor, a taste or a color is performed by a panel of evaluators, each description is made using conceptual descriptors of interest for the domain under study. These descriptors are disambiguated concepts extracted from a knowledge structure corresponding to a taxonomy [1, 2]. The variability of the sets of descriptors provided from one evaluator to another requires a synthesis of the conceptual descriptions to be established, taking into account their diversity [3]. This synthesis can then be used to verify that the effect sought/desired by the industrialist at the initiative

© Springer Nature Switzerland AG 2022
D. Ciucci et al. (Eds.): IPMU 2022, CCIS 1602, pp. 185–196, 2022.
https://doi.org/10.1007/978-3-031-08974-9_15

of this analysis was indeed perceived by the evaluators (e.g. to check if the taste of a product is consistent with a specific target). It is thus necessary to note that the human being, in spite of the subjectivity and the uncertainty which characterize his sensory perception, is positioned as a tool of measurement and decision making in the chain of reliability of the industrialist by the means of the sensory analysis. When it is a panel of experts mastering the solicited sense and the associated controlled vocabulary, the sensory characterization reflects the sole perception of the experienced evaluator. On the other hand, if the evaluators are less experienced, with less trained senses and less familiar with the use of controlled vocabulary, then they will be more inclined, if given the opportunity, to pay attention to the characterizations of other evaluators to refine the expression of their perception. The less experienced will be inspired by the copy of their "elders" for example. The consumers who will buy the perfume, the dish or the paint, and who are therefore the real target of the industrialist, clearly belong to the second category, they are neophyte evaluators sensitive to advertising, to internet influencers or simply to a truth asserted by their neighbor, a proven reference in their eyes. Who among us has not suddenly attributed `blueberry` or `truffle` aromas to a wine after having inquired about its sensory description on the label of the bottle? We wish to instrument the analysis of influence phenomena in this process of qualitative description and to define the models that could be used to measure the influence of an evaluator or a group of evaluators in the sensory domain, the characterization of influencers having obvious economic repercussions.

We will therefore begin by referring to measures of decision-making power derived from game theory. Modelling of influence in a social network assumes that each player has an inclination to say YES or NO which, due to influence of other players, may be different from the final decision of the player. The point of departure of such studies is the concept of the Hoede-Bakker index, which computes the overall decisional 'power' of a player in a social network. The main drawback of the Hoede-Bakker index is that it hides the actual role of the influence function, analyzing only the final decision in terms of success and failure. Several extensions of this work have been proposed to address this issue, in particular by Grabisch and Rusinowska. These measures of decision-making power have been established for decisions where only two antagonistic alternatives are considered (Yes or NO). We then formally define the decision process at stake in sensory analysis: the choice of a semantic summary that synthesizes the descriptions of the sensory analysis panelists. We then propose to transpose the notions of decision power and influence indices when the decision process involves many possible decisions and it is possible to endow this set of decisions with an order relation. Beyond this application to the domain of sensory analysis, this model allows to extend the notions of decision power and influence indices when the decision is not simply boolean, but gradual: an individual can more or less adhere to the collective decision.

The paper is organized as follows: Sect. 2 introduces theoretical notions on which is based our approach as well as notations; it next reports state of the art studies of influence based on the Hoede-Bakker decision power index and its extensions. This index computes the overall decisional power of a player in a social network when the decision is binary; Sect. 3 introduces the decision-making process involved in a collective sensory analysis evaluation to illustrate the concepts of collective decision making and influence in a group; Sect. 4 is our main contribution: it extends the Hoede-Bakker index to non-binary and gradual decisions, when alternatives can be associated with a graded success scale. Section 5 presents the experiment we are building, unfortunately it is still in progress and we did not have time to include it in this article, it is just meant to illustrate our project.

2 Decision-Making Power of an Agent

Different models have been introduced in game theory to represent influence in social networks. We are inspired here by the study of influence based on the Hoede-Bakker decision power index [5] and its extensions. This section is a simple synthesis of what has been written in game theory about the notion of influence; we use the notations and definitions that have become established in this literature. This index allows us to calculate the global decisional power of an actor in a social network. This index has been generalized by the works of Rusinowska and De Swart [6] and then in [4, 7]. The reasons for the existence of influence phenomena, i.e., why an individual changes his decision, are more psychological considerations and are outside the scope of the studies proposed by [8]. All definitions and notations in this section are simply taken from [7] and [8] for the sake of consistency. In these studies, the debate concerns the choice between two options, noted ± 1. The social network considered is composed of a group of n actors $\{a_1, .., a_n\}$. Each actor has an a priori inclination to opt for $+1$ or -1. A vector of inclinations, noted i, is a vector of $+1$ and -1: j^{th} component of i, noted $i_j \in \{-1; +1\}$, represents the inclination of actor a_j. Let $I = \{-1; +1\}^n$ be the set of the 2^n possible inclination vectors.

In this influence model, the basic assumption is that each actor has an a priori inclination, which, under the influence of the other actors, may be different from his final decision. In other words, each vector $i \in I$ of inclinations is transformed into a vector of decisions Bi by a function, denoted B, that models the influence in the social network. The jth coordinate of Bi is $(Bi)_j$, $j \in N := \{1, ..., n\}$ and represents the a posteriori decision of the actor a_j. Then, to each decision vector b is associated a group decision $gd(b) \in \{-1; +1\}$ where $gd : I \rightarrow \{-1; +1\}$ is an aggregation function. The function gd is called the *group decision function* and models the decision of the collective of actors (a majority for example).

An *influence function B* can correspond to a common collective behavior.

For example, in [7] an influence function of the majority type, noted $Maj^{[t]}$ and parameterized by a real t, is introduced. More precisely, for a given vector of inclinations

$$i \in I: Maj^{[t]}(i) = \begin{cases} 1_N & if \quad |i^+| \geq t \\ -1_N & if \quad |i^+| < t \end{cases} \text{ where } i^+ = \{j \in N / i_j = +1\}.$$

The definition of the decision power index of an agent a_j proposed by Hoede-Bakker in [5], where only success with respect to the option $+1$ is considered, was generalized by Rusinowska and De Swart [6] by studying success for both options $+1$ and -1.

Definition 1: Given an influence function B and a group decision function gd, the generalized decision power index of an actor a_j, denoted GHB_{a_j} is defined by:

$$GHB_{a_j}(B, gd) = \frac{1}{2^n}\left(\sum_{i/i_j=+1} gd(Bi) - \sum_{i/i_j=-1} gd(Bi)\right) \tag{1}$$

The main drawback of the Hoede-Bakker index and its generalization is that it masks the actual role of influence. Indeed, it analyses the decision of an actor in terms of success or failure according to whether or not the group's decision coincides with the actor's initial inclination and not with his final decision. Based on this idea, Grabisch and Rusinowska [7] proposed to distinguish the influence function from the group decision function in order to formulate a new index of decision-making power which is based this time on the concordance between the group's decision and the actor's final decision (in this proposal, they also assign a probability of occurrence to each vector of inclinations which we will not develop here):

$$GHB_{a_j}(B, gd) = \frac{1}{2^n}\left(\sum_{i/(Bi)_j=+1} gd(Bi) - \sum_{i/(Bi)_j=-1} gd(Bi)\right) \tag{2}$$

(1) is not better than (2), they are two different views of what success means for a given individual in a collective choice. For example, do we associate ourselves more with the election of a president if the elected candidate corresponds to our choice at the beginning of the campaign ($i_j = gd(Bi)$) or if he is the one whose name is written on the ballot we put in the box ($(Bi)_j = gd(Bi)$)?

Let us now consider how to measure the degree of influence of an agent or a coalition of agents on the others. In general, we say that an agent is influenced by a coalition (a group of agents with the same inclination) if the agent's decision is ultimately different from his original inclination. Indeed, since the agent has changed his opinion and retained the option that he had initially rejected, we can imagine that he has undergone some kind of *external* influence that explains this change of opinion. We assume that this external influence comes from other agents who share the same inclination. In a so-called direct influence, the agent's inclination is therefore different from that of the coalition, but in the end his decision coincides with the coalition's inclination (Grabisch and Rusinowska also define an influence by opposition where the agent who initially shares the coalition's inclination changes his mind to oppose the latter).

The notations that are essential to the definition of the concept of influence in [7] and that will later be used in our model are given below.

The set of inclination vectors for which all agents in a coalition S have the same inclination is defined by:

$$I_S = \{i \in I / \forall k, j \in S[i_k = i_j]\} \tag{3}$$

We note i_S, for $i \in I_S$, the value of i_k, $\forall k \in S$. For any coalition S and any $j \in N$, we define:

$$I_{S \to j} = \{i \in I_S / i_j = -i_S\} \tag{4}$$

$$I^*_{S \to j}(B) = \left\{ i \in I_{S \to j} / (Bi)_j = i_S \right\} \tag{5}$$

$I_{S \to j}$ denotes the inclination vectors for which a direct influence of S on j can be observed (they correspond to decisions for which an influence of the coalition on the agent can be possibly observed since their a priori choices are opposed). $I^*_{S \to j}(B)$ denotes the inclination vectors for which the influence of S on j has potentially been exerted under the assumption of an influence function B (This is a subset of $I_{S \to j}$: only the vectors of $I_{S \to j}$ where the agent has joined the coalition's opinion a posteriori are retained).

We then introduce the index of possibility of direct influence of a coalition S on j:

$$\overline{d}(B, S \to j) = \frac{\left| I^*_{S \to j}(B) \right|}{\left| I_{S \to j} \right|} \tag{6}$$

We can still define an index of certainty of direct influence of a coalition S on j:

$$\underline{d}(B, S \to j) = \frac{\left| \left\{ i \in I^*_{S \to j}(B) / \forall p \notin S [i_p = -i_S] \right\} \right|}{2} \quad \in \left\{ 0, \frac{1}{2}, 1 \right\} \tag{7}$$

Finally, we note the set of *followers* of S:

$$F_B(S) = \left\{ j \in N / \forall i \in I_S \left[(Bi)_j = i_S \right] \right\} \tag{8}$$

3 Decision Model in Sensory Analysis

During a sensory analysis by a panel of evaluators, each description is made using disambiguated concepts extracted from a knowledge structure of interest for the business domain. The variability of the sets of descriptors provided by the evaluators complicates the task of the analyst who must merge them according to the similarity of the concepts in order to hope to obtain a more informative histogram (he must be able to bring together the concepts Raspberry and Strawberry which both evoke the concept Red Berry, a task with a high added cognitive value). The analyst must be able to propose a summary of the collective evaluation that is expressive while being synthetic by managing the abstraction and precision, the redundancy and similarity of the descriptors. To do this, he must manage the semantic relations that bring together or differentiate the concepts and master the knowledge structure from which they are extracted.

In previous work [9], we were interested in the case where this knowledge structure is a taxonomy $T = (\leq, \mathbb{C})$ where $\mathbb{C}$ is the set of concepts (or classes) and $(\leq)$ the associated partial order based upon an abstraction hierarchy (e.g., Raspberry $\leq$ Red Berry) [10]. We have shown how to formally derive the notion of similarity between concepts from this hierarchical structure, the shared and differentiating features between two concepts and the information content of a concept in a given collection [9–11]. We have thus proposed families of functions allowing to model the similarity between two concepts ($sim : \mathbb{C} \times \mathbb{C} \to \mathbb{R}$) [12], then by extension the similarity between two groups of concepts, in other words between two conceptual annotations ($sim : 2^{\mathbb{C}} \times 2^{\mathbb{C}} \to \mathbb{R}$) [9].

Illustration 1. Intuitively, for concepts similarity:

$$sim(\{\texttt{raspberry}\}, \{\texttt{blackberry}\}) \geq sim(\{\texttt{raspberry}\}, \{\texttt{truffle}\})$$

$$sim(\{\texttt{blackberry}\}, \{\texttt{blackcurrant}\}) \geq sim(\{\texttt{blackberry, humus}\})$$

And for group of concepts similarity:

$$sim(\{\texttt{raspberry, blackberry}\}, \{\texttt{strawberry , blackcurrant}\})$$
$$\geq sim(\{\texttt{raspberry, blackberry}\}, \{\texttt{truffle, humus}\})$$

In [13], we then proposed an automated approach to synthesize conceptual descriptions into a single synthetic conceptual description using the taxonomy-induced order relation. The analysis of the semantics of the conceptual descriptors allows managing redundancy, similarity and abstraction. The synthesis takes the most relevant and informative elements by eliminating redundant evocations, and manages the following dilemma: to be too abstract, one loses information; but to be too specific, one loses the synthesis criterion.

Illustration 2. A summary of $\{\texttt{mouse, rat, fieldmouse, ant, fly, mosquito, platypus}\}$ could be $\{\texttt{rodent, insect, platypus}\}$ considering the taxonomy of animals species.

If we now return to the problem of decision in sensory analysis:

– To each of the n evaluators e_i is associated with a set of concepts, *i.e.* a conceptual annotation noted $X_i \in 2^C$.
  - $\hat{X} = (X_1, X_2, .., X_n)$ is the sequence of annotations by n evaluators to be summarized.

Let X be the set of concepts provided by the evaluators, *i.e.*: $X = \bigcup_{i=1}^{n} X_i$.

The objective of Semantic Synthesis is to define a function SS which summarizes the sequence of conceptual annotations $\hat{X}$ into Y:

$$SS : \left(2^C\right)^n \to 2^C$$
$$Y = SS\left(\hat{X}\right) \tag{9}$$

Each set of descriptors $X_i \in 2^C$ proposed by an evaluator e_i will be more or less similar to $Y \in 2^C$. The more X_i is similar to Y, the more e_i will be comforted in the idea that the choice of his descriptors (his decision) has contributed to the collective characterization Y; the more e_i will perceive his decisional power as high.

At this stage, we are thus able to calculate the similarity between the set of descriptors $X_i \in 2^C$ proposed by the evaluator e_i and the summary $Y \in 2^C$ that synthesizes the collective evaluation. The more similar X_i to Y (up to $sim(X_i, Y)$), the greater the decision power of e_i should be. It is this gradual decision process that we will now use to define the notion of influence in sensory analysis.

4   Model of Influence in Sensory Evaluation

In this section, the model of [8] is extended to non-binary and gradual decisions: we extend the equations of [8] when the alternatives are more than two, can be ordered with respect to a similarity relation and then associated with a graded success scale.

We therefore imagine evaluators participating in $|D|$ sensory evaluations $D^{(i)}$, $i = 1..|D|$. The idea is to determine which of these evaluators has the most influence on the group in view of these $|D|$ sensory exercises with the mathematical tools of Sect. 2 transposed to the decision-making process of sensory analysis: "The more similar the conceptual description proposed by e_i to the descriptions retained by the collective, the greater its decision-making power". For each sensory evaluation $D^{(l)}$, each evaluator e_i proposes a conceptual description $D_i^{(l)}$. Then, the evaluations of all the evaluators are brought to the knowledge of each one. After having read the descriptors used by the other evaluators, the evaluator e_i can modify his a priori choice and propose a new a posteriori semantic annotation $BD_i^{(l)}$ for the evaluation $D^{(l)}$ (e.g., some evaluators less familiar with the controlled vocabulary did not find the precise words during their initial characterization, the descriptions of the other evaluators can then allow them to correct some inaccuracies). The semantic summary of the $D_i^{(l)}$ annotations (respectively $BD_i^{(l)}$) is noted $SS(D^{(l)})$ (respectively $SS(BD^{(l)})$) where $D^{(l)}$ (respectively $BD^{(l)}$) is the vector of a priori (respectively a posteriori) evaluations of the n evaluators (a vector whose coordinates are thus elements of $2^{\mathbb{C}}$). This process of sensory qualification is repeated on $|D|$ evaluations by the panel.

In the game-theory model of decision-making power in Sect. 2, the statistics for the coincidence of decisions between an evaluator e_i and the collective are for the 2^n possible inclination vectors, with the influence function B and the collective decision process gd known a priori. Here, we do not try to define the B function a priori, we simply observe the vectors $D^{(l)}$ and $BD^{(l)}$ on the $|D|$ evaluations and the statistics established do not concern all the possible vectors ($2^{\mathbb{C}^n}$), but the $|D|$ observations of the experiment. The semantic synthesis function SS can be seen as playing the role of the aggregation function gd in Sect. 2. It is thus an estimation of the decision power and influence on a case base reduced to the observations that we propose. The following models focus on defining the decision power of an evaluator e_k based on $|D|$ sensory analyses for a given SS annotation synthesis process.

On the other hand, the model in Sect. 2 is concerned with a Boolean decision where the two alternatives are necessarily opposite. Here, the idea that the more similar the conceptual descriptions proposed by e_i are to the descriptions chosen by the collective, the greater its decision-making power, suggests that this gradual aspect must be integrated into the modeling of the power index and the associated influence indices.

We can first introduce a threshold η (η can be interpreted as the radius of a circle with center $SS(BD^{(l)})$) such that if $sim\left(BD_k^{(l)}, SS(BD^{(l)})\right) \geq \eta$ then the final evaluation of e_k coincides with the final characterization of the collective and then count the number of evaluations among the $|D|$ sensory analyses where this constraint is realized. We then define the decision power of e_k as follows[1]:

$$GHB_k(B, SS) = \frac{1}{|D|}\left(\sum_{i=1}^{|D|}\left[sim\left(BD_k^{(i)}, SS\left(BD^{(i)}\right)\right) \geq \eta\right]\right) \quad (10)$$

To take into account the gradual aspect, the following formulation can be preferred:

$$GHB_k(B, SS) = \frac{1}{|D|}\left(\sum_{i=1}^{|D|} sim\left(BD_k^{(i)}, SS\left(BD^{(i)}\right)\right)\right) \quad (11)$$

where the decision-making power of e_k is all the greater as its a posteriori annotations approach the collective annotations retained.

We can then define the sets: I_S, $I_{S\rightarrow j}$ and $I_{S\rightarrow j}^*$ according to the same principle of similarity to the group decision.

Let S be a coalition of evaluators. I_S is the set of a priori conceptual annotation vectors where all evaluators of S agree among the $|D|$ sensory analyses, i.e., they are all inside or outside the "circle" of radius η with centre $SS(D^{(l)})$ around the annotation chosen by the collective (in any case this radius can be chosen different from the radius in Eq. (10)):

$$I_S = \left\{D^{(i)}, i \in 1..|D|/\forall k, j \in S, \left(sim\left(D_k^{(i)}, SS\left(D^{(i)}\right)\right) - \eta\right)\left(sim\left(D_j^{(i)}, SS\left(D^{(i)}\right)\right) - \eta\right) \geq 0\right\} \quad (12)$$

We note for $D^{(i)} \in I_S$ and any $k \in S$, $sign\left(sim\left(D_k^{(i)}, SS\left(D^{(i)}\right)\right) - \eta\right) = s_i$.

It is possible to model the gradation of coalition membership by fuzzyfying I_S. The idea is to model the fact that a synthesis $SS(D^{(i)})$ is more or less unanimous in the coalition S. The degree of membership of $D^{(i)}$ to I_S gives an account of the cohesion of the evaluations around $SS(D^{(i)})$ within the coalition (respectively of the distance of the evaluations to $SS(D^{(i)})$) and can then be defined as follows:

$$\left| \begin{array}{ll} \mu_i = \frac{1}{|S|}\sum_{k\in S} sim\left(D_k^{(i)}, SS(D^{(i)})\right) & if \quad s_i \geq 0 \\ \mu_i = \frac{1}{|S|}\sum_{k\in S}\left(1 - sim\left(D_k^{(i)}, SS(D^{(i)})\right)\right) & if \quad s_i < 0 \end{array} \right.$$

A fuzzy extension after normalization can then be proposed for I_S:

$$\tilde{I}_S = \left\{\left(D^{(i)} \in I_S; \mu_i\right)\right\} \quad (13)$$

If we are interested in the influence of S on e_j, we select among the vectors of I_S those where $D_j^{(i)}$ is not on the same side of the "circle of radius η and center $SS(D^{(i)})$" as the $D_k^{(i)}$, $k \in S$:

$$I_{S\rightarrow j} = \left\{D^{(i)} \in I_S/s_i.\left(sim\left(D_j^{(i)}, SS\left(D^{(i)}\right)\right) - \eta\right) \leq 0\right\} \quad (14)$$

---

[1] $[Constraint] = 1$ if $Constraint$ is true else 0.

One can also quantify the opposition of e_j to S: for a given evaluation $D^{(i)}$ if the conceptual characterizations of the evaluators of S are very similar to (respectively far from, *i.e.,* outside the circle of radius η) $SS(D^{(i)})$, e_j is all the more opposed to S than $D_j^{(i)}$ is different from (respectively similar to) $SS(D^{(i)})$.

This statement allows us to define a degree of membership for $D^{(i)}$ to $I_{S \to j}$:

$$\left| \begin{array}{l} \lambda_i = \mu_i \left(1 - sim\left(D_j^{(i)}, SS(D^{(i)}) \right) \right) \quad if \quad s_i \geq 0 \\ \lambda_i = \mu_i sim\left(D_j^{(i)}, SS(D^{(i)}) \right) \quad if \quad s_i < 0 \end{array} \right.$$

and then after normalization we introduce the fuzzy set:

$$\tilde{I}_{S \to j} = \left\{ \left(D^{(i)} \in \tilde{I}_{S \to j}; \lambda_i \right) \right\} \tag{15}$$

Among the vectors selected in opposition to the coalition S ($I_{S \to j}$), we select those that change the side of the circle (they are outside the circle initially and inside in the end or vice versa) after being possibly influenced by S:

$$I_{S \to j}^* = \left\{ D^{(i)} \in I_{S \to j} / \left(sim\left(D_j^{(i)}, SS(D^{(i)}) \right) - \eta \right) \left(sim\left(BD_j^{(i)}, SS(D^{(i)}) \right) - \eta \right) \leq 0 \right\} \tag{16}$$

We can quantify the "reversal" undergone by e_j under the possible influence of S: the reversal is all the more consequent when $D_j^{(i)}$ is similar to (respectively different from) $SS(D^{(i)})$ and $BD_j^{(i)}$ different from (respectively similar to) $SS(D^{(i)})$. This makes it possible to define a degree of membership for $\tilde{I}_{S \to j}^*$:

$$\left| \begin{array}{l} v_i = \lambda_i sim\left(BD_j^{(i)}, SS(D^{(i)}) \right) \quad if \quad s_i \geq 0 \\ v_i = \lambda_i \left(1 - sim\left(BD_j^{(i)}, SS(D^{(i)}) \right) \right) \quad if \quad s_i < 0 \end{array} \right.$$

And we introduce after normalization:

$$\tilde{I}_{S \to j}^* = \left\{ \left(D^{(i)} \in \tilde{I}_{S \to j}^*; v_i \right) \right\} \tag{17}$$

We then deduce the index of possibility of direct influence of a coalition S on j:

$$\bar{d}(B, S \to j) = \frac{\left| I_{S \to j}^*(B) \right|}{\left| I_{S \to j} \right|} \tag{18}$$

Or in the fuzzy version:

$$\tilde{\bar{d}}(B, S \to j) = \frac{\sum_i v_i}{\sum_i \lambda_i} \tag{19}$$

This index makes it possible to establish the possible influence of any coalition of evaluators on a particular individual. This type of information can be used to predict sensory perceptions based on the characterization of a group of reputable influencers. Influencers can also be used to control collective perception in sensorial marketing. These indexes allow us to answer the question we asked in the introduction about the influence of the sensory description on the label of a bottle of wine. If we are simply interested in the hedonic character of the sensation (pleasant or not), the problem is simpler and can be treated with the tools of Sect. 2. Evaluating the influence through the characterization of a sensory evocation (neutral characteristic descriptors without appreciation judgment) is more complex because there are many possible decisions more or less close to each other.

Beyond this application to the domain of sensory analysis, this model allows to extend the proposals of [7] when the decision is not simply boolean, but gradual: an individual can more or less adhere to the collective decision.

5 Illustrative Example

Verbalizing emotions allows moving from an emotional/reactionary state to a more reasoned state. The experimentation we have implemented consists in proposing to a group of users to verbalize and confront their emotions in front of a series of artworks. The protocol is divided into two phases: an annotation phase and a revision phase. During the first phase, all the users of the group have to express the emotions inspired by the artworks chosen in a priori series. The constraint of this exercise is to choose emotions from a controlled vocabulary represented by Plutchik's wheel of emotions. Plutchik created his "wheel of emotions" to better visualize the intensity and combination of emotions. On the left part of Fig. 1, in the center, we find the most intense emotion (e.g. ecstasy); then in the first crown, the basic emotion (e.g. joy) and finally, in the periphery, the least intense emotion (e.g. serenity). He then defined three levels of possible combinations: primary dyads (combinations of two adjacent basic emotions on the wheel), secondary dyads (combinations of basic emotions within one emotion) and tertiary dyads (combinations of basic emotions within two emotions). We transposed his wheel (left part of Fig. 1) into a multiple inheritance taxonomy structure (a digraph) to be able to calculate the similarities needed for the experiment and to get back to the model described in the article (right part of Fig. 1).

For the experiment, each user for each artwork must fill in at least one emotion (left part of Fig. 2). The second phase can start when all the users of the group have finished filling in the evaluations of the first part. In the second phase (right part of Fig. 2), the same suite of artworks is presented again to each individual in the group. Each individual has the opportunity to revise his initial choices by taking into account the emotions expressed by the other users. In view of the emotions returned by the group and exposed to each individual, each user can give up some of his initial emotions or borrow some of the emotions retained by the others because they seem more relevant.

We did the first tests with a small group of informed users to test the formulas and the feasibility of the experimentation. In these first tests, a user intentionally played the role of a follower or a leader; the test then consisted in automatically finding the role adopted from the model we proposed, but we did not have time to set up an experiment with a network of 'naïve' users.

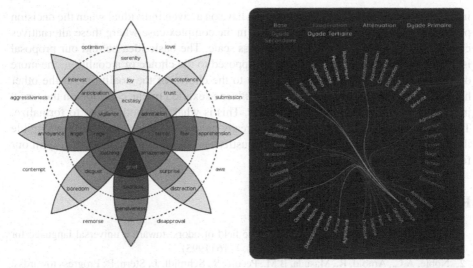

Fig. 1. The Plutchik wheel and its transposition into a graph of relations *(Types of inherited emotions: Base, Exaggeration, Attenuation, Primary Dyad, Secondary Dyad and Tertiary Dyad)* (Color figure online)

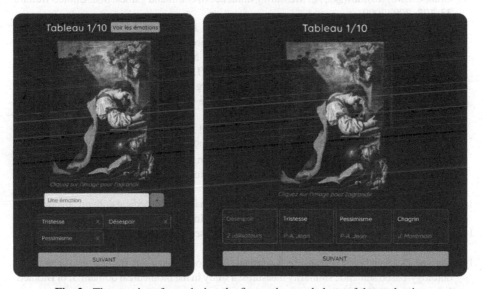

Fig. 2. The user interfaces during the first and second phase of the evaluation

6   Conclusion

We started from Grabisch and Rusinowska's indices of decision-making power and influence defined for a binary decision. We proposed an extension to more sophisticated decisions with n alternatives that can be associated with a graded success scale. Based on Grabisch and Rusinowska's model, our extended fuzzy model allows to measure the influence that a coalition of individuals can have on a given individual when the decision process involves n alternatives, especially in the complex case where these alternatives can be ordered according to graded success scale. The basic idea behind our proposal is that the more an individual is initially opposed to the choice of a coalition, the more difficult it seems to make him/her adhere to the choice of the coalition; on the other hand, if this is the case, it means that the influence exerted by the coalition on the target individual will have been very significant. This is what our model aims to formalize. The sensory analysis on which we have already published seemed to us to be a tangible illustration of the model that we wish to justify and develop more theoretically in our future work.

References

1. Jaubert, J.-N., Tapiero, C., Dore, J.C.: The field of odors: toward a universal language for odor relationships. Perfum. Flavorist **20**(3), 1–16 (1995)
2. Noble, A.C., Arnold, R., Masuda, B.M., Pecore, S., Schmidt, J., Stern, P.: Progress towards a standardized system of wine aroma terminology. Am. J. Enol. Vitic. **35**(2), 107–109 (1984)
3. Johnson, K.E., Mervis, C.B.: Effects of varying levels of expertise on the basic level of categorization. J. Exp. Psychol. Gen. **126**(3), 248 (1997)
4. Grabisch, M., Rusinowska, A.: Measuring influence in command games. Soc. Choice Welfare **33**, 177–209 (2009)
5. Hoede, C., Bakker, R.: A theory of decisional power. J. Math. Sociol. **8**, 309–322 (1982)
6. Rusinowska, A., de Swart, H.: Generalizing and modifying the Hoede-Bakker index. In: de Swart, H., Orłowska, E., Schmidt, G., Roubens, M. (eds.) Theory and Applications of Relational Structures as Knowledge Instruments II. Lecture Notes in Computer Science(), vol. 4342. Springer, Heidelberg (2006). https://doi.org/10.1007/11964810_4
7. Grabisch, M., Rusinowska, A.: A model of influence in a social network. Theor. Decis. **69**, 69–96 (2010)
8. Grabisch, M., Rusinowska, A.: Influence in Social Networks, COGnitive Systems with Interactive Sensors (COGIS2009). France, Paris (2009)
9. Harispe, S., Ranwez, S., Janaqi, S., Montmain, J.: Semantic similarity from natural language and ontology analysis. In: Synthesis Lectures on Human Language Technologies, 254 p. Morgan & Claypool Publishers (2015)
10. Sanchez, D., Batet, M.: Semantic similarity estimation in the biomedical domain: an ontology-based information-theoretic perspective. J. Biomed. Inform. **44**(5), 749–759 (2011)
11. Resnik, P.: Using information content to evaluate semantic similarity in a taxonomy. In: Proceedings of IJCAI-95, Montréal, Canada (1995)
12. Harispe, S., Sanchez, D., Ranwez, S., Janaqi, S., Montmain, J.: A framework for unifying ontology-based semantic similarity measures: a study in the biomedical domain. J. of Biomed. Inform. **48**, 38–53 (2014)
13. Harispe, S., Montmain, J., Medjkoune, M.: Summarizing conceptual descriptions using knowledge representations. In: IEEE Symposium Series on Computational Intelligence for Human-like Intelligence, (SSCI), Athens, Greece (2016)

Study of the Instability of the Sign of the Nonadditivity Index in a Choquet Integral Model

Paul Alain Kaldjob Kaldjob$^{(\boxtimes)}$, Brice Mayag, and Denis Bouyssou

University Paris-Dauphine, University PSL, CNRS, LAMSADE, 75016 Paris, France
{paul-alain.kaldjob-kaldjob,brice.mayag,denis.bouyssou}@dauphine.psl.eu

Abstract. This paper studies the instability of the sign of the nonadditivity index between criteria in a Choquet integral model. Nonadditivity is an essential property of capacities defined on the sets of decision criteria and allows one to flexibly represent the phenomenon of interaction between criteria. In some cases, we show that the sign of the nonadditivity index proposed in the literature depends on arbitrary choice of a numerical representation in the set of all numerical representations compatible with the strict preferential information given by the Decision Maker(DM). This makes its interpretation difficult. We illustrate our results with examples.

Keywords: Instability · Nonadditivity index · Choquet integral model · Numerical representation · Strict preferential information

1 Introduction

In Multiple Criteria Decision Making (MCDM), the theory of value functions aims to assign a real number to each alternative, so that the order on the alternatives induced by these real numbers does not contradict the preferences of the DM. When the preferences of the DM satisfy preferential independence hypothesis, the value assigned to each alternative can be obtained from an additive model [1]. Since this hypothesis is restrictive [9], the Choquet integral model, more general, was popularized by the work of Michel Grabisch [5,6]. It is now considered as a central tool in MCDM when one wants to escape the independence hypothesis [8–10].

When a set of preferential information is not compatible with an additive model, it is common to deduce the existence of interaction between criteria. Interaction among multiple decision criteria can be measured by cardinal probabilistic interaction indices, in particular the Shapley interaction index [7]. More details in the literature on axiomatic properties of some cardinal probabilistic interaction indices are given in [3,8]. In [15], this lack of compatibility with an additive model is simply translated by the notion of nonadditivity index. This article deals with the use of this idea to capture the phenomena of synergy in

© Springer Nature Switzerland AG 2022
D. Ciucci et al. (Eds.): IPMU 2022, CCIS 1602, pp. 197–209, 2022.
https://doi.org/10.1007/978-3-031-08974-9_16

the framework of nonadditivity. In [12], we used the Shapley interaction index to study the interaction between criteria in a Choquet integral model. In particular, in [12] we show that the positive interaction is always possible for any subset of criteria, but could not transpose this result with the negative interaction. Here, we solve this dual problem using the nonadditivity index.

We show that the sign of the nonadditivity index [15] proposed in the literature is not stable in the set of all capacities compatible with strict preferences of DM. Indeed, we prove that from a null nonadditivity index, we can build a strictly positive and a strictly negative nonadditivity indices while remaining in the set of all capacities compatible with strict preferences of DM.

Within the framework of binary alternatives, we show that it is always possible to represent the strict preferences of the DM with a Choquet integral model inducing strictly positive nonadditivity indices, then with another inducing strictly negative nonadditivity indices for all subsets of at least two criteria.

This paper is organized as follows. Section 2 recalls some basic elements on the model of the Choquet integral in MCDM. In Sects. 3 and 4, we present our results. We illustrate each result with an example.

2   Notations and Definitions

2.1   Framework

Let X be a set of alternatives evaluated on an index set of n criteria $N = \{1, 2, \ldots, n\}$ ($n \geq 2$). Throughout this paper we use the notation $A \subseteq N_{\geq 2}$ if and only if $A \subseteq N$ and $|A| \geq 2$. The set of all alternatives X is assumed to be a subset of a Cartesian product $X_1 \times X_2 \times \ldots \times X_n$, where X_i is the set of possible levels on criterion $i \in N$. The criteria are recoded numerically using, for all $i \in N$, a function u_i from X_i into $\mathbb{R}$. Using these functions, we assume that the various recoded criteria are commensurate, so we can use the Choquet integral model [11].

2.2   Choquet Integral

A generalization of criteria weights consists of assigning weights to subsets of criteria. This can be achieved by a capacity [2] defined as a function μ from the power set 2^N into $[0, 1]$ such that:

- $\mu(\emptyset) = 0$,
- $\mu(N) = 1$,
- $\forall S, T \in 2^N, \left[S \subseteq T \implies \mu(S) \leq \mu(T)\right]$ (monotonicity).

For an alternative $x = (x_1, \ldots, x_n) \in X$, the expression of the Choquet integral [7–9] w.r.t. the capacity μ is given by:

$$C_\mu\big(u(x)\big) = \sum_{i=1}^{n} \left[u_{\sigma(i)}(x_{\sigma(i)}) - u_{\sigma(i-1)}(x_{\sigma(i-1)})\right]\mu\big(N_{\sigma(i)}\big),$$

where σ is a permutation on N such that: $N_{\sigma(i)} = \{\sigma(i), \ldots, \sigma(n)\}$, $u_{\sigma(0)}(x_{\sigma(0)}) = 0$ and $u_{\sigma(1)}(x_{\sigma(1)}) \leq u_{\sigma(2)}(x_{\sigma(2)}) \leq \ldots \leq u_{\sigma(n)}(x_{\sigma(n)})$.

In the next subsection, we recall the definition of the nonadditivity index.

2.3 Nonadditivity Index

We work with the nonadditivity index introduced and studied in [15].

Definition 1. *The nonadditivity index w.r.t. a capacity μ is defined by: for all $A \subseteq N_{\geq 2}$,*

$$\eta_A^\mu = \frac{1}{2^{|A|-1} - 1} \sum_{\substack{(B, A\backslash B) \\ \emptyset \subsetneq B \subsetneq A}} \left(\mu(A) - \mu(B) - \mu(A \backslash B) \right) \tag{1}$$

For all $A \subseteq N_{\geq 2}$, for each partition $(B, A \backslash B)$ of A with $\emptyset \subsetneq B \subsetneq A$, we compute the difference $\mu(A) - \left(\mu(B) + \mu(A \backslash B) \right)$. Thus η_A^μ corresponds to the arithmetic mean of these differences.

Remark 1. We have $\eta_{ij}^\mu = \mu_{ij} - \mu_i - \mu_j$, therefore the nonadditivity index coincides with the Shapley interaction index I_{ij}^μ, for pairs $\{i, j\} \subseteq N$.

Remark 2 below gives two equivalent expressions of η_A^μ that we find in [15].

Remark 2. Given a capacity μ on N and $A \subseteq N_{\geq 2}$, Eq. (1) is equivalent to each of Eqs. (2) and (3).

$$\eta_A^\mu = \frac{1}{2^{|A|} - 2} \sum_{\emptyset \subsetneq B \subsetneq A} \left(\mu(A) - \mu(B) - \mu(A \backslash B) \right) \tag{2}$$

$$\eta_A^\mu = \mu(A) - \frac{1}{2^{|A|-1} - 1} \sum_{\emptyset \subsetneq B \subsetneq A} \mu(B) \tag{3}$$

In the following section, we propose the concept of necessary and possible nonadditivity index similar to that of necessary and possible interaction introduced on [14] in the case of 2-additive Choquet integral model.

3 Necessary and Possible Nonadditivity Index on X

3.1 Some Definitions and Notations

The DM compares some alternatives only in terms of strict preference P defined as follow.

Definition 2. *A strict ordinal preference information P on X is given by:*

$$P = \{(a, b) \in X \times X : DM \text{ strictly prefers } a \text{ to } b\}$$

We note $a\,P\,b$ or $(a,b) \in P$. The following definition tests if P is representable by a Choquet integral model.

Definition 3. *A strict ordinal preference information P on X is representable by a Choquet integral model if we can find a capacity μ such that: for all $a,b \in X$,*

$$a\,P\,b \Longrightarrow C_\mu\big(u(a)\big) > C_\mu\big(u(b)\big).$$

We denote by C_{Pref} the set of all capacities compatible with P. The following definition will be central in the rest of this text. It is inspired from [14] where it was given in the case of 2-additive Choquet integral model.

Definition 4. *Let $A \subseteq N_{\geq 2}$ and P a strict ordinal preference information. We say that:*

1. *The nonadditivity for A is possibly positive (resp. null, negative) if there exists $\mu \in C_{Pref}$ such that $\eta_A^\mu > 0$ (resp. $\eta_A^\mu = 0$, $\eta_A^\mu < 0$).*
2. *The nonadditivity for A is necessarily positive (resp. null, negative) if $\eta_A^\mu > 0$ (resp. $\eta_A^\mu = 0$, $\eta_A^\mu < 0$) for all $\mu \in C_{Pref}$.*

The interpretation of sign of the nonadditivity index is difficult in the case of a possible but not necessary, because it depends on the arbitrary choice of a capacity in C_{Pref}. Indeed, the interpretation of the nonadditivity index really makes sense in the case of the necessity. Our results of the next subsection show that null nonadditivity is not necessary.

3.2   Results on X

Proposition 1 shows that from a null nonadditivity index, we can build a strictly positive nonadditivity index while remaining in C_{Pref}.

Proposition 1. *Let P be a strict ordinal preference information on X and $A \subseteq N_{\geq 2}$. Assume that P is representable by a Choquet integral model using a capacity μ for which $\eta_A^\mu = 0$. Then there exists a capacity $\beta^\mu \in C_{Pref}$ such that $\eta_A^{\beta^\mu} > 0$.*

Proof. Let $A \subseteq N_{\geq 2}$, we suppose that P is representable by a Choquet integral model using a capacity μ such that $\eta_A^\mu = 0$. Let us define a function β_ε^μ on power set 2^N into $[0,1]$ by:

$$\beta_\varepsilon^\mu(S) = \begin{cases} \dfrac{1}{1+\varepsilon}(\mu(S)+\varepsilon) & \text{if } A \subseteq S \\[2mm] \dfrac{1}{1+\varepsilon}\mu(S) & \text{otherwise.} \end{cases}$$

where ε is a strictly positive real number to be determined as follows. We show that for all $\varepsilon > 0$, β_ε^μ is a capacity on N.

Let $\varepsilon > 0$. It is obvious that $\beta_\varepsilon^\mu(\emptyset) = 0$ and $\beta_\varepsilon^\mu(N) = 1$. Let $S \subseteq T \subseteq N$.

- If $A \subseteq S$ then $A \subseteq T$ and $\beta_\varepsilon^\mu(T) - \beta_\varepsilon^\mu(S) = \dfrac{1}{1+\varepsilon}(\mu(T) - \mu(S)) \geq 0$ since μ is a capacity on N and $S \subseteq T \subseteq N$.

- If not($A \subseteq S$), then $\beta_\varepsilon^\mu(S) = \dfrac{1}{1+\varepsilon}\mu(S)$. We have $\beta_\varepsilon^\mu(T) = \dfrac{1}{1+\varepsilon}\mu(T)$ (if $A \subseteq T$) or $\beta_\varepsilon^\mu(T) = \dfrac{1}{1+\varepsilon}(\mu(T)+\varepsilon)$, then $\beta_\varepsilon^\mu(T) \geq \dfrac{1}{1+\varepsilon}\mu(T)$ since $\mu(T)+\varepsilon > \mu(T)$. Therefore $\beta_\varepsilon^\mu(T) \geq \dfrac{1}{1+\varepsilon}\mu(T) \geq \dfrac{1}{1+\varepsilon}\mu(S)$ since μ is a capacity on N and $S \subseteq T \subseteq N$, hence $\beta_\varepsilon^\mu(T) \geq \beta_\varepsilon^\mu(S)$.

In both cases, we have $\beta_\varepsilon^\mu(T) \geq \beta_\varepsilon^\mu(S)$. Hence, for all $\varepsilon > 0$, β_ε^μ is a capacity on N.

The capacity β_ε^μ depends of the fact that $A \subseteq N_{\sigma(i)}$ or not, and the Choquet integral $C_{\beta_\varepsilon^\mu}(u(x))$ depends on the sets $N_{\sigma(i)}$, hence for all $x \in X$, we consider the set $\Gamma_{u(x)} = \{i = 1, 2, \ldots, n : A \subseteq N_{\sigma(i)}\}$. For each $x \in X$, this set captures the "pivot" criterion from which A is no longer included in $N_{\sigma(i)}$. We have $1 \in \Gamma_{u(x)}$ for all $x \in X$ since $A \subseteq N = N_{\sigma(1)}$, then $\Gamma_{u(x)} \neq \emptyset$. We then have:

$$C_{\beta_\varepsilon^\mu}(u(x)) = \sum_{i=1}^n \left[u_{\sigma(i)}(x_{\sigma(i)}) - u_{\sigma(i-1)}(x_{\sigma(i-1)}) \right] \beta_\varepsilon^\mu(N_{\sigma(i)})$$

$$= \frac{1}{1+\varepsilon} \sum_{i \notin \Gamma_{u(x)}} \left[u_{\sigma(i)}(x_{\sigma(i)}) - u_{\sigma(i-1)}(x_{\sigma(i-1)}) \right] \mu(N_{\sigma(i)})$$

$$+ \frac{1}{1+\varepsilon} \sum_{i \in \Gamma_{u(x)}} \left[u_{\sigma(i)}(x_{\sigma(i)}) - u_{\sigma(i-1)}(x_{\sigma(i-1)}) \right] \left(\mu(N_{\sigma(i)}) + \varepsilon \right)$$

$$= \frac{1}{1+\varepsilon} \sum_{i \notin \Gamma_{u(x)}} \left[u_{\sigma(i)}(x_{\sigma(i)}) - u_{\sigma(i-1)}(x_{\sigma(i-1)}) \right] \mu(N_{\sigma(i)})$$

$$+ \frac{1}{1+\varepsilon} \sum_{i \in \Gamma_{u(x)}} \left[u_{\sigma(i)}(x_{\sigma(i)}) - u_{\sigma(i-1)}(x_{\sigma(i-1)}) \right] \mu(N_{\sigma(i)})$$

$$+ \frac{1}{1+\varepsilon} \varepsilon \sum_{i \in \Gamma_{u(x)}} \left(u_{\sigma(i)}(x_{\sigma(i)}) - u_{\sigma(i-1)}(x_{\sigma(i-1)}) \right)$$

$$= \frac{1}{1+\varepsilon} \left[C_\mu(u(x)) + \varepsilon v^\sigma(u(x)) \right] \text{ where}$$

$v^\sigma(u(x)) = \sum_{i \in \Gamma_{u(x)}} \left(u_{\sigma(i)}(x_{\sigma(i)}) - u_{\sigma(i-1)}(x_{\sigma(i-1)}) \right)$. For all $(a, b) \in P$, we have

$$C_{\beta_\varepsilon^\mu}(u(a)) - C_{\beta_\varepsilon^\mu}(u(b)) = \frac{1}{1+\varepsilon}\left[\left(C_\mu(u(a)) - C_\mu(u(b)) \right) + \varepsilon \left(v^{\sigma^a}(u(a)) - v^{\sigma^b}(u(b)) \right) \right].$$

ε is such that $C_{\beta_\varepsilon^\mu}\big(u(a)\big) - C_{\beta_\varepsilon^\mu}\big(u(b)\big) > 0$ for all $(a,b) \in P$. We consider the set $\Omega = \{(a,b) \in P : v^{\sigma^a}(u(a)) - v^{\sigma^b}(u(b)) < 0\}$.

- If $\Omega = \emptyset$, then for all $(a,b) \in P$, we have $v^{\sigma^a}(u(a)) - v^{\sigma^b}(u(b)) \geq 0$. Thus for all $(a,b) \in P$, $C_{\beta_\varepsilon^\mu}\big(u(a)\big) - C_{\beta_\varepsilon^\mu}\big(u(b)\big) > 0 \ \forall \varepsilon > 0$.

- If $\Omega \neq \emptyset$, we choose ε such that $0 < \varepsilon < \min\limits_{(a,b)\in\Omega} \left(\dfrac{C_\mu\big(u(b)\big) - C_\mu\big(u(a)\big)}{v^{\sigma^a}(u(a)) - v^{\sigma^b}(u(b))} \right)$ in such a way that $C_{\beta_\varepsilon^\mu}\big(u(a)\big) - C_{\beta_\varepsilon^\mu}\big(u(b)\big) > 0$ for all $(a,b) \in P$.

In both cases we can choose $\varepsilon = \dfrac{1}{2} \min\limits_{(a,b)\in\Omega} \left(\dfrac{C_\mu\big(u(x^j)\big) - C_\mu\big(u(a)\big)}{v^{\sigma^a}(u(a)) - v^{\sigma^b}(u(b))} \right)$ so that $\beta_\varepsilon^\mu \in C_{\mathrm{Pref}}$. Moreover we have:

$$\eta_A^{\beta_\varepsilon^\mu} = \beta_\varepsilon^\mu(A) - \frac{1}{2^{|A|-1}-1} \sum_{\emptyset \neq B \subsetneq A} \beta_\varepsilon^\mu(B)$$

$$= \frac{1}{1+\varepsilon}\left(\varepsilon + \mu(A) - \frac{1}{2^{|A|-1}-1} \sum_{\emptyset \neq B \subsetneq A} \mu(B) \right)$$

$= \dfrac{1}{1+\varepsilon}(\varepsilon + \eta_A^\mu)$. As $\eta_A^\mu = 0$, we have $\eta_A^{\beta_\varepsilon^\mu} = \dfrac{\varepsilon}{1+\varepsilon} > 0$, so the nonadditivity for A is possibly positive. Hence, the nonadditivity is not necessarily null. $\square$

Proposition 2 answers the dual problem. Indeed, it shows that from a null nonadditivity index, we can build a strictly negative nonadditivity index while remaining in C_{Pref}. Note that this dual problem remains open in the case of the necessary and possible interaction with the Shapley interaction index [7].

Proposition 2. *Let P be a strict ordinal preference information on X and $A \subseteq N_{\geq 2}$. Assume that P is representable by a Choquet integral model using a capacity μ for which $\eta_A^\mu = 0$. Then there exists a capacity $\gamma^\mu \in C_{Pref}$ such that $\eta_A^{\gamma^\mu} < 0$.*

Proof. This proof follows the same pattern as proof of Proposition 1, only with things "reversed". Let $A \subseteq N_{\geq 2}$, we suppose that P is representable by a Choquet integral model using a capacity μ such that $\eta_A^\mu = 0$. Let us define a function γ_ε^μ on power set 2^N into $[0,1]$ by:

$$\gamma_\varepsilon^\mu(S) = \begin{cases} \dfrac{1}{1+\varepsilon}(\mu(S) + \varepsilon) & \text{if } S \neq \emptyset \\[2mm] 0 & \text{if } S = \emptyset. \end{cases}$$

where ε is a strictly positive real number to be determined as follows. We show that for all $\varepsilon > 0$, γ_ε^μ is a capacity on N.
Let $\varepsilon > 0$. It is obvious that $\gamma_\varepsilon^\mu(\emptyset) = 0$ and $\gamma_\varepsilon^\mu(N) = 1$. Let $S \subseteq T \subseteq N$.

- If $S = \emptyset$, then $\gamma_\varepsilon^\mu(S) = 0 \leq \gamma_\varepsilon^\mu(T)$.
- If $S \neq \emptyset$, then $\gamma_\varepsilon^\mu(S) = \dfrac{1}{1+\varepsilon}(\mu(S) + \varepsilon) \leq \dfrac{1}{1+\varepsilon}(\mu(T) + \varepsilon) = \gamma_\varepsilon^\mu(T)$.

In the both cases, we have $\gamma_\varepsilon^\mu(T) \geq \gamma_\varepsilon^\mu(S)$. Hence, for all $\varepsilon > 0$, γ_ε^μ is a capacity on N. Besides, we have:

$$C_{\gamma_\varepsilon^\mu}\big(u(x)\big) = \sum_{i=1}^n \left[u_{\sigma(i)}(x_{\sigma(i)}) - u_{\sigma(i-1)}(x_{\sigma(i-1)}) \right] \gamma_\varepsilon^\mu(N_{\sigma(i)})$$

$$= \frac{1}{1+\varepsilon} \sum_{i=1}^n \left[u_{\sigma(i)}(x_{\sigma(i)}) - u_{\sigma(i-1)}(x_{\sigma(i-1)}) \right] \big(\mu(N_{\sigma(i)}) + \varepsilon \big)$$

$$= \frac{1}{1+\varepsilon} \sum_{i=1}^n \left[u_{\sigma(i)}(x_{\sigma(i)}) - u_{\sigma(i-1)}(x_{\sigma(i-1)}) \right] \mu(N_{\sigma(i)})$$

$$+ \frac{\varepsilon}{1+\varepsilon} \sum_{i=1}^n \left[u_{\sigma(i)}(x_{\sigma(i)}) - u_{\sigma(i-1)}(x_{\sigma(i-1)}) \right]$$

Hence, we have $C_{\gamma_\varepsilon^\mu}\big(u(x)\big) = \dfrac{1}{1+\varepsilon}\left[C_\mu\big(u(x)\big) + \varepsilon v^\sigma\big(u(x)\big) \right]$ with

$$v^\sigma\big(u(x)\big) = \sum_{i=1}^n \left[u_{\sigma(i)}(x_{\sigma(i)}) - u_{\sigma(i-1)}(x_{\sigma(i-1)}) \right].$$ For all $(a,b) \in P$, we have

$$C_{\beta_\varepsilon^\mu}\big(u(a)\big) - C_{\beta_\varepsilon^\mu}\big(u(b)\big) = \frac{1}{1+\varepsilon}\left[\Big(C_\mu\big(u(a)\big) - C_\mu\big(u(b)\big) \Big) + \varepsilon\big(v^{\sigma^a}\big(u(a)\big) - \right.$$

$$\left. v^{\sigma^b}\big(u(b)\big) \big) \right].$$

ε is such that $C_{\gamma_\varepsilon^\mu}\big(u(a)\big) - C_{\gamma_\varepsilon^\mu}\big(u(b)\big) > 0$ for all $(a,b) \in P$. We consider the set $\Omega = \{ (a,b) \in P : v^{\sigma^a}\big(u(a)\big) - v^{\sigma^b}\big(u(b)\big) < 0 \}.$

- If $\Omega = \emptyset$, then $v^{\sigma^a}\big(u(a)\big) - v^{\sigma^b}\big(u(b)\big) \geq 0$ for all $(a,b) \in P$. Thus for all $(a,b) \in P$, $C_{\gamma_\varepsilon^\mu}\big(u(a)\big) - C_{\gamma_\varepsilon^\mu}\big(u(b)\big) > 0\ \forall \varepsilon > 0$.

- If $\Omega \neq \emptyset$, we choose ε such that $0 < \varepsilon < \min\limits_{(a,b)\in\Omega} \left(\dfrac{C_\mu\big(u(b)\big) - C_\mu\big(u(a)\big)}{v^{\sigma^a}\big(u(a)\big) - v^{\sigma^b}\big(u(b)\big)} \right)$ in such a way that $C_{\gamma_\varepsilon^\mu}\big(u(a)\big) - C_{\gamma_\varepsilon^\mu}\big(u(b)\big) > 0$ for all $(a,b) \in P$.

So in both cases we can choose $\varepsilon = \dfrac{1}{2} \min\limits_{(a,b)\in\Omega} \left(\dfrac{C_\mu\big(u(b)\big) - C_\mu\big(u(a)\big)}{v^{\sigma^a}\big(u(a)\big) - v^{\sigma^b}(u(b))} \right)$ so that $\gamma_\varepsilon^\mu \in C_{\text{Pref}}$. Moreover we have:

$$\eta_A^{\gamma_\varepsilon^\mu} = \gamma_\varepsilon^\mu(A) - \frac{1}{2^{|A|-1}-1} \sum_{\emptyset \neq B \subsetneq A} \gamma_\varepsilon^\mu(B)$$

$$= \frac{1}{1+\varepsilon}\left[\varepsilon + \mu(A) - \frac{1}{2^{|A|-1}-1} \sum_{\emptyset \neq B \subsetneq A} (\mu(B) + \varepsilon) \right]$$

$$= \frac{1}{1+\varepsilon}\left[\eta_A^\mu + \varepsilon - \frac{1}{2^{|A|-1}-1}\varepsilon(2^{|A|} - 2) \right]$$

$$= \frac{1}{1+\varepsilon}\big(\eta_A^\mu - \varepsilon \big).$$ As $\eta_A^\mu = 0$, we have $\eta_A^{\gamma_\varepsilon^\mu} = \dfrac{-\varepsilon}{1+\varepsilon} < 0$, so the nonadditivity for A is possibly negative. Hence, the nonadditivity is not necessarily null. $\square$

The following example illustrates Propositions 1 and 2.

Example 1. $N = \{1, 2, 3\}$, $X = \{a, b, c, d\}$, $a = (6, 11, 9)$, $b = (6, 13, 7)$, $c = (16, 11, 9)$, $d = (16, 13, 7)$ and $P = \{(d, c), (b, a)\}$.
The strict preference P is representable by the capacity μ (with $\eta_{23}^{\mu} = 0$) given by Table 1 and Choquet integral corresponding is given by Table 2.

Table 1. A capacity $\mu \in C_{\text{Pref}}$ with $\eta_{23}^{\mu} = 0$.

S	$\{1\}$	$\{2\}$	$\{3\}$	$\{1,2\}$	$\{1,3\}$	$\{2,3\}$	$\{1,2,3\}$
$\mu(S)$	0.5	0.5	0	1	0.5	0.5	1

Table 2. Choquet integral corresponding at capacity μ of Table 1.

x	d	c	b	a
$C_{\mu}(u(x))$	14.5	13.5	9.5	8.5

We have $v^{\sigma^d}(d) - v^{\sigma^c}(c) = 7 - 9 = -2 < 0$ and $v^{\sigma^b}(b) - v^{\sigma^a}(a) = 7 - 9 = -2 < 0$ so $\Omega = \{(d, c), (b, a)\}$ and we choose $\varepsilon = \dfrac{1}{2} \min \left(\dfrac{8.5 - 9.5}{7 - 9}, \dfrac{13.5 - 14.5}{7 - 9} \right) = 0.25$. A capacity $\beta_{\varepsilon}^{\mu} \in C_{\text{Pref}}$ such that $\eta_{23}^{\beta_{\varepsilon}} > 0$ and Choquet integral corresponding at $\beta_{\varepsilon}^{\mu}$ are respectively given by Table 3 and Table 4.

Table 3. A capacity $\beta_{\varepsilon}^{\mu} \in C_{\text{Pref}}$ with $\eta_{23}^{\beta_{\varepsilon}} > 0$.

S	$\{1\}$	$\{2\}$	$\{3\}$	$\{1,2\}$	$\{1,3\}$	$\{2,3\}$	$\{1,2,3\}$
$\beta_{\varepsilon}^{\mu}(S)$	0.4	0.4	0	0.8	0.4	0.6	1

Table 4. Choquet integral corresponding at capacity $\beta_{\varepsilon}^{\mu}$ of Table 3.

x	d	c	b	a
$C_{\beta_{\varepsilon}^{\mu}}(u(x))$	13	12.6	9	8.6

Indeed, $\eta_{23}^{\beta_{\varepsilon}^{\mu}} = \dfrac{\varepsilon}{1 + \varepsilon} = \dfrac{0.25}{1 + 0.25} = 0.2 > 0$.

We have $v^{\sigma^d}(d) - v^{\sigma^c}(c) = 16 - 16 = 0$ and $v^{\sigma^b}(b) - v^{\sigma^a}(a) = 13 - 11 = 2 \geq 0$ so $\Omega = \emptyset$ and we can choose any $\varepsilon > 0$. We take $\varepsilon = 1$. A capacity $\gamma_{\varepsilon}^{\mu} \in C_{\text{Pref}}$ such that $\eta_{23}^{\gamma_{\varepsilon}} < 0$ and Choquet integral corresponding $\gamma_{\varepsilon}^{\mu}$ are respectively given by Tables 5 and 6.

Table 5. A capacity $\gamma_\varepsilon^\mu \in C_{\text{Pref}}$ with $\eta_{23}^{\gamma_\varepsilon} < 0$.

S	$\{1\}$	$\{2\}$	$\{3\}$	$\{1,2\}$	$\{1,3\}$	$\{2,3\}$	$\{1,2,3\}$
$\gamma_\varepsilon^\mu(S)$	0.75	0.75	0.5	1	0.75	0.75	1

Table 6. Choquet integral corresponding at capacity γ_ε^μ of Table 5.

x	d	c	b	a
$C_{\gamma_\varepsilon^\mu}(u(x))$	15.25	14.25	11.25	9.75

Indeed, $\eta_{23}^{\gamma_\varepsilon^\mu} = \dfrac{-\varepsilon}{1+\varepsilon} = \dfrac{-1}{1+1} = -0.5 < 0$.

In the next section, we define the set of generalized binary alternatives, then we show that on this set, positive and negative nonadditivity index are always possible for all subsets $A \subseteq N_{\geq 2}$.

4   Necessary and Possible Nonadditivity Index with Generalized Binary Alternatives

4.1   Framework of Binary Alternatives

We assume that the DM can identify two reference levels 0_i and 1_i on each criterion $i \in N$:

- the level 0_i in X_i is considered as a neutral level and we set $u_i(0_i) = 0$,
- the level 1_i in X_i is considered as a good level and we set $u_i(1_i) = 1$.

For all $x = (x_1, \ldots, x_n) \in X$ and $S \subseteq N$, we will sometimes write $u(x)$ as a shorthand for $(u_1(x_1), \ldots, u_n(x_n))$ and we define the alternatives $a_S = (1_S, 0_{-S}) \in X$ such that $a_i = 1_i$ if $i \in S$ and $a_i = 0_i$ otherwise.
We work on the set $\mathcal{B}^g$ which we define as follows.

Definition 5. *We call the set of generalized binary alternatives, the set:*

$$\mathcal{B}^g = a_S = (1_S, 0_{-S}) : S \subseteq N.$$

We add to the strict preference P a binary relation M modeling the monotonicity relations between generalized binary alternatives, and allowing us to ensure the satisfaction of the monotonicity condition: $[S \subseteq T \implies \mu(S) \leq \mu(T)]$.

Definition 6. *For all $a_S, a_T \in \mathcal{B}^g$, $a_S M a_T$ if $[not(a_S P a_T)$ and $S \supseteq T]$.*

In the sequel, we need the following basic definition in graph theory [13].

Definition 7. *There exists a strict cycle in $(P \cup M)$ if there exists elements $x_0, x_1, \ldots, x_r$ of $\mathcal{B}^g$ such that $x_0(P \cup M)x_1(P \cup M)\ldots(P \cup M)x_r(P \cup M)x_0$ and for a least one $i \in \{0, \ldots, r-1\}$, $x_i P x_{i+1}$.*

4.2   Results on Binary Alternatives

In [12] we find a necessary and sufficient condition for a strict ordinal preference information on $\mathcal{B}^g$ to be representable by a Choquet integral model. Under this condition, Proposition 3 below shows that positive nonadditivity is always possible for all subsets $A \subseteq N_{\geq 2}$, in a Choquet integral model. In other words, negative and null nonadditivity are not necessary.

Proposition 3. *Let P be a strict ordinal preference information on $\mathcal{B}^g$ such that $(P \cup M)$ contains no strict cycle. Then there exists a capacity $\mu \in C_{Pref}$ such that $\eta_A^\mu > 0$ for all $A \subseteq N_{\geq 2}$.*

Proof. Assume that $(P \cup M)$ contains no strict cycle, then there exists a partition $\{\mathcal{B}_0, \mathcal{B}_1, \ldots, \mathcal{B}_m\}$ of $\mathcal{B}^g$, build by using a suitable topological sorting on $(P \cup M)$ [4]. We construct a partition $\{\mathcal{B}_0, \mathcal{B}_1, \ldots, \mathcal{B}_m\}$ as follows:

$\mathcal{B}_i = \{x \in \mathcal{B}^g \backslash (\mathcal{B}_0 \cup \ldots \cup \mathcal{B}_{i-1}) : \forall y \in \mathcal{B}^g \backslash (\mathcal{B}_0 \cup \ldots \cup \mathcal{B}_{i-1}), \text{ not } [x(P \cup M)y]\}$, for all $i = 0, 1, 2, \ldots, m$ with $\mathcal{B}_0 \cup \ldots \cup \mathcal{B}_{i-1} = \emptyset$ for $i = 0$. Let us define the capacity $\mu : 2^N \longrightarrow [0, 1]$ as follows.

For all $S \subseteq N$, $\mu(S) = \begin{cases} 0 & \text{if } a_S \in \mathcal{B}_0, \\ (2n)^\ell / (2n)^m & \text{if } a_S \in \mathcal{B}_\ell, \ \ell \in \{1, 2, \ldots, m\}. \end{cases}$

We have $\mu \in C_{\text{Pref}}$. Indeed, if $a_S \, P \, a_T$, then $a_S \in \mathcal{B}_q$ and $a_T \in \mathcal{B}_r$ with $q > r$ (Fig. 1).

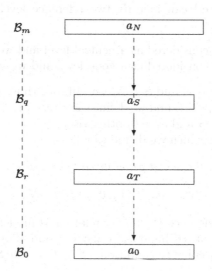

Fig. 1. An illustration of the sets $\mathcal{B}_m$, $\mathcal{B}_q$, $\mathcal{B}_r$ and $\mathcal{B}_0$ such that $m \geq q > r \geq 0$.

Hence $\mu(S) = (2n)^q / (2n)^m$ and $\mu(T) = 0$ (if $r = 0$) or $\mu(T) = (2n)^r / (2n)^m$ (if $r \geq 1$). But $(2n)^q / (2n)^m > \max(0, (2n)^r / (2n)^m)$ since $q > r$, so $\mu(S) > \mu(T)$. Besides, let $A \subseteq N_{\geq 2}$ and $\emptyset \subsetneq B \subsetneq A$. There exists $q, r, s \in \{1, 2, \ldots, m\}$

such that $a_A \in \mathcal{B}_q$, $a_B \in \mathcal{B}_r$ and $a_{A\setminus B} \in \mathcal{B}_s$ with $q > r$ and $q > s$. Hence $\mu(B) = (2n)^r/(2n)^m$, $\mu(A \setminus B) = (2n)^s/(2n)^m$ and $\mu(A) = (2n)^q/(2n)^m = (2n)(2n)^{q-1}/(2n)^m > 2(2n)^{q-1}/(2n)^m = (2n)^{q-1}/(2n)^m + (2n)^{q-1}/(2n)^m \geq (2n)^r/(2n)^m + (2n)^s/(2n)^m$ since $q - 1 \geq r$ and $q - 1 \geq s$. Therefore, we have $\mu(A) - \mu(B) - \mu(A \setminus B) > 0$ for all B such that $\emptyset \subsetneq B \subsetneq A$. Hence, we deduce that $\sum_{\emptyset \subsetneq B \subsetneq A} (\mu(A) - \mu(B) - \mu(A \setminus B)) > 0$, so $\eta_A^\mu > 0$. $\qquad\square$

The following proposition answers the dual question. Indeed, given a strict preference information P on $\mathcal{B}^g$ under the previous conditions, Proposition 4 shows that negative nonadditivity index that is always possible for all subset $A \subseteq N_{\geq 2}$. In other words, positive and null nonadditivity index are not necessary. Note that this dual problem remains open in the case of the necessary and possible interaction with the interaction index [7].

Proposition 4. *Let P be a strict ordinal preference information on $\mathcal{B}^g$ such that $(P \cup M)$ contains no strict cycle. Then there exists a capacity $\mu \in C_{Pref}$ such that $\eta_A^\mu < 0$ for all $A \subseteq N_{\geq 2}$.*

Proof. This proof follows the same pattern as proof of Proposition 3, only a wise choice of capacity. We construct the partition $\{\mathcal{B}_0, \mathcal{B}_1, \ldots, \mathcal{B}_m\}$ as the proof of the Proposition 3 above. Let us define the capacity μ as follows.

$$\text{For all } S \subseteq N, \ \mu(S) = \begin{cases} 0 & \text{if } a_S \in \mathcal{B}_0, \\ \dfrac{\ell + 1}{\ell + 2} & \text{if } a_S \in \mathcal{B}_\ell, \ell \in \{1, 2, \cdots, m - 1\} \\ 1 & \text{if } a_S \in \mathcal{B}_m. \end{cases}$$

Let $a_S, a_T \in \mathcal{B}^g$ such that $a_S \, P \, a_T$. We show that $C_\mu(u(a_S)) > C_\mu(u(a_T))$. Since $a_S, a_T \in \mathcal{B}^g$ and $\{\mathcal{B}_0, \mathcal{B}_1, \ldots, \mathcal{B}_m\}$ is a partition of $\mathcal{B}^g$, then there exists $r, q \in \{0, 1, \ldots, m\}$ such that $a_S \in \mathcal{B}_r$, $a_T \in \mathcal{B}_q$. As $a_S \, P \, a_T$, then $r > q$. We have $C_\mu(u(a_S)) = \mu(S) = \dfrac{r+1}{r+2}$ (if $1 \leq r \leq m - 1$) or $\mu(S) = 1$ (if $r = m$), so

$$C_\mu(u(a_S)) \geq \frac{r+1}{r+2}, \text{ since } 1 \geq \frac{r+1}{r+2}.$$

Moreover, $C_\mu(u(a_T)) = \mu(T) = \dfrac{q+1}{q+2}$ (if $1 \leq q \leq m - 1$) or $\mu(T) = 0$ (if $q = 0$),

then $C_\mu(u(a_T)) \leq \dfrac{q+1}{q+2}$, since $0 \leq \dfrac{q+1}{q+2}$. But $r > q$ therefore $\dfrac{r+1}{r+2} > \dfrac{q+1}{q+2}$,

since the sequence $(f_n)_{n \in \mathbb{N}}$ is strictly increasing, where $f_n = \dfrac{n+1}{n+2} \, \forall n \in \mathbb{N}$. Then $C_\mu(u(a_S)) > C_\mu(u(a_T))$. Then $\mu \in C_{Pref}$.

Let $A \subseteq N_{\geq 2}$ and $\emptyset \subsetneq B \subsetneq A$. Then there exists $q, r, s \in \{1, 2, \ldots, m\}$ such that $a_A \in \mathcal{B}_q$, $a_B \in \mathcal{B}_r$ and $a_{A\setminus B} \in \mathcal{B}_s$ with $q > r$ and $q > s$. Hence $\dfrac{2}{3} \leq \mu(A) \leq 1$, $\dfrac{2}{3} \leq \mu(B) \leq 1$ and $\dfrac{2}{3} \leq \mu(A \setminus B) \leq 1$. Therefore we have,

$\mu(B) + \mu(A \setminus B) \geq \dfrac{2}{3} + \dfrac{2}{3} = \dfrac{4}{3} > 1 \geq \mu(A)$, i.e., $\mu(A) - \mu(B) - \mu(A \setminus B) < 0$ for all $\emptyset \subsetneq B \subsetneq A$. Then $\sum_{\emptyset \subsetneq B \subsetneq A} (\mu(A) - \mu(B) - \mu(A \setminus B)) < 0$, so $\eta_A^\mu < 0$. $\qquad\square$

The following example illustrates Propositions 3 and 4.

Example 2. $N = \{1, 2, 3\}$, $P = \{(a_{23}, a_{12}), (a_2, a_3)\}$.

The binary relation $(P \cup M)$ contains no strict cycle, so P is representable by a Choquet integral model. A suitable topological sorting on $(P \cup M)$ is given by: $\mathcal{B}_0 = \{a_0\}$; $\mathcal{B}_1 = \{a_1, a_3\}$; $\mathcal{B}_2 = \{a_2, a_{13}\}$; $\mathcal{B}_3 = \{a_{12}\}$; $\mathcal{B}_4 = \{a_{23}\}$ and $\mathcal{B}_5 = \{a_{123}\}$. The strict preference P is representable by the following capacities μ and α given by Table 7 and Table 8 respectively.

Table 7. A capacity $\mu \in C_{\text{Pref}}$ and the corresponding nonadditivity index.

A	$\{1\}$	$\{3\}$	$\{2\}$	$\{1,3\}$	$\{1,2\}$	$\{2,3\}$	$\{1,2,3\}$
$6^5 \times \mu(A)$	6	6	6^2	6^2	6^3	6^4	6^5
$6^5 \times \eta_A^\mu$				24	174	1254	7244

Table 8. A capacity $\alpha \in C_{\text{Pref}}$ and the corresponding nonadditivity index.

A	$\{1\}$	$\{3\}$	$\{2\}$	$\{1,3\}$	$\{1,2\}$	$\{2,3\}$	$\{1,2,3\}$
$\alpha(A)$	2/3	2/3	3/4	3/4	4/5	5/6	6/7
cre η_A^α				$-7/12$	$-37/60$	$-7/12$	$-199/315$

We can see that $\forall A \subseteq N_{\geq 2}$, we have $\eta_A^\mu > 0$ and $\eta_A^\alpha < 0$.

5 Conclusion

This article studies the nonadditivity in the Choquet integral model. We make a restriction that the DM only gives strict preferential information. The capacity to represent this strict preference information is not unique and the sign of the nonadditivity index can vary depending on the arbitrary choice of a capacity compatible with the strict preference of DM. Therefore we introduce the concept of necessary and possible nonadditivity similar to that of necessary and possible interaction introduced in [14]. Only necessary nonadditivity are robust since the sign of the nonadditivity index does not vary within the set of all representing capacities.

We prove that neither null, nor positive, or negative nonadditivity is necessary. Thus the sign of nonadditivity index is not stable in the set of all capacities compatible with strict preferences of DM, therefore the interpretation of the nonadditivity index between criteria requires some caution.

In our future research, we will study the case where the ordinal preference information can contain the indifference relation. Moreover, outside the framework of binary alternatives, we will also proposed a linear program allowing to test non stability of the nonadditivity index.

References

1. Bouyssou, D., Pirlot, M.: Conjoint measurement tools for MCDM. In: Greco, S., Ehrgott, M., Figueira, J.R. (eds.) Multiple Criteria Decision Analysis. ISORMS, vol. 233, pp. 97–151. Springer, New York (2016). https://doi.org/10.1007/978-1-4939-3094-4_4
2. Choquet, G.: Theory of capacities. Annales de l'Institut Fourier **5**, 131–295 (1954)
3. Fujimoto, K.: Cardinal-probabilistic interaction indices and their applications: a survey. J. Adv. Comput. Intell. Intell. Inform. **7**(2), 79–85 (2003)
4. Gondran, M., Minoux, M.: Graphes et algorithmes, 3e édition. Eyrolles, Paris (1995)
5. Grabisch, M.: Fuzzy integral in multicriteria decision making. Fuzzy Sets Syst. **69**(3), 279–298 (1995)
6. Grabisch, M.: The application of fuzzy integrals in multicriteria decision making. Eur. J. Oper. Res. **89**(3), 445–456 (1996)
7. Grabisch, M.: k-Order additive discrete fuzzy measures and their representation. Fuzzy Sets Syst. **92**(2), 167–189 (1997)
8. Grabisch, M.: Set Functions, Games and Capacities in Decision Making. Theory and Decision Library C, vol. 46. Springer, Cham (2016). https://doi.org/10.1007/978-3-319-30690-2
9. Grabisch, M., Labreuche, C.: Fuzzy measures and integrals in MCDA. In: Figueira, J., Greco, S., Ehrgott, M. (eds.) Multiple Criteria Decision Analysis. State of the art surveys, number 142 in International Series in Operations Research & Management Science, pp. 563–608. Springer, Berlin (2005)
10. Grabisch, M., Labreuche, C.: A decade of application of the Choquet and Sugeno integrals in multi-criteria decision aid. Ann. Oper. Res. **175**(1), 247–290 (2010)
11. Grabisch, M., Labreuche, C., Vansnick, J.-C.: On the extension of pseudo-Boolean functions for the aggregation of interacting criteria. Eur. J. Oper. Res. **148**, 28–47 (2003)
12. Kaldjob Kaldjob, P.A., Mayag, B., Bouyssou, D.: Necessary and possible interaction between criteria in a general Choquet integral model. In: Information Processing and Management of Uncertainty in Knowledge-Based Systems, pp. 457–466 (2020)
13. Lidl, R., Pilz, G.: Applied Abstract Algebra. Springer, New York (1984). https://doi.org/10.1007/978-1-4615-6465-2
14. Mayag, B., Bouyssou, D.: Necessary and possible interaction between criteria in a 2-additive Choquet integral model. Eur. J. Oper. Res. **283**, 308–320 (2019)
15. Wu, J.Z., Beliakov, G.: Nonadditivity index and capacity identification method in the context of multicriteria decision making. Inf. Sci. **467**, 398–406 (2018)

The *d*-Interaction Index in MCDA

Brice Mayag[1](✉) and Bertrand Tchantcho[2]

[1] University Paris-Dauphine, PSL Research university, LAMSADE, CNRS,
UMR 7243, Place du Maréchal de Lattre de Tassigny, 75775 Paris cedex 16, France
brice.mayag@dauphine.fr
[2] Ecole Normale Supérieure de Yaoundé, MASS Laboratory, BP 47,
Yaoundé, Cameroon

Abstract. In the context of the representation of a preference information, we introduce a new interaction index based on a distance, in order to better model the interactions among criteria. This index, which provides robust interpretations of the interactions, is directly related to the independence property required when an additive model is used as an aggregation function.

Keywords: MCDA · Independence · Interaction · Choquet integral · Kendall distance

1 Introduction

An important aspect in Multiple Criteria Decision Analysis (MCDA) is to represent numerically the preferences of a Decision Maker (DM), by using a suitable aggregation function like the well known weighted sum (additive model). This latter is usually used in real world applications, due to its understandable formula characterized by a weight vector. However, the use of such model needs to respect the mutual preference independence property, generally known as the independence axiom, implying that there is no interaction among criteria. But if these interactions exist, we can model them by a Choquet integral model, which extends the additive model.

The Choquet integral [3,5,7,10] is based on a capacity or a fuzzy measure, defined as a set function allowing weights to all the coalitions of the criteria. From this capacity, an interaction index [12], more precisely a Choquet interaction index, is computed in order to capture the interactions in any subset of criteria. It was proved that the sign of this interaction index is not always stable [11], i.e., given a preference information of the DM, the interaction of a subset could be positive, null or negative, depending on the capacity compatible with these preferences. This leads to some misinterpretations of the interactions among criteria.

In this perspective of finding a convenient and robust interpretation of the sign of the classic Choquet interaction index, we analyze here the link between this index and the independence property. We show that the usual interpretation

D. Ciucci et al. (Eds.): IPMU 2022, CCIS 1602, pp. 210–220, 2022.
https://doi.org/10.1007/978-3-031-08974-9_17

of the null interaction w.r.t. a subset of criteria, as an independence among these criteria, is wrong. Therefore, we introduce a new definition of interaction, based only on a given distance and the preferences of the DM. We prove that this d-interaction remains robust when we need to analyze, from these preferences, the dependencies among criteria.

The paper is organized as follows. The next section presents the definition of the independence property and the basic material we need on the 2-additive Choquet integral. The d-interaction index is introduced and discussed in Sect. 3. We show in Sect. 4 how to compute this new interaction trough two examples. Finally, we give, at the end, some perspectives of this work.

2 Settings

2.1 The Independence Axiom in MCDA

Let X be a set of alternatives evaluated of a finite set of n criteria, $N = \{1, 2, \ldots, n\}$. We assume that X is a Cartesian product $X = X_1 \times X_2 \times \ldots \times X_n$ where X_i refers to a discrete set of attribute i, $i = 1, \ldots, n$, denoted by $X_i = \{c_i^1, c_i^2, \ldots, c_i^k\}$. A marginal utility function $u_i : X_i \to \mathbb{R}_+$, considered as a nondecreasing function, is associated to each attribute X_i, $i = 1, \ldots, n$.

Given a subset S of N, the element $z = (x_S, y_{N \setminus S})$, is an alternative such that $z_i = x_i$ if $i \in S$ and $z_i = y_i$ otherwise. We set, w.l.o.g., that $x_S \in X_S = \prod_{i \in S} X_i$ and $y_{N \setminus S} \in X_{N \setminus S} = \prod_{i \in N \setminus S} X_i$.

In a MCDA decision problem, the DM aims at computing a complete weak order $\succsim_X$ on X. To do so, he provides a preference information $\succsim_{X'}$ on a subset X' of X ($X' \subseteq X$). We consider here that these preferences are translated by two binary relations: a strict preference relation $\succ_{X'}$ and an indifference relation $\sim_{X'}$. The binary relation $\succ_{X'}$ is asymmetric while $\sim_{X'}$ is symmetric.

The additive model F_{add}, equivalent to the weighted sum, is a simple model usually used to represent a preference information $\succsim_X$. It is defined by the following Equation:

$$F_{add}(x_1, \ldots, x_n) = \sum_{i=1}^{n} u_i(x_i), \qquad \forall x = (x_1, \ldots, x_n) \in X \qquad (1)$$

The use of an additive model ensures to respect the following independence axiom, more precisely the mutual preference independence property:

Definition 1 (Mutual preference independence property [1])
  Let $N = \{1, \ldots, n\}$ be a finite set of n criteria, $n \geq 3$. Let $S \subseteq N$ be a subset of N. Let $\succsim_X$ a complete weak order on X.

1. *S is preference independent of $N \setminus S$, w.r.t. $\succsim_X$, if for all $x_S, x_S' \in X_S$, $a_{N \setminus S}, b_{N \setminus S} \in X_{N \setminus S}$,*

$$(x_S, a_{N \setminus S}) \succsim_X (x_S', a_{N \setminus S}) \Rightarrow (x_S, b_{N \setminus S}) \succsim_X (x_S', b_{N \setminus S}) \qquad (2)$$

2. The criteria $1, \ldots, n$ are mutually preference independent if for every subset $S \subseteq N$, S is preference independent of $N \setminus S$.
3. The preference relation $\succsim_X$ is said to satisfy the independence axiom if for every subset $S \subseteq N$, S is preference independent of $N \setminus S$.

Remark 1

- The property of independence is equivalent to the following one:
  $\forall i \in N, \forall z_i, t_i \in X_i, \forall x, y \in X,$

$$(z_i, x_{N-i}) \succsim_X (z_i, y_{N-i}) \Leftrightarrow (t_i, x_{N-i}) \succsim_X (t_i, y_{N-i}) \qquad (3)$$

  Hence an attribute is preferentially independent from all other attributes when changes in the rank ordering of preferences of other attributes does not change the preference order of the attribute $[1, 2, 4]$.
- For any $y_{N \setminus S} \in X_{N \setminus S}$, let us denote by $\succsim_S^{y_{N \setminus S}}$ the binary relation induced in X_S by $\succsim_X$ and defined by the Eq. (4):

$$x_S \succsim_S^{y_{N \setminus S}} x_S' \text{ iff } (x_S, y_{N \setminus S}) \succsim_X (x_S', y_{N \setminus S}) \text{ for all } x_S, x_S' \in X_S \qquad (4)$$

  Using this definition, a preference relation $\succsim_X$ over X satisfies the independence axiom if for all $S \subset N$, for all $y_{N \setminus S} \in X_{N \setminus S}$, the binary relation $\succsim_S^{y_{N \setminus S}}$ is unaffected by the choice of $y_{N \setminus S}$.

Example 1. We consider this classic example introduced in [7]. Four students of a faculty are evaluated on three subjects Mathematics (M), Statistics (S) and Language skills (L). All marks are taken from the same scale, from 0 to 20. The evaluations of these students are given by the table below:

	1: Mathematics (M)	2: Statistics (S)	3: Language (L)
a	16	13	7
b	16	11	9
c	6	13	7
d	6	11	9

- Let us suppose that the Dean of the faculty gives the following preferences:

$$a \succ_X b \text{ and } c \succ_X d \qquad (5)$$

If we set the following values as the marginal utilities functions

1: Mathematics(M)	2: Statistics(S)	3: Language(L)
$u_1(16) = 16$	$u_2(13) = 14$	$u_3(7) = 7$
$u_1(16) = 16$	$u_2(11) = 11$	$u_3(9) = 16$
$u_1(6) = 6$	$u_2(13) = 14$	$u_3(7) = 7$
$u_1(6) = 6$	$u_2(11) = 11$	$u_3(9) = 16$

then the preferences of Eq. (5) are representable by an additive model since

$$F_{add}(a) = u_1(16) + u_2(13) + u_3(7) = 16 + 14 + 7 = 37,$$
$$F_{add}(b) = u_1(16) + u_2(11) + u_3(9) = 16 + 11 + 9 = 36,$$
$$F_{add}(c) = u_1(6) + u_2(13) + u_3(7) = 6 + 14 + 7 = 27,$$
$$F_{add}(d) = u_1(6) + u_2(11) + u_3(9) = 6 + 11 + 9 = 26.$$

Therefore the binary relation $\succ_X$, in this context, satisfies the independence axiom property.

- *Now, let us suppose that the Dean of the faculty gives de the following preferences:*

$$b \succ_X a \ and \ c \succ_X d. \tag{6}$$

By applying the Equation (3) with $i = 1$, it is not difficult to see that these preferences do not satisfy the independence axiom since we have $(16, 11, 9) \succ_X (16, 13, 7)$ and $(6, 13, 7) \succ_X (6, 11, 9)$.

When the independence axiom is not satisfied, we often use a non-additive model in order to represent the preferences of the DM. The Choquet integral model w.r.t. a 2-additive capacity is one of the most popular model used in order to take into account the existence of some interactions among criteria.

2.2   The Choquet Integral Model

The concept of interactions among criteria arises when for a given preference relation, the independence property is not satisfied. As it is more simple and understandable, in MCDA, to interpret the interactions among two criteria, the Choquet integral w.r.t. a 2-additive capacity [7,9,10], also called 2-additive Choquet, was proposed. This aggregation function, considered as a fuzzy integral, is based on the concept of *capacity or fuzzy measure* μ defined as a set function from the powerset of criteria 2^N to $[0,1]$ such that:

1. $\mu(\emptyset) = 0$
2. $\mu(N) = 1$
3. $\forall A, B \in 2^N, \ [A \subseteq B \Rightarrow \mu(A) \leq \mu(B)]$ (monotonicity).

A set function called the *Möbius transform* $m^\mu : 2^N \to \mathbb{R}$, associated to the capacity μ, is defined by

$$m^\mu(T) := \sum_{K \subseteq T} (-1)^{|T \setminus K|} \mu(K), \forall T \in 2^N. \tag{7}$$

A capacity μ on N is said to be *2-additive* if the following two conditions are satisfied:

- For all subset T of N such that $|T| > 2$, $m^\mu(T) = 0$;
- There exists a subset B of N such that $|B| = 2$ and $m^\mu(B) \neq 0$.

Given an alternative $x := (x_1, ..., x_n)$ of X, the expression of the 2-additive Choquet integral of x is given as follows [6]:

$$C_\mu(u_1(x_1), \ldots, u_n(x_n)) = \sum_{i=1}^{n} \phi_i^\mu \, u_i(x_i) - \frac{1}{2} \sum_{\{i,j\} \subseteq N} I_{ij}^\mu \, |u_i(x_i) - u_j(x_j)| \quad (8)$$

where

- $I_{ij}^\mu = \mu(\{i,j\}) - \mu(\{i\}) - (\{j\})$ is the interaction index between the two criteria i and j [5,12];

- $\phi_i^\mu = \sum_{K \subseteq N \setminus i} \frac{(n - |K| - 1)! |K|!}{n!} (\mu(K \cup i) - \mu(K)) = \mu(\{i\}) + \frac{1}{2} \sum_{j \in N, j \neq i} I_{ij}^\mu$

  is defined as the importance of criterion i and it corresponds to the Shapley value of i w.r.t. μ [13].

Equation (8) proves that the 2-additive Choquet integral is a generalization of the weighted sum. Indeed, when there is no interaction among criteria, the Shapley value ϕ_i^μ is the weight associated with the criterion i.

2.3   A Link Between the Independence Axiom and the Null Choquet-interaction Index

Example 2. *Let us assume that the scale* $[0, 20]$ *of the evaluation of the four students, given in Example 1, corresponds to the utility function associated to each subject, i.e.,*

1: Mathematics (M)	2: Statistics (S)	3: Language (L)
$u_1(16) = 16$	$u_2(13) = 13$	$u_3(7) = 7$
$u_1(16) = 16$	$u_2(11) = 11$	$u_3(9) = 16$
$u_1(6) = 6$	$u_2(13) = 13$	$u_3(7) = 7$
$u_1(6) = 6$	$u_2(11) = 11$	$u_3(9) = 16$

We saw above, in Example 1, that the preferences given by the Equation (6) are not representable by an additive model since the independence axiom is not satisfied. The 2-additive capacity μ, *given in Table 1, shows that these preferences are now representable by a 2-additive Choquet integral.*

In Table 2, we can see that the preference information $a \succ_X b \succ_X c \succ_X d$, representable by an additive model in Example 1, is also representable by a 2-additive Choquet integral model in which all the interactions indices are not null.

A misinterpretation of the interaction index might led to believe that in a Choquet model, independence is equivalent to having null interactions. However, as shown in the previous example, preferences may verify the independence

Table 1. $b \succ_X a$ and $c \succ_X d$ are representable by a 2-additive Choquet integral C_μ.

$\mu(\{1\})$	0						
$\mu(\{2\})$	0.25	I_{12}^μ	0	ϕ_1^μ	0.375	$C_\mu(a)$	8.5
$\mu(\{3\})$	0	I_{13}^μ	0.75	ϕ_2^μ	0.25	$C_\mu(b)$	9.5
$\mu(\{1,2\})$	0.25	I_{23}^μ	0	ϕ_3^μ	0.375	$C_\mu(c)$	7.75
$\mu(\{1,3\})$	0.75					$C_\mu(d)$	7.25
$\mu(\{2,3\})$	0.25						

Table 2. $a \succ_X b \succ_X c \succ_X d$ are representable by a 2-additive Choquet integral C_μ.

$\mu(\{1\})$	0.1						
$\mu(\{2\})$	0.5	I_{12}^μ	0.3	ϕ_1^μ	0.3	$C_\mu(a)$	12.7
$\mu(\{3\})$	0.5	I_{13}^μ	0.1	ϕ_2^μ	0.4	$C_\mu(b)$	11.3
$\mu(\{1,2\})$	0.9	I_{23}^μ	-0.5	ϕ_3^μ	0.3	$C_\mu(c)$	9.5
$\mu(\{1,3\})$	0.7					$C_\mu(d)$	8.5
$\mu(\{2,3\})$	0.5						

axiom but been representable by a Choquet model with interactions different from zero. This finding motivates the introduction of a new concept of interaction for which nullity coincides with independence. This is indeed the subject of the next section.

3   A New Definition of the Interaction Index

Definition 2 (The d-interaction index). *Let d be a distance defined on the set of all binary relations and S be a subset of N. The d-interaction index between criteria is defined by:*

$$I^d(S) = \sum_{\{y_{N\setminus S}, z_{N\setminus S}\} \subseteq X_{N\setminus S}} d(\succsim_S^{y_{N\setminus S}}, \succsim_S^{z_{N\setminus S}}) \tag{9}$$

We recall that $\succsim_S^{y_{N\setminus S}}$ and $\succsim_S^{z_{N\setminus S}}$ are binary relations induced in X_S by $\succsim_X$.

The d-interaction index of a coalition S of criteria captures through the distance the effect of criteria of $N \setminus S$ over S. This is a critical observation that makes our contribution different from those who have worked on interactions. Our concept of interaction is unrelated to the preference representation model. We show precisely in the result below that a necessary and sufficient condition under which a preference relation satisfies the independence axiom is that this interaction index is zero for all coalition of criteria.

A particular distance in the literature is the well-known Kendall distance [8]. Given two binary relations R_1 and R_2, the Kendall distance between R_1 and R_2, denoted by $D_K(R_1, R_2)$ is defined as follows:

$$D_K(R_1, R_2) = \left| R_1 \setminus R_2 \right| + \left| R_2 \setminus R_1 \right| \tag{10}$$

Theorem 1. *A preference relation $\succsim_X$ satisfies the independence axiom if and only if $I^d(S) = 0$ for all $S \in 2^N$.*

Proof. Assume that the binary relation $\succsim_X$ over X satisfies the independence axiom. Then, for all $S \in 2^N$, for all $y_{N \setminus S} \in X_{N \setminus S}$, the relation $\succsim_S^{y_{N \setminus S}}$ is unaffected by the choice of $y_{N \setminus S}$, i.e., $\succsim_S^{y_{N \setminus S}} = \succsim_S^{z_{N \setminus S}}$ for all $y_{N \setminus S}$, $z_{N \setminus S}$ belonging to $X_{N \setminus S}$. This implies that $I^d(S) = \displaystyle\sum_{\{y_{N \setminus S}, z_{N \setminus S}\} \subseteq X_{N \setminus S}} d(\succsim_S^{y_{N \setminus S}}, \succsim_S^{z_{N \setminus S}}) = 0$ since d is a distance.

Conversely, if $I^d(S) = 0$, then for all $\{y_{N \setminus S}, z_{N \setminus S}\} \subseteq X_{N \setminus S}$, $d(\succsim_S^{y_{N \setminus S}}, \succsim_S^{z_{N \setminus S}}) = 0$ which means that $\succsim_S^{y_{N \setminus S}} = \succsim_S^{z_{N \setminus S}}$ for all $y_{N \setminus S}$, $z_{N \setminus S} \in X_{N \setminus S}$. Therefore $\succsim_S^{y_{N \setminus S}}$ induced in X_S is not affected by the choice of $y_{N \setminus S}$; thus $\succsim_X$ over X satisfies the independence axiom. ∎

The following result shows that it is enough to show that $I^d(N \setminus i) = 0$ for all $i \in N$, in order to check the independence of $\succsim_X$. This clearly reduces the complexity in verifying the independence axiom. Instead of checking the nullity for 2^n sets, it is enough to check the nullity of the index for the n conditions of criteria of size $n - 1$.

Theorem 2. *Let $\succcurlyeq$ be a preference relation and d a distance over binary relations. A necessary and sufficient condition under which $I^d(S) = 0$ for all $S \in 2^N$, is that, for all $i \in N$, $I^d(N \setminus i) = 0$*

Proof. It is obvious that if $I^d(S) = 0$ for all $S \in 2^N$, then $i \in N$, $I^d(N \setminus i) = 0$.

Conversely, assume that for all $i \in N$, $I^d(N \setminus i) = 0$.

Then, we have $\displaystyle\sum_{\{x_i, y_i\} \subseteq X_i} d(\succsim^{x_i}, \succsim^{y_i}) = 0$, which implies that for all $\{x_i, y_i\} \subseteq X_i$, $d(\succsim^{x_i}, \succsim^{y_i}) = 0$. In other words, $\succsim^{x_i} = \succsim^{y_i}$ for all $x_i, y_i \in X_i$ and therefore, for all $x_i \in X_i$, the binary relation $\succsim_{N \setminus i}^{x_i}$ is unaffected by the choice of x_i in X_i. This clearly means, from Remark 1 that the preference relation $\succsim_X$ satisfies the independence axiom and thanks to Theorem 1, it follows that $I^d(S) = 0$ for all $S \in 2^N$. ∎

Remark 2. *Theorem 2 above also means that a preference relation $\succsim_X$ over X satisfies the independence axiom if and only if $I(N \setminus i) = 0$ for all $i \in N$. Therefore this property, based only on conditions of size $n - 1$, can be used to check the independence of a preference relation over X.*

As seen above, a preference relation satisfies the independence axiom if and only if the d-interaction index satisfies $I^d(S) = 0$ for all subset of criteria S, where d is a distance in the set of binary relation on X. Thanks to Theorem 2, it suffices to check that equality for singletons. In the next section we use the well known Kendall distance to check the independence axiom.

4   Computation of the d-Interaction Index

4.1   Coming Back to the Classic Example

Let us consider the preferences of the Dean on four students, not representable by an additive model and given in Example 1: $b \succ_X a$ and $c \succ_X d$.

We consider the following notations related to the values of attributes given on each criterion:

Criterion 1 : Mathematics $\rightarrow a_1 = 6$ and $a_2 = 16$, with $a_2 \succsim_1 a_1$;
Criterion 2 : Statistic $\quad \rightarrow b_1 = 11$ and $b_2 = 13$, with $b_2 \succsim_2 b_1$;
Criterion 3 : Language $\quad \rightarrow c_1 = 7$ and $c_2 = 9$, with $c_2 \succsim_3 c_1$.

In this context, the corresponding performance matrix is the following:

	1: Mathematics (M)	2: Statistics (S)	3: Language (L)
a	a_2	b_2	c_1
b	a_2	b_1	c_2
c	a_1	b_2	c_1
d	a_1	b_1	c_2

We choose to write an alternative a_i, b_j, c_k $(i, j, k \in \{1, 2\})$ in the simple form $a_i b_j c_k$, and any permutation of $a_i b_j c_k$ will denote the same alternative. The preferences $b \succ_X a$ and $c \succ_X d$ will then be translated by:

$$a_2 b_1 c_2 \succ_X a_2 b_2 c_1 \qquad \text{and} \qquad a_1 b_2 c_1 \succ_X a_1 b_1 c_2$$

Computation of $I^d(23) = \sum_{\{y_{\{1\}}, z_{\{1\}}\} \subseteq X_1} D_K(\succsim_{23}^{y_{\{1\}}}, \succsim_{23}^{z_{\{1\}}}) = D_K(\succsim_{23}^{a_1}, \succsim_{23}^{a_2})$

where $\succsim_{23}^{a_k}$ is the binary relation induced in X_{23} by $\succsim_X$ when the component $a_k \in X_1, k = 1, 2$ is fixed. To determine these binary relations, we look for $i, j, p, q \in \{1, 2\}$ such that for $k \in \{1, 2\}$,

$$b_i c_j \succsim_{23}^{a_k} b_p c_q, \text{ i.e., } a_k b_i c_j \succsim_X a_k b_p c_q. \tag{11}$$

Except the reflexivity property of $\succsim_{23}^{a_k}$, which is not difficult to check, the Equation (11) is valid if and only if

- for $k = 1$, we have $(i \geq 2, j \geq 1)$ and $(1 \geq p, 2 \geq q)$, i.e., $ij \in \{21, 22\}$ and $pq \in \{11, 12\}$, since $\succsim_X$ is transitive and $a_1 b_2 c_1 \succ_X a_1 b_1 c_2$;
- for $k = 2$, we have $(i \geq 1, j \geq 2)$ and $(2 \geq p, 1 \geq q)$, i.e., $ij \in \{12, 22\}$ and $pq \in \{21, 22\}$, since $\succsim_X$ is transitive and $a_2 b_1 c_2 \succ_X a_2 b_2 c_1$.

The two binary relations $\succsim_{23}^{a_1}$ and $\succsim_{23}^{a_2}$ are presented in Table 3, where the valid assertions $b_i c_j \succsim_{23}^{a_k} b_p c_q$, $k = 1, 2$, are marked by "ok". At this stage we can conclude that $I^d(23) > 0$ since $\succsim_{23}^{a_1} \neq \succsim_{23}^{a_2}$ meaning that the DM's preferences do not satisfy the independence axiom.

Table 3. The binary relations $\succsim_{23}^{a_1}$ and $\succsim_{23}^{a_2}$ deduced from the Example 1

$\succsim_{23}^{a_1}, \frown$	b_1c_1	b_1c_2	b_2c_1	b_2c_2
b_1c_1	ok			
b_1c_2		ok		
b_2c_1	ok	ok	ok	
b_2c_2	ok	ok		ok

$\succsim_{23}^{a_2}, \frown$	b_1c_1	b_1c_2	b_2c_1	b_2c_2
b_1c_1	ok			
b_1c_2		ok	ok	ok
b_2c_1			ok	
b_2c_2			ok	ok

4.2   Another Example in Which the Violation of the Independence Axiom Is Not Obvious

Let us consider the following performance matrix where $N = \{1,2,3\}$, $X_1 = \{a_1, a_2, a_3, a_4\} = \{4, 25, 35, 40\}$; $X_2 = \{b_1, b_2, b_3, b_4\} = \{2, 6, 24, 27\}$ and $X_3 = \{c_1, c_2, c_3, c_4\} = \{3, 23, 26, 37\}$ with $a_i \geq a_j, b_i \geq b_j, c_i \geq c_j$ iff $j \geq i$

	Criterion 1	Criterion 2	Criterion 3
x	$a_1 = 4$	$b_3 = 24$	$c_3 = 26$
y	$a_4 = 40$	$b_4 = 27$	$c_2 = 23$
z	$a_3 = 35$	$b_1 = 2$	$c_1 = 3$
t	$a_2 = 25$	$b_2 = 6$	$c_4 = 27$

Using the same notations above (see Sect. 4.1), we assume that the minimal preferences of the decision maker are given as follows:

$$a_1 b_3 c_3 \succ a_4 b_4 c_2 \text{ and } a_3 b_1 c_1 \succ a_2 b_2 c_4$$

Let us check whether the preferences of the DM are independent by computing the d-interaction index. For this purpose we will compute $I^d(N \setminus i)$ for all $i \in N$, i.e., $I^d(23)$, $I^d(13)$ and $I^d(12)$.

Computation of $I^d(13) = \displaystyle\sum_{\{b_i, b_j\} \subseteq X_2} D_K(\succsim_{13}^{b_i}, \succsim_{13}^{b_j})$

Let i, j and p, q belonging to $\{1, 2, 3, 4\}$.

- **Determination of** $\succ_{13}^{b_1}$:

$$a_i c_j \succsim_{13}^{b_1} a_p c_q \Leftrightarrow a_i b_1 c_j \succsim_X a_p b_1 c_q \Leftrightarrow \begin{cases} a_i b_1 c_j \succsim_X a_3 b_1 c_1 \\ \text{and} \\ a_2 b_2 c_4 \succsim_X a_p b_1 c_q \end{cases} \text{iff} \begin{cases} i \geq 3, j \geq 1 \\ \text{and} \\ 2 \geq p, 4 \geq q \end{cases}.$$

- **Determination of** $\succ_{13}^{b_2}$:

$$a_i c_j \succsim_{13}^{b_2} a_p c_q \Leftrightarrow a_i b_2 c_j \succsim_X a_p b_2 c_q \Leftrightarrow \begin{cases} a_i b_2 c_j \succsim_X a_3 b_1 c_1 \\ \text{and} \\ a_2 b_2 c_4 \succsim_X a_p b_2 c_q \end{cases} \text{iff} \begin{cases} i \geq 3, j \geq 1 \\ \text{and} \\ 2 \geq p, 4 \geq q \end{cases}.$$

It follows that $\succ_{13}^{b_2} = \succ_{13}^{b_1}$

- **Determination of** $\succsim_{13}^{b_3}$:

$$a_i c_j \succsim_{13}^{b_3} a_p c_q \Leftrightarrow a_i b_3 c_j \succsim_X a_p b_3 c_q \Leftrightarrow \begin{cases} a_i b_3 c_j \succsim_X a_1 b_3 c_3 \\ \text{and} \\ a_4 b_4 c_2 \succsim_X a_p b_3 c_q \end{cases} \quad \text{iff} \quad \begin{cases} i \geq 1, j \geq 3 \\ \text{and} \\ 4 \geq p, 2 \geq q \end{cases}$$

- **Determination of** $\succsim_{13}^{b_4}$:

$$a_i c_j \succsim_{13}^{b_4} a_p c_q \Leftrightarrow a_i b_4 c_j \succsim_X a_p b_4 c_q \Leftrightarrow \begin{cases} a_i b_4 c_j \succsim_X a_1 b_3 c_3 \\ \text{and} \\ a_4 b_4 c_2 \succsim_X a_p b_4 c_q \end{cases} \quad \Leftrightarrow \quad \begin{cases} i \geq 1, j \geq 3 \\ \text{and} \\ 4 \geq p, 2 \geq q \end{cases}$$

It follows that $\succsim_{13}^{b_4} = \succsim_{13}^{b_3}$.
Therefore, we have
$$I^d(13) = D_K(\succsim_{13}^{b_1}, \succsim_{13}^{b_2}) + D_K(\succsim_{13}^{b_1}, \succsim_{13}^{b_3}) + D_K(\succsim_{13}^{b_1}, \succsim_{13}^{b_4}) + D_K(\succsim_{13}^{b_2}, \succsim_{13}^{b_3}) + D_K(\succsim_{13}^{b_2}, \succsim_{13}^{b_4}) + D_K(\succsim_{13}^{b_3}, \succsim_{13}^{b_4}) = 4 D_K(\succsim_{13}^{b_1}, \succsim_{13}^{b_3})$$
In other to determine $D_K(\succsim_{13}^{b_1}, \succsim_{13}^{b_3})$, let us remark that, in fact, we can identify the binary relations $\succsim_{13}^{b_1}, \succsim_{13}^{b_3}$ as the Cartesian products based on the indices i, j, p, q determined above:

$$\succsim_{13}^{b_1} = \succsim_{13}^{b_2} = \{\{3,4\} \times \{1,2,3,4\}\} \times \{\{1,2\} \times \{1,2,3,4\}\}$$

$$\succsim_{13}^{b_3} = \succsim_{13}^{b_4} = \{\{1,2,3,4\} \times \{3,4\}\} \times \{\{1,2,3,4\} \times \{1,2\}\}$$

Let $i, j, p, q \in \{1,2,3,4\}$. $(ij, pq) \in \succsim_{13}^{b_1} \cap \succsim_{13}^{b_3}$ if and only if

$$\begin{cases} ij \in [\{3,4\} \times \{1,2,3,4\}] \cap [\{1,2,3,4\} \times \{3,4\}] = \{3,4\} \times \{3,4\} \\ \text{and} \\ pq \in [\{1,2\} \times \{1,2,3,4\}] \cap [\{1,2,3,4\} \times \{1,2\}] = \{1,2\} \times \{1,2\} \end{cases}$$

This proves that the cardinality of $\succsim_{13}^{b_1} \cap \succsim_{13}^{b_3}$ is 16.
It follows that

$$I^d(13) = D_K(\succsim_{13}^{b_1}, \succsim_{13}^{b_3}) = \left| \succsim_{13}^{b_1} \setminus \succsim_{13}^{b_3} \right| + \left| \succsim_{13}^{b_3} \setminus \succsim_{13}^{b_1} \right|$$
$$= \left| \succsim_{13}^{b_1} \right| + \left| \succsim_{13}^{b_3} \right| - 2 \left| \succsim_{13}^{b_1} \cap \succsim_{13}^{b_3} \right|$$
$$= 64 + 64 - 2 \times 16$$
$$= 96$$

Hence the preferences of the DM do not satisfy the independence property.

5   Conclusion

We introduced the new concept of d-interaction in MCDA, which is based on a distance and strongly related to the mutual independence axiom. We proved that, in terms of interpretation of the existing of some interactions among criteria, this interaction index is more robust than the Choquet integral interaction index defined by a capacity. In the future works, we will investigate about all the complexity aspects of the computation of d-interaction. For instance, is there exists an algorithm allowing to compute this index in a polynomial time. Furthermore, il will be also interesting to analyze the results when another distance is used, instead of the Kendall distance.

References

1. Bouyssou, D., Marchant, T., Pirlot, M., Perny, P., Tsoukiàs, A., Vincke, P.: Evaluation and Decision Models: A Critical Perspective. Kluwer Academic, Dordrecht (2000)
2. Bouyssou, D., Marchant, T., Pirlot, M., Tsoukiàs, A., Vincke, P.: Evaluation and decision models with multiple criteria: stepping stones for the analyst. In: International Series in Operations Research and Management Science, vol. 86. Boston, 1st edn. (2006)
3. Choquet, G.: Theory of capacities. Annales de l'Institut Fourier **5**, 131–295 (1953)
4. Dyer, J.S.: MAUT - multiattribute utility theory. In: Figueira, J., Greco, S., Ehrgott, M. (eds.) Multiple Criteria Decision Analysis: State of the Art Surveys, pp. 265–285. Springer Verlag, Boston, Dordrecht, London (2005). https://doi.org/10.1007/0-387-23081-5_7
5. Grabisch, M.: k-order additive discrete fuzzy measures and their representation. Fuzzy Sets Syst. **92**, 167–189 (1997)
6. Grabisch, M., Labreuche, Ch.: A decade of application of the Choquet and Sugeno integrals in multi-criteria decision aid. 4OR **6**, 1–44 (2008)
7. Grabisch, M., Labreuche, C.: Fuzzy measures and integrals in MCDA. In: Multiple Criteria Decision Analysis: State of the Art Surveys. ISORMS, vol. 78, pp. 563–604. Springer, New York (2005). https://doi.org/10.1007/0-387-23081-5_14
8. Kendall, M.G.: A new measure of rank correlation. Biometrika **30**(1/2), 81–93 (1938)
9. Mayag, B., Grabisch, M., Labreuche, C.: A characterization of the 2-additive Choquet integral through cardinal information. Fuzzy Sets Syst. **184**(1), 84–105 (2011)
10. Mayag, B., Grabisch, M., Labreuche, C.: A representation of preferences by the Choquet integral with respect to a 2-additive capacity. Theory Dec. **71**(3), 297–324 (2011)
11. Mayag, B., Bouyssou, D.: Necessary and possible interaction between criteria in a 2-additive choquet integral model. Eur. J. Oper. Res. **283**(1), 308–320 (2020)
12. Murofushi, T., Soneda, S.: Techniques for reading fuzzy measures (III): interaction index. In: 9th Fuzzy System Symposium, pp. 693–696, Sapporo, Japan, May 1993. (in Japanese)
13. Shapley, L.S.: A value for n-person games. In: Kuhn, H.W., Tucker, A.W. (eds.) Contributions to the Theory of Games, Vol. II, number 28 in Annals of Mathematics Studies, pp. 307–317. Princeton University Press (1953)

E-Health

E-Health

Analysis of Graphical Causal Models with Discretized Data

Ofir Hanoch[1], Nalan Baştürk[1], Rui Jorge Almeida[1,2(✉)],
and Tesfa Dejenie Habtewold[1]

[1] Department of Quantitative Economics, School of Business and Economics,
Maastricht University, P.O. Box 61, 6200 MD Maastricht, The Netherlands
{o.hanoch,n.basturk,rj.almeida,t.habtewold}@maastrichtuniversity.nl
[2] Department of Data Analytics and Digitilisation,
School of Business and Economics, Maastricht University,
P.O. Box 61, 6200 MD Maastricht, The Netherlands

Abstract. In several fields, sample data are observed at discrete instead
of continuous levels. For example, in psychology an individual's disease
level is typically observed as 'mild', 'moderate' or 'strong', while the
underlying mental disorder intensity is potentially a continuous variable.
Implications of such discretization in linear regression are well-known:
uncertainty increases and estimated causal relations become biased and
inconsistent. For more complex models, implications of discretization are
not theoretically studied. This paper considers an empirical study of complex models where causal relationships are unknown, some variables are
discretized and graphical causal models are used to estimate causal relationships and effects. We study the implications of discretization on the
obtained results using simulations. We show that discretization affects
the correct estimation of causal relations and the uncertainty of obtained
causal relations between discretized variables and non-discretized variables. In addition, we show that discretization influences estimated causal
effects and we relate this influence to the properties of discretized data
and sample size.

Keywords: Causal discovery · Discretized data · Graphical causal
models · Mixed data

1 Introduction

Understanding causal relationships between variables is important in many
applications, such as detecting the causative and influential factors for a disease or understanding the reasons for the good and bad economic conditions.
In conventional statistical models, the researcher starts with an assumption of

We thank David JeanJean for his preliminary research on this project. Baştürk, Hanoch
and Habtewold are financially supported by the Netherlands Organization for Scientific
Research (NWO) under grant number 195.187.

D. Ciucci et al. (Eds.): IPMU 2022, CCIS 1602, pp. 223–234, 2022.
https://doi.org/10.1007/978-3-031-08974-9_18

a causal relationship. Graphical causal models have been proposed to estimate causal relations without making explicit assumptions about part or all of the included variables [10,14]. These models have been applied successfully for causal discovery [3,11]. Recently, graphical causal models have been extended to detect causal relations between mixed (discrete and continuous) variables [5,12,16].

In many applications, discrete variables included in models represent underlying continuous variables. For instance, an individual has the propensity to have a certain disease, but this propensity is not observed. Instead, the observed variable is often discretized based on some disease indicator, such as blood sugar levels or an index for a disorder. In conventional models, such as linear regression, the effects of discretization are studied [2,9]. For linear regression, it is straightforward to show that discretization increases statistical uncertainty and causes biased and inconsistent results for the discretized variable as well as other variables that correlate with the discretized one. For more complex non-linear models, it is shown that discretization can lead to spurious causal effects [15].

In this paper we analyze the estimation properties of graphical causal models in discrete data applications. First, we derive the bias-variance effect of variable discretization in a standard linear regression model. Second, we generate artificial data to study the graphical causal models in terms of their capability to identify the correct relationships and their bias-variance properties. We find that both the estimated causality between variables, in terms of true positive and true negative counts of causal relations, and the bias in estimated causal relations deteriorate with discretization. The discretization effect is more pronounced when more variables in the model are discretized and when we specifically consider the case of false negative relations in the estimates.

2   Graphical Causal Models

Graphical causal models aim to capture the causal structure of multivariate data using a graph structure G, defined as the ordered pair $< V, E >$ where V is a set of vertices, and E is a set of edges which represent the variables [14]. An undirected graph indicates which pairs of vertices in E are correlated. A directed acyclic graph (DAG), on the other hand, represents the causal relations between variables [10]. An example of a DAG for simulated data is given in Fig. 1a. A conventional method to estimate a DAG is the PC algorithm [13] which is extended to mixed data [5]. The PC algorithm for mixed data consists of two steps: (1) the DAG estimation for causal relations and (2) estimation of causal effects based on step (1). The first step of the PC algorithm relies on the correlation matrix between p variables and a sequence of independence tests that identify causal relations. Each element $\rho_{i,j}$ of the correlation matrix is estimated using the sample of N observations:

$$\hat{\rho}_{i,j} = \frac{\sum_{n_1,n_2=1}^{N} k_i(x_{i,n_1}, x_{i,n_2}) k_j(x_{j,n_1}, x_{j,n_2})}{\sqrt{\left(\sum_{n_1,n_2=1}^{N} k_i(x_{i,n_1}, x_{i,n_2})^2\right) \left(\sum_{n_1,n_2=1}^{N} k_j(x_{j,n_1}, x_{j,n_2})^2\right)}}, \quad (1)$$

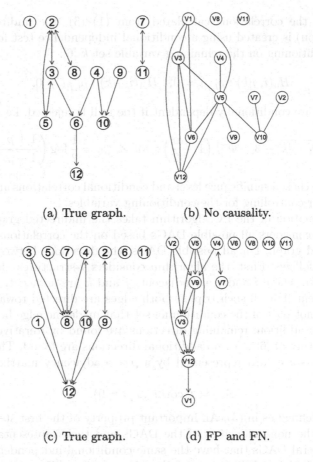

(a) True graph.   (b) No causality.

(c) True graph.   (d) FP and FN.

Fig. 1. True causal relation (left) and estimated causal relation (right) for two simulation cases.

where $k()$ denotes the kernel density estimator. We follow [5] and use a radial basis function for the kernel estimation of continuous variables:

$$k_i(x_{i,n_1}, x_{i,n_2}) = \exp\left(-\frac{(x_{i,n_1} - x_{i,n_2})^2}{2\sigma^2}\right), \tag{2}$$

for discretized variables we use a categorical kernel function:

$$k_i(x_{i,n_1}, x_{i,n_2}) = h_\theta(P_Z(z_{n_1})) \times I(z_{n_1} = z_{n_2}), \tag{3}$$

where $P_Z(z)$ is the probability that a categorical variable Z takes the value z, $h_\theta(z) = (1 - z^\theta)^{\frac{1}{\theta}}$ and $I()$ is the indicator function that takes the value 1 if its argument is true and 0 otherwise. The combination of the continuous and categorical kernels to obtain the correlation values of mixed data is called a kernel alignment [5].

Based on the correlations calculated from (1)–(3), the undirected graph (graph skeleton) is created using a conditional independence test for each variable i, j, conditioning on the remaining variable set k [6]:

$$H_0(i,j|k) : \rho_{i,j|k} = 0, \quad H_A(i,j|k) : \rho_{i,j|k} \neq 0. \tag{4}$$

Variables i, j are conditionally dependent if the null is rejected, i.e.

$$|Z_{i,j|k}|\sqrt{N - |k| - 3} > \Phi^{-1}\left(1 - \frac{\alpha}{2}\right); \text{ for } Z_{i,j|k} = \frac{1}{2}\log\left(\frac{1 + \hat{p}_{i,j|k}}{1 - \hat{p}_{i,j|k}}\right), \tag{5}$$

where α is the chosen significance level and conditional correlations are calculated using (1) after controlling for the conditioning variables.

DAG estimation of the PC algorithm takes the undirected graph obtained from (4) and considers all possible DAGs based on the correlations implied by the undirected graph. For all possible DAGs, the directions between edges are estimated as follows: First, the algorithm considers v-structures, defined as all triplets (i, j, k), where i and j are adjacent, j and k are adjacent, but i and k are not adjacent. For all such triplets both edges are directed towards j if and only if k was not part of the conditioning set that made the edge between i and j drop out. In addition, remaining directions are found by iteratively applying the PC algorithm of [6] until no additional directions are found. The estimated causal relations are then represented by a $p \times p$ adjacency matrix $\hat{A}$ for each possible DAG:

$$\hat{A}_{i,j} = I(\text{cor}(x_i, x_j) \neq 0), \tag{6}$$

where $I()$ is defined as in (3). An important property of the first step of the PC algorithm is the non-uniqueness of the DAGs and their representation in (6): There are several DAGs that have the same conditional independence relationships but the estimated directions of the edges can be different between these DAGs, implying different causal relations. A conventional method is to obtain the best fitting DAG from all alternatives.

The second, relatively straightforward, step of causal effect estimation is the estimation of the effects between variables, assuming the causal relations in (6) are correctly estimated. In this step, total causal effects between variables are obtained using a simple linear regression adjusting for a valid adjustment set. The obtained effects, however, are not unique as the DAGs are not unique. Due to this lack of uniqueness, it is common to report a range of estimates for each causal relation [4].

3   Illustration of Discretization Effects

We first illustrate the problem of discretization in the simple case of a linear regression. In this case we can derive the effects of discretization on parameter estimates, namely their bias, consistency and uncertainty. For this illustration, consider the following data generation process (DGP) where both X_1 and X_2 are normally distributed and $\varepsilon \sim N(0, Q)$:

$$Y = \beta_0 + \beta_1 X_1 + \beta_2 X_2 + \varepsilon, \tag{7}$$

where Y is the $N \times 1$ vector of dependent variable, X_1, X_2 are $N \times 1$ vectors of independent variables and ε is the $N \times 1$ vector of residuals.

Suppose that X_2 is only observed up to a discretized value instead of the underlying continuous variable. We denote the discretized observed value of X_2 as a dummy variable X_3, where $X_3 = x_{i3} = 1$ for observation i if $x_{i2} > 0$ and $X_3 = x_{i3} = 0$ otherwise. The equivalent linear regression of (7) with the discretized variable is:

$$Y = \beta_0 + \beta_1 X_1 + \beta_3 X_3 + \eta, \tag{8}$$

where $\eta = (\beta_2 X_2 - \beta_3 X_3 + \varepsilon)$ is the new error term and it correlates with both X_2 and X_3, while it remains independent from X_1.

The derivation for the estimated coefficients $\hat{\beta}_1$ and $\hat{\beta}_3$ is straightforward:

$$\begin{pmatrix} \hat{\beta}_1 \\ \hat{\beta}_3 \end{pmatrix} = \frac{1}{\tilde{X}} \begin{pmatrix} X_3' X_3 & -X_1' X_3 \\ -X_1' X_3 & X_1' X_1 \end{pmatrix} \begin{pmatrix} X_1' \\ X_3' \end{pmatrix} y$$

$$= \begin{pmatrix} \beta_1 \\ \beta_3 \end{pmatrix} + \frac{1}{\tilde{X}} \begin{pmatrix} X_3' X_3 X_1' \eta - X_1' X_3 X_3' \eta \\ X_1' X_1 X_3' \eta - X_1' X_3 X_1' \eta \end{pmatrix}, \tag{9}$$

where we exclude the intercept β_0 for simplicity, $\tilde{X} = X_1' X_1 X_3' X_3 - (X_1' X_3)^2$ and we use the property that X_1, X_2, X_3 are vectors.[1]

Taking the expectation of (9), we observe that both parameters, (β_1, β_3) are biased. I.e. discretization of variables causes a bias in both the coefficients of continuous variables and the discretized variables:

$$E \begin{pmatrix} \hat{\beta}_1 \\ \hat{\beta}_3 \end{pmatrix} = \begin{pmatrix} \beta_1 \\ \beta_3 \end{pmatrix} + \frac{1}{\tilde{X}} \begin{pmatrix} X_3' X_3 E(X_1' \eta) - X_1' X_3 E(X_3' \eta) \\ X_1' X_1 E(X_3' \eta) - X_1' X_3 E(X_1' \eta) \end{pmatrix}$$

$$= \begin{pmatrix} \beta_1 \\ \beta_3 \end{pmatrix} + \frac{1}{\tilde{X}} \begin{pmatrix} -X_1' X_3 E(X_3' \eta) \\ X_1' X_1 E(X_3' \eta) \end{pmatrix}, \tag{10}$$

which holds from the independence of X_1 and η. Thus, we find both estimators to be biased under the transformation.

We next report the consistency properties of $\hat{\beta}_1$, the effect of continuous parameters and $\hat{\beta}_3$, the effect of discretized variables. An estimator $\hat{\beta}$ for a vector of parameters β is consistent if and only if each component is a consistent estimator of the corresponding component of β, i.e. $\text{plim}(\hat{\beta} - \beta) \longrightarrow 0$. For the models in (7)-(8) we have:

$$\text{plim} \left(\hat{\beta} - \beta \right) = \text{plim} \left(\left(\frac{X'X}{n} \right)^{-1} \right) \text{plim} \begin{pmatrix} \frac{X_1' \eta}{n} \\ \frac{X_3' \eta}{n} \end{pmatrix}, \tag{11}$$

---

[1] Similar conclusions with more involved derivations hold when a constant is included and X_1, X_2, X_3 are matrices.

where plim $\left(\frac{1}{n}X_1'\eta\right) = 0$ as they are independent, whereas plim $\left(\frac{1}{n}X_3'\eta\right) \neq 0$ as they are not independent. Thus plim $\left(\hat{\beta} - \beta\right) \neq 0$, and discretization in linear regression leads to biased and inconsistent estimates.

4   Effects of Discretization in Graphical Causal Models

Graphical causal models have been applied to mixed data recently [5,12,16]. However, so far, the literature does not distinguish the effects of applying graphical causal models on mixed data that occurs as a result of discretization versus mixed data occurring as part of the DGP. The former, mixed data as a result of discretization, implies that the true DGP is not part of the possible models in the set considered by graphical causal models since the underlying, unobserved, continuous variable is not in any one of the graphical model structures.

The effects of the above-mentioned model misspecification, observing discretized data, on the obtained results in graphical causal models are more complex than on the linear regression we illustrate in Sect. 3 and even when the causal relations are estimated by a graphical causal model, the causal estimate is not necessarily unique [7,8,10]. Therefore we analyze the effects of discretization in a detailed simulation study, provided in Sect. 4.1.

Estimation of the graphical causal model in Sect. 2 has two steps, the estimation of the causal relations represented in the adjacency matrix in (6) and estimation of the causal effects between variables. Thus a bias in the obtained results can occur as a result of a bias in estimating the causal relations, i.e. the graph structure, or in the estimated causal effects at the second stage. In order to disentangle these two sources of bias, in Sect. 4.2, we first report the effects of discretization on the estimated adjacency matrices. In Sect. 4.3, we additionally report the effects of discretization on the obtained causal effects.

4.1   Simulation Study Setup

The simulated data setting we consider is similar to [1,6]. For each simulation, we consider a setup with $p = 12$ variables, hence 12 nodes in the graphical causal model and generate the adjacency matrix in (6) such that

$$A_{i,j} \sim \text{Bernoulli}(0.18), \tag{12}$$

where i, j denote the ith row and jth column of the adjacency matrix, o.e. the causal effect from variable i to j. Based on the simulated adjacency matrix, we simulate causal effects from a uniform distribution and store these in matrix $\tilde{A}$ where

$$\tilde{A}_{i,j} = \text{U}(0.1, 1) \times A_{i,j}. \tag{13}$$

Finally, we generate the data matrix as follows:

$$X_n^1 \sim NID(0,1), \forall n \tag{14}$$

$$X_n^i = \sum_{j=1}^{i} \tilde{A}_{i,j} X_n^j + \varepsilon_n^i, \ \varepsilon_n^i \sim NID(0,1), \forall n, \ i = 2, \dots, p, \tag{15}$$

where $n = 1, \ldots, N$ denotes observations. Given the data generating process in (12)–(15), variable $X^i = (X_1^i, \ldots, X_N^i)$ is discretized with a probability p such that $X^i = I(X^i > 0)$ where $I()$ is the indicator function as in (3).

In addition to the setup in [6], we generate data from two sample sizes $N = \{200, 1000\}$ to see the effect of sample size on obtained results and two discretization probabilities $p = \{0.2, 0.8\}$ and replicate each simulation replication 100 times to reduce the effect of simulation noise in obtained results. Where available, we also report the standard deviation of reported results across 100 replications as an illustration of simulation uncertainty. Finally, for the DAG estimation, we consider 9 different kernel specifications which are combinations of means $\theta = \{0.3, 1, 3\}$ and standard deviations $\sigma = \{0.0001, 0.01, 0.1\}$ in order to also asses the effect of different kernel specifications. We use $\alpha = 0.05$ for the conditional dependence tests in (5).

4.2   Effects of Discretization on Estimated Adjacency Matrices

In this section, we present the effects of discretization on the first step of graphical causal model estimation, namely the estimation of the adjacency matrix. A graphical representation of the true adjacency matrix and the estimated adjacency matrix in two simulation examples are given in Fig. 1 where $N = 200, p = 0.2, \theta = 0.3, \sigma = 0.001$. Figures 1a and 1b present a simulation where the estimates detect correlations but no causal directions. Figures 1c and 1d, on the other hand, present a simulation case where there are several false negatives, and where causal relations of variables 8 and 10 with the remaining variables are missed. In addition, causality between variables 1 and 2 is estimated in the opposite direction.

We further report the elements of the confusion matrix with respect to three type of relations: relations between continuous variables, relations between discrete variables and relations between continuous and discretized variables. The reported results are averages from 9 kernel specifications in Sect. 4.1 since the adjacency matrix estimates from different kernel specifications are very similar.

We next summarize the effects of discretization on estimated adjacency matrices from all simulations and parameter specifications and compare these estimates to a baseline where there is no discretization, i.e. $p = 0$ in (15). Table 1 presents the elements of the confusion matrix, namely true positives (TP), false positives (FP), false negatives (FN) and true negatives (TN) obtained from the graphical model estimates of mixed discretized data and continuous data for different simulation settings. Note that, in this paper we consider these measures in relation to counts of causal relations and causal relation counts are calculated as the difference between the true adjacency relations $A_{i,j}$ in (12) and the estimated adjacency relations $\hat{A}_{i,j}$ in (6):

$$\text{TP} = \sum_{i=1}^{p}\sum_{j\neq i} I(\hat{A}_{i,j} = 1, A_{i,j} = 1), \quad \text{FP} = \sum_{i=1}^{p}\sum_{j\neq i} I(\hat{A}_{i,j} = 1, A_{i,j} = 0),$$

$$\text{TN} = \sum_{i=1}^{p}\sum_{j\neq i} I(\hat{A}_{i,j} = 0, A_{i,j} = 0), \quad \text{FN} = \sum_{i=1}^{p}\sum_{j\neq i} I(\hat{A}_{i,j} = 0, A_{i,j} = 1),$$

$$(16)$$

where $I()$ is the indicator function as in (3).

Table 1. Average confusion matrices for mixed data for causal effects between all (all), continuous (cts), discretized (dis) and mixed variables and baseline without discretized variables. Standard deviations in 100 simulations are in parentheses.

Continuous data (baseline)								
	$N = 200$				$N = 1000$			
var. type	TN	FP	FN	TP	TN	FP	FN	TP
all	1043.82	30.87	36.99	76.32	1044.36	13.95	36.45	93.24
	(29.78)	(18.98)	(14.07)	(20.12)	(29.08)	(14.09)	(15.74)	(20.65)
Mixed data $p = 0.2$								
	$N = 200$				$N = 1000$			
var. type	TN	FP	FN	TP	TN	FP	FN	TP
all	964.47	90.33	116.34	16.86	615.99	58.37	464.82	48.82
	(36.81)	(22.11)	(24.21)	(8.63)	(33.95)	(16.74)	(43.05)	(14.87)
cts	658.89	60.90	79.29	10.92	412.95	38.96	324.51	33.58
	(37.38)	(18.28)	(28.48)	(7.86)	(63.08)	(14.54)	(65.11)	(14.80)
dis	14.13	1.53	2.07	0.27	9.51	0.60	6.87	1.02
	(4.83)	(3.18)	(3.43)	(0.96)	(6.06)	(1.76)	(5.35)	(2.49)
mixed	291.45	27.90	34.98	5.67	193.53	18.81	133.44	14.22
	(27.88)	(12.80)	(21.49)	(5.70)	(50.39)	(11.10)	(51.28)	(10.05)
Mixed data $p = 0.8$								
	$N = 200$				$N = 1000$			
var. type	TN	FP	FN	TP	TN	FP	FN	TP
all	1009.39	94.56	71.42	12.63	998.09	84.15	82.72	23.04
	(26.71)	(21.62)	(9.28)	(9.56)	(30.20)	(22.53)	(12.05)	(10.75)
cts.	89.11	8.92	8.45	1.52	88.08	8.43	9.48	2.01
	(13.56)	(6.81)	(11.71)	(3.09)	(13.97)	(7.62)	(10.96)	(3.92)
dis.	426.09	43.67	28.95	5.29	427.47	35.10	31.89	9.54
	(26.22)	(15.29)	(19.98)	(6.01)	(26.92)	(14.60)	(18.00)	(7.97)
mixed	494.19	41.97	34.02	5.82	482.54	40.62	41.35	11.49
	(24.15)	(16.52)	(13.93)	(6.82)	(23.03)	(14.99)	(14.04)	(8.38)

Table 1 shows that both TN and TP decrease with mixed data compared to the baseline. The discretization deteriorates all estimated relationships: relationships between continuous variables, discretized variables and between continuous and discretized variables. The effect of discretization, particularly in terms of true positives, is mitigated with the relatively larger sample size $N = 1000$, but even in this case true positives for all variable types in mixed data are smaller than the baseline. Hence, unlike the linear regression illustration in Sect. 3, we do not find

Table 2. TP, FP, FN, TP rates for mixed data for causal effects between all (all), continuous (cts), discretized (dis) and mixed variables and baseline without discretized variables.

Continuous data (baseline)							
	$N = 200$				$N = 1000$			
var. type	TNR	FPR	FNR	TPR	TNR	FPR	FNR	TPR
all	0.97	0.03	0.33	0.67	0.99	0.01	0.28	0.72
Mixed data $p = 0.2$							
	$N = 200$				$N = 1000$			
var. type	TNR	FPR	FNR	TPR	TNR	FPR	FNR	TPR
all	0.91	0.09	0.87	0.13	0.91	0.09	0.90	0.10
cts	0.92	0.08	0.88	0.12	0.91	0.09	0.91	0.09
dis	0.90	0.10	0.88	0.12	0.94	0.06	0.87	0.13
mixed	0.91	0.09	0.86	0.14	0.91	0.09	0.90	0.10
Mixed data $p = 0.8$							
	$N = 200$				$N = 1000$			
var. type	TNR	FPR	FNR	TPR	TNR	FPR	FNR	TPR
all	0.91	0.09	0.85	0.15	0.92	0.08	0.78	0.22
cts	0.91	0.09	0.85	0.15	0.91	0.09	0.83	0.17
dis	0.91	0.09	0.85	0.15	0.92	0.08	0.77	0.23
mixed	0.92	0.08	0.85	0.15	0.92	0.08	0.78	0.22

an evidence towards consistency for continuous variables. In all comparisons, the standard errors calculated from 100 simulation replications are relatively small, indicating that these findings are robust to simulation noise.

Table 2 presents confusion matrix elements for each simulation experiment as TN, FN, TP, FP rates in order to be able to compare the estimated relations between variables. From a statistical perspective, obtaining true positives is relatively more important in estimating causal relations than obtaining true negatives. Failing in the former by false negatives, missing existing relations, indicates an omitted variable bias. Failing in the latter by false positives, including redundant variables, is generally less problematic as the estimated effect of redundant variables in a statistical model will theoretically be zero. Table 2 shows that, even in the baseline model, true positive rates are much lower than true negative rates, hence the estimated models are likely to have omitted variables bias rather than redundant variables. True positive rates decrease substantially for mixed data compared to the baseline. The higher percentage of discretized variables, $p = 0.8$, leads to a higher deterioration in true positive rates compared to $p = 0.2$. Similar to the conclusions in Table 1, estimated causal relations between discretized, continous as well as mixed variables all deteriorate with discretization.

4.3   Effects of Discretization on Estimated Causal Effects

In this section, we present the effects of discretization on the estimated causal effects as differences between the true causal effects in (13) and estimated causal effects. Recall that causal estimates are not unique in graphical causal models. This could happen e.g. when the causal directions cannot be estimated as in Fig. 1b and each possibility of causality should be considered. The conventional method is to consider the minimum and maximum values of causal relations based on the graph estimates, and report the lower and upper bounds of these estimates [7,8]. For this application, we report the best estimates defined as the estimates that lead to the smallest bias in each simulation.

Table 3. Bias and uncertainty in estimated causal effects

Continuous data (baseline)								
θ	0.3			1			3		
σ	0.001	0.01	0.1	0.001	0.01	0.1	0.001	0.01	0.1
$N = 200$	0.043	0.043	0.043	0.042	0.042	0.042	0.042	0.042	0.042
	(0.012)	(0.012)	(0.012)	(0.012)	(0.012)	(0.012)	(0.011)	(0.011)	(0.011)
$N = 1000$	0.042	0.042	0.042	0.041	0.041	0.042	0.042	0.042	0.042
	(0.012)	(0.012)	(0.012)	(0.011)	(0.011)	(0.011)	(0.011)	(0.011)	(0.011)
Mixed data $p = 0.2$									
θ	0.3			1			3		
σ	0.001	0.01	0.1	0.001	0.01	0.1	0.001	0.01	0.1
$N = 200$	0.058	0.058	0.057	0.052	0.051	0.051	0.045	0.045	0.045
	(0.015)	(0.015)	(0.015)	(0.013)	(0.013)	(0.013)	(0.012)	(0.012)	(0.012)
$N = 1000$	0.051	0.051	0.051	0.049	0.049	0.049	0.044	0.044	0.044
	(0.012)	(0.012)	(0.012)	(0.012)	(0.012)	(0.012)	(0.012)	(0.012)	(0.012)
Mixed data $p = 0.8$									
θ	0.3			1			3		
σ	0.001	0.01	0.1	0.001	0.01	0.1	0.001	0.01	0.1
$N = 200$	0.064	0.064	0.064	0.055	0.054	0.054	0.044	0.044	0.044
	(0.012)	(0.012)	(0.012)	(0.012)	(0.012)	(0.012)	(0.012)	(0.012)	(0.012)
$N = 1000$	0.050	0.049	0.049	0.046	0.046	0.046	0.044	0.044	0.044
	(0.012)	(0.012)	(0.012)	(0.012)	(0.012)	(0.012)	(0.012)	(0.012)	(0.012)

Table 3 presents the minimum bias and standard deviations of minimum bias across 100 simulation replications for each simulation case, $N \in \{200, 1000\}$, $p \in \{0.2, 0.8\}$ and the baseline $p = 0$. We find that the estimated bias and standard deviations are similar across discretized and continuous variables in these simulations. Therefore, as an additional sensitivity analysis, we report the bias results with respect to the kernel parameters (θ, σ) in Table 3.

The comparison of different sample sizes, $N \in \{200, 1000\}$ in Table 3 indicates that sample size is important in the obtained bias. The overall bias in the causal estimates decrease slightly when the sample size increases in all three settings: mixed data case with probabilities of discretized data $p = 0.2$ and $p = 0.8$ as well as continuous data case. The biggest bias reduction with an increased sample size occurs for $p = 0.8$, i.e. when the probability of discretized variables is highest. Comparing different sample sizes and standard deviations from 100 simulations (in parentheses), we do not find a clear uncertainty effect from increasing the number of simulations. This finding is not in line with the classical bias-variance trade-off in statistical models. We conjecture that the stability of the standard deviations across simulation cases and sample sizes can be a result of the non-linearity and non-uniqueness of the graphical causal models, as well as the fact that we report standard deviations between simulations, but not the standard errors for each causal estimation. Obtaining the latter, standard errors for each simulation estimation is not straightforward as they need to be chosen from a non-unique set, as explained in Sect. 4.2.

Table 3 also presents the effects of kernel parameters on obtained biases. For the continuous data case, obtained biases are marginally different for different kernel parameters. For mixed data, however, taking a small θ in both mixed data applications lead to a higher average bias for $N = 200$. For the larger sample size, the differences between these biases are less pronounced. In addition, the kernel standard deviation parameter σ does not seem to affect the obtained results. Thus we conclude that a second important factor for the obtained bias is the kernel parameter θ in case of mixed data, $p \in \{0.2, 0.9\}$, and the relatively small sample size, $N = 200$.

5 Conclusion and Future Work

In this paper we analyze the implications of variable discretization on the estimates of graphical causal model. Since these effects are theoretically complex to analyze, we use a simulation study under different sample size and discretization specifications. We find that discretization affects graphical causal model estimates in several ways. First, the estimated relations in terms of confusion matrices deteriorate with discretization. This deterioration is most visible for the obtained false negative relations, which potentially cause an omitted variable bias and when relatively more variables in the model are discretized. Second, we find that the obtained causal relations have a higher bias compared to the continuous data case and this increase in bias is more visible when the sample size is small and more variables are discretized. Third, particularly for small samples, we find that the parameters of the kernel density in graphical causal models affect the bias deterioration in mixed data compared to continuous data. As future work, we want to expand our study on the effects of variable discretization using other estimation methods for graphical causal models. Furthermore, we want to investigate possible improvements in estimation using interval regression methods.

References

1. Almeida, R.J., Adriaans, G., Shapovalova, Y.: Graphical causal models and imputing missing data: a preliminary study. In: Lesot, M.-J., Vieira, S., Reformat, M.Z., Carvalho, J.P., Wilbik, A., Bouchon-Meunier, B., Yager, R.R. (eds.) IPMU 2020. CCIS, vol. 1237, pp. 485–496. Springer, Cham (2020). https://doi.org/10.1007/978-3-030-50146-4_36
2. Barnwell-Ménard, J.L., Li, Q., Cohen, A.A.: Effects of categorization method, regression type, and variable distribution on the inflation of type-i error rate when categorizing a confounding variable. Stat. Med. **34**(6), 936–949 (2015)
3. Cornelisz, I., Cuijpers, P., Donker, T., van Klaveren, C.: Addressing missing data in randomized clinical trials: a causal inference perspective. PloS one **15**(7), e0234349 (2020)
4. Cui, R., Groot, P., Heskes, T.: Learning causal structure from mixed data with missing values using Gaussian copula models. Stat. Comput. **29**(2), 311–333 (2018). https://doi.org/10.1007/s11222-018-9810-x
5. Handhayani, T., Cussens, J.: Kernel-based approach to handle mixed data for inferring causal graphs. arXiv preprint arXiv:1910.03055 (2019)
6. Kalisch, M., Bühlman, P.: Estimating high-dimensional directed acyclic graphs with the pc-algorithm. J. Mach. Learn. Res. **8**(3), 613–636 (2007)
7. Kalisch, M., Mächler, M., Colombo, D., Maathuis, M.H., Bühlmann, P.: Causal inference using graphical models with the R package pcalg. J. Stat. Softw. **47**(11), 1–26 (2012)
8. Maathuis, M.H., Kalisch, M., Bühlmann, P.: Estimating high-dimensional intervention effects from observational data. Ann. Stat. **37**(6A), 3133–3164 (2009)
9. Maxwell, S.E., Delaney, H.D.: Bivariate median splits and spurious statistical significance. Psychol. Bull. **113**(1), 181 (1993)
10. Pearl, J., Verma, T.S.: A statistical semantics for causation. Stat. Comput. **2**(2), 91–95 (1992)
11. Rohrer, J.M.: Thinking clearly about correlations and causation: graphical causal models for observational data. Adv. Methods Pract. Psychol. Sci. **1**(1), 27–42 (2018)
12. Sokolova, E., Groot, P., Claassen, T., von Rhein, D., Buitelaar, J., Heskes, T.: Causal discovery from medical data: dealing with missing values and a mixture of discrete and continuous data. In: Holmes, J.H., Bellazzi, R., Sacchi, L., Peek, N. (eds.) AIME 2015. LNCS (LNAI), vol. 9105, pp. 177–181. Springer, Cham (2015). https://doi.org/10.1007/978-3-319-19551-3_23
13. Spirtes, P., Glymour, C.: An algorithm for fast recovery of sparse causal graphs. Social Sci. Comput. Rev. **9**(1), 62–72 (1991)
14. Spirtes, P., Glymour, C.N., Scheines, R., Heckerman, D.: Causation, Prediction, and Search. MIT press, Cambridge (2000)
15. Thoresen, M.: Spurious interaction as a result of categorization. BMC Med. Res. Methodol. **19**(1), 1–8 (2019)
16. Zhong, W., et al.: Inferring regulatory networks from mixed observational data using directed acyclic graphs. Front. Genet. **11**, 8 (2020)

Population and Individual Level Meal Response Patterns in Continuous Glucose Data

Danilo Ferreira de Carvalho[1]($\boxtimes$) (iD), Uzay Kaymak[1] (iD), Pieter Van Gorp[2] (iD), and Natal van Riel[3] (iD)

[1] Jheronimus Academy of Data Science, Eindhoven University of Technology, Eindhoven, The Netherlands
d.ferreira.de.carvalho@tue.nl, u.kaymak@ieee.org
[2] Industrial Engineering and Innovation Sciences, Eindhoven University of Technology, Eindhoven, The Netherlands
p.m.e.v.gorp@tue.nl
[3] Department of Biomedical Engineering, Eindhoven University of Technology, Eindhoven, The Netherlands
n.a.w.v.riel@tue.nl

Abstract. Diabetes research has changed with the introduction of wearables that are able to continuously collect physiological data (*e.g.*, blood glucose levels), which has allowed for data-driven solutions. In this context, patients are still expected to self-record events tied to their daily routines (*e.g.*, meals). Since self-recording is prone to errors, automatic detection of meal events could improve the quality of event data and reduce registration burden. In this paper, we investigate the feasibility of meal detection from continuous glucose data by using population level data compared to individual data. We discuss the advantages and disadvantages of both approaches based on a method to identify patterns in time series that can be used to map the characteristics of a glucose signal response to a meal event. Event responses, *i.e.*, subsequences that come right after a recorded event, are identified and fuzzy clustering is used to group different types of them. Our results indicate that both population and individual data give comparable results, which suggests that both could be used interchangeably to develop event identification models.

Keywords: Meal detection · Pattern identification · Fuzzy clustering · Distance profile · Continuous glucose data

1 Introduction

Diabetes is a chronic, lifelong disease that affects millions of individuals [12]. A person with diabetes cannot produce insulin, and for that his/her blood glucose (BG) must be artificially managed using an insulin pump or injections. Many factors play a role with regard to BG variation, in such scenario meals are

© Springer Nature Switzerland AG 2022
D. Ciucci et al. (Eds.): IPMU 2022, CCIS 1602, pp. 235–247, 2022.
https://doi.org/10.1007/978-3-031-08974-9_19

key [18]. This scenario led many researchers to develop solutions able to mitigate the burden that the condition brings to each associated person.

To allow studies to move forward to solutions, data gathered from patients are crucial. Furthermore, data collection becomes a core part of the current daily routine of diabetic patients that aim at providing their own data for such cause. This fact has a big impact not only on people's personal lives, but also on the research efforts done regarding such serious and severe condition [9]. Data generated by patients (*e.g.*, diet, exercises, blood glucose) – if captured by wearables, or self-logged – can serve as basis for data-driven approaches [2,10].

Many approaches try to reduce the burden of asking patients to constantly inform meal events – and the intrinsic error associated to such action – by detecting meals using, by instance, glucose rate of change [5]. This is also a relevant aspect when improving artificial pancreas applications and diabetes related simulators [16,26]. Expanding this scenario to data gathering of events within free-living conditions, other approaches focus on mitigating imperfections (*e.g.*, lack of data) by developing data-driven models able to infer missing events (including meals) based on previously recorded data [8,20].

The work presented in this paper explores the feasibility of identifying patterns in patient gathered glucose data (via continuous glucose monitoring (CGM) devices) that would facilitate the identification of meal events. Different from the aforementioned approaches, which identify direct variations on the incoming BG levels, our method aims at extracting meal response patterns and use them when searching for meal occurrences. Such patterns can be made out of data in a population or individual level, *i.e.*, clustering data generated by a whole group of patients or by a single one. For each level, the existing meal responses can differ, affecting the patterns/clusters found and the meal events to be detected through them. In this paper, we discuss the applicability of such method, as well as the opportunities that each level brings.

The paper is presented as follows. Preliminary concepts are covered in Sect. 2. Details of the proposed method are presented in Sect. 3. Experimental setup and results are presented in Sects. 4 and 5. Finally, conclusions are given in Sect. 6.

2   Preliminaries

Our approach explores CGM data, and in our context such data is taken as a time series to be analyzed together with the associated recorded meal events. In addition, for discovering patterns within the CGM time series, we utilize concepts tied to *shapelets* and *distance profile*.

Definition 1 (Time Series). *A time series $T = t_1, t_2, \ldots, t_m$ is a sequence of m real-valued numbers (usually in temporal order) separated equally.*

In the time series domain, there is a tendency to be interested on the global properties of a series. However, for different problems local properties are more relevant, and specific pieces of a time series T (subsequences) must be handled as the most relevant parts of a bigger time series.

Definition 2 (Subsequence). *A subsequence $T_{i,n}$ of a given time series T of length m is a continuous subset of length $n \leq m$ composed by contiguous positions from T, that is, $T_{i,n} = t_i, t_{i+1}, \ldots, t_{i+n-1}$, where $1 \leq i \leq m - n + 1$.*

For instance, in the classification domain, *shapelets* [24], can be seen as a representation of a class to be identified within a time series. Depending on the level of representation needed, *shapelets* can improve the performance of the commonly used nearest neighbor classification algorithms [7], by reducing the amount of space and time needed during searches. This is due to the fact that each comparison made by the algorithm can rely on just a small subsection of the shapes being searched and compared.

In our paper, the concept behind *shapelets* suits well the problem tackled, as we try to identify response patterns, *i.e.*, specific subsequences, in blood glucose level time series collected from diabetic patients. Such response patterns are by definition limited, as we assume that a response to an event has a pre-determined length, and can be associated to a matching subsequence of equal length. However, such subsequences can carry enough properties for the method to discern if a response to an event happened or not within a time interval.

To query a subsequence – in our case, a pattern – over a time series T, a different representation of T using the information regarding distances to a subsequence Q being queried can be used to identify similarities.

Definition 3 (Distance Profile). *A vector D formed by all Euclidean distances between a given subsequence Q, taken as the query, and all the possible windows of the same length of Q obtained by sliding across a time series T. As an outcome from a sliding window, its cardinality is $|D| = |T| - |Q| + 1$.*

An important remark is that the distance values set in D are the Euclidean distance between the z-normalized subsequences, *i.e.*, both compared subsequence instances are individually normalized, having mean equal to zero, and standard deviation equal to one. This is a transformation that mainly aims at identifying the similarities in the behavior of the compared series (subsequences) while keeping computation efficiency [1].

To compute the distance profile of a time series T with regard to a query Q, the MASS algorithm is used [23]. Such algorithm became an emergent solution for similarity based searches in time series and pattern mining tasks due to its optimized performance [27].

3 Identifying Patterns in CGM Time Series

In this paper, we aim at identifying patterns of blood glucose (BG) responses to daily-living meal events, such as breakfast, lunch, dinner, snacks, and hypo-correction (*e.g.*, sugary drinks like juice or regular soda), and use such patterns to spot similar occurrences in BG signals from CGM data.

For that, a time interval/window size Δ is defined to be used as the length of the responses to the events, *i.e.*, a subsequence of size Δ that comes right after a reported meal event. It is worth mentioning that the amount of responses is equal to the amount of meal events that have a response of length Δ, thus all considered responses can vary in form, but all have the same length.

3.1   Clustering Responses to Establish Patterns

To group response subsequences that hold the same properties, and furthermore shapes, clustering algorithms can be used. When applied to time series, the idea is to find a representation at a lower dimensionality that preserves the original information and describes the original shape of the time series data as closely as possible [21].

A common approach regarding time series clustering is the direct comparison of each series or a (transformed) version of it that suits the problem in hand. Such direct comparison can be done by using a distance metric, commonly Euclidean or a variation of it, and this is particularly useful when dealing with not very long time series [4]. Here, the z-normalized version of the subsequences are used by the distance metric.

To look for similarities (patterns) among the responses found, we perform a clustering step using soft clustering (Fuzzy C-Means [3]) applied to the gathered responses. The center – the mean of each cluster member – of each defined cluster is here used as a pattern, as we assume the cluster center reflects the properties of its associated clustered subsequences. Thus, there will be as many patterns as defined clusters.

The decision for soft clustering lies on the degree of membership that such technique associates to each cluster member. Responses in a BG signal could match different clusters/patterns in certain degree, and this would allow for better insights over the created patterns and their nuances.

3.2   Looking for Matches

With the similar responses separated in clusters, and then the patterns defined, a search is performed in each *testing* day's time series to find the best matches when querying each cluster center. This is done by generating the distance profile [22] of the time series and looking for the smallest distance found. MASS algorithm is used in this step: the queries are the found cluster centers, and the search space (time series) is a set containing all days in the *training* subset.

Definition 4 (Match Index). *Given a computed distance profile D of a time series T regarding a query Q, the index $1 \leq i \leq |T| - |Q|$ from a value $t_i \in T$ associated to the smallest computed distance in D is the index of a match.*

Note that, if i is the index of a match, the subsequence that represents the entire match is $T_{i,|Q|}$, as all compared subsequences have the same length as Q.

The pseudocode in Algorithm 1 presents how the matches are found for each of the days in the *testing* subset. Operations 1, 2, and 3 can be seen as a setup for the matching step. Before iterating over all days, *responses* and *patterns* are already known. By using each pattern as the query Q (Operation 8), we use MASS algorithm to compute the distance profiles regarding Q for each of the days (Operation 9). The index (from T) associated to the smallest distance value computed in D, $i.e.$, the match index, is stored in the 2-D array containing match indices for each day and pattern.

Algorithm 1. Pattern Matching

1: $responses \leftarrow GetResponses(T_{train})$
2: $patterns \leftarrow CreateClusters(responses)$
3: $T_{days} \leftarrow SplitInDays(T_{test})$
4: $M_{days,patterns} \leftarrow$ 2-D array of shape $(len(T_{days}), len(patterns))$
5: **for** $d_i = 1, 2, \ldots, len(T_{days})$ **do**
6:     $T \leftarrow T_{days}[d_i]$
7:     **for** $p_i = 1, 2, \ldots, len(patterns)$ **do**
8:         $Q \leftarrow patterns[p_i]$
9:         $D \leftarrow MASS(Q, T)$
10:         $M_{days,patterns}[d_i][p_i] \leftarrow$ match index from D
11:     **end for**
12: **end for**

An important aspect of Definition 4 is that, although it specifies that the match index is tied to the smallest distance, Algorithm 1 can be used to find the nth smallest match with little modifications. This is possible because the distance profile is a representation of the whole time series T used, thus sorting indices by distance and acquiring the nth index becomes a trivial task. This aspect is particularly useful in the scope of this paper.

4   Experimental Setup

The dataset we use in our experiments is from the $OhioT1DM$ [17], a dataset developed and made publicly available[1] to promote and facilitate research in blood glucose level prediction. Used by many previous researches focusing on diabetes related modeling [10], it contains 12 participants (Type-I diabetic patients), and, for each of them, data from continuous glucose monitoring (measured every 5 min), and also from daily reported events. In this paper, we focus only on meal events, such as breakfast, lunch, dinner, snacks, and hypo-correction. The splitting rules for *training* and *testing* subsets defined by the authors of the dataset were respected. This way, we allow replicability of methods and unbiased comparison of developed models based on such dataset.

Using the *training* subset, BG responses for each existing meal event are collected from the Continuous Glucose Monitoring (CGM) signal. A response interval/window size $\Delta = 2\,h$ (or 24 data points) is used for the experiments.

---

[1] smarthealth.cs.ohio.edu/OhioT1DM-dataset.html.

Table 1. Meal events data used for clustering (*training*), and validation (*testing*).

	Participant ID	540	544	552	559	563	567	570	575	584	588	591	596
Training	Total number of events	73	159	78	150	129	32	136	243	95	221	212	265
	Incomplete BG responses	13	32	14	29	15	11	9	63	18	13	50	82
	Used for clustering	60	127	64	121	114	21	127	180	77	208	162	183
Testing	Total number of events	27	38	21	29	27	0	33	45	23	37	41	54
	Incomplete BG responses	7	7	7	6	4	-	2	16	3	4	6	10
	Used for validation	20	31	14	23	23	-	31	29	20	33	35	44

*Patient 567 has no logged meal events on the *testing* subset.

The information regarding the amount of responses per patient is presented in Table 1. A response is only taken into account if it has the same – and required – amount of data points defined by Δ. This is due to the fact that the similarity between responses is measured by an Euclidean distance based metric, and for that the same amount of data points becomes a requirement [15].

As previously mentioned, for the clustering step using Fuzzy C-Means, the z-normalized version of the responses are clustered. The fuzzifier parameter value in our experimental setup is set to $m = 2$, a value commonly adopted by many researchers when using fuzzy C-Means to incorporate fuzzyness while keeping a decent cluster separation [11]. There are methods that can be applied when looking for optimal values for such parameter [19,25], however, this is not part of the scope and approach described in our work. We vary the amount of clusters from 2 to 8, and then visually inspect the resulting groups. With this, we aim at identifying different core patterns (*e.g.*, ASCENDING, ASCENDING-DESCENDING, DESCENDING-ASCENDING, and DESCENDING). The intention here is to maintain a notable degree of difference between shapes to avoid the search for similar patterns (cluster centers) that would lead to similar – or even the same – matches.

When referring to clusters/patterns generated by using meal responses, two different clustering models will be used and referenced throughout the rest of the paper: *Population*, using meal responses from all patients, *i.e.*, one same model for all participants; and *Individual*, using meal responses per-participant, *i.e.*, one specific model for each participant.

It is worth adding that the number of clusters used still can impact the type of patterns found, specially when dealing with *Individual* models handling very peculiar data. However, this is not in the scope of the proposed study, and thus it is not directly explored in our results.

The implementation of our experiments make use of the fuzzy-c-means[2] and stumpy[3] packages [6,14], both open source projects regarding FCM and distance profile (MASS algorithm) implementations, respectively.

---

[2] http://github.com/omadson/fuzzy-c-means.
[3] http://github.com/TDAmeritrade/stumpy.

5   Results

By making use of the data described in Sect. 4, *Population* and *Individual* models/patterns were defined, and their outcomes are explored in this section.

5.1   Response Patterns

Figure 1 presents clusters associated to the *Population* model. On the 3-clusters scenario (Fig. 1a), the centers of Clusters 0 and 1 fit the ASCENDING and DESCENDING behaviors, respectively, while Cluster 2's has an ASCENDING-DESCENDING mixed behavior. The mixed behavior in Cluster 2's center led us to consider increasing the amount of clusters, in an attempt to split such found pattern, and the result is shown in Fig. 1b. The ASCENDING and DESCENDING behaviors can still be identified (Clusters 2 and 3). However, the mixed one is now split in two centers that share too many similarities, which would lead to a high possibility of similar matches, and this effect escalates over the additional clusters.

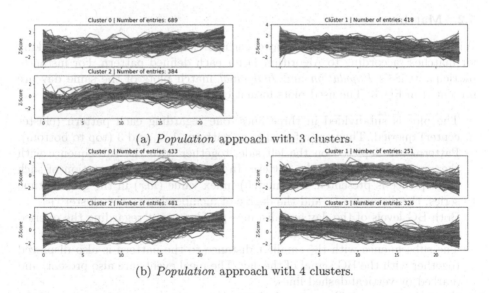

(a) *Population* approach with 3 clusters.

(b) *Population* approach with 4 clusters.

Fig. 1. *Population* approach: clusters of meal responses of all participants. Centers are drawn using strong line and star markers.

The previously used reasoning applies when clustering the meal responses individually, for instance Fig. 2 depicts the clusters generated from the meal responses of the sample patient 588, again for 3- and 4-clusters scenarios. From this, we took 3 clusters as the value for modeling on both *Population* and *Individual* approaches, as it captures enough relevant patterns for our experiments.

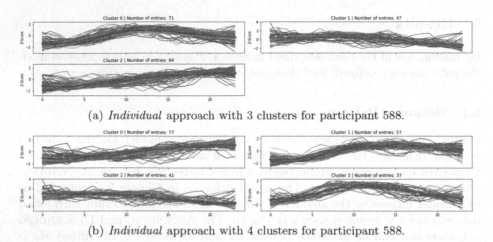

(a) *Individual* approach with 3 clusters for participant 588.

(b) *Individual* approach with 4 clusters for participant 588.

Fig. 2. *Individual* approach: clusters of meal responses for participant 588. Centers are drawn using strong line and star markers.

5.2 Matches

With the patterns in hands, each day of each participant is then scanned, looking for matches (according to Algorithm 1) for each defined pattern. For instance, participant 588's *Population* and *Individual* match plots of the same day are presented in Fig. 3. The used plots format can be described as follows:

– The plot is subdivided in three lines, one regarding each pattern (cluster center) queried. They are referred to as patterns 1, 2, and 3 (top to bottom).
– Patterns are depicted on the left side, together with the subsequence with the smallest distance to the pattern. In addition, extra information of such subsequence is presented: the (match) index value (idx) in the glucose time series, the distance (d), and the degree of membership to the cluster (u).
– Both BG levels of the day and distance profile output regarding the queried pattern are placed on the right.
– The subsequence with the smallest distance to the pattern is also displayed together with the BG signal of the day. The meal events are also present, and marked by vertical dashed lines.
– On the distance profile, we mark the top 10 match indices (the indices of the 10 subsequences with the smallest distances) with triangle shaped markers.

By comparing Fig. 3a and 3b, one can note that for both models, all meal events were spotted by at least one of the top 10 matches of each pattern. To exemplify, we can take the meal events one by one: the one between 06:00 and 09:00 is caught by pattern 3, pattern 1 spots the one close to 12:00, the one a bit after 18:00 by 3, and, finally, the one around 21:00 by pattern 2. Note that, pattern 2 is different from *Population* to *Individual*, which brings up different matches. Depending on the participant, such aspect becomes more apparent.

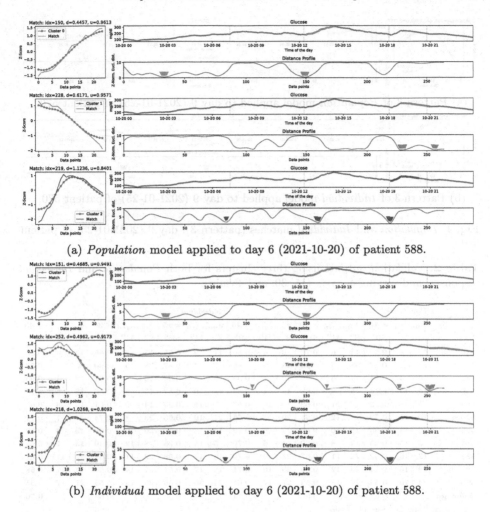

(a) *Population* model applied to day 6 (2021-10-20) of patient 588.

(b) *Individual* model applied to day 6 (2021-10-20) of patient 588.

Fig. 3. *Population* and *Individual* matches: day 6 (2021-10-20) of participant 588.

Figure 4 depicts the results for the participant 570 with focus on pattern 3. The *Individual* pattern 3 matches (in Fig. 4b) were able to spot the event that happened before 12:00, while the *Population* (Fig. 4a) version of the same pattern could not. Individual responses tend to give a more specialized version of the patterns, and this is reflected in the distance profile, as it will give even lower distance values, facilitating the identification of possible matches.

In Table 2, we present validation results. We consider a success when any of the found matches is placed within a success range taking as reference the ground truth, *i.e.*, if the event happened at time t, and a success margin δ_{margin} is used, the match must be within the interval $[t - \delta_{margin}, t + \delta_{margin}]$, where δ_{margin} is set in minutes. Relevant works have shown that there is an expected delay when dealing with meal responses associated to CGM data [5,26], and,

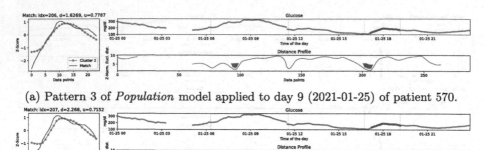

(a) Pattern 3 of *Population* model applied to day 9 (2021-01-25) of patient 570.

(b) Pattern 3 of *Individual* model applied to day 9 (2021-01-25) of patient 570.

Fig. 4. *Population* and *Individual* matches (pattern 3): day 9 (2021-01-25) of patient 570.

Table 2. Validation outputs using three clusters for both *Population* and *Individual* responses, and varying δ_{margin} (15, 30, and 45 min).

	Population									Individual								
	$\delta_{margin} = 15$			$\delta_{margin} = 30$			$\delta_{margin} = 45$			$\delta_{margin} = 15$			$\delta_{margin} = 30$			$\delta_{margin} = 45$		
ID	S[a]	M[b]	SR[c]	S[a]	M[b]	SR[c]	S[a]	M[b]	SR[c]	S[a]	M[b]	SR[c]	S[a]	M[b]	SR[c]	S[a]	M[b]	SR[c]
540	12	8	0.60	14	6	0.70	16	4	0.80	11	9	0.55	13	7	0.65	14	4	0.80
544	27	4	0.87	28	3	0.90	28	3	0.90	27	4	0.87	28	3	0.90	30	1	0.97
552	10	4	0.71	11	3	0.79	13	1	0.93	9	5	0.64	10	4	0.71	13	1	0.93
559	11	12	0.48	15	8	0.65	19	4	0.83	13	10	0.57	15	8	0.65	18	5	0.78
563	5	18	0.22	12	11	0.52	13	10	0.57	8	15	0.35	14	9	0.61	15	8	0.65
570	16	15	0.52	19	12	0.61	20	11	0.65	21	10	0.68	23	8	0.74	24	7	0.77
575	16	13	0.55	19	10	0.65	23	6	0.79	14	15	0.48	20	9	0.69	23	6	0.79
584	5	15	0.25	12	8	0.60	14	6	0.70	5	15	0.25	8	12	0.40	13	7	0.65
588	17	16	0.52	22	11	0.67	23	10	0.70	19	14	0.58	22	11	0.67	22	11	0.67
591	19	16	0.54	23	12	0.66	25	10	0.71	18	17	0.51	21	14	0.60	24	11	0.68
596	24	20	0.55	26	18	0.59	27	17	0.61	24	20	0.55	26	18	0.59	27	17	0.61
Average			0.54			0.67			0.74			0.55			0.66			0.76

[a] Successes, [b] Misses, [c] Success ratio.

based on such information, we used similar values for the margins: 15, 30, and 45 min.

In general, from the values displayed, there is no significant difference between the average success rate when using both *Population* or *Individual* patterns. However, specific participants (*e.g.*, 563, 570) benefit from the individualized use of patterns, being able to successfully spot more meal events.

Moreover, analyzing the success margin (δ_{margin}) used, it is worth noting that there is an expected increase of the average success rate when using a larger margin. However, moving from 15 min to 30, has a higher gain for most of the patients when compared to the change from 30 to 45 min.

6   Conclusion

This work explores a data-driven way of identifying meal events from glucose (CGM) data. For that, we consider the use of two different approaches: patterns from a population of patients, and patterns created individually. In addition, we explore three different delay values during our search for meal events.

As shown in our results, we cannot point any significant success difference between the use of population and individual patterns. Although individual patterns created through clustering are expected to reflect better the nuances of the responses of a person, they demand a large initial set of data from such individual. On the other hand, creating population patterns are easier by nature, as the required data can be collected from multiple people. From this perspective, an existing population model might fit a person with no prior data. There are also participants with very peculiar responses, not fitting well a single general population model. As a future work, we intend to investigate a middle-ground solution: multiple population models, each associated to a specific category/profile.

Alternative distance metrics can be used in order to define the response patterns, which might impact the resulting types of patterns found. This might be also true when finding matches, as the match indices might vary according to different applied distance metrics. To investigate the impact of the aforementioned aspect, a more extensive study over the associated steps is intended to be performed, comparing the results of clustering the BG response time series using different techniques [13]. The comparison can fit well to the scope, steering the next steps of the presented work to new findings, validation steps, and more suitable approaches and analysis.

The approach also opens possibilities regarding the use of the found matches to identify event response windows. As a necessary next step, we should select among the matches found which ones can be labeled as meals. The use of specific events during training, and more narrow search spaces – guided by likelihood [8] – when validating, could also steer the results to a smaller set of matches. With the list of matches in hands, it now turns into a classification problem that makes use of the matches and their features as input for a decision step.

Another important point is the decision to use soft clustering, which is associated to the possibilities this type of clustering technique can offer to further research steps. As there is no optimal number of patterns, it is a difficult task to specify the correct number of clusters to be used. Instead of having crisp established groups, fuzzy ones allow for the use of membership values associated to each cluster member, providing a tool for the identification of members that could belong to multiple clusters, i.e., members that would have influence on the definition of different patterns. The degree of membership also leaves opportunities to: distinguish labelled meal matches from different patterns, or associate a person to a specific patterns profile.

Finally, we focused on creating patterns from responses with pre-defined length, all coming right after a meal. This point could be tackled in an expanded way: not only patterns for one full direct response are considered, but also for different phases in the bodily response to a meal. This is a point that might rely more on domain knowledge, however, would be able to provide insights over the BG behavior over time regarding meal response phases.

Acknowledgment. This publication is part of the project DiaGame (with project number 628.011.027) of the research programme Data2Person which is (partly) financed by the Dutch Research Council (NWO).

References

1. Agrawal, R., Faloutsos, C., Swami, A.: Efficient similarity search in sequence databases. In: Lomet, D.B. (ed.) FODO 1993. LNCS, vol. 730, pp. 69–84. Springer, Heidelberg (1993). https://doi.org/10.1007/3-540-57301-1_5
2. Behera, A.: Use of artificial intelligence for management and identification of complications in diabetes. Clin. Diabetol. **10**(2), 221–225 (2021). https://doi.org/10.5603/DK.a2021.0007
3. Bezdek, J.C., Ehrlich, R., Full, W.: FCM: the fuzzy c-means clustering algorithm. Comput. Geosci. **10**(2–3), 191–203 (1984). https://doi.org/10.1016/0098-3004(84)90020-7
4. Caiado, J., Ann Maharaj, E., D'Urso, P.: Time-series clustering. Handbook of Cluster Analysis, pp. 241–264 (2015). https://doi.org/10.1201/b19706
5. Dassau, E., Bequette, B.W., Buckingham, B.A., Doyle, F.J.: Detection of a meal using continuous glucose monitoring. Diabetes Care **31**(2), 295–300 (2008). https://doi.org/10.2337/dc07-1293
6. Dias, M.L.D.: fuzzy-c-means: An implementation of fuzzy c-means clustering algorithm, September 2021. https://doi.org/10.5281/zenodo.5497844
7. Ding, H., Trajcevski, G., Scheuermann, P., Wang, X., Keogh, E.: Querying and mining of time series data: experimental comparison of representations and distance measures. Proc. VLDB Endow. **1**(2), 1542–1552 (2008). https://doi.org/10.14778/1454159.1454226
8. F. de Carvalho, D., Kaymak, U., Van Gorp, P., van Riel, N.: A Markov model for inferring event types on diabetes patients data. Healthcare Analyt. 100024 (2022). https://doi.org/10.1016/j.health.2022.100024
9. Fagherazzi, G.: Deep digital phenotyping and digital twins for precision health: time to dig deeper. J. Med. Internet Res. **22**(3), e16770 (2020). https://doi.org/10.2196/16770
10. Felizardo, V., Garcia, N.M., Pombo, N., Megdiche, I.: Data-based algorithms and models using diabetics real data for blood glucose and hypoglycaemia prediction - a systematic literature review. Artif. Intell. Med. **118**, 102120 (2021). https://doi.org/10.1016/j.artmed.2021.102120
11. Huang, M., Xia, Z., Wang, H., Zeng, Q., Wang, Q.: The range of the value for the fuzzifier of the fuzzy c-means algorithm. Pattern Recogn. Lett. **33**(16), 2280–2284 (2012). https://doi.org/10.1016/j.patrec.2012.08.014
12. International Diabetes Federation: IDF Diabetes Atlas. Brussels, Belgium: International Diabetes Federation, 10 edn. (2021)
13. Javed, A., Lee, B.S., Rizzo, D.M.: A benchmark study on time series clustering. Mach. Learn. Appl. **1**, 100001 (2020)
14. Law, S.M.: STUMPY: a powerful and scalable Python library for time series data mining. J. Open Source Software **4**(39), 1504 (2019)

15. Liao Warren, T.: Clustering of time series data - a survey. Pattern Recogn. **38**(11), 1857–1874 (2005)
16. Maas, A.H., et al.: A physiology-based model describing heterogeneity in glucose metabolism: the core of the Eindhoven diabetes education simulator (E-DES). J. Diabetes Sci. Technol. **9**(2), 282–292 (2015). https://doi.org/10.1177/1932296814562607
17. Marling, C., Bunescu, R.: The OhioT1DM dataset for blood glucose level prediction: update 2020. In: CEUR Workshop Proceedings, vol. 2675, pp. 71–74 (2020)
18. Nathan, D.M.: The diabetes control and complications trial/epidemiology of diabetes interventions and complications study at 30 years: overview. Diabetes Care **37**(1), 9–16 (2014). https://doi.org/10.2337/dc13-2112
19. Ozkan, I., Turksen, I.B.: Upper and lower values for the level of fuzziness in FCM. Stud. Fuzziness Soft Comput. **215**, 99–112 (2007). https://doi.org/10.1007/978-3-540-71258-9-6
20. Sim, S., Bae, H., Choi, Y.: Likelihood-based multiple imputation by event chain methodology for repair of imperfect event logs with missing data. In: Proceedings - 2019 International Conference on Process Mining, ICPM 2019, pp. 9–16 (2019). https://doi.org/10.1109/ICPM.2019.00013
21. Vlachos, M., Lin, J., Keogh, E., Gunopulos, D.: A wavelet-based anytime algorithm for k-means clustering of time series. In: Workshop on Clustering High Dimensionality Data and Its Applications, at the 3rd SIAM International Conference on Data Mining, pp. 1–3 (2003)
22. Ye, L., Keogh, E.: Time series shapelets: a new primitive for data mining. In: Proceedings of the 15th ACM SIGKDD International Conference on Knowledge Discovery and Data Mining. KDD 2009, New York, NY, USA, pp. 947–956. Association for Computing Machinery (2009). https://doi.org/10.1145/1557019.1557122
23. Yeh, C.C.M., et al.: Matrix profile I: all pairs similarity joins for time series: a unifying view that includes motifs, discords and shapelets. In: 2016 IEEE 16th International Conference on Data Mining (ICDM), pp. 1317–1322 (2016). https://doi.org/10.1109/ICDM.2016.0179
24. Yeh, C.-C.M., Zhu, Y., Ulanova, L., Begum, N., Ding, Y., Dau, H.A., Zimmerman, Z., Silva, D.F., Mueen, A., Keogh, E.: Time series joins, motifs, discords and shapelets: a unifying view that exploits the matrix profile. Data Min. Knowl. Disc. **32**(1), 83–123 (2017). https://doi.org/10.1007/s10618-017-0519-9
25. Yu, J., Cheng, Q., Huang, H.: Analysis of the weighting exponent in the FCM. IEEE Trans. Syst. Man Cybern. B Cybern. **34**(1), 634–639 (2004). https://doi.org/10.1109/TSMCB.2003.810951
26. Zheng, M., Ni, B., Kleinberg, S.: Automated meal detection from continuous glucose monitor data through simulation and explanation. J. Am. Med. Inform. Assoc. **26**(12), 1592–1599 (2019). https://doi.org/10.1093/jamia/ocz159
27. Zhu, Y., et al.: Matrix profile II: exploiting a novel algorithm and GPUs to break the one hundred million barrier for time series motifs and joins. In: 2016 IEEE 16th International Conference on Data Mining (ICDM), pp. 739–748. IEEE (2017). https://doi.org/10.1109/icdm.2016.0085

Analyzing Patient Feedback Data
with Topic Modeling

Jasper Arendsen[1], Emil Rijcken[1,2(✉)], Kalliopi Zervanou[3], Kim Rietjens[4],
Femke Vlems[5], and Uzay Kaymak[1,2]

[1] Jheronimus Academy of Data Science, Den Bosch, The Netherlands
`e.f.g.rijcken@tue.nl`
[2] Eindhoven University of Technology, Eindhoven, The Netherlands
[3] Leiden University, Leiden, The Netherlands
[4] Q-Qonsult Zorg, Utrecht, The Netherlands
[5] Antoni van Leeuwenhoek, Amsterdam, The Netherlands

Abstract. Patient feedback is an increasingly important measure to support quality improvement within healthcare organisations. Until recently, the focus has been on developing mechanisms for collecting patient feedback. However, research into analysis techniques to examine such feedback, especially free-text comments, is limited. The analysis of free-text data requires substantial effort because of the unstructured nature of the responses. As a result, this type of data is often underutilised within healthcare organisations while it contains the most valuable information. This research aims to analyse unstructured patient feedback, collected via a PREM questionnaire, utilising text mining. In particular, the extent to which topics can be extracted from this data is explored. Multiple topic modelling algorithms (LDA, FLSA, FLSA-W, NMF, BTM) are selected based on previous research and the data set characteristics. The applied topic modelling techniques proved to be able to provide a high-level overview of patient experiences. Hence, this research can be considered as one of the first steps towards automated analysis of unstructured patient feedback.

Keywords: Topic modeling · Fuzzy topic models · Patient feedback · Text mining · Information extraction

1 Introduction

Over the years, patient feedback has become an increasingly important outcome measure for healthcare organizations, and it is one of the central pillars that supports quality improvement [8,18]. The methods used to collect this data can be both quantitative and qualitative, and the obtained data can range from individual nurse-patient dialogues to standardized questionnaires [19]. The most common and structured method to collect patient experience data is via a Patient Reported Experience Measure (PREM). The PREM is a nationally coordinated method to measure the patient experience in hospitals [15]. Typically, a PREM contains both quantitative ratings and free-text fields focused on the care provided in the hospital as experienced by the patient. The free-text fields allow the patient to elaborate on the ratings they provided. Also, they can provide information on experiences not covered by the questionnaire. Having such a better understanding of the patient allows hospitals to optimize the care they provide to the wishes of the patients and shift towards a more patient-centred healthcare service [23]. Moreover, understanding patients' specific dissatisfaction can help health professionals and administrators identify and rectify organizational deficiencies before they become costly [21]. The information captured in unstructured text fields may be very valuable for care improvement. However, research into the analysis techniques of free-field feedback is limited and often underutilized within the medical domain [5]; possibly because the analysis of unstructured data is challenging [9]. This research aims to analyze free-text patient feedback utilizing topic modeling. Firstly, we perform a grid search to optimize various topic modelling algorithms. Then, we evaluate the produced topics quantitatively and qualitatively (through domain experts). The domain experts found an in-depth analysis of the topics challenging due to its broad and ambiguous interpretation. Yet, the topics produced by the topic models do provide high-level insights.

The outline of the paper is as follows. In Sect. 2 we discuss the various topic modeling algorithms used in this research. In Sect. 3 we discuss our comparison methodology and data gathering and preprocessing. We present the results from both the quantitative and qualitative evaluation in Sect. 4. Then we discuss our findings in Sect. 5 and conclude the paper in Sect. 6.

2   Topic Modeling

A commonly used text mining method to analyze textual data is topic modeling. Topic modeling extracts hidden topics from a collection of documents. Although various algorithms exist, their output consists of two matrices:

1. $P(W_i|T_k)$ - The probability of word i given topic k,
2. $P(T_k|D_j)$ - The probability of topic k given document j

with:

i word index $i \in \{1, 2, 3, ..., M\}$,
j document index $j \in \{1, 2, 3, ..., N\}$,
k topic index $k \in \{1, 2, 3, ..., C\}$,
M the number of unique words in the data set,
N the number of documents in the data set,
C the number of topics.

Then, the n words with the highest probability per topic are used to represent that topic.

The most popular algorithm is Latent Dirichlet Allocation (LDA), a generative probabilistic model. Documents are represented as random mixtures over latent topics, where a distribution over words characterizes each topic [4]. The document length highly influences LDA's performance; it does not perform well on short texts [11, 26].

In contrast, Non-negative Matrix Factorization (NMF) is known to perform well on short texts [20]. This method projects high-dimensional vectors into a lower-dimensional space. It takes the document-word matrix and represents this in two matrices $\mathbf{U}$ and $\mathbf{V}$. $\mathbf{U}$ consists of the topics found in the documents, and $\mathbf{V}$ consists of the corresponding coefficients representing the weights for those topics. $\mathbf{U}$ and $\mathbf{V}$ are calculated by optimizing the NMF objective function. NMF has fewer parameters than LDA and often distinguishes more realistic topics.

Another algorithm that is designed to deal with short text's sparsity is Biterm Topic Modeling (BTM) [25]. A biterm is defined as an unordered word-pair co-occurring in a short text. This method is based on the assumption that words occurring frequently together belong to the same topic.

Recently, Fuzzy Latent Semantic Analysis (FLSA) [13] was applied to health-care data and showed superior results in comparison to LDA [4]. Just like NMF, FLSA starts with the document-term matrix. It then applies a global term weighting mechanism, after which the representation is projected onto a lower-dimensional space through singular value decomposition. Then, fuzzy C-means clustering [1] is performed on $\mathbf{U}^T$ (thus, documents are being clustered) to find different topics. Inspired by FLSA, FLSA-W works similarly but clusters on $\mathbf{V}^T$ and thus, clusters on words [22]. It outperforms both LDA and FLSA in experimental studies.

3   Analyzing Patient Feedback

We aim to analyse unstructured patient feedback using topic modeling methods. In particular, we explore the extent to which topics can be extracted from this data consisting of Dutch texts. The main steps to achieve this goal are shown in Fig. 1. Below, we explain each step (Fig. 2).

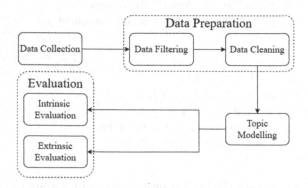

Fig. 1. Overview of the study methodology

3.1 Data Collection

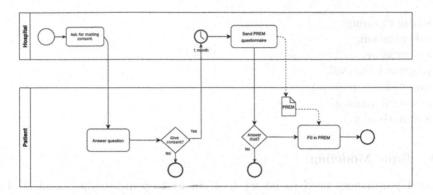

Fig. 2. PREM data collection process.

The data consists of the feedback given in the PREM, collected by the Antoni van Leeuwenhoek (AVL) hospital, a hospital and research institute located in Amsterdam in The Netherlands. There are two types of patients (patients with a malign tumour and patients with a benign tumour) who are asked to fill out different questionnaires because of the difference in treatment. Since most patients have a malign tumour and this type of care is the most important for the AVL, we consider only their feedback in this study. The PREM questionnaire consists of three categories of questions. Firstly, patients are asked for practical data about their treatment. This includes questions about their type of illness, contact, treatment type and whether they participated in a trial. Secondly, patients are asked to rate their satisfaction regarding the provided care overall and for individual healthcare components. The individual components are Interaction, Relationship with healthcare providers, Expertise, Atmosphere, Waiting times, Available information, Aftercare, Facilities, Research, Parking, and others. Lastly, there are two open-ended questions in which the patient can expand on ratings given earlier to indicate what they were and were not satisfied with. In this study, we analyze these two fields with unstructured data separately and refer to them as 'satisfaction data' and 'improvement data'.

3.2 Data Preprocessing

The data preparation phase involves two main steps: (i) data filtering and (ii) data cleaning. Both steps are described in this subsection. Three ways of filtering are applied to the patient feedback data:

1. non-malign patients are removed,
2. missing values are removed.

3. non-Dutch entries are removed[1].

For data cleaning, we have performed the following steps to the filtered data:

1. string cleaning[2]
2. tokenization,
3. lowercasing,
4. stopword removal,[3]
5. punctuation removal,
6. hex-digit removal,
7. lemmatisation[4].

3.3 Topic Modeling

Several parameters need to be set to train and optimize topic models. Two parameters are set for all algorithms: the number of topics and the number of words per topic. Additionally, each algorithm has its own set of hyperparameters to be tuned. The optimal number of topics are five and six, for the satisfaction and improvement data, respectively. They are found by using the elbow method [14,16] for determining an acceptable trade off between maximizing topic coherence (see below) and minimizing the number of topics. This number stays fixed while tuning the other hyperparameters. Furthermore, only the top 10 words per topic are selected in order to reduce the required time effort for the qualitative evaluation by the domain experts. Table 1 shows the optimal- and grid search values for the model-dependent hyperparameters. In this table, the symmetric and asymmetric values in lda's range refer to the Dirichlet priors used [24].

3.4 Evaluation

Evaluation methods of topic model quality can be divided into (i) intrinsic and (ii) extrinsic methods. Intrinsic evaluation methods rely on internal evaluation metrics which directly quantify the task performance, while extrinsic evaluations focus on external evaluation metrics.

Intrinsic Evaluation. Since a topic model's output consists of various topics, each topic containing a collection of words, the quality of a topic model should focus both on the quality of words within each topic (intra-topic quality) and the diversity amongst different topics (inter-topic quality). For the intra-topic quality, we use the C_v coherence metric, which correlates highest to human

---

[1] Filtering is done by applying the Python langdetect algorithm [6]. This algorithm supports over 50 languages and has a precision of 99.77%.

[2] Caused by a different data encoding in the hospital's database.

[3] Stop words are removed from the data by implementing the NLTK Dutch package. [3].

[4] Lemmatisation is applied using the Spacy Dutch Python package [12].

Table 1. Hyperparameter grid search settings and optimal values

Model	Parameter	Range	Optimal value satisfaction	Optimal value improvement
lda	α	[0.01, 0.2, 0.4, 0,6, 0.8, 1, *symmetric*, *asymmetric*]	1	0.4
	β	[0.01, 0.2, 0.4, 0,6, 0.8, 1, *symmetric*, *auto*]	0.4	0.8
	eval_every	[5-40, *steps = 5*]	5	25
	passes	[1-15, *steps = 1*]	14	9
flsa	word_weighting	[idf, probidf]	probidf	probidf
	cluster_method	[fcm, fst-pso, gk]	fst-pso	fst-pso
	svd_factors	[1-5]	4	4
flsa-w	word_weighting	[idf, probidf]	probidf	probidf
	cluster_method	[fcm, fst-pso,gk]	fst-pso	fst-pso
	svd_factors	[1-5]	4	4
nmf	kappa	[0.01, 0.25, 0.5, 0.75, 1, 2]	0.5	1
	eval_every	[5-40, *steps = 5*]	20	15
	passes	[1-15, *steps = 1*]	10	15
btm	α	[0.01, 0.2, 0.4, 0,6, 0.8, 1]	0.8	0.2
	β	[0.01, 0.2, 0.4, 0,6, 0.8, 1]	1.0	0.2
	iterations	[100-800, *steps = 100*]	100	200

judgment amongst all coherence metrics. With c_v, the Normalized Pointwise Mutual Information (2) is calculated for the combination of all the top-n words in a topic. Then, the arithmetic mean is calculated based on all these scores. To calculate the probabilities in Normalized Pointwise Mutual Information, a sliding window of 110 words is being used.

$$PMI(w_i, w_j) = log \frac{P(w_i, w_j) + \epsilon}{P(w_i) \cdot P(w_j)} \tag{1}$$

$$NPMI(w_i, w_j)^\gamma = \frac{PMI}{\sum_{i=1}^{M} \sum_{j=1}^{N} P(W_i, D_j)}^\gamma \tag{2}$$

The coherence score ranges between zero and one, where one means perfect coherence and zero means no coherence whatsoever.

Additionally, the C_{UMass} is considered, which has the second-highest correlation with human judgment. We use the diversity score for the inter-topic quality. We define the following quantities:

W_{unique} the number of unique words in the top-n words of all the topics,
$\quad W_{all}$ the total number of words in all the topics ($n \times C$),
$\quad\quad n$ the number of words per topic,
$\quad\quad C$ the number of topics.

Then, Eq. 3 shows how diversity is calculated.

$$Diversity = \frac{W_{unique}}{W_{all}}. \tag{3}$$

Lastly, we calculate the interpretability score as the product between the coherence (C_v) and diversity score [7], as can be seen in Eq. 4.

$$Interpretability = Coherence(c_v) \times Diversity \tag{4}$$

Extrinsic Evaluation. The domain experts are presented with three questions per topic to measure both the inter- and intra-topic quality for the extrinsic evaluation. For the intra-topic quality, they are asked to rate the topics on their individual quality, similar to [2]. The quality is communicated to the experts as a combination of the coherence, meaningfulness, and interpretability of the top n words per topic with respect to their weights. The quality of the individual topics is measured via an ordinal score [0-3], where 3 represents a 'good/useful' topic, and 0 defines a 'bad/useless' topic. Furthermore, the domain experts are asked to assign each topic with one of the categories extracted from the PREM for the inter-topic quality. For extrinsic evaluations, it is important to take the inter-rater agreement into account because of the variation in human interpretation. Since this research includes nine raters, the Krippendorff's alpha score α is applied to indicate the interrater agreement [17]. Four raters focused on the satisfaction data and five on the improvement data to save time and costs.

4   Results

4.1   Intrinsic Evaluation

The intrinsic metrics, discussed in Sect. 3.4, are presented in Table 2 and 3, where the best values per metric are boldface. The improvement data yields higher coherence scores than the satisfaction data, while diversity scores are comparable for both data sets. FLSA-W and FLSA perform best for both the satisfaction and improvement data. FLSA has the highest coherence scores, whereas FLSA-W produces the most diverse topics. As a result, FLSA-W produces the most interpretable topics for the satisfaction data and FLSA the most interpretable topics for the improvement data. Furthermore, LDA performs much worse than the fuzzy algorithms.

Table 2. Model intrinsic values of the satisfaction data. With C_v and $C_{U_{Mass}}$ coherence score

Model	Satisfaction			
	C_v	C_{UMass}	Diversity	Interpretability
lda	0.490	−2.379	0.780	0.382
flsa	**0.688**	**−2.335**	0.880	**0.605**
flsa-w	0.547	−5.384	**0.920**	0.503
nmf	0.591	−2.590	0.860	0.508
btm	0.518	−3.216	0.820	0.425

Table 3. Model intrinsic values of the improvement data. With C_v and $C_{U_{Mass}}$ coherence score

Model	Improvement			
	C_v	C_{UMass}	Diversity	Interpretability
lda	0.575	−2.134	0.567	0.326
flsa	**0.770**	**−1.449**	0.883	0.680
flsa-w	0.755	−2.269	**0.950**	**0.717**
nmf	0.760	−2.074	0.900	0.684
btm	0.632	−2.263	0.733	0.463

4.2 Extrinsic Evaluation

The Krippendorff's alpha score for the satisfaction data is 0.046, and the alpha score for the improvement evaluation is 0.042. These scores indicate that the results can be interpreted as statistically unrelated. The extrinsic evaluation scores are shown in Table 4. The 'mean' score shows the average ordinal score [0-3] as assigned by the experts. The 'Uniqueness' indicates the average number of labels a topic was assigned to. For this value, a higher value indicates its more challenging to assign a label to a topic, and the ideal value is one. This could mean a topic is hard to interpret. The 'No Category' is used when the human subjects cannot fit the topic into any category. Ideally, this value is as low as possible. Since the variation in scores is relatively small, it seems that quality differences between the topic models are relatively low. Nonetheless, the algorithms generally score higher on the satisfaction data than the improvement data. LDA has the best quality topics for the satisfaction data, whereas FLSA performs best on the improvement data. The differences in performance between the datasets is likely caused by the dataset's characteristics; the texts from the improvement data are longer and more unique, generally, than the satisfaction data.

Table 4. Model extrinsic values

	Satisfaction			Improvement		
Model	Mean	Uniqueness	No category	Mean	Uniqueness	No category
lda	**1.550**	2.300	0.000	1.167	1.467	0.233
flsa	1.300	2.300	0.050	**1.300**	1.533	0.233
flsa-w	1.500	2.350	0.000	0.967	1.300	0.300
nmf	**1.550**	**2.450**	0.000	1.267	1.600	0.133
btm	1.250	1.900	0.000	1.233	**1.700**	0.200

5 Discussion

After evaluating the topics both intrinsically and extrinsically, we find contradicting results between the two. According to the intrinsic evaluation, the
improvement dataset has the highest intra-topic quality, whereas the satisfaction scores best according to the extrinsic evaluation. The results are still rather
preliminary and We need to conduct further experiments to better understand
which evaluation has the most impact. The difference between the two metrics is
likely due to the different characteristics of both datasets: the satisfaction data
has a lower word variability than the improvement data. However, this may also
be caused by the low inter-rater reliability in the extrinsic evaluation. Generally, the domain experts find the quality of the topics relatively low due to their
broad and ambiguous interpretation. In particular, the mean quality is perceived
as relatively low, and the uniqueness and fraction of no categories were relatively
high. Although differences within the extrinsic evaluation were low, dissimilarity
is noticeable concerning the satisfaction and improvement results. Satisfaction
topics are interpreted more broadly due to the average number of categories
selected. Improvement topics are more ambiguous as the fraction of no fitting
topics is relatively high. As a result of the broad and ambiguous interpretation,
the models only allow for a high-level topic overview.

In this research, we have not considered the quantitative scores given by
patients, while these scores are likely to provide valuable insights. Additionally,
the Krippendorff's alpha scores, used for the extrinsic evaluation, are close to
zero, indicating an absence of inter-rater agreement. The low agreement between
the raters affects the reliability of the extrinsic evaluation. Even so, the low alpha
score can be caused by the possible subjective interpretation of the results or
the low number of domain experts [10].

Finally, we use the elbow method to determine the optimal number of topics.
A typical 'elbow' is noticeable for the satisfaction data. Yet, no such pattern
can be found with the improvement data. Hence, a suboptimal number of topics
might have been selected. Consequently, this method is not ideal for determining
the number of clusters for the improvement corpus. Therefore, a more detailed
data analysis should determine the optimal number of topics.

6 Conclusion

In this work, we analyze free-text patient feedback to find relevant information to improve healthcare practices. We have trained/optimized various topic modeling algorithms and evaluated both intrinsically and extrinsically. FLSA and FLSA-W have the highest intrinsic scores, whereas NMF and BTM perform best on the extrinsic evaluation. The methodology used in this paper can be implemented into the hospital's dashboard so that patient feedback is monitored more regularly and adequately. In future work, we plan to include more cohorts of patients to assess the generalizability of our results.

References

1. Bezdek, J.C.: Pattern Recognition with Fuzzy Objective Function Algorithms. Springer, Heidelberg (2013)
2. Bhatia, S., Lau, J.H., Baldwin, T.: An automatic approach for document-level topic model evaluation. In: Proceedings of the 21st Conference on Computational Natural Language Learning (CoNLL 2017), Vancouver, Canada, pp. 206–215. Association for Computational Linguistics, August 2017. https://doi.org/10.18653/v1/K17-1022
3. Bird, S., Klein, E., Loper, E.: Natural Language Processing With Python: Analyzing Text With the Natural Language Toolkit. O'Reilly Media, Inc. (2009)
4. Blei, D.M., Ng, A.Y., Jordan, M.I.: Latent Dirichlet allocation. J. Mach. Learn. Res. **3**, 993–1022 (2003)
5. Cunningham, M., Wells, M.: Qualitative analysis of 6961 free-text comments from the first national cancer patient experience survey in Scotland. BMJ Open **7**(6), e015726 (2017)
6. Danilk, M.: Langdetect. PyPI (2014). https://pypi.org/project/langdetect/
7. Dieng, A.B., Ruiz, F.J., Blei, D.M.: Topic modeling in embedding spaces. Trans. Assoc. Comput. Linguist. **8**, 439–453 (2020)
8. Doyle, C., Lennox, L., Bell, D.: A systematic review of evidence on the links between patient experience and clinical safety and effectiveness. BMJ Open **3**(1) (2013). https://doi.org/10.1136/bmjopen-2012-001570. https://bmjopen.bmj.com/content/3/1/e001570
9. Gleeson, H., Calderon, A., Swami, V., Deighton, J., Wolpert, M., Edbrooke-Childs, J.: Systematic review of approaches to using patient experience data for quality improvement in healthcare settings. BMJ Open **6**(8) (2016). https://doi.org/10.1136/bmjopen-2016-011907. https://bmjopen.bmj.com/content/6/8/e011907
10. Hallgren, K.A.: Computing inter-rater reliability for observational data: an overview and tutorial. Tutor. Quant. Methods Psychol. **8**(1), 23 (2012)
11. Hong, L., Davison, B.D.: Empirical study of topic modeling in Twitter. In: Proceedings of the First Workshop on Social Media Analytics, pp. 80–88 (2010)
12. Honnibal, M., Montani, I.: spaCy 2: natural language understanding With BLoom embeddings, convolutional neural networks and incremental parsing (2017, to appear)
13. Karami, A., Gangopadhyay, A., Zhou, B., Kharrazi, H.: Fuzzy approach topic discovery in health and medical corpora. Int. J. Fuzzy Syst. **20**(4), 1334–1345 (2017). https://doi.org/10.1007/s40815-017-0327-9

14. Khalid, H., Wade, V.: Topic detection from conversational dialogue corpus with parallel Dirichlet allocation model and elbow method. In: Computer Science & Information Technology. AIRCC (2020)

15. Kingsley, C., Patel, S.: Patient-reported outcome measures and patient-reported experience measures. BJA Educ. **17**(4), 137–144 (2017)

16. Kodinariya, T.M., Makwana, P.R.: Review on determining number of cluster in K-means clustering. Int. J. **1**(6), 90–95 (2013)

17. Krippendorff, K.: Content Analysis: An Introduction to Its Methodology. Sage Publications (2018)

18. Kumah, E., Ankomah, S.E., Kesse, F.O.: The impact of patient feedback on clinical practice. Br. J. Hosp. Med. **79**(12), 700–703 (2018)

19. LaVela, S.L., Gallan, A.: Evaluation and measurement of patient experience. Patient Exp. J. **1**(1), 28–36 (2014)

20. Lee, D.D., Seung, H.S.: Learning the parts of objects by non-negative matrix factorization. Nature **401**(6755), 788–791 (1999)

21. Pichert, J.W., Miller, C.S., Hollo, A.H., Gauld-Jaeger, J., Federspiel, C.F., Hickson, G.B.: What health professionals can do to identify and resolve patient dissatisfaction. Joint Comm. J. Qual. Improv. **24**(6), 303–312 (1998). https://doi.org/10.1016/S1070-3241(16)30382-0. https://www.sciencedirect.com/science/article/pii/S1070324116303820

22. Rijcken, E., Scheepers, F., Mosteiro, P., Zervanou, K., Spruit, M., Kaymak, U.: A comparative study of fuzzy topic models and LDA in terms of interpretability. In: Proceedings of the 2021 IEEE Symposium Series on Computational Intelligence (SSCI), pp. 1–8 (2021)

23. Shaller, D.: Patient-Centered Care: What Does It Take? Commonwealth Fund New York (2007)

24. Syed, S., Spruit, M.: Exploring symmetrical and asymmetrical Dirichlet priors for latent Dirichlet allocation. Int. J. Semantic Comput. **12**(03), 399–423 (2018)

25. Yan, X., Guo, J., Lan, Y., Cheng, X.: A biterm topic model for short texts. In: Proceedings of the 22nd International Conference on World Wide Web, pp. 1445–1456 (2013)

26. Zhao, W.X., et al.: Comparing Twitter and traditional media using topic models. In: Clough, P., et al. (eds.) ECIR 2011. LNCS, vol. 6611, pp. 338–349. Springer, Heidelberg (2011). https://doi.org/10.1007/978-3-642-20161-5_34

A Framework for Active Contour Initialization with Application to Liver Segmentation in MRI

Arnau Mir-Fuentes[1,2(✉)] , Arnau Mir[4,5] , Felipe Antunes-Santos[1,2,3] ,
F. Javier Fernandez[1] , and Carlos Lopez-Molina[1,2,3]

[1] Department of Estadística, Informática y Matemáticas,
Universidad Pública de Navarra, 31006 Pamplona, Spain
{arnau.mir,felipe.antunes,fcojavier.fernandez,carlos.lopez}@unavarra.es
[2] NavarraBiomed, Hospital Universitario de Navarra, 31008 Pamplona, Spain
[3] KERMIT, Department of Data Analysis and Mathematical Modelling,
Ghent University, 9000 Ghent, Belgium
[4] Department of Mathematics and Computer Science,
University of the Balearic Islands, 07122 Palma, Spain
arnau.mir@uib.es
[5] Health Research Institute of the Balearic Islands (IdISBa), 07010 Palma, Spain

Abstract. Object segmentation is a prominent low-level task in image processing and computer vision. A technique of special relevance within segmentation algorithms is active contour modeling. An active contour is a closed contour on an image which can be evolved to progressively fit the silhouette of certain area or object. Active contours shall be initialized as a closed contour at some position of the image, further evolving to precisely fit to the silhouette of the object of interest. While the evolution of the contour has been deeply studied in literature [5,11], the study of strategies to define the initial location of the contour is rather absent from it. Typically, such contour is created as a small closed curve around an inner position in the object. However, literature contains no general-purpose algorithms to determine those inner positions, or to quantify their fitness. In fact, such points are frequently set manually by human experts, hence turning the segmentation process into a semi-supervised one. In this work, we present a method to find inner points in relevant object using spatial-tonal fuzzy clustering. Our proposal intends to detect dominant clusters of bright pixels, which are further used to identify candidate points or regions around which active contours can be initialized.

The authors gratefully acknowledge the financial support of the grants PID2019-108392 GB-I00 funded by MCIN/AEI/10.13039/501100011033, as well as that by the Government of Navarra (PC082-083-084 EHGNA). A. Mir acknowledges the financial support of the grant PID2020-113870GB-I00 funded by MCIN/AEI/10.13039/501100011033/.

D. Ciucci et al. (Eds.): IPMU 2022, CCIS 1602, pp. 259–271, 2022.
https://doi.org/10.1007/978-3-031-08974-9_21

Keywords: Hepatic steatosis · Image segmentation · MRI image · Active contour model · Spatial fuzzy c-means · Connected component · Center

1 Introduction

Hepatic steatosis is a common condition caused by the storage of residual fat in the liver. In some cases hepatic steatosis can rapidly lead to liver damage, but early detection and control of the disease can often prevent or even reverse hepatic steatosis with lifestyle changes. While detection can be done with invasive procedures, the ideal case is using non-invasive methods, which are safer, faster and, thanks to the innovations in medical imaging, significantly more accurate. Among non-invasive methods, medical imaging offers a wide range of alternatives with application to the task. Liver segmentation can be achieved with a range of strategies, from thresholding algorithms to neural networks [13,14]. Within such range of alternatives, active contours appear as a prominent one. Active contour models (ACMs), also called snakes, are deformable curves that evolve according to predefined forces. Such forces are normally designed to pull the curve in the normal direction, seeking to minimize an energy functional defined over the space where the curves are defined. Active contour models may be understood as a special case of the general technique of matching a deformable model to an image by means of energy minimization in two dimensions. In this work, ACMs [10] have been chosen over other alternatives for liver segmentation because they do not require training data, they are unsupervised, they have fast convergence and, also, they are robust to noise and topological changes. Interestingly, ACMs allow to control the regularity of objects and their contours adapting the resulting spline to the expected characteristics of the object.

Active contour models were initially introduced by Kass *et al.* in 1988 [10], as part of an effort to apply deformable models in image processing [20]. Among the different follow-up works, the one by Caselles *et al.* [4] is of special relevance. In [4], the authors did not only improve ACMs by giving a more consistent equation for the energy functional, but also gave some proofs on the convergence and stability of ACMs considering viscosity solutions [8]. Despite relevant advances in ACMs, their development has been an ongoing research effort over the past decades. For example, [12] improved the definition of the energy functional equation by adding terms that improve the regularity and speed of convergence.

One of the recurrent problems in the practical use of ACMs is the need to define an initial contour, which is often done manually. In this work, we present different strategies to find a suitable initial curve for ACMs. Note that this task does not only restrict to finding any inner position in a generally-defined object. Also, the selection of the contour shall attempt the point (or the initial contour around it) to have optimal performance for ACMs in terms of convergence and accuracy. The problem we aim at can be stated as follows: Given a grayscale image containing an object of interest, find an initial curve within that object to be used as initial contour in an ACM. The object to be found verifies a local fuzzy property, which is key for our algorithm. In our case, the grayscale image is the MRI visualization of a human torso, the object to be found is the liver and the local fuzzy property is the amount of water at each position. The use of MRI is due to the fact that it comprises a direct relationship between the amount of water and the gray level intensity at each position of the image.

The remainder of this works is divided as follows. In Sect. 2 we present a general framework for ACM initialization, which is instantiated in Sect. 3. Our proposal is put to the test in Sect. 4, while Sect. 5 lists some conclusions.

2 A Framework for Active Contour Model Initialization

In this work, images are taken as mappings $\Omega \mapsto \mathbb{T}$, with Ω representing the set of positions in the image grid and $\mathbb{T}$ represents the tonal palette. The term $\mathbb{I}_\mathbb{T}$ represents the set of images with a certain tonal palette $\mathbb{T}$. For example, $\mathbb{I}_{\{0,1\}}$ represents the set of binary images, while $\mathbb{I}_{[0,1]}$ is the set of real-valued grayscale images. More complex instantiations of $\mathbb{T}$ could be, for example, hyperspectral signatures [15].

Definition 1 (Inclusion of binary images). *Given two binary images I and J, image I is included in image J, $I \subseteq J$, when $I(x,y) \leq J(x,y)$, $\forall(x,y) \in \Omega$.*

Definition 2 (Inclusion of binary and grayscale images). *Given $t \in (0,1)$ a threshold, I a binary image and J a grayscale image, image I is included in image J using threshold t, $I \subseteq_t J$, when $I(x,y) \leq J_t(x,y)$, $\forall(x,y) \in \Omega$, where J_t is the "binarized" J image defined as:* $J_t(x,y) = \begin{cases} 1, & \text{if } J(x,y) \geq t, \\ 0, & \text{otherwise.} \end{cases}$

We propose a novel framework to find the most appropriate center (pixel) around which the initial curve of an ACM can be located. Inspired by the developments of Bezdek *et al.* on edge detection [2], our proposal is based on a four-step procedure in which each phase takes an interpretable goal. The four phases in our proposal are:

- *Object detection*: This is the phase in which visual information is processed to produce an initial estimation of the presence of the object of interest. It is expressed as a mapping $f_{OD} : \mathbb{I}_\mathbb{T} \mapsto \mathbb{I}_{[0,1]}$ which intends to locate the areas of the image in which the object is present. There is no constraints in the tonal palette of the input image (which might be dependent upon specific scenarios), but the output needs to be a value in $[0,1]$. This value is not interpreted as a probability, but as a fuzzy membership degree representing the certainty on the membership of the pixel to the object of interest.
- *Object Selection*: This phase consists of selecting the contiguous area containing the object of interest, out of the information produced at the object detection phase. It can be modeled as a mapping $f_{OS} : \mathbb{I}_{[0,1]} \mapsto \mathbb{I}_{\{0,1\}}$, so that (a) the set of 1-labeled pixels in $f_{OS}(I)$ form a connected component and (b) $f_{OS}(I) \subseteq_t I$, where t is the threshold from which the pixels belong to the

region where the object can be. This represents the fact that (a) one singular object is selected and (b) the object selection phase aims at selecting one the areas identified at the previous phase, not modifying the representation of such area.

- *Object Isolation*: This is the phase in which the morphological priors of the object of interest need to be enforced. Object detection and segmentation in real applications is heavily affected by a list of accepted priors, which can impose restrictions on the object, including shape, regularity, position, etc. The object isolation phase is expressed as a mapping $f_{OI} : \mathbb{I}_{\{0,1\}} \mapsto \mathbb{I}_{\{0,1\}}$ in which the only restrictions to the output are driven by the context of application.

  Note that this phase does not intend to produce a faithfull (or reliable) segmentation of the object. It rather intends to make an approximate detection, in which exactitude shall not be requested. However, regardless of how good the approximation is, we understand that some priors can and should be enforced. For example, an approximate segmentation of a liver might contain lack of precision in the boundaries, but should not contain holes or 1-pixel width linear structures, which are biologically non-viable.

- *Center selection*: This is the phase in which the approximate segmentation is analyzed in order to look for a center around which an initial contour can be set. This can be done on the basis of different inspirations, and is represented as a mapping $f_{CS} : \mathbb{I}_{\{0,1\}} \mapsto \Omega$, with $I(f_{CS}(I)) = 1$ for any binary image I.

Our proposal is divided in four phases which can be combined by function composition. Otherwise said, center detection can be expressed as a function $f_{CD} : \mathbb{I}_\mathbb{T} \mapsto \Omega$, which can be broken down as

$$f_{CD}(I) = f_{CS}(f_{OI}(f_{OS}(f_{OD}(I)))) \,. \tag{1}$$

The following section includes a specific implementation of the proposed framework with application to liver segmentation in MRI imagery.

3   Active Contour Model Initialization for Liver Segmentation in MRI Imagery

In this section we present a specific implementation of the framework in Sect. 2. Specifically, we present an implementation aimed at initializing ACM models for liver segmentation in MRI imagery. This application is relatively frequent in public health and nutrition studies, in which non-invasive liver analysis is a recurrent diagnosis procedure. In this context, layered images from the MRI are analyzed to both (a) quantify the topological characteristics of the liver and (b) measure visual evidence from its state. The workflow for ACM initialization in such context is as presented in Fig. 1.

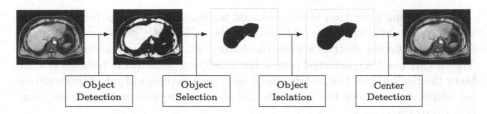

Fig. 1. Schematic representation of the framework presented in Sect. 2, applied to the problem of ACM initialization for liver segmentation. The original image is presented as a colorized image to improve visibility.

3.1 Spatial Fuzzy c-means for Object Detection

Object detection is the phase in which the relevant object has to be discriminated from the remainder of the image. This discrimination need not be precise in any aspect, rather presenting a rough discrimination of the object. Starting a segmentation (ACM) algorithm with a rough approximation to the solution might seem noticeable but is rather common in literature. For example, any context-aware smoothing algorithm (as Anisotropic Diffusion [18,21]) perform an estimation of the presence and strength of object boundaries before modelling tonal diffusion at each iteration. In this case, the estimation shall be good enough to extract the object from the background.

There is little technical requisites to the object detection phase. In fact, simplistic approaches to the task can be made using fuzzy thresholding, which would discriminate bright areas from dark ones according to their tone. The selection of the thresholding algorithm is dependent upon the characteristics of the image. Classical binary thresholding algorithms would provide binary representations of the object. For example, the Otsu method [17] is designed for images with bimodal histograms, while the Rosin method [19] would rather discriminate outliers in a monomodal histogram. However, this approach is not recommended, since it collides with the spirit of the Object Detection phase. Using fuzzy thresholding algorithms [3] does provide a solution to this, since these algorithms typically produce membership information each pixel other than 0 or 1 values for each class. Otherwise said, fuzzy thresholding does not only lead to the discrimination of the pixels in the image, but also to the gradual quantification of such discrimination. A better suited alternative for object detection is, in this regard, clustering algorithms.

The use of clustering algorithms in image binarization lies on a simplistic interpretation: each pixel in the image can be taken as an instance in a dataset [6,16]. Such instance would be composed of the spatial information (pixel location) the tonal information and, potentially, other derived information (texture indicators, etc.). The separation of the instances in such dataset, by means of a clustering algorithm, would be able to group together pixels which are both spatially and tonally close. We believe this ability is crucial in selecting clustering algorithms over thresholding algorithms for object detection.

A clustering algorithm able to account for accounting for spatio-tonal information is the Spatial Fuzzy c-Means algorithm (SFCM), which accommodates the use of different metrics within the framework of the well-studied FCM [1]. The SFCM has been selected over other clustering algorithms because it combines the flexibility of the metric adjustment with two important characteristics: (a) ability to represent the centroids and (b) fuzzy representation of the membership to the classes.

Let $X = \{\mathbf{x}_1, \mathbf{x}_2, \ldots, \mathbf{x}_n\}$ be a set of points over a $k+1$-dimensional Euclidean space $\mathbb{E}$ and $U = \{\mu_1, \ldots, \mu_c\}$ a set of membership functions defined over $\mathbb{E}$. That is, each μ_j is a function:

$$\mu_j : \mathbb{E} \longrightarrow [0, 1],$$

such that:

$$\sum_{j=1}^{c} \mu_j(x) = 1$$

where, given $\mathbf{x} \in \mathbb{E}$, $\mu_j(\mathbf{x})$ represents the membership degree of point $\mathbf{x}$ to cluster C_j.

The goal of the FCM [1] is to minimize the goal function given by:

$$\sum_{j=1}^{c} \sum_{i=1}^{n} \mu_j(x_i)^m \, \|\mathbf{x}_i - \mathbf{v}_j\|^2, \tag{2}$$

where $m > 1$ is a parameter of the algorithm that controls the importance between the distance and the membership function, $\| \cdot \|$ is the Manhattan, Euclidean or Chebyschev norm. $\mathbf{v}_j, j \in \{1, \ldots, c\}$ are the centers of the c clusters, considering the membership function. Such centers are computed as

$$\mathbf{v}_j = \frac{\sum_{\mathbf{x}_i \in C_j} \mu_j(\mathbf{x}_i)\mathbf{x}_i}{\sum_{\mathbf{x}_i \in C_j} \mu_j(\mathbf{x}_i)}, \tag{3}$$

where C_j represents the cluster j,

Given m and the set X, the schema of the algorithm is as follows:

(1) *Initialization:* choose the membership of all the points of set X randomly. For every $\mathbf{x}_i \in X$, $i = 1, \ldots, n$, compute $j = \max_{k=1,\ldots,c} \mu_k(\mathbf{x}_i)$. Next, add the point $\mathbf{x}_i$ to the cluster C_j. In this way, we have calculated the clusters $C_1, \ldots, C_c$. Set the number of iterations n to 1: $n = 1$.

(2) *Computation of the centers:* compute the centers of the algorithm using expression (3). Compute the objective function using expression (2). Let O_i be this value.

(3) *Update of the membership function:* Update the membership function of each point $\mathbf{x}_i \in X$ using this expression:

$$\mu_j(\mathbf{x}_i) = \frac{1}{\left(\sum_{k=1}^{c} \frac{\|\mathbf{x}_i - \mathbf{v}_j\|}{\|\mathbf{x}_i - \mathbf{v}_k\|}\right)^{\frac{2}{m-1}}}, \tag{4}$$

for $i \in \{1, \ldots, n\}$ and $j \in \{1, \ldots, c\}$. Compute the new clusters using the procedure of the first step. Compute the new value of the objective function using expression (2). Let O_{i+1} be this value. Increase the number of iterations n.

(4) *Stopping criterion:* If $|O_i - O_{i+1}| < \epsilon$, where ϵ is the allowed tolerance or if $n \geq n_{\max}$, where $n_{\max}$ is the maximum number of iterations, we stop the process. Otherwise, we go to the step (2) and continue.

While the FCM algorithm has proven valid in a list of applications, we introduce the spatial fuzzy c-means (SFCM) algorithm, which is a generalization of the former. This generalization allows for a better modelling of the distance between elements in the set.

The SFCM is an extension of FCM in which step (3) has two substeps:

(3.1) Update the membership function of each point $\mathbf{x}_i \in X$ using expression (4).

(3.2) Re-update the membership function of each point $\mathbf{x}_i \in X$ using this expression:

$$\mu'_j(\mathbf{x}_i) = \frac{\mu_j(\mathbf{x}_i)^p h_{i,j}^q}{\sum_{k=1}^{c} \mu_j(\mathbf{x}_i)^p h_{i,j}^q},$$

where p and q are parameters that controls the relative importance of the neighbor pixels and $h_{i,j}$ is computed as:

$$h_{i,j} = \sum_{x_j \in NB(x_j)} \mu_i(\mathbf{x}_j),$$

where $NB(\mathbf{x})$ is the set of pixels that belong to the 9×9 square centered at point $\mathbf{x}$. Notice that for $p = 1$ and $q = 0$ the algorithm is exactly the fuzzy c-means algorithm. That is why the spatial fuzzy c-means algorithm are a generalization of the fuzzy c-means algorithm.

It is hence evident why the spatial fuzzy c-means algorithm takes into account the spatial information of the pixels in the substep (3.2).

3.2   Object Selection

Object detection is meant to identify the areas of the image which fit to the expected characteristics of the object of interest. However, it might be the case where more than one region or object is selected. The Object Selection phase shall analyze the image and select the isolated region which is more prone to be the object of interest. The selection of the object might also involve certain restrictions in the size of the object, its shape or proportions, its location or even the roughness of its boundaries.

In the case of liver segmentation, our constraints stem from the basics of MRI technology. In MRI visualization, the regions of a body with greater water indices appear brighter than those with lower water indices. We intend to capitalize on this information to produce an object selection procedure. In this application, we opt out by selecting the largest cluster as region of interest. This idea is based on the fact that the cluster representing the background is larger than the cluster representing the object.

For the experimental results, we restricted to the last strategy. This decision is, still, dependent upon the specific imagery.

3.3   Object Isolation

Once the image has been discriminated into the *object region* and the *background region*, the next step is to find best fitting area among those labeled as *object region*. In our case, we have opted out by selecting the largest connected component of the cluster that contains the object.

To find the largest connected component, we visualize the binary image or the grid as a graph where each pixel represents a separate node of the graph and each node is 8-connected to its neighbors. Next, we apply the Breadth First Search algorithm [7] search for every node of the graph, and find all the nodes connected to the current node with same color value as the current node.

The isolation of the object does not restrict to labeling the object itself. There are semantic constraints that might need to be applied in this phase. For example, since we know that the liver cannot contain holes, any *background* area within the object needs to be relabeled as *object*.

3.4   Center Detection

This phase takes as input a best-possible identification of the object in order to produce an estimation of a center around which an ACM can be initialized. Let $S \subset \Omega$ be the set of positions labeled as *object* in the Object Isolation phase. Let (x_i, y_i) be the positions in S. In order to estimate the center of the object of interest, we take into account different geometric properties of that set, giving rise to the following alternatives to find an object center $(x_c, y_c) \in \Omega$:

- *Centroid*: The center of gravity of the object is taken as the center of gravity. This strategy is problematic in scenarios with non-convex objects, since the center of gravity might in fact fall outside the object itself. The formulation to compute the centroid is:

$$x_c = \frac{\sum_{(x_i, y_i) \in S} x_i}{|S|}, \quad y_c = \frac{\sum_{(x_i, y_i) \in S} y_i}{|S|}.$$

  Note that the centroid can be interpreted as the center of mass of the object S or the arithmetic mean position of all points in S.
- *Maximum distance to the closest point to the boundary:* By analyzing the contour of the region S, we can attempt to find a center which is as far as possible from its boundary. Let $S_b \subseteq S$ be the set of pixels in the contour of S. For every $p_i \in S$, $p_i = (x_i, y_i)$ we first we compute the distance of to S_b,

$$d_{S_b}(p_i) = \min_{p_j \in S_b} m((x_i, y_i), (x_j, y_j)),$$

for some metric m on Ω. Then, the center of the object will be the pixel that maximizes the previous expression:

$$p_c = \text{argmax}_{p_i \in S}\, d_{S_b}(p_i).$$

Topologically, finding this center is equivalent to find the center of the largest ball contained in S, where the ball is defined by the metric m. Hence, it might be of interest in scenarios with round-like objects.

– *Minimum distance to the farthest point to the boundary*: The center expression is similar to the previous expression by changing min to max. First, we need to redefine d_{S_b} as follows:

$$d_{S_b}(p_i) = \max_{p_i \in S_b} m((x_i, y_i), (\tilde{x}_i, \tilde{y}_i)),$$

where m is again some metric on Ω, and the center will be the value that minimizes the previous expression:

$$p_c = \text{argmin}_{p_i \in S}\, d_{S_b}(p_i).$$

Topologically, finding this center is equivalent to find the center of the smallest ball that contains S, where the ball is defined by the metric m.

In terms of vector comparison, a large number of different metrics are available for point-to-point distance measurement. However, as we are modelling solid objects in $\mathbb{R}^2$, we can also consider object-related metrics, such as the geodesic distance [9].

4   Case of Study

In this section, we illustrate some examples of application of our proposal to liver segmentation. Specifically, the images has been taken from a project on non-alcoholic hepatic steatosis, gathered in collaboration with the Clinica Universitaria de Navarra (Pamplona, Spain). Figure 2 shows the visually representable MRI images that we have used in the experimental results. Our goal with this case of study is to illustrate how the different alternatives in the configuration of the workflow have an impact in the final results of both the center detection procedure and the ulterior ACM.

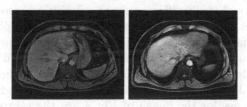

Fig. 2. Original images used in the experimental results in Sect. 4.

In the definition of the algorithm, we have used the proposal in Sect. 3. At the Object Selection phase, we have chosen 0.5 as a threshold, and the parameters in the spatial 2-fuzzy means are $m = 5$ and $p = q = 1$. At the center detection phase, we apply the following three alternatives: (a) The minimum distance to the closest point in the object using the geodesic distance; (b) the minimum distance to the farthest point using the Euclidean distance; and (c) the centroid.

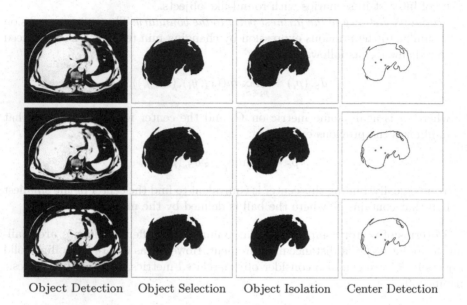

Object Detection   Object Selection   Object Isolation   Center Detection

Fig. 3. In the object detection phase, the Manhattan (upper row), Euclidean (middle row), and Chebyshev (lower row) distances are used for fuzzy clustering. In the center detection phase, we have applied the maximum distance to the closest point using the geodesic distance (red), the minimum distance to the farthest point using the Euclidean distance to the second image (purple) and the centroid (blue). (Color figure online)

Figures 3 and 4 show the results of each of the four phases of the algorithm for the images in Fig. 2. Figure 3 shows the results of the image on the left and Fig. 4, of the image on the right. The last column of these figures shows the center obtained using the three alternatives described above. First, we observe how the distance applied on the clustering does have an effect on the profiling of the object, although such differences are relatively erased after the object selection and object isolation phase. The center detection phase does produce severe differences on the final result of the procedure, which can be directly related to the characteristics of the metric (or centroid) used at such phase.

Overall, ee see that the best strategies for the image of Fig. 3 are the centroid and the minimum distance to the farthest point using the Euclidean distance and the best strategies for the image of Fig. 4 are the centroid and the maximum distance to the closest point using the geodesic distance. Therefore, from this case of study, we can state that there is no optimal strategy for all image, and the parameterization of the workflow needs to adapted to the specific characteristics of the image or image dataset. In order to quantitatively select one parameterization for a whole dataset, a quality measure should be defined for the problem, which is currently absent from literature.

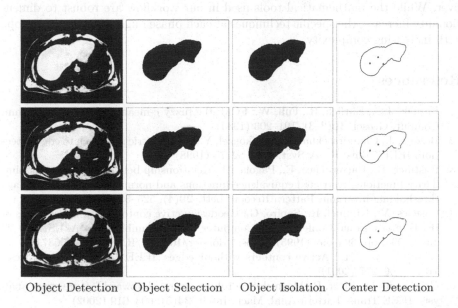

Object Detection   Object Selection   Object Isolation   Center Detection

Fig. 4. In the object detection phase, the Manhattan (upper row), Euclidean (middle row), and Chebyshev (lower row) distances are used for fuzzy clustering. In the center detection phase, we have applied the maximum distance to the closest point using the geodesic distance (red), the minimum distance to the farthest point using the Euclidean distance to the second image (purple) and the centroid (blue). (Color figure online)

5   Conclusions

In this paper we propose a strategy to find initial contours for ACMs, with the final aim of image segmentation in grayscale images. Our proposal is presented as a sequence of four phases represented as functions, so that the workflow materializes as function composition. The four phases in the strategy are:

Object detection. Usage of a spatial 2-fuzzy means algorithm to detect the membership degree of the pixels of the grayscale image.

Object selection. Transformation of the image representing the membership degree of the pixels into a binary image which contains the relevant object.

Object isolation. Application of the restrictions of the relevant object to refine the binary image obtained in the previous phase.

Center selection. Implementation of some strategies to calculate the center of the relevant object around which we will find the initial curve to apply the ACM.

As future work, we intend to extend the worflow to 3D images created as stacked MRI visualizations, so as to perform volumetric segmentations of the liver. While the mathematical tools used in our workflow are robust to dimensionality increase, the specific techniques at each phase might be adapted to cope with increasing complexity.

References

1. Bezdek, J.C., Ehrlich, R., Full, W.: FCM: the fuzzy c-means clustering algorithm. Comput. Geosci. **10**(2–3), 191–203 (1984)
2. Bezdek, J., Chandrasekhar, R., Attikouzel, Y.: A geometric approach to edge detection. IEEE Trans. Fuzzy Syst. **6**(1), 52–75 (1998)
3. Bustince, H., Barrenechea, E., Pagola, M.: Relationship between restricted dissimilarity functions, restricted equivalence functions and normal EN-functions: image thresholding invariant. Pattern Recogn. Lett. **29**(4), 525–536 (2008)
4. Caselles, V., Kimmel, R., Sapiro, G.: Geodesic active contours. In: Proceedings of IEEE International Conference on Computer Vision, Cambridge, MA, USA, 20–23 June 1995, pp. 694–699 (1995). https://doi.org/10.1109/ICCV.1995.466871
5. Chan, T., Vese, L.: Active contours without edges. IEEE Trans. Image Process. **10**(2), 266–277 (2001)
6. Comaniciu, D., Meer, P.: Mean shift: a robust approach toward feature space analysis. IEEE Trans. Pattern Anal. Mach. Intell. **24**(5), 603–619 (2002)
7. Cormen, T.H., Leiserson, C.E., Rivest, R.L., Stein, C.: Introduction to Algorithms, 3rd edn. MIT Press, Cambridge (2009)
8. Crandall, M.G., Ishii, H., Lions, P.L.: Uniqueness of viscosity solutions of Hamilton-Jacobi equations revisited. J. Math. Soc. Jpn. **39**(4), 581–596 (1987)
9. Han, J., Kamber, M., Pei, J. (eds.): Data Mining: Concepts and Techniques. Morgan Kaufmann, Burlington (2012)
10. Kass, M., Witkin, A., Terzopoulos, D.: Snakes: active contour models. Int. J. Comput. Vision **1**(4), 321–331 (1988)
11. Kichenassamy, S., Kumar, A., Olver, P., Tannenbaum, A., Yezzi, A.: Gradient flows and geometric active contour models. In: Proceedings of IEEE International Conference on Computer Vision, Cambridge, MA, USA, 20–23 June 1995, pp. 810–815 (1995). https://doi.org/10.1109/ICCV.1995.466855
12. Li, C., Xu, C., Gui, C., Fox, M.D.: Distance regularized level set evolution and its application to image segmentation. IEEE Trans. on Image Processing **19**(12), 3243–3254 (2010)
13. Li, X., Luo, S., Li, J., et al.: Liver segmentation from CT image using fuzzy clustering and level set. J. Sig. Inf. Process. **4**(03), 36 (2013)
14. Ling, H., Zhou, S.K., Zheng, Y., Georgescu, B., Suehling, M., Comaniciu, D.: Hierarchical, learning-based automatic liver segmentation. In: Proceedings of the IEEE Conference on Computer Vision and Pattern Recognition, pp. 1–8 (2008)
15. Lopez-Maestresalas, A., De Miguel, L., Lopez-Molina, C., Arazuri, S., Bustince, H., Jaren, C.: Hyperspectral imaging using notions from type-2 fuzzy sets. Soft. Comput. **23**(6), 1779–1793 (2018). https://doi.org/10.1007/s00500-018-3208-8

16. Marco-Detchart, C., Lopez-Molina, C., Fernandez, J., Bustince, H.: A gravitational approach to image smoothing. In: Kacprzyk, J., Szmidt, E., Zadrożny, S., Atanassov, K.T., Krawczak, M. (eds.) IWIFSGN/EUSFLAT -2017. AISC, vol. 642, pp. 468–479. Springer, Cham (2018). https://doi.org/10.1007/978-3-319-66824-6_41

17. Otsu, N.: Threshold selection method for gray-level histograms. IEEE Trans. Syst. Man Cybern. **9**(1), 62–66 (1979)

18. Perona, P., Malik, J.: Scale-space and edge detection using anisotropic diffusion. IEEE Trans. Pattern Anal. Mach. Intell. **12**(7), 629–639 (1990)

19. Rosin, P.L.: Unimodal thresholding. Pattern Recogn. **34**(11), 2083–2096 (2001)

20. Terzopoulos, D., Witkin, A., Kass, M.: Constraints on deformable models: recovering 3D shape and nonrigid motion. Artif. Intell. **36**(1), 91–123 (1988)

21. Weickert, J.: Anisotropic Diffusion in Image Processing, ECMI Series. Teubner-Verlag (1998)

Fuzzy Methods in Data Mining and Knowledge Discovery

Improving Text Clustering Using a New Technique for Selecting Trustworthy Content in Social Networks

J. Angel Diaz-Garcia(✉) [iD], Carlos Fernandez-Basso [iD],
Karel Gutiérrez-Batista [iD], M. Dolores Ruiz [iD], and Maria J. Martin-Bautista [iD]

Department of Computer Science and A.I., University of Granada, Granada, Spain
{jagarcia,cjferba,karel,mdruiz,mbautis}@decsai.ugr.es

Abstract. Today's information society has led to the emergence of a large number of applications that generate and consume digital data. Many of these applications are based on social networks, and therefore their information often comes in the form of unstructured text. This text from social media also tends to contain a high level of noise and untrustworthy content. Therefore, having systems capable of dealing with it efficiently is a very relevant issue. In order to verify the trustworthiness of the social media content, it is necessary to analyse and explore social media data by using text mining techniques. One of the most widespread techniques in the field of text mining is text clustering, that allows us to automatically group similar documents into categories. Text clustering is very sensitive to the presence of noise and so in this paper we propose a pre-processing pipeline based on word embedding that allows selecting trustworthy content and discarding noise in a way that improves clustering results. To validate the proposed pipeline, a real use case is provided on a Twitter dataset related to COVID-19.

Keywords: Clustering · Pre-processing · Social media mining

1 Introduction

Nowadays the world in we live, are very influenced by social networks. Every day, we are consuming o generating social networks data. These data, usually come in form of user-generated content, or in other words, unstructured data. Being able to process and analyse the data properly is a very arduous task in which Artificial Intelligence (A.I.) is taking a leading role. Among the A.I. techniques aimed at obtaining relevant information about these unstructured data are supervised techniques such as classification [19], or descriptive techniques such as association rules [16,17] or clustering. Clustering is one of the most widespread A.I. techniques, with great results in various fields of application such as energy [38], health [6], or economics [23]. Due to its potential to obtain hidden groups in data without prior labelling, clustering is also very relevant in social media analysis problems. Some of its most relevant applications have been

© Springer Nature Switzerland AG 2022
D. Ciucci et al. (Eds.): IPMU 2022, CCIS 1602, pp. 275–287, 2022.
https://doi.org/10.1007/978-3-031-08974-9_22

in community analysis [7] or text mining through document clustering [8]. Due to its importance for these sectors, it is necessary to have increasingly reliable clustering techniques.

In this paper, we propose a new text filtering technique to improve clustering results on microblog social networks. Social networks contain a lot of noise and unhelpful or untrustworthy content. This content does not contribute anything, but it is capable of causing traditional data mining techniques to underperform [14]. Therefore, detecting and eliminating it is a relevant task. Our approach is based on selecting trustworthy accounts for a given analysis as well as their content. On the other hand, we eliminate those that are not related to our topic of study, as these accounts have a high probability of generating noise and untrustworthy content. To do this, the system uses on Twitter biographies, as these are typically used to report on professions and interests, so we can create a system that can automatically select which accounts are useful for a topic and which are not in the way a human could. A user on Twitter, to decide whether to believe a certain content or not, would visit the account of the user issuing that tweet and contrast the content of the tweet with the information provided by that person in his or her biography. For example, if a user is reading a tweet about COVID-19, the first thing he or she will do is to check who has issued the tweet. If the author of the tweet provides in his biography information about his profession, such as whether he is a doctor or works in a certain hospital, these will be valuable arguments to give more credibility to the tweet. In this paper, we propose an automatic technique to perform this task of account contrasting, so that we can remove noise and untrustworthy content from the analysis and improve the clustering results. To do this, the system uses the potential of Word Embbedings (W.E.).

Two of the most famous W.E. are Word2Vec [29,30] and FastText [12]. Both are based on the representation of each of the words present in a vocabulary by means of vectors in a vector space so that the distance between words can be operated mathematically. The distance between one word and another will correspond to their semantic relationship. That is, words with a greater vector distance will have less relation and words with a very similar vector distance can be considered synonyms. In this paper, we will use this potential to find similarities between words to guide an automatic and incremental filtering of users in relation to a given topic. The proposed system will be validated in a real problem related to COVID-19, in which we seek to obtain opinion clusters in the social network Twitter.

The paper is organized as follows: Next section focuses on the study of related work. In Sect. 3 we go into detail in the proposed text filtering for content-based. In Sect. 4, we provide the results of the experimentation as well as the parameters used. Finally, in Sect. 5 we examine the conclusions and the future work.

2   Related Works

Regarding the improvement of clustering results, most of the existing studies in the literature are based on the optimization of the algorithm [9,32,35,37]. These approaches are based on improving the selection of the initial parameters of the algorithms such as the initial centroids, the calculation of distances between the centroids and the examples of each of the clusters. While optimisation of algorithms is undoubtedly one of the vital parts of improving today's A.I.-based systems, we should not forget that proper data pre-processing can have even better implications on the final result.

Many of the approaches to pre-processing data for clustering are based on feature selection. By means of this technique, they try to solve one of the biggest problems in text clustering: the sparsity in document-term matrices. Before applying clustering on a document, it must be represented by a matrix in which the rows are the documents and the columns are the terms present in it. Each cell of the matrix will have the frequency of a term in a document. Using this structure a vector representation can be obtained, usually based on TF-IDF. A detailed explanation of the calculation of the TF-IDF and the creation of the vector space for clustering can be studied in [36]. The problem with these matrices is that they are very sparse. The fact that they are sparse is useful for calculating distances and correctly locating clusters, but in many cases this process is heavily biased by words that introduce noise and are not really meaningful. The detection and elimination of these non-useful words in order to achieve a dataset with the best possible characteristics to improve clustering, has been tackled by several A.I. techniques, such as principal component analysis techniques [13], using evolutionary algorithms [2], or even using algorithms based on harmony search [3]. Recently, there is also an attempt to reduce the noise present in the matrices by means W.E. based techniques [10,34].

All of these approaches aim to keep all documents present in the analysis and remove certain features from each of them. Our approach, in contrast to what has been studied in the literature, is to keep all those words but only from those documents that are really useful for a given analysis. In this way, we achieve more cohesive matrices, without having to eliminate characteristics (words), since one of the problems of clustering is usually that it does not have enough words to correctly categorise a document. To select the documents (tweets) that should be deleted, we rely on the trustworthiness of the user issuing them; if a user is trustworthy, we will take the document into account, if not, we will not. The study of the experience or location of trustworthy sources of information, has been approached from multiple perspectives. Some of the most widespread are based on analysing user content and concepts such as engagement (level of interaction with the community) to determine if content is credible or not [4,5]. In our case, we focus on detecting accounts that denote expertise in a sector, and therefore generate useful content related to that sector or domain. In this way, there are also systems such as Cognos [18] or CredSat [1]. These systems offer recommendations of expert users on a topic, making use of lists or again, of the content generated. The ultimate goal of these systems is to retrieve reliable or valuable users, as opposed to our system that seeks to filter a large set of data using the concept of expertise in order to obtain better data structures and improve downstream data mining processes. To the best of our knowledge, this is the first paper that proposes a trustworthy-based filtering approach to reduce the dimensionality of a problem and improve clustering results.

3 Trustworthy Content Selection Pipeline

In this section our proposal is detailed.

3.1 System Architecture

In Fig. 1 we can see a complete scheme of our proposal. The different elements that comprise the proposal shown in the figure will be analysed in detail throughout this section. We start from a database of tweets. We have to bear in mind that from all the metadata that the APIs and tweet databases offer us, we will only keep those relating to the user, such as their username, location, number of followers, favourites, lists, friends, biography and content of the tweet. From all these elements, the really interesting ones are the biography, which we will use for the trustworthy content filter, and the content, which we will use for later stages of data mining, in this case clustering.

The power of biographies lies in the fact that normally are used by people to inform about their professions or interests. therefore, we propose a system capable of automatically detect words related to professions and jobs, and create a list of subject matter experts. At the same time selecting their content as trustworthy for further stages of data mining.

The system is composed of a pre-processing module, a module based on the selection of trustworthy content based on user biographies and finally a layer dedicated to data mining through clustering. Given the importance of these modules in the system, we will see them in detail in the following sections.

3.2 Text Pre-processing

The pre-processing uses common cleaning techniques for user-generated text. This step is applied to the user's biography and to the tweet content. The techniques used are:

1. Twitter domain related cleaning. For this purpose we have eliminated URLs, hashtags, mentions, reserved words from Twitter (RT, FAV...), emojis and smileys.
2. Text mining domain cleaning: This comprises removing numbers, additional spaces and punctuation marks.
3. Turning the text into lowercase letters.
4. Detecting the tweet language. All those tweets using a non-recognised language or from a language other than the one desired by the user are eliminated.
5. Stop words removal.
6. Any empty tweets are removed.
7. Tokenization of the biography and the tweet.

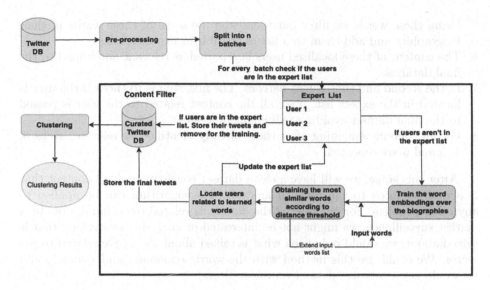

Fig. 1. Data flow and architecture of our proposal.

3.3 Content Filter

For the content filter, we have based it on the premise that if a person has knowledge on a subject, the content that this particular user generates on a topic related to his or her area of expertise will be of high value for the analysis. In contrast, content generated by a user with no knowledge of the topic will generally correspond to noise. As an example, if we look at the COVID-19 use case, it will be more useful to analyse 10 000 tweets created by doctors and researchers than 1 000 000 tweets generated by users who are simply talking about the topic without any real background in it. Therefore, the content filter is based on locating users who are experts in the subject matter and its content. To do this, we will focus on their biographies. The system works as follows (in Fig. 1, we can find the data flow of this filter):

1. The filter method receive as input a list of words related to the topic of analysis, e.g. if we want relevant information about COVID-19 and its symptoms, the initial word list could be *medical, doctor* and *epidemiology.*
2. Using an adaptation of the divide and conquer approach, the dataset resulting from the pre-processing stage is split into different batches.
3. In the first batch, the structures are empty so there are no experts located. Therefore, word embedding is trained on the biographies. The W.E. used for this stage was FastText.
4. On the resulting vector space, we obtain those words that exceed the threshold of 0.6 in similarity over the initial list. These words are added to the search list so that at each iteration the list grows and improves covering more aspects of the problem domain. In this case, in each iteration, the system learns words related to medicine.

5. Using these words we filter out users who use some of these words in their biographies and add them to a list of credible or expert users.
6. The content of these localised users is reported as relevant and added to the final database.
7. In the second and subsequent batches. The first step is to check if the user is located in the expert list, if so, all the content related to the user is passed to the final dataset avoiding redundant processes.
8. For all content remaining after those already identified as relevant, steps 3, 4, 5 and 6 are repeated.

After this stage, we will have a clean dataset consisting only of content that is actually related to the topic of analysis. The algorithm can be applied to any other problem. For example, if the dataset is related to economics or stock market surveillance, we might not be interested in analysing everything that is talked about, we would only need what is talked about by people related to the sector. We could use this method with the words *economist* and *economy* and we would get a cohesive dataset on the subject.

3.4 Clustering

In the final stage of the pipeline we use clustering on the filtered dataset. We must take into account that the level of dimensionality reduction is around 90%, so we will have fewer documents with, also, fewer words. Therefore, we have a more cohesive matrix to apply the K-Means clustering algorithm.

The k-means clustering algorithm is widely known for clustering data using different types of distances [27]. It is based on a calculation of the distance between different data using a number of base centroids that are determined by a parameter.

The main advantages of the k-means method are that it is a simple and fast method [24]. In addition, this algorithm works well with large or small data sets, is efficient and performs well.

Specifically, this clustering algorithm has been widely used for tex mining. Thanks to its advantages such as being able to use different distance measures, to use it with large datasets and to parameterise the number of clusters to extract, it has been used in many applications such as sentiment analysis [26], recommender systems [11] and text analysis [21].

4 Experimentation

For the experimentation, we have proposed a hypothesis test having as a starting hypothesis that the filtering of our dataset improves the clustering result. To demonstrate this, a comparative study has been carried out on the results of the K-Means algorithm. The comparison is made between the case of using the pre-processing seen in Sect. 3.2 together with the content filtering seen in Sect. 3.3 and the case of only using the pre-processing seen in Sect. 3.2, i.e. without filtering the content but pre-processing the text.

4.1 Dataset

The disease caused by the new Coronavirus (Sars-Cov-2) [39], first reported in Wuhan [20] in December 2019, now affects the entire world and is considered one of the largest pandemics in the history of mankind. News and information about the pandemic is generated on a daily basis. Analysing them automatically is an important task. One of the most used channels for disseminating information about COVID-19 is Twitter, so our datasets come from real data from this social network. Currently, the tweet dataset [25] related to COVID-19 includes millions of entries, which makes it a perfect candidate for our use case. Our problem, uses random samples of that dataset comprising a total of 2 626 275 tweets.

4.2 Evaluation Metrics

In order to prove the feasibility of the proposed approach, we have used two automatic metrics: Silhouette coefficient [31], and Davies-Bouldin score [15]. Both metrics will allow the evaluation of the obtained results using our approach and without using it. The reason for using these evaluation techniques is that they are two of the best known, most widely used and complementary evaluation techniques. With the silhouette coefficient we can check how good the separation between clusters is, and with the Davies-Bouldin score, we can check the goodness of the examples within each cluster.

Silhouette Coefficient. The Silhouette coefficient has been widely used in clustering problems because it determines the quality of separation and cohesion of the obtained clusters. One of the main usages of this metric is to obtain the optimal number of clusters for which a clustering algorithm shows a better performance [31].

The coefficient is computed using the mean intra-cluster similarity $a(i)$ and the mean nearest-cluster similarity $b(i)$ for each sample (Eq. 1). The overall Silhouette coefficient is the mean Silhouette coefficient of all samples. The values are in the range [–1,1], being the best results those values that are close to 1.

$$S(i) = \frac{a(i) - b(i)}{max\left\{a(i), b(i)\right\}} \tag{1}$$

Davies-Bouldin Score. Like the Silhouette coefficient, the Davies-Bouldin score allows the evaluation of the results of the clustering algorithms [15]. The Davies-Bouldin score is the average similarity measure of each cluster with its most similar cluster, where similarity is the ratio of within-cluster distances to between-cluster distances (Eq. 3).

$$r_k = \frac{1}{|A_k|} \sum_{x_i \in A_k} d(x_i, C_k) \tag{2}$$

$$DB = \frac{1}{N} \sum_{i}^{N} \sum_{j}^{N} \max_{i \neq j} \frac{r_i + r_j}{d(C_i, C_j)} \tag{3}$$

where r_i and r_j are represented in Eq. 2 (intra-cluster distance), and $d(C_i, C_j)$ is the distance between the centroids C_i and C_j.

It means that clusters that are farther apart and less dispersed will have a better score. The minimum possible value is zero, with lower values indicating better clustering results, unlike the Silhouette coefficient.

4.3   Clustering Results

In this section we present the final results of clustering on the filtered dataset and compare it with the unfiltered dataset. In order to take data so that the results are biased by the number of tweets, a random sample of 10000 tweets was obtained from the original data and the filtered one. So we have 10000 tweets from the filtered dataset and 10000 from the original dataset. In this way we try to mitigate possible biases that might arise from the size of the dataset, since the unfiltered dataset is much larger. These 10000 tweets are obtained randomly for each of the executions, trying to mitigate also that the random choice corresponds to an optimal solution, and favouring the generalisation of the experimentation. So the comparison can be made on the same grounds, and a better result in this case will indicate that the filtering favours the result.

Table 1 shows the average results of 10 runs in terms of Silhouette Score and Davies Bouldin Score. We can see how the results in both metrics are better in the case of filtering the text using our proposal. We must take into account that, although the results may not look good, we are facing a clustering problem in short texts [22]. In these problems, there are few words in each of the documents, in which case, the clustering algorithms do not usually have very high values in terms of fit measures.

Table 1. Average result of the runs on the dataset filtered with our proposal and without filtering.

Metric	Filtered	Unfiltered
Silhouette score	**0,0229**	0,00606
Davies Bouldin	**4,74957**	5,65733

Apart from the goodness of fit measures, to graphically compare the obtained results, we have represented them by means of a t-distributed stochastic neighbour embedding (TSNE) graph [28]. Figure 2 shows the clustering results in the case of filtering the data while Fig. 3 shows the results in the case of no filtering. If we analyse the results we can see that in the first case, we can easily locate 11 clusters, while in the case of not filtering by content, we can only identify 7 clusters.

If we take into account that the final objective of our filtering process is to favour the subsequent stages of data mining, clustering in this case, we see a clear improvement in the case of filtering versus not filtering. In a text mining problem, it is very likely that a large number of clusters will appear, depending on the documents. Specifically in these cases, a battery of tests has been carried out to determine the number of clusters that minimises the error, which is around 26. Therefore, being able to manually locate a greater number of clusters is an indicator that the clustering process is better, and therefore, the preprocessing applied actually works.

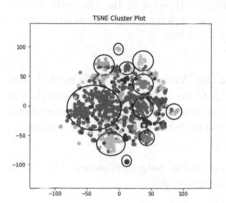

Fig. 2. TSNE plot using the content filter

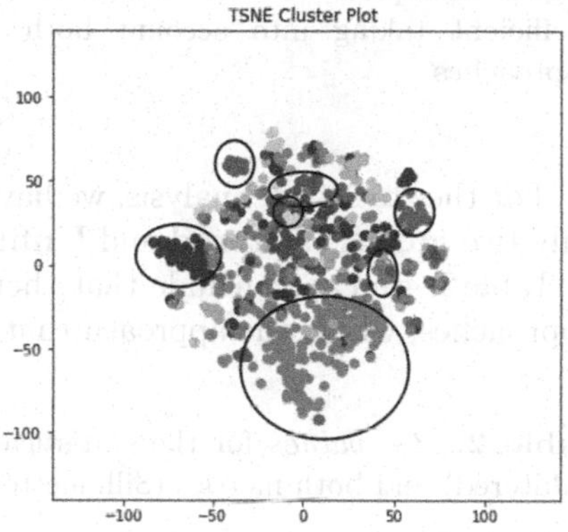

Fig. 3. TSNE plot without filter

For a more detailed analysis and to support the obtained results, we have conducted a statistical analysis to determine whether there are any significant differences among the obtained values for the two approaches (filtered and unfiltered). Figures 4 and 5, depict the boxplots for the Silhouette coefficient and Davies-Bouldin score respectively, regarding the two approaches. At first sight, we can see that the best results are yielded for both metrics when we use our approach.

We can see how the distribution of results, in both cases, is better when using our filtering method. The box is always wider in the case of filtering because the same accounts and content are not always selected. There is a certain random component to word embedding filtering. This causes the result to fluctuate more in the filtering case than in the non-filtering case where it always runs on similar terms. To justify that the improvement is not due to randomness, we have performed numerous runs and statistical tests.

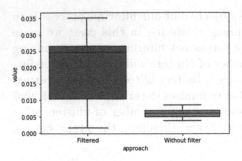

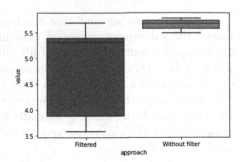

Fig. 4. Boxplot of the Silhouette coefficient taking into account both approaches

Fig. 5. Boxplot of the Davies-Bouldin score taking into account both approaches

For the statistical analysis, we have used the Wilcoxon's test [33] as there are only two groups (**Filtered** and **Unfiltered**). Considering the $p-values$ shown in Table 2, we can conclude that there are significant differences between both approaches, where the approach that applies the filter offers the best results.

Table 2. $P-values$ for the statistical analysis using both approaches (filtered an unfiltered) and both metrics (Silhouette and Davies-Bouldin).

Measure	p-value
Silhouette coefficient	0.02182
Davies-Bouldin score	0.00691

5   Conclusions and Future Work

In this paper we have proposed a useful data filtering technique to improve the textual clustering result on user-generated content in social networks. The values obtained and the statistical tests carried out show that the results are improved compared to not using the proposed filter. These conclusions are also obtained by visualising the clusters generated on a real problem related to COVID-19, where it can be seen how the clusters are of a better quality at a glance.

Is worth to highlight that the system can be very sensitive to lies in the biographies and to possible groups of well-informed users but directed with the intention of misinforming other users or social groups. As future work, a multidimensional solution to mitigate these possible problems of lies should be studied.

Acknowledgment. The research reported in this paper was partially supported by the Andalusian government and the FEDER operative program under the project Big-DataMed (P18-RT-2947 and B-TIC-145-UGR18) and grant PLEC2021-007681 funded by MCIN/AEI/10.13039/501100011033 and by the European Union NextGenerationEU/PRTR. Finally the project is also partially supported by the Spanish Ministry of Education, Culture and Sport (FPU18/00150).

References

1. Abu-Salih, B., Wongthongtham, P., Chan, K.Y., Zhu, D.: CredSat: credibility ranking of users in big social data incorporating semantic analysis and temporal factor. J. Inf. Sci. **45**(2), 259–280 (2019)
2. Abualigah, L.M., Khader, A.T., Al-Betar, M.A.: Unsupervised feature selection technique based on genetic algorithm for improving the text clustering. In: 2016 7th International Conference on Computer Science and Information Technology (CSIT), pp. 1–6. IEEE (2016)
3. Abualigah, L.M., Khader, A.T., AlBetar, M.A., Hanandeh, E.S.: Unsupervised text feature selection technique based on particle swarm optimization algorithm for improving the text clustering. In: 1st EAI International Conference on Computer Science and Engineering, p. 169. European Alliance for Innovation (EAI) (2016)
4. Alrubaian, M., Al-Qurishi, M., Hassan, M.M., Alamri, A.: A credibility analysis system for assessing information on twitter. IEEE Trans. Depend. Secure Comput. **15**(4), 661–674 (2018). https://doi.org/10.1109/TDSC.2016.2602338
5. Alrubaian, M., AL-Qurishi, M., Alrakhami, M., Hassan, M., Alamri, A.: Reputation-based credibility analysis of Twitter social network users: reputation-based credibility analysis of Twitter social network users. Concurrency Comput. Pract. Exp. **29** (2016). https://doi.org/10.1002/cpe.3873
6. Alshabeeb, I.A., Ali, N.G., Naser, S.A., Shakir, W.M.: A clustering algorithm application in Parkinson disease based on k-means method. Comput. Sci. **15**(4), 1005–1014 (2020)
7. Arenas, A., Danon, L., Díaz-Guilera, A., Gleiser, P.M., Guimerá, R.: Community analysis in social networks. Eur. Phys. J. B **38**(2), 373–380 (2004). https://doi.org/10.1140/epjb/e2004-00130-1
8. Arpaci, I., et al.: Analysis of Twitter data using evolutionary clustering during the Covid-19 pandemic. Comput. Mater. Continua **65**(1), 193–204 (2020)
9. Arthur, D., Vassilvitskii, S.: k-means++: the advantages of careful seeding. Technical report, Stanford (2006)
10. Asyaky, M.S., Mandala, R.: Improving the performance of HDBSCAN on short text clustering by using word embedding and UMAP. In: 2021 8th International Conference on Advanced Informatics: Concepts, Theory and Applications (ICAICTA), pp. 1–6 (2021). https://doi.org/10.1109/ICAICTA53211.2021.9640285
11. Berry, M.W., Castellanos, M.: Survey of text mining. Comput. Rev. **45**(9), 548 (2004)
12. Bojanowski, P., Grave, E., Joulin, A., Mikolov, T.: Enriching word vectors with subword information. arXiv preprint arXiv:1607.04606 (2016)
13. Chaudhary, G., Kshirsagar, M.: Enhanced text clustering approach using hierarchical agglomerative clustering with principal components analysis to design document recommendation system. Adv. Res. Comput. Eng. Res. Transcripts Comput. Electr. Electron. Eng. **2**, 1–18 (2021)
14. Dave, R.N.: Characterization and detection of noise in clustering. Pattern Recogn. Lett. **12**(11), 657–664 (1991)
15. Davies, D.L., Bouldin, D.W.: A cluster separation measure. IEEE Trans. Pattern Anal. Mach. Intell. **PAMI-1**(2), 224–227 (1979)

16. Diaz-Garcia, J.A., Fernandez-Basso, C., Ruiz, M.D., Martin-Bautista, M.J.: Mining text patterns over fake and real tweets. In: Lesot, M.-J., et al. (eds.) IPMU 2020. CCIS, vol. 1238, pp. 648–660. Springer, Cham (2020). https://doi.org/10.1007/978-3-030-50143-3_51

17. Diaz-Garcia, J.A., Ruiz, M.D., Martin-Bautista, M.J.: Non-query-based pattern mining and sentiment analysis for massive microblogging online texts. IEEE Access **8**, 78166–78182 (2020). https://doi.org/10.1109/ACCESS.2020.2990461

18. Ghosh, S., Sharma, N., Benevenuto, F., Ganguly, N., Gummadi, K.: Cognos: crowdsourcing search for topic experts in microblogs. In: Proceedings of the 35th International ACM SIGIR Conference on Research and Development in Information Retrieval, pp. 575–590 (2012)

19. Godara, N., Kumar, S.: Twitter sentiment classification using machine learning techniques. Waffen-Und Kostumkunde J. **11**(8), 10–20 (2020)

20. Huang, C., et al.: Clinical features of patients infected with 2019 novel coronavirus in Wuhan, China. Lancet **395**(10223), 497–506 (2020)

21. Jalil, A.M., Hafidi, I., Alami, L., Ensa, K.: Comparative study of clustering algorithms in text mining context (2016)

22. Jin, C., Zhang, S.: Micro-blog short text clustering algorithm based on bootstrapping. In: 2019 12th International Symposium on Computational Intelligence and Design (ISCID), vol. 2, pp. 264–266. IEEE (2019)

23. Jin, Y., Liu, Y., Zhang, W., Zhang, S., Lou, Y.: A novel multi-stage ensemble model with multiple k-means-based selective undersampling: an application in credit scoring. J. Intell. Fuzzy Syst. 1–14 (2021, Preprint)

24. Kodinariya, T.M., Makwana, P.R.: Review on determining number of cluster in k-means clustering. Int. J. **1**(6), 90–95 (2013)

25. Lamsal, R.: Coronavirus (Covid-19) tweets dataset (2020). https://doi.org/10.21227/781w-ef42

26. Li, N., Wu, D.D.: Using text mining and sentiment analysis for online forums hotspot detection and forecast. Decis. Support Syst. **48**(2), 354–368 (2010)

27. Likas, A., Vlassis, N., Verbeek, J.J.: The global k-means clustering algorithm. Pattern Recogn. **36**(2), 451–461 (2003)

28. Maaten, L.v.d., Hinton, G.: Visualizing data using T-SNE. J. Mach. Learn. Res. **9**(Nov), 2579–2605 (2008)

29. Mikolov, T., Chen, K., Corrado, G., Dean, J.: Efficient estimation of word representations in vector space. arXiv preprint arXiv:1301.3781 (2013)

30. Mikolov, T., Sutskever, I., Chen, K., Corrado, G.S., Dean, J.: Distributed representations of words and phrases and their compositionality. Adv. Neural. Inf. Process. Syst. **26**, 3111–3119 (2013)

31. Rousseeuw, P.: Silhouettes: a graphical aid to the interpretation and validation of cluster analysis. J. Comput. Appl. Math. **20**(1), 53–65 (1987)

32. Shi, K., Li, L., He, J., Zhang, N., Liu, H., Song, W.: Improved GA-based text clustering algorithm. In: 2011 4th IEEE International Conference on Broadband Network and Multimedia Technology, pp. 675–679. IEEE (2011)

33. Wilcoxon, F.: Individual comparisons by ranking methods. Biometrics Bull. **1**(6), 80–83 (1945)

34. Xingliang, M., Fangfang, L.: Clustering of short text in micro-blog based on k-means algorithm. In: 2018 IEEE International Conference of Safety Produce Informatization (IICSPI), pp. 812–815 (2018). https://doi.org/10.1109/IICSPI.2018.8690507

35. Yedla, M., Pathakota, S.R., Srinivasa, T.: Enhancing k-means clustering algorithm with improved initial center. Int. J. Comput. Sci. Inf. Technol. **1**(2), 121–125 (2010)

36. Yuan, S., Wenbin, G.: A text clustering algorithm based on simplified cluster hypothesis. In: 2013 2nd International Symposium on Instrumentation and Measurement, Sensor Network and Automation (IMSNA), pp. 412–415 (2013). https://doi.org/10.1109/IMSNA.2013.6743303

37. Zhang, G., Zhang, C., Zhang, H.: Improved k-means algorithm based on density canopy. Knowl.-Based Syst. **145**, 289–297 (2018)

38. Zhang, G., Li, Y., Deng, X.: K-means clustering-based electrical equipment identification for smart building application. Information **11**(1), 27 (2020)

39. Zhou, P., et al.: A pneumonia outbreak associated with a new coronavirus of probable bat origin. Nature **579**(7798), 270–273 (2020)

Fuzzy System-Based Solutions for Traffic Control in Freeway Networks Toward Sustainable Improvement

Mehran Amini[1,2]([✉]), Miklos F. Hatwagner[1,2], and Laszlo T. Koczy[1,2]

[1] Department of Information Technology, Szechenyi Istvan University, Gyor, Hungary
mehran@sze.hu
[2] Department of Telecommunications and Media Informatics,
Budapest University of Technology and Economics, Budapest, Hungary

Abstract. In the scientific community, the topic of traffic control for promoting sustainable transportation in freeway networks is a relatively new field of research that is becoming increasingly relevant. Sustainability is a critical factor in the design and operation of mobility and traffic systems, which impacts the development of freeway traffic control strategies. According to sustainable notions, freeway traffic controllers should be designed to maximize road capacity, minimize vehicle travel delays, and reduce pollution emissions, accidents, and fuel consumption. The problem is full of uncertainty, there is no way to model the whole system analytically, thus a fuzzy modeling approach seems to be not only adequate but necessary. In this study, a Fuzzy Cognitive Map based model (FCM) and a connected simple Fuzzy Inference System (FIS) are presented, as the tools to analyze freeway traffic data with the goal of traffic flow modeling at a macroscopic level, in order to address congestion-related issues as the core of the sustainability improvement strategies. Besides presenting a framework of Fuzzy system-based controllers in freeway traffic, the results of this work indicated that FIS and FCM are capable of realizing traffic control strategies involving the implementation of ramp management policies, controlling vehicle movement within the freeway by mainstream control, and routing vehicles along alternative paths via the execution of suitable route guidance strategy.

Keywords: FCM · FIS · Congestion prediction · Sustainability · Freeway networks

1 Introduction

The expanding number of vehicles has exacerbated traffic congestion, resulting in longer travel times and a decrease in driver confidence in the reliability of traffic services [1]. Moreover, congestion has become a global problem that hinders developing a robust and sustainable transportation infrastructure system. This problem is mainly caused by urbanization, expansion of the number of motor vehicles and associated infrastructure,

© Springer Nature Switzerland AG 2022
D. Ciucci et al. (Eds.): IPMU 2022, CCIS 1602, pp. 288–305, 2022.
https://doi.org/10.1007/978-3-031-08974-9_23

and the growth of ride-share and courier services. This latter becoming even more popular because of the recent pandemic situation. Congestion has been defined based on a variety of approaches [2]. Congestion in the traffic flow state is most commonly defined as when demand for travel exceeds the capacity of a road section, e.g., the freeway. Also, congestion arises when the normal flow of traffic is disrupted by a dense concentration of vehicles, resulting in higher travel time [3]. Despite substantial advancements in information and communication technology, it appears that full utilization of such novel technologies to reduce freeway traffic congestion has not been appropriately obtained [4]. Nevertheless, the rapid growth of personal vehicles has resulted in congestion on a daily basis, both recurrent and non-recurrent, spanning thousands of kilometers worldwide. Therefore, congestions significantly restrict accessible infrastructure capacity during rush hours when needed; consequently, delays, high environmental pollution, and decreased traffic safety occur. Such effects were noted frequently in the event of non-recurring congestions caused by incidents and freeway maintenance [5].

However, many current infrastructures cannot be modified to satisfy ever-increasing traffic volume, mainly due to physical and financial limitations. In this context, the progress of planning and managing tools for traffic systems remains critical in order to maximize the efficiency of the existing freeway network without requiring significant infrastructure improvements [6]. Due to a large portion of the freeway network being incapable of meeting current mobility needs, affecting drivers in the form of congestion, worsened air pollution, and declining safety, various researches have been conducted to advance planning and control techniques for freeway traffic networks. Former researchers were primarily concerned with mitigating congestion problems, but the current global roadmaps for eco-innovation in transportation systems necessitate the fulfillment of much better policies [7]. This requires a reframing of conventional control approaches for a more sustainable perspective because then control purposes cover not only the optimal use of freeway network capacity but the minimization of emission, fuel consumption, accidents, etc. [6].

Computational Intelligence (CI) methods as nature-inspired techniques have been used to address multi-criteria issues in real-world settings [8]. Neural networks, evolutionary computation, and fuzzy systems are the main CI based approaches [9]. Although the majority of existing CI methods to address the problem of vehicle traffic routing and congestion are based on evolutionary computation [10], fuzzy inference-based techniques are also widely applied in traffic-related problems [11]. However, the abilities of the CI techniques concerning sustainable freeway traffic control are relatively neglected. Significant characteristics of sustainable road traffic control mechanisms include traffic flow, dispersion, emission, consumption, and safety models [12]. Accordingly, in this study, flow-based modeling at a macroscopic level is considered to analyze freeway traffic data by employing Fuzzy Cognitive Maps (FCM) and Fuzzy Inference System (FIS) to address congestion-related issues as the core of the sustainability improvement strategies. In addition to providing a generic framework of Fuzzy system-based controllers in freeway traffic, the extended aim of this study is to contribute to the implementation of a sustainable and responsive traffic control and management system. The main contribution of the proposed system is modeling and computing imprecise traffic data at the macroscopic level and mitigating harmful economic, social, and environmental repercussions of congestion in freeway networks.

This work is conducted into five sections. Section 2 introduces traffic flow models with sustainable objectives. An FIS designed for congestion level prediction and an FCM developed for traffic flow simulation after introducing the case study are presented in Sect. 3. In Sect. 4, the results of both fuzzy system-based methods are analyzed, while some conclusions are drawn in Sect. 5.

2 Traffic Flow Models with Sustainable Objectives

Sustainable mobility encompasses a wide variety of issues, from environmental protection to social and economic growth. In this perspective, road transportation is critical for economic growth and social well-being, as it is still the most widely used mode of transport for passengers and commodities. In this view, sustainable transportation aims to meet economic and social needs while also providing a sustainable and accessible service that improves availability and connections for all users. Given the intricacy of the topic, the scientific community has been investigating transportation sustainability-related issues for several decades [13]. Among these problems, reducing traffic congestion becomes the prime common objective of conventional freeway traffic control systems. As previously highlighted, the expansion of freight and passenger mobility systems has contributed to the socio-economic growth of a society. However, it has also led to the spreading of congestion and, as a result, a deterioration of the existing mobility service. Such congestion can manifest itself in various ways, from just forming bottlenecks and increasing travel times to significantly deteriorating the system and bringing vehicular traffic to a halt.

Additionally, as acknowledged by the research conducted in [14], repeated exposure to congestion results in an increase in driver irritation, as drivers view the additional time required to reach their destination as wasted time that could be used for other purposes. Identifying appropriate traffic control actions is a possible approach for expediting the procedure towards resolving congestion and, subsequently, a better sustainable mobility system. Various control measures (Fig. 1) can be employed to manage the flow of traffic on a freeway network. The primary options include ramp management, i.e., ramp metering in conjunction with traffic lights at on-ramps; mainstream control, e.g., variable speed limits, keep-lane directives, lane control, and congestion warnings; and route guiding, i.e., typically, particular indications are displayed at junctions [15].

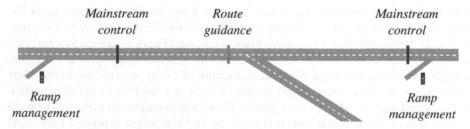

Fig. 1. Various traffic control strategies [15]

To design these control actions, an appropriate modeling approach needs to be defined, not only for describing traffic flow behaviors but also for assessing all sustainability-connected problems. Figure 2 depicts the primary modeling criteria for sustainable freeway traffic control mechanisms, among which the traffic flow-based model is discussed in further detail.

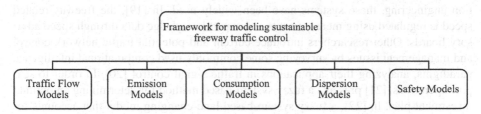

Fig. 2. Modeling framework for sustainable freeway traffic control methods [6]

Traffic flow models were developed in response to the requirement to represent the dynamic behavior of real-world traffic systems mathematically. Apart from evaluation and prediction of the system, traffic flow models can be used to define planning actions, evaluate the effects of new infrastructures or changes to current freeway layouts, and develop, simulate, and evaluate specific control mechanisms. Beginning with work by [16] in the 1950s, a diverse spectrum of traffic flow models with varying features and applications has been developed. Different criteria can be used to classify traffic models [17]. The commonest classification scheme for traffic flow representation is based on their level of detail, with macroscopic, microscopic, and mesoscopic models being the most prevalent.

The most typical and consolidated method of traffic regulation is at the macroscopic level (road-based control measures, i.e., mainstream control, ramp management). Further classification of macroscopic forms is based on the continuous or discrete character of the features representing time and space. Macroscopic discrete models are the most commonly used ones for freeway traffic control because their low degree of detail and discretization enable a low computational complexity, making them particularly well suited for real-time control systems in vast freeway traffic networks. The model's focus in this work is on discrete macroscopic characteristics, emphasizing overall vehicle behavior within hourly time intervals. In addition, instead of employing continuous variables, the related variables are discretized (both spatially and temporally), i.e., freeways are viewed as a collection of sections with fixed lengths, and time is correspondingly divided into distinct intervals [18]. Consequently, these features rely frequently on standard mathematical methodologies, which are regularly incapable of modeling the intricacy of road traffic characteristics and complex interactions among involved parameters. Moreover, these parameters are greatly affected by imprecise and uncertain qualities because of being imposed by constant dynamic behaviors of drivers; therefore, these properties and their varying levels of vagueness need to be included in mathematical reasoning.

3 Fuzzy System-Based Controllers in Transportation

As an essential part of computational intelligence, fuzzy systems use methods and techniques comparable to human observation, reasoning, and decision-making for computing under imprecise conditions. Fuzzy systems lay the groundwork for combining subjective and objective inputs to handle numerical and linguistic data. In the field of transportation engineering, these systems have been widely used. In [19], the freeway-related speed is regulated using measurements and expert knowledge data through speed advisory boards. Other researchers introduce current and potential traffic network control and management issues by surveying some commonly used computational intelligence paradigms, analyzing their applications in traffic signal control [20]. In order to seaport operations, [21] presents a fuzzy system-based method for determining an optimal investment plan. In [22], a fuzzy system-based lane-changing model that accounts for and simulates drivers' socio-demographics was developed to increase the realism of lane-changing operations in work zones.

Even though fuzzy system methods have been used for various traffic-related topics, their application for traffic control in the context of sustainable mobility on freeway networks has been overlooked. However, sustainable mobility is a relatively new area of research that is gaining increasing attention within the scientific community of traffic control experts. In continuation, a Fuzzy Inference System (FIS) and a Fuzzy Cognitive Map (FCM) that were preliminarily designed and developed in [23, 24] respectively, are employed parallelly (see Fig. 3) for traffic control with sustainable mobility purposes in freeway networks at a macroscopic level (road-based control measures, e.g., mainstream control, ramp management). In Fig. 3 a feedback loop is presented as a generic framework of a supervised vehicular traffic system by the proposed fuzzy system-based controllers. In this framework, preventive and uncontrollable inputs are two distinct forms of inputs that influence the characteristics of the system. Preventive inputs are generated from fuzzy controllers that through preassigned actuators, transmitted to the freeway network, e.g., in case of mainstream control, preventive inputs generated by FIS are the number of vehicles that need to be entered in the next segment of the freeway to avoid breaking down of the flow by predefined means such as VMSs. Meanwhile, uncontrollable inputs denote unmanipulable external issues that affect the density of the segment, such as weather-related issues or lane drop caused by accident. Within this context, performance demands are the controllers' computational-related requirements, e.g., time, rule generation, and performance measurements are the key indicators through which evaluating the applicability and efficacy of the control strategy in relation with sustainability-related objectives such as reducing congestion, emission, and travel time is possible.

3.1 The Case Study

The proposed FIS and FCM models were developed using data from the Hungarian freeway networks, wherein their users encounter complicated and dynamic congestion patterns. Apart from other factors, such as the relatively heavy road traffic resulting from Hungary's pivotal placement within Europe's transit system and corridor network [25], this is primarily as a result of an increase in the number of registered vehicles in

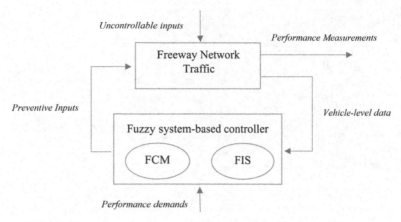

Fig. 3. A framework of Fuzzy system-based controllers in freeway traffic

Hungary, which increased by around 25% between 2010 and 2018 [26]. These issues cause complicated behavior in road traffic, including spatial and temporal changes. The dataset is derived from the Hungarian e-toll system's online transaction processing server, an electronic system maintained by the Hungarian national toll payment services for the country's whole network of freeways and primary roads. This system offers the guidance and support of freeway usage authentication, admittance, levying, and eventually collecting tolls on conventional road sections tollways.

The dataset contains the following independent variables: the freeway's name, the section's name (identifier), the number of e-toll collected over one week in each section (segment) of the 212 freeway sections (links), which latter is used as a proportional criterion of the number of vehicles, the time (per minute), the day, the section's length, and the number of lanes in each section. These links contain a total of 2446 distinct segments. Each segment is between 100 and 18,000 m long. In designing the FIS engine and input and output variables clustering ranges, the entire dataset and freeway sections are analyzed; in the developed FCM to keep the model effective and timely, a sample of 58 segments was chosen, representing the entire set of freeway sections connecting Budapest to the Austrian border.

The majority of road traffic models are designed to describe the behavior of traffic-related variables over a wide range of operations, which recognized locations playing a critical part in the dataset under investigation. This dataset can reflect real-time road traffic behavior based on location. A sample of connections between three segments A, B, and C is shown in Fig. 4. The provided dataset is based on time series; therefore, present traffic circumstances in upstream segments can project future road traffic flow conditions in downstream segments.

Figure 5 shows the causal linkages and correlations between the segments. The first digit on the horizontal axis represents the day, whereas the second and third digits indicate the time (in 24-h format); actual behavior of road traffic flow through time can be seen, demonstrating how traffic flow in the upstream segment might affect the downstream segments. The calculated road traffic flow correlation among segments confirms that A and B correlate 0.03, A and C have a correlation of 0.9, and B and C have a correlation of

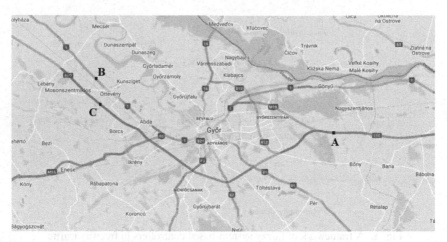

Fig. 4. An example of segments' connections

0.1; using these values, various conclusions, and correlation analyses can be performed to assess the behavior patterns and severity of traffic flow in the freeway.

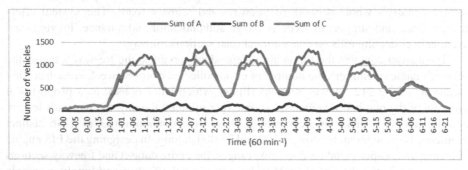

Fig. 5. Causal relations of traffic streamflow on the segments

3.2 The Fuzzy Inference System

Zadeh's original fuzzy set theory [27] has been later used to address a range of industrial and scientific concerns in various technology and science domains. The properties of fuzzy sets, coupled with their possible representation in linguistic terms, provide a computational algorithm for modeling and addressing imprecision and uncertainty-involved problems. Therefore, this work for detecting traffic congestion introduces a fuzzy inference model based on the Mamdani algorithm [28] implemented in MATLAB's Fuzzy Logic Toolbox R2021a. The developed model is designed to analyze and predict the severity of congestion in a freeway network. The model fuzzy inference system's layout in MATLAB with assigned input and output variables is presented (Fig. 6, for further details, see [23]).

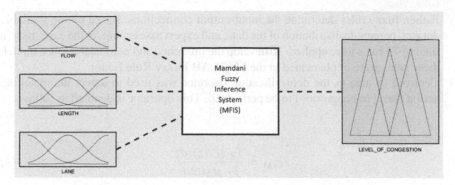

Fig. 6. Schematic representation of the proposed FIS

Specifying the model's input and output parameters is the initial step in developing a fuzzy inference model. The model will use three input parameters (length, number of lanes, and flow) and one output parameter (level of congestion). There are four primary design steps in the proposed Mamdani fuzzy inference algorithm:

1) Determining the numerical ranges for input and output linguistic variables. The following are the input variables:

   • *Flow*, the number of vehicles passing through a specific segment per time unit, which time interval equals 60 min,

$$q = \frac{n}{T} = \frac{n}{\sum_{i=0}^{n} i} \tag{1}$$

   where q is the average number of vehicles (n) that pass a segment during a unit of time (T).

   • *Length* of each segment of freeway networks in kilometers.

   • *Lane*, the number of lanes in each segment.

2) Triangular and trapezoidal membership functions are used for determining the degree of matching of the input and output parameters, as they capture and express the properties of the case study's fuzzy set. Equations 2 and 3 define these triangular and trapezoidal membership functions, respectively:

$$\mu_\Lambda(x) = \begin{cases} 0, & x < \alpha_{min} \\ \frac{x - \alpha_{min}}{\beta - \alpha_{min}}, & x \in (\alpha_{min}, \beta) \\ \frac{\alpha_{max} - x}{\alpha_{max} - \beta}, & x \in (\beta, \alpha_{max}) \\ 0, & x > \alpha_{max} \end{cases} \tag{2}$$

$$\mu_\Lambda(x) = \begin{cases} 0, & x \le \alpha_{min} \\ \frac{x - \alpha_{min}}{\beta_1 - \alpha_{min}}, & x \in (\alpha_{min}, \beta_1) \\ \frac{\alpha_{max} - x}{\alpha_{max} - \beta_2}, & x \in (\beta_2, \alpha_{max}) \\ 0, & x \ge \alpha_{max} \end{cases} \tag{3}$$

3) If-then fuzzy rules determine the input-output connections. Based on the available dataset, percentile distribution of the data, and expert assessment, in the case study, a total of 75 rules were applied. To develop the inference and nonlinear surface model, these rules were implemented in the MATLAB Fuzzy Rule Editor.

4) Centroid of area as the defuzzification operator was used to detect the matching action (level of congestion) to be performed. This operator is denoted as follows:

$$Z_{COA} = \frac{\int_Z \mu_A(z)zdz}{\int_Z \mu_A(z)dz} \tag{4}$$

where z is the fuzzy system output and aggregated output membership function is given as $\mu_A(z)$.

3.3 The Fuzzy Cognitive Map

As a further development of classic cognitive maps [29], Kosko [30] established the notion of the Fuzzy Cognitive Map (FCM) in order to address limitations associated with the binary structure of the original cognitive map model. FCMs combine the idea of cognitive maps and the concept of fuzzy set initially introduced by Zadeh [27], with the additional idea of signed fuzzy effects, forming a special kind of artificial neural network or fuzzy bipolar graph. It features fuzzy nodes or concepts (components) that are used to characterize the non-binary aspects of the modeled system's concepts and their gradual intensities of causal relationships. A simple FCM is schematically illustrated in Fig. 7; linkages and interrelationships between concepts are modeled using weighted arcs.

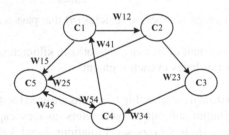

Fig. 7. A basic schematic illustration of FCM [24]

Determining activation values for the concepts related to weight assignments is critical in developing the FCM-based road traffic flow model. The presented model assigns activation values using an inference rule derived from Eqs. (5) and (6) in Table 1. Thus, the proposed integration places a premium on two critical aspects: not only may activation values be computed using the values of the connected concepts and their associated causal weights at each time step, but concepts can also reflect their past values.

Consequently, every freeway segment is signified by a concept whose value is taken as the density ρ of segment i of link m, and the weighted arcs are set to a constant value

Table 1. Involved methods and the proposed inference rule

Author/s	Equation/Method		Application
[31]	$\rho_{m,i}(t+1) =$ $\rho_{m,i}(t) + \frac{T_s}{L_m \lambda_m}[q_{m,i-1}(t) - q_{m,i}(t)]$	(5)	Calculating the density of segment i in link m at different time frames
[32]	$A_i^{(t+1)} = f\left(\sum_{\substack{j=1 \\ i \neq j}}^{n} w_{ji}A_j^{(t)} + A_i^{(t)} \right)$	(6)	Calculating the value of concept C_i at time t, wherein the value of C_i may represent the calculated density in the given segment
[24]	$\rho_{m,i}^{(t+1)} = f\left(\sum_{\substack{j=1 \\ i \neq j}}^{n} \rho_{m,i,j}^{(t)} W_{ij} + \rho_{m,i,i}^{(t)} \right)$	(7)	The proposed inference rule for predicting the density of segment i in link m in $t + 1$ time step by considering the previous density value of the given segment

based on variables $L_m \lambda_m$ as an approximation of the capacity; where L_m denotes the lengths of the segments of link m , and λ_m denotes the quantity of the available lanes in link m. The concepts and the weights are initialized using the aforementioned values. Following that, the system is allowed to interact, and after each iteration, the new state vector is given newly generated values. This procedure will be repeated until the model reaches an equilibrium state by exhibiting a stabilized condition at a fixed numerical boundary (see further details in [24]).

4   Results and Discussion

In this section, further discussion on the advantages of each fuzzy system-based method is deliberated. FIS and FCM are developed in connection with traffic control strategies, i.e., mainstream control, ramp management, route guidance, in freeway networks based on the data analysis at a macroscopic level. In particular, with considering sustainable objectives such as the reduction of the traffic emission, and improving road safety.

4.1 FIS in Congestion Level Prediction

Mainstream control is used to manage the traffic flow of vehicles traveling on the main-line, often by providing suitable indicators to drivers via Variable Message Signs (VMSs) or traffic lights. At a macroscopic level, these widespread control measures aim to homogenize traffic conditions, prevent the formation of recurrent congestion, and reduce the likelihood of vehicle crashes. An additional purpose is to address the emergence of non-recurring congestion problems by boosting the system's efficiency under situations of low capacity [33]. The proposed FIS aims to improve mobility and safety conditions in freeways by suggesting or imposing appropriate speed limits displayed utilizing VMSs. As seen in the preceding section, the level of congestion in each segment is determined using three types of accessible row data: the number of vehicles in a specific time unit, the number of lanes, and the length of the provided segment. All of these input data supplied approximations of the segment's relative capacity to meet recently formed demand, which could result in a change in the Level of Congestion (LOC). The collected findings demonstrate how effective the suggested FIS is at generalizing complicated nonlinear links between congestion levels and other numerical characteristics of the traffic.

The suggested FIS's interdependence between input variables and LOC may be proven by applying a fuzzy control surface in a visual insight view (Fig. 8). It demonstrates the existence of a correlation between LOC and the input variables. The most dramatic change occurs in the LOC when the length is between 4 and 6 km, and the lane count is 2 or 3 (Fig. 8 part I). Additionally, when the length variable is between approximately 1 and 6 km, an intensive reaction (approximately 50% rise) in the LOC occurs in each segment with a rising flow rate of more than 200 vehicles (Fig. 8 part II). Increasing or decreasing the number of lanes has the most significant effect on the LOC. Segments with 3 or 4 lanes will not encounter severe congestion, but raising of the flow rate by 200 vehicles in segments with fewer than two lanes can raise the LOC by more than 50% (Fig. 8 part III).

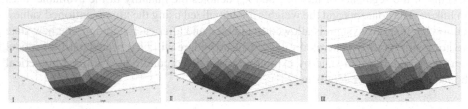

Fig. 8. Rule surface of LOC for length and lane (I), and length and flow (II), and flow and lane (III)

The developed FIS in this study can provide a prediction of congestion severity when input data is inserted. As a sample of the proposed model application from Fig. 9, it can be observed that if real-time input parameter properties are entered as follows:

Flow rate = 253, the segment has two lanes and 5.16 km length, then the LOC would be predicted as 281, which is categorized based on the assigned membership function, for the level of congestion-free.

Congestion measurement is critical for optimizing traffic management and control. The decision-making process that follows in order to create a sustainable transportation system is mainly dependent on current road traffic patterns. Hence, the method used to evaluate the severity of congestion should be realistic enough to enable decision-makers to carry out the necessary steps to alleviate congestion and to build quickly a resilient and sustainable transportation system. Therefore, transportation engineers have identified specific characteristics that are frequently required in a congestion measure [34, 35]. A practical congestion assessment ought to have mainly the following: non-technical individuals should be able to understand and interpret the outcomes of the analysis straightforwardly, give a constant range of possible values, be capable of being utilized for predictive and statistical analysis, and be universally applicable to a variety of road types. Besides all these characteristics, as opposed to conventional methods of traffic detection, the proposed mechanism has a sophisticated discipline known as approximate reasoning [36, 37] through which exact traffic connected properties (e.g., geometric features including junctions, bifurcations, off-ramps, and on-ramps) that can be assigned in both microscopic and mesoscopic types of traffic modeling are sacrificed, to reach significantly low time and computational efforts.

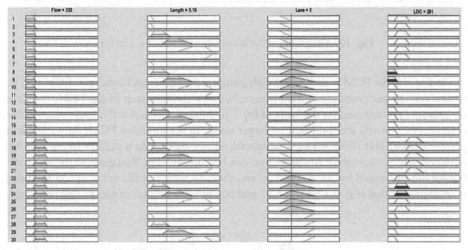

Fig. 9. A sample of the lookup diagram of the fuzzy rules, when: Flow = 253, Length = 5.16, Lane = 2, and the predicted LOC is 281

4.2 FCM in Traffic Flow Simulation

Complex road traffic flow processes are characterized by a variety of interdependent and interrelated elements. Therefore, FCM as a computational intelligence method is presented to address networks of freeways included imprecision and uncertainty. These uncertainties from the macroscopic modeling point of view are mainly connected with road traffic flow, density, and approximate capacity associated variables that can increase the probability of a breakdown and shift the free flow state of traffic to congested flow [18]. In the proposed FCM, segments of each link (freeway) are assigned as the concepts (nodes), where calculated density defines their values. In Fig. 10, a geographical representation of the selected segments is presented.

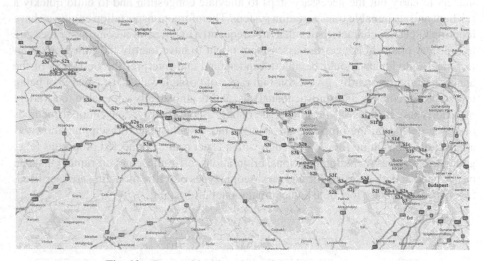

Fig. 10. Geographical locations of the selected segments

In Fig. 11, the FCM is illustrated with initialized weights and concepts. FCM begins to analyze the performance of the process. In every running step of the FCM, the state of concepts is computed on the basis of Eq. 7 in Table 1. Greater activation values in the concepts (segments) are indicated by larger nodes in the modeled FCM; they represent greater density and show stronger activation values that cause a greater impact on the network. Three alternative freeways that can be chosen from Budapest to the Austrian border are illustrated by *S1*, *S2*, and *S3* and their 58 nodes in the network. *S1* includes nine segments that end at segment *ES1* and combines with one of the *S2* segments; *S3*,

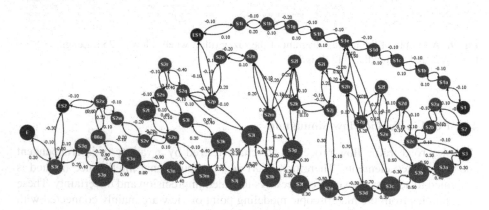

Fig. 11. FCM model of the road traffic flow

as the most chosen route, also has close interaction with the segments in *S2*, which both end at segment *E* as the last Hungarian segment before entering Austrian territory.

To see how alteration in variables properties (i.e., change in the number of lanes or altering in the flow rate) affect system behavior, the FCM traffic flow simulation of the initial states is shown in Table 2 wherein the road segments are defined by nodes, e.g., *S3a*, *S3b*,… and the calculated values indicate the segments' density. This model offers multiple contributions through which the most common traffic control approaches, such as ramp management and route guiding, can be implemented. Accordingly, as one of the common causes of severe LOC, i.e., based on Fig. 8 part III, a lane-drop scenario is simulated in a two-lane segment (*S3h*). However, one of the lanes is reduced, the density declined just marginally, obviously indicating that the remaining lane's density rose considerably and reached severe LOC. In comparison to the initial states, changing in the density values among connected segments, i.e., *S3g* and *S3i* can be seen in Table 3, wherein the density of *S3i* is dropped by 11% and subsequently slight decrease in *S3j*, while in the upstream side *S3g* and *S3f* are escalated by 16% and 8% respectively.

Table 2. Initial simulation result of the traffic flow density in the chosen network

Step	S3	S3a	S3b	S3c	S3d	S3e	S3f	S3g	S3h	S3i	S3j	S3k	S3l	S3m	S3n
0	0	0	0	0	0	0	0	0	0	0	0	0	0	0	0
1	0.50	0.50	0.50	0.50	0.50	0.50	0.50	0.50	0.50	0.50	0.50	0.50	0.50	0.50	0.50
2	0.67	0.71	0.73	0.65	0.71	0.74	0.73	0.74	0.64	0.68	0.80	0.70	0.76	0.66	0.67
3	0.67	0.72	0.76	0.63	0.72	0.77	0.76	0.78	0.63	0.68	0.85	0.72	0.79	0.66	0.66
4	0.67	0.72	0.77	0.63	0.72	0.78	0.77	0.79	0.63	0.67	0.85	0.73	0.80	0.67	0.65
5	0.67	0.72	0.77	0.62	0.72	0.78	0.77	0.79	0.63	0.66	0.85	0.73	0.80	0.67	0.65
6	0.67	0.72	0.77	0.62	0.72	0.78	0.77	0.79	0.63	0.66	0.85	0.73	0.80	0.67	0.65
7	0.67	0.72	0.77	0.62	0.72	0.78	0.77	0.79	0.63	0.66	0.85	0.73	0.80	0.67	0.65

Table 3. Traffic flow density in one lane reduction scenario

Step	S3	S3a	S3b	S3c	S3d	S3e	S3f	S3g	S3h	S3i	S3j	S3k	S3l	S3m	S3n
0	0	0	0	0	0	0	0	0	0	0	0	0	0	0	0
1	0.5	0.5	0.5	0.5	0.5	0.5	0.5	0.5	0.5	0.5	0.5	0.5	0.5	0.5	0.5
2	0.67	0.71	0.73	0.65	0.71	0.74	0.74	0.74	0.62	0.68	0.80	0.70	0.76	0.66	0.67
3	0.67	0.72	0.76	0.63	0.72	0.77	0.78	0.78	0.61	0.68	0.85	0.72	0.79	0.66	0.66
4	0.67	0.72	0.77	0.63	0.72	0.77	0.79	0.79	0.61	0.67	0.83	0.73	0.80	0.67	0.65
5	0.67	0.72	0.77	0.62	0.72	0.79	0.84	0.84	0.61	0.64	0.83	0.73	0.80	0.67	0.65
6	0.67	0.72	0.77	0.62	0.72	0.79	0.84	0.94	0.63	0.62	0.82	0.73	0.80	0.67	0.65
7	0.67	0.72	0.77	0.62	0.72	0.79	0.84	0.94	0.63	0.59	0.82	0.73	0.80	0.67	0.65

The simulations demonstrated the FCM's capabilities as a feasible computational intelligence method, not only at the macroscopic modeling level to investigate the overall behavior of road traffic flow but also to capture the involved features in terms of examining and monitoring meaningful alterations within freeway networks. These characteristics offer valuable information and can contribute to beneficial results related to the traffic control strategies with sustainable objectives, such as prediction and surveillance of the road traffic flow state in complex networks for reducing traffic emissions and improving road safety; the estimation of the influence of new road constructions or comparing the impacts of various development scenarios for planning purposes; predicting the effects of road capacity alteration, e.g., in maintenance purposes; and detecting dynamic congestion patterns and prone error locations for optimizing ramp management and route guidance toward eco-routing [10].

Additionally, as opposed to the obsolete traffic control with fixed strategies that were derived from historical data, current methods, regardless of their unique characteristics, are capable of functioning online based on real-time qualities originating from the road network. The presented FIS and FCM are also able to provide analyses and predictions to feed these traffic control strategies, e.g., mainstream control, ramp management, and route guidance. Moreover, in the classification of the traffic controllers, the local strategies are in the basis of localized data generated by sensors located near the related actuators, while in the global control mechanism, the collected segments data is not considered independently but as an input to analyze the entire freeways network state [15]. Therefore, in the illustrated framework (see Fig. 3), FIS can be proposed as a local traffic controller that can be applied to mainstream control mechanisms and FCM as a global one that is able to compute the dynamics of the whole system in favor of ramp metering and route guiding.

5 Conclusion

The concept of incorporating sustainability considerations into the design of a traffic controller is relatively recent and emerged within the scientific community of traffic control engineers. Moreover, the rapid advancement of road traffic flow modeling necessitates special attention on evaluating the capabilities of various computational intelligence techniques in this field. Therefore, this work proposed FIS and FCM as two computational intelligence methods in analyzing freeway traffic data concerning traffic flow modeling at the macroscopic level for addressing congestion-related issues as the core of the sustainability improvement strategies. While there is no certainty that congestion can be eliminated altogether due to the world's expanding population, these methods are presented to alleviate congestion to a reasonable degree.

This research approach introduced new applications of FIS and FCM to modeling complex freeway networks, with a particular emphasis on practical vehicular traffic congestion control strategies, such as ramp management, mainstream control, and route guidance, with the primary goal of increasing freeway safety and emission reduction. Additionally, by using these methodologies as the primary reason for developing and managing transportation systems, sustainability-related objectives can be improved. It is possible that the FIS and FCM models cannot capture all of the contributions of a macroscopic traffic flow control strategy, according to the problem's complexity, and as such the derived findings may vary from the real state of the freeway traffic. Any estimation technique, however, will inherently include a trade-off between model performance and operating time. In this perspective, methods based on fuzzy systems offer significant advantages in traffic control measures. Additionally, the study's dataset does not include all segments that potentially influence road traffic behavior, but only those that include the e-toll network. It is important to note that by incorporating additional mapping and data, the resolution of the representation of freeway networks can be significantly increased, resulting in more accurate but also more complex FCM models with refined simulation results. Therefore, as a next phase in the research, it would be highly important to take into account the entire involved segments in the freeway networks with a particular emphasis on bottleneck locations, as well as combining FCM with other algorithms such as Dijkstra to develop a real-time route guidance generation method.

Acknowledgements. The authors gratefully acknowledge Gergely Mikulai and Hungarian national toll payment services for their work on providing the original dataset. László T. Kóczy is supported by NKFIH K124055 grants.

References

1. Zhang, K., Batterman, S.: Air pollution and health risks due to vehicle traffic. Sci. Total Environ. **450–451**, 307–316 (2013). https://doi.org/10.1016/j.scitotenv.2013.01.074
2. Ramazani, A., Vahdat-Nejad, H.: A new context-aware approach to traffic congestion estimation. In: 2014 4th International Conference on Computer and Knowledge Engineering (ICCKE), pp. 504–508 (2014)
3. Systematics, C.: Traffic Congestion and Reliability: Trends and Advanced Strategies for Congestion Mitigation. Cambridge Systematics Inc., Cambridge (2005)
4. Faris, H., Yazid, S.: Development of communication technology on VANET with a combination of ad-hoc, cellular and GPS signals as a solution traffic problems. In: 2019 7th International Conference on Information and Communication Technology (ICoICT), pp. 1–9 (2019)
5. Ferrara, A., Sacone, S., Siri, S.: Freeway Traffic Modelling and Control. Springer, Heidelberg (2018). https://doi.org/10.1007/978-3-319-75961-6
6. Pasquale, C., Sacone, S., Siri, S., Ferrara, A.: Traffic control for freeway networks with sustainability-related objectives: Review and future challenges. Annu. Rev. Control. **48**, 312–324 (2019). https://doi.org/10.1016/j.arcontrol.2019.07.002
7. Mavi, R.K., Fathi, A., Saen, R.F., Mavi, N.K.: Eco-innovation in transportation industry: a double frontier common weights analysis with ideal point method for Malmquist productivity index. Res. Cons. Recycl. **147**, 39–48 (2019). ISSN 0921-3449, https://doi.org/10.1016/j.res conrec.2019.04.017
8. Yang, X.-S. (ed.): Nature-Inspired Algorithms and Applied Optimization. SCI, vol. 744. Springer, Cham (2018). https://doi.org/10.1007/978-3-319-67669-2
9. Karaboga, D., Gorkemli, B., Ozturk, C., Karaboga, N.: A comprehensive survey: artificial bee colony (ABC) algorithm and applications. Artif. Intell. Rev. **42**(1), 21–57 (2012). https://doi.org/10.1007/s10462-012-9328-0

10. Jabbarpour, M.R., Zarrabi, H., Khokhar, R.H., Shamshirband, S., Choo, K.-K.: Applications of computational intelligence in vehicle traffic congestion problem: a survey. Soft. Comput. **22**(7), 2299–2320 (2017). https://doi.org/10.1007/s00500-017-2492-z

11. Ai, C., Jia, L., Hong, M., Zhang, C.: Short-term road speed forecasting based on hybrid RBF neural network with the aid of fuzzy system-based techniques in urban traffic flow. IEEE Access **8**, 69461–69470 (2020). https://doi.org/10.1109/ACCESS.2020.2986278

12. Pinto, J.A., et al.: Traffic data in air quality modeling: a review of key variables, improvements in results, open problems and challenges in current research. Atmos. Pollut. Res. **11**(3), 454–468 (2020)

13. Castillo, H., Pitfield, D.E.: ELASTIC-a methodological framework for identifying and selecting sustainable transport indicators. Transp. Res. Part D **15**, 179–188 (2010)

14. Lajunen, T., Parker, D., Summala, H.: Does traffic congestion increase driver aggression? Transp. Res. Part F **2**, 225–236 (1999)

15. Ferrara, A., Sacone, S., Siri, S.: An overview of traffic control schemes for freeway systems. In: Ferrara, A., Sacone, S., Siri, S. (eds.) Freeway Traffic Modelling and Control, pp. 193–234. Springer International Publishing, Cham (2018). https://doi.org/10.1007/978-3-319-759 61-6_8

16. Lighthill, M.J., Whitham, G.B.: On kinematic waves II: a theory of traffic flow on long crowded roads. Proc. Roy. Soc. Lond. A: Math. Phys. Eng. Sci. **229**, 317–345 (1955)

17. Hoogendoorn, S.P., Bovy, P.H.L.: State-of-the-art of vehicular traffic flow modelling. Proc. Inst. Mech. Eng. Part I: J. Syst. Control Eng. **215**, 283–303 (2001)

18. Ferrara, A., Sacone, S., Siri, S.: First-Order Macroscopic Traffic Models. In: Freeway Traffic Modelling and Control. AIC, pp. 47–84. Springer, Cham (2018). https://doi.org/10.1007/978-3-319-75961-6_3

19. Ngo, C.Y., Victor, O.K.L.: Freeway traffic control using fuzzy logic controllers. Inf. Sci. **1**, 59–76 (1994)

20. Zhao, D., Dai, Y., Zhang, Z.: Computational intelligence in urban traffic signal control: a survey. IEEE Trans. Syst. Man Cybern. **42**, 485–494 (2012)

21. John, A., Yang, Z., Riahi, R., Wang, J.: Application of a collaborative modelling and strategic fuzzy decision support system for selecting appropriate resilience strategies for seaport operations. J. Traff. Transp. Eng. (Engl. Ed.) **1**(3), 159–179 (2014). https://doi.org/10.1016/S2095-7564(15)30101-X

22. Li, Q., Qiao, F., Lei, Y.: Socio-demographic impacts on lane-changing response time and distance in work zone with drivers' smart advisory system. J. Traff. Transp. Eng. (Engl. Ed.) **2**(5), 313–326 (2015). https://doi.org/10.1016/j.jtte.2015.08.003

23. Amini, M., Hatwagner, M.F., Mikulai, G.C., Koczy, L.T.: An intelligent traffic congestion detection approach based on fuzzy inference system. In: 2021 IEEE 15th International Symposium on Applied Computational Intelligence and Informatics (SACI), pp. 97–104 (2021). https://doi.org/10.1109/SACI51354.2021.9465637

24. Amini, M., Hatwágner, F.M., Mikulai, G.C., Kóczy, T.L.: Developing a macroscopic model based on fuzzy cognitive map for road traffic flow simulation. Infocommun. J.**13**(3), 14–23 (2021).https://doi.org/10.36244/ICJ.2021.3.2

25. László, F.T., Péter, T.: Hungary's its National report. ITS national report (2018). https://ec.europa.eu/transport/sites/transport/files/2018_hu_its_progress_report_2017.pdf

26. UNECE, E.C.E.: EU transport in figures - Statistical Pocketbook 2020, Number of registered passenger cars in Hungary from 1990 to 2018. European Commission (2020)

27. Zadeh, L.A.: Fuzzy sets. Inf. Control **8**(3), 338–353 (1965). https://doi.org/10.1142/978981 4261302_0021

28. Mamdani, E.H., Assilian, S.: An experiment in linguistic synthesis with a fuzzy logic controller. Int. J. Man. Mach. Stud. **7**(1), 1–13 (1975). https://doi.org/10.1016/S0020-737 3(75)80002-2

29. Axelrod, R.: Structure of Decision: The Cognitive Maps of Political Elites. Princeton University Press, Princeton (1976). https://www.jstor.org/stable/j.ctt13x0vw3
30. Kosko, B.: Fuzzy cognitive maps. Int. J. Man. Mach. Stud. 24(1), 65–75 (1986). https://doi.org/10.1016/S0020-7373(86)80040-2
31. Messmer, A., Papageorgiou, M.: METANET: a macroscopic simulation program for motorway networks. Traffic Eng. Control 31(8–9), 466–470 (1990). https://www.researchgate.net/publication/282285780_METANET_a_macroscopic_simulation_program_for_motorway_networks
32. Stylios, C.D., Groumpos, P.P.: Modeling complex systems using fuzzy cognitive maps. IEEE Trans. Syst. Man Cybern. Part A: Syst. Hum. 34(1), 155–162 (2004). https://doi.org/10.1109/TSMCA.2003.818878
33. Iordanidou, G.-R., Roncoli, C., Papamichail, I., Papageorgiou, M.: Feedback-based mainstream traffic flow control for multiple bottlenecks on motorways. IEEE Trans. Intell. Transp. Syst. 16, 610–621 (2015)
34. Afrin, T., Yodo, N.: A survey of road traffic congestion measures towards a sustainable and resilient transportation system. Sustainability (Switzerland) 12(11), 1–23 (2020)
35. Aftabuzzaman, M.: Measuring traffic congestion—a critical review. In: Proceedings of the 30th Australasian Transport Research Forum (ATRF), Melbourne, Australia, 25–27 September 2007 (2007)
36. Turner, S.M., Lomax, T.J., Levinson, H.S.: Measuring and estimating congestion using travel time-based procedures. Transp. Res. Rec. 1564, 11–19 (1996)
37. Amini, M., Hatwagner, M.F., Mikulai, G.C., Koczy, L.T.: A vehicular traffic congestion predictor system using Mamdani fuzzy inference. Syst. Theor. Control Comput. J. 1(2), 49–57 (2021)

Contextual Sentence Embeddings
for Obtaining Food Recipe Versions

Andrea Morales-Garzón[✉], Juan Gómez-Romero,
and Maria J. Martín-Bautista

Department of Computer Science and Artificial Intelligence,
Universidad de Granada, Granada, Spain
amoralesg@ugr.es, {jgomez,mbautis}@decsai.ugr.es

Abstract. Food and culinary activities related to cooking are present
in our daily lives. The rise of food-related data has led to the term food
computing, which refers to the study and development of computer systems to solve food-related tasks. Despite the large number of food computing systems focused on the collection, recommendation, retrieval, and
creation of recipes, very few have used existing recipes to get adapted versions for user requirements. In this work, we have developed a method for
adapting recipes that suggests food options for substituting their ingredients based on food relations and text similarity. For this purpose, we
employ different deep learning language models based on BERT. These
models incorporate attention mechanisms to extract contextual representations of foods using different strategies to build the word embeddings. We use them to conduct a semantic comparison task for detecting
similar ingredients between the recipe ingredients and a food dataset.
The results show that the method obtains high-quality recipe versions,
thanks to context data, attention mechanisms, and the token representation strategy used for the foods.

Keywords: Food computing · Word embedding · Recipe adaptation

1 Introduction

A healthy diet is essential to prevent risk factors associated with diet. According to the WHO[1], a healthy diet prevents malnutrition and increases protection
against diseases such as obesity, diabetes, or even cancer, which poses extreme
health risks. The WHO plays a great role in encouraging the population to pursue
better lifestyles. It is also responsible for promoting food recommendations by
replacing recipe ingredients with higher-quality ones. The final aim is to guide the
community to nutrient-rich daily diets with greater long-term benefits in their
lifes [23]. Many of these recommendations aim at avoiding the intake of sugar
and ultra-processed products, but there are many other strategies focused on
food substitution. These may be, for example, those that seek to include less fat,

---

[1] https://www.who.int/news-room/fact-sheets/detail/healthy-diet.

© Springer Nature Switzerland AG 2022
D. Ciucci et al. (Eds.): IPMU 2022, CCIS 1602, pp. 306–316, 2022.
https://doi.org/10.1007/978-3-031-08974-9_24

salt, or potassium in food recipes. Also, these recommendations often focus on multiple objectives, such as substitution based on foods from local markets and distributors, organic products, or even considering sustainable and biodegradable packaging in commercial products. Diet and recipe recommendation tasks have been addressed in several computer science problems under the term food computing [18]. Food computing refers to the design and development of computer solutions to food-related problems, including recipe adaptation systems.

We refer to recipe adaptation as the process of getting a version of an existing recipe by changing one or more of its ingredients for other similar ones. These changes can be due to multiple causes and they are not arbitrary. A person may want to make a recipe without having one (or more) of the ingredients available or change specific foods because of their cooking preferences. They also can have dietary restrictions due to medical reasons (e.g., low-fat or sugar-free diets), allergies, intolerances, or ideological/religious reasons. Ingredient modifications are usually determined by food relations, flavor combinations, and even traditional costumes. This task usually involves more than a data source as it requires at least a recipe to adapt and a collection of foods to choose those ingredient substitutes that best suit the recipe. There are multiple advantages to derive recipe versions from an existing one. First, it allows to obtain multiple recipes from existing resources without using text generation tasks. It makes it possible to take profit of available resources, since it is possible to discover many new recipes from a given one. It also provides more flexibility when cooking a recipe, since the customized versions use similar foods that can substitute the original ones. This task, so natural for humans, is hard to do automatically. Recipes contain intrinsic relationships between foods that are difficult to identify. It requires an in-depth study of the existing patterns between foods in recipes. For this, previous works have employed language models such as Word2vec to obtain food data representations [2, 5, 22].

In a previous study, we showed that non-attention-based word embedding models like Word2vec [16], GloVe [24], and Fasttext [3] perform similarly for this task [21]. However, due to recent advances in NLP, attention-based models may outperform those results. For this, we propose to address the recipe adaptation problem using language models based on attention mechanisms. We employ a deep learning language model to learn embedded representations of the foods. Specifically, the strategy suggests similar ingredients for a given one considering food relations in cooking processes. For this, we use learned contextual embeddings generated by BERT to suggest ingredient alternatives for a specific recipe. We implement different strategies for obtaining token and sentence-level vector representations with BERT and SBERT models and analyze which approach performs better. To the best of our knowledge, this is the first food computing approach using context and attention mechanisms for recipe adaptation.

2   Related Work

The analysis of food data has led to great advances and applications of very frequent use among the population. Its relevance for the common good has made it an emerging area of study in both academia and industry [18]. Food computing was first introduced in [7] regarding software systems in the agri-food industry. In [18], the authors repurposed this term to represent food-related tasks that involve acquisition and analysis to get a better understanding of culinary cultures.

Recipe adaptation tasks have not been widely studied, although it has a promising potential for real-world applications. However, it is possible to integrate existing food computing contributions for specific tasks within the problem, for instance, datasets, representation approaches, or relationship extraction algorithms [9,18].

From a more specialized point of view focused on data science, food computing tasks face an initial barrier in almost any of their applications. It is the challenge of combining data from heterogeneous origin and content [15,19]. Information retrieval systems in food computing have addressed this difficulty building a vector from multimodal data to represent recipe objects [20]. Other works have studied relationships between ingredients and recipes to unify data and expand nutritional information from foods to recipe data sources [17,22].

Previous works have used deep learning approaches to encode culinary data for the jointly use of multi-facet attributes in information retrieval systems [19]. In [13], the authors created a large-scale multimodal recipe dataset and used deep learning to obtain an intermediate representation for detecting (recipe, image) pairs. Food computing studies on recipe features allows unifying data and expand nutritional information from ingredients to recipe data sources [22,26]. They also helps to determine intrinsic relationships between ingredients and recipes. Mostly concluded that geographic regions have influence in specific combinations of foods when preparing recipes, considering flavors [1], cuisine styles [10], and pattern trends [11]. Results obtained from recipe analysis can be used to create *"pseudorecipes"* and improve the nutritional features of recipe sets [5]. User preferences and opinions enable to perform popularity analysis and improve recommendation systems performance. Many food computing contributions have studied metrics for recipe recommendation based on the study of ingredients, recipes, and the ratings given by users [8]. They use Word2vec [16] to explore which food groups represent the recipe ingredients and study recipes' trends.

The use of attention mechanisms in neural networks has marked a turning point for many natural language processing tasks. Transformer models first arose in [28], where the authors released a network architecture based on attention mechanisms instead of using convolutions and recurrence. BERT models have achieved significant progress on the state-of-the-art of text similarity tasks by generating high-quality representation vectors [4]. Its architecture has many advantages due to the small number of sequential operations required (i.e., parallelization of training and less computational complexity per layer). For sentence processing with BERT, Sentence-BERT (SBERT) adds a pooling operation to the BERT output to obtain sentence representations [25]. The authors experimented with different pooling strategies while considering attention vectors obtained with the model when operating.

3   Data

The method proposed in this work is a multi-step process that leads to recipe versions. For this reason several data sources are employed, all of them described in English. The final aim is to relate a recipe with a food composition database by connecting its ingredients with equivalent food items. For this, we use a food composition table and an ingredient corpus built from scratch for evaluating the model results. This two sources allow us to detect similarity relations within recipe ingredients and a specialized food resource. We implement a semantic comparison task to relate both sources, thus evaluating the performance of different word representation strategies to build the sentence embedding.

We build an ingredient corpus with the main aim of creating a food collection combining several recipe datasets from different sources. The goal is to have a set as representative and diverse as possible. We use five resources: (1) a dataset created from scratch scrapping the names of the food collection of BBC Foods[2] (2) a collection of the ingredients used in the BBC Recipe dataset, (3) the *Recipe ingredients Dataset* created for a Kaggle competition[3], (4) the *Indian Food 101* dataset with dishes from Indian regions available at Kaggle[4], and (5) the *Foods Recipes* dataset from Archana's Kitchen (www.archanaskitchen.com) also available at Kaggle[5]. We combine the datasets obtaining 104973 ingredients, where 9800 of them are unique. Some of the resources contained typographical errors and misspellings as they are hosted in online blogs managed by their community. We carry out a typical cleaning process by removing symbols, punctuation marks, and nonalphanumeric characters. We remove the content in brackets since it corresponds to additional information for the main ingredient but is not essential. We also singularize the food descriptions for text homogenization.

We use the 2021 version of the Food Composition Database (CoFID). This dataset, available for public use, is maintained by the Public Health Agency (PHE) of the Department of Health and Social Care in England to collect all their available food-related data in a single dataset [14]. Specifically, we use the *Proximates* table, which contains general information on foods from very diverse food groups. This table provides an up-to-date set of 2913 foods currently used in European recipes with a wealth of food-related specialized data. We apply to the food descriptions the same cleaning process described above.

Finally, we use a recipe dataset for testing the proposed method for obtaining recipe versions. For this, we used the Food.com recipe collection, employed in several food computing applications [6, 12].

---

[2] https://www.bbc.co.uk/food.
[3] www.kaggle.com/kaggle/recipe-ingredients-dataset.
[4] https://www.kaggle.com/nehaprabhavalkar/indian-food-101/version/2.
[5] https://www.kaggle.com/sarthak71/food-recipes.

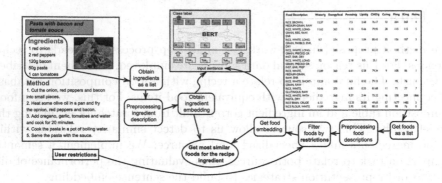

Fig. 1. Overview of the recipe adaptation method

4   Method

To detect similar foods for recipe ingredients in an external dataset we need to find those instances whose similarity to the ingredient is maximal. We propose to treat this problem as a text similarity task, where the food descriptions are the texts sequences between we intend to calculate the similarity degree with. Figure 1 shows an overview of the proposed method.

We first obtain the representation vector for each recipe ingredient. We replicate this procedure for all the foods in the food composition table. After preprocessing and tokenizing the texts descriptions, we calculate their token vector embeddings. Because of its architecture, BERT generates an embedding for each token in the sequence. We obtain the embeddings by extracting the features learned with the model. BERT base-cased version contains 12 hidden layers. Therefore, there are 12 feature vectors per token, with many combinations to obtain a single representation vector for each token in the sequence. Section 5 contains a comparative analysis of different representation strategies to decide which of them is more suitable for the task. The last step is to generate a sentence representation vector by averaging the token embeddings of the sequence. We compare the sentence vectors of both sources pairwise to detect the most similar food for each ingredient. We use the cosine distance function to determine the similarity degree between each pair of descriptions.

We also follow this methodology with SBERT. The main difference resides in the final weighted average with token attention masks when building the token embedding. Token vectors are weighted with their associated attention mask and then averaged to obtain the sentence vector.

5   Experiments

BERT model architecture, with 12 up to 24 hidden layers, provides multiple ways of obtaining token embeddings [28]. The features encoded in each layer differ, and it may contain different information regarding the context [27]. It opens up

Table 1. Comparison of lower bounds for token representation strategies

Representation	Top 1	Top 2	Top 3	Top 4	Top 5	Total
Average all layers	**55.16**	**30.50**	**21.16**	**21.33**	**14.66**	**142.83**
Last layer	36.66	11.50	6.83	9.50	6.66	71.16
Second last layer	38.66	13.00	11.66	8.00	8.66	80.00
Concat 4 last layers	38.66	15.50	11.50	10.33	6.66	82.66

Table 2. Comparison of BERT and SBERT lower bounds

Language model	Top 1	Top 2	Top 3	Top 4	Top 5	Total
BERT	**55.16**	**30.50**	21.16	**21.33**	14.66	**142.83**
SBERT	48.16	27.83	**26.16**	18.50	**16.33**	137

a range of possible combinations for building the vector. We conduct a study to decide which representation is most suitable. We implement four strategies to obtain a token embedding: (1) averaging all BERT hidden layers, (2) concatenate the four last hidden layers, (3) only second last hidden layer, and (4) only the last hidden layer.

One of the challenges of this task is the difficulty in evaluating the results. We cannot guarantee a good candidate for food substitution in the external dataset. This means that it may not be a good option in the food composition database for each recipe ingredient. Also, deciding whether a recipe version is good or not is a very subjective process and depends on many factors, such as cultural and social factors. Therefore, evaluating the results is difficult to automate and mostly depends on annotators and nutrition experts. For this, we designed a lower bound that calculates the syntactic similarity between the recipe ingredient and the best option in the food composition database. The lower bound ranges between 0 and 1, where 0 represents no similarity and 1 is a total syntactic match. This bound helps to determine which implementation shows an overall better performance. We selected the 100 most popular items in the ingredient corpus based on counting repeated corpus instances. We obtained the five most similar options in the food composition database for each ingredient. We calculated the lower bound for these ingredients. Table 1 shows these results. The first column represents the strategy used for calculating the token embeddings. Columns "Top k" represent the lower bound obtained with the ingredients for the k best foods. Note that the maximum lower bound value is 100 as the method is applied to 100 ingredients.

We use the best strategy in Table 1 for comparing BERT and SBERT. We used the same 100 ingredients from the previous experiment. Table 2 shows the results obtained with the lower bound for both language models. The first column represents the language model (i.e., BERT and SBERT). "Top k" columns contain the lower bound for the k best option from the food composition database.

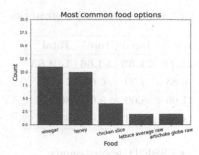
(a) Most common foods for option 1

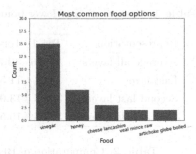
(b) Most common foods for option 2

Fig. 2. Visualization of the most repeated foods obtained with BERT

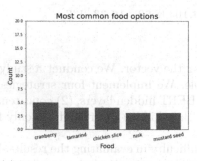
(a) Most common foods for option 1

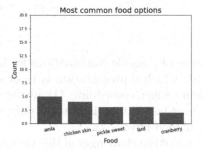
(b) Most common foods for option 2

Fig. 3. Visualization of the most repeated foods obtained with SBERT

6 Discussion

Results from Table 1 indicate that averaging the features from all hidden layers allows to obtain better results. Therefore, there is enriching information encoded in the first layers that help reach similar food substitutes and should be not ignored for our task. Table 2 shows very similar results for BERT and SBERT so we cannot draw clear conclusions about their performances. However, a visual analysis of results with BERT revealed that some foods appear many times for very dissimilar ingredients. We detected those foods that most times appear as the best k options. Figures 2a and b show the most repeated ingredients for the two best options obtained with BERT. It shows that *"Vinegar"* appears 10 and 15 times as first and second-best options respectively in a dataset of 100 items. The same happens to *"Honey"*, appearing 10 and 5 times as first and second options. We think these ingredients correspond to very central vectors in the feature space, thus being more reachable for many options.

We obtained the same visualization for SBERT (see Figs. 3a and b). In this case, we observed that we did not obtain food options more common than others, and there is a wider variety in the results. SBERT generates sentence embeddings from the token embeddings weighted with their corresponding attention

Table 3. Food suggestions for the ingredients of a recipe

Ingredient	Top 1	Top 2	Top 3	Top 4	Top 5
Bacon	Bacon rashers middle grilled (0.8852)	Bacon rashers back grilled (0.8804)	Bacon rashers back tender-sweet grilled (0.8560)	Bacon rashers streaky grilled (0.8501)	Bacon rashers back grilled crispy (0.8477)
Cheddar cheese	Cheese cheddar english (0.9498)	Cheese edam (0.9112)	Cheese ricottum (0.8965)	Cheese camembert (0.8919)	Cheese quark (0.8907)
Onion	Onions raw (0.8301)	Onions baked (0.7599)	Garlic raw (0.7441)	Chutney tomato (0.7381)	Potato ring (0.7201)
Salad dressing	Salad cream (0.8284)	Spinach dried (0.7005)	Salami snack (0.6977)	Tamarind (0.6908)	Marjoram dried (0.6889)

masks. Therefore, the attention values of the network contribute to building the representation vector. This helps build a high-quality vector to distinguish better between food descriptions with a more efficient distribution of feature space.

Generic food composition tables are a good solution for proposing foods for daily use. Table 3 shows the five best options found for each of the ingredients in a recipe. The first column contains the ingredient description, and the rest of the columns correspond to the best substitutes in the food composition database. Results show that our method finds promising alternatives, given the high similarity and suitability of the options regarding the original ingredient. This provides a reliable methodology for detecting food equivalents in both sources, thus enabling data fusion for taking advantage of food properties in specialized sources. In this way, we could use extra data to enrich recipe versions by including dietary restrictions. Note that suggestions for the ingredient "Bacon" (first row) are long food descriptions and still reached by our method. It illustrates that our method generates representation vectors capable of distinguishing the main ingredient in a description. Suggestions for "*onion*" are also noteworthy, as the best option achieved is the raw version of the food. Therefore, the method detects as the first option the ingredient without food processing followed by cooked versions in the second instance. Suggestions for "*Salad dressing*" show how the method detects different but still suitable options, thus capturing the semantics of the original ingredient description.

7 Conclusions and Future Work

This work exposes the high value contained in text descriptions. The method is capable of obtaining great similar ingredients without the need of adding extra features. This is possible through the richness of the language and the intrinsic information and relationships contained in recipe texts. The features encoded in BERT hidden layers include sufficient information to detect high-quality semantic relations between the sources. This is due to the strong capability of BERT to get sequence contextual representations for determining similarity relationships. Attention mechanisms play a fundamental role enabling to obtain valuable representation vectors. With our method we can obtain recipe versions in an automated and widespread way, thus taking advantage of existing resources and with no need to create new recipes from scratch. This method is able to combine two datasets from different sources. This has great implications because it means that we can connect the data and use them jointly.

This study opens several lines of research to pursue to improve the results. The next task to undertake consists of an extensive evaluation of the results. The most straightforward solution is a user evaluation of adapted recipes, in which we plan to compare the results with other sentence-based models. Developing an evaluation framework for recipe adaptation is compelling since it implies dealing with subjectivity, food tastiness, and cultural influences. This task remains a challenging task to pursue in the near future. Also, the quality of food suggestions obtained with our method has a strong relation to the data sets used. We suggest extending the food composition table to consider foods for specific and restrictive diets. Our method does not consider the users' opinions when generating a recipe. A more comprehensive validation based on the reviews of users and experts would allow us to detect specific shortcomings of our model. We believe in the potential of this avenue since it could provide external knowledge regarding tastes and ingredient trends. We also intend to adapt the selection of ingredients by considering the rest of the information in the recipe (such as the title, quantities and preparation steps), while modifying them accordingly.

Acknowledgements. This project is partially supported by the Andalusian government and the FEDER operative program under the project BigDataMed (P18-RT-2947 and B-TIC-145-UGR18). It is also supported by the Department of Economic Transformation, Industry, Knowledge and Universities of the Junta de Andalucía and the program of research initiation for master students of the University of Granada.

References

1. Ahn, Y.Y., Ahnert, S.E., Bagrow, J.P., Barabási, A.L.: Flavor network and the principles of food pairing. Sci. Rep. **1**(1), 1–7 (2011)
2. Altossar, J.: food2vec-augmented-cooking-machine intelligence. Jaan Altossar's blog (2015). Accessed 17 December 2015
3. Bojanowski, P., Grave, E., Joulin, A., Mikolov, T.: Enriching word vectors with subword information. Trans. Assoc. Comput. Linguist. **5**, 135–146 (2017)

4. Cer, D., Diab, M., Agirre, E., Lopez-Gazpio, I., Specia, L.: Semeval-2017 task 1: semantic textual similarity-multilingual and cross-lingual focused evaluation. arXiv preprint arXiv:1708.00055 (2017)

5. Chen, M., Jia, X., Gorbonos, E., Hong, C.T., Yu, X., Liu, Y.: Eating healthier: exploring nutrition information for healthier recipe recommendation. Inf. Process. Manag. 102051 (2019)

6. Fujita, J., Sato, M., Nobuhara, H.: Model for cooking recipe generation using reinforcement learning. In: 2021 IEEE 37th International Conference on Data Engineering Workshops (ICDEW), pp. 1–4. IEEE (2021)

7. Harper, C., Siller, M.: OpenAG: a globally distributed network of food computing. IEEE Pervasive Comput. **14**(4), 24–27 (2015). https://doi.org/10.1109/MPRV. 2015.72

8. Harvey, M., Ludwig, B., Elsweiler, D.: You are what you eat: learning user tastes for rating prediction. In: Kurland, O., Lewenstein, M., Porat, E. (eds.) SPIRE 2013. LNCS, vol. 8214, pp. 153–164. Springer, Cham (2013). https://doi.org/10. 1007/978-3-319-02432-5_19

9. Jiang, S., Min, W.: Food computing for multimedia. In: Proceedings of the 28th ACM International Conference on Multimedia. MM 2020, pp. 4782–4784. Association for Computing Machinery, New York (2020). https://doi.org/10.1145/ 3394171.3418544

10. Kazama, M., Sugimoto, M., Hosokawa, C., Matsushima, K., Varshney, L.R., Ishikawa, Y.: A neural network system for transformation of regional cuisine style. Front. ICT **5**, 14 (2018)

11. Kim, K.J., Chung, C.H.: Tell me what you eat, and i will tell you where you come from: a data science approach for global recipe data on the web. IEEE Access **4**, 8199–8211 (2016)

12. Majumder, B.P., Li, S., Ni, J., McAuley, J.: Generating personalized recipes from historical user preferences. arXiv preprint arXiv:1909.00105 (2019)

13. Marin, J., et al.: Recipe1m+: a dataset for learning cross-modal embeddings for cooking recipes and food images. IEEE Trans. Pattern Anal. Mach. Intell. **43**(1), 187–203 (2019)

14. McCance, R.A., Widdowson, E.M.: McCance and Widdowson's the composition of foods. Roy/ Soc. Chem. (2014)

15. Metwally, A.A., Leong, A.K., Desai, A., Nagarjuna, A., Perelman, D., Snyder, M.: Learning personal food preferences via food logs embedding. arXiv preprint arXiv:2110.15498 (2021)

16. Mikolov, T., Chen, K., Corrado, G., Dean, J.: Efficient estimation of word representations in vector space. arXiv preprint arXiv:1301.3781 (2013)

17. Min, W., Jiang, S., Jain, R.C.: Food recommendation: framework, existing solutions, and challenges. IEEE Trans. Multimedia **22**, 2659–2671 (2020)

18. Min, W., Jiang, S., Liu, L., Rui, Y., Jain, R.: A survey on food computing. ACM Comput. Surv. (CSUR) **52**(5), 1–36 (2019)

19. Min, W., Jiang, S., Sang, J., Wang, H., Liu, X., Herranz, L.: being a supercook: joint food attributes and multimodal content modeling for recipe retrieval and exploration. IEEE Trans. Multimedia **19**(5), 1100–1113 (2016)

20. Min, W., Jiang, S., Wang, S., Sang, J., Mei, S.: A delicious recipe analysis framework for exploring multi-modal recipes with various attributes. In: Proceedings of the 25th ACM International Conference on Multimedia, pp. 402–410 (2017)

21. Morales-Garzón, A., Gómez-Romero, J., Martin-Bautista, M.J.: A word embedding-based method for unsupervised adaptation of cooking recipes. IEEE Access **9**, 27389–27404 (2021)

22. Morales-Garzón, A., Gómez-Romero, J., Martin-Bautista, M.J.: A word embedding model for mapping food composition databases using fuzzy logic. In: Lesot, M.-J., et al. (eds.) IPMU 2020. CCIS, vol. 1238, pp. 635–647. Springer, Cham (2020). https://doi.org/10.1007/978-3-030-50143-3_50
23. World Health Organization et al.: Healthy diet. Technical report, World Health Organization. Regional Office for the Eastern Mediterranean (2019)
24. Pennington, J., Socher, R., Manning, C.D.: Glove: global vectors for word representation. In: Empirical Methods in Natural Language Processing (EMNLP), pp. 1532–1543 (2014). http://www.aclweb.org/anthology/D14-1162
25. Reimers, N., Gurevych, I.: Sentence-Bert: sentence embeddings using Siamese Bert-networks. arXiv preprint arXiv:1908.10084 (2019)
26. Su, H., Lin, T.W., Li, C.T., Shan, M.K., Chang, J.: Automatic recipe cuisine classification by ingredients. In: Proceedings of the 2014 ACM International Joint Conference on Pervasive and Ubiquitous Computing: Adjunct Publication, pp. 565–570 (2014)
27. Toneva, M., Wehbe, L.: Interpreting and improving natural-language processing (in machines) with natural language-processing (in the brain). arXiv preprint arXiv:1905.11833 (2019)
28. Vaswani, A., et al.: Attention is all you need. In: Advances in Neural Information Processing Systems, pp. 5998–6008 (2017)

A Fuzzy-Based Approach
for Cyberbullying Analysis

J. Angel Diaz-Garcia[1]([⊠]) [iD], Carlos Fernandez-Basso[1] [iD],
Jesica Gómez-Sánchez[2] [iD], Karel Gutiérrez-Batista[1] [iD], M. Dolores Ruiz[1] [iD],
and Maria J. Martin-Bautista[1] [iD]

[1] Department of Computer Science and A.I., University of Granada, Granada, Spain
{jagarcia,cjferba,karel,mdruiz,mbautis}@decsai.ugr.es
[2] Department of Developmental and Educational Psychology, University of Granada,
Granada, Spain
gomezjs@ugr.es

Abstract. Social networks and new technologies are present today in
almost all aspects of social relations. These technologies, which have so
many advantages in the daily lives of young and older people, can become
a double-edged sword. One of the most negative connotations created by
the incursion of social networks and new technologies in the lives of children and adolescents is cyberbullying. Cyberbullying is the act of harassment using social networks, causing severe emotional problems for the
victim and, in the worst-case scenario, even leading to suicide. Therefore,
detecting and preventing this misuse of social networks and the internet
is a task of significant interest for society. Considering the overlapping of
different types of cyberbullying within the same document, in this paper,
we propose the use of fuzzy logic for the study and modelling of cyberbullying from an explainable point of view. On a dataset containing five
different types of cyberbullying, we demonstrate that fuzzy modelling is
a promising technique.

Keywords: Fuzzy logic · Cyberbullying · Social media mining ·
Education

1 Introduction

The development and assimilation of new technologies and social networks in our
lives have meant that many of our social relations are conducted via online social
networks. These social relationships are not always positive, as the anonymity,
immediacy and total interconnections between users can be the perfect breeding
ground for abuses such as cyberbullying. Cyberbullying is a problem in both
children and adolescents, and there is common agreement in defining it as inten-
tional, violent, cruel and repetitive behaviour against peers. This problem, which
is pervasive in today's society, is even listed by the US Center for Disease Control
and Prevention (CDC) as a severe public health threat [1]. This problem can

D. Ciucci et al. (Eds.): IPMU 2022, CCIS 1602, pp. 317–328, 2022.
https://doi.org/10.1007/978-3-031-08974-9_25

cause serious emotional and psychological problems and, in the worst cases, can even lead to suicide in those who suffer cyberbullying [5]. Therefore, detecting cyberbullying early and efficiently can help save lives or achieve a better quality of mental health for adolescents and children.

According to [14], cyberbullying can be categorised into a few categories such as flaming, harassment, denigration, impersonation, outing, boycott and cyberstalking. In the vast majority of these categories of bullying, especially in the more severe ones, such as flaming and harassment, cyberbullying occurs in texts, through comments or messages exchanged between the bully and the victim. These messages are usually based on insults and attacks based on racism, sexuality, ethnicity, or social status. These messages are, in fact, user-generated content in an unstructured way so that they can be analysed automatically by Artificial Intelligence (AI) systems based on natural language processing (NLP) and text mining. According to [7] NLP is a collection of computational techniques for automatic analysis and representation of human languages. Moreover, text mining systems use data mining techniques on unstructured text to obtain value from it [26] or structures that other algorithms can easily handle.

In this paper, we propose a study of cyberbullying based on NLP and Data Mining techniques such as fuzzy text clustering and vector representation models. The initial premise is that several types of cyberbullying can appear in the same message, and modelling them with a degree of membership can offer great potential. In order to validate these assumptions in terms of our results, we have relied on a validation by experts in psychology. The major contributions of this paper to state of the art are:

- We offer a starting point for the fuzzy study of cyberbullying. It is very much in line with the nature of the problem, as the same message can contain attacks of different types or characteristics.
- An automatic approach for obtaining degrees of fuzzy membership to one or many types of cyberbullying from sentence embeddings is proposed.
- As far as we know, it is the first time that fuzzy clustering has been applied to model and attempt to categorise and detect cyberbullying from an unsupervised point of view, with promising results.

The rest of the paper is organised as follows. The following section is a detailed state-of-the-art study of the detection and study of cyberbullying using Artificial Intelligence (AI) techniques. Section 3 is devoted to the fuzzy analysis and modelling of cyberbullying. In Sect. 4, the experimental process and interpretation of results are presented. Section 5 is dedicated to the future work that this paper opens up, as well as to the conclusions.

2   Related Works

Typically, the works present in the literature on AI techniques to detect cyberbullying are based on supervised classification algorithms [11]. In [24] Rosa et al. review state of the art, in which all the papers analysed are based on classification. All these works focus on training models with cyberbullying or non-cyberbullying labels and then using them to detect possible cases of cyberbullying in social networks. Also, some works focus on looking at the type of cyberbullying [8,16]. These techniques, with artificial neural networks being the current state of the art [4,23], offer excellent results from an accuracy point of view but have two problems. On the one hand, it requires large labelled datasets to be trained. On the other hand, these are black-box models in which it is difficult to understand and explain the reason for the detection. If we consider that these models often have to be used by professionals who do not necessarily have computer skills, such as psychologists or teachers, it is necessary to propose explainable, descriptive and unsupervised approaches.

In [2], Abou El-Seoud et al. catalogue a total of 19 articles and conference contributions dealing with cyberbullying from an unsupervised or semi-supervised point of view. The works mentioned in the review try to solve the challenge of the need for large labelled databases. In this review and the literature in general, deep-learning models based on autoencoders [28,29] are again very present. These models seek to generate a feature space related to cyberbullying through neural networks that can infer new examples. Also in line with neural networks is the work proposed by Di Capula et al. [10]. In this work, they explore a type of model based on a self-organising map (SOM) [12], a type of neural network that is suitable for unsupervised modelling. In [10], documents (tweets) with similar characteristics are grouped hierarchically throughout the training, generating maps for documents with and without cyberbullying. Although the paper indicates that it is an unsupervised approach, it is true that, in part, they use manual labelling to construct the maps. From a purely unsupervised point of view, we find the works [21,27]. In [21] it is employed the K-Means clustering algorithm to categorise the documents arriving in a data stream into two clusters (abusive and polite). In [27], another of the most widespread branches of unsupervised learning, association rules, is explored. [27] offers a descriptive approach to cyberbullying-related patterns in the Malay language. To do this, the authors use the Apriori algorithm.

If we focus on the detection and fuzzy modelling of cyberbullying, all the articles we find are once again based on classification. In [22] Rosa et al. propose the use of fuzzy fingerprints for the classification and detection of cyberbullying. This technique is based on fuzzy modelling of the essential characteristics that identify one group or another (cyberbullying or not cyberbullying) and, in the case of a new example, discern which vector of characteristics is closer. In this case, it is taken into account that different types of cyberbullying can appear in one message. The model is trained on a labelled dataset from a question and answer website. In [3] authors study cyberbullying from the point of view of multi-class classification using multinomial Naïve Bayes. This point of view aligns with the nature of the problem since the bully usually attacks using different arguments such as racism, sexual harassment or shaming. Once classified, fuzzy rules determine the "strength of bully". At this point, fuzzy logic is applied at an endpoint after classification.

These previous works use fuzzy logic to improve the classification or labelling process. Our approach is based on a prior study of messages from a fuzzy point of view to improve the results. We obtain their essential characteristics by generating a feature vector in a fuzzy way that considers the whole set of data according to each class of cyberbullying. Then we model the messages taking into account that different types of attacks may appear with different strengths (more significant or lower degree of belonging) in the same message. As far as we know, this work is the first paper that studies cyberbullying in that fuzzy way. It is also the first paper that presents the use of fuzzy clustering in the problem of cyberbullying, laying the foundations for a future explainable AI tool on the problem.

3   Fuzzy Analysis of Cyberbullying

As mentioned, the objective of this work is to use fuzzy logic for cyberbullying analysis in textual data from social networks (specifically from Twitter) to facilitate and make more flexible the cyberbullying analysis in social networks. This section presents all the details of our proposal. Figure 1 shows the workflow followed to process, extract, and analyse cyberbullying from the textual data. We will explain each component of our approach in detail in the following sections.

3.1   Text Preprocessing

The preprocessing step is a crucial phase for almost all downstream NLP tasks. It cleans the textual data by applying some filters to represent data properly for later analysis. The filters applied were: Convert the texts to lower case, remove punctuations, remove words containing digits, remove stopwords, remove leading and trailing white spaces and finally, remove URLs. It can be noticed that we do not use advanced preprocessing techniques such as stemming, lemmatization, n-grams analysis, etc., because, as we will explain in Sect. 3.2, we use a pre-trained language model to encode the texts, and we need it as original as possible.

3.2   Sentence Embeddings

Text representation for NLP has caught the attention of researchers for several years. As is well-known, data mining techniques can not understand the raw text. That is why the proper text representation technique directly affects the performance of the used algorithm.

Generally speaking, the text representation process is a basic step for all NLP tasks, and its main goal is to convert words into numbers for machines to understand the natural language. In the state-of-the-art, we can find several approaches for text representation such as *One-Hot encoding, Bag-of-words representation (BOW), CountVectorizer*, and *Tf-idf term weighting*.

In the last decade, we have witnessed many groundbreaking techniques for text representation, thanks to deep learning algorithms. These representations

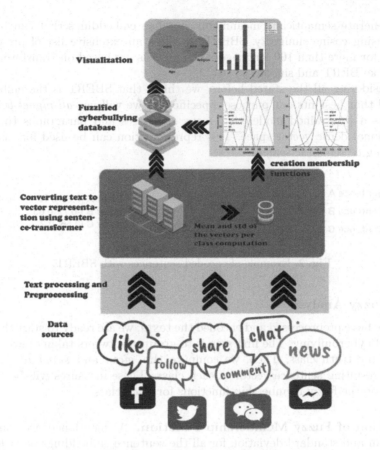

Fig. 1. Workflow of the entire proposal process.

are all based on word embeddings and are responsible for the latest break-throughs in NLP in recent years.

Word embeddings allow representing each word by a featured dense vector of fixed length. Each vector contains the semantic meaning of each word for a specific context. Word embeddings can be divided into two main groups: **static** and **contextualized**. Static word embeddings are the oldest type of word embedding, where the embeddings are generated from a large corpus, but a word used in different contexts always has the same representation. Example of static pre-trained word embeddings are word2vec [15], GloVe [17] and fastText [6]. On the opposite, contextualized word embeddings enable capturing the different meanings of a word taking into account the context. BERT [9], ELMo [18] and OpenAI GPT [19] are examples of contextualized pre-trained word embeddings.

In this work, we use a generalization of this idea that employs sentences instead of words. In particular, we use Sentence-BERT (SBERT) [20] for text representation. SBERT allows computing dense vector representations for sentences (see Fig. 2), paragraphs, and images. One of the main objectives of SBERT is to generate semantically meaningful sentence embeddings that can be compared using cosine similarity. SBERT provides an extensive list of pre-trained models for more than 100 languages. The models are based on transformer networks like BERT and similar.

Considering all the stated before, we think that SBERT is the embedding method that best fits our purpose. Specifically, we will use *all-mpnet-base-v2*[1], which is a pre-trained model that maps sentences and paragraphs to a 768-dimensional dense vector space. This representation can be used for tasks like clustering or semantic search.

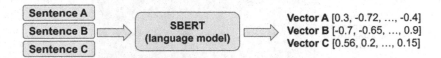

Fig. 2. Example of encoded sentences with SBERT.

3.3   Fuzzy Analysis

Once we have preprocessed and encoded the texts, we are ready to start the fuzzy study of cyberbullying. The first step is to group the tweets taking into account the ground-truth class. Then, we compute the *mean* and *standard deviation* for the resulting sentence vector of each class. These measures will be used to determine the fuzzy membership functions for each class.

Definition of Fuzzy Membership Function. As mentioned, we computed the mean and standard deviation for all the sentence embeddings of each class. We must remember that a 768-dimensional vector represents both the text representation and the computed measures. Considering this, we need to reduce the current dimension of the data to a smaller one. In our cases, we want to define the membership functions for each class. For that, we transform all the vectors into unidimensional data using the Uniform Manifold Approximation and Projection (UMAP), which is a recent technique for dimensionality reduction [13].

Finally, to define the fuzzy membership functions (Eq. 1) of each class, we have used the mean (m) and the standard deviation (d) in one dimension corresponding to each group. We must highlight that the definition process of the membership functions is entirely automatic.

$$\mu_{m,d}(x) = \begin{cases} 1 \; if \; x \leq min\_value \; or \; x \geq max\_value \; for \; triangular \; and \; Gaussian \; functions \\ gaussmf(x,m,d*0.3) \; for \; Gaussian \; function \\ trimf(x,d*0.5,m,m+(d*0.5)) \; for \; triangular \; function \end{cases} \quad (1)$$

where x is the unidimensional representation of the sentence embedding vectors, and min_value and max_value are the minimum and maximum values, respectively, after transforming to one dimension and normalising each sentence embedding (each document).

---

[1] https://huggingface.co/sentence-transformers/all-mpnet-base-v2.

4 Experimentation

This section describes the experimentation carried out on the dataset of tweets about cyberbullying.

4.1 Dataset

The dataset used was collected by the Twitter API and classified into different categories in [25]. The dataset comprises 46017 tweets with full text and the category corresponding to their classification. These categories can be seen in Fig. 3.

We must remark that this dataset does not represent a realistic dataset, as in the real world, the non-cyberbullying instances are the majority compared to the total number of cyberbullying instances. In our case, we have chosen this dataset because its characteristics make it suitable for our purpose of analysing the overlapping between the different kinds of cyberbullying.

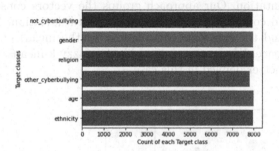

Fig. 3. Class distribution.

4.2 Experiment Setup

To demonstrate the feasibility of our proposal, we have divided the experimentation into two phases. First, we have tried fuzzy clustering, particularly fuzzy k-means, a trivial technique that we will use to compare our approach. Figure 4 depicts the tests carried out with fuzzy k-means and their different parameterisations. In this case, the values of *Fuzzy partition coefficient metric* are low, and the centroids are practically identical for all the classes (the crosses overlap), making the analysis very difficult.

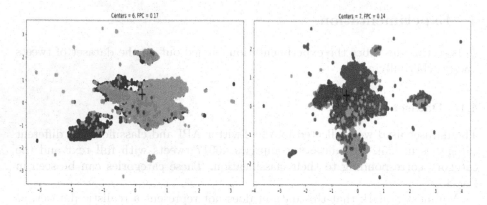

Fig. 4. Results for fuzzy clustering.

Since results obtained with fuzzy k-means were not good, we have applied our fuzzy approach to model and analyse cyberbullying in the second phase of the experimentation. Our approach groups the vectors considering ground-truth classes and computes the mean and standard deviation. Figure 5 depicts all the tweets and the centroids (represented by the mean) of each class in a 2-dimensional space. We can see that, unlike for fuzzy k-means, the centroids of the groups are better defined in this case.

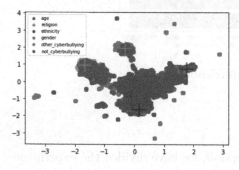

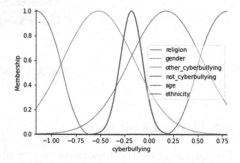

Fig. 5. Tweets and centroids per class in a 2-dimensional space.

Fig. 6. Membership functions defined through our proposal using gaussian bell functions.

Finally, we create the membership functions using the mean and standard deviation in a 1-dimensional space. The membership function can be created with two different functions: triangular and Gaussian. In Fig. 6, we have only shown the Gaussian function due to space restriction. These functions allow us to analyse cyberbullying from a fuzzy point of view, and now each tweet can be labelled with more than one class, probably with different degrees of membership.

4.3   Analysis of Our Fuzzy Approach

After this process, an analysis of the results obtained has been carried out. Different improvements can be seen concerning the interpretability and knowledge associated with the dataset. In Fig. 7 we can notice how the distribution of the classes has changed concerning the original dataset (see Fig. 3). Our proposal allows us to have a degree of membership for each category, and now we can have tweets in several categories with different strengths.

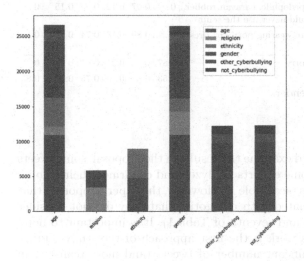

Fig. 7. Count of tweets per category with some degree of membership to any other category.

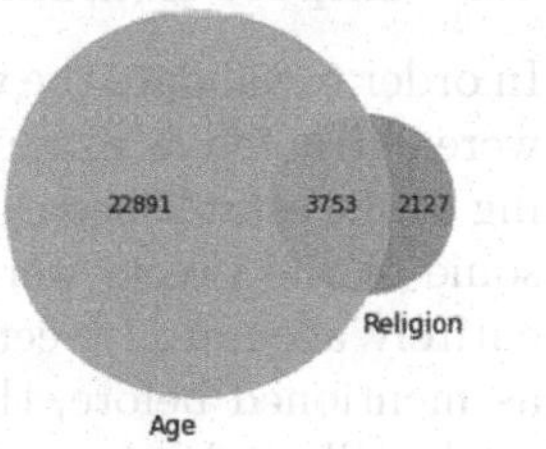

Fig. 8. Elements of categories **age**, **religion**, and the intersection of both sets according to their degrees of membership.

Furthermore, thanks to this process, we have been able to represent the data's knowledge and nature better, as a text does not have to belong to only one category. In other words, we could have a tweet about bullying that deals with a person's **religion** and also with their **age**. This relationship between elements of the database can be seen in Fig. 8. In this figure, we have data with some degree of membership to **age** or **religion**, and we also have elements that share both categories (3753 tweets).

Although the examples showcased in this work can be considered slightly straightforward, we must highlight that the proposed approach constitutes a starting point for cyberbullying analysis from a fuzzy perspective. The approach enables a broad spectrum of queries that facilitate the analysis of cyberbullying on textual data. For example, a user could filter texts with a 0.4 degree of membership to **religion** and a 0.6 degree of membership to **age**.

Table 1. Tweets after computing the degree of membership for each class[a].

Tweet	A	R	E	G	Other	Not
You leave me here! And no joke I will find you and rape and then kill you!	0.11	0	0	0.97	0.13	0
Fuck you bitch. Can't text a nigger back. I did text your dumb ass back. Faggot	0	0	0.87	0.79	0.85	0
Slaggg on the phone plzzzz I beening picked on fuck offf you dumb ass dirty crab of a whore shut ya mouth don't ever call my mate a nigger	0	0	0.12	0.87	0.17	0
Your prophet was a rapist, murderer, pedophile, caravan robber, slave trader, bigot and sexist. God would never use the scum	0	0.97	0.12	0	0.15	0
I hate you. You are Muslim but you are wearing porn girl dress. Go to hell bloody idiot	0	0.89	0.13	0.74	0	0
You look like a gypsy. Sorry I'm not sorry	0.87	0	0.2	0.05	0	0
Snorlax: I choose you!	0.53	0	0	0.75	0.87	0.85

[a]Age, Religion, Ethnicity, and Gender.

4.4 Expert Evaluation

In order to validate the work and examine the results of the proposal, some tweets were extracted as examples. Some experts analyzed and confirmed their belonging to the established categories (see Table 1). However, the experts reported that some of the tweets were associated with categories that they did not consider entirely accurate (tweets sixth and seventh of Table 1). It is important to note, as mentioned before, that this work is the first approach of the study. Future work will analyze a more significant number of tweets (and more realistic) in greater depth to improve the tool's performance. Expert recommendations will be considered in this future step.

5   Conclusions and Future Work

This work has aimed to show a preliminary solution for modelling and analysing cyberbullying from a fuzzy point of view using sentence embeddings. The presented approach improves the explainability of the analysis of cyberbullying when different types of cyberbullying are overlapped within the same document. This approach also makes it possible to generate more knowledge about classes such as those that allow classifying texts related to cyberbullying since the same text can belong to different classes with different degrees of membership.

For future research, we propose to use this novel processing method to create a decision support system to detect and resolve cases of bullying in different environments such as class groups, photos or social networks. Moreover, thanks to the great explainability and interpretability provided by fuzzy labels, we plan to improve the explainability of the data and the processes applied to them.

Acknowledgment. The research reported in this paper was partially supported by the Andalusian government and the FEDER operative program under the project Big-DataMed (P18-RT-2947 and B-TIC-145-UGR18) and grant PLEC2021-007681 funded by MCIN/AEI/10.13039/501100011033 and by the European Union NextGenerationEU/PRTR. Finally, the project is also partially supported by the Spanish Ministry of Education, Culture and Sport (FPU18/00150). In addition, this work has been partially supported by the Ministry of Universities through the EU-funded Margarita Salas programme - NextGenerationEU.

References

1. Centers for disease control and prevention: technology and youth: protecting your child from electronic aggression. Injury Prevention and Control: Violence Prevention (2009). Accessed 24 Apr 2013
2. Abou El-Seoud, S., Farag, N., McKee, G.: A review on non-supervised approaches for cyberbullying detection. Int. J. Eng. Pedagog. **10**(4), 25–34 (2020)
3. Akhter, A., Acharjee, U.K., Polash, M.M.A.: Cyber bullying detection and classification using multinomial Naïve Bayes and fuzzy logic. Int. J. Math. Sci. Comput **5**, 1–12 (2019)
4. Al-Ajlan, M.A., Ykhlef, M.: Deep learning algorithm for cyberbullying detection. Int. J. Adv. Comput. Sci. Appl. **9**(9), 199–205 (2018)
5. Bauman, S., Toomey, R.B., Walker, J.L.: Associations among bullying, cyberbullying, and suicide in high school students. J. Adolesc. **36**(2), 341–350 (2013)
6. Bojanowski, P., Grave, E., Joulin, A., Mikolov, T.: Enriching word vectors with subword information. Trans. Assoc. Comput. Linguist. **5**, 135–146 (2017)
7. Chowdhary, K.: Natural language processing. In: Fundamentals of Artificial Intelligence, pp. 603–649 (2020)
8. Dadvar, M., Eckert, K.: Cyberbullying detection in social networks using deep learning based models. In: Song, M., Song, I.-Y., Kotsis, G., Tjoa, A.M., Khalil, I. (eds.) DaWaK 2020. LNCS, vol. 12393, pp. 245–255. Springer, Cham (2020). https://doi.org/10.1007/978-3-030-59065-9_20
9. Devlin, J., Chang, M.W., Lee, K., Toutanova, K.: BERT: Pre-training of deep bidirectional transformers for language understanding. In: Proceedings of the 2019 Conference of the North American Chapter of the Association for Computational Linguistics: Human Language Technologies, Volume 1 (Long and Short Papers). Association for Computational Linguistics, Minneapolis, Minnesota (Jun 2019)
10. Di Capua, M., Di Nardo, E., Petrosino, A.: Unsupervised cyber bullying detection in social networks. In: 2016 23rd International Conference on Pattern Recognition (ICPR), pp. 432–437. IEEE (2016)
11. Huang, Q., Singh, V.K., Atrey, P.K.: Cyber bullying detection using social and textual analysis. In: Proceedings of the 3rd International Workshop on Socially-Aware Multimedia, pp. 3–6 (2014)
12. Kohonen, T.: The self-organizing map. Proc. IEEE **78**(9), 1464–1480 (1990)
13. McInnes, L., Healy, J., Melville, J.: Umap: uniform manifold approximation and projection for dimension reduction. arXiv preprint arXiv:1802.03426 (2018)
14. McLoughlin, C., Burgess, J.: Texting, sexting and social networking among australian youth and the need for cyber safety education. In: Proceedings of AARE International Education Research Conference. Australian Association for Research in Education, Melbourne (2009)
15. Mikolov, T., Chen, K., Corrado, G., Dean, J.: Efficient estimation of word representations in vector space. In: Bengio, Y., LeCun, Y. (eds.) 1st International Conference on Learning Representations. ICLR 2013, Scottsdale, Arizona, USA, 2–4 May 2013, Workshop Track Proceedings (2013)

16. Paul, S., Saha, S.: CyberBert: Bert for cyberbullying identification. Multimedia Syst. 1–8 (2020)
17. Pennington, J., Socher, R., Manning, C.D.: GloVe: global vectors for word representation. In: Proceedings of the 2014 Conference on Empirical Methods in Natural Language Processing (EMNLP), pp. 1532–1543 (2014)
18. Peters, M.E., et al.: Deep contextualized word representations. In: Proceedings of the 2018 Conference of the North American Chapter of the Association for Computational Linguistics: Human Language Technologies (Long Papers), vol. 1, pp. 2227–2237. Association for Computational Linguistics, New Orleans, June 2018
19. Radford, A., Narasimhan, K., Salimans, T., Sutskever, I.: Improving language understanding by generative pre-training (2018)
20. Reimers, N., Gurevych, I.: Sentence-Bert: Sentence embeddings using Siamese Bert-networks. In: Proceedings of the 2019 Conference on Empirical Methods in Natural Language Processing. Association for Computational Linguistics, November 2019
21. Romsaiyud, W., na Nakornphanom, K., Prasertsilp, P., Nurarak, P., Konglerd, P.: Automated cyberbullying detection using clustering appearance patterns. In: 2017 9th International Conference on Knowledge and Smart Technology (KST), pp. 242–247 (2017)
22. Rosa, H., Carvalho, J.P., Calado, P., Martins, B., Ribeiro, R., Coheur, L.: Using fuzzy fingerprints for cyberbullying detection in social networks. In: 2018 IEEE International Conference on Fuzzy Systems (FUZZ-IEEE), pp. 1–7. IEEE (2018)
23. Rosa, H., Matos, D., Ribeiro, R., Coheur, L., Carvalho, J.P.: A "deeper" look at detecting cyberbullying in social networks. In: 2018 International Joint Conference on Neural Networks (IJCNN), pp. 1–8. IEEE (2018)
24. Rosa, H., et al.: Automatic cyberbullying detection: a systematic review. Comput. Hum. Behav. 93, 333–345 (2019)
25. Wang, J., Fu, K., Lu, C.T.: SosNet: a graph convolutional network approach to fine-grained cyberbullying detection. In: 2020 IEEE International Conference on Big Data (Big Data), pp. 1699–1708. IEEE (2020)
26. Wang, L.L., Lo, K.: Text mining approaches for dealing with the rapidly expanding literature on Covid-19. Brief. Bioinform. 22(2), 781–799 (2021)
27. Zainol, Z., Wani, S., Nohuddin, P., Noormanshah, W., Marzukhi, S.: Association analysis of cyberbullying on social media using apriori algorithm. Int. J. Eng. Technol. 7(4.29), 72–75 (2018)
28. Zhao, R., Mao, K.: Cyberbullying detection based on semantic-enhanced marginalized denoising auto-encoder. IEEE Trans. Affect. Comput. 8(3), 328–339 (2016)
29. Zhao, Z., Gao, M., Luo, F., Zhang, Y., Xiong, Q.: LSHWE: improving similarity-based word embedding with locality sensitive hashing for cyberbullying detection. In: 2020 International Joint Conference on Neural Networks (IJCNN), pp. 1–8 (2020)

Flexible Division Queries Based on RL-Instances

Patricia Córdoba-Hidalgo ⓘ, Nicolás Marín ⓘ, and Daniel Sánchez(✉) ⓘ

Department Computer Science and Artificial Intelligence,
University of Granada, Granada, Spain
patriciacorhid@gmail.com, {nicm,daniel}@decsai.ugr.es

Abstract. In this paper we discuss on performing flexible division queries in crisp relational databases using gradual restrictions on the values of attributes. Our approach is based on using fuzzy sets for representing the restrictions and a recently proposed approach for computing and representing the result, based on the paradigm of Representations by Levels. We show some examples of usual division query patterns with SQL based on the NOT EXISTS operator.

Keywords: Flexible querying · Relational division · Fuzzy restrictions · Representations by levels

1 Introduction

Starting from crisp tables in relational databases, we consider flexible queries involving fuzzy restrictions on attribute values. The graduality in the conditions leads to graduality in the relation instances obtained, which have been represented by conjunctive fuzzy sets of tuples in the literature that we call *fuzzy instances*, with the corresponding extension of Relational Algebra operators. See a summary for instance in [2,3].

An alternative has been proposed in [3] based on the Representation by Levels (RL) Theory [9]. In this theory, graduality is represented using a finite set of levels in [0, 1], and mathematical objects and their operations and predicates are extended by considering the crisp versions in each level independently. When using RLs, the result of a flexible query is what we call an *RL-instance*, which is an assignment of crisp relational instances to levels. In the spirit of RLs, the extension of crisp Relational Algebra operations is performed by applying

Authors appear in alphabetical order. All authors have participated in conceptualization and research tasks. P. Córdoba-Hidalgo carried out most of the implementation tasks under the supervision of her advisors N. Marín and D. Sánchez.

This work has been partially supported by the Andalusian Government and the European Regional Development Fund - ERDF (Fondo Europeo de Desarrollo Regional - FEDER) under grants *B-TIC-145-UGR18* and *P18-RT-1765, BIGDATAMED. Analisis de Datos en Medicina: de las Historias Clínicas al Big Data.*

D. Ciucci et al. (Eds.): IPMU 2022, CCIS 1602, pp. 329–340, 2022.
https://doi.org/10.1007/978-3-031-08974-9_26

them independently to the corresponding crisp versions in each level of the RL-instances involved [3].

Remarkably, this new proposal keeps all the properties of the conventional Relational Algebra, particularly guaranteeing that all equivalent query patterns keep being so in the gradual case, something that cannot be guaranteed with fuzzy approaches, which is the main gain of our approach. As a consequence, since basic operators of Relational Algebra are part of SQL, it is possible to perform queries using the mentioned approach through the usual Relational Algebra query patterns, as shown in [3].

In this paper we go beyond the basic approach to division shown in [3], providing additional SQL patterns for implementing flexible division queries using RL-instances. The problem of extending division queries to fuzzy databases has been widely dealt with in the literature, see for example [1,2,6–8,10]. We focus here on relational databases and show that the use of RL-instances allows us to take advantage of the NOT EXISTS operator of SQL to solve division queries affected by graduality, beyond the well known query pattern based on the formulation of division in terms of basic Relational Algebra operators. We show that flexible division queries can be performed using the usual crisp SQL patterns based on this operator, avoiding the necessity to develop complex specific tools and the corresponding cost in training the users.

2   Preliminaries

2.1   Relational Databases and Division Queries

In relational databases, *relations* (also called *tables*) are the basic data structures. A relation is comprised of a *schema*, which is a set of attributes with associated domains, and a *relation instance*[1]. In this paper $R(A_1 : D_1, \ldots, A_n : D_n)$ denotes the schema of a relation R with attributes A_i having associated domains D_i (sometimes domains are omitted in the notation); every tuple in the actual instance of R is an element of the Cartesian product $D_1 \times \cdots \times D_n$. Given a tuple t, we will denote by $t[A_i]$ the value of attribute A_i in t; in general, we will denote by $t[A_1, \ldots, A_i]$ the i-tuple formed by the values of attributes from A_1 to A_i in tuple t. We shall use capital letters for relations and the corresponding small letters for denoting their instances.

Relational Algebra is a well known functional query language for relational databases. Every operator in Relational Algebra takes as input one or two relations and gives a relation as output. Tables 1 and 2 show the basic and non-basic operators, respectively, as given in [3][2]. Non-basic relational operators can be derived from the basic operators in Table 1 as shown in Table 2. I_R represents the set of all possible instances of relation R.

---

[1] We shall use the term *instance* to refer to *relation instances* from now on.

[2] In this paper, logical expressions are those expressions of first order logic based on the use of conjunction, disjunction, negation, and comparison operators in the domains of attributes, as commonly employed in relational database queries.

Table 1. Basic Relational Algebra operators. $R(A_1 : D_1, \ldots, A_n : D_n)$, $T_1(A_1 : D_1, \ldots, A_i : D_i)$ with $1 \leq i \leq n$, $S_1(B_1 : D_{n+1}, \ldots, B_m : D_{n+m})$, $T_2(A_1 : D_1, \ldots, A_n : D_n, B_1 : D_{n+1}, \ldots, B_m : D_{n+m})$, $S_2(A_1 : D_1, \ldots, A_n : D_n)$ are schemata. Note that $I_R = I_{S_2}$. Θ is a logical expression defined on the attributes of R.

Operator	Definition	
Selection	$\sigma_\Theta : I_R \rightarrow I_R$	$\sigma_\Theta(r) = \{t \in r \mid t \text{ satisfies } \Theta\}$
Projection	$\pi_{A_1, \ldots, A_i} : I_R \rightarrow I_{T_1}$	$\pi_{A_1, \ldots, A_i}(r) = \{t[A_1, \ldots, A_i] \mid t \in r\}$
Cartesian product	$\times : I_R \times I_{S_1} \rightarrow I_{T_2}$	$r \times s = \{t \in D_1 \times \ldots \times D_{n+m} \mid t[A_1, \ldots, A_n] \in r \wedge t[B_1, \ldots, B_m] \in s\}$
Union	$\cup : I_R \times I_{S_2} \rightarrow I_R$	$r \cup s = \{t \mid t \in r \vee t \in s\}$
Difference	$\setminus : I_R \times I_{S_2} \rightarrow I_R$	$r \setminus s = \{t \mid t \in r \wedge t \notin s\}$

In this paper we are concerned with the division operator, whose semantics can be expressed in terms of set inclusion as follows: let $R(A_1, \ldots, A_n)$ and $S(A_i, \ldots, A_n)$ be two relations with $2 \leq i \leq n$. Then the scheme of $r \div s$ is $(A_1, \ldots, A_{i-1})$, and its instance is computed from r and s as:

$$r \div s = \{t[A_1, \ldots, A_{i-1}] \mid t \in r \wedge s \subseteq \pi_{A_i, \ldots, A_n}(\sigma_{A_1 = t[A_1] \wedge \cdots \wedge A_{i-1} = t[A_{i-1}]}(r))\} \quad (1)$$

This semantics is the basis for the classical approaches to extend the division operator to the fuzzy case, as we will see in the next section.

2.2   Fuzzy Instances and Flexible Division Queries

As we said in the introduction, we are concerned with flexible queries in relational databases, corresponding to queries involving conditions containing fuzzy restrictions on the values of attributes. The result of such queries are conjunctive fuzzy sets of tuples that we call *fuzzy (relational) instances*. There are several approaches to the extension of Relational Algebra for performing flexible querying [2] which, using the same notation of the crisp case above, have been described in [3]. Particularly, the division operator is implemented by means of a fuzzy inclusion indicator, following the semantics given by Eq. (1), as follows:

$$\mu_{r \div s}(t) = Inc(s, \pi_{A_i, \ldots, A_n}(\sigma_{A_1 = t[A_1] \wedge \cdots \wedge A_{i-1} = t[A_{i-1}]}(r))) \quad (2)$$

where r and s are fuzzy instances of the schemata R and S employed in Eq. (1), Inc being a fuzzy inclusion index. A study of different sets of axioms required for such indicators, as well as some proposals that satisfy them, can be found for instance in [4].

3   RL-Instances and RL-Relational Algebra

In [3] we propose an alternative to fuzzy operations for flexible querying. Our proposal relies on the idea of *RL-systems* [9], which are systems that allow fuzzy

Table 2. Non-basic Relational Algebra operators. $R(A_1 : D_1, \ldots, A_n : D_n)$, $S_1(B_1 : D_{n+1}, \ldots, B_m : D_{n+m})$, $T_1(A_1 : D_1, \ldots, A_n : D_n, B_1 : D_{n+1}, \ldots, B_m : D_{n+m})$, $S_2(A_i : D_i, \ldots, A_n : D_n[, B_1 : D_{n+1}, \ldots, B_m : D_{n+m}])$ with $1 \le i \le n$, $T_2(A_1 : D_1, \ldots, A_n : D_n[, B_1 : D_{n+1}, \ldots, B_m : D_{n+m}])$, $S_3(A_i : D_i, \ldots, A_n : D_n)$ with $1 < i \le n$, $T_3(A_1 : D_1, \ldots, A_{i-1} : D_{i-1})$ with $1 < i \le n$, and $S_4(A_1 : D_1, \ldots, A_n : D_n)$ are schemata. Note that $I_R = I_{S_4}$. Θ is a logical expression defined on the attributes of T_1.

Operator	Definition	
Θ-join	$\bowtie_\Theta : I_R \times I_{S_1} \to I_{T_1}$	$r \bowtie_\Theta s := \sigma_\Theta(r \times s)$
Natural join	$\bowtie : I_R \times I_{S_2} \to I_{T_2}$	$r \bowtie s := \pi_{A_1, \ldots, A_n[, B_1, \ldots, B_m]}$ $(\sigma_{(R.A_i = S_2.A_i) \wedge \ldots \wedge (R.A_n = S_2.A_n)}(r \times s))$
Division	$\div : I_R \times I_{S_3} \to I_{T_3}$	$r \div s := \pi_{A_1, \ldots, A_{i-1}}(r) \backslash$ $\pi_{A_1, \ldots, A_{i-1}}((\pi_{A_1, \ldots, A_{i-1}}(r) \times s) \backslash r)$
Intersection	$\cap : I_R \times I_{S_4} \to I_R$	$r \cap s := r \backslash (r \backslash s)$

sets as input and output, but do not use Fuzzy Set Theory for internally representing and operating with graduality. Instead, the Theory of Representation by Levels is employed, akin to the proposal in [5], which is based on the following ideas:

– Graduality is represented using a finite set of levels $\Lambda = \{\alpha_1, \ldots, \alpha_m\} \subset (0, 1]$ with $1 = \alpha_1 > \alpha_2 > \cdots > \alpha_m > \alpha_{m+1} = 0$. Each level represents a degree of "relaxation" or allowance in the definition of a property, 1 meaning "maximum restriction" and 0 meaning "no restriction at all". Semantics of intermediate values are given by their distance to these extremes. Usual choices are sets of equidistant levels with 10 or 100 levels.
– A gradual set defined on X is represented as an RL-set, which is an assignment of crisp sets to levels of the form $\rho : \Lambda \to \{0, 1\}^X$. This assignment is extended to all levels in $(0, 1]$ as follows: $\rho(\alpha) = \rho(\alpha_i) \,\forall \alpha_i > \alpha > \alpha_{i+1}$. There is no restriction on the crisp sets in each levels, particularly the nestedness typical of α-cuts of fuzzy sets is not imposed.
– Graduality can be extended to other kinds of objects and notions such as elements (particularly numbers), predicates, truth values, etc. In each case, the gradual version of a crisp kind of object is an assignment of an object of that kind to each level.
– Operations between RL-objects are performed in each level independently, performing the corresponding crisp operation on the crisp versions of the objects in the same level.
– Summaries of properties of RL-objects are provided as the probability that the property holds in a level taken at random in $(0, 1]$. Particularly, the membership of an element to an RL-set is computed as the probability that the element is in a level taken at random in $(0, 1]$. Note that the amount of levels considered determines the precision employed in the membership function (e.g., 10 and 100 equidistant levels correspond to a precision of 1 and 2 decimals, respectively).

On this basis, an RL-instance is defined in [3] as follows:

Definition 1 ([3]). *Let $R(A_1, \ldots, A_n)$ be a schema. An RL-instance is a pair (Λ_r, ρ_r), where*

- $\Lambda_r = \{\alpha_1, \ldots, \alpha_m\}$ *is a set of levels, with $1 = \alpha_1 > \ldots > \alpha_m > \alpha_{m+1} = 0$,*
- *and $\rho_r : \Lambda_r \to I_R$ is a function that assigns a specific relation instance to each level in Λ_r.*

Similarly, the notion of RL-logical expression is introduced:

Definition 2 ([3]). *Let $\{A_1, \ldots, A_n\}$ be a set of attributes. An RL-logical expression defined on $\{A_1, \ldots, A_n\}$ is a pair $(\Lambda_\Theta, \rho_\Theta)$ where*

- $\Lambda_\Theta = \{\alpha_1, \ldots, \alpha_m\}$ *is a set of levels, with $1 = \alpha_1 > \ldots > \alpha_m > \alpha_{m+1} = 0$,*
- ρ_Θ *is a function that assigns a specific logical expression defined on attributes $\{A_1, \ldots, A_n\}$ to each level in Λ_Θ.*

Using these definitions, all Relational Algebra operators are extended to RL-Relational Algebra operators by performing the usual crisp operations in each level independently, using in each level the corresponding crisp versions of both instances and logical expressions. Contrary to the case of fuzzy instances, the application of equivalent query patterns to RL-instances always yields the same result [3]. These results can be used for flexible querying as follows:

- Fuzzy sets are employed for building RL-logical expressions with an approximation given by the amount of equidistant levels employed.
- The application on crisp instances of RL-Relational Algebra operators based on such RL-logical expressions yield RL-instances as result.
- In addition to RL-instances, fuzzy summaries in the form of fuzzy instances can be provided. Each tuple is assigned the probability that a tuple is in the crisp instance of a level α taken at random, which can be obtained in the case of equidistant levels as the ratio between the number of levels where the tuple appear and the number of levels employed.

4 Gradual Division on RL-Instances

In addition to keeping all properties of crisp Relational Algebra, the use of RL-instances for flexible querying allows for extending the full power of SQL queries to the gradual case. Remarkably, as we shall show, flexible division queries can be friendly performed using SQL and the well-known NOT EXISTS operator. This proposal goes beyond that in [3], which is based on the expression of the division operator in terms of basic Relational Algebra operator shown in Table 2.

4.1 RL-Systems with SQL

One way to use RL-systems based on RL-instances in a relational database is by using a table with a predefined and equidistant set of α values. This table will allow us to convert any crisp instance into an equivalent RL-instance, by just

M_ID	M_TITLE	M_DURATION
M1	They Die by Dawn	51
M2	The Nightmare Before Christmas	76
M3	Primer	77
M4	Borat	84
M5	Reservoir Dogs	99
M6	Ed Wood	125
M7	The Great Dictator	125
M8	The Shining	146
M9	Django Unchained	165
M10	LOTR1	178

(b) Media

C_ID	C_NAME	AGE
C1	Daniel Marín	20
C2	Nicolás Sánchez	40
C3	M. Amparo Pons	59
C4	Olga Vila	70
C5	Miguel Berzal	72

(a) Customers

C_ID	M_ID	V_DATE	V_RATING
C1	M1	03/15/21	70
C1	M2	02/17/20	80
C1	M3	07/10/21	80
C1	M4	09/12/19	60
C1	M5	10/01/20	60
C1	M6	03/03/21	75
C2	M1	03/10/21	90
C2	M2	05/21/20	100
C2	M3	07/27/19	90
C2	M5	09/17/19	55
C3	M1	12/11/18	40
C3	M7	11/23/20	85
C4	M2	02/10/18	75
C5	M1	01/11/21	60

(c) Views

Fig. 1. An example crisp database. *Customers* is a view created on a table containing the birthdate of the customer, from which the attribute AGE is obtained.

performing the Cartesian product of the crisp instance with the aforementioned auxiliary table [3]. In our examples, run on an Oracle DBMS[3], we will use a table `lambda` with a single attribute `alpha` and ten tuples with values from 0.1 to 1.0. for the sake of space.

In our system, we have also defined a class `FT` that allows us to represent trapezoidal values and labels on $\mathbb{R}$ and that, among its functionalities, incorporates a method that allows checking if a crisp value belongs to a certain α-cut of a trapezoid. To illustrate how an RL-instance is obtained using a fuzzy restriction given by one of these trapezoids, consider the `customers` relation shown in Fig. 1(a). The following code builds an RL-instance of adult customers:

```
SELECT alpha, c_name
FROM lambda, customers
WHERE FT('ADULT').matches(age,alpha)='T'
ORDER BY alpha DESC, c_name ASC;
```

The Cartesian product of `lambda` and `customers` generates an RL-instance in which all customers appear at all levels (i.e., they are associated with all the values of α in `lambda`). The condition expressed in the `WHERE` clause removes customers from some of these levels according to membership of the customer's age to the corresponding α-cut. The `ADULT` label is defined in Fig. 2.

---

[3] www.oracle.com.

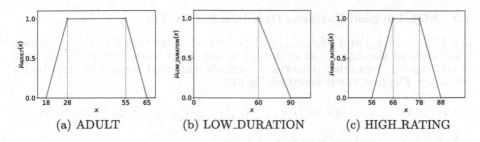

Fig. 2. Semantics for linguistic terms

ALPHA	P_NAME
1	Nicolás Sánchez
0.9	Nicolás Sánchez
0.8	Nicolás Sánchez
0.7	Nicolás Sánchez
0.6	M. Amparo Pons
0.6	Nicolás Sánchez
0.5	M. Amparo Pons
0.5	Nicolás Sánchez
0.4	M. Amparo Pons
0.4	Nicolás Sánchez
0.3	M. Amparo Pons
0.3	Nicolás Sánchez
0.2	Daniel Marín
0.2	M. Amparo Pons
0.2	Nicolás Sánchez
0.1	Daniel Marín
0.1	M. Amparo Pons
0.1	Nicolás Sánchez

(a) RL-instance

MU	P_NAME
1	Nicolás Sánchez
0.6	M. Amparo Pons
0.2	Daniel Marín

(b) Fuzzy summary

Fig. 3. Adult customers

The resulting RL-instance (which can be seen in Fig. 3(a)) can be transformed into a fuzzy instance by calculating the ratio of levels in which each tuple appears, using the following SQL code (the result is shown in Fig. 3(b)):

```
WITH rl_p AS (
  SELECT *
  FROM lambda, customers
  WHERE FT('ADULT').matches(age,alpha)='T')
SELECT COUNT(alpha)/10 mu, c_name
FROM rl_p
GROUP BY c_id, c_name
ORDER BY mu DESC, c_name ASC;
```

In general, as it can be seen in the examples above, the query that yields the fuzzy summary is just a simple grouping extension of the query that produces the corresponding RL-instance. For the sake of space, we shall use directly the second one in the examples of the remaining of this paper.

4.2   SQL Division Patterns Based on NOT EXISTS

Let $R(A_1,\dots,A_n)$ and $S(A_i,\dots,A_n)$ be two relations with $2 \le i \le n$. A pure translation of the expression in Table 2 is not usual when performing division in crisp tables. Two other patterns are usually employed because of their better efficiency. The first one is based on Eq. (3):

$$r \div s = \{t[A_1,\dots,A_{i-1}] \mid t \in r \wedge \left(s \backslash \pi_{A_1,\dots,A_{i-1}}(\sigma_{A_1=t[A_1]\wedge\cdots\wedge A_{i-1}=t[A_{i-1}]}(r))\right) = \emptyset\}, \quad (3)$$

which can be translated into SQL as follows:

```
SELECT A_1 , ... , A_i_1
FROM r t1
WHERE NOT EXISTS (
  SELECT *
  FROM s
  MINUS
  SELECT A_i , ... , A_n
  FROM r t2
  WHERE t1.A_1 = t2.A_1 , ... , t1.A_i_1=t2.A_i_1);
```

A second pattern is based on Eq. (4):

$$r \div s = \{t_1[A_1,\dots,A_{i-1}] \mid r(t_1) \wedge \not\exists t_2(s(t_2) \wedge \not\exists t_3(r(t_3)\wedge$$
$$\wedge t_3[A_1] = t_1[A_1] \wedge \cdots \wedge t_3[A_{i-1}] = t_1[A_{i-1}]\wedge \quad (4)$$
$$\wedge t_3[A_i] = t_2[A_i] \wedge \cdots \wedge t_3[A_n] = t_2[A_n]))\}$$

where $r(t)$ holds iff $t \in r$. This pattern can be translated into SQL as follows:

```
SELECT A_1 , ... , A_i_1
FROM r t1
WHERE NOT EXISTS (
  SELECT *
  FROM s t2
  WHERE NOT EXISTS (
    SELECT *
    FROM r t3
    WHERE t3.A_1 = t1.A_1 , ... , t3.A_i_1=t1.A_i_1
      AND t3.A_i = t2.A_i , ... , t3.A_n=t2.A_n ));
```

In the next subsection, we will show that these two patterns can be friendly used to solve flexible division queries using RL-instances.

4.3   SQL Division Patterns for Flexible Querying with RL-Instances

Consider the relational database in Fig. 1. A flexible division query might be *Find those customers who have viewed all the low duration media* (Query 1). Using the label **low duration** to constrain the value of the media's duration produces graduality in the divisor and makes the division query a flexible query.

Using the approach described in Sect. 4.1 and the division pattern of Eq. 3, we can solve Query 1 through the following SQL code:

```
WITH rl_c AS (
    SELECT *
    FROM lambda, customers),
  rl_m AS (
    SELECT *
    FROM lambda, media
    WHERE FT('LOW DURATION').matches(m_duration,alpha)='T'),
  rl_v AS (
    SELECT *
    FROM lambda, views
)
SELECT COUNT(alpha)/10 mu, c_name
FROM rl_c
WHERE NOT EXISTS (
  SELECT m_id
  FROM rl_m
  WHERE rl_m.alpha=rl_c.alpha
  MINUS
  SELECT m_id
  FROM rl_v
  WHERE rl_v.c_id=rl_c.c_id AND rl_v.alpha=rl_c.alpha)
GROUP BY c_id, c_name
ORDER BY mu DESC;
```

As can be seen, in the with clause we define the RL-instances for the query using the database crisp tables. Among them, the rl_m table contains the RL-instance with the gradual divisor generated by using the low_duration label (the definition of this label can be seen in Fig. 2(b)). The query below the with clause uses the pattern of Eq. 3 to solve the division and compute the fuzzy summary. Note that the pattern coincides with the one that would be used in a crisp environment, with the single addition of a restriction in the WHERE clauses that imposes equality at each α level. The resulting fuzzy summary and the corresponding RL-instance are shown in Figs. 4(a) and 5(a), respectively.

Alternatively, but with the *same* result, we can apply the pattern of Eq. 4 to solve Query 1 through the following SQL code:

```
WITH rl_c AS (
    SELECT *
    FROM lambda, customers),
  rl_m AS (
    SELECT *
    FROM lambda, media
    WHERE FT('LOW DURATION').matches(m_duration,alpha)='T'),
  rl_v AS (
    SELECT *
    FROM lambda, views
)
SELECT COUNT(alpha)/10 mu, c_name
FROM rl_c
WHERE NOT EXISTS (
```

MU	P_NAME
1	Daniel Marín
0.8	Nicolás Sánchez
0.6	M. Amparo Pons
0.6	Miguel Berzal

(a) Query 1

MU	P_NAME
0.8	Nicolás Sánchez
0.2	D. Marín
0.2	M. Amparo Pons

(b) Query 2

MU	P_NAME
0.2	Daniel Marín

(c) Query 3

Fig. 4. Fuzzy summaries for the three division queries

ALPHA	P_NAME
1	Daniel Marín
1	M. Amparo Pons
1	Miguel Berzal
1	Nicolás Sánchez
0.9	Daniel Marín
0.9	M. Amparo Pons
0.9	Miguel Berzal
0.9	Nicolás Sánchez
0.8	Daniel Marín
0.8	M. Amparo Pons
0.8	Miguel Berzal
0.8	Nicolás Sánchez
0.7	Daniel Marín
0.7	M. Amparo Pons
0.7	Miguel Berzal
0.7	Nicolás Sánchez
0.6	Daniel Marín
0.6	M. Amparo Pons
0.6	Miguel Berzal
0.6	Nicolás Sánchez
0.5	Daniel Marín
0.5	M. Amparo Pons
0.5	Miguel Berzal
0.5	Nicolás Sánchez
0.4	Daniel Marín
0.4	Nicolás Sánchez
0.3	Daniel Marín
0.3	Nicolás Sánchez
0.2	Daniel Marín
0.1	Daniel Marín

(a) Query 1

ALPHA	P_NAME
1	Nicolás Sánchez
0.9	Nicolás Sánchez
0.8	Nicolás Sánchez
0.7	Nicolás Sánchez
0.6	M. Amparo Pons
0.6	Nicolás Sánchez
0.5	M. Amparo Pons
0.5	Nicolás Sánchez
0.4	Nicolás Sánchez
0.3	Nicolás Sánchez
0.2	Daniel Marín
0.1	Daniel Marín

(b) Query 2

ALPHA	P_NAME
0.2	Daniel Marín
0.1	Daniel Marín

(c) Query 3

Fig. 5. RL-instances for the three division queries

```
SELECT *
FROM rl_m
WHERE rl_m.alpha=rl_c.alpha
  AND NOT EXISTS (
    SELECT *
    FROM rl_v
    WHERE rl_v.c_id=rl_c.c_id AND rl_v.alpha=rl_c.alpha
      AND rl_m.m_id = rl_v.m_id))
GROUP BY c_id, c_name
ORDER BY mu DESC;
```

Again, the pattern coincides with the one that would be used in a crisp environment, with the aforementioned single addition regarding equality at each α level.

Using this approach, it is easy to consider divisions where graduality also affects the candidates to be part of the quotient. For example, we could consider the query *Find those adult customers who have viewed all the low duration media* (Query 2), which would be solved with a code similar to the previous one, by only changing the definition of the RL-instance for customers as follows:

```
WITH rl_c AS (
    SELECT *
    FROM lambda , customers
    WHERE FT('ADULT').matches(age,alpha)='T'),
 ...
)
 ...
```

The generated RL-instance is depicted in Fig. 5(b) and the corresponding fuzzy summary is shown in Fig. 4(b).

Finally, adding a restriction that affects the dividend by constraining the relationship between each candidate and each element of the divisor does not pose much difficulty either. For example, the query *Find those adult customers who have viewed all the low duration media, having assigned them a high rating in the view* (Query 3) could be solved again with the same code, by only changing the definition of the RL-instance for `views` in the `with` clause:

```
WITH rl_c AS (
    SELECT *
    FROM lambda , customers
    WHERE FT('ADULT').matches(age,alpha)='T'),
  rl_m AS (
    SELECT *
    FROM lambda , media
    WHERE FT('LOW DURATION').matches(m_duration,alpha)='T'),
  rl_v AS (
    SELECT *
    FROM lambda , views
    WHERE FT('HIGH RATING').matches(v_rating,alpha)='T'
)
 ...
```

The generated RL-instance is depicted in Fig. 5(c) and the corresponding fuzzy summary is shown in Fig. 4(c). The semantics for HIGH RATING is shown in Fig. 2(c).

5   Conclusions

The resolution of flexible division queries in relational databases can be performed in a simple and direct way through the usual query patterns employing the NOT EXISTS operator, thanks to the use of RL-instances based on the Theory of Representation by Levels. As we have shown in this paper, this holds for queries involving fuzzy restrictions affecting the divisor, the candidate, and/or the dividend.

Following the approach in [3], it suffices to create a table `lambda` with a predefined set of equidistant α values in $(0, 1]$, according to the degrees of discernibility required in the results. Though the theoretical complexity of this solution is the same as that of the conventional case, the space and time required to solve the query is multiplied by a factor given by the number of levels considered, which can be a problem in large databases. A solution to this problem can be achieved by the use of parallelizing solutions similar to those employed in the modern NewSQL systems, since the computation of queries can be performed in each level independently. This will be an object of future research.

The semantics of the fuzzy summaries of our division queries comes from the analysis of the RL-instance obtained as quotient, assigning to each crisp tuple a degree representing the probability of finding it at a given level of the mentioned RL-instance. That is, the degree α assigned to a tuple in the final fuzzy summary tells us how probable is to find this tuple in the crisp instance obtained by carrying out the conventional relational division in a level taken at random from $(0, 1]$.

As explained in [3], the use of RL-systems keeps all the properties of the crisp case, particularly the Boolean properties of set operations, because the relation between fuzzy inputs and outputs is not functionally expressible in general. As a consequence, we conjecture (and will study) that the fuzzy summaries of division queries we obtain will be different in general from those provided by the usual fuzzy inclusion indicators based on functional expressions like those in [4].

References

1. Bosc, P., Pivert, O.: On diverse approaches to bipolar division operators. Int. J. Intell. Syst. **26**(10), 911–929 (2011)
2. Bosc, P., Pivert, O., Rocacher, D.: Tolerant division queries and possibilistic database querying. Fuzzy Sets Syst. **160**(15), 2120–2140 (2009)
3. Córdoba-Hidalgo, P., Marín, N., Sánchez, D.: RL-instances: an alternative to conjunctive fuzzy sets of tuples for flexible querying in relational databases. Fuzzy Sets Syst. https://doi.org/10.1016/j.fss.2022.03.022
4. Cornelis, C., der Donck, C.V., Kerre, E.E.: Sinha-Dougherty approach to the fuzzification of set inclusion revisited. Fuzzy Sets Syst. **134**(2), 283–295 (2003)
5. Dubois, D., Prade, H.: Gradual elements in a fuzzy set. Soft Comput. **12**, 165–175 (2008)
6. Dubois, D., Prade, H.: Semantics of quotient operators in fuzzy relational databases. Fuzzy Sets Syst. **78**(1), 89–93 (1996)
7. Galindo, J., Medina, J.M., Garrido, M.C.: Fuzzy division in fuzzy relational databases: an approach. Fuzzy Sets Syst. **121**(3), 471–490 (2001)
8. Marín, N., Molina, C., Pons, O., Vila, M.A.: An approach to solve division-like queries in fuzzy object databases. Fuzzy Sets Syst. **196**, 47–68 (2012)
9. Sánchez, D., Delgado, M., Vila, M.A., Chamorro-Martínez, J.: On a non-nested level-based representation of fuzziness. Fuzzy Sets Syst. **192**(1), 159–175 (2012)
10. Yager, R.R.: Fuzzy quotient operators for fuzzy relational databases. In: Proceedings IFES 1991, pp. 289–296 (1991)

Soft Computing and Artificial Intelligence Techniques in Image Processing

Image Segmentation Losses with Modules Expressing a Relationship Between Predictions

Petr Hurtik[ID], Vojtech Molek[✉][ID], and Hana Zámečníková

University of Ostrava, Centre of Excellence IT4Innovations, Institute for Research
and Applications of Fuzzy Modeling, 30. dubna 22, Ostrava, Czech Republic
{petr.hurtik,vojtech.molek,hana.zamecnikova}@osu.cz
http://gitlab.com/irafm-ai/segmentation-loss-hack

Abstract. We focus on semantic image segmentation with the usage
of deep neural networks and give emphasis on the loss functions used
for training the networks. Considering region-based losses, Dice loss, and
Tversky loss, we propose two independent modules that easily modify the
loss functions to take into account the relationship between the class predictions and increase the slope of the gradient. The first module expresses
the ambiguity between classes and the second module utilizes a differentiable soft argmax function. Each of the modules is used before the
standard loss is computed and remains untouched. In the benchmark,
we involved two neural network architectures with two different backbones, selected two loss functions, and examined separately two scenarios for softmax and sigmoid top activation functions. In the experiment,
we demonstrate the usefulness of our modules by improving the IOU
and F1 coefficients on the test dataset for all scenarios tested. Moreover,
the usage of the modules decreases overfitting. The proposed modules are
easy to integrate into existing solutions and add near-zero computational
overhead.

Keywords: Image segmentation · Smooth loss function · Soft argmax

1 Introduction

Using deep neural networks for image processing has led to the automation of
many routine tasks. Generally, we can divide image processing tasks into three
main categories: *image classification*, *image segmentation*, and *object detection*.
In this study, we focus on image segmentation. The goal of the methods in the
category is to assign a probability vector to each pixel, where each vector value
indicates the probability that a pixel belongs to a certain class. In contrast to
instance segmentation [11], a subtask of object detection, it does not distinguish
between individual object instances of the same class. The use of image segmentation can be found in autonomous driving to identify lanes on a road [33],
medical imaging to find particular parts of organs [10,42], remote sensing to

© Springer Nature Switzerland AG 2022
D. Ciucci et al. (Eds.): IPMU 2022, CCIS 1602, pp. 343–354, 2022.
https://doi.org/10.1007/978-3-031-08974-9_27

estimate the area of rivers, trees, etc. [13,16], and in modern industrial defectoscopy [9,18].

Obtaining a well-performing deep neural network model consists of two critical processes: selecting a *suitable architecture* and *training pipeline*. An *architecture* defines a sequence of layers that transform the input into the desired output. The current research focuses on developing an optimal architecture, given a certain problem. We are witnessing the appearance of new, bigger, and more powerful schemes, among the most famous are U-Net [27], Feature Pyramid Network (FPN) [21], DeepLab [6], HR-Net [34], and Segmentation Transformer (OCR) [39]. These schemes are coupled in their encoding phase with backbones that serve as powerful feature extractors, where the most frequently used are ResNet [40], ResNeSt [41], EfficientNet [32], and Swin [23], to name a few. The second direction of research is focused on the *training pipeline*, emphasizing the role of data and the setting of training. Here are developed new augmentation and regularization techniques [7,12], loss functions, or principles of working with training data such as Meta Pseudo Labels [26] or Noisy Student [35].

Our study follows the second research direction and focuses on the loss functions used to train deep neural networks for image segmentation. We present two simple modules (modifications) of known region-based loss functions (Dice loss [25] and Tversky loss [29]) that change the derivative slope and deal with the relationship between class predictions, leading to higher precision while the model, data, and training scheme remain fixed. Such a combination of loss functions and the proposed modules is *novel* and is not described in the literature.

2 Related Work and Preliminaries

The loss function is the key ingredient in neural network training. It describes how the difference between the ground truth and the produced prediction is computed during the training phase. A model can achieve significantly different performance depending on the used loss function, see [17]. Loss functions can be divided into distribution-based loss functions, region-based ones, and their combination, where in semantic segmentation, the last two mentioned are involved most frequently. In this study, we will focus on region-based loss functions in our benchmark.

Distribution-based losses are those known from the classification task, such as cross-entropy, Focal loss [22], or Kullback-Leibler divergence [20]. Their disadvantage is that an incorrectly segmented small object has only a negligible impact on a global loss value, and the network might ignore small areas. Moreover, the network tends to optimize toward objects with a bigger area as the losses are size-dependent. The solution to the issue is to use class weighting [14,24], balancing [36], or decrease the loss produced by correct predictions and increase the loss of incorrect predictions [22], which can lead to overfitting of noisy labels.

Region-based losses include Dice loss [25], Lovasz-Softmax loss [2], Jaccard loss [38], or Tversky loss [29], to name a few. Their advantage is size independence, i.e., a neural network does not tend to ignore tiny objects. The last group, combined losses, includes a simple aggregation of distribution-based and area-based losses (such as adding cross-entropy to Dice loss), Focal Tversky loss [1] – applying 'Focal function' to Tversky loss, or Combo loss [31] that weights modified cross-entropy with Dice loss. These combinations are the most general solution and can be safely used for the image segmentation problem without knowing the specifics of the data.

In the following text, we describe our modules for two region-based loss representatives – Dice and Tversky loss. Let $\mathbf{y}$, $\mathbf{y'}$ be a label and a prediction, respectively. Then, Dice loss is defined as

$$\mathcal{L}_D(\mathbf{y}, \mathbf{y'}) = 1 - \frac{2|\mathbf{y} \cap \mathbf{y'}| + s}{|\mathbf{y}| + |\mathbf{y'}| + s}, \tag{1}$$

where $| \cdot |$ denotes the cardinality of a set and $s \in [0, 1]$ is a smoothing constant that provides a similar effect as label smoothing [30]. Note that the original Dice loss [25] did not include the smoothing constant s, which may help to avoid overfitting noisy labels. Dice loss is derived from Dice index [8] (the loss is an inverse value of the index) originally used as a two-sample similarity measurement in plant sociology. Despite its similarity to Jaccard index, Dice loss does not satisfy the triangle inequality. The second representative, Tversky loss, can be considered a generalization of Dice loss, defined as follows:

$$\mathcal{L}_T(\mathbf{y}, \mathbf{y'}) = 1 - \frac{|\mathbf{y} \cap \mathbf{y'}| + s}{|\mathbf{y} \cap \mathbf{y'}| + \alpha|(1 - \mathbf{y}) \cap \mathbf{y'}| + \beta|\mathbf{y} \cap (1 - \mathbf{y'})| + s}, \tag{2}$$

where $\alpha + \beta = 1$ are weighting constants. The proper setting of α and β is problem dependant and helps to control the importance of false negative and false positive. For example, in the optical quality assessment, it is usually crucial to catch all damaged product parts, and it is acceptable to predict some good parts as faulty. In such a case, $\beta > \alpha$.

3   The Modules

The motivation of our modules is the different requirements during training and inference. During training, we want to obtain the real-valued class probabilities that are used in the loss function to compute the gradient. During the inference, however, we are interested in the highest probability class, computed by argmax operation. Example: the information that a pixel is recognized as ['background', 'car'] with probability [0.4, 0.6] is necessary for the optimizer to determine how the weights should be updated to obtain the actual true probabilities, e.g., [0, 1]. The gradient for such a prediction is different for, e.g., prediction [0.3, 0.7]. During the inference, both predictions of [0.4, 0.6] and [0.3, 0.7] yield 'car', as the highest class probability is the second vector element in both cases. The only rule is to have the probability of the correct class higher than the probability of the other class. Our goal is to reflect such behavior needed in inference in the training part. To achieve that, we propose two modules that are coupled with a loss function. Note that in one time moment, only one of the modules can be used.

Let a neural network $\mathcal{N}$ realizing image segmentation be $\mathcal{N} : \mathbb{R}^{w \times h \times 3} \to \mathbb{R}^{w \times h \times c}$, where c indicates the number of resulting classes. Furthermore, let $\mathbf{y}$, $\mathbf{y}'$ be a label with a prediction, and $\mathcal{L} : \mathbb{R}^{w \times h \times c} \times \mathbb{R}^{w \times h \times c} \to \mathbb{R}_0^+$ be a loss function if the following holds:

1. $\mathcal{L}(\mathbf{y}, \mathbf{y}') = 0$ iff $\mathbf{y} = \mathbf{y}'$, otherwise $\mathcal{L}(\mathbf{y}, \mathbf{y}') > 0$,
2. for two predictions $\mathbf{y}', \mathbf{y}''$, where $p_{\mathbf{y}'} \leq p_{\mathbf{y}''}$:

$$\mathcal{L}(\mathbf{y}, \mathbf{y}'') \leq \mathcal{L}(\mathbf{y}, \mathbf{y}') \text{ if } p_{\mathbf{y}} = 1,$$
$$\mathcal{L}(\mathbf{y}, \mathbf{y}'') \geq \mathcal{L}(\mathbf{y}, \mathbf{y}') \text{ if } p_{\mathbf{y}} = 0,$$

3. $\mathcal{L}$ is differentiable.

The symbol $p_{(.)}$ denotes the value of $p \in [0, 1]$ in connection with certain prediction (in case of $c = 2$, e.g., the prediction of $[1 - p, p]$). Note that the third condition is necessary to perform the Backpropagation algorithm [28] that computes gradient. The third condition prevents the argmax function from being used inside a loss function.

3.1   Module 1

We consider a segmentation task into $c = 2$ classes. It means that the goal is to assign to each of the pixel in an image a probability of being/not being a certain object. Note that the sum of the probabilities for a certain pixel can be different from 1. Practical usage is, e.g., segmentation of a car vs. background, iris vs. background, or segmenting a fault in a pruduct such as scratch. We define the module as $\mathcal{M}_1 : \mathbb{R}^{w \times h \times 2} \to \mathbb{R}^{w \times h \times 1}$ where

$$\mathcal{M}_1(\mathbf{y}) = 0.5(1 + \mathbf{y}_{\cdot, \cdot, 1} - \mathbf{y}_{\cdot, \cdot, 0}) \tag{3}$$

and the operations are performed element-wise. $\mathbf{y}_{\cdot, \cdot, 0}$ usually denotes the probability of being an object and $\mathbf{y}_{\cdot, \cdot, 0}$ not being the object. With this modification, a loss function $\mathcal{L}^{M_1}$ is computed as

$$\mathcal{L}^{M_1}(\mathbf{y}, \mathbf{y}') = \mathcal{L}(\mathcal{M}_1(\mathbf{y}), \mathcal{M}_1(\mathbf{y}')). \tag{4}$$

The proposed module has two interesting properties. Firstly, if $\mathbf{y}', \mathbf{y}$ are tensors whose elements $y'_{m,n,l}$, $y_{m,n,l} \in \{0, 1\}$ and if $y_{m,n,0} + y'_{m,n,1} = 1$ holds for arbitrary m, n then it coincides with argmax. If $y'_{m,n,l}$, $y_{m,n,l} \in [0, 1]$, then with fuzzy argmax. This is valid when the top activation function is softmax. Secondly, if $y_{m,n,0} + y'_{m,n,1} \neq 1$ holds for arbitrary m, n, then it expresses unambiguity between the two classes. That may occur when the sigmoid is used as the top activation function.

An illustration of the behavior is shown in Fig. 1 for Dice loss and in Fig. 2 for Tversky loss. For simplicity, we consider only a single pixel with the label of $[0, 1]$ and the prediction with $[1 - p, p]$ where $p \in [0, 1]$. The enriched loss functions produce a nonlinear response and put a bigger emphasis on bad predictions. That is similar to Focal loss [22] used in image classification that puts emphasis on incorrect classification and suppresses the loss of near correct classification.

3.2 Module 2

The second module $\mathcal{M}_2 : \mathbb{R}^{w \times h \times c} \rightarrow \mathbb{R}^{w \times h \times 1}$ relies on performing soft argmax, i.e., differentiable softmax. It was proposed by Yi et al. [37] in the task of key-point position estimation and it is based on a numerical trick [5] that allows to construct smooth functions, and we modify it to be used for image segmentation. It is defined as

$$\mathcal{M}_2(\mathbf{y}) = \sum_{i=1}^{c} \mathbf{a}_{.,.,i} \tag{5}$$

with

$$\mathbf{a} = \mathbf{p} \frac{e^{\omega \mathbf{y}}}{\sum_{i=1}^{c} e^{\omega \mathbf{y}_{.,.,i}}}, \tag{6}$$

where c determines the number of classes, operations are defined element-wise, ω is a positive constant ([37] recommends $\omega = 10$), and $\mathbf{p} = \{0, 1, \ldots, c-1\}$.

The application of (5) to a loss function is identical to the scheme given by (4), i.e., as

$$\mathcal{L}^{M_2}(\mathbf{y}, \mathbf{y}') = \mathcal{L}\left(\mathcal{M}_2(\mathbf{y}), \mathcal{M}_2(\mathbf{y}')\right). \tag{7}$$

In contrast to the standard argmax which is not differentiable, soft argmax can control the smoothness of the transition between the crisp values via ω parameter. With $\omega \approx \infty$ (in practice for, e.g., $\omega = 1 \cdot 10^3$), it coincides with the standard argmax. It becomes problematic when a batch of samples produces nearly perfect or nearly worst predictions; in both cases, the derivative is close to zero, and therefore the weight updates are negligible, and the training slows down considerably.

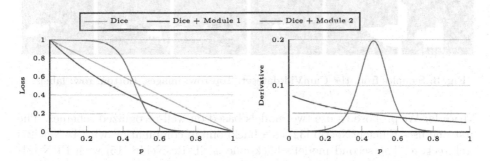

Fig. 1. Illustration of progression of Dice loss (left) and its derivation (right) for a single pixel with $c = 2$, the label of $[0, 1]$, and the prediction of $[1 - p, p]$ where $p \in [0, 1]$. For Module 2, we set $\omega = 10$.

4 Benchmark

Data: The experimental verification is performed on the CamVid dataset [3], which consists of pixel-level annotated images captured from a car. The dataset includes 367 training and 233 testing images, each image with a size of 480×360

px. Note that the number of images may seem low for a deep learning approach, but due to the dense annotation, it includes 63.4M and 40.2M annotated pixels. Each pixel is assigned to one of 32 semantic classes. For our purposes, we created a binary problem by extracting the 'car' semantic class and assigning the class 'other' to all remaining pixels. The visualization is shown in Fig. 3.

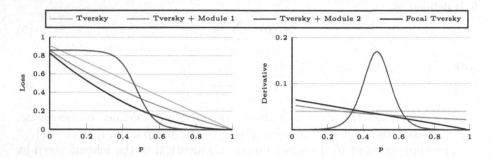

Fig. 2. Illustration of progression of Tversky loss (left) its derivation (right) for a single pixel with $c = 2$, the label of $[0, 1]$, and the prediction of $[1 - p, p]$ where $p \in [0, 1]$. The parameters of Tversky loss are set to $\alpha = 0.4$, $\beta = 0.6$, $s = 0.1$. For Tversky + Module 2, we set $\omega = 10$ and for Focal Tversky $\gamma = 2$.

Fig. 3. Samples from the CamVid dataset. Top row: images. Bottom row: labels.

Models: We propose to use two models based on well-recognized schemes. The first one uses EfficientNetB0 [32] as a backbone in combination with U-Net [27] architecture. The second model's backbone is SE-ResNet18 [15] with FPN [21] architecture. The backbone serves as a feature extractor from the input image, then the features are used by the U-Net/FPN by projecting them back into the input image resolution and transforming them into pixel-level predictions. We have selected the respective combinations due to their proven effectiveness in many winning solutions on the Kaggle competition website and their accessible implementation in a well-established library[1].

Methods: We involve Dice loss and Tversky loss in (1) and (2) form with the setting of $s = 1$ for Dice loss and $\alpha = 0.4, \beta = 0.6, s = 0.1$ for Tversky loss.

---

[1] http://github.com/qubvel/segmentation_models.

We also consider their enriched versions according to (4) for Module 1 and (7) for Module 2. Module 2 is used with $\omega = 10$. For numerical comparison of performance, we also add Focal Tversky loss [1] with $\gamma = 2$. The loss functions are coupled with the defined models separately using sigmoid and softmax as the top activation function. At the end of the training, we measure Intersect Over Union (IOU) and the harmonic mean of the precision and recall (F_1) scores on both training and testing datasets.

Training Specification: Each model is trained from scratch for 200 epochs with images padded up to 480×384px and a batch size of 4. The optimizer is Adam [19] with an initial learning rate of $1 \cdot 10^{-4}$ and is decreased by a factor of 0.1 in the epochs 120, 140, and 160. For data augmentation, we use Albumentations [4] and the augmentations of random brightness, contrast, motion blur, perspective distortion, and Gaussian noise. After 200 epochs of training, the test set is evaluated. Furthermore, we train each model five times and compute the mean.

Results: In total, we conducted 140 full trainings (200 epochs each) and according to the results in Table 1, we can claim that both proposed modules lead in all tested scenarios to higher IOU and F_1 coefficients than the standard version of the losses. The results also demonstrate that the score improvement is higher for the two modules than for the Focal version of Tversky loss and that Module 2 yields slightly better scores than Module 1 in most cases. The hypothesis that loss functions enriched by Module 2 will fail during the start of the training or will be unable to fine-tune the model due to close to zero gradient was not confirmed. The welcomed side effect of both modules is also the reduction of overfitting compared to the original version of the loss functions.

    Ablation study: To reveal the negative impact of the wrong selection of ω used in Module 2, we measured IOU coefficient on the test dataset for models trained with various ω. The used architecture is U-Net + EfficientNetB0 and the training scheme is identical to that described in 'Training specification' paragraph. Graph 4 visualizes the results from which we can conclude that the valid setting is for $\omega \in [3, 80]$ and outside this interval, the training does not work at all.

5   Discussion

The proposed modules are similar to focal loss in the sense that they both produce a higher loss value for bad predictions and a relatively smaller loss value for good predictions and thus prevent overfitting. Such behavior is undesirable when a dataset contains a large number of noisy labels. The difference between the proposed module and focal loss lies in the shape of the loss function itself. While focal loss forms a convex shape, the proposed modules create a concave-convex shape. This produces a different derivative, where the maximum value of a derivative is for the case of the highest uncertainty, i.e., when the probabilities of the classes are equal. It makes the modules unique, has a positive impact on neural network performance, and also yields much higher precision than the combination of a generic loss function with focal loss.

Table 1. The results for train score, test score, and the overfitting value. The bold value indicates a better result. The higher the IOU and F₁ scores, the better. The lower the value of the train test, the better.

U-Net + EfficientNetB0

Setting			Train · 100		Test · 100		Train − Test	
Loss	Module	Activation	IOU	F_1	IOU	F_1	IOU	F_1
Dice	——	Softmax	94.57	96.85	86.50	90.54	8.07	6.31
Dice	Module 1	Softmax	93.83	96.43	87.30	**93.83**	**6.53**	**2.60**
Dice	Module 2	Softmax	94.61	96.92	**87.49**	91.57	7.12	5.35
Tversky	——	Softmax	94.62	96.93	87.17	91.15	7.45	5.78
Tversky	Module 1	Softmax	92.58	96.86	**87.42**	91.43	**5.16**	5.43
Tversky	Module 2	Softmax	94.60	96.87	87.13	**91.69**	7.47	5.18
Focal Tversky	——	Softmax	92.67	95.58	85.67	90.79	7.00	**4.79**
Dice	——	Sigmoid	94.51	96.81	87.01	91.04	7.50	5.77
Dice	Module 1	Sigmoid	94.00	96.52	87.49	91.60	**6.51**	**4.92**
Dice	Module 2	Sigmoid	94.91	97.18	**87.81**	**91.86**	7.10	5.32
Tversky	——	Sigmoid	94.48	94.75	86.90	90.94	7.58	3.81
Tversky	Module 1	Sigmoid	94.59	96.86	87.46	91.43	7.13	5.43
Tversky	Module 2	Sigmoid	94.58	96.85	**87.56**	**91.55**	**7.02**	**5.30**
Focal Tversky	——	Sigmoid	93.18	95.19	85.08	89.46	8.10	5.73

FPN + SE-ResNet18

Setting			Train · 100		Test · 100		Train − Test	
Loss	Module	Activation	IOU	F_1	IOU	F_1	IOU	F_1
Dice	——	Softmax	94.66	97.00	86.14	90.36	8.52	6.64
Dice	Module 1	Softmax	94.48	97.04	86.54	90.88	7.94	6.16
Dice	Module 2	Softmax	94.96	97.31	**87.64**	**91.68**	**7.32**	**5.63**
Tversky	——	Softmax	94.70	97.03	85.76	89.97	8.94	7.06
Tversky	Module 1	Softmax	94.50	97.09	86.23	90.52	8.34	6.57
Tversky	Module 2	Softmax	94.60	96.87	**87.13**	**91.69**	7.47	**5.18**
Focal Tversky	——	Softmax	91.86	95.01	84.92	89.34	**6.94**	5.67
Dice	——	Sigmoid	94.44	96.88	85.68	89.13	8.76	7.75
Dice	Module 1	Sigmoid	94.61	97.11	86.92	91.13	7.69	5.98
Dice	Module 2	Sigmoid	94.91	97.18	**87.81**	**91.86**	7.10	5.32
Tversky	——	Sigmoid	94.42	96.82	85.32	89.64	9.10	7.18
Tversky	Module 1	Sigmoid	94.76	97.20	86.59	90.86	8.17	6.34
Tversky	Module 2	Sigmoid	94.58	96.85	**87.56**	**91.55**	7.02	**5.30**
Focal Tversky	——	Sigmoid	91.56	95.08	84.56	89.05	**7.00**	6.03

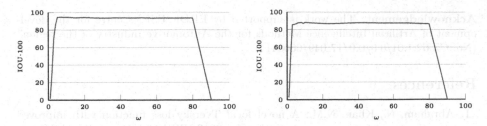

Fig. 4. The impact of various setting of ω in Module 2 on IOU coefficient. Left: evaluation on train dataset, right: evaluation on test dataset.

The proposed modules improve training performance, test precision, and reduce overfitting, but still, have several limitations. The first module is numerically stable and produces robust results, but it works only on two-class segmentation and in the current form cannot be applied to the n-class segmentation. We do not see the possibility of a direct generalization of the first module for n-classes segmentation at the moment, thus it is an open issue. The main limitation is the small number of operations available because the loss function must be differentiable. The second module is not restricted to the two classes only and yields better results, but there is a theoretical threat of training instability due to the nonzero gradient for predictions that are too good or too bad. Furthermore, as we have shown in the ablation study, a wrong choice of ω value can completely ruin the training procedure, ending with close to zero IOU. Due to this fact, engineers should keep ω in the recommended range of $\omega \in [3, 80]$.

6   Summary

In the area of deep learning, there are different requirements for predictions produced by a neural network. During training, the goal is to get as good predictions as possible and not to put too much effort into the separation of classes. During inference, it is enough to obtain the highest probability of the correct class disregarding the absolute value, the correct class after argmax operation is crucial. We are focusing on unifying these different requirements for both training and inference and therefore propose two independent modules that enrich the existing region-based loss functions for image segmentation. Modules are performed before the original loss function is computed and increase the model's performance. The main idea relies on involving the relationship between the predicted class probabilities, and, as a side effect, it increases the slope of the gradient. In the benchmark, we have shown that this idea leads to a performance increase in all tested scenarios while the computation overhead is close to zero and is easy to integrate into existing training pipelines. The source code is available online at http://gitlab.com/irafm-ai/segmentation-loss-hack.

Acknowledgement. The work is supported by ERDF/ESF "Centre for the development of Artificial Intelligence Methods for the Automotive Industry of the region" (No. CZ.02.1.01/0.0/0.0/17_049/0008414).

References

1. Abraham, N., Khan, N.M.: A novel focal Tversky loss function with improved attention u-net for lesion segmentation. In: 2019 IEEE 16th International Symposium on Biomedical Imaging (ISBI 2019), pp. 683–687. IEEE (2019)
2. Berman, M., Triki, A.R., Blaschko, M.B.: The lovász-softmax loss: a tractable surrogate for the optimization of the intersection-over-union measure in neural networks. In: Proceedings of the IEEE Conference on Computer Vision and Pattern Recognition, pp. 4413–4421 (2018)
3. Brostow, G.J., Shotton, J., Fauqueur, J., Cipolla, R.: Segmentation and recognition using structure from motion point clouds. In: Forsyth, D., Torr, P., Zisserman, A. (eds.) ECCV 2008. LNCS, vol. 5302, pp. 44–57. Springer, Heidelberg (2008). https://doi.org/10.1007/978-3-540-88682-2_5
4. Buslaev, A., Iglovikov, V.I., Khvedchenya, E., Parinov, A., Druzhinin, M., Kalinin, A.A.: Albumentations: fast and flexible image augmentations. Information **11**(2), 125 (2020)
5. Chapelle, O., Wu, M.: Gradient descent optimization of smoothed information retrieval metrics. Inf. Retrieval **13**(3), 216–235 (2010)
6. Chen, L.C., Papandreou, G., Schroff, F., Adam, H.: Rethinking atrous convolution for semantic image segmentation. arXiv preprint arXiv:1706.05587 (2017)
7. Cubuk, E.D., Zoph, B., Mane, D., Vasudevan, V., Le, Q.V.: Autoaugment: learning augmentation strategies from data. In: Proceedings of the IEEE/CVF Conference on Computer Vision and Pattern Recognition, pp. 113–123 (2019)
8. Dice, L.R.: Measures of the amount of ecologic association between species. Ecology **26**(3), 297–302 (1945)
9. Golodov, V., Maltseva, A.: Approach to weld segmentation and defect classification in radiographic images of pipe welds. NDT E Int. **127**, 102597 (2021)
10. Hatamizadeh, A., et al.: Unetr: transformers for 3d medical image segmentation. In: Proceedings of the IEEE/CVF Winter Conference on Applications of Computer Vision, pp. 574–584 (2022)
11. He, K., Gkioxari, G., Dollar, P., Girshick, R.: Mask R-CNN. IEEE Trans. Pattern Anal. Mach. Intell. **42**(2), 386–397 (2020)
12. He, Z., Xie, L., Chen, X., Zhang, Y., Wang, Y., Tian, Q.: Data augmentation revisited: rethinking the distribution gap between clean and augmented data. arXiv preprint arXiv:1909.09148 (2019)
13. Heidler, K., Mou, L., Baumhoer, C., Dietz, A., Zhu, X.X.: HED-UNet: combined segmentation and edge detection for monitoring the Antarctic coastline. IEEE Trans. Geosci. Remote Sens. **60**, 1–14 (2021)
14. Ho, Y., Wookey, S.: The real-world-weight cross-entropy loss function: modeling the costs of mislabeling. IEEE Access **8**, 4806–4813 (2019)
15. Hu, J., Shen, L., Sun, G.: Squeeze-and-excitation networks. In: Proceedings of the IEEE Conference on Computer Vision and Pattern Recognition, pp. 7132–7141 (2018)
16. Hua, Y., Marcos, D., Mou, L., Zhu, X.X., Tuia, D.: Semantic segmentation of remote sensing images with sparse annotations. IEEE Geosci. Remote Sens. Lett. **19**, 1–5 (2021)

17. Jadon, S.: A survey of loss functions for semantic segmentation. In: 2020 IEEE Conference on Computational Intelligence in Bioinformatics and Computational Biology (CIBCB), pp. 1–7. IEEE (2020)
18. Jang, J., et al.: Residual neural network-based fully convolutional network for microstructure segmentation. Sci. Technol. Weld. Joining **25**(4), 282–289 (2020)
19. Kingma, D.P., Ba, J.: Adam: a method for stochastic optimization. arXiv preprint arXiv:1412.6980 (2014)
20. Kullback, S., Leibler, R.A.: On information and sufficiency. Ann. Math. Stat. **22**(1), 79–86 (1951)
21. Lin, T.Y., Dollár, P., Girshick, R., He, K., Hariharan, B., Belongie, S.: Feature pyramid networks for object detection. In: Proceedings of the IEEE Conference on Computer Vision and Pattern Recognition, pp. 2117–2125 (2017)
22. Lin, T.Y., Goyal, P., Girshick, R., He, K., Dollár, P.: Focal loss for dense object detection. In: Proceedings of the IEEE International Conference on Computer Vision, pp. 2980–2988 (2017)
23. Liu, Z., et al.: SWIN transformer: hierarchical vision transformer using shifted windows. In: International Conference on Computer Vision (ICCV) (2021)
24. Long, J., Shelhamer, E., Darrell, T.: Fully convolutional networks for semantic segmentation. In: Proceedings of the IEEE Conference on Computer Vision and Pattern Recognition, pp. 3431–3440 (2015)
25. Milletari, F., Navab, N., Ahmadi, S.A.: V-NET: Fully convolutional neural networks for volumetric medical image segmentation. In: 2016 Fourth International Conference on 3D Vision (3DV), pp. 565–571. IEEE (2016)
26. Pham, H., Dai, Z., Xie, Q., Le, Q.V.: Meta pseudo labels. In: Proceedings of the IEEE/CVF Conference on Computer Vision and Pattern Recognition, pp. 11557–11568 (2021)
27. Ronneberger, O.: Invited talk: U-Net convolutional networks for biomedical image segmentation. In: Bildverarbeitung für die Medizin 2017. I, pp. 3–3. Springer, Heidelberg (2017). https://doi.org/10.1007/978-3-662-54345-0_3
28. Rumelhart, D.E., Hinton, G.E., Williams, R.J.: Learning representations by back-propagating errors. Nature **323**(6088), 533–536 (1986)
29. Salehi, S.S.M., Erdogmus, D., Gholipour, A.: Tversky loss function for image segmentation using 3D fully convolutional deep networks. In: Wang, Q., Shi, Y., Suk, H.-I., Suzuki, K. (eds.) MLMI 2017. LNCS, vol. 10541, pp. 379–387. Springer, Cham (2017). https://doi.org/10.1007/978-3-319-67389-9_44
30. Szegedy, C., Vanhoucke, V., Ioffe, S., Shlens, J., Wojna, Z.: Rethinking the inception architecture for computer vision. In: Proceedings of the IEEE Conference on Computer Vision and Pattern Recognition, pp. 2818–2826 (2016)
31. Taghanaki, S.A., et al.: Combo loss: handling input and output imbalance in multi-organ segmentation. Comput. Med. Imaging Graph. **75**, 24–33 (2019)
32. Tan, M., Le, Q.: Efficientnet: rethinking model scaling for convolutional neural networks. In: International Conference on Machine Learning, pp. 6105–6114. PMLR (2019)
33. Tang, J., Li, S., Liu, P.: A review of lane detection methods based on deep learning. Pattern Recogn. **111**, 107623 (2021)
34. Wang, J., et al.: Deep high-resolution representation learning for visual recognition. IEEE Trans. Pattern Anal. Mach. Intell. **43**, 3349–3364 (2020)
35. Xie, Q., Luong, M.T., Hovy, E., Le, Q.V.: Self-training with noisy student improves imagenet classification. In: Proceedings of the IEEE/CVF Conference on Computer Vision and Pattern Recognition, pp. 10687–10698 (2020)

36. Xie, S., Tu, Z.: Holistically-nested edge detection. In: Proceedings of the IEEE International Conference on Computer Vision, pp. 1395–1403 (2015)
37. Yi, K.M., Trulls, E., Lepetit, V., Fua, P.: LIFT: learned invariant feature transform. In: Leibe, B., Matas, J., Sebe, N., Welling, M. (eds.) ECCV 2016. LNCS, vol. 9910, pp. 467–483. Springer, Cham (2016). https://doi.org/10.1007/978-3-319-46466-4_28
38. Yuan, Y., Chao, M., Lo, Y.C.: Automatic skin lesion segmentation using deep fully convolutional networks with Jaccard distance. IEEE Trans. Med. Imaging **36**(9), 1876–1886 (2017)
39. Yuan, Y., Chen, X., Chen, X., Wang, J.: Segmentation transformer: object-contextual representations for semantic segmentation. In: European Conference on Computer Vision (ECCV), vol. 1 (2021)
40. Zagoruyko, S., Komodakis, N.: Wide residual networks. arXiv preprint arXiv:1605.07146 (2016)
41. Zhang, H., et al.: ResNeSt: split-attention networks. arXiv preprint arXiv:2004.08955 (2020)
42. Zhou, Z., Rahman Siddiquee, M.M., Tajbakhsh, N., Liang, J.: UNet++: a nested U-Net architecture for medical image segmentation. In: Stoyanov, D., et al. (eds.) DLMIA/ML-CDS -2018. LNCS, vol. 11045, pp. 3–11. Springer, Cham (2018). https://doi.org/10.1007/978-3-030-00889-5_1

Fuzzy Clustering to Encode Contextual Information in Artistic Image Classification

Javier Fumanal-Idocin[1]([envelope]) [iD], Zdenko Takáč[2] [iD], Ľubomíra Horanská[2] [iD],
Humberto Bustince[1] [iD], and Oscar Cordon[3] [iD]

[1] Public University of Navarra, Pamplona, Spain
{javier.fumanal,bustince}@unavarra.com
[2] Institute of Information Engineering, Automation and Mathematics,
Slovak University of Technology in Bratislava, Bratislava, Slovakia
{zdenko.takac,lubomira.horanska}@stuba.sk
[3] DaSCI Research Institute and DECSAI, University of Granada, Granada, Spain
ocordon@decsai.ugr.es

Abstract. Automatic art analysis comprises of utilizing diverse processing methods to classify and categorize works of art. When working with this kind of pictures, we have to take under consideration different considerations compared to classical picture handling, since works of art alter definitely depending on the creator, the scene delineated or their aesthetic fashion. This extra data improves the visual signals gotten from the images and can lead to better performance. However, this information needs to be modeled and embed alongside the visual features of the image. This is often performed utilizing deep learning models, but they are expensive to train. In this paper we utilize the Fuzzy C-Means algorithm to create a embedding strategy based on fuzzy memberships to extract relevant information from the clusters present in the contextual information. We extend an existing state-of-the-art art classification system utilizing this strategy to get a new version that presents similar results without training additional deep learning models.

Keywords: Clustering · Image classification · Fuzzy C Means · Representation learning

1 Introduction

The digitalization of various artworks and collections all around the world has utilized well known methods of computer vision and image processing on creative information [5]. One of the most encouraging themes toward this path is the programmed examination of compositions. These strategies are applied in tasks

Javier Fumanal Idocin and Humberto Bustince's research has been supported by the project PID2019-108392GB I00 (AEI/10.13039/501100011033). Zdenko Takáč and Ľubomíra Horanská's research has been supported by the grant VEGA 1/0267/21.

D. Ciucci et al. (Eds.): IPMU 2022, CCIS 1602, pp. 355–366, 2022.
https://doi.org/10.1007/978-3-031-08974-9_28

generally performed on most of the galleries and museums. For example, author verification [14], style investigation [25] or restauration [47].

Automatic art analysis examination can be performed utilizing hand-created features [10,39], or automatically extracted features using deep-learning [22,41]. Utilizing a Convolutional Neural Network to extract visual features from a picture is extremely well known and popular [3,4,35]. But for the instance of imaginative and creative pictures, human experts perform their examination using not only visual cues. They also rely on their insight on the chronicled setting, other artworks, materials, and so on [27].

There are numerous ways in which this information can be used. One of the most popular ones is a knowledge graph [18,19]. An knowledge graph models the connection between various ideas and attributions utilizing the construction of an network [12,40]. Networks are a successful way to model interactions [29] and they have been utilized to tackle a heap of issues in various subject matters, from computer science [8,16,30], to biology [31] and the social sciences [2,17]. The network connections can be utilized to capture all the information connected with a painting that trascends the visual cues, like the author or the artistic style. The most popular way to use this information is to develop continuous space representation from the nodes in the graph [21,28]. These sort of representation are popular since they can work with conventional AI approaches [20].

The combination of visual and contextual features stands for a promising direction in which to perform automatic artistic images analysis [18,19,44]. However, even though visual features have been widely studied in the context of deep learning, there is not a straightforward procedure to model the contextual information associated with each image. The different possibilities of representation and fusion of this information can result in a wide range of different performances, as some modelizations can be more suitable for one task than others.

The mix of visual and context oriented elements represents a promising direction in which to perform automatic artistic images analysis [18,19,44]. However, despite the fact that visual elements have been broadly considered using deep learning, there is certainly not a direct methodology to show the relevant data related with each picture. He different possibilities of representation and fusion of this information can result in a wide range of different performances, as some modelizations can be more appropriate for one errand than others.

The aim of this paper is to propose different methods of representing the contextual embeddings from different works of art utilizing different traditional strategies, less complex and quicker than the development of a deep learning model. In order to so, we shall study different representation space obtained, and the various clusters obtained utilizing the Fuzzy C-Means [7]. Thus, we can develop an implanting strategy where each aspect is an embedding method where each dimension is a fuzzy membership to each of the relevant groups found in the original representation space.

The rest of the paper goes as follows: in Sect. 2 we displayed some of the previous concepts required to understand this work. In Sect. 3 we show our new method to obtain contextual embeddings from textual annotations using the Fuzzy C-Means clustering. Subsequently, in Sect. 4 we display our proposed framework for artistic image classification using contextual embeddings. In Sect. 5 we show the results of our experimentation and of other classification frameworks compared to ours. Finally, in Sect. 6 we give our final conclusions and future lines for this work.

2 Background

In this section we revise some previous works regarding knowledge graphs, the Fuzzy C-Means clustering algorithm and context aware embeddings.

2.1 Knowledge Graphs

Knowledge graphs are a form of knowledge representation in which each concept is modeled as a node that is related to others with different relationships that are represented as edges. Knowledge graphs have been very popular in fuzzy literature because of fuzzy cognitive maps, and they have been instrumental in the development on a many systems.

There are different strategies in which a knowledge graph can be constructed and exploit. One possibility is to form hierarchies of concepts detected in images, or exploiting semantic similarities between different concepts or to use external knowledge bases.

Once the knowledge graph is built, there are different ways in which it can be exploited. One possibility is to feed it directly to a graph neural network. It is also possible to extract embeddings using node2vec, that uses random walks to create representations of each node that balance local proximity and homophily. Sometimes, it is also possible to use inference methods directly in the graph.

2.2 Fuzzy C-Means

The Fuzzy C-Means (FCM) is a well known fuzzy clustering algorithm, in which each element is assigned not only to one group, but rather, presents a membership to each of the groups considered [7].

The FCM aims to minimize the corresponding objective function:

$$\arg\min \sum_{i=1}^{n} \sum_{j=1}^{c} w_{ij}^m \|\mathbf{x}_i - \mathbf{c}_j\|^2 \tag{1}$$

where n is the number of observations, m is a constant, $\mathbf{c}$ is a cluster and c is the number of different clusters, and $\mathbf{x}$ is an observation. Finally, w_{ij}^m is the membership of the $i-th$ particle to the $j-th$ cluster, that follows this expression:

$$w_{ij} = \frac{1}{\sum_{k=1}^{c} \left(\frac{\|\mathbf{x}_i - \mathbf{c}_j\|}{\|\mathbf{x}_i - \mathbf{c}_k\|} \right)^{\frac{2}{m-1}}} \tag{2}$$

The algorithm assigns randomly a coefficient for each observation to each cluster. Then, computes the centroid for each cluster, and the computes each membership again. It repeats this process until it has converged.

2.3 Multi-task Learning

Similarly to transfer learning, the idea of multi-task learning (MTL) is that features can be useful for more tasks than they were originally intended to [43]. MTL is based on training for more than one task at a time, so that the resulting features will generalize better [11,13].

MTL can be performed using hard parameter [38] and soft parameter sharing [36]. In the first case, the parameters for all the tasks are shared until the last layer, while in the latter one, each task has its own set of parameters, but they are encouraged to stay similar using different regularization methods. MTL is a popular deep-learning approach that has been successfully applied in different environments [48].

2.4 Representation Learning

Representation learning consist of automatically extracting and computing features suitable for machine learning tasks from unstructured data like text, video or image [6]. Deep learning is one of the most popular fields in which feature learning is performed. Convolutional neural networks have been massively popular tools to embed images and video into vector spaces [15,23,26,32] as well as text [33].

Just as image, video and text, networks can also be embed into vector space using deep learning models [21,37,46]. The deep learning models can be combined with other classical methods in text, using TF-IDF or latent Dirichlet allocation [9,24].

3 Fuzzy Representations for Contextual Information Using Fuzzy C-Means

In this section we present our new proposal to substitute and enhance the contextual information extraction in two steps: first, we discuss how can we obtain bag of words representations from each contextual image, and how can we use the Fuzzy C-Means algorithm to extract the relevant information from the chosen feature extraction method.

3.1 Encoding Context Using Bag of Words

Contextual information in artistic images is usually encoded using textual representation (Fig. 2). Textual representation can be encoded in different ways. The most classical one is the Bag of Words (BoW) in which each phrase is represented as a vector of numbers in the [0, 1] where each position is associated with one word.

This representation is usually computed using the term frequency-inverse document frequency (TF-IDF) metric represent each word [1]. TF-IDF has been very popular in other artistic image processing tasks [19] and in text processing tasks in general [34, 45].

However, this representation presents some issues for our application. The vocabulary of these kind of contextual annotations is usually full of personal names and words that are present in very few samples. We are interested in a representation that scales well with the number of features, so that the vocabulary can be easily fixed. It can also be problematic that the TF-IDF ponders most words that are frequent in a description but absent in others. The idea is that those words are most discriminative than others, but this is not necessarily true in our case. "Landscape", for example, is a word that appears in many descriptions but is also very discriminative. This also happens with other words such as "Portrait".

In order to solve these problems, we have opted to use only the top k most common words and a standard BoW representation. In this representation each contextual information is coded as a vector of size k in which each position n is 1 if the n-th most common word is present in the text. We shall fine-tune the k value empirically.

3.2   Extracting Relevant Context Information Using Fuzzy C-Means

The idea of using FCM for this task is that the space formed using our embedding method can be a faithful representation of the original domain, bur not useful to solve the task at hand. Since we are interested in using these features to discriminate between classes, we are more interested in the topology of the representation obtained and the groups that are naturally present in them.

We expect that these groups should agglomerate categories that are not mutually exclusive. For example, in the case of artistic representation, style and year can be very correlated, because of artistic movements. Sometimes artists just don't follow their contemporary trends. There are many more possible examples in this case: landscapes can be together but belong to different authors, etc.

The FCM is the most suitable clustering algorithm for this task, since we intend to express the membership to different, not mutually exclusive groups. For each observation, the FCM returns a fuzzy membership degree for each of the pertinent groups. We can use this information to better characterize each observation with respect to the rest of them: each feature will correspond to one of the different relevant groups found using the FCM algorithm, instead of just encoding the original contextual information.

4 Artistic Image Classification Framework

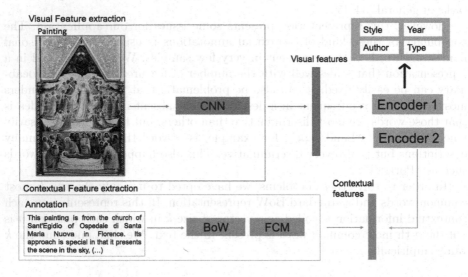

Fig. 1. Scheme of the proposed classification framework.

Our proposed framework consists of two different parts. On one hand, we compute the contextual embeddings in two step process:

1. Compute the BoW encoding for each annotation.
2. Compute the FCM over the BoW encodings in order to obtain the fuzzy memberships to each of the clusters founds.

On the other hand, we use a ResNet 50 [42] to compute visual features for each image. This ResNet is trained in a multi-task setting, so that it must learn at the same time style, year, author and type of each painting. It also trained to learn a reconstruction of the contextual embeddings computed alongside the classification problem. In order to do so, we have two "final" layers: one encoder that transforms the final feature vector of the network into the contextual features, and another one that performs the classification. These encoders are single full connected layers with a Rectified Linear Unit activation function.

The loss for each class c is the standard cross-entropy:

$$l_c(y, \hat{y}) = -\frac{1}{n} \sum_{i=1}^{n} y_{ci} \log \hat{y}_{ci} + (1 - y_{ci}) \log(1 - \hat{y}_{ci}) \tag{3}$$

Given r, the final embedding obtained from the ResNet, m the number of clusters obtained with the FCM, the loss function for the reconstruction of the fuzzy memberships vector is the Smooth L1:

$$\delta_{emb}(i,j) = \begin{cases} \frac{1}{2}(i-j), & \text{if } |i-j| \leq 1 \\ |i-j| - \frac{1}{2}, & \text{otherwise} \end{cases} \tag{4}$$

$$l_{emb} = \sum_{i=1}^{n}\sum_{j=1}^{m} \delta_{emb}(w_{ij}, r_{ij}) \tag{5}$$

A visual scheme of our proposed framework is displayed in Fig. 1.

5 Experimentation

"Tiepolo painted this altarpiece during his stay in Germany. Because of the damp climate, he could only work on the frescoes in the Wurzburg Residenz in the spring and summer. So in the fall and winter he had to concentrate on painting in oil on canvas. He produced some fantastic and exotically beautiful works in which the religious subject seems merely a pretext for eye-catching, showy images, but he himself was genuinely religious. The style of the age, however, meant that even religious topics often became theatrical"

Fig. 2. An example of a Semart painting alongside its contextual information.

In this section we have computed the results with our proposed method, using the visual embeddings from the ResNet and the context aware embeddings enhanced using the Fuzzy C-means algorithm. We have also compared our solution with other proposals that use both visual and contextual embeddings.

For our experimentation we have used the SemArt dataset [19]. This dataset consists of 21,384 painting images, from which 19,244 are used for training, 1,069 for validation and 1,069 for test. Each painting has associates an artistic comment, alongside the following attributes: Author, Title, Date, Technique, Type, School and Timeframe.

In this experimentation three different classification tasks are proposed:

- Type: each painting is classified according to 10 different common types of paintings: portrait, landscape, religious, etc.
- School: each painting is identified with different schools of art: Italian, Dutch, French, Spanish, etc. There are a total of 25 classes of this kind.
- Timeframe: The attribute Timeframe, which corresponds to periods of 50 years evenly distributed between 801 and 1900, is used to classify each painting according to its creation date. We consider only the timeframes where at least 10 are present. This corresponds to 18 classes.
- Author: corresponds to the author of each paintings. We consider a total of 350 painters, that comprise the set of authors with more than 10 paintings in the dataset.

Our first experimental study was finding the optimal number of dimensions to use in the BoW embedding. Since we are interested in using the FCM on the computed embeddings, we are interested in finding a representation that clearly shows cluster of observations. We found the optimal number using visual inspection with the PCA algorithm, as the results were straightforward enough to interpret (Fig. 5).

After the ideal number was stablished as $k = 5$, we computed the FCM for different numbers of clusters, and we computed the silhouette index in each case. We use the "elbow rule" to choose the optimal number of clusters, which we established as 13 (Fig. 5d).

Once we have stablished the dimensionality of the BoW vectors and the number of clusters for the FCM, we can train our model using the computed contextual embeddings. We show the results obtained using our method in Table 1. In order to check the importance of the Bow and FCM parameters we also trained our proposed framework using a bigger number of words for the BoW model and a bigger number of clusters for the FCM.

Besides our method, we have also shown the results obtained with other classification methods:

1. The ResNet50, ResNet152 and the VGG16 using their correspondent pre-trained weights. We adapt the last layer to match the number of target classes. These solutions consider only the visual information for each image.
2. The ResNet50, ResNet152 and the VGG16 fine-tuning their weights. We also adapt the last layer to match the number of target classes. These solutions consider only the visual information for each image.
3. The ResNet50 precomputed weights with information captured from contextual annotations using node2vec representations using a Knowledge graph [18].

From the results obtained, we can conclude that pre-trained models using only visual features performed the worst. When taking into account contextual features using both FCM or KGM, the performance improved substantially for all classes. Comparing the context-aware proposals, the FCM-based frameworks performed better than their KGM counterparts in the "Type" and "Author" classes, where they also obtained the vest result overall. The best results for the two other classes were obtained using a fine-tuned ResNet50 model.

Table 1. Classification results for the different attributes on SemArt Dataset.

Method	Type	School	TF	Author
VGG16 pre-trained	0.706	0.502	0.418	0.482
ResNet50 pre-trained	0.726	0.557	0.456	0.500
VGG16 fine-tuned	0.768	0.616	0.559	0.520
ResNet50 fine-tuned	0.765	**0.655**	**0.604**	0.515
ResNet50 pre-trained+KGM	0.786	0.647	0.597	0.548
ResNet50 pre-trained+FCM_{5-15}	0.778	0.625	0.591	**0.564**
ResNet50 pre-trained+$FCM_{100-150}$	**0.793**	0.630	0.586	0.559

6   Conclusions and Future Lines

In this work we have presented a new method to extract features from the contextual annotations of a dataset of artistic images. We have shown the classification framework used, that uses a fine-tuned ResNet 50 in a multi-task environment. This network learns to solve a classification problem and to reconstruct the features extracted from the contextual image for each image, which helps the network generalize better, as it cannot rely only on visual cues to classify each sample.

In order to construct the contextual representations, we extract the k most common words and we construct use a bag of words method. Then, we use Fuzzy C-Means clustering on these features to obtain a fuzzy membership for each of the natural clusters formed in this representation. In this way, the network is forced to learn a representation for each sample that is useful to solve the classification problem, but it is also a faithful representation of the different coalitions present in the contextual information embeddings.

We have compared our proposal with other similar classification frameworks. We found that contextual works using a knowledge graph surpass our performance, but are considerably more expensive to compute. We also found that our framework performed better than others using only visual features. Future lines of our research shall study the use of fuzzy linguistic variables to characterize the images in order to find a more expressive space in which to represent some of the images characteristics and attributes.

References

1. Aizawa, A.: An information-theoretic perspective of TF-IDF measures. Inf. Process. Manage. **39**(1), 45–65 (2003)

2. Akça, S., Akbulut, M.: Social network analysis of mythology field. Library Hi Tech (2021)
3. Aloysius, N., Geetha, M.: A review on deep convolutional neural networks. In: 2017 International Conference on Communication and Signal Processing (ICCSP), pp. 0588–0592. IEEE (2017)
4. Anwar, S.M., Majid, M., Qayyum, A., Awais, M., Alnowami, M., Khan, M.K.: Medical image analysis using convolutional neural networks: a review. J. Med. Syst. **42**(11), 1–13 (2018)
5. Barni, M., Pelagotti, A., Piva, A.: Image processing for the analysis and conservation of paintings: opportunities and challenges. IEEE Signal Process. Mag. **22**(5), 141–144 (2005)
6. Bengio, Y., Courville, A., Vincent, P.: Representation learning: a review and new perspectives. IEEE Trans. Pattern Anal. Mach. Intell. **35**(8), 1798–1828 (2013)
7. Bezdek, J.C., Ehrlich, R., Full, W.: FCM: the fuzzy c-means clustering algorithm. Comput. Geosci. **10**(2–3), 191–203 (1984)
8. Blondel, V.D., Guillaume, J.L., Lambiotte, R., Lefebvre, E.: Fast unfolding of communities in large networks. J. Stat. Mech. Theory Exper. **2008**(10) (2008)
9. Bounabi, M., Moutaouakil, K., Satori, K.: Text classification using fuzzy TF-IDF and machine learning models. In: BDIoT 2019 (2019)
10. Carneiro, G., da Silva, N.P., Del Bue, A., Costeira, J.P.: Artistic image classification: an analysis on the PRINTART database. In: Fitzgibbon, A., Lazebnik, S., Perona, P., Sato, Y., Schmid, C. (eds.) ECCV 2012. LNCS, vol. 7575, pp. 143–157. Springer, Heidelberg (2012). https://doi.org/10.1007/978-3-642-33765-9_11
11. Caruana, R.: Multitask learning. Machine Learn. **28**(1), 41–75 (1997)
12. Chen, X., Jia, S., Xiang, Y.: A review: knowledge reasoning over knowledge graph. Expert Syst. Appl. **141**, 112948 (2020)
13. Collobert, R., Weston, J.: A unified architecture for natural language processing: Deep neural networks with multitask learning. In: Proceedings of the 25th International Conference on Machine Learning, pp. 160–167 (2008)
14. Crowley, E.J., Zisserman, A.: The art of detection. In: Hua, G., Jégou, H. (eds.) ECCV 2016. LNCS, vol. 9913, pp. 721–737. Springer, Cham (2016). https://doi.org/10.1007/978-3-319-46604-0_50
15. Dosovitskiy, A., Springenberg, J.T., Riedmiller, M., Brox, T.: Discriminative unsupervised feature learning with convolutional neural networks. Adv. Neural. Inf. Process. Syst. **27**, 766–774 (2014)
16. Fumanal-Idocin, J., Alonso-Betanzos, A., Cordón, O., Bustince, H., Minárová, M.: Community detection and social network analysis based on the Italian wars of the 15th century. Futur. Gener. Comput. Syst. **113**, 25–40 (2020)
17. Fumanal-Idocin, J., Cordón, O., Dimuro, G., Minárová, M., Bustince, H.: The concept of semantic value in social network analysis: an application to comparative mythology. arXiv preprint arXiv:2109.08023 (2021)
18. Garcia, N., Renoust, B., Nakashima, Y.: Context-aware embeddings for automatic art analysis. In: Proceedings of the 2019 on International Conference on Multimedia Retrieval, pp. 25–33 (2019)
19. Garcia, N., Vogiatzis, G.: How to read paintings: semantic art understanding with multi-modal retrieval. In: Leal-Taixé, L., Roth, S. (eds.) ECCV 2018. LNCS, vol. 11130, pp. 676–691. Springer, Cham (2019). https://doi.org/10.1007/978-3-030-11012-3_52
20. Grohe, M.: word2vec, node2vec, graph2vec, x2vec: towards a theory of vector embeddings of structured data. In: Proceedings of the 39th ACM SIGMOD-SIGACT-SIGAI Symposium on Principles of Database Systems, pp. 1–16 (2020)

21. Grover, A., Leskovec, J.: node2vec: Scalable feature learning for networks. In: Proceedings of the 22nd ACM SIGKDD International Conference on Knowledge Discovery and Data Mining, pp. 855–864 (2016)

22. Guo, B., Hao, P.: Analysis of artistic styles in oil painting using deep-learning features. In: 2020 IEEE International Conference on Multimedia & Expo Workshops (ICMEW), pp. 1–4. IEEE (2020)

23. Jing, L., Tian, Y.: Self-supervised visual feature learning with deep neural networks: a survey. IEEE Trans. Pattern Anal. Mach. Intell. **43**, 4037–4058 (2021)

24. Kim, D., Seo, D., Cho, S., Kang, P.: Multi-co-training for document classification using various document representations: TF-IDF, LDA, and doc2vec. Inf. Sci. **477**, 15–29 (2019)

25. Lecoutre, A., Negrevergne, B., Yger, F.: Recognizing art style automatically in painting with deep learning. In: Asian Conference on Machine Learning, pp. 327–342. PMLR (2017)

26. LeCun, Y., Bengio, Y., Hinton, G.: Deep learning. Nature **521**(7553), 436–444 (2015)

27. Lombardi, T.E.: The classification of style in fine-art painting. Pace University (2005)

28. Mikolov, T., Chen, K., Corrado, G., Dean, J.: Efficient estimation of word representations in vector space. arXiv preprint arXiv:1301.3781 (2013)

29. Newman, M.: Networks. Oxford University Press, Oxford (2018)

30. Newman, M.E.: Modularity and community structure in networks. Proc. Natl. Acad. Sci. **103**(23), 8577–8582 (2006)

31. Palla, G., Derényi, I., Farkas, I., Vicsek, T.: Uncovering the overlapping community structure of complex networks in nature and society. Nature **435**(7043), 814 (2005)

32. Pathak, D., Krähenbühl, P., Donahue, J., Darrell, T., Efros, A.A.: Context encoders: Feature learning by inpainting. In:2016 IEEE Conference on Computer Vision and Pattern Recognition (CVPR), pp. 2536–2544 (2016)

33. Poria, S., Cambria, E., Gelbukh, A.: Deep convolutional neural network textual features and multiple kernel learning for utterance-level multimodal sentiment analysis. In: EMNLP (2015)

34. Qaiser, S., Ali, R.: Text mining: use of TF-IDF to examine the relevance of words to documents. Int. J. Comput. Appl. **181**(1), 25–29 (2018)

35. Rawat, W., Wang, Z.: Deep convolutional neural networks for image classification: a comprehensive review. Neural Comput. **29**(9), 2352–2449 (2017)

36. Ruder, S.: An overview of multi-task learning in deep neural networks. arXiv preprint arXiv:1706.05098 (2017)

37. Sarlin, P.E., DeTone, D., Malisiewicz, T., Rabinovich, A.: Superglue: Learning feature matching with graph neural networks. In: 2020 IEEE/CVF Conference on Computer Vision and Pattern Recognition (CVPR), pp. 4937–4946 (2020)

38. Sener, O., Koltun, V.: Multi-task learning as multi-objective optimization. arXiv preprint arXiv:1810.04650 (2018)

39. Shamir, L., Macura, T., Orlov, N., Eckley, D.M., Goldberg, I.G.: Impressionism, expressionism, surrealism: automated recognition of painters and schools of art. ACM Trans. Appl. Percept. (TAP) **7**(2), 1–17 (2010)

40. Taber, R.: Knowledge processing with fuzzy cognitive maps. Expert Syst. Appl. **2**(1), 83–87 (1991)

41. Tan, W.R., Chan, C.S., Aguirre, H.E., Tanaka, K.: Ceci n'est pas une pipe: a deep convolutional network for fine-art paintings classification. In: 2016 IEEE International Conference on Image Processing (ICIP), pp. 3703–3707. IEEE (2016)

42. Targ, S., Almeida, D., Lyman, K.: Resnet in resnet: Generalizing residual architectures. arXiv preprint arXiv:1603.08029 (2016)
43. Torrey, L., Shavlik, J.: Transfer learning. In: Handbook of Research on Machine Learning Applications and Trends: Algorithms, Methods, and Techniques, pp. 242–264. IGI global (2010)
44. Vaigh, C.B.E., Garcia, N., Renoust, B., Chu, C., Nakashima, Y., Nagahara, H.: Gcnboost: artwork classification by label propagation through a knowledge graph. arXiv preprint arXiv:2105.11852 (2021)
45. Vijayarani, S., Ilamathi, M.J., Nithya, M., et al.: Preprocessing techniques for text mining-an overview. Int. J. Comput. Sci. Commun. Networks 5(1), 7–16 (2015)
46. Wang, H., Zhang, F., Zhao, M., Li, W., Xie, X., Guo, M.: Multi-task feature learning for knowledge graph enhanced recommendation. In: The World Wide Web Conference (2019)
47. Zeng, Y., Gong, Y., Zeng, X.: Controllable digital restoration of ancient paintings using convolutional neural network and nearest neighbor. Pattern Recogn. Lett. **133**, 158–164 (2020)
48. Zhang, Y., Yang, Q.: An overview of multi-task learning. Natl. Sci. Rev. **5**(1), 30–43 (2018)

New Aggregation Strategies in Color Edge Detection with HSV Images

Pablo A. Flores-Vidal[1]([⊠]) [iD], Daniel Gómez[2] [iD], Javier Castro[2] [iD],
and Javier Montero[3] [iD]

[1] Dpto. de Producción animal, Facultad de Veterinaria, Universidad Complutense de
Madrid, Av. Puerta de Hierro s/n, 28040 Madrid, Spain
pflores@ucm.es
[2] Dpto. de Estadística y Ciencia de los datos, Facultad de Estudios Estadísticos,
Universidad Complutense de Madrid, Av. Puerta de Hierro s/n, 28040 Madrid, Spain
[3] Dpto. de Estadística e Investigación Operativa, Facultad de Ciencias Matemáticas,
Universidad Complutense de Madrid, Plaza de las Ciencias 3, 28040 Madrid, Spain

Abstract. Most edge detection algorithms only deal with grayscale images, while their use with color images remains an open problem. This paper explores different approaches to aggregating color information from RGB and HSV images for edge extraction purposes through the usage of the Canny algorithm. The Berkeley's image data set is used to evaluate the performance of the different aggregation methods. Precision, Recall and F-score are computed. Better performance of aggregations with HSV channels than with RGB's was found. This article also shows that depending on the type of image used -RGB or HSV-, some methodologies are more appropriate than others.

Keywords: Color edge detection · HSV · Hexcone model · RGB · Pre-aggregation · Post-aggregation · Canny

1 Introduction

McAndrew [21] argued that 'for human beings, colour provides one of the most important descriptors of the world around us'. However, despite the undoubtful importance of color, traditionally, grayscale images have been more widely used when it comes to Image Processing (IP). This bears particular significance in relation to the edge detection problem, where dealing with color images introduces some complications.

The RGB color space is roughly based on human vision as the human eye has three different cone cells, one to capture red luminosity, another to capture green, and the last one to capture blue. As a result, only 'three numerical components are necessary and sufficient to describe a color' as Bogumil has indicated [3]. From this point of view, the RGB color space should be ideal for capturing all the color information.

Supported by Government of Spain, grant PGC2018-096509-B-I00.

HSV was created by graphic designers to mimic the artist process of creating colors [25]. HSV is made up of three channels, with the color information captured in one single channel: Hue -'H'-, while the Saturation (S) is placed in a different channel. The higher the saturation -'S'-, the purer the color. A third channel contains the brightness information: the Value -'V'-. The higher is the value, the whiter the color. This color space has recently found new applications as skin detection and face detection [17,23,24].

Edge detection (ED) is considered one of the main techniques inside Image Processing field (IP) [15,16,18]. Its importance lies in the fact that most higher-level techniques make use of it. Most of the ED literature works with grey images, as without color the algorithms are usually easier to create and perform. Contrarily, color edge detection is more complex to deal with than grayscale's [10], but it is of great importance as most images contain color information. One reason for this greater complexity is that the distance/dissimilarity between pixels luminosity in one dimension is easier to compute than in the multi-dimensional case. Therefore, the main problem that arises in color ED is how to measure the distance between colors. Due to this problem, the most common method is first converting the color image into one single channel (grayscale image), which can be considered an oversimplification that causes loss of information. Two main methods for dealing with this have been proposed in the literature [27]: individual channel [8] and vector-based's [9,26,30]. The first approach consists of extracting edges for each channel separately, which brings the problem of choosing the right aggregation for blending the different channels. This is especially intricate in the case of HSV images due to the different nature of the channels. Another motivation for this research is that color ED with HSV images has been less common in the literature than RGB's.

The proposal of this paper shows that applying different color ED algorithms over HSV images making use of aggregation operators inspired by Yager's [29] overtakes some RGB based approaches. In this paper, 4 different methodologies for aggregating the RGB and HSV channels are proposed. The rest of this paper is organized as follows: The following Sect. 2 explains some preliminary information including IP basic concepts, color edge detection, HSV images generation from RGB images, and relevant Aggregation Operators concepts. Then, the different approaches for aggregating RGB and HSV channels are presented in Sect. 3. Finally, Sect. 4 and Sect. 5 present the comparison, its results, and the conclusions of this research.

2   Preliminaries

2.1   Digital Image Processing

Let us denote a digital image by I, where each point (i, j) represents a pixel with $i = 1, \ldots, n$ and $j = 1, \ldots, m$. Therefore, $n \times m$ is the size of an image and the number of pixel inside it. A $k = 1, \ldots, \tilde{k}$ index is needed for expressing the number of channels in the image. Let us denote by $I_{i,j}^k$ the spectral information associated with each pixel (i, j) at channel k. The value range of this information is dependent on the digital image type that is being considered:

- *Binary map* (I_{bin}) where $I_{i,j} \in \{0, 255\}$ (or $\{0, 1\}$).
- *Grayscale image* (I_{gray}) where $I_{i,j} \in \{0, 1, \ldots, 255\}$.
- *Soft image* or *normalized grayscale image* (I_{soft}) where $I_{i,j} \in [0, 1]$.
- *RGB image* (I_{RGB}) where $I_{i,j} \in \{0, 1, \ldots, 255\}^3$. (R=*Red*; G=*Green* and B=*Blue*. In this paper, the channels are referred as both, $I_{RGB}^R, I_{RGB}^G, I_{RGB}^B$ or as the more simplified version I_R, I_G, I_B.
- *HSV image* (I_{HSV}) composed by three channels $(I_{HSV}^H, I_{HSV}^S, I_{HSV}^V)$ being H=*Hue*; S=*Saturation* and V=*Value*), $I_{i,j}^S, I_{i,j}^V \in [0, 1]$ and for $I_{i,j}^H$ there are two possible definitions: $I_{i,j}^H \in [0^o, 360^o]$ or $I_{i,j}^H \in [0, 1]$ -the one used in this paper-. The first one can easily be obtained through multiplying the second one by 360 and changing the scale to degrees. In this paper the HSV channels are also referred to as I_H, I_S, I_V. Moreover, for the use inside a formula by means of numeric index a third kind of expressions are employed: $I_{HSV}^1, I_{HSV}^2, I_{HSV}^3$ where 1 stands for Hue, 2 for Saturation, and 3 for Value. See Subsect. 2.3 for more information about HSV images.

2.2 Color Edge Detection

From a mathematical point of view, an edge detection algorithm is a function that converts a digital image into a binary image. The literature on the topic offers two main approaches to color edge detection:

1. *Individual channel*: The edges are extracted for each channel. This is the approach this paper employs.
2. *Vector-based approach*: An aggregation function is applied: For example, a median filter [26], a range operator [30], or other statistical aggregation methods [7]. This approach was employed in [9].

2.3 HSV and Other Color Models

HSV was created for mimicking the process of a painter that starts by choosing a hue -a color-, and then adds some white to it to give more light, or some black to darken it. Hue is any pure color, and can be represented as a point placed in a disk -or in an hexagon-. This color ranges according to its saturation from the purest color -maximum saturation that is placed over the disk or hexagon- to white or gray -minimum saturation that is situated in the center of the circle or hexagon-. 'Value' is the third dimension and represents the grade in which this color is non-black -0 = 'black' and 1 = 'white'-. A regular cone or a similar figure is usually employed to represent this color model -see Fig. 1-. The less black, the higher the value is.

The literature has proposed color spaces different from RGB and HSV to study the dissimilarities between colors, as in the case of YUV and its variant

YCoCg [13], CIELab [19], CYMK, among others. One algorithm that transforms RGB into an HSV image is known as 'Hexcone model' [25]:

After this algorithm is applied, an HSV image is obtained, with $I_H \in [0,1]$, $I_S \in [0,1]$ and $I_V \in [0,1]$. This paper is concerned with RGB and HSV color spaces (see Fig. 1), and bothe are employed by means of individual-channel approach (multi-channel was employed in [9]).

2.4   Aggregation Operators

Aggregation Operators (AO) are one of the most important disciplines in information sciences since they are part of knowledge acquisition. The process of aggregating the information is a key tool for most knowledge-based systems.

Definition 1. *A function $A : [0,1]^n \to [0,1]$ is said to be an n-ary aggregation function if the following conditions hold:*

(A1) *A is increasing in each argument: for each $i \in \{1, \ldots, n\}$, if $x_i \leq y$, then $A(x_1, \ldots, x_i, \ldots, x_n) \leq A(x_1, \ldots, x_{i-1}, y, x_{i+1}, \ldots, x_n)$;*
(A2) *A satisfies the boundary conditions: $A(0, \ldots, 0) = 0$ and $A(1, \ldots, 1) = 1$.*

AO have been employed in different disciplines and applications, IP being one of them [2,4,5]. In this paper, the use of AO is justified by the fact that the value of each channel is related to the likeliness of a given pixel to be an edge, or what the literature has called *edginess*. Then, there is a natural connection between the boundary conditions of AO and the potential edginess of a certain pixel. In this sense, the complete lack of edginess can be associated to the concept of minimal boundary. Conversely, the pixel is an edge when the supreme boundary is reached. Another desired propriety is monotonicity, as for a certain pixel an increment in the value of any channel means a higher likeliness for the pixel to be an edge.

The classical definition of AO can be naturally extended by means of replacing the unit interval $[0,1]$ with a more general lattice, which in the fuzzy area is traditionally assumed to be a complete lattice [14]. Some of these operators allow giving priorization to some type of data against others. This can be done, for example, dealing with prioritized information, as it happens with the prioritized aggregation operators proposed by Yager in [29], which were generalized by Rojas et al. [22] In the latter, the generalization consisted in the use of general weights acting over the hierarchy and internal aggregation operators that can differ from the minimum, which was employed by Yager [29].

Definition 2. *The Yager-inspired hierarchical prioritized aggregation operator is defined as:*

$$V(y_1, \cdots, y_n) = \sum_{i=1}^{n} w_i \prod_{k=1}^{i} y_{n-k+1} \qquad (1)$$

Where 'V' stands for vertical, as every hierarchized set of data is placed in a different box inside a vertical structure that shows the hierarchy between the clusters 'y_i', and 'n' is the number of different clusters (for further information see [22]). The *prioritization* of clusters (or 'categories' in Yager's words) enables the assignment of a different importance for each one (this could also be done employing weighted operators) and the use of 'a kind of importance weight in which the importance of a lower priority criteria will be based on its satisfaction to the higher priority criteria' [28]. For this paper, and within the HSV context, the y_1 = 'value' channel is in the top box, followed by the y_2 = 'saturation' channel in the middle, with -the 'least important'- the y_3 ='Hue' channel at the bottom. Moreover, in order to satisfy the conditions of Definition 1, $\sum_{i=1}^{n} w_i = 1$ and $w_i \in [0,1]$.

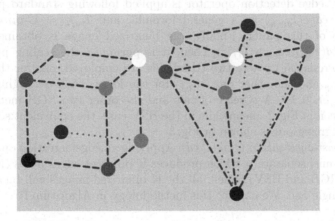

Fig. 1. A visual representation of RGB and HSV -Hexcone model- color spaces. (Color figure online)

3   Aggregating Channels in Color Edge Detection

This paper proposes 7 different methodologies for aggregating the channels of RGB (3 methods) and HSV (4 methods) images. Canny algorithm [6] is used as ED algorithm, and it uses individual channel approach[1].

1. *Crispy pre-aggregation (A method)*: This method aggregates the different channels into one single channel, normally employing the mean. After this aggregation, an edge detection algorithm is applied over the resulting grayscale image. This method can be regarded as the classic one, as it is the most common procedure employed in the literature.

$$I_{gray} = \frac{I_R + I_G + I_B}{3} \qquad (2)$$

---

[1] Multichannel approach was employed in [9].

When aggregating the three different channels of an HSV image, the specific nature of each channel has to be taken into account. For it, this paper proposes what we have termed *Yager aggregation* of channels, which is inspired in [29], employed in [22] and explained in Sect. 2.4. The different weights -w_1, w_2, w_3- are applied hierarchically as shown below:

$$Yager(I_{HSV}, w_1, w_2, w_3) = \sum_{i=1}^{3} w_i \prod_{k=1}^{i} I_{HSV}^{4-k} \tag{3}$$

which is an I_{gray} image, that can also be expressed as:

$$I_{gray} = Yager(I_{HSV}, w_1, w_2, w_3) = w_1 \cdot I_V + w_2 \cdot I_V \cdot I_S + w_3 \cdot I_V \cdot I_S \cdot I_H \tag{4}$$

After, an edge detection operator is applied following standard procedures: $I_{soft} = edge(I_{gray})$ as a general formula, and $I_{soft} = Canny(I_{gray})$ in the case of this paper. Finally, the binarized image is obtained: $I_{bin} = threshold(I_{soft}, \alpha_{sup}, \alpha_{inf})$[2]. Then, it is perfomed the scaling phase using the hysteresis ([6]) where two trhesholds are employed, one for the superior limit α_{sup} and the other one -α_{inf}- for the lower, both α ranging from 0% to 100% or $[0, 1]$-. A scheme of this and the other two RGB methodologies can be seen in Fig. 2, meanwhile in the HSV case, the equivalent scheme with Yager aggregations is shown in Fig. 3.

2. *Crispy post-aggregation (B method)*: Applying an edge detection operator over each channel separately, which produces $\tilde{K}$ different edges maps ($\tilde{K}$=3 in the case of RGB and HSV). Then, all the $\tilde{K}$ binarized images will be aggregated into a single one. We can see this methodology in Algorithm 1:

Algorithm 1. Crispy post-aggregation (B method)

1: **procedure** (Crispy post-aggregation of $\tilde{k}$ channels)
2:    **for** $k = 1, \ldots, \tilde{k}$ **do**
3:       $I_{soft}^k = edge(I_{RGB}^k$ or $I_{HSV}^k)$
4:       $I_{bin}^k = threshold(I_{soft}^k, \alpha)$
5:    $I_{bin} = \Theta(I_{bin}^1, \ldots, I_{bin}^{\tilde{k}})$

In the case of HSV images, the three channels differ in their nature, as pointed out in Sect. 2. Therefore, in order to adapt the I_{soft}^k images, these three images resulted as it is shown below.

1. $I_{soft}^{YagerV} = edge(w_1 \cdot I_V)$
2. $I_{soft}^{YagerS} = edge(w_2 \cdot I_V \cdot I_S)$
3. $I_{soft}^{YagerH} = edge(w_3 \cdot I_V \cdot I_H \cdot I_S)$

---

[2] A detailed explanation of the different phases of edge detection can be found in [11,12].

At the next step, the binarized images of each channel are created after applying a threshold (see 4th line of Algorithm 1). Finally, they are aggregated (5th line of Algorithm 1). The maximum was employed as aggregation function - $\Theta() = Max()$-: $I_{bin} = \max(I_{bin}^1, \ldots, I_{bin}^{\tilde{k}})$

3. *Fuzzy post-aggregation (Methodology C)*: In this case the aggregation function is applied over the soft image corresponding to each color channel. At the last step of the Algorithm 2 the binarized image is produced following a soft approach that we have called "fuzzy" approach. One more algorithm -Algorithm 3- has been developed going beyond what was done in [10]. In this new approach, the aggregation function is applied before combining the edge information -edginess- of two different directions, horizontal and vertical.

Algorithm 2. 'Fuzzy' post-aggregation (Method C1)

1: **procedure** ('Fuzzy' aggregation of channels)
2:     **for** $k = 1, \ldots, \tilde{k}$ **do**
3:         $I_{soft}^k = edge(I_{RGB}^k \text{ or } I_{HSV}^k)$
4:     $I_{soft} = \Theta(I_{soft}^1, \ldots, I_{soft}^{\tilde{k}})$
5:     $I_{bin} = threshold(I_{soft}, \alpha)$

Algorithm 3. 'Fuzzy' post-aggregation with direction (Method C2)

1: **procedure** ('Fuzzy' aggregation of channels directions)
2:     **for** $k = 1, \ldots, \tilde{k}$ **do**
3:         $I_{soft_x}^k = edge_x(I_{RGB}^k \text{ or } I_{HSV}^k)$
4:         $I_{soft_y}^k = edge_y(I_{RGB}^k \text{ or } I_{HSV}^k)$
5:     $I_{soft_x} = \Theta(I_{soft_x}^1, \ldots, I_{soft_x}^{\tilde{k}})$
6:     $I_{soft_y} = \Theta(I_{soft_y}^1, \ldots, I_{soft_y}^{\tilde{k}})$
7:     $I_{soft} = sqrt(I_{soft_x}^2 + I_{soft_y}^2)$
8:     $I_{bin} = threshold(I_{soft}, \alpha)$

---

4   Comparisons and Results

To test the ED quality of the different methods, the first 50 images of Berkeley's images data set (BSDS500) [20] were used. This sample (from '100075.jpg' to '16052.jpg') was divided into training's and test, which allowed learning the best paremeters. Following a cross-validation method, they were divided into 5 blocks of 10 images each, which resulted in 10 folds containing 30 images for training and 20 for the test. Performing cross-validation was another improvement compared to what was done in previous research [10]. The comparisons were conducted for RGB and HSV images. The RGB version was the original one included in Berkeley's dataset, while the HSV version was created from

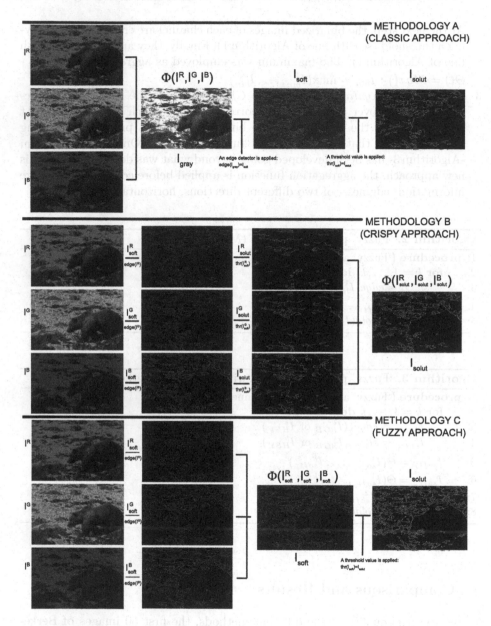

Fig. 2. Scheme of methodologies A, B and C with RGB images.

the RGB's employing the $rgb2hsv(I_{RGB})$ function of *Matlab2020b*, which transforms an RGB image into HSV's applying the hexcone model [25]. Different weights were allowed ($w_1 = \{0.4, 0.5, ..., 0.9, 1\}$ and $w_2 = \{0, 0.1, ..., 0.5\} = w_3$), following two restrictions: first, the boundary conditions of Definition 1 were respected, and secondly, it was considered that w_1 -the 'value' weight- had to

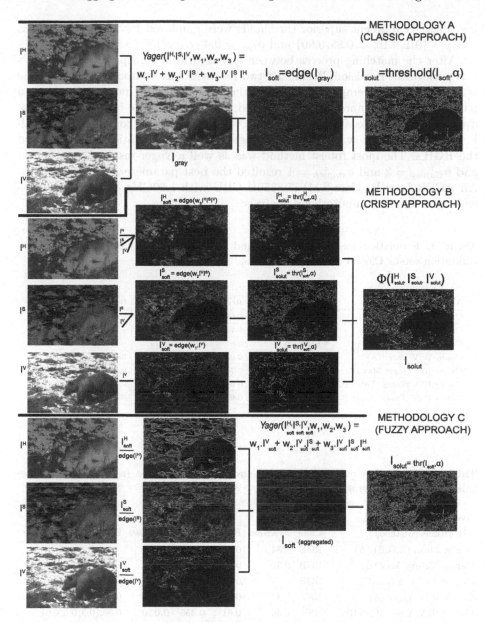

Fig. 3. Scheme of methodologies A, B and C with HSV images.

be at least equal to the other two, i.e., $w_1 \geq w_2$ and $w_1 \geq w_3$. In total, in this paper are being evaluated 50 different color images and 2 different color spaces (RGB and HSV), with 3 and 4 methods respectively, with 3 Gaussian smoothing values ($\sigma_{smooth} = \{0.6, 1, 2\}$), with 24 different weigths combinations and 2 different Gaussian sigma values: $\sigma_{Canny} = \{1, 2\}$). Moreover, for the scaling phase 17 different superior thresholds were employed for hysteresis process ($\alpha_{sup} = \{0.1, 0.15, ..., 0.85, 0.90\}$ and $\alpha_{low} = 0.4 \cdot \alpha_{sup}$ [3]).

After the matching process between the algorithm's outputs with human's was performed, Precision, Recall and two F-scores were computed. The first one is focused on the average human, while the other is for the most similar human (this was another improvement compared to what was done in [10]). Then, the average Prec and Recall for the 10 folds were employed to compute the overall F. Tables 1 and 2 show that the HSV Yager-inspired aggregations outperformed the RGB's. The most robust method was as well a Yager-inspired aggregation and $\sigma_{Canny} = 2$ and $\sigma_{smooth} = 1$ resulted the best parameters for all methods. An ACER Intel(R) Core(TM) i7-8550U CPU with 1.80GHz and 2.00 GHz was employed for the computational analysis.

Table 1. Evaluation measures for HSV and RGB algorithms (average of 10 cross-validation sets) - Closest human.

Closest human						
Algorithms (Method)	Color	Precision	Recall	F	Std. Dev	Best threshold and w_i
Canny Mean (R, G, B) (A)	RGB	0.46	0.68	0.550	0.019	–
Canny 'Crispy' Max (B)	RGB	0.47	0.70	0.559	0.039	–
Canny 'Fuzzy' Mean (C)	RGB	0.46	0.67	0.548	0.029	–
Canny HSV Yager (A)	HSV	0.48	0.68	0.566	**0.012**	0.50;{0.5, 0.1, 0.4}
Canny HSV Yager Max (B)	HSV	0.47	0.70	0.563	0.019	0.50;{1.0, 0.0, 0.0}
Canny HSV 'Fuzzy' Yager (C1)	HSV	0.48	0.67	0.563	0.033	0.25;{0.5, 0.3, 0.2}
Canny HSV 'Fuzzy' Yager (C2)	HSV	0.48	0.69	**0.567**	0.023	0.50;{0.5, 0.5, 0.0}

Table 2. Evaluation measures for HSV and RGB algorithms (average of 10 cross-validation sets) - Average human

Average human						
Algorithms (Method)	Color	Precision	Recall	F	Std. Dev	Best threshold and w_i
Canny Mean (R,G,B) (A)	RGB	0.34	0.61	0.440	0.014	–
Canny 'Crispy' Max (B)	RGB	0.37	0.58	0.450	0.022	–
Canny 'Fuzzy' Mean (C)	RGB	0.35	0.58	0.440	0.022	–
Canny HSV Yager (A)	HSV	0.37	0.59	0.456	0.013	0.50;{0.5, 0.1, 0.4}
Canny HSV Yager Max (B)	HSV	0.36	0.61	0.456	**0.010**	0.50;{0.5, 0.5, 0.0}
Canny HSV 'Fuzzy' Yager (C1)	HSV	0.37	0.60	0.454	0.019	0.25;{0.5, 0.3, 0.2}
Canny HSV 'Fuzzy' Yager (C2)	HSV	0.37	0.59	**0.457**	0.021	0.50;{0.5, 0.2, 0.3}

---

[3] This relationship between thresholds had been discovered in previous researches.

5 Conclusions

This paper's results showed the potential of Yager-inspired aggregations with HSV channels, which showed to be closer to human vision than its equivalent RGB aggregations. This was tested for Canny's. These Yager-inspired aggregations proved to be an efficient approach for Color edge detection. Moreover, the value's weight (w_1) was proof to be the most relevant for ED performance.

Future research could continue exploring these aggregation methods -or equivalent ones- with other color spaces such as HSL -which is similar to HSV-, CYMK, CIELAB, or Super 8 [9]. As well, other interesting possibility would be comparing the proposed HSV aggregations with other HSV aggregations proposed in the literature. Finally, the approach based on mixing different channels from different color spaces using deep learning techniques seems like a promising and natural next step.

Acknowledgments. This research has been partially supported by the Government of Spain, grant PGC2018-096509-B-I00.

For conducting this research, the code created by Kermit Research Unit has been helpful [1].

References

1. de Baets, B., López-Molina, C.: The kermit image toolkit (kitt), ghent university. www.kermitimagetoolkit.net (2016)
2. Beliakov, G., Bustince, H., Paternain, D.: Image reduction using means on discrete product lattices. IEEE Trans. Image Process. **21**(3), 1070–1083 (2011)
3. Bogumil, S.: Color image edge detection and segmentation: a comparison of the vector angle and the Euclidean distance color similarity measures. Ph.D. thesis, University of Waterloo (1999)
4. Bouchon-Meunier, B.: Aggregation and fusion of imperfect information, vol. 12. Physica (2013)
5. Bustince, H., Fernández, J., Kolesárová, A., Mesiar, R.: Generation of linear orders for intervals by means of aggregation functions. Fuzzy Sets Syst. **220**, 69–77 (2013)
6. Canny, J.: A computational approach to edge detection. IEEE Trans. Pattern Anal. Mach. Intell. **PAMI-8**(6), 679–698 (1986). https://doi.org/10.1109/TPAMI.1986.4767851
7. Dutta, S.: A color edge detection algorithm in RGB color space, pp. 337–340 (2009)
8. Flores-Vidal, P.A., Gómez, D., Castro, J., Montero, J.: The different importance of each color in edge detection. In: Developments of Artificial Intelligence Technologies in Computation and Robotics - Proceedings of the 14th International FLINS Conference (FLINS2020), pp. 931–938 (2020)
9. Flores-Vidal, P.A., Gómez, D., Castro, J., Montero, J.: A new approach to color edge detection by means of transforming RGB images into an 8-dimension color space. In: Proceedings of the EEE 14th International Conference on Intelligent Systems and Knowledge Engineering (ISKE 2919), pp. 1140–1147 (2020)
10. Flores-Vidal, P.A., Gómez, D., Villarino, G., Castro, J., Montero, J.: A new approach to color edge detection. In: Atlantis Studies in Uncertainty Modelling, 2019 Conference of the International Fuzzy Systems Association and the European Society for Fuzzy Logic and Technology (EUSFLAT 2019), pp. 376–384 (2019)

11. Flores-Vidal, P.A., Olaso, P., Gómez, D., Guada, C.: A new edge detection method based on global evaluation using fuzzy clustering. Soft. Comput. **23**(6), 1809–1821 (2018). https://doi.org/10.1007/s00500-018-3540-z

12. Flores Vidal, P.A., Villarino, G., Gómez, D., Montero, J.: A new edge detection method based on global evaluation using supervised classification algorithms. Int. J. Comput. Intell. Syst. **12**(1), 367–378 (2019)

13. Gnanatheja, R., Reddy, T.S.: YCoCg color image edge detection. Int. J. Eng. Res. Appl. **2**(2), 152–156 (2012)

14. Goguen, J.A.: L-fuzzy sets. J. Math. Anal. Appl. **18**(1), 145–174 (1967)

15. González, R.C., Woods, R.E.: Digital Image Processing, 3rd edn. (2008)

16. Guada, C., Gómez, D., Rodríguez, J.T., Yáñez, J., Montero, J.: Classifying image analysis techniques from their output. Int. J. Comput. Intell. Syst. **9**, 43–68 (2016). https://doi.org/10.1080/18756891.2016.1180819

17. Lee, D., Wang, J., Plataniotis, K.N.: Contribution of skin color cue in face detection applications. In: Celebi, M.E., Smolka, B. (eds.) Advances in Low-Level Color Image Processing. LNCVB, vol. 11, pp. 367–407. Springer, Dordrecht (2014). https://doi.org/10.1007/978-94-007-7584-8_12

18. López-Molina, C.: The breakdown structure of edge detection: analysis of individual components and revisit of the overall structure. Ph.D. thesis (2012)

19. Macedo-Cruz, A., Pajares, G., Santos, M., Villegas-Romero, I.: Digital image sensor-based assessment of the status of oat (avena sativa l.) crops after frost damage. Sensors **11**(6), 6015–6036 (2011)

20. Martin, D., Fowlkes, C., Tal, D., Malik, J.: A database of human segmented natural images y its application to evaluating segmentation algorithms y measuring ecological statistics. In: Proceedings of the IEEE International Conference on Computer Vision, vol. 2, pp. 416–423 (2001)

21. McAndrew, A.: An introduction to digital image processing with matlab notes for SCM2511 image processing. Sch. Comput. Sci. Math. Victoria Univ. Technol. **264**(1), 1–264 (2004)

22. Rojas, K., Gómez, D., Montero, J., Rodríguez, J.T., Valdivia Barrios, A., Paiva, F.: Development of child's home environment indexes based on consistent families of aggregation operators with prioritized hierarchical information. Fuzzy Sets Syst. **241**, 41–60 (2014)

23. Sandeep, K., Rajagopalan, A.: Human face detection in cluttered color images using skin color, edge information. In: ICVGIP (2002)

24. Shaik, K.B., Ganesan, P., Kalist, V., Sathish, B., Jenitha, J.M.M.: Comparative study of skin color detection and segmentation in HSV and YCBCR color space. Procedia Comput. Sci. **57**, 41–48 (2015)

25. Smith, A.R.: Color gamut transform pairs. In: ACM SIGGRAPH Computer Graphics, vol. 12, no. 3, pp. 12–19 (1978)

26. Trahanias, P.E., Venetsanopoulos, A.N.: Color edge detection using vector order statistics. IEEE Trans. Image Process. **2**(2), 259–264 (1993)

27. Turhan, H.I., Sahin, G., Erkmen, A.M.: Comparing color edge detection techniques (unpublished)

28. Yager, R.R.: Modeling prioritized multicriteria decision making. IEEE Trans. Syst. Man Cybernet. Part B (Cybernetics) **34**(6), 2396–2404 (2004)

29. Yager, R.R.: Prioritized aggregation operators. Int. J. Approximate Reasoning **48**(1), 263–274 (2008)

30. Yang, Y.: Colour edge detection and segmentation using vector analysis. University of Toronto (1996)

Representing Vietnamese Traditional Dances and Handling Inconsistent Information

Salem Benferhat[1,2]([⊠]), Zied Bouraoui[1,2], Truong-Thanh Ma[1,2], and Karim Tabia[1,2]

[1] CRIL CNRS and Artois University, Lens, France
{benferhat,bouraoui,ma,tabia}@cril.fr
[2] Centre de Recherche en Informartique de Lens, Université d'Artois and CNRS UMR 8188, Rue Jean Souvraz, 62307 Lens, France

Abstract. The first part of this paper proposes an application based upon a Vietnamese traditional Dance (VTD) ontology used for storing, searching, and reasoning with VTD knowledge and multimedia data available on the Web. We use lightweight ontologies, with prioritized assertional bases, to represent expert knowledge. We focus on the detection of movements in the ethnic VTD, where prioritized assertional bases contain motion present in the data. In general, the automated classification of the movements leads to inconsistency problems. In the second part of this paper, we propose a practical inconsistency-tolerant method for totally ordered information that goes beyond the so-called non-defeated inference.

1 Introduction

Southeast Asia is one of the most assertive growing regions in the world with natural and cultural resources. Specially, the cultural foundation regarding dance domain plays an important role in community life, it always brings the historical and cultural knowledge to the adjacent generation. The intangible culture heritages (ICHs) of the ethnic group dance are quite difficult to identify the exact values and great significance.

This paper describes an ontology-based modelling of Vietnamese Thai Dances movements. A ontology plays an important role in the representation of information processing systems and also is one of the formalism of describing entities, the properties and relationships. Especially, it brings a mission of transshipping information from multimedia data (such as raw Video/images) to computer. Correspondingly, utilizing ontology for preserving ICH is a completely appropriate selection because the traditional dances is recorded in raw videos which are complicated and enigmatic for storing heterogeneous information blocks. Different ontologies have been proposed in the literature to represent traditional dances [14–16,24,25]. This paper first proposes an ontology, built in the same spirit as those proposed in [23,24], and oriented towards the representation of

D. Ciucci et al. (Eds.): IPMU 2022, CCIS 1602, pp. 379–393, 2022.
https://doi.org/10.1007/978-3-031-08974-9_30

the so-called "CaUoc" movement in the ethnic Vietnamese Thai dances. We use lightweight ontologies to represent expert knowledge on "CaUoc" movements. We also propose a tool for detecting typical postures and orientations extracted from raw videos of VTDs. We lastly use prioritized assertional bases to encode factual concepts present in video data.

Many problems, such as classification, induction and knowledge base completion, need practical solutions for handling conflicting information. In our case study, inconsistency may occur when some automatically detected movements, from video data, contradict the knowledge encoded in the ontology about "CaUoc" movements. The problem of inconsistency management has received considerable attention in the literature in Artificial Intelligence and Databases. Different propositions have been studied in various knowledge representation formats (e.g., propositional logic [5], databases [6,12,13,19,20], and description logics [2,7,9,26], etc.). Such proposals are often based on the notion of a repair (usually defined as a maximally consistent subset of initial knowledge base). Querying all possible repairs often lead to an inconsistency-tolerant semantic with a high computational complexity. One way to circumvent the computational complexity is to select a single consistent subset that would replace the inconsistent knowledge base for query processing. Among these methods, we find the so-called IAR-semantics method [18], which consists in considering the intersection of repairs. Other approaches have been proposed for selecting a single preferred consistent sub-base and where the available information is totally or partially ordered [3,4]. In this paper, we also propose to select a single consistent sub-base, in the spirit of the non-defeated repair method proposed in [4] for totally ordered knowledge bases.

The rest of this paper is organised as follows. The first part of the paper contains a case study on ontology-based classification of Vietnamese traditional dance movements videos. The second part of the paper proposes a discussion on a minimal set of properties for practical inconsistency-tolerant methods. It also identify a practical inconsistency-tolerant method for prioritized information. It is based on an extension of the non-defeated semantics and which is appropriate for our application.

2   An Ontology-based Modelling for Ethnic Vietnamese Thai Dances

Vietnam has fifty-four ethnic groups living and working on the entire territory, in which, Thai community is one of the ethnic groups existing the large number of the traditional dances. In order to understand explicitly with respect to Ethnic Vietnamese Thai dance (EVTD), in this paper, we concentrated on analyzing and determining the fundamental movement features which is one of the important foundation to determine the particular EVTDs. Representing the details of each motion in the EVTDs is quite essential for considering the fundamental movement characteristics. It is divided in five prime classifications: Orientation, Arm Posture, Leg Posture, Sitting Posture, Standing Posture.

Orientation it is one of the most significant characteristics in the sense that the motions, postures and gestures of VTDs are described explicitly through the orientations. They are split in eight orientations as Fig. 1, including from orientation 1 to orientation 8. Orientation 1 is the direction of the dancer opposite to spectator. Each EVTD movement is represented in each part of body, an evidence as the left leg is in orientation 8 and move to orientation 2 or the right shoulder is in orientation 4.

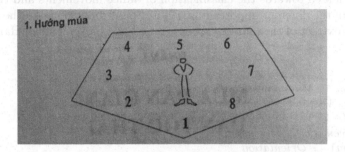

Fig. 1. The Orientation in EVTD

Arm Posture. Most of arm postures are concentrated on depicting life activities in Thai community. It is divided in five primary postures: VN-Thai-Thế-1-Arm, VN-Thai-Thế-2-Arm, VN-Thai-Thế-3-Arm, VN-Thai-Thế-4-Arm, VN-Thai-Thế-5-Arm. They are grouped into two distinct classes (ontology concepts): open-arm posture and close-arm posture. Many configurations are available: In the first posture (1) the two arms are parallel in orientation 1; the next posture (2), left arm is in orientation 7 and right arm is in orientation 3, height of arms is about shoulder's height; etc.

Leg Posture. There are five significant leg postures to represent for EVTD movements. It includes VN-Thai-Thế-1-Leg, VN-Thai-Thế-2-Leg, VN-Thai-Thế-3-Leg, VN-Thai-Thế-4-Leg, VN-Thai-Thế-5-Leg. There are many descriptions: Firstly (1), right foot is in orientation 2 and left foot is on orientation 8, heels of foot together; Secondly (2), left foot is in orientation 8 as VN-Thai-The-1-Leg and right foot is forwarded a step and being in orientation 2; etc.

Sitting Posture. Regarding sitting posture is divided in two postures, it consists of VN-Thai-Thế-1-Sitting, VN-Thai-Thế-2-Sitting. Description as follows: (1) sitting on heels of foot, keeping the foots closely.(2) sitting on a heel of the foot and another foot is as VN-Thai-The-1-Leg.

Standing Posture. There are three standing posture in EVTD: VN-Thai-Thế-5-Standing, VN-Thai-Thế-2-Standing, VN-Thai-Thế-4-Standing. It concentrates primarily on describing the angle of leg when the dances is standing. For example, (1) one leg is in orientation 1 and another leg is to create an angle 25 °C at knee between thigh and calf. (2) one leg is in orientation 1 and another leg is lifted forwardly to create an angle 45 °C at knee between thigh and calf, the pillar is in the backward leg. (3) one leg is in orientation 1 and another leg is lifted backwardly to create an angle 45°C at knee between thigh and calf, the pillar is in the forward leg.

Ontology Moddeling. After collecting knowledge from dance domain experts, we selected the significant features to reconstruct the schema for EVTD movements and shown in Fig. 2. Our ontology is inspired by the work done in [23, 24] but it is oriented towards the classification of dance movements and the characterization of the different important attributes described above. In the following, we give an extract of the ontology about some basis movements of Handkerchief Dance.

$Orientation \sqsubseteq VietnameseThaiDance$
$Posture \sqsubseteq VietnameseThaiDance$
$Movements \sqsubseteq VietnameseThaiDance$
$CaUocMovement \sqsubseteq Movements$
$Orientation1 \sqsubseteq Orientation$
$Orientation3 \sqsubseteq Orientation$
$ArmPosture \sqsubseteq Posture$
$LegPosture \sqsubseteq Posture$
$VNThaiThe1Arm \sqsubseteq ArmPosture$
$VNThaiThe2Arm \sqsubseteq ArmPosture$
$Orientation1 \sqsubseteq CaUocMovement$
$Orientation3 \sqsubseteq CaUocMovement$
$Orientation7 \sqsubseteq CaUocMovement$
$VNThaiThe1Arm \sqsubseteq CaUocMovement$
$VNThaiThe2Arm \sqsubseteq CaUocMovement$
$VNThaiThe3Arm \sqsubseteq \neg CaUocMovement$
$VNThaiThe4Arm \sqsubseteq \neg CaUocMovement$
$VNThaiThe5Arm \sqsubseteq \neg CaUocMovement$
$Orientation4 \sqsubseteq \neg CaUocMovement$
$Orientation1 \sqsubseteq \neg VNThaiThe3Arm$
$\exists isDirectionOf \sqsubseteq Orientation$
$\exists isDirectionOf^- \sqsubseteq Movements$
$\exists isPostureOf \sqsubseteq Posture$
$\exists isPostureOf^- \sqsubseteq Movements$
$\exists ConfidenceOfOrt \sqsubseteq Orientation$
$\exists ConfidenceOfOrt^- \sqsubseteq IntegerType$
$\exists CombineWith \sqsubseteq Orientation$
$\exists CombineWith^- \sqsubseteq Posture$
...

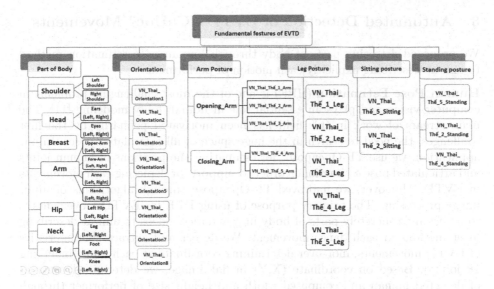

Fig. 2. The basis feature schema for EVTD "CaUoc" movements

Our aim is to have a repository served for searching, reasoning and query-answering each motion of each movement that may be represented by the fundamental features. Our ontology is mainly oriented towards the classification of movements associated with traditional Vietnamese CuAoac dances. It is therefore complementary to work that focuses on the definition of ontologies (for instance given in [16,23]), on traditional dances in a general way In the following, we describe how we automatically detect movements (Fig. 3).

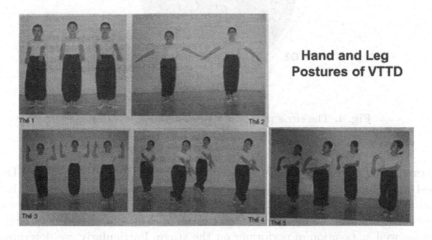

Fig. 3. The hand and leg posture in EVTD

3   Automated Detection of EVTD "CaUoc" Movements

We proposed detecting parts of body through human pose estimation method to aim at describing explicitly each motion of performers.

Human Pose Estimation (HPE) is one of the most challenging problems in computer vision and plays an essential role in human body modeling [11]. The main stream of work in this field has been motivated primarily by the first challenges, the need to search in the huge space of all articulated poses. In our application, we use TF-Openpose[1] for estimating the positions of human joints and articulated pose estimation in order to support for depicting each movements in EVTD. Moreover, we improved TF-Openpose through algorithms of input image processing. The primary purpose of using HPE for EVTD movements is to determine concretely parts of body in raw dance videos to aim at describing most motions in each dance movement. We detect and extract characteristics of EVTD movements, moreover determining coordinate at each joint (including 18 joints). Based on coordinate (X,Y) in flat image, we determined centroid of detected human and computed width and height size of performer through surrounding circle as showed in Fig. 4.

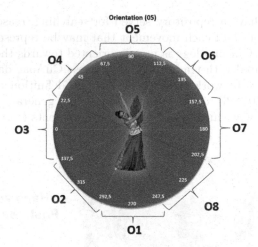

Fig. 4. The circle proposed for determining orientations

Determining Features of EVTDs Movements. The analysis of EVTDs is based upon four main terms: movements-phrases, movement primitives, basic dance postures and dance orientations. In order to define automatically for each orientation, we propose an orientation circle as in Fig. 4 for describing parts of body as well as position of performer on the stage. Particularly, we determined

---

[1] https://github.com/evalsocket/tf-openpose.

the direction of each part based upon the angle of two vectors: $\vec{x}$ and $\overrightarrow{ort}$, where $\vec{x}$ is an abscissa vector and $\overrightarrow{ort}$ is a vector of the interested part of body.

At the position of the parts will be computed for each angle between two distinct vectors ($\vec{x}$ and $\overrightarrow{ort}$) through the following Cosine formulas:

$$cos\theta = \frac{(\vec{x} \cdot \overrightarrow{ort})}{(|\vec{x}| \cdot |\overrightarrow{ort}|)}$$

$$angle\theta = arccosine\frac{(\vec{x} \cdot \overrightarrow{ort})}{(|\vec{x}| \cdot |\overrightarrow{ort}|)}$$

Figure 5 illustrates explicitly how to compute angle between joint-to-joint to determine orientations. Depending on angle θ at each joint-to-joint that we could assemble straightforwardly the description of the motions in each frame.

Following this method, we detect automatically the significant features served for classifying the EVTD's raw videos (low revolution videos). Nevertheless, we realized, in detectable process, several inconsistencies of features when query-answering into ontology-based model of ETVD movement. The contradictions are generated from recognizing and detecting automatically the characteristics properties. In the following, we present an extract from the obtained ontology assertions.

$CaUocMovement("F1")$
$CaUocMovement("F2")$
$CaUocMovement("F3")$
$Orientation1("O1")$
$Orientation3("O3")$
$Orientation2("O2")$
$VNThaiThe1Arm("Arm1")$
$VNThaiThe2Arm("Arm2")$
$VNThaiThe3Arm('Arm3")$
$isDirectionOf("O1", "F1")$
$isDirectionOf("O3", "F1")$
$isPostureOf("Arm1", "F2")$
$isPostureOf("Arm3", "F1")$
$ConfidenceOfOrt("O1", "50")$
$ConfidenceOfOrt("O2", "60")$
$CombineWith("O1", "Arm1")$
$CombineWith("O1", "Arm3")$

We could consider two scenarios of conflicts given a frame analysis as follows: The first scenario, the arm posture analysis of frame F1 of CaUoc movement produces VNThaiThe3Arm, while the role are *"orientation 1 of CaUoc movement, we could not exist VNThaiThe3Arm"*. Therefore, there is a conflict regarding F1 between orientation 1 and VNThaiThe3Arm posture. In the other hand, considering the same frame, it could not represent two orientations together into one frame regarding a performer. Particularly, the orientation analysis of a frame produces two possibilities: orientation 1 with a confident score A1 and orientation 2 with a confident score A2. Appearing of this case because the angles between two orientations could not be accurate completely when they are belonged to neighbouring of the boundary of two of those orientation regions.

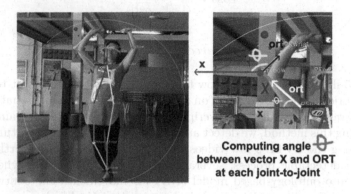

Fig. 5. Computing angle θ at each joint-to-joint

Classifying Vietnamese traditional dance movements may then raise the problem of conflicting information. This is particularly true when different experts or automated tools annotate a same dance video. Given the size of the data to be processed, it is important to have practical and efficient methods for managing prioritized and inconsistent data. This is detailed in next section.

4   Practical Inconsistency-Tolerant Strategies

This sections deals with problem of inconsistency in knowledge bases. In the literature, various inconsistency-tolerant semantics have been proposed to allow for meaningful reasoning with inconsistent knowledge bases (e.g., [2, 8, 9, 22, 27]). Most of these semantics consist in replacing the inconsistent knowledge base by one or more of its consistent (maximal or not) sub-bases. From these consistent sub-bases, different inference (or query answering) mechanisms have been defined depending on whether one considers all the set of consistent sub-bases, a single consistent sub-base or even the intersection of the maximally consistent sub-bases. Considering all maximally consistent sub-bases often leads to sound semantics because no arbitrary choice is made between the different consistent sub-bases. However, the negative point is the computational complexity which is often very high. Considering a single consistent sub-base (maximal or not) can lead to adventurous semantics (when randomly selecting a single consistent sub-base), or to cautious semantics (when taking the intersection of all maximally consistent sub-bases). The advantage of choosing only one consistent sub-base is its low computational complexity, which makes its semantics suitable for querying large knowledge bases. The different proposed semantics have been studied both in the case of flat bases (where all the assertions have the same level of importance) and in the case where the knowledge bases are totally or partially ordered.

In this paper, we consider a description logic (DL) knowledge base $\mathcal{K} = <\mathcal{T}, \mathcal{A}>$ where assertions have different priority levels. We assume that the TBox content $\mathcal{T}$ is correct, stable and cannot be questionable. However, in case of inconsistency, the assertions from the ABox $\mathcal{A}$ can be questionable and ignored. We view the prioritized ABox $\mathcal{A}$ as a well ordered partition simply represented by $(\mathcal{A}_1, \mathcal{A}_2, ..., \mathcal{A}_n)$. When there is no ambiguity, we write $f \in \mathcal{A}$ to denote that there exists $i \in \{1, ..., n\}$ such that $f \in \mathcal{A}_i$. We also assume that an assertion $f \in \mathcal{A}$ is present only in one $\mathcal{A}_i$. The stratum $\mathcal{A}_1$ contains the most important assertions in $\mathcal{A}$, while the stratum $\mathcal{A}_n$ contains the least important assertions in $\mathcal{A}$.

More generally, an assertion $f \in \mathcal{A}$ is considered as having more priority than an assertion $g \in \mathcal{A}$ (denoted by $f \succ g$) if $\exists i \in \{1, ..., n\}$, and $\exists j \in \{1, ..., n\}$ such that $f \in \mathcal{A}_i, g \in \mathcal{A}_j$ and $i < j$. Assertions belonging to a same layer $\mathcal{A}_i$ are considered as equally reliable.

Let us introduce the following notions, needed to determine the status of each assertion on the basis of its priority level in the ABox and whether it is involved or not in some conflicts.

- A conflict $\mathcal{C}$ is a subset of $\mathcal{A}$ that is minimally inconsistent with $\mathcal{T}$. Namely, the knowledge sub-base $< \mathcal{T}, \mathcal{C} >$ is inconsistent, and for each $\mathcal{C}'$ that is strictly included in $\mathcal{C}$, we have $< \mathcal{T}, \mathcal{C} >$ is consistent. Let $\mathcal{E}(\mathcal{A})$ be the set of all conflicts in $\langle \mathcal{T}, \mathcal{A} \rangle$.
- An assertion $f \in \mathcal{A}$ is said to be innocent if for any conflict $\mathcal{C} \subseteq \mathcal{A}$ of $\langle \mathcal{T}, \mathcal{A} \rangle$, $f \notin \mathcal{C}$. Said differently, an assertion f is said to be innocent if it is not involved in any conflict. Innocent elements are also called free elements in the context of propositional logic [5].

In the following, we denote $free(\mathcal{A})$ (or $IAR(\mathcal{A})$) the set of innocent or free elements.

We now discuss a set of minimal properties for a practical inconsistency-tolerant strategy, denoted by Δ. $\Delta(\mathcal{A})$, a set of assertions, denotes the result of applying Δ on $\mathcal{A}$.

The following are basic (minimal) properties for choosing Δ:

(P1) $\Delta(\mathcal{A})$ is consistent with $\mathcal{T}$
(P2) Computing $\Delta(\mathcal{A})$ is tractable
(P3) $\Delta(\mathcal{A}) \subseteq \mathcal{A}_1 \cup ... \cup \mathcal{A}_n$
(P4) if f is an innocent assertion in $\mathcal{A}$, then $f \in \Delta(\mathcal{A})$

Property **(P1)** states that the result of an inconsistency-tolerant strategy is a consistent set of assertions with $\mathcal{T}$. **(P3)** means that $\Delta(\mathcal{A})$ is only composed of elements of $\mathcal{A}$, while **(P4)** requires that $\Delta(\mathcal{A})$ should contain innocent assertions.

The following property is a weaker form of **(P3)**:

(P5) $\Delta(\mathcal{A}) \subseteq cl(\mathcal{A}_1 \cup ... \cup \mathcal{A}_n)$,

where $cl(.)$ represents the deductive closure of a set of assertions, defined by considering $\mathcal{T}_p$ the set of all the positive inclusion axioms of $\mathcal{T}$. The deductive closure of $\mathcal{A}$ with respect to $\mathcal{T}$ is given by:
$cl(\mathcal{A}) = \{B(a) \mid \langle \mathcal{T}_p, \mathcal{A} \rangle \models B(a) \text{ s.t. } B \text{ is a concept in } \mathcal{T}, a \text{ is an individual of } \mathcal{A}\}$
    $\cup \{R(a, b) \mid \langle \mathcal{T}_p, \mathcal{A} \rangle \models R(a, b) \text{ s.t. } R \text{ is a role in } \mathcal{T}, a \text{ and b are individuals of } \mathcal{A}\}$.
Here $\models$ is a standard DL inference relation.

The spirit of the **(P3)** and **(P5)** properties is that to resolve conflicts, we assume that certain assertions are false and should be ignored. These properties **(P3)** and **(P5)** do not allow replacing these false assertions with other assertions that are not explicitly or implicitly present in the knowledge base.

The properties **(P1–P4)** do not guarantee the maximality of the set $\Delta(\mathcal{A})$. It may therefore be useful to identify other additional properties that involve conflicting assertions. A natural idea is to say that if two assertions $\{f, g\}$ are in conflict then one of the two elements must be part of the repair $\Delta(\mathcal{A})$. This property is not acceptable because it is not compatible with **(P1)**. Indeed, if $\mathcal{E}(\mathcal{A})$ contains the three following conflicts $\{f, g\}$, $\{f, h\}$ and $\{g, h\}$, then it is not possible to keep, in a consistent way, more than one of the three assertions f, g, h.

Let us introduce the concept of doubtful assertions:

– An assertion $f \in \mathcal{A}$ is said to be doubtful if f is conflicting with some assertion of $\mathcal{A}$.

The following presents a new property regarding doubtful assertions.

(P6) if $f \in \mathcal{A}$ is doubtful and f is consistent with $\Delta(\mathcal{A})$ then $f \in \Delta(\mathcal{A})$.

This property **(P6)** implies maximality (together with **(P1)**–**(P5)**) of $\Delta(\mathcal{A})$. Let $\mathcal{A}$ be a flat assertion base. Assume $\Delta(\mathcal{A})$ satisfies **(P1)**–**(P6)**. Then one can easily check that $\Delta(\mathcal{A})$ is a maximally consistent sub-base of $\mathcal{A}$. Indeed, the consistency of $\Delta(\mathcal{A})$ is guaranteed by **(P1)**, while the maximality of $\Delta(\mathcal{A})$ is guaranteed by **(P6)**.

In what follow, we focus on ontology languages where computing conflicts in done in polynomial time with respect to the size of ABox [21] (data complexity). This assumption is needed to satisfy **(P2)**. Examples of such ontology languages are $DL - Lite_R$ [1,10].

Let us now discuss these properties with respect to some inconsistency-tolerant strategies defined for flat and prioritized knowledge bases. As it is said previously, the main ingredient in these strategies is the concept of assertional repairs. An assertional repair is usually defined a maximally sub-base of $\mathcal{A}$ which is consistent with $\mathcal{T}$. For sake of simplicity, in the rest of this paper, we will use, in certain strategies, the term repair to designate a consistent sub-bases whether it is maximal or not.

In practice, an inconsistent base $\langle \mathcal{T}, \mathcal{A} \rangle$ may have several maximal repairs. The well-known AR-Semantics[2] and its variant cardinality-based CAR[3] fail to satisfy (P3) since they cannot be efficiently represented by a unique repair. Taking the intersection of all repairs leads to the well-known IAR- semantics [17,18]. In fact, using IAR semantics simply comes down to only consider the set of innocent assertions, also called free elements. Assume that computing conflicts of $\langle \mathcal{T}, \mathcal{A} \rangle$ is tractable. Then $\Delta(\mathcal{A}) = IAR(\mathcal{A})$ satisfies (P1)–(P5). However, it obviously fails to satisfies (P6), as it is well-known that $IAR(\mathcal{A})$ is not a maximally consistent sub-base of $\mathcal{A}$.

In [4], different strategies have been proposed to select one non-maximal prefer sub-base: Possibilistic repair $(\pi(\mathcal{A}))$, Linear-based repair $(l(\mathcal{A}))$, Non-defeated repair $(nd(\mathcal{A}))$ etc. All these inconsistency-tolerant strategies compute $\Delta(\mathcal{A})$ in a polynomial time with respect to the size of ABox. Hence, they statisfy (P2).

The possibilistic strategy is the most cautious one, in the sense that $\pi(\mathcal{A})$ is included in $l(\mathcal{A})$ an $nd(\mathcal{A})$ etc. It is defined by:

$$\pi(\mathcal{A}) = \mathcal{A}_1 \cup \mathcal{A}_2 \cup \cdots \cup \mathcal{A}_i$$

such that $\pi(\mathcal{A})$ is consistent but $\pi(\mathcal{A}) \cup \mathcal{A}_{i+1}$ is inconsistent. If $\mathcal{A}_1$ is inconsistent then $\pi(\mathcal{A}) = \emptyset$. If $\mathcal{A}$ is consistent then $\pi(\mathcal{A}) = \mathcal{A}$ (the whole Abox).

The possibilistic strategy is so cautious that it may drop some innocent assertions. Hence, it fails to satisfy the above properties and more precisely (P4) and (P6).

When the knowledge base is flat, the sub-bases $l(\mathcal{A})$, $nd(\mathcal{A})$ and $lcd(\mathcal{A})$ collapse with the set of innocent assertions. However, when the knowledge base is prioritized, $l(\mathcal{A})$ fails to recover the whole set of innocent assertions. Hence, the linear-based strategy does not satisfy (P4).

Lastly, the non-defeated strategy is quite a reasonable strategy, since the core of its definition is based on innocent assertions. The non-defeated repair $nd(\mathcal{A})$ is defined by:

$$nd(\mathcal{A}) = free(\mathcal{A}_1) \cup free(\mathcal{A}_1 \cup \mathcal{A}_2) \cup \cdots \cup free(\mathcal{A}_1 \cup \cdots \cup \mathcal{A}_n),$$

where we recall that $free(X)$ (or $IAR(X)$) contains free or innocent assertions of X. The non-defeated repair $nd(\mathcal{A})$ is an iterative application of the IAR-semantics over the nested sets $\mathcal{A}_1, \mathcal{A}_1 \cup \mathcal{A}_2, ..., \mathcal{A}_1 \cup \cdots \cup \mathcal{A}_n)$.

As for IAR-semantics, the non-defeated semantics satisfies (P1)–P5 but it fails to satisfy (P6). The following proposition provides a situation where $nd(\mathcal{A})$ satisfies (P1)–(P6):

---

[2] Where a query is considered as valid if it follows from each repair.
[3] Where for query-answering only the largest repairs are considered.

Proposition 1. *Let $\mathcal{A}$ be a prioritized assertion base where the priority relation between assertions is strict. Namely, $\forall f \in \mathcal{A}$ and $\forall g \in \mathcal{A}$ such that $f \neq g$ we have either $f > g$ or $g > f$. Then the non-defeated repairs $\Delta(\mathcal{A}) = nd(\mathcal{A})$ satisfies* **(P1)–(P6)**.

Note that in the case where the order between the assertions is strict and total then the non-defeated repair is equal to the linear-based repair. The non-defeated repair is built progressively, assertion by assertion, from the highest priority to the lowest priority. Initially the set $nd(\mathcal{A})$ is initialized to an empty set. Then, at each step if an assertion is consistent with the base $nd(\mathcal{A})$, then it is added to $nd(\mathcal{A})$. Otherwise the assertion is ignored and the next lower priority assertion is considered. Clearly, the obtained set of assertions is a maximal consistent subbase of $\mathcal{A}$ (hence it satisfies **P6**). The other properties are satisfied because the non-defeated repair satisfies them.

On the basis of the above proposition, one can propose a new strategy that satisfies all desirable properties **(P1)–(P6)** presented above. This strategy derives more assertions than IAR-semantics, possibilistic semantics and also non-defeated semantics (but remains incomparable with the linear-based semantics). Our aim is also to compute a unique repair, denoted by $ndw(\mathcal{A})$, where $nd(\mathcal{A})$ is first applied. Then the set $D(\mathcal{A}) = \mathcal{A}/nd(\mathcal{A})$ is computed and ordered in a strict way (and that extends the prioritized ordering defined on $\mathcal{A}/nd(\mathcal{A})$). The definition of new criteria, which would make it possible to order the contradictory assertions, is left for future work. The idea is then to split each of the strata of the $\mathcal{A}$ so that the elements which are not in $nd(\mathcal{A})$ are strictly ordered. More presicely, let $\mathcal{A} = (\mathcal{A}_1, \mathcal{A}_2, ..., \mathcal{A}_n)$ be the the prioritized ABox. Let us define a new stratification (a well oredered partition) of the Abox as follows:

$$D(\mathcal{A}) = (\mathcal{A}'_0, \mathcal{A}'_{11}, ..., \mathcal{A}'_{1m1}, \mathcal{A}'_{21}, ..., \mathcal{A}'_{2m2}, ..., \mathcal{A}'_{n1}, ..., \mathcal{A}'_{nmn}).$$

where $\mathcal{A}'_0 = nd(\mathcal{A})$ and for each $i = 1, ..n$ we have :

- $\mathcal{A}_i/nd(\mathcal{A}) = \mathcal{A}'_{i1} \cup ...\mathcal{A}'_{imi}$ (where mi is the size of $\mathcal{A}_i$), and
- each stratum $\mathcal{A}'_{ij}$ is a singleton.

Lastly, the non-defeated repair is applied on $D(\mathcal{A})$. More precisely, the new strategy defines the following repair:

$$ndw(\mathcal{A}) = nd(\mathcal{D}(\mathcal{A})).$$

One can easily check that $IAR(\mathcal{A}) \subseteq ndw(\mathcal{A})$, $\pi(\mathcal{A}) \subseteq ndw(\mathcal{A})$ and $nd(\mathcal{A}) \subseteq ndw(\mathcal{A})$. The following proposition shows that this strategy satisfies **(P1)–(P6)**:

Proposition 2. *Let $\mathcal{A}$ be a prioritized assertion base. $D(\mathcal{A})$ be the new stratification of $\mathcal{A}$ defined above. Then $\Delta(\mathcal{A}) = ndw(\mathcal{A})$ satisfies* **(P1)–(P6)**.

The proof is immediate. The consistency of $ndw(\mathcal{A}$ (i.e. **(P1)**), as well as the tractability property (i.e. **(P2)**), are satisfied, because the non-defeated repair also satisfied them. The satisfaction of **(P3)**, **(P5)** is obvious. The property

(**P4**) is ensured by the fact that the set $nd(\mathcal{A})$ is consistent and is considered as the highest priority stratum of $\mathcal{D}(\mathcal{A})$. Therefore, $nd(\mathcal{A})$ is included in $ndw(\mathcal{A})$. Hence innocent assertions, as well as non-defeated assertions, are included in $ndw(\mathcal{A})$. Lastly, (**P6**) is satisfied thanks to Proposition 1.

5   Conclusion

This paper proposed an ontology-based modeling for Vietnamese traditional Dance (VTD) movements with an automated detection of such movements from raw videos. In general, the classification of the movements introduces inconsistency on the knowledge base. We reviewed basic properties for handling inconsistency and we proposed a practical method that extends the non-defeated repair. A future work is to define additional criteria to decide between conflicting assertions. This will make it easier to define the preferred repair; or at least get a reasonable number of preferred repairs.

Acknowledgements. The first part of this work (on representing Vietnamese Traditional Dances) has received support from the European Project H2020 Marie Sklodowska-Curie Actions (MSCA), Research and Innovation Staff Exchange (RISE): Aniage project (High Dimensional Heterogeneous Data based Animation Techniques for Southeast Asian Intangible Cultural Heritage Digital Content), project number 691215.

References

1. Artale, A., Calvanese, D., Kontchakov, R., Zakharyaschev, M.: The dl-lite family and relations. J. Artif. Intell. Res. **36**, 1–69 (2009)
2. Baget, J., Benferhat, S., Bouraoui, Z., Croitoru, M., Mugnier, M., Papini, O., Rocher, S., Tabia, K.: A general modifier-based framework for inconsistency-tolerant query answering. In: KR, Cape Town, South Africa. pp. 513–516 (2016)
3. Belabbes, S., Benferhat, S., Chomicki, J.: Handling inconsistency in partially pre-ordered ontologies: the elect method. J. Logic Comput. **31**(5), 1356–1388 (2021)
4. Benferhat, S., Bouraoui, Z., Tabia, K.: How to select one preferred assertional-based repair from inconsistent and prioritized dl-lite knowledge bases? In: Proceedings of the Twenty-Fourth International Joint Conference on Artificial Intelligence, IJCAI, 2015, pp. 1450–1456 (2015)
5. Benferhat, S., Dubois, D., Prade, H.: Some Syntactic Approaches to the Handling of Inconsistent Knowledge Bases : a Comparative Study. Part 2 : the Prioritized Case, vol. 24, pp. 473–511. Physica-Verlag, Heidelberg (1998)
6. Bertossi, L.E.: Database Repairing and Consistent Query Answering. Morgan & Claypool Publishers, Synthesis Lectures on Data Management (2011)
7. Bienvenu, M., Bourgaux, C.: Querying and repairing inconsistent prioritized knowledge bases: complexity analysis and links with abstract argumentation. In: Calvanese, D., Erdem, E., Thielscher, M. (eds.) International Conference on Principles of Knowledge Representation and Reasoning, KR 2020, Rhodes, Greece, 2020, pp. 141–151 (2020)

8. Bienvenu, M., Bourgaux, C., Goasdoué, F.: Query-driven repairing of inconsistent DL-Lite knowledge bases. In: IJCAI, New York, USA, pp. 957–964 (2016)
9. Bienvenu, M., Bourgaux, C., Goasdoué, F.: Computing and explaining query answers over inconsistent DL-Lite knowledge bases. J. Artif. Intell. Res. **64**, 563–644 (2019)
10. Calvanese, D., De Giacomo, G., Lembo, D., Lenzerini, M., Rosati, R.: Tractable reasoning and efficient query answering in description logics: the DL-Lite family. J. Autom. Reasoning **39**(3), 385–429 (2007)
11. Cao, Z., Simon, T., Wei, S., Sheikh, Y.: Realtime multi-person 2D pose estimation using part affinity fields. CoRR abs/1611.08050 (2016)
12. Chomicki, J.: Consistent query answering: five easy pieces. In: Schwentick, T., Suciu, D. (eds.) ICDT 2007. LNCS, vol. 4353, pp. 1–17. Springer, Heidelberg (2006). https://doi.org/10.1007/11965893_1
13. Du, J., Qi, G., Shen, Y.: Weight-based consistent query answering over inconsistent SHIQ knowledge bases. Knowl. Inf. Syst. **34**(2), 335–371 (2013)
14. El Raheb, K., Ioannidis, Y.: A labanotation based ontology for representing dance movement. In: Efthimiou, E., Kouroupetroglou, G., Fotinea, S.E. (eds.) Gesture and Sign Language in Human-Computer Interaction and Embodied Communication, pp. 106–117. Springer, Berlin Heidelberg, Berlin, Heidelberg (2012). https://doi.org/10.1007/978-3-642-34182-3_10
15. El Raheb, K., Papapetrou, N., Katifori, V., Ioannidis, Y.: Balonse: ballet ontology for annotating and searching video performances. In: Proceedings of the 3rd International Symposium on Movement and Computing. Association for Computing Machinery, New York, NY, USA (2016)
16. Kalita, D., Deka, D.: Ontology for preserving the knowledge base of traditional dances (otd). Electron. Libr. **38**, 785–803 (2020)
17. Lembo, D., Lenzerini, M., Rosati, R., Ruzzi, M., Savo, D.F.: Inconsistency-tolerant semantics for description logics. In: Hitzler, P., Lukasiewicz, T. (eds.) RR 2010. LNCS, vol. 6333, pp. 103–117. Springer, Heidelberg (2010). https://doi.org/10.1007/978-3-642-15918-3_9
18. Lembo, D., Lenzerini, M., Rosati, R., Ruzzi, M., Savo, D.F.: Inconsistency-tolerant query answering in ontology-based data access. J. Web Sem. **33**, 3–29 (2015). https://doi.org/10.1016/j.websem.2015.04.002
19. Livshits, E., Kimelfeld, B.: Counting and enumerating (preferred) database repairs. In: SIGMOD, pp. 289–301 (2017)
20. Livshits, E., Kochirgan, R., Tsur, S., Ilyas, I.F., Kimelfeld, B., Roy, S.: Properties of inconsistency measures for databases. In: SIGMOD, pp. 1182–1194 (2021)
21. Lukasiewicz, T., Malizia, E., Vaicenavičius, A.: Complexity of inconsistency-tolerant query answering in Datalog+/- under cardinality-based repairs. Proc. AAAI Conf. Artif. Intell. **33**(01), 2962–2969 (2019)
22. Lukasiewicz, T., Martinez, M.V., Simari, G.I.: Inconsistency handling in datalog+/- ontologies. In: 20th European Conference on Artificial Intelligence ECAI, 2012, pp. 558–563 (2012)
23. Ma, T., Benferhat, S., Bouraoui, Z., Do, T., Nguyen, H.: Developing application based upon an ontology-based modelling of vietnamese traditional dances. In: Addison, A.C., Thwaites, H. (eds.) 3rd Digital Heritage International Congress, held jointly with 24th International Conference on Virtual Systems & Multimedia, DigitalHERITAGE/VSMM 2018, San Francisco, CA, USA, 26–30 October 2018, pp. 1–7. IEEE (2018)

24. Ma, T., Benferhat, S., Bouraoui, Z., Tabia, K., Do, T., Nguyen, H.: An ontology-based modelling of vietnamese traditional dances (S). In: Pereira, Ó.M. (ed.) The 30th International Conference on Software Engineering and Knowledge Engineering, Hotel Pullman, Redwood City, California, USA, 1–3 July 2018, pp. 64–67. KSI Research Inc. and Knowledge Systems Institute Graduate School (2018)

25. Silva, D., Rocha Souza, R.: Knowledge organization systems for the representation of multimedia resources on the web: a comparative analysis. Knowl. Organ **47**, 300–319 (2020)

26. Trivela, D., Stoilos, G., Vassalos, V.: Querying expressive DL ontologies under the ICAR semantics. In: 31st International Workshop on Description Logics, USA (2018)

27. Tsalapati, E., Stoilos, G., Stamou, G., Koletsos, G.: Efficient query answering over expressive inconsistent description logics. In: Proceedings of the International Joint Conference on Artificial Intelligence IJCAI, New York, USA, pp. 1279–1285 (2016)

Laplace Operator in Connection to Underlying Space Structure

Hana Zámečníková[(✉)] and Irina Perfilieva[iD]

IRAFM, University of Ostrava, 30. dubna 22, 701 03 Ostrava, Czech Republic
{hana.zamecnikova,irina.perfilieva}@osu.cz
http://www.osu.eu/

Abstract. Laplace operator is a diverse concept throughout natural sciences. It appears in many research areas and every such area defines it accordingly based on underlying domain and plans on follow-up applications. This operator attracts a lot of attention e.g. in signal and image processing applications [6]. However, signals, in general, can be defined not only on Euclidean domains such as regular grids (in case of images). There are cases when underlying space is considered to be e.g. a non-regular graph or even a manifold, but the Laplace operator is still closely bound to the space structure. Therefore, we investigated this operator from point of view of spaces, where distance may not be explicitly defined and thus is being replaced by more general, so-called, proximity. Our goal was to find such a representation, that would be simple for computations but at the same time applicable to more general domains, possibly to spaces without a notion of a classic distance. In this article, we will mention some of the various ways in which this operator can be introduced in relation to the corresponding space. Also, we will introduce the formula for the Laplace operator in the space whose structure is determined by a fuzzy partition [8]. And we will investigate the properties of this kind of representation in parallelisms to standard well-known versions.

Keywords: Laplace operator · Proximity · Fuzzy transform

1 Introduction

Laplace operator (or Laplacian) is a fundamental concept in natural sciences. It can be introduced in many ways (some of them we will discuss in following section), but the important idea behind all definitions is that *Laplace operator provides a deviation in values of a given function from a local average in their neighborhood*. Therefore, it is essential when concerning wave or heat propagation [2], and any form of diffusion processes. One can also observe in literature the significance of Laplacian in geometry processing, since its eigensystem [11,15] enables to characterize surface parametrization [1]. In this area, this operator can be used in a variety of problems such as surface smoothing [22] or vector field decomposition [23]. Applications in differential geometry [4] or mesh

© Springer Nature Switzerland AG 2022
D. Ciucci et al. (Eds.): IPMU 2022, CCIS 1602, pp. 394–404, 2022.
https://doi.org/10.1007/978-3-031-08974-9_31

processing [17] can be also mentioned, where Laplacian acts well for studying surfaces, because it admits isometry invariance (meaning if we move the surface in a way that does not change point-to-point distances along the surface then the Laplacian is preserved).

However, its utilization goes far beyond just geometry processing [18] and mentioned applications in physics. It is a very powerful tool that allows all different kinds of computations. Laplacian gets a lot of attention when it comes to image processing [7,10,13] (or any signal processing in general). Because of its ability to reveal local changes in function values, Laplacian can detect so-called keypoints in image [24], which are, as its name implies, crucial points in the image, important for a deeper understanding of the image itself and also for further processing. One can also observe that Laplacian finds its usage in machine learning [21], not mentioning graph theory [1], differential equations area [2] or even the connection to discrete dynamical systems [20].

In this paper, we start with an overview of the different ways how the Laplacian can be defined in, both, a continuous and discrete setting. We introduce some well-known formulas so that the reader can understand the story behind that led at the end to our approach (Sect. 2). In this part, we also touch up on the possible space structure. However, as was mentioned at the beginning, not all spaces are equipped with a classic distance. Therefore, this section will be followed by an explanation of the concept of proximity and we also provide here a brief introduction to notions of the fuzzy theory that are necessary for our approach, like fuzzy partition or fuzzy transform (Sect. 3). Last, but not least, in Sect. 4, we introduce the discrete variant of the fuzzy-transform-based Laplace operator and clarify its properties that reflect the well-known standard properties of Laplacian.

2 Various Versions of Definition

Most of the key concepts can be introduced in a few different ways, you can use functional dependent definition, or you can introduce term axiomatically. When it comes to Laplacian, you can also involve inner product and adjoint operators [6]. A lot of notions also have a continuous and respective discrete version, usually because the continuous one is useful to build up the theory behind and the discrete one finds its usage in applications.

As we indicated, there are many different definitions that can be used to express the Laplacian. For instance, it can be introduced as a sum of unmixed partial derivatives or divergence of the gradient (these two are most common especially in Euclidean space), it can be written in terms of differential forms, we can think of it as a trace of Hessian or we can also connect it to random walks, where all these mentioned possibilities concern the smooth setting. There is even a nonlocal approach, which allows one to define Laplacian as a kind of weighted sum (or integral in continuous case). Most of these definitions apply not just in Euclidean space but one can use almost all of them directly also in curved domains.

So let us take a look at some of these definitions in detail. It is worth mentioning the following common ones, that either influenced our research, or that are closely connected to it.

2.1 Sum of Partial Derivatives

When it comes to Euclidean space, Laplacian is frequently defined as a second order differential operator acting on functions in a way that it assigns the sum of all unmixed second partial derivatives along coordinate axes in the Cartesian coordinates. Let us recall this particular definition.

Assume f to be a twice differentiable real function $f : \mathbb{R}^n \to \mathbb{R}$, then Laplace operator Δ in $\mathbb{R}^n$ is defined as follows

$$\Delta f = \sum_{i=1}^{n} \frac{\partial^2}{\partial x_i^2} f.$$

This formula can be translated to some domains that admit an amount of curvature. It is not applicable to any manifold in general, but in the case of the Riemannian manifold, it is possible to introduce an analogical definition with some adjustments, by incorporating a few additional terms like Riemannian metric or tangent spaces. A reader interested in this problematics can find details in [5].

2.2 Divergence of the Gradient

Now, let us focus on defining the Laplace operator by a divergence of a gradient. As the name Laplace *operator* implies, it is an operator, meaning, it takes a function and assigns it a new function, in this case, defined as a composition of two operators - the divergence and the gradient. So it is a tool which takes a function as an argument and converts it to another new function, where the function value of the new one at each point is equal to the divergence of the gradient

$$f \mapsto \mathrm{div}(\mathrm{grad}(f)).$$

This idea then projects also to curved spaces. Again we can think of it as a composition of two operators, one of which is a differential type - the gradient and the second one is a kind of aggregation - the divergence. As a recommendation for further details, we advise the reader to observe e.g. [5].

2.3 Nonlocal Approach

Moving to another kind of definition, which takes advantage of the previous 'gradient-divergence' definition and considers the notion from a different side, commonly used in the context of image processing. When the gradient is mentioned, one can assume the concept of a vector of partial derivatives. Every such derivative can be seen as a ratio of change in function values over the change in values of arguments. In the nonlocal case, the ratio changes into multiplication, what can be explained in a way that the inverse denominator of the derivative becomes its so-called *weight*. The concept of the weight and its importance will be explained later on. However, when 'gradient' is computed, then the only thing that remains is to apply 'divergence' and by this, to aggregate the result, here, in a form of integral. So, the *nonlocal Laplace operator* is defined as follows [6]

$$\Delta_w f(x) = \int_a^b w(x,y)(f(y) - f(x))dy. \tag{1}$$

This transition is crucial for all so-called nonlocal-type operators. In this case one does not incorporate the limit transition, instead, assume the multiplication with an evaluated-bond weight function w, which represents the relation between considered points. This semantic difference also corresponds to our approach (which will be introduced in Sect. 4).

Since we have already investigated the continuous case in our previous research [25], and when it comes to applications, the discrete setting is needed anyway, let us now introduce a discrete counterpart of the latter mentioned definition.

2.4 Discrete Laplacian

The discrete Laplace operator can be seen as an analog of the continuous Laplace operator, defined so that it has meaning on a graph or a discrete grid [16]. Considering image processing, an image, as a key object of attention here, is usually represented as a set of pixels arranged into a grid structure. Alternatively, such an image can be expressed in the form of a graph, where each pixel is identified with a vertex. Edges of such originated graph then represent similarities between pixels, resulting in a standard grid structure.

Besides the image processing point of view, the graph is a useful concept here, since it provides a model of underlying discrete space in general.

Let us therefore recall basic definitions. Assume a weighted graph $G = (V, E, w)$ with a finite set of vertices $V = \{v_1, \ldots, v_n\}$ and a finite set of weighted edges E ($E \subset V \times V$). Two vertices u and v are connected by edge (u, v) and the weight of such edge is then $w(u, v)$. The weights are assigned by using a function $w : V \times V \to \mathbb{R}_+^0$, which is symmetric ($w(u, v) = w(v, u), \forall (u, v) \in E$), and $w(u, v) = 0$ if $(u, v) \notin E$. Notation $u \sim v$ stands for two adjacent vertices u and v, meaning there exists en edge between u and v.

Let $H(V)$ denote the Hilbert space of real-valued functions on the vertices V of the graph. The space $H(V)$ is endowed with the usual inner product $\langle f, h \rangle_{H(V)} = \sum_{v \in V} f(v)h(v)$, where $f, h : V \to \mathbb{R}$. Analogously, let $H(E)$ be the space of real-valued functions defined on the edges of a graph G. This space is endowed with the inner product $\langle F, H \rangle_{H(E)} = \sum_{(u,v) \in E} F(u,v)H(u,v) = \sum_{u \in V} \sum_{v \sim u} F(u,v)H(u,v)$, where $F, H : E \to \mathbb{R}$ are two functions of $H(E)$.

Representation of the Laplacian. Let G be a weighted graph, and let f be a function of $H(V)$. The *difference operator* of f, noted $d : H(V) \to H(E)$, is defined on an edge $(u, v) \in E$ by

$$(df)(u, v) = \sqrt{w(u, v)}\,(f(v) - f(u)). \tag{2}$$

Note the parallelism with the 'divergence-gradient' definition. When one observes the latter formula (2), it can be considered as a directional derivative, or the weighted deviation, and therefore as a form of a gradient, which we aggregate later on. Formally, we can incorporate the difference operator and its adjoint in the following way.

The *adjoint of the difference operator*, noted $d^* : H(E) \to H(V)$, is a linear operator defined by

$$\langle df, H \rangle_{H(E)} = \langle f, d^* H \rangle_{H(V)}, \tag{3}$$

for any functions $H \in H(E)$ and $f \in H(V)$.

Laplace operator of a function $f \in H(V)$, noted $\Delta_w : H(V) \to H(V)$, is identified [12] by

$$\Delta_w f = -\frac{1}{2} d^*(df). \tag{4}$$

However, since the adjoint operator d^* at each vertex $u \in V$ can be expressed [12] as follows

$$(d^* H)(u) = \sum_{v \sim u} \sqrt{w(u, v)}\,(H(v, u) - H(u, v)), \tag{5}$$

we can substitute this representation into formula (4), resulting in

$$\begin{aligned}
\Delta_w f(u) &= -\frac{1}{2} d^*(df) = -\frac{1}{2} \sum_{v \sim u} \sqrt{w(u, v)}(df(v, u) - df(u, v)) \\
&= -\frac{1}{2} \sum_{v \sim u} \sqrt{w(u, v)} \left(\sqrt{w(u, v)}\,(f(u) - f(v)) - \sqrt{w(u, v)}\,(f(v) - f(u)) \right) \\
&= -\sum_{v \sim u} w(u, v)(f(u) - f(v)),
\end{aligned}$$

and therefore state the following.

Proposition 1. *Considering weighted graph, the Laplace operator of $f \in H(V)$, at a vertex $u \in V$, can be computed by*

$$\Delta_w f(u) = -\sum_{v \sim u} w(u, v)(f(u) - f(v)). \tag{6}$$

Note, in the graph theory literature, it is customary to define the Laplace operator with the opposite sign [3], but since we would like to keep consistency, we assume the Laplacian in the form (6).

3   Structure of the Underlying Space, Proximity

In the continuous setting, the structure of the space determines the computation of the gradient. When it comes to the discrete case, there is an especially strong connection, where space structure is given by a weighted graph, including connectivity and also edge weights which represent kind of evaluation of these connections.

In connection to weights, the first question that usually comes up is *how one can choose the proper weight function while applying Laplacian?* In literature, there are plenty of options, usually based on specific applications, e.g. [12].

As we pointed out several times, there are spaces where distance is not implicitly defined but it is still necessary to find out how close some objects are in a more general way. For this reason, one can use the so-called *measure of proximity*, which is a function, which tells us how close are two points, meaning, the closer they are, the larger is their proximity.

The specific proximity function we are using is based on a generating basic function of fuzzy partition. It is natural to set the weights this way because membership degrees provide a good intuition about the relationship between points.

However, the general concept of proximity can be introduced in the following way.

Definition 1. *Let function $w : \Omega \times \Omega \to \mathbb{R}$ be non-negative $(0 \leq w(x, y) < \infty)$ and symmetric $(w(x, y) = w(y, x))$, then w is called a measure of proximity. A pair (Ω, w) defines a proximity space [26].*

Note, the proximity space (Ω, w) can be also introduced with the reference to a positive (distance) measure $\tilde{d}$, $0 < \tilde{d}(x, y) \leq \infty$, between two points x and y, e.g. defining

$$w(x, y) = \tilde{d}^{-2}(x, y). \tag{7}$$

As we referred, our later on ideas are based on such one, where proximity is determined by the *fuzzy partition* $A_1, \ldots, A_n$. With this purpose in mind, we now recall the basic notions regarding this topic.

3.1   Fuzzy Partition

Definition 2. *Fuzzy sets $A_1, \ldots, A_n$, establish a fuzzy partition of a real interval $[a, b]$ with nodes $x_1 < \cdots < x_n$ if for $k = 1, \ldots, n$ holds*

1. *$A_k : [a, b] \to [0, 1]$, $A_k(x_k) = 1$, $A_k(x) > 0$ if $x \in (x_{k-1}, x_{k+1})$*
2. *$A_k(x) = 0$ if $x \notin (x_{k-1}, x_{k+1})$, where $x_0 = a$ and $x_{n+1} = b$*
3. *$A_k(x)$ is continuous*
4. *$A_k(x)$, for $k = 2, \ldots, n$, strictly increases on $[x_{k-1}, x_k]$ and $A_k(x)$ strictly decreases on $[x_k, x_{k+1}]$ for $k = 1, \ldots, n - 1$*
5. *$\forall x \in [a, b]$ holds Ruspini condition*

$$\sum_{k=1}^{n} A_k(x) = 1. \tag{8}$$

The membership functions $A_1, \ldots, A_n$ are called basic functions [8].

Definition 3. *The fuzzy partition $A_1, \ldots, A_n$, for $n \geq 2$ is h-uniform if nodes $x_0 < \cdots < x_{n+1}$ are h-equidistant, i.e. for all $k = 1, \ldots, n+1$, $x_k = x_{k-1} + h$, where $h = (b-a)/(n+1)$ and the following additional properties are fulfilled [8].*

1. *for all $k = 1, \ldots, n$ and for all $x \in [0, h]$, $A_k(x_k - x) = A_k(x_k + x)$,*
2. *for all $k = 2, \ldots, n$ and for all $x \in [x_{k-1}, x_{k+1}]$, $A_k(x) = A_{k-1}(x - h)$.*

Definition 4. *If the fuzzy partition $A_1, \ldots, A_n$ of $[a, b]$ is h-uniform, then there exists an even function $A_0 : [-1, 1] \to [0, 1]$, such that for all $k = 1, \ldots, n$*

$$A_k(x) = A_0\left(\frac{x - x_k}{h}\right), \quad x \in [x_{k-1}, x_{k+1}].$$

A_0 is called a generating function of uniform fuzzy partition [8].

Corollary 1. *Generating function A_0 produces infinitely many rescaled functions $A_H : \mathbb{R} \to [0, 1]$ with the scale factor $H > 0$, so that*

$$A_H(x) = A_0\left(\frac{x}{H}\right).$$

A (h,H)-uniform partition of $\mathbb{R}$ is then a collection of translations $\{A_H(x - k \cdot h), k \in \mathbb{Z}\}$ [10].

3.2 Fuzzy Transform

Direct Fuzzy transform or *F-transform* is a result of weighted linear integral transformation of a continuous function with weights determined by basic functions.

Definition 5. *Let $A_1, \ldots, A_n$ be basic functions which form a fuzzy partition of $[a, b]$ and f be any function from $C([a, b])$. We say that n-tuple of real numbers $F[f] = (F_1[f], \ldots, F_n[f])$ given by*

$$F_k[f] = \frac{\int_a^b f(x) A_k(x) dx}{\int_a^b A_k(x) dx}, \quad k = 1, \ldots, n, \tag{9}$$

is the direct integral F-transform of f with respect to $A_1, \ldots, A_n$ [10].

F-transform establishes a correspondence between a set of continuous functions on $[a, b]$ and the set of n-dimensional vectors. *Inverse F-transform* then converts an n-dimensional vector of components $(F_1[f], \ldots, F_n[f])$ into another continuous function

$$\hat{f}(x) = \sum_{k=1}^{n} F_k A_k(x), \tag{10}$$

which approximates the original one [10].

3.3 Discrete F-transform

As we mentioned, major part of applications do not consider continuous function as an input. Therefore, let us now modify the previous definition [9].

Assume that the function f is defined only on a finite set $P = \{x_1, \ldots, x_p\} \subseteq [a, b]$.

Definition 6. *The domain P of the function f is sufficiently dense with respect to the fixed fuzzy partition if for each fuzzy set A_k, $k = 1, \ldots, n$, from the fuzzy partition, there is an element $x_j \in P$ belonging to the support of A_k. Formally*

$$(\forall k)(\exists j) A_k(x_j) > 0.$$

Definition 7. *Let $A_1, \ldots, A_n$ be a fuzzy partition of $[a, b]$ and a function f be defined on the set $P = \{x_1, \ldots, x_p\} \subseteq [a, b]$ that is sufficiently dense with respect to $A_1, \ldots, A_n$. We say that the n-tuple of real numbers $F[f] = (F_1[f], \ldots, F_n[f])$ is a discrete F-transform of f with respect to $A_1, \ldots, A_n$ if*

$$F_k[f] = \frac{\sum_{j=1}^{p} f(x_j) A_k(x_j)}{\sum_{j=1}^{p} A_k(x_j)}. \tag{11}$$

Note here that according this definition we strictly distinguish points and nodes, meaning not every point from P has to be considered as a node.

4 Discrete FT-Based Laplacian

Now we have all the necessary tools for introducing the discrete F-transform-based Laplace operator. Supposing that the underlying space is endowed with the fuzzy partition, the first step is to characterize the model of the space with proximity using the weighted graph, which means to specify the weights of the graph. They will be assigned on the basis of the basic functions of the fuzzy partition.

Let us assume that graph $G = (V, E, w)$ does not contain any self-loops, i.e. $u \not\sim u$, and any isolated vertices, meaning graph G is connected. Let weights of edges be defined as follows

$$w(u, v) = A_u(v) = A(u, v). \tag{12}$$

In addition, each weight is multiplied by normalization coefficient. Then, based on Eq. (6), we can define discrete Laplace operator by its linear action on vertex-based functions

$$\Delta f(u) = F_u[f] - f(u). \tag{13}$$

Symbol $F_u[f]$ stands for F-transform component computed at node u. It is worth mentioning that in our approach we do not distinguish between nodes and points. Meaning, every point (vertex) is considered as a node of the constructing fuzzy partition, therefore we have a component for each $u \in V$ and the latter equation is well-defined. Note, when one observes the expression on the right-hand side of the equality, this is nothing more than a difference between an average value of the function f at vertex u (provided by F-transform) and the actual function value at the same vertex. Therefore it tells us how the function differs from its average value (in some small neighborhood), which fully corresponds with the key idea behind the Laplace operator.

Also, semantically the weight function in our approach corresponds to the concept of weights proposed in literature since the F-transform admits the evaluation of the connections by the degrees of membership functions of the fuzzy partition that indicates the proximity between points.

4.1 Properties

With this representation (13), let us consider again the graph G (with the same assumptions as above). Then (motivated by a smooth Laplace operator [14]) the discrete Laplacian based on F-transform admits the following properties

1. Δ has constant functions in the kernel

$$\Delta f(u) = 0 \text{ iff } f(u) = \text{const.}$$

   Intuitively it makes sense since the constant function does not differ in function values throughout the whole domain. Moreover, in $\mathbb{R}^n$ this holds also for all linear functions at each interior vertex, assuming that f is a linear function on the plane, point-sampled at vertices of the graph G.
   In parallelism to other definitions, this one is pretty obvious. When consider e.g. 'sum of partial derivatives' definition, constant or linear functions have both second partial derivatives equal to zero, which in summation results again in zero value of Laplacian.
2. Δ is a symmetric operator, whenever weights are symmetric ($w(u,v) = w(v,u)$).
3. Δ has local support. Since weights are defined by basic functions and the latter have local support. Moreover, weights are assigned only to two adjacent vertices, i.e. those connected with an edge. Since the formula of Laplacian takes into account just neighbor vertices, i.e. $A(u,v) = 0$ if $u \nsim v$, and the sum is taken just over vertices that are adjacent to the current vertex, function value at other vertices does not affect the value of Laplacian.
4. Weights are positive, which immediately follows from the definition of basic functions of fuzzy partition, $A(u,v) \geq 0$. Moreover, in a neighbourhood of current vertex u, there should always exist such a vertex v, $v \sim u$: $A(u,v) > 0$.
5. Operator $-\Delta$ is positive semi-definite

$$(-\Delta f, f) \geq 0.$$

The last property is important especially when it comes to the eigensystem of the Laplace operator because it ensures that the eigenvalues of the operator are real and eigenvectors orthogonal.

In the above a reader could notice that discrete nonlocal Laplace operator (6) can be seen as a kind of discretization of its continuous counterpart (1). As shown by the work of Belkin [19], it can be proven that under certain conditions, the discretization of the Laplace operator converges to its smooth version. In the case of FT-based Laplacian, this is an open issue, which we plan to investigate in our upcoming research.

5   Conclusion

This contribution reminds well-known definitions of the Laplace operator in connection to underlying space. Besides that, a discrete representation of the Laplace operator in a space with a fuzzy partition is proposed and analyzed. It is based on the theory of fuzzy transform, where the weight assignment is done with membership degrees of respective points into the fuzzy set, that forms a fuzzy partition. The values corresponding to membership degrees (=weights) nicely represent proximity between points, meaning the larger the weight value of the connection, the closer these points are. The resulting definition of the Laplace operator is then dependent on the difference between the F-transform component of the function and an actual function value at each point. To this end, we mentioned certain properties of this operator, where each of them is a known structural property of the standard Laplacian.

Acknowledgements. The authors thank the reviewers for their valuable comments and suggestions to improve the quality of the paper.

References

1. Fiedler, M.: Geometry of the Laplacian. Linear Algebra Appl. **403**(5), 409–413 (2005)
2. Styer, D.F.: The geometrical significance of the Laplacian. Am. J. Phys. **83**, 992–997 (2015)
3. Biyikoglu, T., Leydold, J., Stadler, P.F.: Laplacian Eigenvectors of Graphs, 1st edn. Springer, Heidelberg (2007). https://doi.org/10.1007/978-3-540-73510-6
4. Urakawa, H.: Spectral Geometry Of The Laplacian: Spectral Analysis And Differential Geometry Of The Laplacian, 2nd edn. World Scientific, New Jersey (2017)
5. Rosenberg, S.: The Laplacian on a Riemannian Manifold: An Introduction to Analysis on Manifolds, 2nd edn. Cambridge University Press, Cambridge (1997)
6. Gilboa, G., Osher, S.: Nonlocal operators with applications to image processing. Multiscale Model. Simul. **3**, 1005–1028 (2008)
7. Gilboa, G.: Nonlocal linear image regularization and supervised segmentation. Multiscale Model. Simul. **6**(2), 595–630 (2007)
8. Perfiljeva, I.: Fuzzy transforms: theory and applications. Fuzzy Sets Syst. **8**, 993–1023 (2006)
9. Perfiljeva, I., Daňková, M., Bede, B.: Towards a higher degree F-transform. Fuzzy Sets Syst. **180**, 3–19 (2011)
10. Perfiljeva, I., Vlašánek, P.: Total variation with nonlocal FT-Laplacian for Patch-based Inpainting. Soft Comput. **23**, 1833–1841 (2018)

11. Belkin, M., Niyogi, P.: Laplacian eigenmaps for dimensionality reduction and data representation. Neural Comput. **6**(15), 1373–1396 (2003)
12. Elmoataz, A., Lézoray, O., Bougleux, S.: Nonlocal discrete regularization on weighted graphs: a framework for image and manifold processing. IEEE Trans. Image Process. **17**, 1047–1060 (2008)
13. Lezoray, O., Ta, V.T., Elmoataz, A.: Nonlocal graph regularization for image colorization. In: 19th International Conference on Pattern Recognition, Tampa, FL, pp. 1–4 (2008)
14. Wardetzky, M., Mathur, S., Kälberer, F., Grinspun, E.: Discrete Laplace operators: no free lunch. In: Proceedings of the Fifth Eurographics Symposium on Geometry Processing, Goslar, DEU, pp. 33–37 (2007)
15. Belkin, M., Niyogi, P.: Laplacian eigenmaps and spectral techniques for embedding and clustering. In: Proceedings of the 14th International Conference on Neural Information Processing Systems: Natural and Synthetic, Vancouver, British Columbia, Canada, pp. 585–591 (2001)
16. Burago, D., Ivanov, S., Kurylev, J.: A graph discretization of the Laplace-Beltrami operator. J. Spectral Theory **4**, 675–714 (2014)
17. Sorkine, O.: Laplacian Mesh Processing. In: Eurographics 2005 - State of the Art Reports, The Eurographics Association, Dublin, pp. 53–70 (2005)
18. Styer, D.: The geometrical significance of the Laplacian. Am. J. Phys. **83**, 992–997 (2015)
19. Belkin, M., Niyogi, P.: Towards a theoretical foundation for Laplacian-based manifold methods. J. Comput. Syst. Sci. **74**(8), 1289–1308 (2008)
20. Giunti, B., Perri, V.: Dynamical systems on graphs through the signless Laplacian matrix. Ricerche di Matematica **67**(2), 533–547 (2017). https://doi.org/10.1007/s11587-017-0326-z
21. Shaham, U., Stanton, K. P., Li, H., Nadler B., Basri R., Kluger, Y.: SpectralNet: spectral clustering using deep neural networks. ArXiv (2018)
22. Vollmer, J., Mencl, R., Müller, H.: Improved Laplacian smoothing of noisy surface meshes. Comput. Graph. Forum **18**(3), 131–138 (1999)
23. Campos, D., Pérez-de la Rosa, M., Bory-Reyes, J.: Generalized Laplacian decomposition of vector fields on fractal surfaces. J. Math. Anal. Appl. **499**(2), 125038 (2021)
24. Perfiljeva, I., Adamczyk, D.: Representative keypoints of images: a new selection criterion. In: 14th International FLINS Conference 2020: Developments of Artificial Intelligence Technologies in Computation and Robotics 2020, Cologne, Germany, pp. 939–946. Singapore, World Scientific (2020). https://doi.org/10.1142/9789811223334_0113
25. Zámečníková, H., Perfilieva, I.: Nonlocal Laplace operator in a space with the fuzzy partition. In: Lesot, M.J., et al. (eds.) IPMU 2020. CCIS, vol. 1239, pp. 295–303. Springer, Cham (2020). https://doi.org/10.1007/978-3-030-50153-2_22
26. Perfiljeva, I., Zámečníková, H., Valášek, R.: Nonlocal Laplace operator in image processing. In: The 14th International FLINS Conference on Robotics and Artificial Intelligence: Developments of Artificial Intelligence Technologies in Computation and Robotics 2020, vol. 12, pp. 956–963. World Scientific, Cologne (2020). https://doi.org/10.1142/9789811223334_0115

Noise Reduction as an Inverse Problem in F-Transform Modelling

Jiří Janeček[1](✉) and Irina Perfilieva[2]

[1] Department of Mathematics, Faculty of Science, University of Ostrava,
30. dubna 22, 701 03 Ostrava, Czech Republic
Jiri.Janecek@osu.cz

[2] Institute for Research and Applications of Fuzzy Modeling,
University of Ostrava, 30. dubna 22, 701 03 Ostrava, Czech Republic
Irina.Perfilieva@osu.cz
http://prf.osu.eu/kma, http://ifm.osu.eu

Abstract. In this paper, we discuss a special type of fuzzy partitioned space generated by a fuzzy set that is used to enrich the data domain with a notion of closeness. We utilize this notion to sketch the solution to the denoising problem in the discrete, now only 1-D setting, where the Nyquist-Shannon-Kotelnikov sampling theorem in not applicable. The finite-dimensional space with closeness is described by a closeness matrix that transforms discrete one-dimensional signals (considered as functions defined on the space and identified with high-dimensional vectors) into a lower-dimensional vectors. On the basis of this and the corresponding pseudo-inverse transformation, we characterize the signal denoising problem as a type of inverse problem. This opens a new perspective on discrete data processing involving algebraic tools and singular value matrix decomposition. As there are many degrees of freedom in initializing parameters of the chosen model, we restrict ourselves on some special cases. The link between the generating function of the fuzzy partition and a fundamental subspace of the closeness matrix is expressed in terms of Euclidean orthogonality. The theoretical background as well as solutions in particular settings are illustrated by numerical examples.

Keywords: Closeness · Fuzzy partition · Denoising

Introduction

This paper was motivated by our previous research on spaces with closeness [2–4] as well as by the paper [5]. In the latter, the authors considered the task of reconstruction of a continuous function from its F-transform components. In analogy with the Nyquist-Shannon-Kotelnikov sampling theorem (NSK), they proposed a condition on (i) how densely the noisy signal must be sampled in order to reconstruct it, and (ii) how Gaussian noise can be removed by the reconstruction. This condition depends on two major parameters that will also

© Springer Nature Switzerland AG 2022
D. Ciucci et al. (Eds.): IPMU 2022, CCIS 1602, pp. 405–417, 2022.
https://doi.org/10.1007/978-3-031-08974-9_32

play an important role in this paper – the width of a generating function and the distance between equidistant nodes of the fuzzy partition that underlies the F-transform.

In this contribution, we would like to show that a similar result can be obtained in a space with a weaker structure than that of a Hilbert space. Therefore, we utilize the notion of closeness to sketch the solution to the denoising problem in the purely discrete, yet 1-D setting where NSK in inapplicable. We believe that this approach can be further used in image processing (in reconstruction and denoising of images that are 2-D discrete signals). Closeness space is weaker than a metric space and more suitable for data having the so called graph structure, i.e. for data (images, etc.) characterized by its weighted neighbourhoods.

In Sect. 1 we introduce all necessary concepts and give general definitions. In Sect. 2, we describe the initial setting in which the subsequent problems arise with the emphasis on properties of the data domain and the relation between the concepts from the previous section. In Sect. 3, we recall the technique of singular value decomposition and apply it to the closeness matrix. Section 4 characterizes the link between the generating function of the fuzzy partition and a fundamental subspace of the closeness matrix in terms of Euclidean orthogonality. Section 5 relates denoising of a 1-D signal with a fundamental subspace of the closeness matrix and by that, characterizes an inverse problem. We indicate a partial solution to this problem and illustrate it in two numerical examples. Section 6 follows our assumption on the noise and signal separability and indicates that procedure proposed in Sect. 5 can be used to decompose the space. Section 7 suggests how to extend the applicability of the procedures above to the 2-D case.

1   Preliminaries

In this section, we give basic definitions from relevant areas (linear algebra, fuzzy set theory, notion of closeness, signal and additive noise).

Throughout the manuscript, we will stick to the convention that vectors are column vectors (i.e. matrices of the order $m \times 1$) while matrices of the order $1 \times n$ are row vectors that are always denoted by the transposition symbol T. Moreover, $A_j = A(\cdot, j)$ will denote the j-th column of a matrix A (column vector) while $A^i = A(i, \cdot)$ will denote the i-th row of a matrix A (row vector). To denote a subspace of a vector space, we will use the double inclusion symbol $\subseteq\subseteq$.

We will use the standard notions of three (out of four) fundamental subspaces defined in matrix calculus. Their list is given below.

Definition 1 (Matrix Range). *Let $A \in \mathbb{R}^{m \times n}$ be a matrix, then its* range *(or right column space, commonly referred to as just a* column space*) is a linear subspace of $\mathbb{R}^m$ spanned by all columns of A. We will denote it by* range A.

$$\text{range } A = \left\{ \sum_{j=1}^{n} c_j A_j \,\middle|\, c_j \in \mathbb{R} \right\} \subseteq\subseteq \mathbb{R}^m$$

Definition 2 (Matrix Row Space). *Let $A \in \mathbb{R}^{m \times n}$ be a matrix, then its* row *space (or left column space) is a range of its transpose, $A^\top$, i.e. it is linear subspace of $\mathbb{R}^n$ spanned by all transposed rows of A. We will denote it by* row A.

$$\text{row } A = \text{range } A^\top = \left\{ \sum_{j=1}^{m} c_j \left(A^\top \right)_j \,\middle|\, c_j \in \mathbb{R} \right\} = \left\{ \sum_{j=1}^{m} c_j A^{j\top} \,\middle|\, c_j \in \mathbb{R} \right\} \subseteq \subseteq \mathbb{R}^n$$

Definition 3 (Matrix Null Space). *Let $A \in \mathbb{R}^{m \times n}$ be a matrix, then its* right null space *(or kernel), commonly referred to as just a* null space*) is a linear subspace of $\mathbb{R}^n$ containing all vectors from $\mathbb{R}^n$ that A maps to the zero vector of $\mathbb{R}^m$. We will denote it by* null A.

$$\text{null } A = \{ x \in \mathbb{R}^n \mid Ax = 0 \} \subseteq \subseteq \mathbb{R}^n$$

Definition 4 (Fuzzy set). *Let U be a set, then any function $K \colon U \to [0,1]$ is a* fuzzy set*. We say that K is a* fuzzy subset *of U which is denoted by $K \subsetneq U$. The domain U is called the* universe*. We say that the point $x \in U$ is* covered *by the fuzzy set K if $K(x) > 0$, i.e. if the functional value (called* membership degree*) is positive. The set of all points covered by K is called* support *of K and denoted by* Supp K.

$$\forall K \subsetneq U \colon \text{ Supp } K = \{ x \in U \mid K(x) > 0 \}$$

As any fuzzy set is a function, it can be described either using a formal expression or by writing the corresponding pairs of samples.

We will denote by U^2 the Cartesian product of a set U with itself, i.e. $U^2 = U \times U = \{ (x,y) \mid x, y \in U \}$. Below, we describe closeness in case of finite sets.

Definition 5 (Closeness). *Let $U = \{ x_i \mid i = 1, \ldots, n \}$ be a set, $n \in \mathbb{N}$, then* closeness *on U is any symmetric, non-negative function $w \colon U^2 \to \mathbb{R}$.*

Definition 6 (Space with Closeness). *Let U be a finite set and w be closeness on U, then the pair (U, w) is called a* space with closeness *or simply* closeness space.

Definition 7 (Closeness Matrix). *Let $n \in \mathbb{N}$, $U = \{ x_i \mid i = 1, \ldots, n \}$ be a set and $w \colon U^2 \to \mathbb{R}$ be closeness on U, then a matrix $W \in \mathbb{R}^{n \times n}$ given by*

$$\forall i, j = 1, \ldots, n \colon W(i,j) = w_{ij} = W_j[i] = W^i[j] \;\; = \;\; w(x_i, x_j),$$

is called closeness matrix.

The preceding definition states that every finite space with closeness can be uniquely characterized using a closeness matrix (as this matrix stores all values of closeness) and vice versa: every closeness matrix (a symmetric matrix with non-negative entries) associated with a fixed set determines a unique closeness space.

Definition 8 (Generating Function). *Let $A_0 \subsetneq \mathbb{Z}$ be a fuzzy set with a finite support of the form*

$$\operatorname{Supp} A_0 = [-h, h] \cap \mathbb{Z} = \{-h, -h+1, \ldots, -1, 0, 1, \ldots, h\} \quad \text{for some} \quad h \in \mathbb{N}.$$

We call it a generating function (or kernel) *of width $2h$.*

Definition 9 (Generated Fuzzy Partition with Nodes). *Let $X = \{x_j \mid j = 1, \ldots, L\} \subset \mathbb{R}$ be a finite set and Y be its non-empty subset, s.t. $0 < |Y| = l \leq L = |X| < +\infty$ and $x_1 \leq x_2 \leq \ldots \leq x_L$. Let $I_Y \subseteq \{1, \ldots, L\}$ denote the indices of points in Y, so that $Y = \{x_j \in X \mid j \in I_Y\}$, and let $A_0 \subsetneq \mathbb{Z}$ be a generating function of width $2h$. Then* fuzzy partition of the universe X with nodes Y generated by A_0 *is given by the set of fuzzy sets (called* basic functions*) $\{A_t \mid t \in I_Y\}$, s.t.*

1. $\forall t \in I_Y \colon A_t \colon X \to [0, 1]$,
2. $\forall t \in I_Y \ \forall x_j \in X \colon A_t(x_j) = A_0(j - t)$,
3. $\displaystyle\bigcup_{t \in I_Y} \operatorname{Supp} A_t = X$.

For each $t \in I_Y$, A_t is basic function *associated with the node x_t.*

Following the definition above, we see that for the generating function A_0, it holds that $A_0(k) > 0 \Leftrightarrow |k| \leq h$ and that every node has the same membership degree in the associated, generated basic function, i.e. for every node index $t \in I_Y$ it holds that $A_t(x_t) = A_0(0)$ is a positive constant. Condition 3. means that

$$\bigcup_{t \in I_Y} \{x \in X \mid A_t(x) > 0\} = X,$$

i.e. that every point of the universe X is covered by at least one basic function. This condition determines the minimal value of the half-width parameter h, i.e.

$$h \geq h_{\min} \quad \text{for} \quad h_{\min} = \min\left\{ h \in \mathbb{N} \ \middle| \ \bigcup_{t \in I_Y} [t - h, t + h] \cap \mathbb{Z} \supseteq \{1, \ldots, L\} \right\}. \quad (1)$$

Definition 10 (Noise w.r.t. Generated Fuzzy Partition). *Let a set $X = \{x_j \mid j = 1, \ldots, L\} \subset \mathbb{R}$ be finite, s.t. $x_1 \leq x_2 \leq \ldots \leq x_L$, Y be its non-empty subset indexed by I_Y and $A_0 \subsetneq \mathbb{Z}$ be a generating function of width $2h$, then $v \colon X \to \mathbb{R}$ is an* additive *noise w.r.t. fuzzy partition generated by A_0 if it fulfills $\forall t \in I_Y \colon \sum_{j \in J(t)} A_0(j - t)v(x_j) = 0$ where $J(t) = \{j = -h + t, \ldots, h + t \mid x_j \in X\}$. We say that $u \colon X \to \mathbb{R}$ is a* noisy signal *on the universe X if it can be represented as $u = u' + v$ where $u' \colon X \to \mathbb{R}$ and $v \neq 0$ is a noise as defined above.*

Signal and noise describing functions u and v are identified with vectors in $\mathbb{R}^L$: $\forall j = 1, \ldots, L \colon u_j = u(x_j)$, $v_j = v(x_j)$.

We assume that the noise can be separated from the clear signals in the following way: the space $\mathbb{R}^L$ of all functions on the 1-D universe (noise mixed with signals) can be decomposed into two subsets $\langle N \rangle$ and $\langle S \rangle$, s.t. $\mathbb{R}^L = \langle N \cup S \rangle$

and $N \cap S = \emptyset$, in the other words, $\mathbb{R}^L = \langle N \rangle \oplus \langle S \rangle$ is their direct sum. The linear spans of sets of vectors N and S, $\langle N \rangle$ and $\langle S \rangle$, denote the subspace of noise and the subspace of clear signals, respectively.

Above, we defined noise w.r.t. fuzzy partition and below, we work with noise w.r.t. closeness. These two notions will be naturally unified in Sect. 2.

Definition 11 (Denoising). *The closeness matrix W denoises the noisy signal $\overline{u} = u + v$ with noise v iff $W(u + v) = Wu$.*

2 Assumptions on the Data

Let us have a finite set of points on $\mathbb{R}$:

$$X = \{x_j \,|\, j = 1, \ldots, L\}, \tag{2}$$

its non-empty subset

$$Y = \{x_j \in X \,|\, j \in I_Y\} \subseteq X, \quad 0 < |Y| = |I_Y| = l \leq L = |X| < +\infty,$$

and let us have a generating function $A_0 \subsetneq \mathbb{Z}$ of a sufficient width satisfying the Eq. (1). Following the Definition 9, the universe X with nodes Y has a structure of a fuzzy-partitioned space where its fuzzy partition is generated by A_0.

One of the most distinguishable features of this paper is the assumption that instead of working with the closeness on the whole universe X, given by a function on X^2, we work only with its restriction on $Y \times X$ – the values of closeness on $(X \setminus Y) \times X$ are undefined. Let

$$w(x_t, x_j) = A_t(x_j), \quad \text{where } t \in I_Y, x_j \in X,$$

describe the closeness of the henceforth specified closeness space. Its values are stored in the closeness matrix $W \in \mathbb{R}^{l \times L}$. It means that by fixing values of the generating function A_0 at the beginning, we set all basic functions (and their particular values) and we simultaneously set values of the closeness-describing function $w \colon Y \times X \to \mathbb{R}$. Hence, it holds:

$$\forall t \in I_Y, j \in \{1, \ldots, L\}\colon W(t, j) = w(x_t, x_j) = A_t(x_j) = A_0(j - t). \tag{3}$$

Note that all l rows of the closeness matrix W are indexed by I_Y, i.e. w.r.t. original indices in the universe X and not from 1 to l. This simplifying convention is respected throughout the manuscript as well as the notation of the universe X, it subset of nodes Y, their cardinalities (L and l, respectively) and index sets ($\{1, \ldots, L\}$ and I_Y, respectively) and $W \in \mathbb{R}^{l \times L}$ will below always denote a closeness matrix.

3   Singular Value Decomposition of the Closeness Matrix

Assume that we found a singular value decomposition (commonly abbreviated by SVD, formalized e.g. in [1]) of the closeness matrix in the form $W = PSZ^\top$, where the matrices $P \in \mathbb{R}^{l \times l}$ and $Z \in \mathbb{R}^{L \times L}$ are orthogonal and $S \in \mathbb{R}^{l \times L}$ is diagonal ($\forall i, j\colon i \neq j \Rightarrow s_{ij} = 0$) with singular values $\forall i = 1, \ldots, l\colon \sigma_i = s_{ii}$ on its diagonal, s.t. $\sigma_1 \geq \sigma_2 \geq \cdots \geq \sigma_k > \sigma_{k+1} = \cdots = \sigma_l = 0$. For $0 < k = \operatorname{rank} W \leq l$, we have

$$\operatorname{range} W = \left\{ \sum_{j=1}^{L} c_j W_j \,\middle|\, c_j \in \mathbb{R} \right\} = \operatorname{range} PSZ^\top = \left\{ \sum_{i=1}^{k} c_i P_i \,\middle|\, c_i \in \mathbb{R} \right\} \subseteq \mathbb{R}^l,$$

i.e. the range of W is spanned by the left singular vectors corresponding to its positive singular values. Of course, if $l = k$, $\operatorname{range} W = \mathbb{R}^l$. As W always contains a non-zero element, $k \neq 0$. Next, null space of W is spanned by right singular vectors corresponding to zero singular values:

$$\operatorname{null} W = \left\{ v \in \mathbb{R}^L \mid Wv = 0 \right\} = \left\{ \sum_{i=k+1}^{L} c_i Z_i \,\middle|\, c_i \in \mathbb{R} \right\} \subseteq \mathbb{R}^L.$$

And lastly, the row space of W is spanned by right singular vectors of W corresponding to positive singular values:

$$\operatorname{row} W = \operatorname{range} W^\top = \left\{ \sum_{t \in I_Y} c_t \left(W^\top \right)_t \,\middle|\, c_t \in \mathbb{R} \right\} = \left\{ \sum_{t \in I_Y} c_t {W^t}^\top \,\middle|\, c_t \in \mathbb{R} \right\}$$

$$= \operatorname{row} PSZ^\top = \operatorname{range} ZS^\top P^\top = \left\{ \sum_{i=1}^{k} c_i Z_i \,\middle|\, c_i \in \mathbb{R} \right\} \subseteq \mathbb{R}^L.$$

Lemma 1. *The null space and the row space of the closeness matrix $W \in \mathbb{R}^{l \times L}$ are orthogonal complements w.r.t. the standard inner product on $\mathbb{R}^L$, $\langle x, y \rangle = x^\top y$.*

Proof. The matrix Z is orthogonal.   □

This orthogonality holds true in general for any matrix in $\mathbb{R}^{\cdot \times L}$.

4   Orthogonal Characterization

Let us use a compact notation for vectors with components as index functions, e.g.

$$[A_0(j - t)]_{j=1}^{L} = \begin{bmatrix} A_0(1 - t) \\ \vdots \\ A_0(L - t) \end{bmatrix}.$$

Theorem 1. *The space spanned by a set of vectors given by a generating function A_0 in the form*

$$\left\{ [A_0(j-t)]_{j=1}^{L} \,\middle|\, t \in I_Y \right\}, \tag{4}$$

is the orthogonal complement to null W *w.r.t. the standard inner product on* $\mathbb{R}^L$ *where the closeness matrix* $W \in \mathbb{R}^{l \times L}$ *is determined by A_0 in accordance with the Eq. (3).*

Proof. Using the same SVD as above, $W = PSZ^\top$, let us first express the t-indexed row of the closeness matrix W as $W^t = P^t SZ^\top$. Then, we can express its arbitrary element by

$$w_{tj} = W^t[j] = P^t \left(SZ^\top\right)_j = P^t \left[\sigma_i z_{ji}\right]_{i=1}^{l} = \sum_{i=1}^{l} p_{ti}\sigma_i z_{ji} = P^t S \left(Z^\top\right)_j.$$

Following Lemma 1 and recalling (3), we can express the orthogonality of the null space in the following form:

$$(\text{null } W)^{\perp} = \text{row } W = \left\{ \sum_{t \in I_Y} c_t W^{t\top} \,\middle|\, c_t \in \mathbb{R} \right\} = \left\{ \sum_{t \in I_Y} c_t \left(P^t SZ^\top\right)^\top \,\middle|\, c_t \in \mathbb{R} \right\}$$

$$= \left\{ \sum_{t \in I_Y} c_t \left[P^t S \left(Z^\top\right)_j\right]_{j=1}^{L} \,\middle|\, c_t \in \mathbb{R} \right\} = \left\{ \sum_{t \in I_Y} c_t [w_{tj}]_{j=1}^{L} \,\middle|\, c_t \in \mathbb{R} \right\}$$

$$= \left\{ \sum_{t \in I_Y} c_t [A_t(x_j)]_{j=1}^{L} \,\middle|\, c_t \in \mathbb{R} \right\} = \left\{ \sum_{t \in I_Y} c_t [A_0(j-t)]_{j=1}^{L} \,\middle|\, c_t \in \mathbb{R} \right\}$$

$$= \left\langle \left\{ [A_0(j-t)]_{j=1}^{L} \,\middle|\, t \in I_Y \right\} \right\rangle.$$

$\square$

Lemma 2. $\forall y, v' \in \mathbb{R}^L : Wy = Wv' \Leftrightarrow \forall x \subset \left\{ [A_0(j-t)]_{j=1}^{L} \,\middle|\, t \in I_Y \right\} : x^\top y = x^\top v'$.

Proof.

$$\forall v' \in \mathbb{R}^L \, \forall v \in \text{null } W \, \forall x \in \text{row } W : x^\top (v' + v) = x^\top v',$$

which implies that

$$\forall y, v' \in \mathbb{R}^L \, \forall x \in \text{row } W : x^\top y = x^\top v' \Leftrightarrow x^\top (v' - y) = 0$$
$$\Leftrightarrow v' - y \in \text{null } W \Leftrightarrow W(v' - y) = 0 = Wv' - Wy,$$

which proves that $Wy = Wv'$ iff y and v' have the same inner product value with any element of the row space of W which is spanned by the set (4). $\square$

5 Inverse Denoising Problem

Inverse problem generally consists in finding parameters of a given model that agrees with the data (either just with the observed outputs, or with both inputs and outputs). We work with a model introduced in Sect. 2. We model closeness within data domain X given by the Eq. (2) and indexed in an order[1] $x_1 \leq \ldots \leq x_L$, using a space with closeness (X, w) that is fully characterized by the closeness matrix W. This means that the inverse problem consist in finding the values of all entries of this matrix. This is equivalent with finding the values of a generating function A_0. At the beginning, we choose its width $2h$ and node indices I_Y.

Consider a one-dimensional signal (e.g. a scalar-valued time series[2]) u that spreads over the space X (which represents e.g. time points) and that describes a certain phenomenon. Then noise is a type of distortion arising between the source of this phenomenon and the observer (or recorder of a certain quality produced by the phenomenon) that measures the noisy signal. Assume that the additive noise $v\colon X \to \mathbb{R}$ is represented by a linear combination of vectors from the linearly independent set $N = \{n_1, \ldots\}$, i.e. $v \in \langle N \rangle$, and we know one of its elements, say n_1. Therefore, the noise can be characterized only partially[3].

Our aim is to analyze the parameters of this model (all entries of W) in order to remove this noise, i.e. we would like to endow the given data with such closeness that the signal is denoised[4] (hence the name of this section).

The full inverse problem would be given by a set of b's for which we would want to analyze W and a corresponding set of u's, s.t. the noises v's are removed by W that models the signal transformation. In this contribution, however, we focus only on the denoising part (noise removal). Important parameters of W are foremostly the width of the generating function and the node distribution within the universe – this fixes the number of rows and the number of non-zero entries of W leading to a particular number of values of A_0 to be computed.

We propose to assume that the noise forms a part of the null space of the closeness matrix because by that and following the previous section, we get $v \in \langle N \rangle$ & $\langle N \rangle \subseteq \text{null}\, W \Rightarrow v \in \text{null}\, W$ and $W(u + v) = Wu + Wv = b$ & $Wv = 0 \Rightarrow Wu = b$. Hence the original signal u can be reconstructed[5] by solving $Wu = b$. Therefore, we must firstly create W.

---

[1] X can be naturally connected with a topology that admits its homeomorphic embedding into $\mathbb{R}$.

[2] Such as list of registered values of light frequencies produced by an ideal light source that creates only one frequency at a time.

[3] A particular occurrence of this phenomenon can be considered on a cassette player. The signal is stored in the tape but the device produces sound also based on the volume level set by a continuous rotation of volume button (dial knob that does not switch only between a few possible, distinguishable levels). We measure the produced signal but we would like to ignore the unknown volume setting – we consider it as a constant noise in the measured signal represented by a constant vector n_1.

[4] In the sense of the Definition 11.

[5] Under sufficient conditions, reconstruction of u is equivalent with denoising $u + v$.

We know that $n_1 \in N \subseteq \text{null}\,W$. The set (4) spans $(\text{null}\,W)^\perp$. As $n_1 \in \text{null}\,W$, we have $Wn_1 = 0$ which means, recalling the Eq. (3), that

$$\forall t \in I_Y : W^t n_1 = \sum_{j=1}^{L} w_{tj} n_1[j] = \sum_{j=1}^{L} A_0(j-t) n_1[j] = 0.$$

In case of too many unknowns, we propose to choose other element(s) of N and based on that, to compute W and A_0. To conclude, we solve the inverse problem given noise and we propose a procedure of its removal w.r.t. A_0.

5.1 Special Case $h = 1$

Assume that L is odd, $I_Y = \{2, 4, \ldots, L-1\}$, $l = \frac{L-1}{2}$ and $h = 1$, i.e. the set of node indices is equidistant ($\forall i = 2, \ldots, l-1 : t_{i+1} = t_i + 2$) end every point between two neighbouring nodes is covered by the two basic functions corresponding to these nodes.

If we are given a noise element $n_1 \in \mathbb{R}^L$, then the inverse denoising problem has the form $W(u+v) = Wu$ where W is to be determined solving $Wn_1 = 0$. Based on the initial setting for node indices I_Y and generating function width $2h$ we get

$$Wn_1 = \begin{bmatrix} A_0(-1) & A_0(0) & A_0(1) & 0 & 0 & 0 & \cdots & 0 \\ 0 & 0 & A_0(-1) & A_0(0) & A_0(1) & 0 & \cdots & 0 \\ \vdots & & & \ddots & \ddots & \ddots & & \vdots \\ \vdots & & & & \ddots & \ddots & \ddots & 0 \\ 0 & \cdots & & & 0 & A_0(-1) & A_0(0) & A_0(1) \end{bmatrix} \begin{bmatrix} n_1[1] \\ n_1[2] \\ \vdots \\ \vdots \\ n_1[L] \end{bmatrix} = 0.$$

We have l equations of 3 unknowns. Adding more vectors would yield to even more equations. We can see that for this initial setting, one vector in N and $l = 3$ nodes make this case the best applicable to only universes consisting of $L = 7$ points. This is illustrated by the following example.

Example 1. Let $L = 7$, $I_Y = \{2, 4, 6\}$, $l = 3$ and $h = 1$. We assume an oscillating noise (w.l.o.g. positive on nodes and negative otherwise, a type of 1-D salt-and-pepper) that admits all unknown values to be positive:

$$n_1 = \begin{bmatrix} -1 & 1 & -1 & 1 & -1 & 1 & -1 \end{bmatrix}^\top .$$

We need to find W generated by A_0, s.t. $Wn_1 = 0$, and because this system is homogeneous, the solution A_0 can be arbitrarily scaled, so we assume that $A_0(0) = 1$. Hence we solve

$$Wn_1 = \begin{bmatrix} A_0(-1) & 1 & A_0(1) & 0 & 0 & 0 & 0 \\ 0 & 0 & A_0(-1) & 1 & A_0(1) & 0 & 0 \\ 0 & 0 & 0 & 0 & A_0(-1) & 1 & A_0(1) \end{bmatrix} \begin{bmatrix} -1 \\ 1 \\ -1 \\ 1 \\ -1 \\ 1 \\ -1 \end{bmatrix} = \begin{bmatrix} 0 \\ 0 \\ 0 \end{bmatrix} .$$

This system has infinitely many solutions satisfying $A_0(-1) + A_0(1) = 1$ which, in order to create a generating function and closeness, must be restricted by $0 \le A_0(-1), A_0(1) \le 1$. A symmetric solution is given by $A_0(-1) = A_0(1) = \frac{1}{2}$.

To demonstrate the denoising property of W, recall that n_1 is a noise, and hence is any of its multiples, e.g. $3n_1$, so we have $W(3n_1) = 0$. Conversely, consider a constant signal $u = \begin{bmatrix} 1\,1\,1\,1\,1\,1\,1 \end{bmatrix}^T$ which is aggregated by W to

$$Wu = \begin{bmatrix} \frac{1}{2} & 1 & \frac{1}{2} & 0 & 0 & 0 & 0 \\ 0 & 0 & \frac{1}{2} & 1 & \frac{1}{2} & 0 & 0 \\ 0 & 0 & 0 & 0 & \frac{1}{2} & 1 & \frac{1}{2} \end{bmatrix} \begin{bmatrix} 1 \\ 1 \\ 1 \\ 1 \\ 1 \\ 1 \\ 1 \end{bmatrix} = \begin{bmatrix} 2 \\ 2 \\ 2 \end{bmatrix}.$$

If we measured a noisy value $u + 3n_1 = \begin{bmatrix} -2\,4\,-2\,4\,-2\,4\,-2 \end{bmatrix}^T$, we could use the newly created closeness matrix W to aggregate it to

$$W(u + 3n_1) = \begin{bmatrix} \frac{1}{2} & 1 & \frac{1}{2} & 0 & 0 & 0 & 0 \\ 0 & 0 & \frac{1}{2} & 1 & \frac{1}{2} & 0 & 0 \\ 0 & 0 & 0 & 0 & \frac{1}{2} & 1 & \frac{1}{2} \end{bmatrix} \begin{bmatrix} -2 \\ 4 \\ -2 \\ 4 \\ -2 \\ 4 \\ -2 \end{bmatrix} = \begin{bmatrix} 2 \\ 2 \\ 2 \end{bmatrix}.$$

As we can see, the impact of the noise is completely obliterated.

5.2   Special Case $h = 2$

In this subsection, we demonstrate the correctness of this theory on a bigger example where another element of N is added. Assume that $L \equiv 1 \pmod 4$, $I_Y = \{3, 7, \ldots, L - 2\}$, $l = \frac{L-1}{4}$ and $h = 2$. Consider again an equidistant set of node indices ($\forall i = 2, \ldots, l-1\colon t_{i+1} = t_i + 4$) where only the middle point between two neighbouring nodes is covered by both corresponding basic functions.

Example 2. Let $L = 9$, $I_Y = \{3, 7\}$, $l = 2$ and $h = 2$. We assume a known noise element: $n_1 = \begin{bmatrix} 1\,1\,-3\,2\,3\,3\,0\,-1\,-1 \end{bmatrix}^T$. We need to find W generated by A_0, s.t. $Wn_1 = 0$, and we assume that $A_0(0) = 1$, hence we solve $Wn_1 = 0$, i.e.

$$\begin{bmatrix} A_0(-2) & A_0(-1) & 1 & A_0(1) & A_0(2) & 0 & 0 & 0 & 0 \\ 0 & 0 & 0 & 0 & A_0(-2) & A_0(-1) & 1 & A_0(1) & A_0(2) \end{bmatrix} \begin{bmatrix} 1 \\ 1 \\ -3 \\ 2 \\ 3 \\ 3 \\ 0 \\ -1 \\ -1 \end{bmatrix} = \begin{bmatrix} 0 \\ 0 \end{bmatrix}.$$

This system of 2 equations with 4 unknowns has infinitely many solutions satisfying $A_0(1) = 10A_0(-2) + 10A_0(-1) - 3$ and $A_0(2) = -7A_0(-2) - 7A_0(-1) + 3$. Let us choose another noise element: $n_2 = \begin{bmatrix} 1 & 4 & 0 & -1 & -1 & -1 & 0 & -1 & 1 \end{bmatrix}^T$. Hence we solve $W[n_1, n_2] = 0$, i.e.

$$\begin{bmatrix} A_0(-2) & A_0(-1) & 1 & A_0(1) & A_0(2) & 0 & 0 & 0 & 0 \\ 0 & 0 & 0 & 0 & A_0(-2) & A_0(-1) & 1 & A_0(1) & A_0(2) \end{bmatrix} \begin{bmatrix} 1 & 1 \\ 1 & 4 \\ -3 & 0 \\ 2 & -1 \\ 3 & -1 \\ 3 & -1 \\ 0 & 0 \\ -1 & -1 \\ -1 & 1 \end{bmatrix} = \begin{bmatrix} 0 & 0 \\ 0 & 0 \end{bmatrix}.$$

This system of 4 equations with 4 unknowns has a unique solution: $A_0(-2) = \frac{1}{9}$, $A_0(-1) = \frac{2}{9}$, $A_0(1) = \frac{1}{3}$, $A_0(2) = \frac{2}{3}$. Denoising properties of the closeness matrix generated by this newly found generating function can be demonstrated in the same way as above, i.e. $\forall a, b \in \mathbb{R} \colon W(u + an_1 + bn_2) = Wu$.

6   Inverse Signal and Noise Separation Problem

Consider again a finite universe (domain of a given data set) X over which a one-dimensional signal u spreads. We would like to endow X with a notion of closeness described by a closeness matrix $W \in \mathbb{R}^{l \times L}$ generated by a generating function $A_0 \subsetneq \mathbb{Z}$ as above.

Now, consider an orthogonal matrix $Z \in \mathbb{R}^{L \times L}$. Its columns

$$Z' = \{Z_i \mid i = 1, \dots, L\},$$

form a basis of $\mathbb{R}^L$, the space of all noisy signals that can be registered[6] on X. Our aim is to decompose Z' into two subsets N and S ($Z' = N \cup S$, $N \cap S = \emptyset$) that demarcate the subspace of noise $\langle N \rangle$ and the subspace of clear signals $\langle S \rangle$, respectively.

Let $k = \operatorname{rank} W$ denote the rank of the closeness matrix that is to be found. Following the fact that any element of $\operatorname{null} W$ is mapped on zero, we utilize this denoising property by requiring that $\langle N \rangle = \operatorname{null} W$. Then $|N| = L - k$ and $|S| = k$. Given N, we would like to create W satisfying the above.

Lemma 3. *Let A_0 be a generating function of a width $2h > 0$, $L \equiv 1 \pmod{2h}$ and $I_Y = \{h + 1, 3h + 1, 5h + 1, \dots, L - h\} = \{t_1, \dots, t_l\} \subset \{1, \dots, L\}$, $|I_Y| = l = \frac{L-1}{2h}$, be an equidistant set of node indices, s.t. only the middle point between two neighbouring nodes is covered by the two corresponding basic functions, i.e. $\forall i = 2, \dots, l-1 \colon t_{i+1} = t_i + 2h$. Let W be generated by A_0, then $k = \operatorname{rank} W = l$.*

---

[6] The points of the universe are indexed in the same way the components of these vectors are.

Proof. Each node-indexed column of W contains exactly one positive entry that is located on the coordinate with the same index and hence no row can be obtained as a linear combination of the others, i.e. all rows of W are linearly independent.   □

Under the assumptions of Lemma 3, $|N| = L - l$ and $|S| = l$. Moreover, $\langle S \rangle = \text{row } W$. Therefore, to decompose Z', we only need to determine W. To find the membership degrees of A_0, we use the proposition of the preceding section – assuming that we know a noise element, we can compute the degrees directly, or if there are too many unknowns, we choose another, linearly independent noise element to compute the degrees within the interval $[0, 1]$.

Example 2 illustrates the fact that for determining A_0 and W we needed only two vectors from the noise subspace (and hence from the closeness matrix null space) that is seven-dimensional subspace of $\mathbb{R}^9$ in which all signals (noise mixed with clear signal) live and hence the complete description of N is not always necessary. The two-dimensional subspace, $\text{row } W$, should coincide with $\langle S \rangle$ and all others elements of Z' should belong to N where only n_1 and n_2 were used to find it. These vectors (in the particular Example 2) are not orthonormal but the procedure and subsequent reasoning would be the same if $n_1, n_2 \in Z'$.

7   Application to Image Processing

In this section, we sketch a possible extension of the propositions above to the 2-D case, i.e. an application in image processing.

Consider that gray-scale image consisting of $M \times Q$ pixels is a function $u \colon M_1 \times Q_1 \to \mathbb{R}$ where $M_1 = \{1, \ldots, M\}$ and $Q_1 = \{1, \ldots, Q\}$. By stacking all its columns into a single vector, we can represent it in the 1-D style as a vector $u \in \mathbb{R}^{M \cdot Q}$.

In order to keep the 2-D connectivity, we first endow the set $M_1 \times Q_1$ by a 2-D–styled closeness that will be later transformed in the same manner as the image was transformed into a vector to enable the closeness matrix to emerge.

The 2-D–styled generating function can be defined as

Definition 12 (2-D Generating Function). *Let $A_0 \subsetneq \mathbb{Z}^2$ be a fuzzy set with a finite support of the form*

$$\text{Supp } A_0 = [-h, h]^2 \cap \mathbb{Z}^2$$
$$= \{(-h, h), (-h, h-1), \ldots, (-h, -h), (-h+1, h), \ldots, (-h+1, -h), \ldots,$$
$$(-1, h), \ldots, (-1, -h), (0, h), \ldots, (0, -h), \ldots, (h, -h)\},$$

for some $h \in \mathbb{N}$. We call it a 2-D *generating function of width* $2h$.

Following the procedure above, we create a generated fuzzy partition, s.t.

$$\forall (t_1, t_2) \in I_Y \; \forall (x_{j_1}, y_{j_2}) \in X \colon A_{(t_1, t_2)}(x_{j_1}, y_{j_2}) = A_0(j_1 - t_1, j_2 - t_2),$$

where the nodes are indexed by $I_Y \subset M_1 \times Q_1 = X$, $0 < |I_Y| = l \leq L = M \cdot Q < +\infty$. Each node $(t_1, t_2) \in I_Y$ is associated with a basic function $A_{(t_1,t_2)} \colon X \to [0,1]$.

Then, the closeness matrix $W \in \mathbb{R}^{l \times L}$ is created, s.t. each of its rows corresponds to one node (nodes are ordered lexicographically in I_Y which agrees with stacking columns of the image into a vector, the transformation that the universe X undergoes). After that, the denoising or noise separation procedure following the corresponding 2-D assumptions about the noise, is the same. How the obtained results can be transformed back from the 1-D case (created by reshaping an image into a vector) to the original 2-D case is subject to the further research.

Conclusions

We introduced a space with closeness and showed its connection with fuzzy partition with nodes generated by a fuzzy set. W.r.t. that, we characterized a type of noise using algebraic properties (stemming from SVD) of fundamental subspaces of the closeness matrix. We proposed how to remove this type of noise from 1-D discrete signals as an inverse problem solution. We restricted ourselves to some special cases where we demonstrated the denoising (including removal of salt-and-pepper) on particular examples. At the end, we sketched the extension of our propositions to the 2-D case.

Acknowledgements. The support of the project SGS20/PřF-MF/2022 of the University of Ostrava is kindly announced.

References

1. Golub, G.H., Van Loan, C.F.: Matrix Computations. Johns Hopkins University Press, Baltimore (1996)
2. Janeček, J., Perfilieva, I.: Dimensionality reduction and its F-transform representation. In: Atlantis Studies in Uncertainty Modelling, Atlantis Press (2019)
3. Janeček, J., Perfilieva, I.: Three methods of data analysis in a space with closeness. In: Developments of Artificial Intelligence Technologies in Computation and Robotics, pp. 947–955 (2020). https://doi.org/10.1142/9789811223334_0114
4. Janeček, J., Perfiljeva, I.: Laplacian singular values. In: Joint Proceedings of the 19th World Congress of the International Fuzzy Systems Association (IFSA), the 12th Conference of the European Society for Fuzzy Logic and Technology (EUSFLAT), and the 11th International Summer School on Aggregation Operators (AGOP), pp. 142–146. Atlantis Press (2021)
5. Perfilieva, I., Holčapek, M., Kreinovich, V.: A new reconstruction from the F-transform components. Fuzzy Sets Syst. **288**, 3–25 (2016). https://doi.org/10.1016/j.fss.2015.10.003. ISSN 0165-0114

Selection of Keypoints in 2D Images Using F-Transform

Irina Perfilieva[✉] ⓘ and David Adamczyk ⓘ

Institute for Research and Applications of Fuzzy Modeling,
Centre of Excellence IT4Innovations, University of Ostrava,
30. dubna 22, Ostrava, Czech Republic
{irina.perfilieva,david.adamczyk}@osu.cz

Abstract. We focus on a new fast and robust algorithm for selecting keypoints in 2D images using the following techniques: image regularization, selection of spaces with closeness, and design of the corresponding graph Laplacians. Then, the representative keypoints are local extrema in the image after the Laplacian operator is applied. The convolution kernels, used for regularization, are extracted from the uniform partition of the image domain, and the graph Laplacian is constructed using the theory of F^0-transforms. Empirically, we show that sequences of F-transform kernels that correspond to different regularization levels share the property that they do not introduce new local extrema into the image under convolution. This justifies the computation of keypoints as points where local extrema are reached and allows them to be classified according to the values of the local extrema.

We show that the extracted key points are representative in the sense that they allow a good approximate reconstruction of the original image from the calculated components of the F-transform taken from different convolutions. In addition, we show that the proposed algorithm is resistant to Gaussian noise.

1 Introduction

The proposed contribution is focused on a new fast and robust algorithm of image feature extraction in the form of representative keypoints. This problem closely relates to the data processing by neural networks, the extraction of stable features that are invariant under various geometric transformations. Our motivation is supported by the overview of the explainability of artificial intelligence [1], the authors of which pointed out that "The sophistication of AI-powered systems has lately increased to such an extent that almost no human intervention is required for their design and deployment. When decisions derived from such systems ultimately affect human life (as in, e.g., medicine, law, or defense), there is an emerging need for understanding how such decisions are furnished by AI methods." They propose to link explainability to understanding how the data is processed by the model, which means how the results obtained contribute to solving the imposed problem. This forces us to focus on feature extraction because

D. Ciucci et al. (Eds.): IPMU 2022, CCIS 1602, pp. 418–430, 2022.
https://doi.org/10.1007/978-3-031-08974-9_33

it is the first step in every machine learning algorithm. Depending on the feature extractors, we distinguish between local, semi-local, and nonlocal features. The character of their extraction (by means of convolution) depends on the receptive field of the corresponding kernel. The latter is connected with the semantics of the extracted value and by this, contributes to the explainability. Therefore, to increase the level of explainability, we shall consciously select convolution kernels. In image processing tasks, features are associated with keypoints and their descriptors. In early works, both were identified on local areas of the image that correspond to its content (not related to the background). Thus, the processing was computationally consuming and depending on many space-environment conditions: illumination, position, resolution, etc. In the seminal work, [9], it was shown that invariant local features (Harris corners and their rotation invariant descriptors) can contribute to solving general problems of image recognition. In detail, an invariant image description in a neighborhood of a point can be described by the set of its convolutions with Gaussian derivatives. Because the Harris corner detector is sensitive to changes in image scale, it was expanded in [4] to the descriptor known in the literature as SIFT to achieve scale and rotation invariance and to provide reliable matching over a wide range of affine distortions. The publication of SIFT has inspired many modifications: SURF, PCA-SIFT, GLOH, Gauss-SIFT, etc., (see [3] and references therein), aimed at improving its efficiency in various senses: reliability, computation time, etc. However, the main stages and their semantics have been preserved. Our contribution to this topic is as follows: we use the basic methodology of SIFT and its modifications, but use nontraditional kernels derived from the theory of F-transforms [5]. This allows you to simplify the scaling and selection of keypoints, as well as reduce their number and enhance the robustness.

2 Space with a Fuzzy Partition

In this section, we introduce a space that plays an important role in our research. A space with a fuzzy partition is considered as a space with a proximity (closeness) relation, which is a weak version of a metric space. Our goal is to show that the diffusion (heat conduction) equation in (2) can be extended to spaces with closeness, where the concepts of derivatives are adapted to nonlocal cases.

Let us first recall the basic definitions. As we indicated at the beginning, our goal is to extend the Laplace operators to those that take into account the specifics of spaces with fuzzy partitions. For this reason, in the following sections, we recall the basic concepts on this topic.

2.1 Fuzzy Partition

The notion of fuzzy partition has been evolved in the theory of fuzzy sets being adjusted to various requests to a space structure. The closest form to that which we use in this paper, it has been introduced in [5]. Below, we give a relaxed version of this notion and then show how it relates to the standard one.

Definition 1: Fuzzy sets $A_1, \ldots, A_n : [a, b] \to \mathbb{R}$, establish a *fuzzy h-partition* of the real interval $[a, b]$ with nodes $a = x_1 < \ldots < x_n = b$, where $n \geq 3$, if for all $k = 1, \ldots, n$, the following conditions are valid (we assume $x_0 = a$, $x_{n+1} = b$):

1. $A_k(x_k) = 1$, $A_k(x) > 0$ if $x \in (x_k - h, x_k + h)$;
2. $A_k(x) = 0$ if $x \notin (x_k - h, x_k + h)$;
3. $A_k(x)$ is continuous,
4. $A_k(x)$, for $k = 2, \ldots, n$, strictly increases on $[x_k - h, x_k]$ and $A_k(x)$ strictly decreases on $[x_k, x_k + h]$ for $k = 1, \ldots, n - 1$,
5. $\forall x \in [a, b]$, $\sum_{k=1}^{n} A_k(x) > 0$.

The membership functions $A_1, \ldots, A_n$ are called *basic functions* [5].

Definition 2:

1. A continuous bell-shaped even function $A_0 : [-1, 1] \to [0, 1]$, such that $A_0(0) = 1$, $A_0(-1) = A_0(1) = 0$, is called a *generating function*.
2. Fuzzy h-partition $A_1, \ldots, A_n$, $n \geq 3$, of $[a, b]$ with nodes $a = x_1 < \ldots < x_n = b$, is *generated by* the h-scaled $A_0^h : [-h, h] \to [0, 1]$, where $A_0^h(x) = A_0(\frac{x}{h})$, if for all $k = 1, \ldots, n$,

$$A_k(x) = \begin{cases} A_0^h(x - x_k), & x \in [x_k - h, x_k + h], \\ 0, & \text{otherwise.} \end{cases}$$

3. Fuzzy h-partition $A_1, \ldots, A_n$, $n \geq 3$, of $[a, b]$, generated by the h-scaled $A_0^h : [-h, h] \to [0, 1]$, with nodes $a = x_1 < \ldots < x_n = b$, is called *h- uniform* if nodes are h-equidistant, i.e. for all $k = 2, \ldots, n$, $x_k - x_{k-1} = h$.

In the sequel, we will work with the triangular generating function $\hat{A}_0 : [-1, 1] \to [0, 1]$, such that

$$\hat{A}_0(x) = 1 - |x|. \tag{1}$$

The corresponding fuzzy h-partitions will be called *triangular*.

Proposition: Assume that the basic functions $\hat{A}_1^h, \ldots, \hat{A}_n^h$ and $\hat{A}_1^{2h}, \ldots, \hat{A}_n^{2h}$, establish triangular fuzzy h- and $2h$-partitions with the same nodes $a = x_1 < \ldots < x_n = b$. Let moreover, the fuzzy partition $\hat{A}_1^h, \ldots, \hat{A}_n^h$ be an h-uniform. Then, for $k = 3, \ldots, n - 2$,

$$\hat{A}_k^{2h} = \hat{A}_k^h + \frac{1}{2}\hat{A}_{k-1}^h + \frac{1}{2}\hat{A}_{k+1}^h. \tag{2}$$

2.2   Discrete Universe and Its Fuzzy Partition

From the point of view of signal processing, we assume that the domain of the corresponding functions is finite, i.e., finitely sampled in $\mathbb{R}$, and the functions are identified with high-dimensional vectors of their values at the selected samples in the discretized domain. Moreover, we assume that the domain and the range of all considered functions are equipped with the corresponding relations of closeness.

The best formal model of all these assumptions is a weighted graph $G = (V, E, w)$ where $V = \{v_1, \ldots, v_\ell\}$ is a finite set of vertices, and E ($E \subset V \times V$) is a set of weighted edges so that $w : E \to \mathbb{R}_+$. The edge $e = (v_i, v_j)$ connects two vertices v_i and v_j, and then the weight of e is $w(v_i, v_j)$ or just w_{ij}. Weights are set using the function $w : V \times V \to \mathbb{R}_+$, which is symmetric ($w_{ij} = w_{ji}, \forall 1 \leq i, j \leq \ell$), non-negative ($w_{ij} \geq 0$) and $w_{ij} = 0$ if $(v_i, v_j) \notin E$. The notation $v_i \sim v_j$ denotes two adjacent vertices v_i and v_j with an existing edge connecting them.

Let $H(V)$ denote the Hilbert space of real-valued functions on the set of vertices V of the graph, where if $f, h \in H(V)$ and $f, h : V \to \mathbb{R}$, then the inner product $\langle f, h \rangle_{H(V)} = \sum_{v \in V} f(v)h(v)$. Similarly, $H(E)$ denotes the space of real-valued functions defined on the set E of edges of a graph G. This space has the inner product $\langle F, H \rangle_{H(E)} = \sum_{(u,v) \in E} F(u,v)H(u,v) = \sum_{u \in V} \sum_{v \sim u} F(u,v)H(u,v)$, where $F, H : E \to \mathbb{R}$ are two functions on $H(E)$.

We assume that the set of vertices V is identified with the set of indices $V = \{1, \ldots, \ell\}$ and that $[1, \ell]$ is h-uniform fuzzy partitioned with normalized basic functions $A_1^h, \ldots A_\ell^h$, so that $A_k^h(x) = A_h(x - k)/h$, $k = 1, \ldots, \ell$, $A_h(x) = A_0(x/h)$ and A_0 is the generating function.

Definition 3: A weighted graph $G = (V, E, w)$ is fuzzy weighted, if $V = \{1, \ldots, \ell\}$, $A_1^h, \ldots A_\ell^h$ is an h-uniform fuzzy partition, generated by A_0, and $w_{ij} = A_i^h(j)$, $i, j = 1, \ldots, \ell$. The fuzzy weighted graph $G = (V, E, w)$, corresponding to the h-uniform fuzzy partition, will be denoted $G_h = (V, E, A_h)$.

3   Discrete (Non-local) Laplace Operator

In this section, we aim to develop elements of functional analysis on spaces with closeness to be able to introduce an operator with properties similar to the Laplacian. We recall the definition of a (non-local) Laplace operator as a differential operator given by the divergence of the gradient of a function (see [2]). On spaces with closeness, the generalized version known as Laplace-Beltrami operator is used. The definition is based on the self-adjoint property, which leads us to definitions of the corresponding Hilbert spaces.

Let $G = (V, E, w)$ be a weighted graph model of a space with closeness, and let $f : V \to \mathbb{R}$ be a real-valued function. Let $H(V)$ denote the Hilbert space of real-valued functions on V, such that if $f, h \in H(V)$ and $f, h : V \to \mathbb{R}$, then the inner product $\langle f, h \rangle_{H(V)} = \sum_{v \in V} f(v)h(v)$. Similarly, $H(E)$ denotes the space of real-valued functions defined on the set E of edges of a graph

G. This space has the inner product $\langle F, H \rangle_{H(E)} = \sum_{(u,v) \in E} F(u,v)H(u,v) = \sum_{u \in V} \sum_{v \sim u} F(u,v)H(u,v)$, where $F, H : E \to \mathbb{R}$ are two functions on $H(E)$.

The difference operator $d : H(V) \to H(E)$ of f, is defined on $(u,v) \in E$ by

$$(df)(u,v) = \sqrt{w(u,v)}\,(f(v) - f(u)). \tag{3}$$

The directional derivative of f, at vertex $v \in V$, along the edge $e = (u,v)$, is defined as:

$$\partial_v f(u) = (df)(u,v). \tag{4}$$

The adjoint to the difference operator $d^* : H(E) \to H(V)$, is a linear operator defined by:

$$\langle df, H \rangle_{H(E)} = \langle f, d^* H \rangle_{H(V)}, \tag{5}$$

for any function $H \in H(E)$ and function $f \in H(V)$.

Proposition 1: The adjoint operator d^* can be expressed at a vertex $u \in V$ by the following formula:

$$(d^* H)(u) = \sum_{v \sim u} \sqrt{w(u,v)}\,(H(v,u) - H(u,v)). \tag{6}$$

The divergence operator, defined by $-d^*$, measures the network outflow of a function in $H(E)$, at each vertex of the graph.

The weighted gradient operator of $f \in H(V)$, at vertex $u \in V$, $\forall (u, v_i) \in E$, is a column vector:

$$\nabla_w f(u) = (\partial_v f(u) : v \sim u)^T = (\partial_{v_1} f(u), \ldots, \partial_{v_k} f(u))^T.$$

The weighted Laplace operator $\Delta_w : H(V) \to H(V)$, is defined by:

$$\Delta_w f = -\frac{1}{2} d^*(df). \tag{7}$$

Proposition 2 [2]: The weighted Laplace operator Δ_w at $f \in H(V)$ acts as follows:

$$(\Delta_w f)(u) = -\sum_{v \sim u} w(u,v)(f(v) - f(u)).$$

This Laplace operator is linear and corresponds to the graph Laplacian.

Proposition 3 [7]: Let $G_H = (V, E, w_H)$ be a fuzzy weighted graph, corresponding to the 1-uniform fuzzy partition of $V = \{1, \ldots \ell\}$. Then, the weighted Laplace operator Δ_H at $f \in H(V)$ acts as follows:

$$(\Delta_H f)(i) = -\sum_{i \sim j} A_i^H(j)(f(j) - f(i)) = f(i) - F^H[f]_i,$$

where $F^h[f]_i$, $i = 1, \ldots, \ell$, is the i-th discrete F-transform component of f, cf. [5].

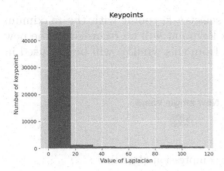
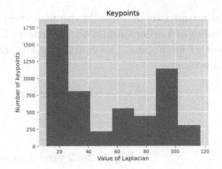

Fig. 1. Left: distribution of keypoints and their Laplacian values. **Right:** distribution of keypoints after filtering the first bin

4 Keypoints Localization

Our proposed technique of keypoint detection is based on our previous work [6] and [8] where we propose an algorithm for keypoint detection on 1D time series. Above of our technique for keypoint detection on 1D time series or 2D images are built on detection extreme values of Laplacians. Due to the different spatial organization of the analyzed objects (time series vs. images), we found simpler solutions to the problem raised. For example, to exclude extrema close to each other (this can be seen especially on images with noise), we leave only representative keypoints by removing them from the histogram to partition our set of keypoints by extrema values. We leave only those representative keypoints, which gives the best semantic correlation with the characteristic of this particular extremum. Below, we give the histograms before and after they have been bin removed Fig. 1. That set of keypoints is further filtered by our technique for keypoint compression and keypoint filtering (Fig. 2).

Algorithm 1. Keypoint localization

$img_{gray} = convertTograyscale(Image_{RGB})$
$img_{FT5} = computeFT5(img_{gray})$
$laplacian = img_{FT5} - img_{gray}$
$keypoints_{ALL} = findExtrema(laplacian)$
$keypoints = removeBin(keypoints_{ALL}, 1)$

4.1 Keypoint Compression

During a detailed analysis of the distribution of keypoints, we found that a small subset of keypoints share a particular area. Technique for keypoint compression is the first step of our filtering algorithm. The idea of compression is based on iterating over FT image with a small window of 2×2 pixels. When the window is found two or more keypoints, then we select a keypoint with the maximum value of Laplacian and the position of this keypoint will be recorded in the new set of keypoints. The remaining keypoints from this window will be not used in the future.

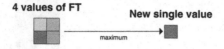

Fig. 2. 2×2 pixel window in compression algorithm

Algorithm 2. Keypoint compression

$compressionMatrix \leftarrow Matrix(img_{gray}.shape)$
for all $window_{ij} \in img_{gray}$ **do**
$\quad compressionMatrix_{ij} \leftarrow Maximum(window_{ij})$
end for

4.2 Keypoint Filtering

We assume that the keypoint M is a corner or edge point on the image. In previous step, we obtain a 3 sets of keypoints $K1, K2, K3$ where $K3 <= K2 <= K1$. The largest set of keypoints $K1$ originates from the first step of our algorithm by computing Laplacian. Let the keypoint M belong to $K1$ and it is in one of 1–6 zones. We found his coordinates (x_M, y_M) and we assign a label to this keypoint. The result of this step is that we assign a label to each of the keypoints in the set $K1$. The first step of compression is performed with an iterating window of 2×2 pixels. If in the current position the new keypoint with maximum value is found, then we assign a new label (x_M, y_M) for this keypoint. Thus, some of the pixels have a new label after the first compressions were performed and we assign them to $K2$ set. Now we can repeat the computation of Laplacian and keypoint detection on the set $K2$. Let the keypoint P belongs to $K2$ and be assigned to one of 1–6 zones. If P are assigned to the pixel with label then continue to another P. If P is assigned to a pixel without labels, then we search the area of 8 neighborhood pixels until we find a pixel with the label. If we find a pixel with a label, then we assign that label to the keypoint P otherwise the keypoint P will be removed from $K2$. Thus, all keypoints from $K2$ have labels. Repeating the previous steps will be ensured that all keypoints from $K3$ will have a label.

Algorithm 3. Keypoint filtration

$K_1 \leftarrow keypointLocalization(image_{gray})$
for all $M \in K1$ **do**
   if $M \in Zone_1, ..., Zone_6$ **then**
      $Label_{K_1} \leftarrow (x_M, y_M)$ ▷ Assign keypoints coordinations as label
   end if
end for
$K_2 \leftarrow keypointCompression(K_1)$
for all $M \in K_1$ **do**
   if $M \in K_2$ **then**
      $Label_{K_2} \leftarrow Label_M$
   end if
end for
$K_3 \leftarrow keypointCompression(K_2)$
for all $M \in K_2$ **do**
   if $M \in K_3$ **then**
      $Label_{K_3} \leftarrow Label_M$
   end if
end for

4.3   Reconstruction

The set of keypoints provides information about the features of the image. The best keypoints representation gives us a high-quality reconstruction of the original image. The reconstruction based on every keypoint position has high computational complexity. When we divide the image into two parts by bounding box, we can perform the reconstruction faster. To demonstrate the effectiveness of the proposed keypoint representation, we first show that an initial image can be reconstructed from FT based image. The iterative reconstruction algorithm has a large computational complexity with an increasing number of keypoints. That algorithm will be more computationally efficient when we divide the image into two parts and the computation will be performed for two parts separately. The positions of representative keypoints provide a boundary to the bounding box which divides the image into two parts. The first part of the image, which is the background, has a low number of details. This part will be processed by FT15. The second part of the image has a high number of details, therefore will be processed by FT3 (FT5). The algorithm of the reconstruction performs Inverse Fuzzy Transform on two parts of the image. Outputs will be merged into one image.

5   Experiments with Images

To examine our keypoint detection algorithm, we performed a series of experiments. In the first place, we were an experiment with a simple one-time filtration Fig. 3 and two-time filtration Fig. 4. The input image destined to the experiments was modified by adding one 2×2 pixel black rectangle for stability verification of our algorithm. The second verification of our algorithm was added as random Gaussian noise to the image Fig. 6.

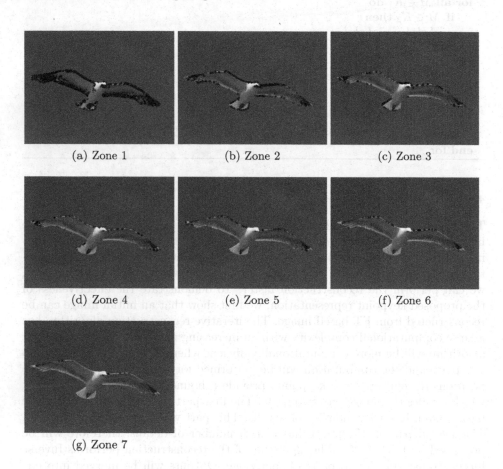

(a) Zone 1   (b) Zone 2   (c) Zone 3

(d) Zone 4   (e) Zone 5   (f) Zone 6

(g) Zone 7

Fig. 3. Keypoints and their zones obtained from histogram after filtering

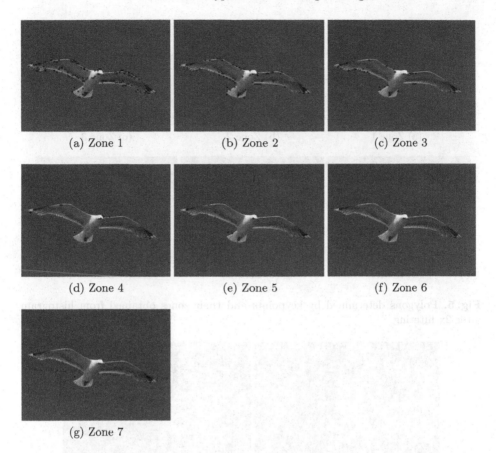

(a) Zone 1   (b) Zone 2   (c) Zone 3

(d) Zone 4   (e) Zone 5   (f) Zone 6

(g) Zone 7

Fig. 4. Keypoints and their zones obtained from histogram after 2x filtering

Figure 5 shows polygons determined by keypoints positions. The polygons were computed by the ConvexHull algorithm. The area of the particular polygon depends on the particular zone. The polygons for Zone 1 and Zone 2 correspond to the current object on the image and we assume that the representation of keypoints can be useful for describing the object.

The last experiment was the reconstruction of our image from sets of keypoints, where was realized experiment with FT3 Fig. 7 and FT5 Fig. 8 for a particular bounding box.

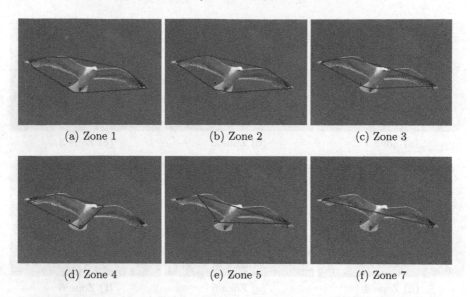

(a) Zone 1   (b) Zone 2   (c) Zone 3

(d) Zone 4   (e) Zone 5   (f) Zone 7

Fig. 5. Polygons determined by keypoints and their zones obtained from histogram after 2x filtering

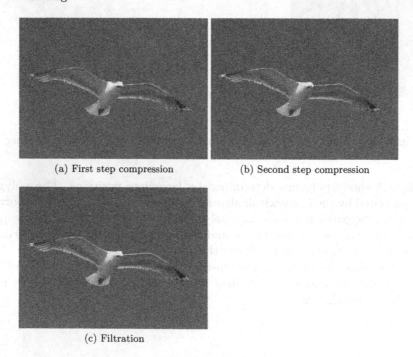

(a) First step compression   (b) Second step compression

(c) Filtration

Fig. 6. Images with Gaussian Noise

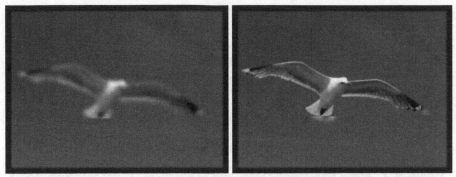

(a) FT 3, Filtration stage 1, bounding box   (b) FT 3, Filtration stage 2, bounding box
size: 70x354   size: 104x340

Fig. 7. Reconstruction with FT3

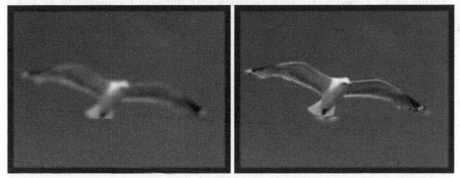

(a) FT 5, Filtration stage 1, bounding box   (b) FT 5, Filtration stage 2, bounding box
size: 70x354   size: 104x340

Fig. 8. Reconstruction with FT5

6 Conclusion

We are focused on a new fast and robust algorithm of image/signal feature extraction in the form of representative keypoints. We have contributed to this topic by showing that the use of nontraditional kernels derived from the theory of F-transforms [5] leads to simplified algorithms and comparable efficiency in the selection of keypoints. Moreover, we reduced their number and enhanced the robustness. This has been shown at the theoretical and experimental levels.

In future work we would like to focus on generalization our algorithm to 3D objects. And we will elevate our algorithm to perform tasks like classification or segmentation.

Acknowledgment. The work was supported by the grant project SGS18/PrF-MF/2021 (Ostrava University) is kindly announced.

References

1. Arrieta, A.B., et al.: Explainable artificial intelligence (XAI): concepts, taxonomies, opportunities and challenges toward responsible AI. Inf. Fusion **58**, 82–115 (2020)
2. Elmoataz, A., Lezoray, O., Bougleux, S.: Nonlocal discrete regularization on weighted graphs: a framework for image and manifold processing. IEEE Trans. Image Process. **17**(7), 1047–1060 (2008)
3. Lindeberg, T.: Image matching using generalized scale-space interest points. J. Math. Imaging Vision **52**(1), 3–36 (2015)
4. Lowe, D.G.: Distinctive image features from scale-invariant keypoints. Int. J. Comput. Vision **60**(2), 91–110 (2004)
5. Perfilieva, I.: Fuzzy transforms: theory and applications. Fuzzy Sets Syst. **157**(8), 993–1023 (2006)
6. Perfilieva, I., Adamczyk, D.: Features as keypoints and how fuzzy transforms retrieve them. In: Rojas, I., Joya, G., Català, A. (eds.) IWANN 2021. LNCS, vol. 12862, pp. 14–27. Springer, Cham (2021). https://doi.org/10.1007/978-3-030-85099-9_2
7. Perfilieva, I., Vlašánek, P.: Total variation with nonlocal FT-Laplacian for patch-based inpainting. Soft. Comput. **23**(6), 1833–1841 (2019)
8. Perfilievaa, I., Adamczyka, D., Perfilieva, I.: Fuzzy partitions and multi-scale representation of 1D data. In: 19th World Congress of the International Fuzzy Systems Association (IFSA), 12th Conference of the European Society for Fuzzy Logic and Technology (EUSFLAT), and 11th International Summer School on Aggregation Operators (AGOP), pp. 153–157. Atlantis Press (2021)
9. Schmid, C., Mohr, R.: Local grayvalue invariants for image retrieval. IEEE Trans. Pattern Anal. Mach. Intell. **19**(5), 530–535 (1997)

3D Shapes Classification Using Intermediate Parts Representation

Jan Hula[1,2]([✉])(iD), David Mojzisek[2](iD), and David Adamczyk[2](iD)

[1] Czech Technical University in Prague, Prague, Czechia
[2] CE IT4I - IRAFM, University of Ostrava, Ostrava, Czechia
{jan.hula,david.mojzisek,david.adamczyk}@osu.cz

Abstract. We describe a novel approach for 3D shape classification which classifies the shape based on a graph of its parts. To segment out the parts of a given object, we train a shape segmentation network to mimic the segments obtained from an offline co-segmentation method. Using the predicted segments, our approach constructs a spatial graph of the parts which reflects the spatial relations between them. The graph of parts is finally classified by a Tensor Field Network - a type of a graph neural network which is designed to be equivariant to rotations and translations. Therefore, the classification of the spatial graph of parts is not influenced by the choice of the coordinate frame. We also introduce a data augmentation method which is particularly suitable to our setting. A preliminary experimental results show that our method is competitive with the standard approach which does not detect parts as an intermediate step. The intermediate representation of parts makes the whole model more interpretable.

Keywords: Geometry processing · Graph neural networks · Co-segmentation · Shape classification

1 Introduction

3D point cloud classification is a classical task in geometric data processing. We are given a set of 3D coordinates of points lying on the surface of a given object and should predict a class of the object as an output. To each point, we may also associate a feature vector that may contain information such as the color of a surface at that point or a normal vector that is perpendicular to the tangent plane at the given point. In the past, manually designed feature descriptors have been used to obtain a representation of an object, which was then classified by a simple machine learning algorithm [7].

Recently, the classification of point clouds is being done almost exclusively with neural networks. Similar to convolutional neural networks, which are popular in image processing, neural networks for point cloud processing contain localized filters that aggregate information from points in a small neighborhood of a given point. As with most approaches that use neural networks, the resulting model is not interpretable.

© Springer Nature Switzerland AG 2022
D. Ciucci et al. (Eds.): IPMU 2022, CCIS 1602, pp. 431–442, 2022.
https://doi.org/10.1007/978-3-031-08974-9_34

In this contribution, we describe a novel approach for the classification of 3D objects. Our approach works by segmenting an object represented as a point cloud into multiple interpretable parts and then classifying the spatial graph which reflects how these parts are related geometrically and topologically.

The whole approach consists of three phases. In the first phase, we jointly segment all objects in a particular class into interpretable parts. This problem, usually called co-segmentation, is fully unsupervised and is only used to create data for a supervised segmentation model.

In the second phase, the supervised segmentation model is trained to mimic the output of the unsupervised co-segmentation. The difference between co-segmentation and supervised segmentation is that co-segmentation segments each object jointly with all other objects in a given class. It can not segment an object separately without the context of other objects and therefore is computationally expensive. Its purpose is to discover parts which are shared among many objects of the class. The supervised model, on the other hand, processes each point cloud independently and therefore can be used online.

In the third phase, the segmented parts from the supervised segmentation model are used to create a spatial graph of parts, which is then classified by the final classification model. The classification model is designed to be invariant with respect to rotations and translations and therefore we do not require that the point cloud is aligned to a canonical coordinate system. We also propose a suitable data augmentation that fits well into our approach. Finally, we compare our approach to a standard neural network for point cloud classification and show that it is competitive. In summary, our contribution lies in the following:

- We describe a novel approach for point cloud classification which is more interpretable than methods that do not use parts of objects as an intermediate representation.
- We propose an augmentation that fits well with a model that classifies an object based on its parts.

The rest of the paper has the following structure: In Sect. 2 we describe the related work. Section 3 contains a summary of the approach. Section 4 provides a background for graph neural networks. Section 5 describes the method for obtaining labels for supervised segmentation using unsupervised co-segmentation. Section 6 describes PointNet++ which is used for supervised segmentation. Section 7 describes the construction of the graph of parts and its classification using the Tensor Field Network. Section 8 describes our data augmentation method. Experiments are described in Sect. 9 and we conclude with discussion and the conclusion in Sects. 10 and 11, respectively.

2   Related Work

We are not aware of previous work that would classify a point cloud by constructing a graph of object parts and then classifying this graph. The most similar to our approach is an early work by Huber et al. [7] where the classification is also based on object parts. Our approach represents a new iteration of this idea with new tools such as graph neural networks. Currently, the classification of point clouds is most often approached by using various types of graph neural networks that process the graph created by connecting points lying close to each other.

PointNet [14] was one of the first neural networks trained for the classification of point clouds and later was improved by *PointNet++* [15] which is used in our experiments. There exist more recent types of neural networks with higher accuracy such as [13, 18, 20]. We use PointNet++ because its popularity and availability of reliable implementation. The mentioned architectures could be used for classification or for segmentation.

Apart from the work mentioned above, our research is also related to the work on co-segmentation and equivariant graph neural networks. 3D object segmentation is a popular task in geometry processing. Its goal is to decompose the object into recognizable parts. Most often, it is approached on an object-by-object basis so that each object is segmented separately. The segmentation is often achieved using a neural network trained on a dataset with manually annotated segmentation or using manually designed heuristics that segment the object based on purely geometric properties such as convexity [4, 10].

Huang et al. [6] popularized an approach called co-segmentation which segments all objects in a given class jointly by leveraging the fact that these objects could be decomposed into similar parts. The co-segmentation procedure described in Sect. 5 uses their approach. Recently, co-segmentation was also approached using neural networks [2, 22] but in our experience, these networks do not produce reliable results.

Lastly, our classification model is a Tensor Field Network (TFN) introduced by Thomas et al. [19]. It solves the problem of invariance to rotations and translations. Many other neural networks working on a similar principle as TFN were later introduced. A notable example is *SE(3)-Transformer* [3] which adds an attention mechanism to TFN and was used in the ground-breaking approach for the prediction of protein folding called AlphaFold2 [9].

3   Description of the Approach

The presented approach consists of three optimization phases. In the first phase, we jointly optimize the segmentation of all point clouds within one class. This process, often called co-segmentation, makes use of the fact that objects from the same class typically share a lot of similar parts and these are obtained as individual segments by the co-segmentation process. We describe the objective function for co-segmentation in Sect. 5. The drawback of most co-segmentation methods is that they can not be used online at a test time.

For this reason, the second optimization phase consists of training a standard segmentation network Pointnet++ [15], to mimic the results from the co-segmentation. This phase is described in Sect. 6. The trained network can be used to segment each new object independently and therefore could be used at a test time.

The third and final optimization phase consists of training the classification model which takes as input a spatial graph of the parts of the object and predicts the class of the corresponding object. We use a Tensor Field Network [19], which is a neural network designed to be equivariant/invariant with respect to translations and rotations. The construction of the spatial graph and the classification network are described in Sect. 7.

4   Graph Neural Networks

Alongside Convolutional Neural Networks (CNNs) suitable for regular grid-like data and recurrent neural networks or transformers for sequential data, Graph Neural Networks (GNNs) are an established way to process graph-structured data and learn meaningful representations from them. In this section, we will briefly introduce GNNs to later explain the variants used in our work.

A graph is a data structure consisting of a collection of entities (nodes) and relations between them (edges). In 3D computer vision, points in a point cloud can be represented as nodes and are related by their closeness, which can be reflected as an edge. The nodes of a graph can carry additional meta-information in the form of a fixed-size feature vector. In the case of a point cloud, it can be color encoding or a normal vector.

The basic variant of a GNN is a Graph Convolutional Network (GCN). It works by applying the same transformation to each node of the graph. This transformation combines the feature vector of a given node with the aggregated feature vectors of its neighbors. The weights of that transformation are learned. In the usual setup, multiple steps of this convolution with different parameters are applied, resembling the layers in the classical neural network.

More formally: Let $G = (V, E)$ be a graph with a feature vector x_v assigned to each node. Furthermore, let $N(v)$ denote the neighborhood of a node v, that is, the set of all nodes adjacent to v. In each propagation step $k = 1, 2, 3 \ldots$, a new feature vector $h_v^{(k)}$ (s.t. $h_v^{(0)} = x_v$) is computed as follows [21]:

$$a_v^{(k)} = \mathsf{AGGREGATE}^{(k)} \left(\left\{ h_u^{(k-1)} \mid u \in N(v) \right\} \right)$$

$$h_v^{(k)} = \mathsf{COMBINE}^{(k)} \left(h_v^{(k)-1}, a_v^{(k)} \right)$$

Often these two steps are integrated and the aggregation is done over $N(v) \cup \{v\}$. In this way, the new feature vector also contains useful information from the neighborhood of a given node.

A simple node update rule used in the basic Graph Convolutional Network model (GCN) can have the following form [12]:

$$h_u^{(k)} = \sigma \left(W^{(k)} \sum_{v \in N(u) \cup \{u\}} \frac{h_v^{(k-1)}}{\sqrt{|N(u)||N(v)|}} \right)$$

The matrix $W^{(k)} \in \mathbb{R}^{dim(h_u^k) \times dim(h_u^{k-1})}$ contains trainable parameters, the square root in the denominator scales the vectors according to a degree of the respective node, and σ is the application of a nonlinear activation function (such as ReLU). We stress that $W^{(k)}$ is shared by all nodes in layer k but may differ between layers, allowing successive change of the size of the feature vector.

At the end, we get the final feature vector for each node. These features can be aggregated, and a final vector representing the whole graph can be used for classification. The main advantage of a GNN is that it can be applied to a graph of any size and the order of nodes does not matter since the graph convolution is applied locally and commonly used aggregation operators (mean, max, ...) are invariant to the argument permutation.

GNN has become a popular choice for the processing of many types of graph data. Alongside point clouds, they can find their usage, for example, in the domain of automated reasoning, since logical expressions can be naturally described by directed acyclic graphs [8].

5   Discovery of Parts of the Object

We assume that the dataset in question does not have segmentation labels assigned, and therefore we use a co-segmentation method to segment the objects in a given class in a consistent way. Concretely, we use a simplified version of a co-segmentation method developed by Huang et al. [6]. Here, we provide a high-level description of their method. For more details, see the original publication.

The basic idea of co-segmentation is to segment all objects within one class jointly in a consistent way. By consistent segmentation of objects A and B, we mean that there has to exist a mapping from a segmentation of the object A to a segmentation of the object B that maps each segment to a similar segment (measured by a similarity function) and at the same time respects the neighborhood structure of the segments.

The method of Huang et al. works by first decomposing the object to a set of patches P using a normalized cuts method [5]. The final segments are obtained by making a partition of the set P which means that each final segment consists of one or more patches of P merged together. Using randomized cuts [5], the method then finds a large number of possible partitions that are grouped together to form an overcomplete set I of possible segments. To state it explicitly, elements in I may share one or more patches from P. Additionally, every element in I is assigned a heuristic score w that reflects the quality of the segment.

The final step of the co-segmentation method is to select a nonoverlapping subset S_i of I_i for each object i with the constraint that there should exist a consistent mapping between segmentations of an arbitrary pair of objects from a given class. This goal is achieved by setting up a discrete optimization problem whose objective is to maximize the quality and consistency of the segmentation. For a pair of two objects whose sets of possible segmentations are denoted by I_1 and I_2, the optimization problem has the following form:

$$\underset{S_1 \subset I_1, S_2 \subset I_2}{\operatorname{argmax}} \; Seg(S_1) + Seg(S_2) + Consistency(S_1, S_2) \qquad (1)$$

where $Seg(S_i)$ denotes the score for the segmentation S_i of object i and $Consistency(S_i, S_j)$ measures the consistency of two segmentations of objects i and j. $Seg(S_i)$ is computed by summing the weights w_k over all segments $s_k \in S_i$ and the consistency score is computed by:

$$Consistency(S_i, S_j) = \sum_{ij \in \{12,21\}} \max_{M_{ij}} Consistency(M_{ij})$$

where $M_{ij} \subset S_i \times S_j$ is a directed mapping from S_i to S_j. The score measures the quality of the best possible mapping between the two segmentations. This quality reflects whether individual segments are assigned to segments of similar shape and whether the neighborhood structure of the segments is being preserved.

The objective function 1 is expressed as an integer quadratic program where the optimization variables range over possible segmentations and their mappings with a constraint that every patch in P_i should be covered by exactly one segment.

Setting up the whole objective as a maximization over all possible pairs of objects leads to unreasonable computational requirements and therefore the authors relax the problem to a *linear program* which is optimized using a block coordinate descent procedure. The block coordinate descent iteratively optimizes subsets of variables corresponding to a pair of objects.

At the end of the optimization procedure, we obtain a segmentation for every object and a correspondence between segments of distinct objects. Using this procedure, we obtain labels for a supervised segmentation described in the next section.

6   Supervised Segmentation

For supervised segmentation, we use PointNet++ [15] which is trained on the labels obtained from co-segmentation. It is based on a simpler version called PointNet, which was one of the first neural networks designed to work directly on raw point cloud data [14]. The point cloud is a set of points described by their 3D coordinates. In practice, the dimensionality of the input is often increased by adding more features, such as surface normals. Unlike voxel or other representations, points in a point cloud are unordered. PointNet takes the input feature of each point and applies the same transformations to it. These transformations map the input feature vector into a 64-dimensional feature space with a shared MLP. For the direct classification task, the feature of each point is further transformed to a 1024-dimensional space with another MLP, and a symmetric function (max pooling) is applied to obtain a single global feature vector output that is then fed to a soft-max classifier.

For segmentation, the 64-dimensional feature vector of each point is concatenated with the global feature vector of a given point cloud. Another MLP is used to first decrease the dimension from 1088 to 128 and finally from 128 to the number of segmentation classes.

PointNet also transforms the input to normalize the pose of the point cloud. That should make the output of the network invariant to input transformations such as rotation and translation. Since the MLPs with shared weights are applied independently to each point, and a symmetric function is used to get the final feature vector, PointNet is also invariant to the permutation of points.

In our work, we use a more advanced architecture called Pointnet++. The difference is that Pointnet++ exploits local structures obtained by using a distance metric with the input point cloud. PointNet++ is a graph neural network applying PointNet recursively on a nested partitioning of the input point cloud. This change to the original architecture allows PointNet++ to better capture local structures.

7   Classification of the Graph of Object Parts

7.1   Constructing the Graph of Object Parts

Using the prediction of the supervised segmentation model, we construct a graph of the object parts of each point cloud. This graph is constructed in the following way:

1. Separate points in the point cloud according to the assigned label.
2. Points assigned to the same class may belong to disconnected components of the object as depicted in Fig. 1. To obtain the disconnected subparts, we cluster each separated part into possibly multiple clusters by assigning points v_i and v_j to the same cluster if $||v_i - v_j|| < \epsilon_1$ (in our case $\epsilon_1 = 0.1$). Each cluster corresponds to a node in the final graph.
3. Each point in a given cluster has a label distribution assigned by the supervised segmentation model. Compute an average of this distribution over all points within the cluster and assign this averaged distribution as a feature vector to the node corresponding to this cluster.
4. Assign a position to node j by averaging positions of all points in the cluster corresponding to node j.
5. Create an edge between two nodes i and j if there exist a point v_i in cluster corresponding to node i and a point v_j in the cluster corresponding to node j such that $||v_i - v_j|| < \epsilon_2$ (in our case $\epsilon_2 = 0.1$).

At the end of this procedure, we obtain a spatial graph for the object in which every node represents a part of the object. An example of such a graph is depicted in Fig. 1. Each node in this graph is summarised by a feature vector and 3D position. The last step of our pipeline is to classify this graph. The model for the classification and its properties are described in the next section.

7.2   Tensor Field Networks

Tensor Field Network (TFN) is a type of graph neural network that is designed to be equivariant to rigid transformations of the given point cloud. This means the

Fig. 1. The figure depicts a graph representation for the three selected object. The color of a point corresponds to its segmentation label. If more than one group of the same label is present, we separate them and calculate their centroids. If two groups are touching, we connect their centroids by an edge to form a final graph. (Color figure online)

feature representation of every point in a given layer changes in a predictable way when the point cloud is rotated or translated. Actually, the network is automatically invariant to translations because the feature vector of every point is created in a relative coordinate frame with the given point as an origin. Therefore, the only interesting equivariance is with respect to rotations. The rotation equivariance of the whole network is achieved by a composition of rotation equivariant parts. It can be shown that the rotation equivariance is preserved by addition, function composition, and formation of tensor products. The update rule of one layer of the TFN has the following form:

$$f_i^{l+1} = \sum_{j \in Neigh(i)} g(x_j - x_i) \otimes f_j^l,$$

where f_i^{l+1} is a feature representation of node i in layer $l + 1$. $Neigh(i)$ are nodes connected to node i according to the spatial graph whose construction was described in Sect. 7.1. Summation is used here for aggregation. The aggregated values are formed by a tensor product between the feature vector from the previous layer f_j^l and a learnable filter g which is parametrized by a vector that points from node i to node j (x_i is a 3D coordinate of node i). Therefore, if the filter g and a feature vector f_j^l are both equivariant, the new feature vector f_i^{l+1} will also be equivariant. The equivariance of the filter g is achieved by decomposing it into radial and angular components. The radial component contains learnable weight, and the angular component is based on spherical harmonics (which are equivariant filters) and contains no learnable parameters. The TFN consists of several such layers which are interleaved with equivariant nonlinearities. For classification the feature vectors of the final layer are aggregated using a symmetric operation such as addition, maximum or mean and the aggregated feature vector is processed by an MLP. For a more detailed description, see the original publication [19].

8 Data Augmentations and Noise Injection

Data augmentation and noise injection are popular techniques for making machine learning models more robust to variations of the input and less prone to overfitting. Data augmentation techniques enrich the training dataset with variations of the training examples and the noise injection techniques inject noise to various parts of the model. To give a concrete example, data augmentation for a dataset of images may create new images by rotating every image in the training dataset by a random angle from a prespecified range. The noise injection may, for example, set the activation value of randomly chosen neurons to zero (a technique called Dropout [17]).

In our case, we do not require typical augmentations such as random rotations, scaling, and translations because our classification model is designed to be equivariant/invariant to these transformations as mentioned in Sect. 7. We augment the training dataset by injecting noise to the spatial graph constructed from the object parts. The noise consists of randomly removing parts from the spatial graph of parts with a small probability p (in our experiments $p = 0.05$) and adding Gaussian noise to the positions of the nodes within this graph ($\sigma = 0.01$, $\mu = 0$). The intention behind the removal of random nodes from the graph is to make the network more robust to faulty segmentation results, and the intention behind the addition of Gaussian noise is to make the network more robust to variations of parts.

9 Experiments

For our experiments, we used the publicly available Pytorch Geometric version of the ShapeNet dataset [1]. The dataset contains about 17 000 examples of 3D objects represented by point clouds. Those examples are divided into 16 shape categories (Motorbike, Airplane, Guitar, Pistol, Mug, ...). The first step of our pipeline is a co-segmentation, which creates labels of parts in each category for the supervised segmentation. In the case of ShapeNet, this step could be omitted since the dataset contains segmentation labels. We decided to use co-segmentation because we wanted to test our method in a situation where segmentation labels are not available.

Our goal is to show that the proposed approach is competitive with an approach which classifies the point cloud directly without the intermediate representation of parts. We compare the two approaches in a small data regime. Concretely, for each class, we select 50 samples for training and 100 samples for testing. We do not perform tuning of hyper-parameters and therefore do not use a validation set. The reason we choose to compare the approaches in a small data regime is that the co-segmentation method is computationally expensive and in real applications it may be unreasonable to collect hundreds of training samples per class. The baseline we compare with is an implementation of PointNet++ trained for point cloud classification provided by Pytorch Geometric library[1].

---

[1] https://github.com/pyg-team/pytorch_geometric/tree/master/examples.

For the supervised segmentation, which is part of our approach, we also use an implementation of PointNet++ provided in the same library. For the classification of the graph of parts, we use the implementation of Tensor Filed Network provided at the repository maintained by the authors[2].

For our approach, we train two models: PointNet++ for segmentation and Tensor Field Network for graph classification. For the baseline, we train only PointNet++ for classification. The training of all networks is performed for 150 epochs with a batch size of 32 samples. We use Adam [11] optimizer with a learning rate set to 10^{-5} (Table 1).

Table 1. The results of our experiments - the comparison between PointNet++ and our approach in the point cloud classification task.

Approach	Accuracy (%)
PointNet++	92.5
Our approach	93.7

10   Discussion

In this section, we discuss the drawbacks of our approach and the possible direction of future work. The main drawback of our approach in its current form is that in our experiments, the improvement over the baseline approach is not significant enough to compensate for the additional complexity. We view this as a temporary issue. Our approach significantly departs from the standard approaches which were incrementally improved by many researchers. The fact that our model is competitive in the basic version described in this paper is a promising sign of there being a space for improvement. Another drawback is that the supervised segmentation network is not equivariant/invariant with respect to rotations and translations by design. Therefore, the segmentation may fail if the object is represented in an unexpected coordinate frame. This can be mitigated by applying random rotations during training, but this diminishes the benefits of the classification architecture which is equivariant/invariant by design.

In future work, we plan to focus on the improvement of the classification by using better features for the segmented parts. Currently, the features are distributions of labels (which correspond to different parts) averaged over all points in a given segment. Adding either features which reflect the geometry or are learned together with the classification should further boost the accuracy of the approach. We also plan to use a segmentation neural network that is equivariant/invariant by design such as a *DiffusionNet* [16]. Lastly, we plan to conduct a more extensive set of experiments using data obtained from 3D scans.

---

[2] https://github.com/e3nn/e3nn.

11 Conclusion

In this contribution, we presented a novel approach for the classification of 3D objects represented as point clouds. It works by segmenting the object into representative parts, constructing a spatial graph of these parts and classifying the resulting graph. The classification model is equivariant/invariant with respect to rotations and translations by design and, therefore, is not dependent on the coordinate frame with respect to which is the spatial graph represented. To obtain the representative parts, we run a co-segmentation method on all training examples of a given class and then train a segmentation model to mimic the results from the co-segmentation. We also described a data augmentation technique that fits well to our setting. Our approach is competitive with a standard neural network for point cloud classification, which does not use the intermediate representation of parts. The intermediate representation of parts increases the interpretability of our approach.

Acknowledgement. The results were supported by the Ministry of Education, Youth and Sports within the dedicated program ERC CZ under the project POSTMAN no. LL1902. This scientific article is part of the RICAIP project that has received funding from the European Union's Horizon 2020 research and innovation programme under grant agreement No. 857306.

References

1. Chang, A.X., et al.: ShapeNet: an information-rich 3D model repository. arXiv preprint arXiv:1512.03012 (2015)
2. Chen, Z., Yin, K., Fisher, M., Chaudhuri, S., Zhang, H.: BAE-NET: branched autoencoder for shape co-segmentation. In: 2019 IEEE/CVF International Conference on Computer Vision (ICCV), pp. 8489–8498 (2019)
3. Fuchs, F., Worrall, D., Fischer, V., Welling, M.: Se (3)-transformers: 3D roto-translation equivariant attention networks. Adv. Neural. Inf. Process. Syst. **33**, 1970–1981 (2020)
4. Gadelha, M., et al.: Label-efficient learning on point clouds using approximate convex decompositions. arXiv, abs/2003.13834 (2020)
5. Golovinskiy, A., Funkhouser, T.: Randomized cuts for 3D mesh analysis. In: ACM SIGGRAPH Asia 2008 Papers, pp. 1–12 (2008)
6. Huang, Q., Koltun, V., Guibas, L.: Joint shape segmentation with linear programming. In: Proceedings of the 2011 SIGGRAPH Asia Conference, pp. 1–12 (2011)
7. Huber, D., Kapuria, A., Donamukkala, R., Hebert, M.: Parts-based 3D object classification. In: Proceedings of the 2004 IEEE Computer Society Conference on Computer Vision and Pattern Recognition. CVPR 2004, vol. 2, p. II-II. IEEE (2004)
8. Hula, J., Mojžíšek, D., Janota, M.: Graph neural networks for scheduling of SMT solvers. In: 2021 IEEE 33rd International Conference on Tools with Artificial Intelligence (ICTAI), pp. 447–451. IEEE (2021)
9. Jumper, J.M., et al.: Highly accurate protein structure prediction with alphafold. Nature **596**, 583–589 (2021)

10. Van Kaick, O., Fish, N., Kleiman, Y., Asafi, S., Cohen-Or, D.: Shape segmentation by approximate convexity analysis. ACM Trans. Graph. (TOG) **34**(1), 1–11 (2014)
11. Kingma, D.P., Ba, J.: Adam: a method for stochastic optimization. arXiv preprint arXiv:1412.6980 (2014)
12. Kipf, T.N., Welling, M.: Semi-supervised classification with graph convolutional networks. arXiv preprint arXiv:1609.02907 (2016)
13. Ma, X., Qin, C., You, H., Ran, H., Fu, Y.: Rethinking network design and local geometry in point cloud: a simple residual MLP framework. arXiv preprint arXiv:2202.07123 (2022)
14. Qi, C.R., Su, H., Mo, K., Guibas, L.J.: PointNet: deep learning on point sets for 3D classification and segmentation. In: Proceedings of the IEEE Conference on Computer Vision and Pattern Recognition, pp. 652–660 (2017)
15. Qi, C.R., Yi, L., Su, H., Guibas, L.J.: Pointnet++: deep hierarchical feature learning on point sets in a metric space. In: Advances in Neural Information Processing Systems, vol. 30 (2017)
16. Sharp, N., Attaiki, S., Crane, K., Ovsjanikov, M.: DiffusionNet: discretization agnostic learning on surfaces. arXiv preprint arXiv:2012.00888 (2020)
17. Srivastava, N., Hinton, G., Krizhevsky, A., Sutskever, I., Salakhutdinov, R.: Dropout: a simple way to prevent neural networks from overfitting. J. Mach. Learn. Res. **15**(1), 1929–1958 (2014)
18. Thomas, H., Qi, C.R., Deschaud, J.-E., Marcotegui, B., Goulette, F., Guibas, L.J.: KPConv: flexible and deformable convolution for point clouds. In: Proceedings of the IEEE/CVF International Conference on Computer Vision, pp. 6411–6420 (2019)
19. Thomas, N., et al.: Tensor field networks: rotation-and translation-equivariant neural networks for 3D point clouds. arXiv preprint arXiv:1802.08219 (2018)
20. Wang, Y., Sun, Y., Liu, Z., Sarma, S.E., Bronstein, M.M., Solomon, J.M.: Dynamic graph CNN for learning on point clouds. ACM Trans. Graph. (TOG) **38**(5), 1–12 (2019)
21. Xu, K., Hu, W., Leskovec, J., Jegelka, S.: How powerful are graph neural networks? In: ICLR (2019)
22. Zhu, C., Xu, K., Chaudhuri, S., Yi, L., Guibas, L.J., Zhang, H.: AdacoSeg: adaptive shape co-segmentation with group consistency loss. In: 2020 IEEE/CVF Conference on Computer Vision and Pattern Recognition (CVPR), pp. 8540–8549 (2020)

Content-Aware Image Smoothing Based on Fuzzy Clustering

Felipe Antunes-Santos[1,2,3](✉) , Carlos Lopez-Molina[1,2,3] ,
Arnau Mir-Fuentes[1,3] , Maite Mendioroz[3] , and Bernard De Baets[2]

[1] Department of Estadistica, Informatica y Matematicas,
Universidad Pública de Navarra, 31006 Pamplona, Spain
{felipe.antunes,carlos.lopez,arnau.mir}@unavarra.es
[2] KERMIT, Department of Data Analysis and Mathematical Modelling,
Ghent University, 9000 Ghent, Belgium
bernard.debaets@ugent.be
[3] NavarraBiomed,
Hospital Universitario de Navarra, 31008 Pamplona, Spain
tmendioi@navarra.es

Abstract. Literature contains a large variety of content-aware smoothing methods. As opposed to classical smoothing methods, content-aware ones intend to regularize the image while avoiding the loss of relevant visual information. In this work, we propose a novel approach to content-aware image smoothing based on fuzzy clustering, specifically the Spatial Fuzzy c-Means (SFCM) algorithm. We develop the proposal and put it to the test in the context of automatic analysis of immunohistochemistry imagery for neural tissue analysis.

Keywords: Image smoothing · Fuzzy clustering · Progressive supranuclear palsy

1 Introduction

Image regularisation is one of the most basic tasks in computer vision. Initially, its goal was to produce a regularised, a.k.a. smooth, version of the original signal, hence preventing problems due to noise or contamination in the signal. Despite large improvements in understanding image smoothing, e.g. the use of Gaussian filters and the creation of the Gaussian Scale Space [16,17], it was soon evident that smoothing brought undesired distortions to images. Specifically, such distortions included the removal of small objects, as well as the blurring of certain image artefacts, mainly object boundaries. It then became evident that it was necessary to have smoothing techniques that, while regularising the image

The authors gratefully acknowledge the financial support of the Spanish Ministry of Science (Project PID2019-108392GB-I00 AEI/FEDER, UE), as well as the funding from the European Union's H2020 research and innovation programme under Marie Sklodowska-Curie Grant Agreement Number 801586.

D. Ciucci et al. (Eds.): IPMU 2022, CCIS 1602, pp. 443–454, 2022.
https://doi.org/10.1007/978-3-031-08974-9_35

content, would not entail any of the drawbacks. In this context, two main families of smoothing techniques can be discriminated: content-unaware (CUS) and content-aware smoothing (CAS) techniques. The former apply the same smoothing operation across the image, while the latter adapt the smoothing operation locally to avoid object removal and boundary blurring.

Content-unaware techniques have historically been based on Gaussian smoothing, especially since the theoretical results by Babaud *et al.* [1]. Large efforts were devoted to understand Gaussian smoothing, develop theories as the anisotropic Gaussian filters or the Gaussian Scale-Space [13,17]. However, the panorama in content-aware smoothing is significantly richer. Early attempts are due to Saint-Marc [24], who presented a Gaussian-based smoothing technique in which the standard deviation of the Gaussian kernel applied at each pixel is dependent upon local characteristics. A more elaborate proposal was that by Perona and Malik, who presented a discrete schema for the so-called Anisotropic Diffusion model [23]. These pioneering efforts were continued to produce continuous schemas, as well as to customise the smoothing behaviour [26,28]. For example, Weickert presented Coherence-Enhancing Anisotropic Diffusion [18,27], which not only aimed at preserving structurally strong objects, but also at improving the visibility of visual structures. Other authors elaborated on theories different from anisotropic diffusion. Examples are bilateral filtering [25], aimed at content-aware smoothing using principles from spatio-tonal filtering, and Mean Shift [8], which incorporates notions from clustering and multivariate analysis. Despite the variability in both inspirations and specific implementations, many of the proposals for content-aware smoothing can be studied under the prism of unifying theories [3, 4].

Among the inspirations for content-aware smoothing, a very promising trend is that based on pixel clustering. The underlying idea behind this trend is as simple as powerful: the pixels in an image can be seen as n-dimensional feature vectors comprising (a) their position and (b) their (possibly multivalued) tone. Hence, an image becomes a point cloud in either 3D (for grayscale images), 5D (for most colour images), or event larger spaces (for, e.g., multispectral images). By performing clustering in such point cloud, pixels will be grouped according to spatial and tonal similarity. Otherwise said, the tone at each pixel will be influenced by the tones in pixels that are both spatially and tonally close. This shall achieve intra-object regularisation (since tones within an object will be grouped to a single tone) while avoiding object boundary blurring (since nearby pixels will have low influence on each other if they are tonally different). The main representative for clustering-based content-aware smoothing is Mean Shift, as presented by Comaniciu and Meer [8]. However, other clustering techniques are equally valid, e.g. gravitational clustering [20,31]. Literature contains, to the best of our knowledge, no proposals for content-aware smoothing using fuzzy clustering. This is relatively surprising, given the significant impact of fuzzy clustering techniques in both fuzzy set theory and machine learning.

In this work, we propose a novel method for content-aware smoothing using Spatial Fuzzy c-Means (SFCM). Our proposal is put to the test in the context of medical image processing, a field of particular interest for image smoothing given the proneness of such images to noise and external contamination.

The remainder of this work is organised as follows. Section 2 presents the general notions of clustering-based content-aware smoothing. Then, Sect. 3 depicts our proposal for image smoothing using Spatial Fuzzy c-Means. Our proposal is experimentally tested in Sect. 4. Finally, Sect. 5 lists some general conclusions and future work.

2   Clustering-Based Content-Aware Smoothing

Among the inspirations for content-aware smoothing, a relevant trend is that inspired by multidimensional clustering. This trend is based on the interpretation of images as datasets to be analysed from a spatio-tonal perspective. Let I : $\Omega \mapsto \mathbb{T}$ be an image, with $\Omega = \{1, ..., N\} \times \{1, ..., M\}$ representing the set of positions and $\mathbb{T}$ representing the tonal palette. The image I can be understood as a dataset in which each pixel becomes an instance $p \in \{1, ..., N\} \times \{1, ..., M\} \times \mathbb{T}$. Consider a pixel $I(i, j) = t \in \mathbb{T}$ with $t = (t_1, ..., t_k)$ a given tone; the instance corresponding to this pixel would be $(i, j, t_1, ..., t_k)$.

The idea behind clustering-based content-aware smoothing is that clustering pixels in the spatio-tonal space will group pixels that are both similar in tone and similar in position. In this manner, pixels that are both spatially and tonally similar will be grouped to similar tones. This shall produce inter-object regularisation. Also, pixels that are spatially similar, but tonally different, shall not be grouped together, preventing inter-object blurring. Hence, the evolution of instances in the clustering process is meant to produce content-aware smoothing.

Nevertheless, the underlying inspiration of these smoothing techniques needs to render into functional algorithms, and such algorithms are dependent upon the specific clustering method. There are two main technical issues to be faced in clustering-based smoothing techniques. Firstly, we need to build an image from the dataset, at each stage of the clustering process, since the result of the process at each stage should be an image. Secondly, since clustering methods are normally based on comparison measures (often, on metrics), it is necessary to design comparison measures able to produce meaningful results in the spatio-tonal universe.

The generation of images at each stage of the clustering process is heavily dependent upon the specific body of knowledge generated in the process itself. In clustering methods that properly evolve the instances in the dataset (e.g., mean shift or gravitational clustering), each instance p in the image will be modified iteratively. This shall affect the tonal information in p, but also the spatial information in it. Otherwise said, the positions shall no longer fit the original pixel grid in Ω. Hence, when using such clustering methods, positional information is normally reset back to the original values after each iteration in the clustering process. The situation is different for clustering methods that evolve a set of centroids, leaving the instances unaltered. In such cases, it is necessary to work with the membership of each pixel to each of the centroids, regardless of whether such membership is expressed as probability, a fuzzy membership degree or any other numerical representation. Since each centroid is an element in $\mathbb{R}^+ \times \mathbb{R}^+ \times \mathbb{T}$, it does represent a tone. Hence, at each iteration of the clustering process, an image can be reconstructed through linear combination of the tones at each centroid, using as weighting coefficients the *memberships* to such centroids. More image-creating strategies could be designed for clustering methods that shall not fit in any of the previous two descriptions. Still, it is evident that the use of clustering methods for content-aware smoothing requires not only the creation of a dataset incorporating spatio-tonal information. Also, it requires the design of a strategy to create an image from the body of knowledge generated in the clustering itself.

As for the design of comparison operators for the spatio-temporal universe, there is no predefined solution. This is mainly because it is extremely dependent upon the tonal palette $\mathbb{T}$, which can vary from grayscale tones (scalar values in $\mathbb{R}^+$ or $\mathbb{N}^+$) to hyperspectral signatures (vectors in $(\mathbb{R}^+)^{256}$ or $(\mathbb{R}^+)^{512}$). However, it is typical to produce a metric from the convex combination of two metrics: one in the spatial universe ($\mathbb{R}^+ \times \mathbb{R}^+$), to account for spatial similarity, and another one in the tonal universe ($\mathbb{T}$), to account for the tonal one. Still, the weights in the convex combination must be adjusted to ensure the representativeness of both spatial and tonal information in the spatio-tonal metric. It is remarkable that the design of comparison measures able to work on a spatio-tonal universe is recurrent in image processing literature. For example, it was a key in the evolution of Baddeley's delta metric [2] from binary images to gray-scale ones [9].

It is relevant to mention that clustering, at final stages, can also be used for both segmentation and hierarchical segmentation. Literature contains different successful examples, such as the graph-based hierarchical clustering method by Felzebschwalb and Huttenlocher [11] and the FCM-based segmentation based by Yang et al. [30]. In these segmentation procedures there is no need to produce intermediate images as the clustering evolves, since the only required result is the distribution of pixels in the final partition. However, they do require spatio-tonal comparison measures for the clustering.

An example of the performance of CUS and clustering-based CAS can be seen in Fig. 1. In this figure we display a colour image, together with the result of a Gaussian smoothing with $\sigma = 2$. We can also observe the image at its initial state, together with its state after the 10th and 100th iterations of the gravitational clustering procedure in [20].

3   Content-Aware Smoothing Based on Spatial Fuzzy c-Means

The goal of a content-aware smoothing algorithm is to homogenize image regions while maximally preserving the information on regions of interest, which in this case are mainly edges [19]. However, edges are (a) some of the most sensitive areas when performing image smoothing and (b) a naturally color-wise imprecise zone [5]. In order to preserve information about objects on a image, and given

(a) Original image   (b) Gaussian smoothing   (c) Gravitational smoothing (10th, 100th its.)

Fig. 1. Example of image smoothing performed with content-aware and content-unaware techniques. (a) The original image; (b) the result with content-unaware smoothing (Gaussian smoothing, $\sigma = 2$); (c) the smoothing output from the 10th and 100th iteration of gravitational (content-aware) smoothing, respectively. For CUS, (b) shows the blurring of the edges and loss of information. For CAS, (c) shows the progressive blending of tones within objects, while preserving their boundaries.

their border properties, fuzzy set theory comes as a natural alternative. We have hence focused on an existing fuzzy clustering algorithm which embodies the expected needs in our approach: the Spatial Fuzzy c-Means (SFCM) [7].

The SFCM is an unsupervised n-dimensional clustering algorithm primarily designed for data clustering [7]. Through the image-to-dataset conversion, any image can be seen as a dataset on which clustering can be applied. The SFCM will receive and image I as input and will output a set of N cluster centroids $\mathbf{C} = (C_1, C_2, C_3, ..., C_N)$, with $C_i \in \Omega \times \mathbb{T}$. However, the result of a smoothing procedure is an image, which means that the information produced at each iteration needs to be used to produce a progressively smoother version of the image. Multiple approaches can be taken to map the clusters back to image, but the process mainly comprises of two phases: pixel-to-cluster assignment and pixel colour definition.

The first phase consists of assigning the pixels to each of the clusters, so as to determine which cluster or clusters (that is, which class centroids) will be used in the determination of the colour at each pixel. A list of alternatives is available. In a simplistic approach, each pixel can be assigned to the cluster to which it has the highest membership degree. Also, a combination of the membership degrees to the clusters can be used to perform the cluster assignment, in a procedure similar to Generalized Mixture Functions [10]. A third alternative is to keep the partial membership to each of the clusters, so as to represent that the pixel is not completely assigned to any of them.

At the second phase, the cluster information at each pixel is used to produce a tonal value for each pixel $p \in \Omega$. Surely, this phase depends on the decision made in the former phase. If each pixel is uniquely associated to one cluster, its tone in the smooth image will be that represented in the tonal part of the cluster it is assigned to. If the pixels are considered to have multiple partial membership degrees to all clusters, the tone value can be obtained from the weighted combination of the membership degrees and the tonal information of the clusters.

Although membership degrees shall not be understood as weights to be operated with, we can use the membership functions for each cluster to materialise such weighted combination. Let I be an image and let $\mathbf{C}_i = \{C_{i,1}, \ldots, C_{i,N}\}$ be the set of class centroids at some iteration of the clustering. From each centroid $C_{i,j}$, we can compute a membership function $\mu_i : \Omega \mapsto [0,1]$ representing the membership degree of each pixel to the i-th cluster. The tone of a pixel $p \in \Omega$ in the i-th smooth image I'_i can be computed as

$$I'_i(p) = \sum_{j=1}^{N} \mu_j(p) \cdot C_{i,j} \,, \tag{1}$$

where $C_{i,j}$ represents the tonal information of j-th cluster at the i-th iteration.

The overall workflow of the proposal is shown in Fig. 2. From now onward, this algorithm will be referred to as CAS-SFCM.

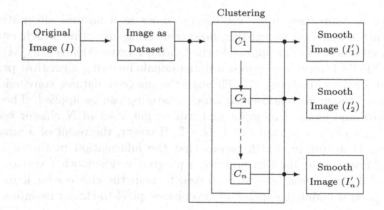

Fig. 2. Schematic representation of content-aware smoothing based on Spatial Fuzzy c-Means (SFCM). After converting an image I into a dataset, the clustering procedure iteratively produces cluster centers $\mathbf{C}_i$. In order for those centers to be converted into a progressively smoother image, the information from the centers is combined with the data in the original image, as covered in Sect. 3.

The result of the CAS-SFCM is expected to be an edge-preserving, smoothen version of the image, since the pixels will take colors progressively evolving toward clustering centroids. As such centroids are influenced by groups of (spatio-tonally) similar pixels, they shall evolve to represent highly populated regions in the spatio-tonal space. By applying Eq. (1) in the reconstruction of the images from the clustering data, pixels shall evolve towards the colors represented by such centroids, hence producing a context-aware smoothing behaviour. This shall be put to the test in the upcoming section.

4   Experimental Results

In order to test the applicability of our proposal to realistic scenarios, we put it to the test in a dataset of electronic microscopy for neural tissue analysis. Specifically, we use a dataset of neural biopsies from patients affected by Progressive Supranuclear Palsy (PSP). These images are used to determine how the presence of free protein (in this case, Tau protein) at different areas of the brain affects the degeneration of the brain functionality. The dataset is composed of 188 high resolution images of all brain regions from 14 PSP-affected patients. For better visualisation, patches of the original images will analysed on this paper. From now onward, the PSP image dataset will be referred to as PSP dataset.

4.1   Progressive Supranuclear Palsy

Neurosciences, as well as neurology, is heavily hampered by the fact that few invasive studies can be used to audit the state of a diseased organ. Invasive study methods can permanently damage neural tissues and, are unsuitable for the study of many processes related to neurodegeneration [29]. In this context, scientists have developed strategies for neural tissue analysis that do not involve invasive techniques. For Progressive Supranuclear Palsy, for instance, mislocalized Tau protein is the main study object to understand the condition [14].

The Tau protein is present in all humans in a natural manner. It has the essential function of structurally stabilising the neuron's microtubes and to regulate some biological processes [22]. For patients with PSP, an anomaly causes the Tau protein to detach from its original place, which in time will cause neural death [12]. The detached Tau protein also act as a catalyst for the degeneration process. Thus, Tau protein is a key biomarker for PSP and several studies have been performed to relate the quantity, location and form of the free Tau protein to the impact of the disease in a patient [22]. Most of these studies use imaging techniques to either manually or automatically segment the Tau protein and identify its form [6,15]. However, segmenting, identifying and quantifying the Tau protein are all challenging tasks, as the imaging conditions vary greatly in the few data that exist. One of these conditions is the visual aspect of the Tau protein after immunohistochemistry processing. After tinting the neural tissue, Tau protein tends to take a characteristic colour. Figure 3 contains examples of Tau protein deposits, which are identifiable by the brownish-tone it takes in comparison to other artefacts.

Considering the existing challenges in defining the edges of Tau protein areas due to the tonality variation, a smoother version of the images, in which the tonality change issue is addressed, would greatly improve efficiency and results of future studies, both for automated and non-automated approaches. In addition, given that Tau protein is analysed by experts as regions and not by pixel, the spatio-tonal clustering comes as a natural approach, following the hypothesis that a pixel that belongs to a Tau region is usually near other pixels that belong to the same Tau region.

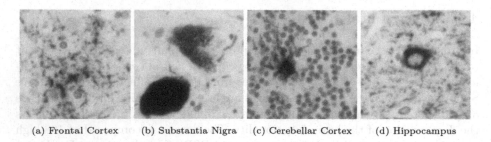

(a) Frontal Cortex   (b) Substantia Nigra   (c) Cerebellar Cortex   (d) Hippocampus

Fig. 3. Patches of biopsies after being tinted to highlight Tau protein, which is identifiable by its brownish tone. Patches are taken from different brain regions, as indicated individually for each column.

4.2   Experimental Configuration

This section intends to illustrate the specific configuration of our proposal for its application to a realistic dataset. The PSP dataset is used for illustration purposes.

The first relevant decision is related to the tonal palette of the images, which needs to be coordinated with the decision on the metrics to be used in the SFCM. In this case study, we select the CIELab colour space. The main reason is that distances yielded by the Euclidean metric on CIELab tones are consistent with tonal dissimilarities in human perception. Thus, the PSP dataset will contain 5-dimensional instances comprising L, a and b colour space components, as well as the pixel coordinates.

A second decision of interest is related to the construction of the smooth images from the centroids. In this case, we use the strategy in Eq. (1) so as to combine as much information as possible from all clusters. Figure 4 contains a visual representation of the clusters in an image from the PSP dataset. Specifically, we observe the three clusters generated from the 100th iteration of the clustering process. Such visual representations can be used to inspect the actual fitting of the clusters to the different regions or artefacts in the image.

4.3   Result Evaluation

Ideally, content-aware smoothing should be evaluated in a quantitative way. Certain quantitative measure or strategy should evaluate the results in terms of intra-region smoothing and inter-region contrast enhancement. However, such measures or strategies are absent from the literature. On the one hand, there is no possibility to generate ground truth, since it is unclear which is the optimal final state of a content-aware smoothing procedure. Hence, comparison-based strategies are discarded. On the other hand, standalone image quality strategies (such as BRISQUE [21]) attempt to measure how good an image looks to the Human Visual System, which is not really the goal of content-aware smoothing. Alternative options can be built around the intelligent use of contrast, homogeneity and luminance quantifications. However, the use of multiple metrics or

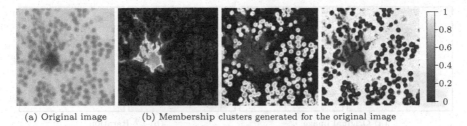

(a) Original image   (b) Membership clusters generated for the original image

Fig. 4. Visual representation of the membership functions modelled after configuring the clustering process with three clusters. (a) The original image on which the smoothing will be performed; (b) the visual representation of the three clusters generated by the clustering for (a). We can observe that each cluster in (b) mainly contains one object area from (a), specifically the tau protein (leftmost cluster), the blueish nuclei (middle cluster) and the background (rightmost cluster).

quantifiers would generate a complex measurement strategy and, as long as all individual strategies are weak by themselves, we might reach a questionable result evaluation.

The indirect evaluation of the CAS-SFCM would be the more natural approach to take, as content-aware smoothing is normally performed to improve the image quality for object segmentation. However, the use of a method or segmentation schema would add another layer of parameters to the CAS-SFCM evaluation. Also, if this indirect strategy is applied, the evaluation would be performed on the output of the method with the smoother images applied and not on the images themselves.

Considering the above-mentioned factors, the direct and indirect evaluations of the CAS-SFCM in our view are unsuitable in the current state of literature. We have hence opted for a visual evaluation of the results. In this manner, we do not intend to identify the best performing set of parameters, or to compare the results of our proposal to that by other proposals in literature. Instead, we intend to illustrate how our proposal actually leads to interesting results in the context of applications.

The first question relates to the ability of our algorithm to perform both intra-region regularisation and inter-region contrast enhancement. In order to illustrate this fact, we present in Fig. 5 the line-based analysis of the evolution of an image path from the PSP dataset. The figure contains the state of the image patch in its original state, after the 1st, 10th and 100th iterations, together with the plot-based representation of one of its rows (highlighted in orange). This plot-based representation displays the red, green and blue channel of the selected row. Although the variation in the image patch itself is subtle, we can observe how the signal evolves in all channels. This evolution is seen in two different aspects. Firstly, we observe a reduction in the rugosity of the quasi-flat areas of the image. Secondly, we see how the contrast at the transition points (between the background and the Tau protein deposit) are sharpened progressively.

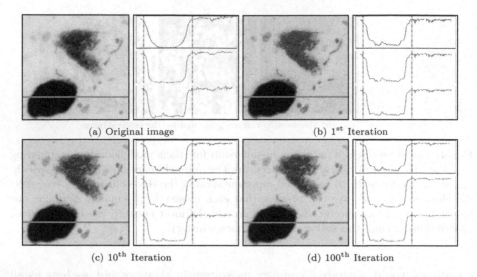

(a) Original image   (b) 1st Iteration

(c) 10th Iteration   (d) 100th Iteration

Fig. 5. Visual representation of the state of one row in an image patch at different iterations of the smoothing procedure. The figure displays the state of the image patch in its original form, and after the 1st, 10th and 100th iterations, as well as the state of the row marked in orange (individually for each RGB component). We can observe both the intra-region regularisation (reduction of the rugosity inside the objects) and the inter-region contrast enhancement (sharper tonal changes at object boundaries). The vertical grey lines are for better visualisation. (Color figure online)

A detailed view of this fact can be seen in Fig. 6. In this figure we present a detailed patch of the image to observe how the regularisation of the image is not only noticeable for individual components, as seen in Fig. 5, but also for multivalued tones. Figure 6 displays the original patch (Fig. 6(a)) and the state of the patch after 1st, 10th and 100th iterations (Fig. 6(b)–(c)). We observe in Fig. 6(a) how deposits normally feature smooth and gradual boundaries. However, with content-aware smoothing (as in Fig. 6(b)) the boundary is sharpened.

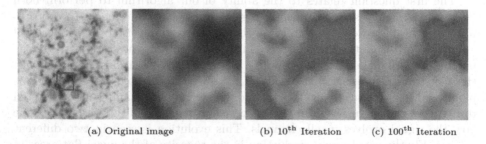

(a) Original image   (b) 10th Iteration   (c) 100th Iteration

Fig. 6. Zoomed-in analysis of the preservation of boundaries and the smoothing behaviour for an image from the PSP dataset. The patch is displayed in both its original state and its state after the 10th and 100th iterations.

5   Conclusions

In this paper, we proposed a content-aware image smoothing approach for neural tissue images based on fuzzy clustering. Our proposal is inspired by the problems and limitations that experts have when segmenting regions of interest from neural tissue imagery to perform studies on neurodegenerative diseases. Specifically, the approach is divided in three phases, being: (1) turning images into clusterable data; (2) clustering the data; and (3) generating a smooth version of the input image from the clustering output. For the experimentation and results obtained, our proposal is analysed by the prism of tonality stability, colour distribution and expert feedback, as a direct or indirect analysis is unsuitable for the context. The three analyses yielded positive feedback, showing meaningful and useful improvements through the use of the CAS-SFCM. Nevertheless, a natural possibility for future work is to delve into means to quantify the results attained by the CAS-SFCM. The comparison of our proposal with other smoothing procedures is another possibility of future work.

References

1. Babaud, J., Witkin, A.P., Baudin, M., Duda, R.O.: Uniqueness of the Gaussian kernel for scale-space filtering. IEEE Trans. Pattern Anal. Mach. Intell. **8**(1), 26–33 (1986)
2. Baddeley, A.J.: Errors in binary images and an L^p version of the Hausdorff metric. Nieuw Archief voor Wiskunde **10**, 157–183 (1992)
3. Barash, D.: A fundamental relationship between bilateral filtering, adaptive smoothing, and the nonlinear diffusion equation. IEEE Trans. Pattern Anal. Mach. Intell. **24**, 844–847 (2002)
4. Barash, D., Comaniciu, D.: A common framework for nonlinear diffusion, adaptive smoothing, bilateral filtering and mean shift. Image Vis. Comput. **22**(1), 73–81 (2004)
5. Bonnet, A.: On the regularity of edges in image segmentation. In: Annales de l'Institut Henri Poincaré C, Analyse non linéaire, vol. 13, pp. 485–528 (1996)
6. Borroni, B., Gardoni, F., Parnetti, L., et al.: Pattern of Tau forms in CSF is altered in progressive supranuclear palsy. Neurobiol. Aging **30**(1), 34–40 (2009)
7. Chuang, K.S., Tzeng, H.L., Chen, S., et al.: Fuzzy c-means clustering with spatial information for image segmentation. Comput. Med. Imaging Graph. **30**(1), 9–15 (2006)
8. Comaniciu, D., Meer, P.: Mean shift: a robust approach toward feature space analysis. IEEE Trans. Pattern Anal. Mach. Intell. **24**(5), 603–619 (2002)
9. Coquin, D., Bolon, P.: Application of Baddeley's distance to dissimilarity measurement between gray scale images. Pattern Recogn. Lett. **22**(14), 1483–1502 (2001)
10. Farias, A.D.S., Santiago, R.H., Bedregal, B.: Some properties of generalized mixture functions. In: 2016 IEEE International Conference on Fuzzy Systems, pp. 288–293 (2016)
11. Felzenszwalb, P.F., Huttenlocher, D.P.: Efficient graph-based image segmentation. Int. J. Comput. Vision **59**(2), 167–181 (2004)
12. Flament, S., Delacourte, A., Verny, M., et al.: Abnormal tau proteins in progressive supranuclear palsy. Acta Neuropathol. **81**(6), 591–596 (1991)

13. Florack, L.: Image Structure, vol. 10. Springer, Dordrecht (1997). https://doi.org/10.1007/978-94-015-8845-4
14. Iglesias-Rey, S., Antunes-Santos, F., Hagemann, C., et al.: Unsupervised cell segmentation and labelling in neural tissue images. Appl. Sci. **11**(9), 3733 (2021)
15. Kuiperij, H.B., Verbeek, M.M.: Diagnosis of progressive supranuclear palsy: can measurement of tau forms help? Neurobiol. Aging **33**(1), 204.e17-204.e18 (2012)
16. Lindeberg, T.: Generalized Gaussian scale-space axiomatics comprising linear scale-space, affine scale-space and spatio-temporal scale-space. Technical report, KTH (Royal Institute of Technology) (2011)
17. Lindeberg, T.: Scale-space theory in computer vision. Ph.D. thesis, KTH (Royal Institute of Technology) (1991)
18. Lopez-Molina, C., Galar, M., Bustince, H., De Baets, B.: On the impact of anisotropic diffusion on edge detection. Pattern Recogn. **47**(1), 270–281 (2014)
19. Madhulatha, T.S.: An overview on clustering methods. IOSR J. Eng. **2**(4), 719–725 (2012)
20. Marco-Detchart, C., Lopez-Molina, C., Fernandez, J., Bustince, H.: A gravitational approach to image smoothing. In: Kacprzyk, J., Szmidt, E., Zadrożny, S., Atanassov, K.T., Krawczak, M. (eds.) IWIFSGN/EUSFLAT -2017. AISC, vol. 642, pp. 468–479. Springer, Cham (2018). https://doi.org/10.1007/978-3-319-66824-6_41
21. Mittal, A., Moorthy, A.K., Bovik, A.C.: No-reference image quality assessment in the spatial domain. IEEE Trans. Image Process. **21**(12), 4695–4708 (2012)
22. Nizynski, B., Dzwolak, W., Nieznanski, K.: Amyloidogenesis of Tau protein. Protein Sci. **26**(11), 2126–2150 (2017)
23. Perona, P., Malik, J.: Scale-space and edge detection using anisotropic diffusion. IEEE Trans. Pattern Anal. Mach. Intell. **12**(7), 629–639 (1990)
24. Saint-Marc, P., Chen, J.S., Medioni, G.: Adaptive smoothing: a general tool for early vision. IEEE Trans. Pattern Anal. Mach. Intell. **13**(6), 514–529 (1991)
25. Tomasi, C., Manduchi, R.: Bilateral filtering for gray and color images. In: Proceedings of the IEEE International Conference on Computer Vision, pp. 838–846 (1998)
26. Weickert, J.: Anisotropic Diffusion in Image Processing. ECMI Series. Teubner-Verlag (1998)
27. Weickert, J.: Coherence-enhancing diffusion filtering. Int. J. Comput. Vision **31**(2–3), 111–127 (1999)
28. Weickert, J.: Nonlinear diffusion scale-spaces: from the continuous to the discrete setting. In: Berger, M.O., Deriche, R., Herlin, I., Jaffré, J., Morel, J.M. (eds.) ICAOS 1996. LNCIS, vol. 219, pp. 111–118. Springer, Berlin Heidelberg (1996). https://doi.org/10.1007/3-540-76076-8_123
29. Werner, C.T., Williams, C.J., Fermelia, M.R., et al.: Circuit mechanisms of neurodegenerative diseases: a new frontier with miniature fluorescence microscopy. Front. Neurosci. 1174 (2019)
30. Yang, Z., Chung, F.L., Shitong, W.: Robust fuzzy clustering-based image segmentation. Appl. Soft Comput. **9**(1), 80–84 (2009)
31. Yung, H.C., Lai, H.S.: Segmentation of color images based on the gravitational clustering concept. Opt. Eng. **37**(3), 989–1000 (1998)

Soft Methods in Statistics and Data Analysis

A Probabilistic Tree Model to Analyze Fuzzy Rating Data

Antonio Calcagnì[1]([⊠])[iD] and Luigi Lombardi[2]

[1] DPSS, University of Padova, Padova, Italy
antonio.calcagni@unipd.it
[2] DIPSCO, University of Trento, Trento, Italy
luigi.lombardi@unitn.it

Abstract. In this contribution we provide initial findings to the problem of modeling fuzzy rating responses in a psychometric modeling context. In particular, we study a probabilistic tree model with the aim of representing the stage-wise mechanisms of direct fuzzy rating scales. A Multinomial model coupled with a mixture of Binomial distributions is adopted to model the parameters of LR-type fuzzy responses whereas a binary decision tree is used for the stage-wise rating mechanism. Parameter estimation is performed via marginal maximum likelihood approach whereas the characteristics of the proposed model are evaluated by means of an application to a real dataset.

Keywords: Fuzzy rating data · Probabilistic tree model · Direct fuzzy rating scale · Triangular fuzzy numbers

1 Introduction

Rating data are ubiquitous across many disciplines that deal with the measurement of human attitudes, opinions, and sociodemographic constructs. In these cases, as the measurement process involves cognitive actors as the primary source of information, the collected data are often affected by fuzziness or imprecision. Fuzziness in rating data has multiple origins, which go from the semantic aspects of the questions/items being rated to the decision uncertainty that affects the rater response process [2]. By and large, the differences along this continuum might reflect the differences between the ontic and epistemic viewpoint on fuzzy statistics [6]. To give an example of what is intended with fuzziness as decision uncertainty, consider the case where a rater is presented with a question/item "I am satisfied with my current work" and a five-point scale ranging from "strongly disagree" to "strongly agree". In order to provide a response - which corresponds to mark one of the five labels/levels of the scale - a rater behaves according to a sequential process, the first step of which consists in

---

A. Calcagnì is member of the INdAM research group GNCS (National Institute of Advanced Mathematics; National Group of Scientific Computing).

D. Ciucci et al. (Eds.): IPMU 2022, CCIS 1602, pp. 457–468, 2022.
https://doi.org/10.1007/978-3-031-08974-9_36

the opinion formation stage in which cognitive and affective information about the item being rated - i.e., job satisfaction - are retrieved and integrated until the decision stage is triggered (second step). This includes the selection stage, where the set of response choices is pruned to obtain the final rating response, for example "strongly agree". Decision uncertainty arises from the conflicting demands of the opinion formation stage (first step), which requires the integration of often conflicting cognitive and affective information (for instance, a work problem with the boss might increase the probability of answering the item negatively) [10]. Stated in this way, fuzziness does not reflect an ontic property of the item being rated, rather it originates from the cognitive demands underlying the response process, namely the epistemic state of the rater.

Over the recent years, a number of fuzzy rating scales have been proposed to quantify fuzziness from rating data, including both direct/indirect fuzzy rating scales and fuzzy conversion scales (for an extensive review, see [3]. In addition, see [12,16,17] for further developments on this topic). In its most typical implementation, a (direct) fuzzy rating scale allows the rater to provide his/her response by adopting a stage-wise procedure [8,13]. To exemplify, consider the following five-point scale: (1) "strongly disagree", (2) "disagree", (3) "neither agree nor disagree", (4) "agree", (5) "strongly agree". First, the rater marks his/her choice on the scale (e.g., "agree") and then he/she extends the previous choice by marking another point both on the left (e.g., "disagree") and right (e.g., "strongly agree") sides. Finally, the marks are integrated to form a triangular fuzzy number where the core of the set is linked to the first mark whereas the support of the set is linked to the left and right extensions. Figure 1 shows a graphical representation of such a procedure.

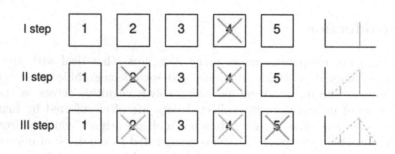

Fig. 1. Example of direct fuzzy rating scale with five-point levels along with the resulting triangular fuzzy response.

Fuzzy rating data can be analyzed either by means of standard statistical approaches or by adopting fuzzy statistical methods devoted to this purpose. In the first case, fuzzy numbers need to be turned into crisp numbers in advance through a defuzzification procedure whereas in the second case fuzzy numbers are used as is. Several fuzzy statistical methods are available nowadays (for a recent review, see [5]). However, as for many statistical models, they are quite general and, in the case of fuzzy rating data, these models do not offer a thorough formal account of the mechanism underlying the fuzzy rating process.

In this contribution, we introduce a novel statistical model to analyse LR-type triangular fuzzy data $\text{Trg}(c, l, r)$. The aim is to provide a tailor-made statistical model which mimics the stage-wise response process of direct fuzzy rating scales as those developed by [8] and [13]. In particular, such a model would be of great interest for those who are interested in studying the relationships among fuzzy rating responses and other variables (e.g., covariates) from the perspective of the mechanisms at the origin of fuzzy responses (i.e., the three-stage response mechanism). The remainder of this short paper is organized as follows. Section 2 describes the model along with the estimation procedure. Section 3 describes the results of a real case study used to assess the features of the proposed model. Finally, Sect. 4 concludes this contribution by providing final remarks and suggestions for future extensions.

2   Model

Let $\tilde{\mathbf{y}} = ((c_1, l_1, r_1), \ldots, (c_i, l_i, r_i), \ldots, (c_I, l_I, r_I))$ be a $I \times 1$ sample of triangular fuzzy numbers represented using the LR parameterization. In this context, $c_i \in \{1, \ldots, M\}$ is the core of the fuzzy number and represents the first step of the stage-wise rating process, $l_i \in \{0, \ldots, M-1\}$ is the left spread of the fuzzy number and codifies the second step of the rating process, whereas $r_i \in \{0, \ldots, M-1\}$ is the right spread of the fuzzy number and codifies the last step of the rating process (M is the number of levels of the rating scale). The magnitude of l_i and r_i quantifies the fuzziness of the rating process. It is straightforward to notice that the data encapsulate two types of uncertainty, one related to the sampling mechanism (i.e., randomness) and one related to the response process (i.e., the decision uncertainty expressed in terms of fuzziness). We assume that fuzziness results from the interplay among different components such as the characteristics of the item/question being assessed (e.g., the easiness, with higher values being associated to less difficult items in terms of response process), the characteristics of the rater (e.g., his/her ability to respond the item), and further contextual factors like social desirability, faking or cheating. For the sake of simplicity, as in the traditional Rasch modeling framework [11], we shall consider the first two components only, namely the item $\alpha \in \mathbb{R}$ and the rater's ability $\eta_i \in \mathbb{R}$. Under the stage-wise mechanism depicted in Fig. 1, the probability of a fuzzy response can be factorized as follows:

$$\mathbb{P}(Y_i = (c, l, r)|\eta_i; \boldsymbol{\theta}) = \mathbb{P}(C_i = c|\eta_i; \boldsymbol{\theta}) \tag{1}$$

$$\cdot \Big[\xi_i \mathbb{P}(L_i = l|C_i, \eta_i; \boldsymbol{\theta})\mathbb{P}(R_i = r|C_i, \eta_i; \boldsymbol{\theta}) \tag{2}$$

$$+ (1 - \xi_i)\mathbb{P}(L_i = 0|C_i, \eta_i; \boldsymbol{\theta})\mathbb{P}(R_i = 0|C_i, \eta_i; \boldsymbol{\theta})\Big]$$

where (1) indicates the probability model for the first step of the rating process, (2) represents the second and third steps of the rating process, $\boldsymbol{\theta}$ is a real vector of parameters which governs the behavior of the model (to be specified later), whereas $\xi \in [0,1]$ controls the mixture component of the model. Note that (i) conditionally on C_i, L_i and R_i are independent (i.e., $L_i \perp\!\!\!\perp R_i$, for all $i \in \{1,\ldots,I\}$), (ii) the mixture component (2) allows for disentangling those situations involving a certain level of decision uncertainty (i.e., $\xi_i > 0$) from those situations with no decision uncertainty (i.e., $\xi_i = 0$). In what follows, we will describe all the terms involved by Eqs. (1)–(2) in more details.

2.1 About the Probabilistic Term (1)

To instantiate the first term of the joint probabilistic model, we use the Rasch-tree model which is part of the family of IRTrees [1,11]. Among other advantages, they offer a simple and effective statistical representation of rating responses in terms of conditional binary trees [1,9]. Figure 2 shows two examples of IRTree for modeling rating responses.

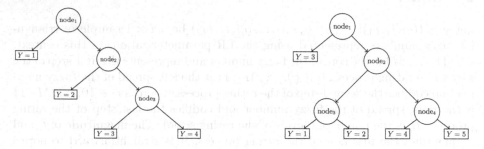

Fig. 2. Examples of IRTree models for modeling response processes in rating scales.

More formally, we set:

$$\mathbb{P}(C_i = c|\eta_i; \boldsymbol{\theta}) = Multinom(\mathbf{c}^\dagger; 1, \boldsymbol{\pi}_i^y) \tag{1.1}$$

where $\mathbf{c}_i^\dagger \in \{0,1\}^M$ is the event $C_i = c$ represented as a Boolean vector via the indicator function $\mathcal{I}(C_i^\dagger = c)$. Note that, in light of the mapping between Multinomial and Categorical random variables [14], the outcomes of C_i can be rewritten using a dummy vector with M elements, all of which are zero except for the entry $C_i = c$. For example, the event $C_i = 3$ can be rewritten as $c^\dagger = (0,0,1,0,0)$. The $M \times 1$ vector of probabilities $\boldsymbol{\pi}_i^y$ is defined according to a user-defined IRTree model as follows:

$$\pi_{im}^y = \prod_{n=1}^{N} \left(\frac{\exp(\eta_i + \alpha_n)t_{mn}}{1 + \exp(\eta_i + \alpha_n)} \right)^{\delta_{mn}} \quad m \in \{1,\ldots,M\} \tag{1.2}$$

$$\eta_i \sim \mathcal{N}(\eta; \mu_i, \sigma_\eta^2) \tag{1.3}$$

where t_{mn} is an entry of the mapping matrix $\mathbf{T}_{M \times N}$, which indicate how each response category (in rows) is associated to each node (in columns) of the tree. For the right-most tree in Fig. 2, the mapping matrix is as follows:

$$\mathbf{T}_{5 \times 4} = \begin{bmatrix} 1 & 0 & 0 & \text{NA} \\ 1 & 0 & 1 & \text{NA} \\ 0 & \text{NA} & \text{NA} & \text{NA} \\ 1 & 1 & \text{NA} & 0 \\ 1 & 1 & \text{NA} & 1 \end{bmatrix}$$

with $N = M - 1$ being the number of nodes. As $t_{mn} \in \{0, 1\}$, $t_{mn} = 1$ indicates that the m-th category of response involves the node n, $t_{mn} = 0$ indicates that the m-th category of response does not involve the node n, whereas $t_{mn} = \text{NA}$ indicates that the m-th category of response is not connected to the n-th node at all. The term δ_{mn} is defined as follows: $\delta_{mn} = 0$ if $t_{mn} = \text{NA}$ and $\delta_{mn} = 1$ otherwise. The rater's ability η_i is a random quantity from a Normal distribution with mean $\mu_i \in \mathbb{R}$ and variance $\sigma_\eta^2 \in \mathbb{R}^+$. Usually, $\mu_i = 0$ for most applications, although it can be rewritten as a linear combinations of K variables $\mu_i = \mathbf{x}_i \boldsymbol{\beta}$ to account for the effect of external covariates. Finally, the parameter $\alpha_n \in \mathbb{R}$ expresses the easiness of choosing the n-node of the tree. In general, we may have as many α's as the number of nodes or, more simply, a single α for all the nodes [1].

2.2 About the Probabilistic Term (2)

The second term of the model is a mixture distribution representing the last two stages of the fuzzy rating process. Conditioned on the first stage $C_i = c$, the final response might be affected by decision uncertainty at some degrees - a case in which $\xi_i > 0$ - or, conversely, it might be free of fuzziness. To exemplify the idea behind this representation, consider once again the right-most decision tree in Fig. 2. We expect that a higher degree of decision uncertainty entails a higher difficulty level to navigate the tree structure, which in turn increases all the response probabilities π_i^y. Conversely, a lower degree of decision uncertainty implies a lower difficulty to go through the tree nodes, which in turn decreases the probability to activate contiguous responses. This suggests to use π_i^y in the definition of (2). In particular, we define the mixture probability ξ_i in terms of the normalized Shannon entropy:

$$\xi_i = - \left(\sum_{m=1}^{M} \pi_{im}^y \ln \pi_{im}^y \right) \Big/ \ln M \tag{2.1}$$

and set the mixture components to be Binomial as follows:

$$\mathbb{P}(L_i = l | C_i, \eta_i; \boldsymbol{\theta}) = \mathcal{B}in(l; \ C_i - 1, \pi_i^s) \tag{2.2}$$

$$\mathbb{P}(R_i = r | C_i, \eta_i; \boldsymbol{\theta}) = \mathcal{B}in(r; \ M - C_i, 1 - \pi_i^s) \tag{2.3}$$

$$\mathbb{P}(L_i = 0 | C_i, \eta_i; \boldsymbol{\theta}) = \mathcal{B}in(l; \ C_i - 1, 0) \tag{2.4}$$

$$\mathbb{P}(R_i = 0 | C_i, \eta_i; \boldsymbol{\theta}) = \mathcal{B}in(r; \ M - C_i, 0) \tag{2.5}$$

where (2.4)–(2.5) are degenerate distribution with mass one on the element zero of the support [4]. The parameter π_i^s is the probability to activate lower response categories and it is defined as follows:

$$\pi_i^s = \sum_{m \in \{1,\dots,M\}\setminus C_i} \pi_{im}^y \Big/ (1 - \pi_{i,m=c_i}^y) \tag{2.6}$$

under the convention that $\pi_i^s = 0$ if $m = c_i$ and where $\pi_{i,m=c_i}^y$ is the probability of the current response $C_i = c$. Note that the normalized Shannon entropy increases as $\boldsymbol{\pi}^y$ gets uniform and decreases as $\boldsymbol{\pi}^y$ becomes degenerate for a single element of $\{1,\dots,M\}$. This property makes the entropy measure suitable to quantify varying levels of decision uncertainty in the rating process.

2.3 Sampling Schema

In short, the proposed model can be rewritten in terms of the underlying sampling process as follows:

$$\eta_i \sim \mathcal{N}(\eta; \mathbf{x}_i \boldsymbol{\beta}, \sigma_\eta^2)$$

$$C_i^\dagger | \eta_i \sim \mathcal{M}ultinom(c^\dagger; 1, \boldsymbol{\pi}_i^y(\boldsymbol{\alpha}, \eta_i))$$

$$Z_i | \eta_i \sim \mathcal{B}in(z; 1, \xi_i(\boldsymbol{\alpha}, \eta_i))$$

$$Z_i = 1 \begin{cases} L_i | C_i, \eta_i \sim \mathcal{B}in(l; C_i - 1, \pi_i^s(\boldsymbol{\alpha}, \eta_i)) \\ R_i | C_i, \eta_i \sim \mathcal{B}in(l; M - C_i, 1 - \pi_i^s(\boldsymbol{\alpha}, \eta_i)) \end{cases} \tag{3}$$

$$Z_i = 0 \begin{cases} L_i | C_i, \eta_i \sim \mathcal{B}in(l; C_i - 1, 0) \\ R_i | C_i, \eta_i \sim \mathcal{B}in(l; M - C_i, 0) \end{cases}$$

where $C_i = \mathcal{I}(C_i^\dagger)$, $\boldsymbol{\pi}_i^y(\boldsymbol{\alpha}, \eta_i)$ is defined via Eq. (1.2), $\xi_i(\boldsymbol{\alpha}, \eta_i)$ is defined via Eq. (2.1), whereas $\pi_i^s(\boldsymbol{\alpha}, \eta_i)$ is defined according to Eq. (2.6).

According to the stage-wise representation of the rating response process, model (3) is self-consistent in the manner through which the fuzzy data $\tilde{\mathbf{y}}$ are modeled. Indeed, given an IRTree structure according to which the rating process is supposed to behave, a particular instance of $\boldsymbol{\theta} = \{\boldsymbol{\alpha}, \boldsymbol{\beta}, \sigma_\eta^2\}$ gives rise to a cascade computations from the input to the output $\{\hat{c}, \hat{l}, \hat{r}\}$ through the model equations. As a result, external information in terms of explaining variables or covariates can be plugged-in to the model through the model parameters only and there is no way to link them to the outcome variable directly.

Finally, the probability of a fuzzy response is as follows:

$$
\begin{aligned}
\mathbb{P}(Y_i =(c_i, l_i, r_i)|\eta_i; \boldsymbol{\theta}) = \; & \pi^y_{i, m=c_i} \\
\times \; & \left[\xi_i \binom{c_i - 1}{l_i} \binom{M - c_i}{r_i} (\pi^s_i)^{l_i + M - c_i - r_i} \cdot (1 - \pi^s_i)^{r_i + c_i - l_i - 1} \right. \\
& \left. + (1 - \xi_i) \binom{c_i - 1}{l_i} \binom{M - c_i}{r_i} 0^{l_i + r_i} \cdot 1^{M - r_i - l_i - 1} \right] \\
\times \; & \frac{1}{\sigma_\eta \sqrt{2\pi}} \exp\left(-\frac{1}{2\sigma_\eta^2}(\eta_i - \mathbf{x}_i \boldsymbol{\beta})^2 \right)
\end{aligned}
\tag{4}
$$

where $\pi^y_{i, m=c_i}$ indicates the probability of the response $C_i = c$.

2.4 Parameter Estimation

Model (4) implies the following parameters $\boldsymbol{\theta} = \{\boldsymbol{\alpha}, \boldsymbol{\beta}, \sigma_\eta^2\} \subset \mathbb{R}^N \times \mathbb{R}^K \times \mathbb{R}_+$. Since the model uses a logistic function to determine $\boldsymbol{\pi}^y$, we can further simplify the parameter estimation by restricting the parameter space in a subset of reals, for instance by means of the following constraints: $|(\boldsymbol{\alpha}, \boldsymbol{\beta})|^T 1_{N+K} \leq 5$ and $\sigma_\eta \in (0, 3.5]$. They are justified by the simple fact that the logistic curve increases quickly only in a small subset of its domain. The model parameters can be estimated by maximizing the marginal likelihood function, which is obtained by integrating out the random terms $\eta_1, \ldots, \eta_I$ from the full likelihood function [15]. This requires the computation of the following marginal probability distribution:

$$
\begin{aligned}
\mathbb{P}(Y_i = (c_i, l_i, r_i); \boldsymbol{\theta}) = \; & \int_{\mathbb{R}} \mathbb{P}(Y_i = (c_i, l_i, r_i)|\eta_i; \boldsymbol{\alpha}) f_{\eta_i}(\eta; \mathbf{x}_i \boldsymbol{\beta}, \sigma_\eta^2) \, d\eta \\
\propto \; & \int_{\mathbb{R}} \pi^y_{i, m=c_i} \left[\xi_i \left((\pi^s_i)^{l_i + M - c_i - r_i} \cdot (1 - \pi^s_i)^{r_i + c_i - l_i - 1} \right. \right. \\
& \left. \left. - 0^{l_i + r_i} \cdot 1^{M - r_i - l_i - 1} \right) + 0^{l_i + r_i} \cdot 1^{M - r_i - l_i - 1} \right] \\
& \times \exp\left(-\frac{1}{2\sigma_\eta^2}(\eta_i - \mathbf{x}_i \boldsymbol{\beta})^2 \right) \, d\eta \\
\propto \; & \int_{\mathbb{R}} h(c_i, l_i, r_i, \boldsymbol{\alpha}, \eta_i) \exp\left(-\frac{1}{2\sigma_\eta^2}(\eta_i - \mathbf{x}_i \boldsymbol{\beta})^2 \right) \, d\eta
\end{aligned}
\tag{5}
$$

where the integral can be solved numerically via the Gauss-Hermite quadrature. By the change of variable $d_i = \sigma_\eta \sqrt{2}\eta_i + \mathbf{x}_i \boldsymbol{\beta}$, the integral is approximated as follows:

$$
\begin{aligned}
\mathbb{P}(Y_i = (c_i, l_i, r_i); \boldsymbol{\theta}) \propto \; & \int_{\mathbb{R}} h(c_i, l_i, r_i, \boldsymbol{\alpha}, d_i) \exp\left(-\frac{1}{2\sigma_\eta^2} d_i^2 \right) \, dd_i \\
\approx \; & \frac{1}{\sqrt{\pi}} \sum_{h=1}^{H} h(c_i, l_i, r_i, \boldsymbol{\alpha}, \sigma_\eta \sqrt{2}\gamma_h + \mathbf{x}_i \boldsymbol{\beta}) \, \omega_h
\end{aligned}
\tag{6}
$$

where $\gamma_1, \ldots, \gamma_H$ and $\omega_1, \ldots, \omega_H$ are the nodes and weights of the quadrature to be computed numerically for a fixed H [7]. Finally, the log-likelihood function:

$$\ln \mathcal{L}(\boldsymbol{\theta}) \propto \sum_{i=1}^{I} \ln \left(\sum_{h=1}^{H} h(c_i, l_i, r_i, \boldsymbol{\alpha}, \sigma_\eta \sqrt{2}\gamma_h + \mathbf{x}_i\boldsymbol{\beta}) \, \omega_h \right) \tag{7}$$

can be maximized numerically via either the Broyden-Fletcher-Goldfarb-Shanno (BFGS) or the Augmented Lagrangian (AUGLAG) algorithms. Note that in the first case the variance parameter has to be transformed to lie into the real line (e.g., via exp function) whereas in the second case the constraints $|(\boldsymbol{\alpha}, \boldsymbol{\beta})|^T \mathbf{1} \leq 5$ and $\sigma_\eta \in (0, 3.5]$ can be directly plugged in to the optimization routine.

3 Application

In this section we illustrate the characteristics of the proposed model by means of an application to a real dataset. In particular, data refers to a survey administered to $n = 69$ young drivers in Trentino region (north-est of Italy). Of these, 45% were women with mean age of 18.23 years. All participants were young drivers with an average of driving experience of 12 months since receipt of their driver's license. About 74% of them drove frequently during the week, 26% drove once a week. Participants were asked to self-assess their reckless-driving behavior (RDB) along with a short version of the Driving Anger Scale (DAS), adopted to evaluate the driving anger provoked by someone else's behaviors. Ratings were collected using a four-point direct fuzzy rating scale (see Fig. 1). For both scales, higher categories indicate higher scores on RDB and DAS items, respectively. To simplify the interpretation of the results, the items of the Driving Anger Scale were aggregated to form a crisp total score. In the next data analysis, the fuzzy variable RDB was used as response variable whereas the DAS total score was used as crisp predictor.

Table 1. Application: models for the RDB fuzzy rating data. Note that model M3 is the best model according to the lowest BIC criterion.

Model	Covariates	No. of parameters	$\ln \mathcal{L}(\boldsymbol{\theta})$	BIC
M1: linear tree	-	2	-161.15	330.767
M2: linear tree	sex	3	-157.855	328.412
M3: linear tree	sex, DAS	4	-155.268	327.472
M4: linear tree	sex, DAS, sex:DAS	5	-155.253	331.676
M5: nested tree	sex, DAS	5	-158.937	339.044

Table 2. Application: parameter estimation and standard errors for model M3.

Parameter	Estimate	Std. Error
α	-1.248	0.09
β_{sex}	0.408	0.119
β_{das}	1.284	0.093
σ_η	0.005	19.947

Three models (M1-M4) with a linear decision tree (see Fig. 2, leftmost panel) and an additional model (M5) with a nested decision tree structure (see Fig. 2, rightmost panel) were run on RDB. The models varied in terms of covariates (see Table 1). In particular, model M1 involved no covariates and a common α parameter for all the $N = 4$ nodes of the decision tree. On the contrary, models M2-M4 differed from M1 just in terms of covariates, with M4 including the interaction term sex:DAS. Finally, model M5 differed from M3 as this uses a different decision tree with a nested structure (see Fig. 2, rightmost panel). The final model was chosen according to the fitting measure $BIC = -2\ln\mathcal{L}(\boldsymbol{\theta}) + p\ln I$, with p being the number of parameters implied by the model. The best model is that achieving the lowest BIC, in this case M3. Table 2 reports the estimated parameters whereas Figs. 3–4 show the marginal effects for the chosen model. As it includes the categorical covariate sex, the parameters α codify the intercept of the model across all the nodes, which in this case is the coefficient for the level sex=F when DAS=0.

Overall, when DAS=0, participants in the group sex=F showed a stronger tendency to choose lower response categories ($\hat{\alpha} = -1.248$, $\sigma_{\hat{\alpha}} = 0.09$) if compared to participants in the group sex=M ($\hat{\beta}_{\text{sex}} = 0.408$, $\sigma_{\hat{\beta}_{\text{sex}}} = 0.119$). Similarly, DAS was positively associated to RDB ($\hat{\beta}_{\text{das}} = 1.284$, $\sigma_{\hat{\beta}_{\text{das}}} = 0.093$) and acted by increasing the tendency to activate the last nodes of the decision tree (see Fig. 3, first row). With regards to the parameter ξ, participants in the group sex=M showed a higher probability to activate the spread components of the fuzzy response across all the levels of DAS as opposed to participants in the group sex=F (see Fig. 4). Thus, all in all, the results suggest that driving anger increased the levels of decision uncertainty, with male participants showing a larger fuzziness if compared to female participants.

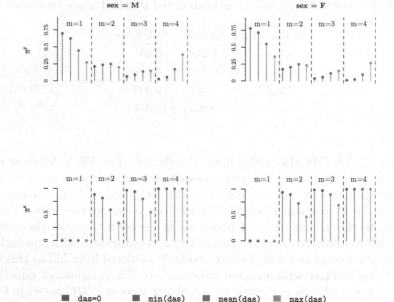

Fig. 3. Application: marginal effects computed over four reference values of das (das=0, das=min, das=mean, das=max) and for both sex=M and sex=F. The effects are computed for the response probability π^y (first row) and for the probability to activate a lower response π^s (second row). Note that $m = 1, \ldots, m = 4$ indicate the response categories of the rating scale.

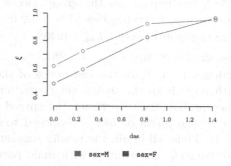

Fig. 4. Application: probability ξ to activate the mixture components (marginal effect) computed as a function of das and for both sex=M and sex=F.

4   Conclusions

In this contribution we have described a new statistical model for fuzzy rating responses that are collected by means of direct fuzzy rating scales. With the aim of representing the stage-wise decision process underlying a rating response, the model revolves around the adoption of a conditional representation where a

Multinomial tree component is coupled with a mixture of Binomial distributions to represent the fuzziness of rating responses. A nice advantage of the proposed method is its ability to deal with LR-type triangular fuzzy data in terms of the stage-wise mechanisms supposed to drive the unobserved rating response process. However, as for any statistical model, it has some limitations. In particular, the current version of the model does not take into account the shape of LR-type fuzzy numbers and it cannot be used in a multivariate context (i.e., the model works with a single outcome variable per time). Further investigations might consider these limitations more explicitly, for instance by means of additional simulation studies. The results of this contribution should be considered as initial findings to the problem of analysing fuzzy responses in a psychometric modeling context.

References

1. Boeck, P.D., Partchev, I.: IRTrees: tree-based item response models of the GLMM family. J. Stat. Soft. **48**(Code Snippet 1) (2012). https://doi.org/10.18637/jss. v048.c01
2. Calcagnì, A.: fIRTree: an item response theory modeling of fuzzy rating data. In: Joint Proceedings of the 19th World Congress of the International Fuzzy Systems Association (IFSA), the 12th Conference of the European Society for Fuzzy Logic and Technology (EUSFLAT), and the 11th International Summer School on Aggregation Operators (AGOP), pp. 471–477. Atlantis Press (2021)
3. Calcagnì, A., Cao, N., Rubaltelli, E., Lombardi, L.: A psychometric modeling approach to fuzzy rating data. Fuzzy Sets Syst. (2022)
4. Cao, J., Yao, W.: Semiparametric mixture of binomial regression with a degenerate component. Statistica Sinica 27–46 (2012)
5. Couso, I., Borgelt, C., Hullermeier, E., Kruse, R.: Fuzzy sets in data analysis: from statistical foundations to machine learning. IEEE Comput. Intell. Mag. **14**(1), 31–44 (2019)
6. Couso, I., Dubois, D.: Statistical reasoning with set-valued information: Ontic vs. epistemic views. Int. J. Approximate Reasoning **55**(7), 1502–1518 (2014). https:// doi.org/10.1016/j.ijar.2013.07.002
7. Golub, G.H., Welsch, J.H.: Calculation of gauss quadrature rules. Math. Comput. **23**(106), 221–230 (1969)
8. Hesketh, T., Pryor, R., Hesketh, B.: An application of a computerized fuzzy graphic rating scale to the psychological measurement of individual differences. Int. J. Man-Mach. Stud. **29**(1), 21–35 (1988)
9. Jeon, M., De Boeck, P.: A generalized item response tree model for psychological assessments. Behav. Res. Methods **48**(3), 1070–1085 (2015). https://doi.org/10. 3758/s13428-015-0631-y
10. Leary, M.R., Kowalski, R.M.: Impression management: a literature review and two-component model. Psychol. Bull. **107**(1), 34 (1990)
11. van der Linden, W.J.: Handbook of Item Response Theory: Volume 1: Models. CRC Press, Boca Raton (2016)
12. Lubiano, M.A., García-Izquierdo, A.L., Gil, M.Á.: Fuzzy rating scales: does internal consistency of a measurement scale benefit from coping with imprecision and individual differences in psychological rating? Inf. Sci. **550**, 91–108 (2021)

13. Lubiano, M.A., de Sáa, S.d.l.R., Montenegro, M., Sinova, B., Gil, M.Á.: Descriptive analysis of responses to items in questionnaires. why not using a fuzzy rating scale? Inf. Sci. **360**, 131–148 (2016)
14. Murphy, K.P.: Machine Learning: a Probabilistic Perspective. MIT Press, Cambridge (2012)
15. Pawitan, Y.: In All Likelihood: Statistical Modelling and Inference using Likelihood. Oxford University Press, Oxford (2001)
16. Rosa de Sáa, S.d.l., Gil, M.Á., García, M.T.L., Lubiano, M.A.: Fuzzy rating vs. fuzzy conversion scales: an empirical comparison through the MSE. In: Kruse, R., Berthold, M., Moewes, C., Gil, M., Grzegorzewski, P., Hryniewicz, O. (eds.) Synergies of Soft Computing and Statistics for Intelligent Data Analysis. Advances in Intelligent Systems and Computing, vol. 190. Springer, Berlin (2013). https://doi.org/10.1007/978-3-642-33042-1_15
17. de Sáa, S.d.l.R., Gil, M.Á., González-Rodríguez, G., López, M.T., Lubiano, M.A.: Fuzzy rating scale-based questionnaires and their statistical analysis. IEEE Trans. Fuzzy Syst. **23**(1), 111–126 (2014)

Distance Metrics for Evaluating the Use of Exogenous Data in Load Forecasting

Ramón Christen[1]([✉])(iD), Luca Mazzola[1](iD), Alexander Denzler[1],
and Edy Portmann[2](iD)

[1] Computer Science and Information Technology, HSLU - Lucerne University of
Applied Sciences and Arts, Suurstoffi 1, 6343 Rotkruez, Switzerland
{ramon.christen,luca.mazzola,alexander.denzler}@hslu.ch
[2] Human-IST Institute, University of Fribourg, Bd de Pérolles 90, 1700 Fribourg,
Switzerland
edy.portman@unifr.ch

Abstract. Similarity metrics measure distance to a compared time series. It allows for a classification and dependency search. These metrics are used for the selection of additional time series in forecasting, which involves advanced information on the target time series, known as exogenous data. Several studies demonstrate an accuracy gain by including such data that represents the environment for the target time series. Yet, robust significance analysis represents a key prerequisite for the correct context identification towards and accurate time series forecast.

For this reason, this article presents current similarity metrics and demonstrates significant aspects in aligning time series. The concept of dynamic comparison and its importance will be discussed as a basis for a robust perceptual significance analysis. By employing a pair of exemplary load and exogenous time series, alignment capabilities of promising distance metrics were tested, thus demonstrating gaps for an effective perceptual computing methodology. Finally, this paper examines prerequisites necessary for a new robust significance analysis methodology on three characteristic exogenous and target time-series combinations.

Keywords: Time series · Similarity metric · Alignment · Time warping · Feature extraction · Perceptual computing · Robustness

1 Introduction

Exogenous data are additional information for a time series. It allows for the improvement of forecast accuracy by providing extra information about the target time series behaviour. Numerous publications highlight this advantage through tested examples of large numbers of exogenous variables that showed an increase in forecast accuracy in the domain of power consumption and/or production. In addition, the increasing interconnection of Internet of Things (IoT) devices, handhelds and social media platforms offers new sources for exogenous

© Springer Nature Switzerland AG 2022
D. Ciucci et al. (Eds.): IPMU 2022, CCIS 1602, pp. 469–482, 2022.
https://doi.org/10.1007/978-3-031-08974-9_37

data that might be worthwhile to consider for enhanced forecast accuracy [5]. Yet, two aspects are necessary for a useful inclusion of this data in forecasting: a good knowledge and understanding of the behaviour, but also a change in the computational method, such as in the perceptual computing approach [6]. In fact it is necessary to know precisely which characteristics influence the target time series and which information is appropriate in the forecast.

Time series usually show volatile and complex characteristics, far away from strict and simple rules. This also applies for the influence of external variables on time series, which is not constant on a same level. While some parts have a significant impact on the target variable, others appear to have a weak influence. In worst case, an external variable provides only a short sequence describing a specific characteristic of the target time series while it has no remarkable influence for most of the time. The room temperature for instance shows many dependencies on external variables. Some of them appear to have a linear influence (such as the outside air temperature), others present a more complex effect. So the influence of sunlight. Depending on the solar irradiation angle and windows state, it changes the room temperature directly proportional to the difference with the outside temperature. Both variables have a varying influence that is neither of a constant strength nor following a strict periodic scheme. Calendar data show this behaviour even more obvious. While calendars typically provide only weak related information to room temperature, they may indicate an increased room occupation that directly affects the room temperature. This means, it is indispensable to know the conditions (how and when) under which these exogenous data influence the behaviour of the target variable. This applies for all remarkable characteristics in target variable including sudden changes, trends, seasonality as well as slow changes or long lasting effects.

As result, the evaluation of considering exogenous data needs to take account of the influence of data on target time series. State-of-the-art forecast approaches measure the significance and influence, respectively, of an exogenous variable by similarity metrics while comparing it with the target time series. Based on the value reported, the variable will be considered or not. The higher the similarity between exogenous data and the target variable is, the more useful seems the information to be. This similarity evaluation usually encloses the whole time series or forecast sequence. Yet, it neglects local, short-term characteristics providing potentially valuable information for forecast. In most cases, exogenous variables show correlated features to the target variable. Anyway, they can appear only in a different time instant and/or present a different shape. As such, similar feature detection requires robustness, as shown by [17] with respect to average forecasting.

Some distance measures applied in similarity metrics also have the ability to provide information on feature level, meaning, which signal parts might be of relevance for the target variable. These similarity metrics might provide a reasonable influence estimation approach for our goal: a section related individual inclusion of exogenous data. With this focus, we evaluate existing similarity metrics and discuss prerequisites for a new significance analysis methodology allowing for quantifying the relevance of exogenous data in load forecasting. The following sections discuss the fundamentals of similarity metric and differentiates lock-step and dynamic comparison methods. Section 3 describes the concept of elastic metrics and introduces exemplary exogenous and power load time series for a comparison of promising state-of-the-art approaches. As result, the Sect. 4 discusses on three exemplary cases relevant aspects for aligning time series and evaluating the influence of exogenous data. Finally, in Sect. 5, this work points out the next steps towards a new analysis method before it concludes the contribution in Sect. 6.

2   Exogenous Data Evaluation Metrics

The benefit of including exogenous data in load forecasting depends on the provided information about distinctive features in target time series. Valuable information reveals the presence and the shape of distinct parts in target time series. Present forecast methods quantify the value of exogenous data by means of simple similarity metrics. Thereby, the similarity is a proportional measure for the potential benefit of exogenous data.

In best case, information in exogenous data, namely the feature, is present before its effect on target variable. However, exogenous variable can also be synchronous with target variable or can even have a delay to them. Similar applies for amplitude. In practice, time series typically have different amplitude scaling. This implies constant but also dynamic amplitude shifts and scaling differences and changes respectively. These differences lead to varying Euclidean distances and make it difficult to achieve reasonable and comparable similarity indications. As result, time series comparison normally require a normalisation of the amplitude scaling as well as an offset harmonisation by a cancellation. Relative metrics, such as correlation coefficients, are resilient against amplitude warping and allow for comparing time series with different amplitude scaling/shift effects [8].

An extensive review on warping capable similarity measuring methods (here called elastic) provided a collection of several approaches listed in Table 1. It comprises methods based on the Dynamic Time Warping (DTW) mechanism and the Edit Distance (ED) as well as multiple other approaches. The further evaluation of these methods focus on the main concepts with respect to a qualitative exogenous data influence estimation and feature detection. Thus, the **bold** entries in Table 1 are the ones fully considered in our in-depth analysis. This selection excludes special derivations of the DTW and ED and methods that seem less promising for the target use case.

Time series transformation such as Fourier or Wavelet allow to detect and extract distinctive features [1,13]. Representing time series in alternative domains allow for making features more obvious, yet still distorted in time and amplitude. This helps to detect valuable key features in time series indeed. However, such transformations do not provide any distance measure nor similarity metric but are part of distance measure approaches. As a result, transformations

Table 1. Summary of elastic distance measure approaches

Elastic measures	
Dynamic Time Warping Methods	
DTW	**Dynamic Time Warping**
DDTW	**Derivative DTW**
IDTW	Incremental DTW
WDTW	Weighted DTW
Edit Distance Methods	
EDR	**Edit Distance on Real Sequence**
ERP	**Edit Distance with Real Penalty**
LCSS	Longest Common Subsequence
TWED	Time Warping Edit Distance
Other Methods	
ALoT	**Alignment of Textures**
AMSS	**Angular Metric for Shape Similarity**
FSTE	Fast Time Series Evaluation
MSM	**Move-Split-Merge**
MVM	**Minimal Variance Matching**
SpADe	Spatial Assembling Distance
Swale	Sequence Weighted Alignment Model

like Fourier or Wavelet are not suitable for evaluating the value of exogenous data in forecasting and are therefore out of scope for this research and not present in Table 1. Nevertheless, they might be considered for the development of a new significance metric.

In brief, measuring and quantifying influences between time series deals with two main challenges: a) **time-** and b) **amplitude-warping** between compared time series. Even if the shape of two time series is congruent, they may still show a time or amplitude warping. Matching characteristic features may be present in differing shapes that show a stretch, compression, drift or a constant shift to each other. And all these differences may appear in variable strength. In contrast to constant differences, such variations are hard to compensate when aiming for meaningful feature matching. The capability of coping with those hard aspects is the central pint of elastic measure methods. As a result, an accurate value estimation of exogenous data addresses elastic measures with capability to detect similar and warped features in compared time series.

Advantage of Dynamic vs. Lock-step Comparison

Time series comparison typically implies point wise distance evaluation - applicable as **lock-step** or **dynamic** comparison. As the name implies, lock-step

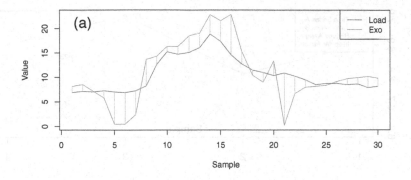

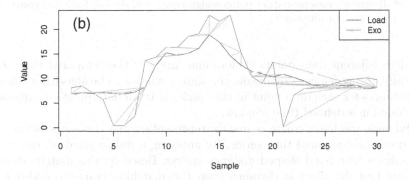

Fig. 1. Lock-step (a) vs. dynamic (b) comparison method: (a) compares time series data with a fix time shift whereas (b) allows flexible time shift in finding shortest distance. As result, in (b), data points of one time series may have shortest distances to multiple data points of the compared time series.

methods compare data points with a fix time distance to each other. This forces a constant time shift between compared time series. Literature usually presents the lock-step method as a time synchronised comparison that suggest comparing only data points at the same instant of time, as illustrated in Fig. 1(a) (hides possible constant shift compensation). It defines the (amplitude) distance of two fix assigned data points and leaves out all other data points, even if they have a shorter distance to each other. That means, for each point of time, this comparison considers the distance from the data point in time series A to one and only one of the time series B. In this way, it retains a fix time shift between data points of the compared time series. The lock-step comparison is suitable for similarity measures with strict time relations such as for detecting a distinct pattern in a time series that exactly matches a desired amplitude and time behaviour.

In contrast, dynamic comparison allows some time flexibility in data point matching by assigning data points of different points in time to each other, as shown in Fig. 1(b). In addition to a fix point-to-point comparison, it also

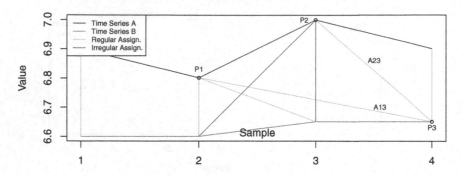

Fig. 2. Constraint for **distance selection in dynamic comparison**. Following distance selection (data point assignment) after *A13* requires continuing in time series. The red distances cross preceding point assignments and do not fulfil the constraints. Hence, those are not allowed.

considers adjacent data points (before and after) of the compared time series. Typically, this is achieved by means of a sliding window. The algorithm calculate the distances for each data point in time series A to all data points in time series B included in a defined time window.

Dynamic methods create a matrix representing the distances between all data points of compared time series. By applying a sliding window, this results in a Sakoe-Chiba band shaped *distance matrix*. Based on this matrix, dynamic methods find the shortest distance from the matching start- to end-point in compliance with the rule that each point assignment does not cross preceding ones. Figure 2 demonstrates this constraint for the assignment of *P2* to *P3*. The point *P3* is the last data point of time series *B* inside the window and defines to *P1* the last selected distance *A13*. The next possible assignment for this data point is *P2* with the distance *A23*. Yet, point *P2* has no assignment up to the last selected distance *A13* and might be assigned to all points of time series *B*. But, the constraint only allows an assignment to point *P3* as this is the only one that is not back in time to the last assigned data point of time series *B*. In contrast to the red highlighted distances, solely distance *A23* does not cross previous assignments like *A13*. For this reason, also for point *P2*, the assignment *A23* is the only one that fulfills the constraints.

This method allows to stretch or compress the time relations of compared time series and is key component to allow dynamic comparison. For this reason, dynamic methods are suitable for comparing time distorted series while lock-step approaches benefit locating strict parameterised features.

3 Evaluation of Elastic Metrics

Elastic metrics provide significant advantages in similarity measure compared to static metrics. As shown in previous sections, the main concept of elastic distances relies on a one-to-many data point mapping that results in a distance

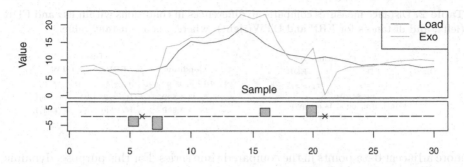

Fig. 3. Exemplary time series applied to discussed elastic metrics. The Load series present two main peaks and a small decrease towards the end. The exogenous series (Exo) is a distorted copy of the load series, by overlapping with white noise. It has a large deviation in four data points and two missing data points. The inferior parts shows (1) the levels of distortions with a boxplot and (2) missing data point by crosses. (Color figure online)

matrix for revealing shortest distances. In more detail, a table aggregates the distances of every data point to a row of data points of the comparing time series. The following subsections discuss fundamental elastic metric approaches that are able suppressing time and/or amplitude shifts in comparing time series. Based on two exemplary time series with remarkable but distorted similarities they show traits of key approaches by comparing basic distance matrices.

The exemplary load time series (red) in Fig. 3 has three main characteristic: two rather flat peaks and a slightly decreasing tail. The exogenous time series (green) is a copy of the load time series with amplitude and time distortions. In amplitude, this time series has three modifications compared to the original. a) All data points have a randomised deviation of max ±0.5. b) Four data points present a large deviation between 20% to 50% of the peak value. The boxplot in Fig. 3 shows the deviations in size and position. c) Two data points marked with a cross simulate missing samples with a *null*. In addition, exogenous time series has a positive time warping in the middle and a negative at the endings. The warping follows a cosine whereas time stretching implies a value duplication and the compression a deletion of some data points.

State-of-the-Art Elastic Similarity Metrics

Originally developed for time normalisation in the research area of spoken word recognition, the DTW algorithm [14] presents a warping method that obtained attention in various data-science applications. In recent years, numerous variants have been developed addressing problems in aligning time and/or amplitude warped time series for comparison - with one common: all rely on a distance matrix.

Contrary to lock-step metrics, dynamic approaches allow multiple assignments for data matching. In fact, they additionally evaluate distances to one or

Table 2. Distance measures comparison: differences in conditions within ED and EDR (left), and distances for ERP and DTW(right), where g is a constant value.

Dist.	Condition ED	EDR	Condition	Distance ERP	DTW
0	$r_i == s_i$	$\|r_i - s_i\| \leq \epsilon$	r_i and s_i is no gap	$\|r_i - s_i\|$	$\|r_i - s_i\|$
1	otherwise*		s_i is a gap	$\|r_i - g\|$	$\|r_i - s_{(i-1)}\|$
	(*incl. if r_i or s_i is a gap)		r_i is a gap	$\|s_i - g\|$	$\|s_i - r_{(i-1)}\|$

more adjacent data points in the compared time series. For this purpose, dynamic approaches also consider multidimensional aspects (such as module and rotation in a vector) to calculate a distance matrix for all data points.

The main difference between the analysed approaches lies in the data used in the distance matrix and the way of optimisation, particularly the definition of the matching path through the matrix. The contrasted approaches in Fig. 4 measure the distance between data points differently. This ranges from simple Manhattan (L1) to more complex distances as in the ALignment of Textures (ALoT) approach [13] that includes multiple parameters in the distance calculation. Inspired by the ED which is used to quantify the dissimilarity of two strings, Chen et al. applied a threshold on this distance to make ED applicable as dissimilarity measure for time series. In this way, the proposed Edit Distance on Real Sequence (EDR) approach [3] fills the distance matrix with transformed dissimilarities into binary values using a threshold ϵ. Thus, the EDR approach does not fulfil the triangle inequality and therefore it is not a metric. For this reason the Edit Distance with Real Penalty (ERP) distance drops the threshold and applies the L1-norm [2]. Additionally, it introduces a constant for evaluating gaps. According to [2], Table 2 presents the differences for computing the distance $dist(r_i, s_i)$ with respect to DTW.

With additional constraints, the Correlation Optimized Warping (COW) algorithm can be seen as special case of DTW. It reduces the search space for optimal warping and employs correlation coefficient as optimisation criterion [16]. The COW was developed for aligning chromatographic (peak) profiles. Other further developments of the DTW are the Derivative Dynamic Time Warping (DDTW) that quantifies the distance based on time series's deviation instead absolute values [9]. The Incremental Dynamic Time Warping (IDTW) by Khan et al. providing accurate comparison between incomplete and complete sequences [10] and Weighted Dynamic Time Warping (WDTW) proposed by Jeong et al. that penalises points with higher phase difference [7].

The Minimum Variance Matching (MVM) method combines the straight distance calculation from DTW with partial match of the query time series to subsequence of the target time series from Longest Common Subsequence (LCSS) [11]. Similarly, the Move-Split-Merge (MSM) method deviates significantly from ERP and DTW. It replaces the *insert* and *delete* operations by combinations of *move* and *split* operations. In this way, the MSM considers the current and adjacent values for insert and delete operations and satisfies the triangular inequality [15].

Similar to EDR, the ALoT method [13] uses binary values for employing the hamming distance on time series. For this purpose, it applies a texture extraction first, by substituting values not inferior to the current one with 'zero' and setting to 'one' all the others. Other approaches, such as Angular Metric for Shape Similarity (AMSS), DDTW and Spacial Assembling Distance (SpADe), rely on continuous values. AMSS uses the orientation of vectors set by two successive values. In this way, it defines the distance on the derivation of time series as the DDTW. Yet, the AMSS method relies on cosine distance between vectors [12]. This takes time and amplitude warping into account, as the SpADe also does. However, the SpADe computes warping path on the distance of *local pattern matches* - a vector of six different features describing a match of two subsequences according to a threshold [4].

In Fig. 4 we present a comparison of some of the methods from the paragraphs above. The red 'REF' line shows the modulated time warping. Notably, no algorithm can accurately replicate the optimal assignment. All methods use the same window length and DTW's backtrack algorithm. And parameters were set to default values according to literature. The comparison omits the SpADe method as we avoided the path finding implementation with the proposed Dijkstra's algorithm.

The figure shows larger deviations in the proposed alignment from the modulated warping for the methods DTW, DDTW and ALoT. In contrast, the ERP method matches the modulated time warping promisingly. Yet, finding high-value features in exogenous data for forecasting, presents other difficulties in addition to aligning time series with similar shape. Some of them will be introduced and discussed in the next chapter.

4   Evaluation of Time Series Alignment Methods

Time and amplitude warping is a main topic in time-series alignment and is crucial for pointing out essential information from exogenous data. A proper alignment is precondition for recognising influences on the target, due to the fact that key parts can assume different shapes in exogenous time series. Steps, spikes but also large variance in periods may be informative regarding the load time series. In Fig. 5 we show three possible variants of exogenous time series indicating a distinctive behaviour in target time series: in a) time shifted triangular series with identical shape for an energy consumption that straight follows the behaviour of exogenous data. b) A preceding spike followed by a significant change in the target variable with a certain time shift and c) a step in exogenous data causing a short term change.

Identical shapes in exogenous and target time series, meaning strong correlating sub-sequences in exogenous and target time series, may appear in the power consumption of cooling or heating units following fluctuating ambient temperatures. This is the ideal case that allows for accurate alignment results. Secondly, exogenous data with a constant time shift (a) provide the most valuable data for forecasting. On the downside, indications in exogenous data also occur in

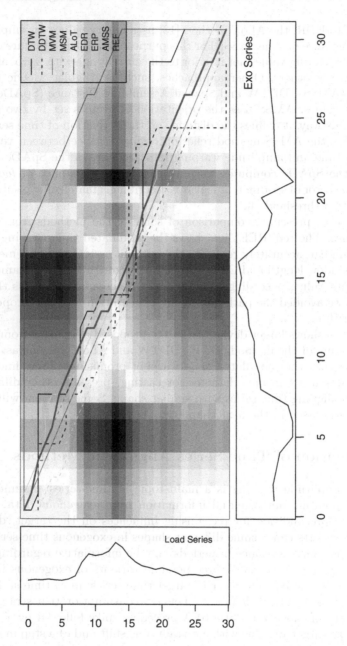

Fig. 4. A comparison of state-of-the-art dynamic warping methods. As reference, the matrix shows the L1 distance in increasingly saturated grey scale.

different shapes than target data. Two non-identical variations are shown in the other figures. A spike (b) or a single step (c) in exogenous data may point to a specific behaviour in target time series as well. This can be an opening and clos-

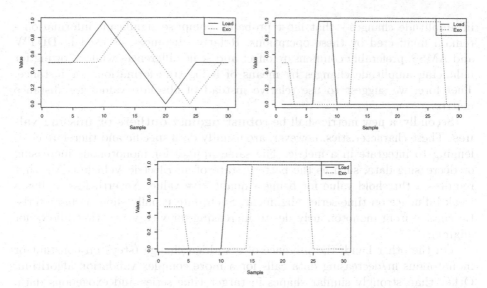

Fig. 5. Possible time warped matching features. (a) matching time warped triangular signal for load and exo series. (b) large rect. load shape triggered by an exo spike. (c) large rect. load shape triggered by a single change in exo signal.

ing of a door or an arrival of goods, representing a spike and step respectively, that triggers an increased power consumption.

These examples reveal two relevant aspects: 1. A feature providing high value information does not necessarily need a high correlation with the target time series sub-sequence it points to. 2. Recognising valuable features in exogenous time series requires serious alignments even if its shape shows a low similarity to the target time series. Both of them are of high relevance for applied similarity metrics in order to evaluate the influence of exogenous data on time series. As shown in Fig. 4, none of the investigated alignment methods is able to accurately replicate the optimal matching. Thus, in the next section we discuss aspects to be considered for developing an appropriate alignment metric.

5   Towards a Significance Analysis Methodology

Evaluating the influence of exogenous data on target time series requires a robust and dynamic metric. In fact, the measure needs to combine **robust alignment with flexibility in terms of amplitude warping**. This is mandatory as the evaluation compares different time series, more specifically, recordings with different sampling and amplitude rates. A car or handheld device provides its location probably in higher rates than the power consumption sampling. In the same way, the amplitude can also differ significantly by offset, scale or large varying amplitude changes. Indeed, adding and multiplying constant values can correct predominantly amplitude warping in offset and scale. Yet, large variations in amplitude changes - that most probably comprise significant information - remain unaffected by these operations. Relative distances, as used in DDTW and AMSS, preferably compensate offset and scale differences while allowing for balancing amplitude changes by means of a log transformation, for instance. Therefore, we suggest to use relative instead of absolute values for distance matrix.

Secondly, a new metric shall be **robust against outliers or missing values**. These characteristics, however, are usually data specific and therefore challenging to integrate in a metric. The same applies for monotonous increasing or decreasing data, such as the battery state of an Electric Vehicle (EV), that requires a threshold value for being a qualitative value. Nevertheless, it has a weak influence on time-series alignment. Such data usually show a characteristic change from monotonously de- to increasing, or vice versa, that allows for aligning.

On the other hand, the presence of a sudden change (step) on constant or monotonous in/decreasing data calls for a more complex validation algorithm. Other than strongly similar shapes in target time series and exogenous data, monotonous changing data require different values than amplitudes for distance calculation.

The similarity metrics we studied and analysed in this work do not meet the requirements for a proper significance analysis. A serious methodology presents two main characteristics: robust alignment capability and sensitivity in counting and delimiting key aspects. On top of it, fulfilling the triangle inequality will allow its usage as a regular metric instead of only as distance measure. The development of such a methodology is the next step in our research agenda.

6   Conclusions

Exogenous data provide promising added value for increasing time series forecast as shown in [5]. The use of these information, however, imply an appropriate feature selection in advance. This usually relies on similarity metrics comparing the behaviour of time series with exogenous information. Although several metric take time and/or amplitude warping into account, their primarily aim is measuring shape instead of looking at valuable feature matching.

In this article we discuss different approaches for similarity metrics and reveal hurdles and difficulties in comparing time series. Thereby, it clarifies the difference between Lock-Step and Dynamic comparison and shows that primarily time warping causes challenges, when aligning time series properly. For this reason, it focuses on elastic metrics, for which it reviews state-of-the-art approaches and compares the performance of promising approaches on exemplary distorted load and exogenous time series. Aiming at the alignment of exogenous data on power time series, we discuss essential requirements with three typical dissimilarities: 1. time shifted series, 2. spike and 3. step functions in leading time series. By means of these examples, we show a lack of suitable distance metrics for measuring the influence of exogenous data on target time series. Thus, our proposal is to develop a metric relying on delta values - as in DDTW and AMSS - for distance evaluation that primarily stresses similar features. Accordingly, our next step will be to take into account the identified missing aspects towards a better suited significance analysis approach for evaluating the influences and relevance of exogenous data on forecasting time series.

Acknowledgment. Thanks to Ben Haymond for his help in the language editing of this abstract.

References

1. Abanda, A., Mori, U., Lozano, J.A.: A review on distance based time series classification. arXiv:1806.04509 [cs, stat], June 2018. http://arxiv.org/abs/1806.04509, arXiv: 1806.04509
2. Chen, L.: On the marriage of Lp-norms and edit distance. In: Proceedings of the 30th VLDB Conference, p. 12. Toronto, Canada, (2004)
3. Chen, L., Özsu, M.T., Oria, V.: Robust and fast similarity search for moving object trajectories. In: Proceedings of the 2005 ACM SIGMOD International Conference on Management of Data - SIGMOD 2005, p. 491. ACM Press, Baltimore, Maryland (2005). https://doi.org/10.1145/1066157.1066213, http://portal.acm.org/citation.cfm?doid=1066157.1066213
4. Chen, Y., Nascimento, M., Ooi, B., Tung, A.: SpADe: on shape-based pattern detection in streaming time series. In: Proceedings - International Conference on Data Engineering, pp. 786–795, May 2007. https://doi.org/10.1109/ICDE.2007.367924
5. Christen, R., Mazzola, L., Denzler, A., Portmann, E.: Exogenous data for load forecasting: a review. In: Proceedings of the 12th International Joint Conference on Computational Intelligence. pp. 489–500. SCITEPRESS - Science and Technology Publications, Budapest, Hungary (2020). https://doi.org/10.5220/0010213204890500, https://www.scitepress.org/DigitalLibrary/Link.aspx?doi=10.5220/0010213204890500
6. Henson, C., Sheth, A., Thirunarayan, K.: Semantic perception: converting sensory observations to abstractions. IEEE Internet Comput. **16**(2), 26–34 (2012). https://doi.org/10.1109/MIC.2012.20
7. Jeong, Y.S., Jeong, M.K., Omitaomu, O.A.: Weighted dynamic time warping for time series classification. Pattern Recogn. **44**(9), 2231–2240 (2011). https://doi.org/10.1016/j.patcog.2010.09.022. https://linkinghub.elsevier.com/retrieve/pii/S003132031000484X
8. Jiang, G., Wang, W., Zhang, W.: A novel distance measure for time series: maximum shifting correlation distance. Pattern Recogn. Lett. **117**, 58–65 (Jan 2019). https://doi.org/10.1016/j.patrec.2018.11.013. https://www.sciencedirect.com/science/article/pii/S0167865518308985
9. Keogh, E.J., Pazzani, M.J.: Derivative dynamic time warping. In: Proceedings of the 2001 SIAM International Conference on Data Mining, pp. 1–11. Society for Industrial and Applied Mathematics, April 2001. https://doi.org/10.1137/1.9781611972719.1. https://epubs.siam.org/doi/10.1137/1.9781611972719.1
10. Khan, N.M., Lin, S., Guan, L., Guo, B.: A visual evaluation framework for in-home physical rehabilitation. In: 2014 IEEE International Symposium on Multimedia, pp. 237–240, December 2014. https://doi.org/10.1109/ISM.2014.21

11. Latecki, L.J., Megalooikonomou, V., Wang, Q., Lakaemper, R., Ratanamahatana, C.A., Keogh, E.: Elastic partial matching of time series. In: Jorge, A.M., Torgo, L., Brazdil, P., Camacho, R., Gama, J. (eds.) PKDD 2005. LNCS (LNAI), vol. 3721, pp. 577–584. Springer, Heidelberg (2005). https://doi.org/10.1007/11564126_60
12. Nakamura, T., Taki, K., Nomiya, H., Seki, K., Uehara, K.: A shape-based similarity measure for time series data with ensemble learning. Pattern Anal. Appl. **16**(4), 535–548 (2013). https://doi.org/10.1007/s10044-011-0262-6. http://link.springer.com/10.1007/s10044-011-0262-6
13. Oğul, H.: ALoT: a time-series similarity measure based on alignment of textures. In: Yin, H., Camacho, D., Novais, P., Tallón-Ballesteros, A.J. (eds.) IDEAL 2018. LNCS, vol. 11314, pp. 576–585. Springer, Cham (2018). https://doi.org/10.1007/978-3-030-03493-1_60
14. Sakoe, H., Chiba, S.: Dynamic programming algorithm optimization for spoken word recognition. IEEE Trans. Acoustics Speech Sig. Process. **26**(1), 43–49 (1978). https://doi.org/10.1109/TASSP.1978.1163055
15. Stefan, A., Athitsos, V., Das, G.: The Move-Split-Merge Metric for Time Series. IEEE Trans. Knowl. Data Eng. **25**(6), 1425–1438 (2013). https://doi.org/10.1109/TKDE.2012.88
16. Tomasi, G., van den Berg, F., Andersson, C.: Correlation optimized warping and dynamic time warping as preprocessing methods for chromatographic data. J. Chemometrics **18**(5), 231–241 (2004). https://doi.org/10.1002/cem.859, https://onlinelibrary.wiley.com/doi/abs/10.1002/cem.859
17. Zhao, Y.: The robustness of forecast combination in unstable environments: a Monte Carlo study of advanced algorithms. Empirical Econ. **61**(1), 173–199 (2021). https://doi.org/10.1007/s00181-020-01864-w

On the Role of the Considered Measure in a Quantile-Based Graded Version of Stochastic Dominance

Raúl Pérez-Fernández[✉] and Juan Baz

Department of Statistics and O.R. and Mathematics Didactics, University of Oviedo, Oviedo, Spain
{perezfernandez,bazjuan}@uniovi.es

Abstract. Stochastic dominance is a popular stochastic order for the comparison of random variables. Even though it carries many intuitive properties, stochastic dominance does not result in a complete order relation, even when restricted to certain parametric families of probability distributions. In this paper, we explore a graded version of stochastic dominance measuring the degree in which a random variable stochastically dominates another random variable. This graded version of stochastic dominance allows to quantify the degree in which a random variable stochastically dominates another random variable by measuring the subset of the unit interval on which the quantile function of the first random variable is pointwisely greater than or equal to the quantile function of the second random variable. Particular attention is paid to the choice of measure, exploring its influence on four different location-scale distribution families: Gaussian, uniform, Cauchy and exponential.

Keywords: Stochastic order · Stochastic dominance · Graded relation

1 Introduction

Stochastic orders [3,13,17] are a popular tool in probability and statistics for the comparison of random variables. There is little doubt that the most prominent stochastic order is that of stochastic dominance [7], which has been successfully applied in several fields of application (see, e.g., [8]). We say that a random variable stochastically dominates another random variable in case the cumulative distribution function of the first random variable is pointwisely smaller than or equal to the cumulative distribution function of the second random variable. This implies that the first random variable is taking greater values with a greater probability.

This research has been partially supported by the Spanish Ministry of Science and Technology (TIN-2017-87600-P). The authors are grateful to Inés Couso for valuable suggestions and comments that helped improving the quality of the manuscript.

© Springer Nature Switzerland AG 2022
D. Ciucci et al. (Eds.): IPMU 2022, CCIS 1602, pp. 483–493, 2022.
https://doi.org/10.1007/978-3-031-08974-9_38

As intuitive as it is, stochastic dominance is not a complete relation on the set of random variables, i.e., it does not allow to compare all pairs of random variables. For instance, there exists no Gaussian-distributed random variable that stochastically dominates another Gaussian-distributed random variable with a different standard deviation. For this reason, other stochastic orders have been considered such as the comparison of expected utilities [12], statistical preference [5,6] and probabilistic preference [10]. Here, we follow the direction started in [1,2] in which a graded version of stochastic dominance is proposed by measuring the proportion of the unit interval in which the quantile function of a random variable is pointwisely greater than or equal to the quantile function of another random variable. The goal of this paper is to analyse the influence of the choice of measure, by exploring four prominent location-scale distribution families.

The remainder of the paper is structured as follows. Section 2 recalls the basic concepts on stochastic dominance that are necessary for the correct understanding of the paper and presents the notion of graded stochastic dominance. This graded stochastic dominance is studied within a location-scale distribution family in Sect. 3. The case of four prominent location-scale distribution families are studied in Sect. 4. The paper ends with some conclusions and open problems in Sect. 5.

2   Stochastic Dominance

The notion of stochastic dominance has been deeply studied for the comparison of random variables. A random variable X is said to stochastically dominate another random variable Y if the cumulative distribution function F_X of X is pointwisely smaller than or equal to the cumulative distribution function F_Y of Y. This implies that X tends to take larger values than Y.

Definition 1. *Given two random variables X and Y, X is said to stochastically dominate Y, denoted by $X \succeq_{FSD} Y$, if $F_X(x) \leq F_Y(x)$ for any $x \in \mathbb{R}$.*

The left-hand side of Fig. 1 illustrates stochastic dominance for two Gaussian-distributed random variables. Note that the cumulative distribution function F_X of the random variable $X \rightsquigarrow N(0,1)$ is pointwisely greater than the cumulative distribution function F_Y of the random variable $Y \rightsquigarrow N(1,1)$, therefore $Y \succeq_{FSD} X$.

It must be remarked that the stochastic dominance relation may equivalently be defined in terms of the survival functions $\overline{F_X} = 1 - F_X$ and $\overline{F_Y} = 1 - F_Y$ as $X \succeq_{FSD} Y$, if $\overline{F_X}(x) \geq \overline{F_Y}(x)$ for any $x \in \mathbb{R}$. This characterization is oftentimes preferred over the original definition since it associates greater random variables in terms of stochastic dominance with pointwisely greater survival functions (rather than smaller cumulative distribution functions).

Another equivalent, yet less prominent, characterization of stochastic dominance in terms of the quantile functions shall be of interest to this paper. We

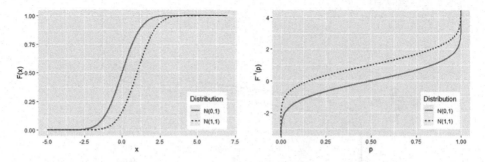

Fig. 1. Illustration of stochastic dominance between two Gaussian-distributed random variables in terms of cumulative distribution functions (left) and in terms of quantile functions (right).

recall that the quantile function $F_X^{-1} :]0,1[\to \mathbb{R}$ of a random variable X is the function defined as

$$F_X^{-1}(p) = \inf\{x \in \mathbb{R} \mid p \le F_X(x)\}.$$

If the cumulative distribution function F_X of X is continuous and strictly increasing, then the quantile function F_X^{-1} is the actual inverse of F_X. As shown in [17], a random variable X stochastically dominates another random variable Y if and only if the quantile function F_X^{-1} of X is pointwisely greater than or equal to the quantile function F_Y^{-1} of Y.

Proposition 1. *Given two random variables X and Y, it holds that $X \succeq_{FSD} Y$ if and only if $F_X^{-1}(p) \ge F_Y^{-1}(p)$ for any $p \in]0,1[$.*

The right-hand side of Fig. 1 illustrates this characterization of stochastic dominance for two Gaussian-distributed random variables. Note that the quantile function F_X^{-1} of the random variable $X \rightsquigarrow N(0,1)$ is pointwisely smaller than the quantile function F_Y^{-1} of the random variable $Y \rightsquigarrow N(1,1)$, therefore $Y \succeq_{FSD} X$.

Stochastic dominance is known to be an order, i.e., it is a reflexive, transitive and antisymmetric binary relation. Unfortunately, it is only a partial order relation (i.e., not a complete order relation) and does not allow for the comparison of many pairs of random variables for which the cumulative distribution functions intersect. For instance, there exists no Gaussian-distributed random variable that stochastically dominates another Gaussian-distributed random variable with a different standard deviation. For this reason and with the goal of allowing for the comparison of more random variables, other stochastic orders refining that of stochastic dominance have been proposed in the literature [11].

Figure 2 shows the quantile functions of three Gaussian-distributed random variables: $X \rightsquigarrow N(0,1)$, $Y \rightsquigarrow N(1,1)$ and $Z \rightsquigarrow N(0,2)$. It can be seen that it does not hold that $X \succeq_{FSD} Z$, neither does it hold that $Y \succeq_{FSD} Z$. However, it can be seen from Fig. 2 that Y is in some sense closer to stochastically dominating Z than X is.

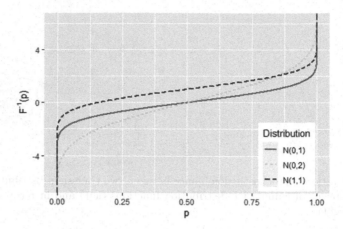

Fig. 2. Comparison of the quantile functions of three Gaussian-distributed random variables.

This intuition may be formalized by means of a graded relation representing the degree in which a random variable stochastically dominates another random variable. In this case, one could say that the degree in which Y stochastically dominates Z is greater than the degree in which X stochastically dominates Z. A first approach to formalize so, already explored in [15], consists on measuring the proportion of support on which one cumulative distribution function is above the other one. However, this resulted in several problems, mainly related to the comparison of random variables with unbounded support. With the intention of avoiding such problems, the approach considered in this work aligns with the direction initiated in [1,2], where instead of measuring the proportion of support on which one cumulative distribution function is above the other one, we measure the proportion of the unit interval on which one quantile function is above the other one. For the purpose of defining such version of graded stochastic dominance, we need to fix a non-null and finite measure $\nu : \Sigma_\nu \to [0,\infty[$ on the unit interval $]0,1[$, where Σ_ν is a σ-algebra with respect to which the quantile functions are measurable. Notice that, since the quantile functions are left-continuous, they are Borel-measurable, thus it is quite natural to choose the Borel σ-algebra, here denoted by $\Sigma_\mathbb{B}$.

Definition 2. *Consider two random variables X and Y and a non-null and finite measure $\nu : \Sigma_\mathbb{B} \to [0,\infty[$ on the unit interval. The degree in which X stochastically dominates Y, denoted by $R_\nu(X,Y)$, is defined as*

$$R_\nu(X,Y) = \frac{\nu(\{p \in \,]0,1[\,|\; F_X^{-1}(p) \geq F_Y^{-1}(p)\})}{\nu(]0,1[)}.$$

A natural choice for ν is the Lebesgue measure ν_L on $]0,1[$, which actually returns the γ-index presented in [1,2] (letting aside small technicalities for quantile functions that intersect on a subset with positive Lebesgue measure).

Different choices for ν may be used for modelling a more optimistic/pessimistic approach to the comparison of random variables by focusing more strongly on the right/left tails of the distributions. For instance, this optimistic/pessimistic approach plays an important role when considering stochastic orders in the context of investment decisions, where some investors might prefer a riskier/safer investment plan. We will return to the role of the choice of measure on the comparison of random variables in the upcoming sections.

An important result connects the stochastic dominance relation and the degree of stochastic dominance.

Proposition 2. *Consider two random variables X and Y and a non-null and finite measure $\nu : \Sigma_{\mathbb{B}} \to [0, \infty[$ on the unit interval.*

(i) If $X \succeq_{FSD} Y$, then $R_\nu(X, Y) = 1$.
(ii) If $\nu(A) > 0$ for any nonempty open set $A \in \Sigma_{\mathbb{B}}$, then $X \succeq_{FSD} Y$ if and only if $R_\nu(X, Y) = 1$.

Proof. Suppose $X \succeq_{FSD} Y$. It trivially follows that

$$\{p \in]0, 1[| \ F_X^{-1}(p) \geq F_Y^{-1}(p)\} =]0, 1[,$$

therefore, $R_\nu(X, Y) = \frac{\nu(]0,1[)}{\nu(]0,1[)} = 1$.

Suppose $R_\nu(X, Y) = 1$. It follows that $\nu(\{p \in]0, 1[| \ F_X^{-1}(p) < F_Y^{-1}(p)\}) = 0$. We need to prove that $\{p \in]0, 1[| \ F_X^{-1}(p) < F_Y^{-1}(p)\} = \emptyset$. Suppose that there exists $q \in]0, 1[$ such that $F_X^{-1}(q) < F_Y^{-1}(q)$. Since both quantile functions are left-continuous, there exists $\varepsilon > 0$ such that $F_X^{-1}(q') < F_Y^{-1}(q')$ for any $q' \in]q - \varepsilon, q[$. The contradiction follows from the fact that $\nu(A) > 0$ for any nonempty open set A. It is concluded that $X \succeq_{FSD} Y$. ∎

3   Graded Stochastic Dominance for a Location-Scale Distribution Family

The location-scale distribution family generated by a cumulative distribution function F_0 is the set of all distribution functions F for which there exists a location parameter $\mu \in \mathbb{R}$ and a scale parameter $\lambda \in \mathbb{R}+$ such that the distribution function F may be expressed as $F(x) = F_0(\frac{x-\mu}{\lambda})$ for any $x \in \mathbb{R}$. The notation $F_{\mu,\lambda}$ is then used for referring to such distribution function. In terms of quantile functions, it holds that $F_{\mu,\lambda}^{-1}(p) = \mu + \lambda F_0^{-1}(p)$. Therefore, it is straightforward to verify that $F_{\mu,\lambda}^{-1}(p) \geq F_{\mu',\lambda'}^{-1}(p)$ if

$$(\mu - \mu') \geq (\lambda' - \lambda)F_0^{-1}(p).$$

The result above implies that, for random variables whose distributions belong to the same location-scale distribution family and have the same scale parameter, stochastic dominance holds if and only if there is a non-negative location shift. More formally, $F_{\mu,\lambda}^{-1}(p) \geq F_{\mu',\lambda}^{-1}(p)$ for any $p \in]0, 1[$ holds if and only if

$\mu \geq \mu'$. Furthermore, if F_0 is continuous and strictly increasing, stochastic dominance may only hold in a location-scale distribution family if both compared distributions have the same scale parameter, i.e., $F_{\mu,\lambda}^{-1}(p) \geq F_{\mu',\lambda'}^{-1}(p)$ for any $p \in]0,1[$ holds if and only if $\lambda = \lambda'$ and $\mu \geq \mu'$. Additionally, two different cumulative quantile functions $F_{\mu,\lambda}^{-1}$ and $F_{\mu',\lambda'}^{-1}$ from a location-scale distribution family may only intersect if $\lambda \neq \lambda'$ and $F_0^{-1}(p) = \frac{\mu-\mu'}{\lambda'-\lambda}$. If F_0 is continuous, then two different cumulative quantile functions $F_{\mu,\lambda}^{-1}$ and $F_{\mu',\lambda'}^{-1}$ (with $\lambda \neq \lambda'$) in a location-scale distribution family may only intersect at $p_0 = F_0\left(\frac{\mu-\mu'}{\lambda'-\lambda}\right)$.

Interestingly, the properties above lead to the following result for continuous random variables.

Proposition 3. *Consider two random variables X and Y from the location-scale family generated by a continuous distribution F_0 and a non-null and finite measure $\nu : \Sigma_{\mathbb{B}} \to [0, \infty[$ on the unit interval. It follows that:*

(i) If $\lambda_X = \lambda_Y$ and $\mu_X \geq \mu_Y$, then $R_\nu(X,Y) = 1$;
(ii) If $\lambda_X = \lambda_Y$ and $\mu_X < \mu_Y$, then $R_\nu(X,Y) = 0$;
(iii) If $\lambda_X \neq \lambda_Y$, then

$$R_\nu(X,Y) \in \left\{0, \nu\left(\left]0, F_0\left(\frac{\mu_X - \mu_Y}{\sigma_Y - \sigma_X}\right)\right]\right), \nu\left(\left[F_0\left(\frac{\mu_X - \mu_Y}{\sigma_Y - \sigma_X}\right), 1\right[\right), 1\right\}.$$

(iv) If $\lambda_X \neq \lambda_Y$ and F_0 is strictly increasing, then

$$R_\nu(X,Y) \in \left\{\nu\left(\left]0, F_0\left(\frac{\mu_X - \mu_Y}{\sigma_Y - \sigma_X}\right)\right]\right), \nu\left(\left[F_0\left(\frac{\mu_X - \mu_Y}{\sigma_Y - \sigma_X}\right), 1\right[\right)\right\}.$$

The Gaussian distribution family is a prototypical example of location-scale distribution family generated by the standard Gaussian distribution (where λ is simply denoted by σ for historical reasons). As an illustrative example, we provide several heatmaps for illustrating the degree of stochastic dominance of a standard Gaussian distribution over any other Gaussian distribution with parameters $\mu \in [-5, 5]$ and $\sigma \in]0, 4]$ for different measures. Figure 3 considers the Lebesgue measure, whereas Fig. 4 shows how the value of $R_\nu(N(0,1), N(\mu,\sigma))$ changes for four different measures. Considering a measure that gives more importance to the centers of the distributions (as that on the top-left subfigure of Fig. 4) increases the relevance of the location parameter, basically imposing that $R_\nu(N(0,1), N(\mu,\sigma))$ is close to one for $\mu < 0$ and close to zero for $\mu > 0$. Analogously, considering a measure that gives more importance to the extremes of the distributions (as that on the top-right subfigure of Fig. 4) increases the relevance of the scale parameter, basically imposing that $R_\nu(N(0,1), N(\mu,\sigma))$ is close to 0.5 for values of σ that differ from one. Finally, the bottom-left and bottom-right subfigures of Fig. 4 show how the symmetric pattern is lost by considering measures that give more importance to the left/right-hand side of the unit interval. It should be remarked that the four considered measures are just simple examples of measures weighing differently the left and right halves of the unit interval. In particular, since the beta distribution family allows to construct a large variety of measures with a very different behaviour, all four considered measures were obtained from a beta distribution $B(\alpha, \beta)$ as follows:

$$\nu(A) = \int_A f(x)\, dx,$$

where f is the density function of the considered beta distribution. Note that ν is assured to be a measure due to the Radon-Nikodym theorem [14, 16].

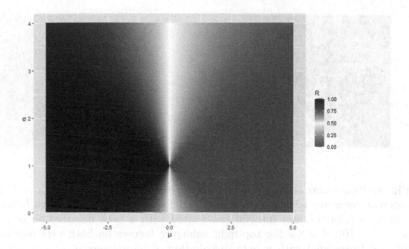

Fig. 3. Illustration of $R_{\nu_L}(N(0,1), N(\mu, \sigma))$ for different values of μ and σ.

4   ROC Curves for Different Measures

Consider the problem in which we have a sample of each of two random variables X and Y with unknown distribution, yet it is known that both random variables belong to the same location-scale distribution family and that one of the random variables stochastically dominates the other one. The goal is to identify which of the random variables stochastically dominates the other one by considering a decision rule based on the estimation $\widehat{R_\nu}$ of $R_\nu(X, Y)$ obtained by considering the quantile functions associated with the empirical cumulative distribution functions. More specifically, we are considering a decision rule of the type $X \succeq_{\mathrm{FSD}} Y$ if $\widehat{R_\nu} \geq c$ and $Y \succeq_{\mathrm{FSD}} X$ if $\widehat{R_\nu} < c$ for a certain $c \in [0, 1]$. Note the asymmetric role played by the first and second samples, especially when $c \neq 0.5$.

For evaluating the obtained family of decision rules for different measures ν, we run an experiment for simulating the distribution of $\widehat{R_\nu}$ by independently sampling from two random variables within the same location-scale distribution family. Within each of the four considered location-scale distribution families,

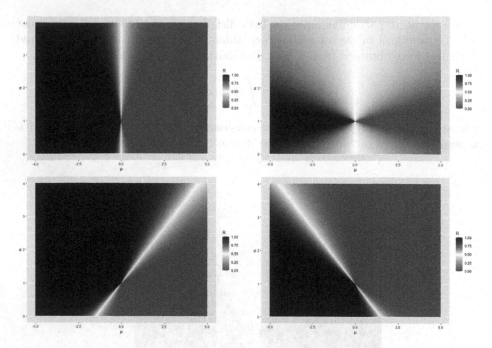

Fig. 4. Illustration of $R_\nu(N(0,1), N(\mu,\sigma))$ for different values of μ and σ for four different measures ν. The measure ν used for the top-left subfigure focuses on the central part of the unit interval (associated with a beta distribution of parameters $\alpha = \beta = 10$); that for the top-right subfigure focuses on both extremes of the unit interval (associated with a beta distribution of parameters $\alpha = \beta = 0.1$); that for the bottom-left subfigure focuses slightly on the left-hand side of the unit interval (associated with a beta distribution of parameters $\alpha = 1$ and $\beta = 10$); that for the bottom-right subfigure focuses strongly on the right-hand side of the unit interval (associated with a beta distribution of parameters $\alpha = 10$ and $\beta = 1$).

the standard distribution is compared against the same standard distribution shifted 0.1 units to the right. All samples are considered of equal size $n = 50$ and the number of repetitions is set to 10000. The results are illustrated in Fig. 5 by means of the Receiver Operating Characteristic curve (ROC curve) obtained when representing the True Positive Ratio (TPR) against the False Positive Ratio (FPR) for all threshold values c.

It is interesting to note the different behaviour of the five measures for the four different location-scale distribution families, as illustrated in Fig. 5. For instance, for the Gaussian distribution family, the Lebesgue measure seems to be the best possible measure. However, for the uniform distribution family, where the minimum and maximum statistics are linked to the maximum likelihood estimates of the extremes of the support of the uniform distribution, all three measures increasing the importance of the extremes lead to a better ROC curve than the Lebesgue measure. For the Cauchy distribution family, where the fat

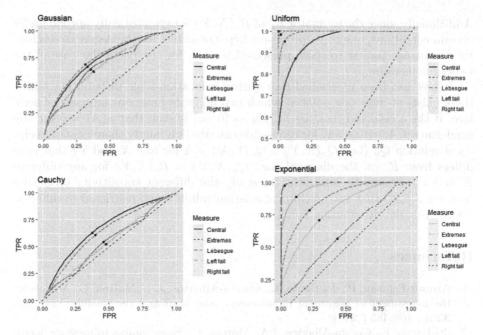

Fig. 5. Illustration of the ROC curves associated with five different measures ν (Central: measure associated with a beta distribution of parameters $\alpha = \beta = 10$; Extremes: measure associated with a beta distribution of parameters $\alpha = \beta = 0.1$; Lebesgue: Lebesgue measure; Left tail: measure associated with a beta distribution of parameters $\alpha = 1$ and $\beta = 10$; and Right tail: measure associated with a beta distribution of parameters $\alpha = 10$ and $\beta = 1$) for four different location-scale distribution families (top-left subfigure: Gaussian; top-right subfigure: uniform, bottom-left subfigure: Cauchy; and bottom-right subfigure: exponential). The black circles represent the cases in which the threshold parameter is set to $c = 0.5$.

tails often result in outliers, the measure that reduces the importance of the extremes leads to the best ROC curve. Finally, the most interesting example arises in the exponential location-scale distribution family, where the left tail easily identifies the support of the random variables (thus leading to the best ROC curve) and the right tail is heavily influenced by the maximum obtained values that typically show high variability (thus leading to the worst ROC curve).

5 Conclusions and Open Problems

In this paper, we have explored the importance of considering different measures within the graded version of stochastic dominance presented in [1,2]. In particular, four prominent location-scale distribution families have been analysed. Future work will consider different families of probability distributions, including continuous distribution families that are not a location-scale distribution family and discrete distribution families such as the binomial distribution.

Additionally, since the estimator $\widehat{R_\nu}$ of $R_\nu(X,Y)$ is asymptotically normal under certain conditions, one could propose a hypothesis test for the degree of stochastic dominance similar to those presented in [1,2] for the particular case of the Lebesgue measure.

It has also been mentioned that within a location-scale distribution family there exists at most one point at which two quantile functions intersect. Therefore, if the considered measure $\nu : \Sigma_\mathbb{B} \to [0,\infty[$ is such that $\nu(A) = 0$ for any singleton set A, within any location-scale distribution family there exists a reciprocal relation Q_ν (i.e., $Q_\nu(X,Y) + Q_\nu(Y,X) = 1$ for any X and Y) that only differs from R_ν on the diagonal (i.e., $Q_\nu(X,Y) = R_\nu(X,Y)$ for any different X and Y). It may be interesting to study the different transitivity properties (see, e.g., [4,9]) that such reciprocal relation fulfills when restricted to different (location-scale) distribution families.

References

1. Álvarez-Esteban, P., del Barrio, E., Cuesta-Albertos, J.A., Matrán, C.: Models for the assessment of treatment improvement: the ideal and the feasible. Stat. Sci. **32**(3), 469–485 (2017)
2. del Barrio, E., Cuesta-Albertos, J.A., Matrán, C.: Some indices to measure departures from stochastic order. arXiv preprint arXiv:1804.02905 (2018)
3. Belzunce, F., Riquelme, C.M., Mulero, J.: An Introduction to Stochastic Orders. Academic Press, Cambridge (2015)
4. De Baets, B., De Meyer, H., De Schuymer, B.: Cyclic evaluation of transitivity of reciprocal relations. Soc. Choice Welfare **26**(2), 217–238 (2006)
5. De Schuymer, B., De Meyer, H., De Baets, B.: A fuzzy approach to stochastic dominance of random variables. In: Bilgiç, T., De Baets, B., Kaynak, O. (eds.) IFSA 2003. LNCS, vol. 2715, pp. 253–260. Springer, Heidelberg (2003). https://doi.org/10.1007/3-540-44967-1_30
6. De Schuymer, B., De Meyer, H., De Baets, B., Jenei, S.: On the cycle-transitivity of the dice model. Theor. Decis. **54**(3), 261–285 (2003)
7. Lehmann, E.: Ordered families of distributions. Ann. Math. Stat. **26**, 399–419 (1955)
8. Levy, H.: Stochastic Dominance: Investment Decision Making under Uncertainty. Kluwer Academic Publishers, Boston (1998)
9. Monjardet, B.: A generalisation of probabilistic consistency: linearity conditions for valued preference relations. In: Kacprzyk, J., Roubens, M. (eds.) Non-Conventional Preference Relations in Decision Making, pp. 36–53. Springer, Heidelberg (1988)
10. Montes, I., Montes, S., De Baets, B.: Multivariate winning probabilities. Fuzzy Sets Syst. **362**, 129–143 (2019)
11. Montes, I., Rademaker, M., Pérez-Fernández, R., De Baets, B.: A correspondence between voting procedures and stochastic orderings. Eur. J. Oper. Res. **285**(3), 977–987 (2020)
12. Morgenstern, O., Von Neumann, J.: Theory of Games and Economic Behavior. Princeton University Press, Princeton (1953)
13. Müller, A., Stoyan, D.: Comparison Methods for Stochastic Models and Risks. Wiley, Hoboken (2002)

14. Nikodym, O.: Sur une généralisation des intégrales de MJ Radon. Fundam. Math. **15**(1), 131–179 (1930)
15. Pérez-Fernández, R., Baz, J., Díaz, I., Montes, S.: On a graded version of stochastic dominance. In: 19th World Congress of the International Fuzzy Systems Association (IFSA), 12th Conference of the European Society for Fuzzy Logic and Technology (EUSFLAT), and 11th International Summer School on Aggregation Operators (AGOP), pp. 494–500. Atlantis Press (2021)
16. Radon, J.: Theorie und anwendungen der absolut additiven mengenfunktionen. Hölder, Wien (1913)
17. Shaked, M., Shanthikumar, J.: Stochastic Orders. Springer Science & Business Media, New York (2007)

Bootstrapped Kolmogorov-Smirnov Test for Epistemic Fuzzy Data

Przemyslaw Grzegorzewski[1,2] and Maciej Romaniuk[1,3]

[1] Systems Research Institute, Polish Academy of Sciences, Newelska 6,
01-447 Warsaw, Poland
pgrzeg@ibspan.waw.pl
[2] Faculty of Mathematics and Information Science, Warsaw University of
Technology, Koszykowa 75, 00-662 Warsaw, Poland
[3] Warsaw School of Information Technology, Newelska 6, 01-447 Warsaw, Poland
mroman@ibspan.waw.pl

Abstract. A typical two-sample problem we deal with in statistical reasoning is to conclude whether the distributions of these samples differ significantly. In other words, we want to know if the given two samples come from the same distribution. To cope with this task one can utilize the Kolmogorov-Smirnov test which is probably the most popular goodness-of-fit two-sample test for real-valued observations. However, quite often the results of an experiment cannot be observed precisely and their imprecise outcomes are recorded as fuzzy numbers. Hence the main goal of this paper is to generalize the Kolmogorov-Smirnov test for fuzzy data. We propose a new version of this goodness-of-fit test based on the so-called epistemic bootstrap and designed for the fuzzy random samples analysis.

Keywords: Bootstrap · Empirical distribution function · Epistemic fuzzy data · Goodness-of-fit test · Kolmogorov-smirnov test

1 Introduction

The problem of comparing two samples has been addressed by many researchers and practitioners in various fields, wherever we would like to find evidence of whether the distribution functions of the populations behind these samples are identical or not. Many situations that require two-sample comparisons arise in biomedical problems, climate dynamics, engineering, stock prices analysis, etc. Usually, we suspect that the population distributions have different locations. Sometimes the distribution functions may differ in scale. It may also happen that we are interested in simultaneously detecting either location or scale differences between two populations. In each case, one has to apply a statistical test suitable for the considered alternative. However, if we suspect that differences in shape, as well as differences in location and/or scale, are possible, it is suggested to use a test against all types of differences that may exist between distributions.

© Springer Nature Switzerland AG 2022
D. Ciucci et al. (Eds.): IPMU 2022, CCIS 1602, pp. 494–507, 2022.
https://doi.org/10.1007/978-3-031-08974-9_39

One of the most famous ways to assess whether there are significant differences of any kind between distributions is the two-sample Kolmogorov-Smirnov test [13,20].

Unfortunately, quite often the results of an experiment cannot be observed precisely and we are provided only with their imprecise outcomes. It is caused by some general measurement problems, unreliable measuring tool, or other deficiencies in data recording. In these cases, imprecise observations can be effectively modeled and processed as fuzzy numbers. Since the corresponding fuzzy data are considered as disjunctive sets representing incomplete information they are called *epistemic* as opposed to *ontic* fuzzy sets that refer to essentially fuzzy data (see [2]).

Suppose now that we are faced with two samples of imprecise data modeled by fuzzy numbers and we would like to assess whether there are any differences between the parent population distributions. A suitable generalization of the Kolmogorov-Smirnov test would be desirable for this purpose. Several attempts to solve this problem have already appeared in the literature. Destercke, Strauss [3] and Grzegorzewski [6,7] proposed a generalization of the Kolmogorov-Smirnov test for the interval-valued data but their tests (unlike classical ones) do not lead necessarily to a binary decision (i.e. either reject or do not reject the null hypothesis). Unfortunately, no strict decision can be made for highly imprecise data. Hesamian, Chachi [11] and Hesamian, Taheri [12] considered the Kolmogorov-Smirnov test in a fuzzy environment but their tests produce a fuzzy p-value which is too often inconclusive, as it was shown by many experiments. This is the reason that the need for an effective Kolmogorov-Smirnov test for fuzzy data is still valid.

In this contribution, we propose a new generalization of the Kolmogorov-Smirnov test for fuzzy data. Our method is based on the so-called epistemic bootstrap (see [9,10]) which results in real-valued outcomes. This approach provides a binary decision of rejection/acceptance of the null hypothesis under study. This very feature turns out to be highly desired by practitioners. Moreover, a simulation study shows good statistical properties, which suggests that it will be met with great interest.

The paper is organized as follows. In Sect. 2 we recall the classical two-sample Kolmogorov-Smirnov test. In Sect. 3 we introduce basic concepts related to fuzzy data and random fuzzy numbers. The epistemic bootstrap is described in Sect. 4. In Sect. 5 we propose four versions of the two-sample Kolmogorov-Smirnov test for fuzzy data. Then, in Sect. 6 some results of the simulation study performed to examine statistical properties of the suggested test generalizations are given. The final conclusions are summarized in Sect. 7.

2   The Kolmogorov-Smirnov Test

Let $\mathbb{X} = (X_1, \ldots, X_n)$ and $\mathbb{Y} = (Y_1, \ldots, Y_m)$ denote two independent random samples drawn from populations with unknown cumulative distribution functions (c.d.f.) F and G, respectively. To check whether both samples come from the same distribution we verify the null hypothesis

$$H_0 : F(t) = G(t) \quad \text{for all} \quad t \in \mathbb{R}, \tag{1}$$

against the alternative $H_1 : F(t) \neq G(t)$ for some $t \in \mathbb{R}$ that their distributions differ. Among various goodness-of-fit tests one can apply in this problem the most often used is the **Kolmogorov-Smirnov** test (KS-test) with the following test statistic

$$D_{n,m} = \sup_{t \in \mathbb{R}} |\widehat{F}_n(t) - \widehat{G}_m(t)|, \tag{2}$$

where $\widehat{F}_n(t) = \frac{1}{n}\sum_{i=1}^{n} \mathbb{1}(X_i \leqslant t)$ and $\widehat{G}_m(t) = \frac{1}{m}\sum_{i=1}^{m} \mathbb{1}(Y_i \leqslant t)$ are the empirical distribution functions (e.d.f.) based on the first and the second sample, respectively. Usefulness of the test statistic (2) is motivated by two general facts. Firstly, for any fixed $t \in \mathbb{R}$ the e.d.f. $\widehat{F}_n(t)$ is a consistent estimator of $F(t)$. Moreover, by the Glivenko-Cantelli lemma $\widehat{F}_n$ converges to F with probability 1. Secondly, the null distribution of $D_{n,m}$ does not depend on unknown distributions F and G as long as they are continuous.

Therefore, $D_{n,m}$ might be applied in our goodness-of-fit testing problem, since if the null hypothesis (1) holds than the distance between $\widehat{F}_n(t)$ and $\widehat{G}_m(t)$ should be small for each t. Conversely, large differences between $\widehat{F}_n(t)$ and $\widehat{G}_m(t)$ indicate that our samples $\mathbb{X}$ and $\mathbb{Y}$ are not identically distributed, so we should reject H_0. A final decision whether to reject or accept H_0 can be taken with respect to the p-value

$$p = \mathbb{P}_{H_0}(D_{n,m} \geqslant d), \tag{3}$$

where d is the actual value of $D_{n,m}$ and the desired probability might be obtained using the following approximation (see [20])

$$\lim_{n,m \to \infty} \mathbb{P}_{H_0}\left(\sqrt{\frac{nm}{n+m}} D_{n,m} \leqslant d\right) = 1 - 2\sum_{i=1}^{\infty} (-1)^{i-1} e^{-2i^2 d^2}. \tag{4}$$

3 Epistemic Fuzzy Data

In many situations where the experiment outcomes are real-valued, we actually have to face their imprecise or vague perceptions. This involves the need to adopt an appropriate mathematical model of imprecise data. Secondly, statistical tools which were originally designed for real-valued observations have to be modified adequately to cope with imprecise data. Let us start from data modeling where fuzzy set theory appears useful. Indeed, a natural counterpart of the real-valued outcomes $\mathbf{x} = (x_1, \ldots, x_n)$, where $x_i \in \mathbb{R}$, is a sequence of fuzzy numbers $\widetilde{\mathbf{x}} = (\widetilde{x}_1, \ldots, \widetilde{x}_n)$, where each $\widetilde{x}_i$, $i = 1, \ldots, n$, is a fuzzy number.

Definition 1. *A mapping $\widetilde{x} : \mathbb{R} \to [0,1]$ is a **fuzzy number** if its α-cuts $(\widetilde{x})_\alpha$, i.e.*

$$(\widetilde{x})_\alpha = \begin{cases} \{x \in \mathbb{R} : \widetilde{x} \geqslant \alpha\} & \text{if} \quad \alpha \in (0,1], \\ cl\{x \in \mathbb{R} : \widetilde{x} > 0\} & \text{if} \quad \alpha = 0, \end{cases}$$

where cl stands for the closure, are nonempty compact intervals for all $\alpha \in [0,1]$.

Further on the family of all fuzzy numbers will be denoted by $\mathbb{F}(\mathbb{R})$.

Although fuzzy numbers may assume different forms, several families of somehow regular shapes are usually used in practice. A family of fuzzy numbers which is used most often is the one of **trapezoidal fuzzy numbers** of the following form

$$\widetilde{x}(x) = \begin{cases} \frac{x-a}{b-a} & \text{if } a < x \leqslant b, \\ 1 & \text{if } b \leqslant x \leqslant c, \\ \frac{d-x}{d-c} & \text{if } c \leqslant x < d, \\ 0 & \text{otherwise}, \end{cases} \tag{5}$$

where $a, b, c, d \in \mathbb{R}$ such that $a \leqslant b \leqslant c \leqslant d$. Since each trapezoidal fuzzy number is completely characterized by four reals, then if $\widetilde{x}$ is given by (5), we have $\widetilde{x} = \text{Tra}(a, b, c, d)$. If $b = c$ then $\widetilde{x}$ is said to be a **triangular fuzzy number**. Restricting to triangular or trapezoidal fuzzy numbers is generally motivated both by their simplicity in calculations and their natural interpretation. Moreover, fuzzy numbers of other types are quite often approximated by the trapezoidal fuzzy numbers. For more information we refer the reader to [1].

When imprecise data appear as the outcomes of a random experiment, we may treat them as fuzzy values generated by fuzzy-valued random variables [14,15].

Definition 2. *Given a probability space* $(\Omega, \mathcal{F}, P)$, *a mapping* $\widetilde{X} : \Omega \to \mathbb{F}(\mathbb{R})$ *is said to be a **fuzzy random variable** (f.r.v.) if for each* $\alpha \in [0,1]$ $(\inf \widetilde{X}_\alpha) :$ $\Omega \to \mathbb{R}$ *and* $(\sup \widetilde{X}_\alpha) : \Omega \to \mathbb{R}$ *are real-valued random variables on* $(\Omega, \mathcal{F}, P)$.

Thus a fuzzy random variable $\widetilde{X}$ might be considered as a *fuzzy perception* of a usual random variable X, called the *original* of $\widetilde{X}$, which remains unknown. Similarly, a **fuzzy random sample** $\widetilde{\mathbb{X}} = (\widetilde{X}_1, \ldots, \widetilde{X}_n)$ might be treated as a fuzzy perception of a usual real-valued random sample.

Following the Extension Principle any statistic $T = T(\mathbb{X})$ can be extended to a fuzzy statistic $\widetilde{T} = \widetilde{T}(\widetilde{\mathbb{X}})$. Given a realization $\widetilde{\mathbb{x}}$ of a fuzzy random sample $\widetilde{\mathbb{X}}$, a corresponding value of the fuzzy statistic $\widetilde{T}$ is a fuzzy set $\widetilde{T}(\widetilde{\mathbb{x}})$ described by the following family of its α-cuts

$$\widetilde{T}_\alpha = \left(\widetilde{T}(\widetilde{\mathbb{x}})\right)_\alpha = \{T(x_1, \ldots, x_n) : x_1 \in (\widetilde{x}_1)_\alpha, \ldots, x_n \in (\widetilde{x}_n)_\alpha\}.$$

If T is regular enough (monotone, continuous, etc.) then α-cuts $\widetilde{T}_\alpha$ are intervals, so for obtaining $\widetilde{T}_\alpha$ it suffices to find their borders $\inf \widetilde{T}_\alpha$ and $\sup \widetilde{T}_\alpha$. Consequently, calculations for fuzzy case can be reduced to calculations on intervals. Sometimes it works smoothly, like the determination of the average of a fuzzy sample $\widetilde{x}_1, \ldots, \widetilde{x}_n$ which is obtained immediately using the Minkowski sum and the scalar product. Indeed, the desired fuzzy sample average is a fuzzy number with the following α-cuts

$$\left(\frac{1}{n} \sum_{i=1}^{n} \widetilde{x}_i\right)_\alpha = \left[\frac{1}{n} \sum_{i=1}^{n} \inf(\widetilde{x}_i)_\alpha, \frac{1}{n} \sum_{i=1}^{n} \sup(\widetilde{x}_i)_\alpha\right].$$

However, calculations are not always so straightforward as shown above. For instance, the sample variance of a fuzzy sample is a fuzzy number defined as follows

$$\widetilde{s}^2 = \frac{1}{n-1} \sum_{i=1}^{n} \left[\widetilde{x}_i - \left(\frac{1}{n} \sum_{i=1}^{n} \widetilde{x}_i \right) \right]^2.$$

Unfortunately, the upper border computation for $(\widetilde{s}^2)_\alpha = [\inf(\widetilde{s}^2)_\alpha, \sup(\widetilde{s}^2)_\alpha]$ is in general the NP-hard problem. There are algorithms that compute this border in a more effective time ($O(n \log n)$ or even $O(n)$) but they impose certain restrictions on a fuzzy sample [17].

Moreover, even if obtaining $\widetilde{T}(\widetilde{x}_1, \ldots, \widetilde{x}_n)$ involves no calculation problems a final solution can be completely useless for practitioners, e.g., if α-cuts of $\widetilde{T}$ are too wide, then point estimates appear "too fuzzy", confidence intervals are too wide and fuzzy tests are usually inconclusive. Obviously, such procedures cannot satisfy the potential users. Therefore, researchers are inclined to construct statistical procedures delivering "more precise" final decisions. In parametric models, several approaches to improve estimation based on imprecise data were discussed in [4]. However, it seems that nonparametric methods are much more appropriate for imprecise data. Here we find motivations for a universal nonparametric technique, called the **epistemic bootstrap** [9], which could be helpful when the existing methods do not work or do not give satisfactory results.

4   Epistemic Bootstrap

Suppose instead of the real-valued sample $(x_1, \ldots, x_n)$ we have only its imprecise perception modeled by a fuzzy sample $(\widetilde{x}_1, \ldots, \widetilde{x}_n)$. It means that a fuzzy set $\widetilde{x}_i$ contains the actual real-valued realization of the i-th observation but we do not know where it is actually located. On the other hand, the membership function of $\widetilde{x}_i$ attributes to each point the possibility that it is the true realization of X_i.

Thus, following the general bootstrap idea of resampling from the initial fuzzy sample $\widetilde{x}_1, \ldots, \widetilde{x}_n \in \mathbb{F}(\mathbb{R})$ we take into account the degree of possibility that x_i^* is the true outcome of the experiment. To generate a single real-valued element x_i^* of a bootstrap sample we need just one loop with two steps: firstly, we generate an α-cut, and secondly, we draw randomly a real value from this α-cut. Following these steps n times we get the entire bootstrap sample $x_1^*, \ldots, x_n^*$. Next, repeating such resampling for several (say, $B \geqslant 1$) α-cuts for each fuzzy set, we obtain a multiplicity of bootstrap samples, i.e. $(x_{1j}^*, \ldots, x_{nj}^*)$, where $j = 1, \ldots, B$. This general idea of drawing bootstrap samples from epistemic fuzzy data is shown in Algorithm 1 (see [9]).

Algorithm 1 (Epistemic fuzzy bootstrap – standard approach).
Require: Initial fuzzy sample $\widetilde{x}_1, \ldots, \widetilde{x}_n \in \mathbb{F}(\mathbb{R})$.
Ensure: B bootstrap samples.
  for $j = 1$ to B **do**
    for $i = 1$ to n **do**

Generate randomly a real number α_{ij} from the uniform distribution on the unit interval $[0, 1]$.

Generate randomly a real number x^*_{ij} from the uniform distribution on the α-cut $(\widetilde{x}_i)_{\alpha_{ij}}$.

   end for
end for
return bootstrap samples $\mathbf{x}^*_j = (x^*_{1j}, \ldots, x^*_{nj})$, where $j = 1, \ldots, B$.

Suppose our statistical inference is based on a statistic $T = T(\mathbf{x})$. Because in the "bootstrap world" we have B bootstrap samples, hence we obtain respective B values $T(\mathbf{x}^*_1), \ldots, T(\mathbf{x}^*_B)$ of T which are then usually aggregated by simple averaging, i.e. $T^* = \frac{1}{B} \sum_{j=1}^{B} T(\mathbf{x}^*_1)$.

Algorithm 1 proved its usefulness in some areas of statistical inference, like point estimation or hypothesis testing [9,10]. To improve its efficiency and to reduce some errors that may appear in estimation, a new version of the epistemic bootstrap algorithm, described as the antithetic approach, was proposed in [10] and is shown as Algorithm 2.

Algorithm 2 (Epistemic fuzzy bootstrap – antithetic approach).
Require: Initial fuzzy sample $\widetilde{x}_1, \ldots, \widetilde{x}_n \in \mathbb{F}(\mathbb{R})$.
Ensure: B bootstrap samples.
   for $j = 1$ to B **do**
      for $i = 1$ to n **do**
         Generate randomly a real number α_{ij} from the uniform distribution on the unit interval $[0, 1]$.
         Generate randomly two real numbers: x'_{ij} from the uniform distribution on the α-cut $(\widetilde{x}_i)_{\alpha_{ij}}$, and x''_{ij} from the uniform distribution on the $1 - \alpha$-cut $(\widetilde{x}_i)_{1-\alpha_{ij}}$.
         Let $x^*_{ij} = \frac{1}{2}(x'_{ij} + x''_{ij})$.
      end for
   end for
   return bootstrap samples $\mathbf{x}^*_j = (x^*_{1j}, \ldots, x^*_{nj})$, where $j = 1, \ldots, B$.

Below we show that the epistemic bootstrap may be successfully applied to generalize the classical Kolmogorov-Smirnov test for fuzzy samples.

5   Bootstrapped Kolmogorov-Smirnov Test

Suppose that we observe two independent random samples $\mathbb{X} = (X_1, \ldots, X_n)$ and $\mathbb{Y} = (Y_1, \ldots, Y_m)$ from populations with unknown c.d.f. F and G, respectively, but instead of precise real-valued measurements we have only their imprecise perceptions modeled by two fuzzy samples $\mathbf{x} = (\widetilde{x}_1, \ldots, \widetilde{x}_n)$ and $\mathbf{y} = (\widetilde{y}_1, \ldots, \widetilde{y}_m)$. Our goal is still to test the null hypothesis $H_0 : F = G$. As we cannot apply the classical Kolmogorov-Smirnov goodness-of-fit test for fuzzy samples, we propose its bootstrapped version to cope with this challenge.

Let us consider fuzzy samples $\mathbf{x} = (\widetilde{x}_1, \ldots, \widetilde{x}_n)$ and $\mathbf{y} = (\widetilde{y}_1, \ldots, \widetilde{y}_m)$ as the initial input samples for the epistemic bootstrap described in Sect. 4. Thus, following Algorithm 1 (or Algorithm 2) we generate B bootstrap samples $\mathbf{x}_j^* = (x_{1j}^*, \ldots, x_{nj}^*)$ and $\mathbf{y}_j^* = (y_{1j}^*, \ldots, y_{mj}^*)$, where $j = 1, \ldots, B$. Since each bootstrap sample generates its own e.d.f.'s, we obtain two sequences: $\widehat{F}_n^{*[1]}(t), \ldots, \widehat{F}_n^{*[B]}(t)$ and $\widehat{G}_m^{*[1]}(t), \ldots, \widehat{G}_m^{*[B]}(t)$, where for $j = 1, \ldots, B$

$$\widehat{F}_n^{*[j]}(t) = \frac{1}{n} \sum_{i=1}^n \mathbb{1}(x_{ij}^* \leqslant t) \quad \text{and} \quad \widehat{G}_m^{*[j]}(t) = \frac{1}{n} \sum_{i=1}^m \mathbb{1}(y_{ij}^* \leqslant t). \tag{6}$$

Further on sequences $\widehat{F}_n^{*[1]}(t), \ldots, \widehat{F}_n^{*[B]}(t)$ and $\widehat{G}_m^{*[1]}(t), \ldots, \widehat{G}_m^{*[B]}(t)$ might be utilized in different manners to verify the null hypothesis H_0. This way we obtain a few variants of the bootstrapped Kolmogorov-Smirnov test denoted as BKS with a number indicating a particular variant.

BKS1
Given $\widehat{F}_n^{*[1]}(t), \ldots, \widehat{F}_n^{*[B]}(t)$ and $\widehat{G}_m^{*[1]}(t), \ldots, \widehat{G}_m^{*[B]}(t)$ we can determine the following B classical KS test statistics $d_{n,m}^{*[1]}, \ldots, d_{n,m}^{*[B]}$, where

$$d_{n,m}^{*[j]} = \sup_{t \in \mathbb{R}} \left| \widehat{F}_n^{*[j]}(t) - \widehat{G}_m^{*[j]}(t) \right|. \tag{7}$$

Next, following (3), we calculate the corresponding p-values

$$p^{*[j]} = \mathbb{P}_{H_0} \left(D_{n,m} \geqslant d_{n,m}^{*[j]} \right), \tag{8}$$

where $j = 1, \ldots, B$. Then, to make a decision we have to aggregate p-values $p^{*[1]}, \ldots, p^{*[B]}$. Several approaches for combining p-values were suggested in the literature but most of them assume independence between tests which is usually not our case. Hence, we recommend the Simes method [19] which is robust to dependencies among the combined tests.

BKS2
We can consider another resampling to choose pairs of the bootstrap samples required for the KS test statistics computation. Namely, we fix a number of pairs K and for each $k = 1, \ldots, K$ we generate randomly (with replacement) one e.d.f from $\widehat{F}_n^{*[1]}(t), \ldots, \widehat{F}_n^{*[B]}(t)$ and denote it as $\widehat{F}_n^{**[k]}(t)$. Similarly, we generate one e.d.f from $\widehat{G}_m^{*[1]}(t), \ldots, \widehat{G}_m^{*[B]}(t)$ and denote it as $\widehat{G}_m^{**[k]}(t)$. Then we compute

$$d_{n,m}^{**[k]} = \sup_{t \in \mathbb{R}} \left| \widehat{F}_n^{**[k]}(t) - \widehat{G}_m^{**[k]}(t) \right|, p^{**[k]} = \mathbb{P}_{H_0} \left(D_{n,m} \geqslant d_{n,m}^{**[k]} \right), \tag{9}$$

for $k = 1, \ldots, K$. As before, to make a decision we have to aggregate the available p-values $p^{**[1]}, \ldots, p^{**[K]}$.

Here one can ask a natural question why instead of considering all possible pairs of e.d.f. we generate K such pairs. Obviously, if B is small considering all $\binom{B}{2}$ pairs does not cause trouble. But usually, B is rather big and hence we consider only some pairs as it is a typical practice in permutation tests.

BKS3
The main drawback of the two previous approaches is the need for the p-values aggregation which usually provokes discussions among statisticians. Therefore the third proposed method involves the aggregation of statistics obtained for the bootstrap samples, which is in some sense a common practice in bootstrap.

Thus, after generating $\widehat{F}_n^{*[1]}(t), \ldots, \widehat{F}_n^{*[B]}(t)$ and $\widehat{G}_m^{*[1]}(t), \ldots, \widehat{G}_m^{*[B]}(t)$ we determine the bootstrap e.d.f. $\widehat{F}_n^*(t)$ and $\widehat{G}_m^*(t)$, respectively, such that

$$\widehat{F}_n^*(t) = \frac{1}{B} \sum_{i=1}^{B} \widehat{F}_n^{*[i]}(t) \quad \text{and} \quad \widehat{G}_m^*(t) = \frac{1}{B} \sum_{i=1}^{B} \widehat{G}_m^{*[i]}(t). \tag{10}$$

Then, following (2) we calculate a value of the KS test statistic

$$d_{n,m}^{***} = \sup_{t \in \mathbb{R}} |\widehat{F}_n^*(t) - \widehat{G}_m^*(t)|, \tag{11}$$

and the corresponding p-value

$$p^{***} = \mathbb{P}_{H_0}(D_{n,m} \geqslant d_{n,m}^{***}), \tag{12}$$

which is unique and might be applied directly to make a final decision.

BKS4
Another approach that avoids p-values aggregation works as follows: we aggregate the KS test statistics (7) obtained for the subsequent bootstrap samples

$$d_{n,m}^* = \frac{1}{B} \sum_{j=1}^{B} d_{n,m}^{*[j]} \tag{13}$$

which leads to the unique p-value

$$p^* = \mathbb{P}_{H_0}(D_{n,m} \geqslant d_{n,m}^*). \tag{14}$$

6   Simulation Study

In numerical experiments we restrict our attention to trapezoidal fuzzy numbers since many simulations indicate that the shape of the membership function scarcely affects statistical conclusions (see [16]). By (5), to obtain a fuzzy number $\widetilde{x} = \text{Tra}(a, b, c, d)$ we have to generate four real numbers $a \leqslant b \leqslant c \leqslant d$. It is done according to the following formulas

$$a = X - S^l - C^l, \ b = X - C^l, \ c = X + C^r, \ d = X + C^r + S^r, \tag{15}$$

where X is a random variable corresponding to the "true" population distribution of which fuzzy perception is $\widetilde{x}$, while C^l, C^r, S^l and S^r denote some random variables applied for modeling this fuzzy value. All random variables are generated independently from the distributions shown in Table 1. The notation applied there is self-explanatory, e.g., if $\widetilde{x}_i \in \mathbb{F}_{(N,U,U)}$, then X is simulated from the standard normal distribution $N(0,1)$, while C^l and C^r are generated from the uniform distribution on the interval $(0, 0.6)$, and S^l, S^r from the uniform distribution on $(0, 0.8)$. In the same manner, $Exp(1)$ denotes the exponential distribution with the parameter 1, $\Gamma(2, 2)$ stands for the gamma distribution with the shape and scale parameter equal to 2, $\beta(2, 5)$ for the beta distribution with parameters 2 and 5, etc. This way of simulating fuzzy random variables is widely used in the literature [5, 8, 18].

Table 1. Scenarios for simulating fuzzy random variables.

Type	X	C^l, C^r	S^l, S^r
$\mathbb{F}_{(N,U,U)}$	$N(0, 1)$	$U(0, 0.6)$	$U(0, 0.8)$
$\mathbb{F}_{(E,U,U)}$	$Exp(1)$	$U(0, 0.6)$	$U(0, 1.2)$
$\mathbb{F}_{(N,U,U,1)}$	$N(4, \sqrt{8})$	$U(0, 0.6)$	$U(0, 0.8)$
$\mathbb{F}_{(\Gamma,U,U)}$	$\Gamma(2, 2)$	$U(0, 0.6)$	$U(0, 0.8)$
$\mathbb{F}_{(\Gamma,U,U,1)}$	$\Gamma\left(\frac{16}{5}, \frac{5}{56}\right)$	$U(0, 0.6)$	$U(0, 0.8)$
$\mathbb{F}_{(\beta,U,U)}$	$\beta(2, 5)$	$U(0, 0.6)$	$U(0, 0.8)$
$\mathbb{F}_{(U,U,U)}$	$U(\frac{4-\sqrt{15}}{14}, \frac{4+\sqrt{15}}{14})$	$U(0, 0.6)$	$U(0, 0.8)$
$\mathbb{F}_{(U,U,U,1)}$	$U\left(-\sqrt{3}, \sqrt{3}\right)$	$U(0, 0.6)$	$U(0, 0.8)$

We start from juxtaposing our BKS test and its primary crisp two-sample KS test. Unlike situations encountered in practice, when the original distribution is unknown, here we have the underlying distribution of random variable X at our disposal. Thus given real-valued samples $\mathbb{X} = (X_1, \ldots, X_n)$ and $\mathbb{Y} = (Y_1, \ldots, Y_m)$ we perform the classical KS test. Next we use the real-valued samples to generate fuzzy samples $\widetilde{x}$ and $\widetilde{y}$ according to (15). Then B epistemic bootstrap samples x^* and y^* are created from these initial fuzzy samples using both the standard (denoted as std) and antithetic ($anti$) method, i.e. following Algorithm 1 or Algorithm 2, respectively. Finally, we perform all four variants of the BKS test for each pair of the bootstrap samples x^* and y^*. In the case of BKS1 and BKS2 tests, the Simes method is applied to aggregate p-values.

In our experiments we consider the relatively small ($n = m = 10$) or moderate ($n = m = 100$) sample sizes. To curb undesired random effects each experiment was repeated 10000 times and the results were averaged. Some exemplary p-values obtained for all BKS test variants (when $B = 10$), as well as the p-values for the crisp KS test, are given in Fig. 1a and 2a (for the small and moderate sample size, respectively). More specifically, the curves show the simulated p-values of these tests as a function of the deterministic shift added to the second sample Y. Because of this increasing shift, these p-values should decrease. And the closer the respective curve of the considered BKS test to its counterpart for the KS test (i.e. our benchmark), the better. In Fig. 1b and 2b we illustrate a

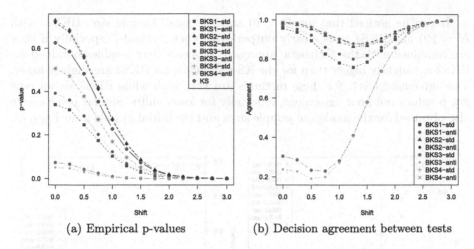

Fig. 1. Simulated results for $X \sim \mathbb{F}_{(N,U,U)}$ and $n = m = 10$ (as a function of the shift added to Y).

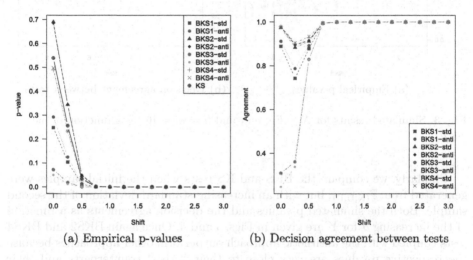

Fig. 2. Simulated results for $X \sim \mathbb{F}_{(N,U,U)}$ and $n = m = 100$ (as a function of the shift added to Y).

decision agreement between tests calculated as a frequency of situations when both BKS and KS tests simultaneously reject or accept the null hypothesis (at the significance level 0.05). The higher value of the decision agreement indicates the bigger consistency between the considered BKS test and our benchmark (i.e. the "crisp" KS test).

It can be noticed that for $B = 10$ and the small sample size, BKS2 (with $K = 10$) and BKS4 tests clearly outperform other methods, especially if they are combined with the antithetic approach. It seems that p-values obtained for BKS2 are slightly bigger than for the KS test, while for BKS4 are slightly lower. The agreement levels for these methods are very high while BKS3 shows very low p-values and poor agreement, especially for lower shifts. Similar results were also obtained for the moderate sample sizes and the initial sample from $\mathbb{F}_{(E,U,U)}$.

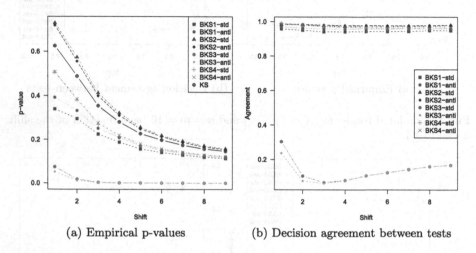

(a) Empirical p-values   (b) Decision agreement between tests

Fig. 3. Simulated results for $X \sim \mathbb{F}_{(N,U,U)}$ and $n = m = 10$ (as a function of increasing σ).

Similarly, we compare the BKS and KS tests when the initial samples were generated from $\mathbb{F}_{(N,U,U)}$, but with an increasing standard deviation of the second sample. Both the simulated p-values and the decision agreements as a function of the increasing σ for Y are given in Figs. 3 and 4. Once again BKS2 and BKS4 tests coupled with the antithetic approach outperform other approaches because the respective p-values are very close to their "crisp" counterparts and their decision agreements levels are high.

Finally, we consider situations when X and Y come from totally different distributions, although chosen so that they have identical expected values and variances but different shapes (e.g., the normal distribution against the gamma distribution). Then the decision agreement between each variant of the BSK test and KS test for the small and moderate sample sizes is computed.

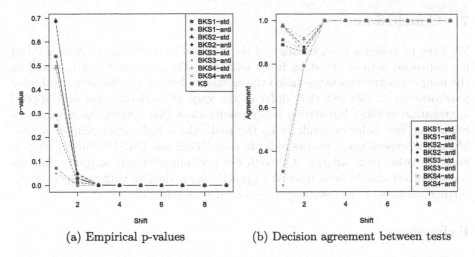

(a) Empirical p-values   (b) Decision agreement between tests

Fig. 4. Simulated results for $X \sim \mathbb{F}_{(N,U,U)}$ and $n = m = 100$ (as a function of increasing σ).

Results obtained for several pairs are given in Table 2. It is seen that BKS2 and BKS4 tests (especially with the antithetic approach) reveal very high agreement with the KS test for all of the considered cases (even about 98–99%, not less than 88%). On the other hand, BKS3 usually behaves rather poorly (not more than 70% for the decision agreement, even about 10–30% in some cases).

Table 2. Decision agreement between BKS and KS tests.

	BKS1		BKS2		BKS3		BKS4	
	Std	Anti	Std	Anti	Std	Anti	Std	Anti
$\mathbb{F}_{(N,U,U,1)}/\mathbb{F}_{(\Gamma,U,U)}$								
$n = 10$	0.9807	0.9871	0.9916	0.9927	0.1216	0.0892	0.9891	0.9896
$n = 100$	0.8681	0.8921	0.9381	0.9453	0.2178	0.2161	0.9501	0.9537
$\mathbb{F}_{(\beta,U,U)}/\mathbb{F}_{(U,U,U)}$								
$n = 10$	0.8728	0.8990	0.9627	0.9676	0.7112	0.6139	0.9847	0.9845
$n = 100$	0.3683	0.7260	0.7878	0.8830	0.5978	0.6043	0.8958	0.8974
$\mathbb{F}_{(N,U,U)}/\mathbb{F}_{(U,U,U,1)}$								
$n = 10$	0.9573	0.9698	0.9856	0.9876	0.2853	0.2250	0.9889	0.9885
$n = 100$	0.8518	0.8690	0.9339	0.9410	0.2168	0.1752	0.9419	0.9477
$\mathbb{F}_{(\beta,U,U)}/\mathbb{F}_{(\Gamma,U,U,1)}$								
$n = 10$	0.8693	0.8968	0.9659	0.9719	0.7052	0.6098	0.9872	0.9871
$n = 100$	0.3346	0.7497	0.8071	0.9246	0.6003	0.6060	0.9405	0.9425

7 Conclusions

We have proposed a generalization of the two-sample Kolmogorov-Smirnov test for imprecise data modeled by fuzzy numbers. The suggested test is based on the nonparametric technique called the epistemic bootstrap. We have considered four variants of this test that differ in the way of randomization and applied aggregation method. Simulation study results show that despite the data imprecision our test behaves similarly to the real-valued Kolmogorov-Smirnov test. Moreover, experimental results indicate that BKS2 and BKS4 definitely outperform the other test variants. Although our preliminary study suggests that the proposed test should be of interest to practitioners dealing with imprecise data, further analysis of the test properties would be advisable.

References

1. Ban, A., Coroianu, L., Grzegorzewski, P.: Fuzzy Numbers: Approximations. Ranking and Applications. Polish Academy of Sciences, Warsaw (2015)
2. Couso, I., Dubois, D.: Statistical reasoning with set-valued information: Ontic vs. epistemic views. Int. J. Approximate Reasoning **55**(7), 1502–1518 (2014). https://doi.org/10.1016/j.ijar.2013.07.002
3. Destercke, S., Strauss, O.: Kolmogorov-Smirnov test for interval data. In: Laurent, A., Strauss, O., Bouchon-Meunier, B., Yager, R.R. (eds.) IPMU 2014. CCIS, vol. 444, pp. 416–425. Springer, Cham (2014). https://doi.org/10.1007/978-3-319-08852-5_43
4. Grzegorzewski, P., Gołowska, J.: In search of a precise estimator based on imprecise data. In: Joint Proceedings of IFSA-EUSFLAT-AGOP 2021 Conferences, pp. 530–537. Atlantis Press (2021)
5. Grzegorzewski, P., Hryniewicz, O., Romaniuk, M.: Flexible resampling for fuzzy data. Int. J. Appl. Math. Comput. Sci. **30**, 281–297 (2020)
6. Grzegorzewski, P.: The Kolmogorov goodness-of-fit test for interval-valued data. In: 2017 IEEE International Conference on Fuzzy Systems (FUZZ-IEEE), pp. 1–6 (2017). https://doi.org/10.1109/FUZZ-IEEE.2017.8015557
7. Grzegorzewski, P.: The Kolmogorov-Smirnov goodness-of-fit test for interval-valued data. In: Gil, E., Gil, E., Gil, J., Gil, M.Á. (eds.) The Mathematics of the Uncertain. SSDC, vol. 142, pp. 615–627. Springer, Cham (2018). https://doi.org/10.1007/978-3-319-73848-2_57
8. Grzegorzewski, P., Hryniewicz, O., Romaniuk, M.: Flexible bootstrap based on the canonical representation of fuzzy numbers. In: Proceedings of EUSFLAT 2019. Atlantis Press (2019). https://doi.org/10.2991/eusflat-19.2019.68
9. Grzegorzewski, P., Romaniuk, M.: Epistemic bootstrap for fuzzy data. In: Joint Proceedings of IFSA-EUSFLAT-AGOP 2021 Conferences, pp. 538–545. Atlantis Press (2021)
10. Grzegorzewski, P., Romaniuk, M.: Bootstrap methods for epistemic fuzzy data. International Journal of Applied Mathematics and Computer Science (accepted) (2022)
11. Hesamian, G., Chachi, J.: Two-sample Kolmogorov-Smirnov fuzzy test for fuzzy random variables. Stat. Pap. **56**, 61–82 (2013)
12. Hesamian, G., Taheri, S.M.: Fuzzy empirical distribution function: properties and application. Kybernetika **49**, 962–982 (2013)

13. Kolmogorov, A.: Sulla determinazione empirica di una legge di distribuzione. Giornale dell'Istituto Italiano degli Attuari **4**, 83–91 (1933)
14. Kruse, R.: The strong law of large numbers for fuzzy random variables. Inf. Sci. **28**, 233–241 (1982)
15. Kwakernaak, H.: Fuzzy random variables, part I: Definitions and theorems. Inf. Sci. **15**, 1–15 (1978)
16. Lubiano, M.A., Salas, A., Gil, M.A.: A hypothesis testing-based discussion on the sensitivity of means of fuzzy data with respect to data shape. Fuzzy Sets Syst. **328**, 54–69 (2017). https://doi.org/10.1016/j.fss.2016.10.015
17. Nguyen, H., Kreinovich, V., Wu, B., G., X.: Computing Statistics under Interval and Fuzzy Uncertainty. Springer (2012)
18. Romaniuk, M., Hryniewicz, O.: Discrete and smoothed resampling methods for interval-valued fuzzy numbers. IEEE Trans. Fuzzy Syst. **29**, 599–611 (2021). https://doi.org/10.1109/TFUZZ.2019.2957253
19. Simes, R.J.: An improved Bonferroni procedure for multiple tests of significance. Biometrika **73**(3), 751–754 (1986)
20. Smirnov, N.: Estimate of deviation between empirical distribution functions in two independent samples. Bull. Moscow Univ. **2**, 3–16 (1933)

Connections Between Granular Counts and Twofold Fuzzy Sets

Corrado Mencar[1]([✉]) [iD] and Didier Dubois[2] [iD]

[1] Department of Computer Science, University of Bari Aldo Moro, Bari, Italy
corrado.mencar@uniba.it
[2] IRIT-CNRS, University Paul Sabatier, Toulouse, France
dubois@irit.fr

Abstract. Twofold fuzzy sets model ill-known collections of objects, where, for each element of a given domain, both necessity and possibility degrees of membership are specified. It is proved that the cardinality of a twofold fuzzy set is equivalent to possibilistic granular count when the twofold fuzzy set is derived from a possibilistic assignment table. This result sheds light on the connections between twofold fuzzy sets and granular counts: in particular, it is possible to take interactivity into account when aggregating twofold fuzzy sets derived from a possibilistic assignment table.

1 Introduction

Ill-known sets are ordinary sets to which it is not known whether some elements belong or not. When the uncertainty of an ill-known set F, defined on a domain X, is due to incomplete information, it can be formalized as a possibility distribution over the powerset of X. However, such a distribution could be hard to manage; therefore, an approximate representation may be more convenient. Twofold fuzzy sets are mathematical structures modeling ill-known sets of objects; in essence, a twofold fuzzy set $\mathcal{F}$ represents the lower and upper approximation of the ill-known set F by a pair of fuzzy sets, namely an *interior* fuzzy set, which can be viewed as a set of elements which (more or less) certainly belong to F, and a *closure* fuzzy set of elements that (more or less) possibly belong to F [3]. Twofold fuzzy sets extend the idea of interval sets [10] by introducing fuzziness, and resemble fuzzy set intervals [5], or interval-valued fuzzy sets, but twofold fuzzy sets are consistent with Possibility Theory.

More recently, the connection between twofold fuzzy sets and the granular counting of uncertain data was suggested as possible development [9]. In essence, granular counting refers to the idea of counting the number of observations that pertain to a referent, when observations lack complete information; therefore, their referent is partially unknown. This work starts from this idea and shows that the granular count of a referent coincides with the cardinality of a twofold fuzzy set properly defined on a collection of observations. The advantages of connecting the two theories are twofold: on one hand, the theory of

© Springer Nature Switzerland AG 2022
D. Ciucci et al. (Eds.): IPMU 2022, CCIS 1602, pp. 508–519, 2022.
https://doi.org/10.1007/978-3-031-08974-9_41

granular counting can be enriched with the algebra of twofold fuzzy sets that enable approximate reasoning involving ill-known concepts; on the other hand, the algorithmic properties of granular counting and other theoretical properties can be exploited within models based on twofold fuzzy sets.

In the next two sections, the main concepts of twofold fuzzy sets and granular counting, which are instrumental to the presented work, are briefly presented and formalized; in Sect. 4 it is proved that granular counting, which is a property derived from some basic assumptions on observations and referents, coincides with the cardinality of a twofold fuzzy set as defined by Dubois and Prade [3]. Section 5 reports an approach to take interactivity into account when aggregating twofold fuzzy sets.

2   Twofold Fuzzy Sets

Let X be a referential set and $F \subseteq X$. We assume that F is *ill-known*, i.e., there exists at least one element $x \in X$ for which it is not known whether x belongs to F or not. This situation can be formalized by a possibility distribution π_F defined on the powerset of X, namely $\pi_F : 2^X \mapsto [0,1]$, where $\pi_F(A)$ evaluates the possibility that a subset $A \subseteq X$ is the unknown F.

Being defined on the powerset of X, the possibility distribution π_F could be hard to manage; therefore, an approximate representation could be more convenient. This is accomplished by introducing two fuzzy sets, I and C, such that I represents the extent to which it is certain that an element x belongs to F, while C represents the extent to which it is possible that an element x belongs to F. Formally:

$$\mu_I(x) = \inf_{A : x \notin A} (1 - \pi_F(A)) \tag{1}$$

and

$$\mu_C(x) = \sup_{A : x \in A} \pi_F(A) \tag{2}$$

It is easy to check that, due to the normality of π_F, whenever $\mu_C(x) < 1$, then $\mu_I(x) = 0$. Thus, the reverse problem can be considered: given fuzzy sets I and C, is it possible to recover the original possibility distribution π_F? In order to approach this problem, twofold fuzzy sets are introduced as follows.

Definition 1. *Let X be a Universe of Discourse. A twofold fuzzy set is a pair $\mathcal{F} = (I, C)$ of fuzzy sets on X, respectively called* interior *and* closure *of $\mathcal{F}$, iff I is included in the core of C, i.e.*

$$\forall x \in X : \mu_C(x) < 1 \rightarrow \mu_I(x) = 0 \tag{3}$$

Given a twofold fuzzy set $\mathcal{F}$, it is not generally possible to recover the original possibility distribution π_F because the fuzzy sets I and C are only an approximate representation of the information contained in π_F; yet, it is possible to estimate the largest possibility distribution π^* (in the sense that $\forall A \in 2^X : \pi^*(A) \geq \pi(A)$ for any π satisfying Eqs. (1) and (2)) as follows [4]:

$$\pi^*(A) = \min \left\{ \inf_{x \in A} \mu_C(x), \inf_{x \notin A} (1 - \mu_I(x)) \right\} \tag{4}$$

Set-theoretic operations on twofold fuzzy sets have been defined as follows [3]:

$$\bar{\mathcal{F}} = (\bar{C}, \bar{I})$$
$$\mathcal{F}_1 \cup \mathcal{F}_2 = (I_1 \cup I_2, C_1 \cup C_2)$$
$$\mathcal{F}_1 \cap \mathcal{F}_2 = (I_1 \cap I_2, C_1 \cap C_2)$$

where $\mathcal{F}_1 = (I_1, C_1)$ and $\mathcal{F}_2 = (I_2, C_2)$. These set-theoretic operations are consistent with the corresponding operations on interval-valued fuzzy sets, by viewing the membership function of a twofold fuzzy set as $\mu_{\mathcal{F}}(x) = [\mu_I(x), \mu_C(x)]$. However, it is worth noticing that the compositionality of the union and intersection operations presupposes that the two possibility distributions underlying $\mathcal{F}_1$ and $\mathcal{F}_2$ are non-interactive.

In this work, we are interested in the study of the cardinality of a twofold fuzzy set, because it will be put in relation with granular counting.[1] Given a fuzzy set A, the possibility of the event "A has at least n elements" can be estimated by

$$\Pi\left(|A| \geq n\right) = \begin{cases} 0, & \text{if } \nexists \alpha : |A_\alpha| \geq n \\ \sup\left\{\alpha : |A_\alpha| \geq n\right\}, & \text{otherwise} \end{cases}$$

for any fuzzy set A, where A_α is the α-cut of A, namely $A_\alpha = \{x : \mu_A(x) \geq \alpha\}$, $\alpha \in]0, 1]$. For the ill-known set F, the possibility that it has exactly n elements is given by the conjunction of two conditions:(i) the number of certain elements should be at most n (otherwise it is impossible that F has n elements); and (ii) the number of possible elements should be at least n (because if it is possible that F has $m > n$ elements, then it is possible it has n elements). This leads to the following definition (see also [2] for further insights):

Definition 2. *Let $\mathcal{F} = (I, C)$ a twofold fuzzy set on X. The cardinality $\|\mathcal{F}\|$ of $\mathcal{F}$ is defined as the fuzzy set*

$$\mu_{\|\mathcal{F}\|}(n) = \min\left\{1 - \Pi\left(|I| \geq n + 1\right), \Pi\left(|C| \geq n\right)\right\} \tag{5}$$

3   Granular Count

An example involving ill-known sets consists of a finite set O of m items (or *observations*), which refer to some attributes (or *referents*) in a domain R.[2]

---

[1] Henceforth, we assume the referential X to be finite.

[2] From now on, the notation used in [9] will be used, in order to facilitate cross reading.

Table 1. Example of possibilistic assignment table, where $O = \{o_1, \ldots, o_6\}$ and $R = \{r_1, r_2, r_3\}$. The j-th row represents a possibility distribution π_{o_j} on R.

	r_1	r_2	r_3
o_1	1	0.3	0.5
o_2	0.8	1	0.6
o_3	1	0	0
o_4	0.8	0.9	1
o_5	1	0.8	1
o_6	0.2	0.5	1

A reference function (or selection function) assigning each observation to a referent

$$\rho : O \mapsto R$$

is assumed to exist, but it is unknown. The reference function formalizes the assumption that an observation cannot refer to more than one referent, and each observation refers to a referent. (We will call this as *functionality* assumption.)

A concrete example of referents and observations can be found in some applications in Bioinformatics, where it is necessary to estimate the *gene expression* in a cell or tissue, i.e., the process by which the information encoded in a gene is used to direct the assembly of a protein molecule. Current technologies (called Next Generation Sequencing, or NGS) estimate gene expression by reading fragments of messenger RNA (mRNA) and then matching these fragments to a database of known genes. However, this matching is uncertain because the fragments (also called "read") do not carry enough information for unequivocal association to genes. Therefore, while each read (which stands for an observation) is actually the product of the expression of a single gene (which stands for a referent), this referent is not known but from a possibility distribution, where grades of possibility are obtained from matching degrees as well as technological and biological criteria [1].

The uncertainty on the values referred to by each observation is formalized by a (normal) possibility distribution for each $o \in O$, i.e.

$$\pi_o : R \mapsto [0, 1]$$

such that $\exists r \in R : \pi_o(r) = 1$. The possibility distributions of all observations can be arranged in a *possibilistic assignment table*, as exemplified in Table 1. It is further assumed that the knowledge of the actual referent of an observation does not affect the knowledge of the referent of any other observation. (This will be called *non-interactivity* assumption.)

The uncertainty represented by the possibility distributions π_o propagates to the uncertainty on the reference function, which is defined by a possibility degree $\pi(\rho) \in [0, 1]$. This possibility degree is determined by the joint possibility of each observation o referring to $\rho(o) \in R$. Based on the non-interactivity assumption, it is possible to define the possibility degree as

$$\pi(\rho) = \min_{o \in O} \pi_o(\rho(o)) \tag{6}$$

As a consequence, the possibility measure of a set $P \subseteq R^O$ of reference functions is determined as

$$\Pi(P) = \max_{\rho \in P} \pi(\rho) \tag{7}$$

Sets of reference functions are instrumental for evaluating the possibility of complex and possibly interactive events involving sets of observations and/or referents. As an example, let P_r^n be the set of all reference functions where referent r is referred to by n observations, $n \leq m$. Formally:

$$P_r^n = \left\{ \rho \in R^O : \left| \rho^{-1}(r) \right| = n \right\} \tag{8}$$

Then, the following theorem holds [8]:

Theorem 1. *The possibility measure of P_r^n is determined by*

$$\Pi(P_r^n) = \max_{O_n \subseteq O} \min \left\{ \min_{o \in O_n} \pi_o(r), \min_{o \notin O_n} \max_{r' \neq r} \pi_o(r') \right\} \tag{9}$$

where $O_n \subseteq O$ is any subset of $n \in \mathbb{N}$ observations, i.e. $|O_n| = n$.[3]

Intuitively, the measure $\Pi(P_r^n)$ evaluates the possibility that there exists a set of n observations that refer to r and all the remaining observations refer to a different referent. The direct application of the granular counting formula (9) leads to a procedure that has exponential time complexity. Nevertheless, a quadratic-time algorithm was defined for exact granular counting and a linear-time algorithm was proposed for approximate granular counting [9]. Furthermore, on the one hand, certified error bounds for approximate counting were identified [7]; on the other hand, an incremental exact granular counting algorithm was devised to deal with large data sets [6].

4   Twofold Fuzzy Sets from Possibilistic Assignment Tables

A possibilistic assignment table can be read both horizontally and vertically. In fact, while each row expresses the distribution of the more or less possible referents of an observation, each column represent the possibility degrees that each observation refers to a specific referent; also, the remaining columns express the possibility that each observation refers to some other referents. In a nutshell, the columns of a possibilistic assignment table represent inclusion and exclusion possibilities of each observation to each referent: this naturally leads to the idea of twofold fuzzy set, as showed in the following.

---

[3] The conventions $\min \emptyset = 1$ and $\max \emptyset = 0$ are followed throughout the paper.

Proposition 1. *Let O be a set of m observations, R be a finite set of referents; let $\pi_o : R \mapsto [0,1]$ be a possibility distribution for each $o \in O$, and $r \in R$. The fuzzy sets C_r of objects that possibly have referent r, and I_r of objects that have r as their sure referent, defined as*

$$\mu_{C_r}(o) = \pi_o(r) \quad and \quad \mu_{I_r}(o) = 1 - \max_{r' \neq r} \pi_o(r')$$

form a twofold fuzzy set $\mathcal{F}_r = (I_r, C_r)$ describing the ill-known set O_r of observations with referent r.

Example 1. Given the possibilistic assignment table in Table 1, and referent r_1, the twofold fuzzy set $\mathcal{F}_{r_1} = (I_{r_1}, C_{r_1})$ is represented in Table 2. Indeed. C_{r_1} represents the set of observations more or less possibly referring to r_1, while I_{r_1} is the set of observation more or less certainly referring to observation r_1. As can be seen from the tables, o_1 is possibly referring to r_1 but without high certainty because there is a non-negligible possibility that o_1 could refer to r_3 (or, to a lesser extent, to r_2). On the other hand, the possibility that o_2 refers to r_1 is not maximal, therefore this referent is not certain at all. Furthermore, o_3 is certainly referring to r_1 (and therefore, possibly) because there are no further alternatives. On the other hand, o_5 is possibly referring to r_1 but this is not certain at all because o_5 could possibly refer to r_3 (or to r_2, though to a lesser extent).

Table 2. The twofold fuzzy set $\mathcal{F}_1$ derived from r_1 as in Table 1.

	C_{r_1}	I_{r_1}
o_1	1	0.5
o_2	0.8	0
o_3	1	1
o_4	0.8	0
o_5	1	0
o_6	0.2	0

The cardinality of a twofold fuzzy set is defined as in (5); on the other hand, the granular count formula (9) is a property derived by the functionality and non-interactivity assumptions. They actually coincide, as proved in the following:

Theorem 2. *Let O be a set of m observations, R be a finite set of referents; let $\pi_o : R \mapsto [0,1]$ be a possibility distribution for each $o \in O$, and $r \in R$. Let $\mathcal{F}_r = (I_r, C_r)$ be the twofold fuzzy set describing the ill-known set O_r of observations with referent r. Then, for each $n \in \mathbb{N}$:*

$$\mu_{\|\mathcal{F}_r\|}(n) = \Pi\left(P_r^n\right)$$

Proof (sketch). From Theorem 1:

$$\Pi\left(P_r^n\right) = \max_{O_n \subseteq O} \min \left\{ \min_{o \in O_n} \pi_o\left(r\right), \min_{o \notin O_n} \max_{r' \neq r} \pi_o\left(r'\right) \right\}$$

Because of the distributive law of max over min and further considerations:

$$\Pi\left(P_r^n\right) = \min \left\{ \max_{O_n \subseteq O} \min_{o \in O_n} \pi_o\left(r\right), \max_{O_n \subseteq O} \min_{o \notin O_n} \max_{r' \neq r} \pi_o\left(r'\right) \right\}$$

$$= \min\left\{A, B\right\}$$

where:

$$A = \max_{O_n \subseteq O} \min_{o \in O_n} \pi_o\left(r\right)$$

$$B = \max_{O_n \subseteq O} \min_{o \notin O_n} \max_{r' \neq r} \pi_o\left(r'\right)$$

In order to prove the theorem, it is sufficient to prove that $A = \Pi\left(|C_r| \geq n\right)$ and $B = 1 - \Pi\left(|I_r| \geq n+1\right)$.

(A) Let $[C_r]_\alpha$ the α-cut of C_r; then its cardinality $|[C_r]_\alpha| \geq n$ iff there are at least n observations $o_1, o_2, \ldots, o_n$ such that $\alpha \leq \mu_{C_r}\left(o_i\right) = \pi_{o_i}\left(r\right)$ or, equivalently, there exists a subset O_n of n observations such that $\min_{o \in O_n} \pi_o\left(r\right) \geq \alpha$. By varying the subsets of observations (provided they all have cardinality n) the value of $\min_{o \in O_n} \pi_o\left(r\right)$ changes accordingly; therefore, the value $A = \max_{O_n \subseteq O} \min_{o \in O_n} \pi_o\left(r\right)$ is the maximally attainable possibility degree of n observations referring to r: for any $\alpha > A$ there is not any subset of n observations such that all of them have a possibility degree greater than or equal to α. Furthermore, it is easy to show that, for $n' > n$,

$$\max_{O_{n'} \subseteq O} \min_{o \in O_{n'}} \pi_o\left(r\right) = \max_{O_{n'} \subseteq O} \min \left\{ \min_{o \in O_n} \pi_o\left(r\right), \min_{o \in O_{n'} \backslash O_n} \pi_o\left(r\right) \right\} \leq A$$

because $\min_{o \in O_n} \pi_o\left(r\right) \leq A$. As a consequence, $A = \sup\left\{\alpha : |[C_r]_\alpha| \geq n\right\} = \Pi\left(|C_r| \geq n\right)$.

(B) In order to prove that $B = 1 - \Pi\left(|I_r| \geq n+1\right)$, it can be observed that the equality is equivalent to proving the equality

$$\Pi\left(|I_r| \geq n+1\right) = \min_{O_n \subseteq O} \max_{o' \notin O_n} \mu_{I_r}\left(o'\right) \tag{10}$$

Let $\bar{\alpha} = \min_{O_n \subseteq O} \max_{o' \notin O_n} \mu_{I_r}\left(o'\right)$; then

$$\forall O_n \subseteq O : \max_{o' \notin O_n} \mu_{I_r}\left(o'\right) \geq \bar{\alpha}$$

which implies

$$\forall O_n \subseteq O \exists o' \notin O_n : \mu_{I_r}\left(o'\right) \geq \bar{\alpha} \tag{11}$$

We prove the existence of a set $\bar{O}_n$ of n observations such that $\min_{o \in \bar{O}_n} \mu_{I_r}\left(o\right) \geq \bar{\alpha}$ by *reductio ad absurdum*. We suppose that such set does not exist, i.e., $\forall O_n \subseteq O : \min_{o \in O_n} \mu_{I_r}\left(o\right) < \bar{\alpha}$, which is equivalent to asserting

$$\forall O_n \subseteq O \exists o \in O_n : \mu_{I_r}\left(o\right) < \bar{\alpha} \tag{12}$$

Let $O_n \subseteq O$; in force of (12), there exists $o \in O_n$ such that $\mu_{I_r}(o) < \bar{\alpha}$ while, in force of (11), there exists $o' \notin O_n$ such that $\mu_{I_r}(o') \geq \bar{\alpha}$. It is therefore possible to "swap" o and o' by defining $O'_n = (O \setminus \{o\}) \cup \{o'\}$; the existence of o' and o guarantees that the cardinality of O'_n is still n. Two cases can occur on O'_n:

1. all elements of O'_n have membership $\geq \bar{\alpha}$, thus contradicting the hypothesis, or
2. there still exists an element in O'_n such that $\mu_{I_r}(o) < \bar{\alpha}$. In such a case, (11) still guarantees the existence of an element $o'' \notin O'_n$ such that $\mu_{I_r}(o'') \geq \bar{\alpha}$. By repeating the "swap" of elements, both cases can be considered again. Eventually, in up to n repetitions, a set $\bar{O}_n$ is constructed such that $\forall o \in \bar{O}_n : \mu_{I_r}(o) \geq \bar{\alpha}$, thus contradicting the absurd hypothesis (12).

Given such $\bar{O}_n$, it is possible to apply (11) again, which assures the existence of an element $o' \notin \bar{O}_n$ such that $\mu_{I_r}(o') \geq \bar{\alpha}$. Therefore, there exists a set $O_{n+1} = \bar{O}_n \cup \{o'\}$ with $n + 1$ elements, such that $\min_{o \in O_{n+1}} \mu_{I_r}(o) \geq \bar{\alpha}$, i.e., $O_{n+1} \subseteq [I_r]_{\bar{\alpha}}$, which implies $|[I_r]_{\bar{\alpha}}| \geq n + 1$.

By definition of $\bar{\alpha}$, there exists $O_n^{\min} \subseteq O$ such that $\max_{o' \notin O_n^{\min}} \mu_{I_r}(o') = \bar{\alpha}$, which implies that $\forall o' \notin O_n^{\min} : \mu_{I_r}(o') \leq \bar{\alpha}$. Thus, any subset O_{n+1} of $n + 1$ elements must intersect $O \setminus O_n^{\min}$, therefore $\min_{o \in O_{n+1}} \mu_{I_r}(o) \leq \bar{\alpha}$. As a consequence, $\bar{\alpha} = \sup \{\alpha : |[I_r]_\alpha| \geq n + 1\} = \Pi(|I_r| \geq n + 1)$, thus proving (10).

$\square$

It is worth noticing the similarity between Eqs. (9) and (4); thanks to Theorem 2 it is possible to claim that:

$$\mu_{\|\mathcal{F}\|}(n) = \max_{O_n \subseteq O} \pi^*(O_n) \qquad (13)$$

that is, the possibility that the cardinality of the ill-known set F is n, corresponds to the maximum possibility of any subset of n elements in O, according to the largest possibility distribution π^* induced by $\mathcal{F}$.

5   Interactivity-Aware Operations on Twofold Fuzzy Sets

The granular counting formula (9) takes into account the interactivity of referents. In fact, a summation operator can be defined on the granular counts of the referents in R and it can be proved that the sum of all granular counts of all referents equals the number of observations (a precise number) [8]. To prove this property, the possibility measure of some appropriate sets of reference functions is computed. A similar approach could be adopted to take interactivity of twofold fuzzy sets into account, when these represent ill-known sets of observations. Let

$$P_{o \to r} = \{\rho \in R^O : \rho(o) = r\}$$

then, the following proposition is true:

Proposition 2. $\Pi(P_{o \to r}) = \pi_o(r)$

Proof. By definition:

$$\Pi\left(P_{o \to r}\right) = \max_{\rho \in P_{o \to r}} \pi\left(\rho\right)$$

$$= \max_{\rho \in P_{o \to r}} \min_{o' \in O} \pi_{o'}\left(\rho\left(o'\right)\right)$$

$$= \max_{\rho \in P_{o \to r}} \min \left\{ \min_{o' \in O \setminus \{o\}} \pi_{o'}\left(\rho\left(o'\right)\right), \pi_o\left(r\right) \right\}$$

$$= \min \left\{ \max_{\rho \in P_{o \to r}} \min_{o' \in O \setminus \{o\}} \pi_{o'}\left(\rho\left(o'\right)\right), \pi_o\left(r\right) \right\}$$

Because of normality of possibility distributions, there exists a reference function ρ^* such that $\rho^*\left(o\right) = r$ (thus $\rho^* \in P_{o \to r}$) and $\rho^*\left(o'\right) = r'$ where $\pi_{o'}\left(r'\right) = 1$ for all $o' \neq o$. Therefore:

$$\Pi\left(P_{o \to r}\right) = \min\left\{1, \pi_o\left(r\right)\right\}$$

$$= \pi_o\left(r\right)$$

$\square$

With similar arguments, for the set

$$P_{o \not\to r} = \left\{\rho \in R^O : \rho\left(o\right) \neq r\right\} = \overline{P_{o \to r}}$$

it is easy to check that:

$$\Pi\left(P_{o \not\to r}\right) = \max_{r' \neq r} \pi_o\left(r'\right)$$

As a consequence, $N\left(P_{o \to r}\right) = 1 - \Pi\left(P_{o \not\to r}\right) = \max_{r' \neq r} \pi_o\left(r'\right)$. Thus, the following property holds:

Proposition 3. *Let $\mathcal{F}_r = \left(I_r, C_r\right)$ a twofold fuzzy set resulting from referent r as in Proposition 1. Then:*

$$\mu_{C_r}\left(o\right) = \Pi\left(P_{o \to r}\right)$$

and

$$\mu_{I_r}\left(o\right) = N\left(P_{o \to r}\right)$$

It is now possible to consider the more complex case of the set of all reference functions that map an observation o to one referent in a given set. Namely, given $S \subseteq R$, the following set is defined:

$$P_{o \to S} = \left\{\rho \in R^O : \rho\left(o\right) \in S\right\} = \bigcup_{r \in S} P_{o \to r}$$

Thus, the set $P_{o \to S}$ represents the disjunction of elementary events, each one consisting of o referring to a referent in S. It is easy to observe that the pair $\left(I_S, C_S\right)$ defined as

$$\mu_{C_S}\left(o\right) = \Pi\left(P_{o \to S}\right)$$

and

$$\mu_{I_S}(o) = N(P_{o \to S})$$

forms a twofold fuzzy set, which is denoted by $\mathcal{F}_S$. It is therefore possible to take interactivity into account when computing the union of two twofold fuzzy sets as follows:

Definition 3. *Given two twofold fuzzy sets $\mathcal{F}_{S_1}$ and $\mathcal{F}_{S_2}$ as derived from a possibilistic assignment table with $S_1, S_2 \subseteq R$, the twofold fuzzy set $\mathcal{F}_{S_1} \uplus \mathcal{F}_{S_2}$ is defined as*

$$\mathcal{F}_{S_1} \uplus \mathcal{F}_{S_2} = (I_{S_1 \cup S_2}, C_{S_1 \cup S_2})$$

where

$$\mu_{C_{S_1 \cup S_2}}(o) = \Pi(P_{o \to S_1 \cup S_2})$$

and

$$\mu_{I_{S_1 \cup S_2}}(o) = N(P_{o \to S_1 \cup S_2})$$

The following theorem holds:

Theorem 3. *The closure and interior of $\mathcal{F}_{S_1} \uplus \mathcal{F}_{S_2}$ are*

$$\mu_{C_{S_1 \cup S_2}}(o) = \max_{r \in S_1 \cup S_2} \pi_o(r)$$

and

$$\mu_{I_{S_1 \cup S_2}}(o) = 1 - \max_{r \notin S_1 \cup S_2} \pi_o(r)$$

Proof. It is straightforward to observe that

$$P_{o \to S_1 \cup S_2} = P_{o \to S_1} \cup P_{o \to S_2}$$

therefore

$$\begin{aligned}
\mu_{C_{S_1 \cup S_2}}(o) &= \Pi(P_{o \to S_1 \cup S_2}) = \Pi(P_{o \to S_1} \cup P_{o \to S_2}) \\
&= \max\{\Pi(P_{o \to S_1}), \Pi(P_{o \to S_2})\} \\
&= \max_{r \in S_1 \cup S_2} \pi_o(r)
\end{aligned}$$

Similarly,

$$\mu_{I_{S_1 \cup S_2}}(o) = N(P_{o \to S_1 \cup S_2}) = 1 - \Pi\left(\overline{P_{o \to S_1 \cup S_2}}\right)$$

Since

$$\overline{P_{o \to S_1 \cup S_2}} = \bigcup_{r \notin S_1 \cup S_2} P_{o \to r}$$

then

$$\begin{aligned}
\mu_{I_{S_1 \cup S_2}}(o) &= 1 - \Pi\left(\bigcup_{r \notin S_1 \cup S_2} P_{o \to r}\right) \\
&= 1 - \max_{r \notin S_1 \cup S_2} \pi_o(r)
\end{aligned}$$

$\square$

Table 3. Union of two twofold fuzzy sets according to standard union $\cup$ and interactivity-aware union $\uplus$. Differences are marked by ">".

	$\mathcal{F}_{r_2}$		$\mathcal{F}_{r_3}$		$\mathcal{F}_{r_2} \cup \mathcal{F}_{r_3}$		$\mathcal{F}_{r_2} \uplus \mathcal{F}_{r_3}$	
	C	I	C	I	C	I	C	I
o_1	0.3	0	0.5	0	0.5	0	0.5	0
o_2	1	0.2	0.6	0	1	0.2	1	0.2
o_3	0	0	0	0	0	0	0	0
> o_4	0.9	0	1	0.1	1	0.1	1	0.2
o_5	0.8	0	1	0	1	0	1	0
> o_6	0.5	0	1	0.5	1	0.5	1	0.8

In a similar way, it is possible to take interactivity into account when computing intersection and complement of twofold fuzzy sets.

Example 2. Table 3 compares the results of applying the standard union operator $\cup$ and the interactivity-aware union $\uplus$ for the union of the twofold fuzzy sets derived from referents r_2 and r_3 of the possibilistic assignment table reported in Table 1. The difference can be noted only in the interior fuzzy sets because the closure fuzzy sets are determined in the same way. Among the differences, we highlight the case of o_6: from the table we see that it is not necessary that o_6 belongs to O_{r_2} (the ill-known set of observations described by $\mathcal{F}_{r_2}$) because it is fully possible that o_6 refers to r_3, and it is necessary with degree 0.5 that o_6 belongs to O_{r_3} (the ill-known set described by $\mathcal{F}_{r_3}$) because there is a possibility of 0.5 that o_6 refers to r_2. According to the standard union, the necessity that o_6 belongs to the union of O_{r_2} and O_{r_3} is *at least* 0.5; in fact, the interactivity-aware union yields a value of necessity equal to 0.8. This value is a consequence of interactivity of r_2 and r_3: while the necessity that o_6 refers to one of the two referents is low when taken individually, the necessity that o_6 refers to one of the two referents when taken together is much higher, because the possibility that o_6 refers to r_1 is very small (0.2).

6 Conclusion

Twofold fuzzy sets can be profitably used to represent the ill-known collection of observations that pertain to a referent (or to a set of referents). We have proved that the cardinality of these twofold fuzzy sets coincides with the granular count of observations, the latter being a property derived from two very weak assumptions: functionality (each observation refers to one referent only) and non-interactivity (each observation does not affect the others).

Granular count is derived by considering the possibility measure of sets of reference functions, which formalize the above-mentioned assumptions. Following the same path, it is possible to take interactivity into account when computing aggregations of twofold fuzzy sets derived from a possibilistic assignment table. The use of these interactivity-aware operations avoids to relax the interpretation of closure and interior fuzzy sets as upper and lower bounds of possibility and necessity measures, respectively, which is necessary when standard set-theoretic operations are adopted. However, the computation of these set operations requires the availability of the possibilistic assignment table, which could affect performance in the case of large datasets.

Acknowledgments. The research is partially supported by Ministero dello Sviluppo Economico (MISE) under grant F/190030/01-03/X44 "LIFT". C.M. is member of the INdAM Research group GNCS and CILA (Centro Interdipartimentale di Logica e Applicazioni).

References

1. Consiglio, A., et al.: A fuzzy method for RNA-Seq differential expression analysis in presence of multireads. BMC Bioinform. **17**, 95–110 (2016)
2. Dubois, D., Prade, H.: Fuzzy cardinality and the modeling of imprecise quantification. Fuzzy Sets Syst. **16**(3), 199–230 (1985)
3. Dubois, D., Prade, H.: Twofold fuzzy sets and rough sets-some issues in knowledge representation. Fuzzy Sets Syst. **23**(1), 3–18 (1987)
4. Dubois, D., Prade, H.: Incomplete conjunctive information. Comput. Math. Appl. **15**(10), 797–810 (1988)
5. Yao, J., et al.: Towards more adequate representation of uncertainty: from intervals to set intervals, with the possible addition of probabilities and certainty degrees. In: 2008 IEEE International Conference on Fuzzy Systems (IEEE World Congress on Computational Intelligence), pp. 983–990. IEEE, June 2008
6. Mencar, C.: An incremental algorithm for granular counting with possibility theory. In: IEEE International Conference on Fuzzy Systems, pp. 1–7, Glasgow, United Kingdom. IEEE, July 2020
7. Mencar, C.: Possibilistic bounds for granular counting. In: Lesot, M.-J., et al. (eds.) Information Processing and Management of Uncertainty in Knowledge-Based Systems - IPMU 2020. Communications in Computer and Information Science, vol. 1239, pp. 27–40. Springer, Cham (2020). https://doi.org/10.1007/978-3-030-50153-2_3
8. Mencar, C.: Possibilistic granular count: derivation and extension to granular sum. In: Joint Proceedings of the 19th World Congress of the International Fuzzy Systems Association (IFSA), the 12th Conference of the European Society for Fuzzy Logic and Technology (EUSFLAT), and the 11th International Summer School on Aggregation Operators AG, pp. 486–493. Atlantis Press (2021)
9. Mencar, C., Pedrycz, W.: Granular counting of uncertain data. Fuzzy Sets Syst. **387**, 108–126 (2020)
10. Yao, Y.: Interval sets and interval-set algebras. In: 2009 8th IEEE International Conference on Cognitive Informatics, pp. 307–314. IEEE, June 2009

Testing Independence with Fuzzy Data

Przemyslaw Grzegorzewski[1,2]([✉]) [iD]

[1] Systems Research Institute, Polish Academy of Sciences,
Newelska 6, 01-447 Warsaw, Poland
pgrzeg@ibspan.waw.pl
[2] Faculty of Mathematics and Information Science, Warsaw University
of Technology, Koszykowa 75, 00-662 Warsaw, Poland

Abstract. The question of whether two random variables describing two features under study are independent arises in many statistical studies and practical applications. Many parametric and nonparametric tests of bivariate independence can be found in the literature, like the chi-square test for contingency tables, tests based on Pearson's, Kendall's, Spearman's correlation coefficients and so on. The problem of testing independence becomes much more difficult when the available data are imprecise, incomplete or vague. Although fuzzy modeling provide appropriate tools for dealing with uncertain data, some limitations of fuzzy random variables inhibit the straightforward generalization of the classical tests of independence into fuzzy framework. At the same time, this situation imposes the quest for new solutions that may be applied in fuzzy environment. In this contribution we propose a new permutation test of independence for fuzzy data based on the so-called distance covariance.

Keywords: Fuzzy data · Fuzzy numbers · Permutation test · Random fuzzy numbers · Test of independence

1 Introduction

In many practical applications, when random samples describing two features X and Y are observed, we are interested in finding out whether X and Y are somehow associated. It is important, since if they are related then knowledge about a value of X or Y provides some information about a value of the latter variable, even if it does not tell us this value exactly. Otherwise, if the knowledge about X and Y does not bring us any information about the latter variable than we conclude that both variables are independent.

Many tests of the bivariate independence have been proposed. If one assumes multivariate normality of the joint distribution function the problem of testing that the covariance matrix equals zero is equivalent to independence testing. When data reveal a monotone dependence then tests based on Kendall's tau, Spearman's rho or the Goodman-Kruskal coefficients perform well (see, e.g., [5]). For categorized data structured in contingency table the famous chi-square test of independence is usually applied.

© Springer Nature Switzerland AG 2022
D. Ciucci et al. (Eds.): IPMU 2022, CCIS 1602, pp. 520–531, 2022.
https://doi.org/10.1007/978-3-031-08974-9_42

The problem of testing independence becomes much more difficult in the case of imprecise phenomena. In particular, the lack of universally accepted total ranking between fuzzy sets used for modeling imprecise data eliminates tests based on ranks. Therefore, a few contributions on testing independence with fuzzy data appeared in the literature. A fuzzy version of the chi-square test of independence in a two-way contingency table was considered by Hryniewicz [11] or Taheri et al. [23]. Hryniewicz [12] considered also a fuzzy version of the Goodman-Kruskal test for categorical data when the values of the response variable are imprecise while observations of the explanatory variable are crisp (see also Taheri and Hesamian [22] for situations when variables of interest are categorized by linguistic terms). Moreover, Gil et al. [6] discussed the strength of association between fuzzy variables in terms of the extended determination coefficient, which is equivalent to the "affine independence". A similar and yet a separate problem is the topic of fuzzy regression analysis, especially concerning a possibilistic approach or fuzzy least squares – for some recent works on a fuzzy regression we refer the reader to [4].

In this paper we propose a new method for testing independence with fuzzy data based on the concept of the distance covariance and distance correlation introduced by Székely, Rizzo and Bakirov [20]. Applying the distance correlation Bakirov et al. [1] designed a new nonparametric approach for testing the joint independence of two or more random vectors in arbitrary dimension. We show that tailoring the distance covariance to fuzzy data analysis, we are able to construct a permutation test suitable for verification of independence with fuzzy data. It appears that our test is quite universal since it works well against various alternatives and types of dependence.

The paper is organized as follows: in Sect. 2 and Sect. 3 we recall basic information related to fuzzy data and fuzzy random variables, respectively. In Sect. 4 we describe the concept of the distance covariance and distance correlation which is afterwards applied in testing independence with fuzzy data. A new test is proposed in Sect. 5, while some numerical experiments and examples are reported in Sect. 6.

2   Fuzzy Data

Real numbers are typical outputs of many experiments. In the presence of imprecise results we need a tool for describing such observations. A common way for modeling imprecise values is to apply fuzzy numbers. A **fuzzy number** is identified by a mapping $\widetilde{A} : \mathbb{R} \to [0,1]$, called a membership function, such that its α-cuts

$$\widetilde{A}_\alpha = \begin{cases} \{x \in \mathbb{R} : \widetilde{A}(x) \geqslant \alpha\} & \text{if } \alpha \in (0,1], \\ cl\{x \in \mathbb{R} : \widetilde{A}(x) > 0\} & \text{if } \alpha = 0, \end{cases} \tag{1}$$

are nonempty compact intervals for each $\alpha \in [0,1]$, where cl denotes the closure operator. Thus a fuzzy number is completely characterized either by its membership function $\widetilde{A}(x)$ or by a family of its α-cuts $\{\widetilde{A}_\alpha\}_{\alpha \in [0,1]}$. Two α-cuts are of special interest: $\tilde{A}_1 = \text{core}(\tilde{A})$, called the **core**, which contains all values fully compatible with the concept described by $\tilde{A}$ and $\tilde{A}_0 = \text{supp}(\tilde{A})$, called the **support**, which contains values compatible to some extent with $\tilde{A}$.

Membership functions may assume different shapes but the most often used fuzzy numbers are the **trapezoidal fuzzy numbers** (or fuzzy intervals) with membership functions of the form

$$\tilde{A}(x) = \begin{cases} \frac{x-a_1}{a_2-a_1} & \text{if } a_1 \leqslant x < a_2, \\ 1 & \text{if } a_2 \leqslant x \leqslant a_3, \\ \frac{a_4-x}{a_4-a_3} & \text{if } a_3 < x \leqslant a_4, \\ 0 & \text{otherwise,} \end{cases} \tag{2}$$

where $a_1, a_2, a_3, a_4 \in \mathbb{R}$ such that $a_1 \leqslant a_2 \leqslant a_3 \leqslant a_4$. It's easily seen that a trapezoidal fuzzy number is completely described by its support and core, so a trapezoidal fuzzy number given by (2) is often denoted as $\tilde{A} = \text{Tra}(a_1, a_2, a_3, a_4)$. If $a_2 = a_3$ then $\tilde{A}$ is said to be a **triangular fuzzy number**; if $a_1 = a_2$ and $a_3 = a_4$ we have the so-called **interval** (or rectangular) fuzzy number. The families of all fuzzy numbers, trapezoidal fuzzy numbers, triangular fuzzy number and interval fuzzy numbers will be denoted by $\mathbb{F}(\mathbb{R})$, $\mathbb{F}^T(\mathbb{R})$, $\mathbb{F}^\Delta(\mathbb{R})$ and $\mathbb{F}^I(\mathbb{R})$, respectively, where $\mathbb{F}^\Delta(\mathbb{R}) \subset \mathbb{F}^T(\mathbb{R}) \subset \mathbb{F}(\mathbb{R})$ and $\mathbb{F}^I(\mathbb{R}) \subset \mathbb{F}^T(\mathbb{R}) \subset \mathbb{F}(\mathbb{R})$.

Basic arithmetic operations in $\mathbb{F}(\mathbb{R})$ are defined with α-cut-wise operations on intervals. Thus the sum of $\tilde{A} \in \mathbb{F}(\mathbb{R})$ and $\tilde{B} \in \mathbb{F}(\mathbb{R})$ is given by the Minkowski addition of the corresponding α-cuts, i.e. for all $\alpha \in [0,1]$ we have

$$(\tilde{A} + \tilde{B})_\alpha = \left[\inf \tilde{A}_\alpha + \inf \tilde{B}_\alpha, \sup \tilde{A}_\alpha + \sup \tilde{B}_\alpha\right].$$

Similarly, the product of a fuzzy number $\tilde{A} \in \mathbb{F}(\mathbb{R})$ by a scalar $\theta \in \mathbb{R}$ is defined by the Minkowski scalar product for intervals, i.e. for all $\alpha \in [0,1]$

$$(\theta \cdot \tilde{A})_\alpha = \left[\min\{\theta \inf \tilde{A}_\alpha, \theta \sup \tilde{A}_\alpha\}, \max\{\theta \inf \tilde{A}_\alpha, \theta \sup \tilde{A}_\alpha\}\right].$$

This type of calculation is much easier for trapezoidal fuzzy numbers. Indeed, if $\tilde{A} = \text{Tra}(a_1, a_2, a_3, a_4)$, $\tilde{B} = \text{Tra}(b_1, b_2, b_3, b_4)$ and $\theta \in \mathbb{R}$ then

$$\tilde{A} + \tilde{B} = \text{Tra}(a_1 + b_1, a_2 + b_2, a_3 + b_3, a_4 + b_4), \tag{3}$$

$$\theta \cdot \tilde{A} = \begin{cases} \text{Tra}(\theta \cdot a_1, \theta \cdot a_2, \theta \cdot a_3, \theta \cdot a_4) & \text{if } \theta \geqslant 0, \\ \text{Tra}(\theta \cdot a_4, \theta \cdot a_3, \theta \cdot a_2, \theta \cdot a_1) & \text{if } \theta < 0. \end{cases} \tag{4}$$

Unfortunately, $(\mathbb{F}(\mathbb{R}), +, \cdot)$ has not linear but a semilinear structure since in general $\tilde{A} + (-1 \cdot \tilde{A}) \neq \mathbb{1}_{\{0\}}$. The Minkowski-based difference does not satisfy, in general, the addition/subtraction property, i.e. $(\tilde{A} + (-1 \cdot \tilde{B})) + \tilde{B} \neq \tilde{A}$. To overcome this problem the so-called Hukuhara difference [13] has been proposed but the Hukuhara difference does not always exist. To avoid subtraction problems in statistical reasoning with fuzzy observations an alternative distance-based approach was developed (see [3]). Obviously, one can define many metrics in

$\mathbb{F}(\mathbb{R})$ but the most often used in statistical context is the distance proposed by Gil et al. [6] and Trutschnig et al. [24], defined for any $A, B \in \mathbb{F}(\mathbb{R})$ as follows

$$D_\theta^\lambda(\widetilde{A}, \widetilde{B}) = \left(\int_0^1 [(\operatorname{mid} \widetilde{A}_\alpha - \operatorname{mid} \widetilde{B}_\alpha)^2 + \theta \cdot (\operatorname{spr} \widetilde{A}_\alpha - \operatorname{spr} \widetilde{B}_\alpha)^2] d\lambda(\alpha) \right)^{1/2}, \quad (5)$$

where λ is a normalized measure associated with a continuous distribution on $[0, 1]$, θ is a positive constant and

$$\operatorname{mid} \widetilde{A}_\alpha = \frac{1}{2}(\inf \widetilde{A}_\alpha + \sup \widetilde{A}_\alpha), \qquad \operatorname{spr} \widetilde{A}_\alpha = \frac{1}{2}(\sup \widetilde{A}_\alpha - \inf \widetilde{A}_\alpha) \quad (6)$$

denote the mid-point and the radius of the α-cut $\widetilde{A}_\alpha$, respectively. Here λ allows to weight the α-cut's influence, while by θ we may weight the impact of the difference between spreads of the α-cuts (difference in vagueness) and the distance between their mid-points (difference in location). The most common choice of λ is the Lebesgue measure on $[0, 1]$ and $\theta = 1$ or $\theta = \frac{1}{3}$. Fortunately, whatever (λ, θ) is chosen D_θ^λ is an L^2-type metric in $\mathbb{F}(\mathbb{R})$ and it is invariant to translations and rotations. Moreover, $(\mathbb{F}(\mathbb{R}), D_\theta^\lambda)$ is a separable metric space and for each fixed λ all metrics D_θ^λ are topologically equivalent. For more details on fuzzy numbers we refer the reader to [2].

3   Fuzzy Random Variables

In classical statistics we deal with crisp experimental results modeled usually by real-valued random variables. When the outputs of an experiment are imprecise they are often satisfactorily described by fuzzy numbers. Hence, for the purposes of statistical inference we need a model which allows to grasp both aspects of uncertainty that appear in such data, i.e. fuzziness (connected with data imprecision) and randomness (associated with data generation mechanism). To handle such data Puri and Ralescu [18] introduced the concept of a **fuzzy random variable**, sometimes also called a **random fuzzy number**.

Definition 1. *Let $(\Omega, \mathcal{A}, P)$ be a probability space. A mapping $\widetilde{X} : \Omega \to \mathbb{F}(\mathbb{R})$ is a **fuzzy random variable (random fuzzy number)** if for all $\alpha \in [0, 1]$ the α-level function is a compact random interval.*

It can be shown that $\widetilde{X}$ is a fuzzy random variable if and only if $\widetilde{X}$ is a Borel measurable function w.r.t. the Borel σ-field generated by the topology induced by D_θ^λ.

To describe a fuzzy random variable in a concise form its location and dispersion parameters are defined. The Aumann-type mean of a fuzzy random variable $\widetilde{X}$ is the fuzzy number $E(\widetilde{X}) \in \mathbb{F}(\mathbb{R})$ such that for each $\alpha \in [0, 1]$ the α-cut $\left(E(\widetilde{X})\right)_\alpha$ is equal to the Aumann integral of $\widetilde{X}_\alpha$, i.e. (cf. [18])

$$\left(E(\widetilde{X})\right)_\alpha = [\mathbb{E}(\operatorname{mid} \widetilde{X}_\alpha) - \mathbb{E}(\operatorname{spr} \widetilde{X}_\alpha), \mathbb{E}(\operatorname{mid} \widetilde{X}_\alpha) + \mathbb{E}(\operatorname{spr} \widetilde{X}_\alpha)]. \quad (7)$$

The D_θ^λ-Fréchet-type variance $V(\widetilde{X})$ is a non-negative real number such that (cf. [14])

$$V(\widetilde{X}) = \mathbb{E}\left(\left[D_\theta^\lambda(\widetilde{X}, E(\widetilde{X}))\right]^2\right) = \int_0^1 \text{Var}(\text{mid }\widetilde{X}_\alpha)d\lambda(\alpha) + \theta \int_0^1 \text{Var}(\text{spr }\widetilde{X}_\alpha)d\lambda(\alpha).$$
(8)

Consequently, a **fuzzy random sample** (**fuzzy sample**, in brief) is a finite sequence of independent random fuzzy numbers from the same distribution. Given such fuzzy sample $\mathbb{X} = (\widetilde{X}_1, \ldots, \widetilde{X}_n)$ we can determine its various characteristics, like the average $\overline{\widetilde{X}} \in \mathbb{F}(\mathbb{R})$ defined by the following α-cuts

$$\overline{\widetilde{X}}_\alpha = \left[\frac{1}{n}\sum_{i=1}^n \text{mid}\,(\widetilde{X}_i)_\alpha - \frac{1}{n}\sum_{i=1}^n \text{spr}\,(\widetilde{X}_i)_\alpha, \frac{1}{n}\sum_{i=1}^n \text{mid}\,(\widetilde{X}_i)_\alpha + \frac{1}{n}\sum_{i=1}^n \text{spr}\,(\widetilde{X}_i)_\alpha\right],$$
(9)

where $\alpha \in [0, 1]$, or the sample variance $S^2 \in \mathbb{R}$ given by

$$S^2 = \frac{1}{n-1}\sum_{i=1}^n D_\theta^\lambda(\widetilde{X}_i, \overline{\widetilde{X}})^2.$$
(10)

Although (9) and (10) estimate nicely (7) and (8), respectively (see [19]), no one should think too easily that transferring well-known facts from classical statistics to the fuzzy domain is straightforward and obvious. Firstly, one should be aware of problems with subtraction and division of fuzzy numbers, mentioned in Sect. 2, which lead to the conclusion that avoiding these operations whenever it is possible is strongly recommended. Another source of potential problems is the absence of suitable models for the distribution of fuzzy random variables which makes the statistical reasoning really hard. Hence nonparametric methodology seem to be a natural solution in the absence of parametric models. Unfortunately, since fuzzy numbers are not linearly ordered, the lack of a commonly accepted total ranking precludes the application of the rank methods so popular in nonparametric statistics. Finally, there are not yet satisfying Central Limit Theorems for fuzzy random variables that can be applied directly in decision making. All these disadvantages impede the straightforward generalization of traditional methods applied successfully in statistics with precise data. Thus, to make inference with fuzzy samples we need innovative solutions and a non-standard reasoning. Some successful attempts in hypotheses testing with fuzzy data are associated with the idea of the bootstrap (see, e.g. [7,15,17]) or permutation tests (see, e.g. [8–10]). The last methodology is applied also in this contribution.

4   Distance Covariance and Distance Correlation

Recently Székely, Rizzo and Bakirov [20] introduced a new measure of association between random variables, called the **distance correlation**. Suppose X and Y are random variables of arbitrary dimensions (say in $\mathbb{R}^p$ and $\mathbb{R}^q$, respectively) from distributions with finite first moments. Let ξ_X and ξ_Y denote the character-istic functions of X and Y, respectively, and let $\xi_{X,Y}$ be the joint characteristic function of X and Y.

Definition 2. *(cf. [20]) The **distance covariance** between random variables X and Y with finite first moments is the nonnegative number $\mathcal{V}(X,Y)$ defined as follows*

$$\mathcal{V}^2(X,Y) = \|\xi_{X,Y}(t,s) - \xi_X(t)\xi_Y(s)\|^2, \tag{11}$$

where $\|\cdot\|$ stands for the norm in the L^2 space of functions in $\mathbb{R}^{p+q}$.

By analogy, we may define the distance variance $\mathcal{V}(X)$ as the square root of

$$\mathcal{V}^2(X) = \mathcal{V}^2(X,X) = \|\xi_{X,X}(t,s) - \xi_X(t)\xi_X(s)\|^2.$$

Definition 3. *(cf. [20]) The **distance correlation** between random variables X and Y with finite first moments is the nonnegative number $\mathcal{R}(X,Y)$ defined as a square root of*

$$\mathcal{R}^2(X,Y) = \begin{cases} \frac{\mathcal{V}^2(X,Y)}{\sqrt{\mathcal{V}^2(X)\mathcal{V}^2(Y)}} & \text{if } \mathcal{V}^2(X)\mathcal{V}^2(Y) > 0, \\ 0 & \text{if } \mathcal{V}^2(X)\mathcal{V}^2(Y) = 0. \end{cases} \tag{12}$$

So defined distance correlation has properties analogous to the well-known product-moment correlation. In particular, it satisfies $0 \leqslant \mathcal{R} \leqslant 1$, but unlike the classical correlation $\mathcal{R} = 0$ only if the random vectors are independent. Moreover, for the bivariate normal distributions $\mathcal{R}$ is a function of the classical correlation coefficient $\rho(X,Y)$ and

$$\mathcal{R}(X,Y) \leqslant |\rho(X,Y)|,$$

with equality if $\rho(X,Y) = \pm 1$. For more information on the distance correlation we refer the reader to [20].

Now let us move into the statistical ground, i.e. to situation where a random sample $(\mathbb{X},\mathbb{Y}) = (X_1,Y_1),\ldots,(X_n,Y_n)$ from the joint distribution of random variables X in $\mathbb{R}^p$ and Y in $\mathbb{R}^q$ is observed. Following [20] we define

$$A_{ij} = a_{ij} - \bar{a}_{i\cdot} - \bar{a}_{\cdot j} + \bar{a}_{\cdot\cdot}, \tag{13}$$

where

$$a_{ij} = \|X_i - X_j\|_p, \quad \bar{a}_{i\cdot} = \frac{1}{n}\sum_{j=1}^n a_{ij}, \quad \bar{a}_{\cdot j} = \frac{1}{n}\sum_{i=1}^n a_{ij}, \quad \bar{a}_{\cdot\cdot} = \frac{1}{n^2}\sum_{i=1}^n\sum_{j=1}^n a_{ij} \tag{14}$$

for $i,j = 1,\ldots,n$ and $\|\cdot\|$ denoting the Euclidean norm in $\mathbb{R}^p$. Similarly, we define $b_{ij} = \|Y_i - Y_j\|_q$ and other analogues of (14) for Y's to obtain values

$$B_{ij} = b_{ij} - \bar{b}_{i\cdot} - \bar{b}_{\cdot j} + \bar{b}_{\cdot\cdot} \tag{15}$$

required for empirical counterparts of the notions showed in Definition 2 and Definition 3.

Definition 4. *(cf. [20]) The **empirical distance covariance** is the nonnegative number* $\mathcal{V}_n(\mathbb{X}, \mathbb{Y})$ *defined as follows*

$$\mathcal{V}_n^2(\mathbb{X}, \mathbb{Y}) = \frac{1}{n^2} \sum_{i=1}^{n} \sum_{j=1}^{n} A_{ij} B_{ij}. \tag{16}$$

It can be shown that $\mathcal{V}_n^2(\mathbb{X}, \mathbb{Y}) \geqslant 0$. For the proof we refer to [20]. Obviously,

$$\mathcal{V}_n^2(\mathbb{X}) = \mathcal{V}_n^2(\mathbb{X}, \mathbb{X}) = \frac{1}{n^2} \sum_{i=1}^{n} \sum_{j=1}^{n} A_{ij}^2.$$

Definition 5. *(cf. [20]) The **empirical distance correlation** is the nonnegative number* $\mathcal{R}_n(\mathbb{X}, \mathbb{Y})$ *defined as a square root of*

$$\mathcal{R}_n^2(\mathbb{X}, \mathbb{Y}) = \begin{cases} \frac{\mathcal{V}_n^2(\mathbb{X}, \mathbb{Y})}{\sqrt{\mathcal{V}_n^2(\mathbb{X}) \mathcal{V}_n^2(\mathbb{Y})}} & if \quad \mathcal{V}_n^2(\mathbb{X}) \mathcal{V}_n^2(\mathbb{Y}) > 0, \\ 0 & if \quad \mathcal{V}_n^2(\mathbb{X}) \mathcal{V}_n^2(\mathbb{Y}) = 0. \end{cases} \tag{17}$$

So defined empirical distance correlation $\mathcal{R}_n(\mathbb{X}, \mathbb{Y})$ appears to be a useful empirical measure of association between random variables that can be applied, for instance, to check whether two features under study are independent. The asymptotic distribution of $\mathcal{V}_n^2(\mathbb{X}, \mathbb{Y})$ is a quadratic form of centered Gaussian random variables with coefficients that depend on the distributions of X and Y.

In statistical inference we generally consider that the dimensions of random variables are fixed and then we investigate the effect of a sample size on the inference results. In [21] Székely et al. considered the situation where the dimensions of random vectors are large relative to the sample size. With the help of a modified distance correlation they obtained a test statistic that has an asymptotic (with respect to dimension) Student's t-distribution under independence.

Moreover, Lyons [16] extended the distance covariance to all separable Hilbert spaces which makes it applicable to functional data. This, in turn, opened the way to applications in fuzzy environment. And here's what we're going to suggest in the next section on how to utilize the generalized distance correlation coefficient for testing hypotheses based on fuzzy data.

5   Testing Independence with Fuzzy Data

Now suppose that we have no longer a typical real-valued sample but we have to perform the desired statistical reasoning with imprecise data. To be more strict, we have a fuzzy random sample $(\widetilde{\mathbb{X}}, \widetilde{\mathbb{Y}}) = (\widetilde{X}_1, \widetilde{Y}_1), \ldots, (\widetilde{X}_n, \widetilde{Y}_n)$, where $\widetilde{X}_i$ and $\widetilde{Y}_i$, for $i = 1, \ldots, n$, are random fuzzy numbers. Given such fuzzy sample we define the counterparts of some notions defined in Sect. 4.

Definition 6. *The empirical distance covariance between fuzzy samples* $\widetilde{\mathbb{X}}$ *and* $\widetilde{\mathbb{Y}}$ *with respect to the distance* D_θ^λ *is the nonnegative number* $\mathcal{V}_{D_\theta^\lambda}(\widetilde{\mathbb{X}}, \widetilde{\mathbb{Y}})$ *defined as follows*

$$\mathcal{V}_{D_\theta^\lambda}^2(\widetilde{\mathbb{X}}, \widetilde{\mathbb{Y}}) = \frac{1}{n^2} \sum_{i=1}^n \sum_{j=1}^n G_{ij} H_{ij}, \tag{18}$$

where

$$G_{ij} = D_\theta^\lambda(\widetilde{X}_i, \widetilde{X}_j) - \frac{1}{n} \sum_{j=1}^n D_\theta^\lambda(\widetilde{X}_i, \widetilde{X}_j) - \frac{1}{n} \sum_{i=1}^n D_\theta^\lambda(\widetilde{X}_i, \widetilde{X}_j) \tag{19}$$

$$+ \frac{1}{n^2} \sum_{i=1}^n \sum_{j=1}^n D_\theta^\lambda(\widetilde{X}_i, \widetilde{X}_j),$$

$$H_{ij} = D_\theta^\lambda(\widetilde{Y}_i, \widetilde{Y}_j) - \frac{1}{n} \sum_{j=1}^n D_\theta^\lambda(\widetilde{Y}_i, \widetilde{Y}_j) - \frac{1}{n} \sum_{i=1}^n D_\theta^\lambda(\widetilde{Y}_i, \widetilde{Y}_j) \tag{20}$$

$$+ \frac{1}{n^2} \sum_{i=1}^n \sum_{j=1}^n D_\theta^\lambda(\widetilde{Y}_i, \widetilde{Y}_j).$$

Our goal is to apply the empirical distance covariance between fuzzy samples defined above to construct a test for independence. As it happens in many other situations (see, e.g. [8–10]), since the distributions of our fuzzy samples are unknown, the proposed test based on (18) will be implemented as a permutation test.

Let $\widetilde{X} : \Omega \to \mathbb{F}(\mathbb{R})$ and $\widetilde{Y} : \Omega \to \mathbb{F}(\mathbb{R})$. Then $\widetilde{Z} = (\widetilde{X}, \widetilde{Y})$ is a random vector which assumes values in $\mathbb{F}(\mathbb{R})^2$. Let $\widetilde{z}$ denote n independent observations of $\widetilde{Z}$ which form our fuzzy sample. It can be considered as a data matrix with two rows and n columns, i.e.

$$\widetilde{z} = \widetilde{z}(\widetilde{\mathbb{x}}, \widetilde{\mathbb{y}}) = \begin{bmatrix} \widetilde{x}_1 & \widetilde{x}_2 & \dots & \widetilde{x}_n \\ \widetilde{y}_1 & \widetilde{y}_2 & \dots & \widetilde{y}_n \end{bmatrix}. \tag{21}$$

Let ν_X and ν_Y denote labels of $\widetilde{\mathbb{x}}$ sample (first row) and $\widetilde{\mathbb{y}}$ sample (second row), respectively. If $\widetilde{X}$ and $\widetilde{Y}$ are dependent, the samples $\widetilde{\mathbb{x}}$ and $\widetilde{\mathbb{y}}$ must be paired and the ordering of labels ν_Y cannot be changed independently of ν_X. Under independence, the samples $\widetilde{\mathbb{x}}$ and $\widetilde{\mathbb{y}}$ need not be matched. Any permutation of the labels of $\widetilde{\mathbb{x}}$ or $\widetilde{\mathbb{y}}$ sample generates a permutation replicate. The permutation test procedure for independence permutes the indices corresponding to one of the samples, so it is not necessary to permute both ν_X and ν_Y.

Let $T = T(\widetilde{\mathbb{X}}, \widetilde{\mathbb{Y}})$ be a test statistic for testing independence. For a fixed distance D_θ^λ our test statistic based on the empirical distance covariance between fuzzy samples is defined by

$$T(\widetilde{\mathbb{X}}, \widetilde{\mathbb{Y}}) = n \cdot \mathcal{V}_{D_\theta^\lambda}^2(\widetilde{\mathbb{X}}, \widetilde{\mathbb{Y}}). \tag{22}$$

Then our permutation test of independence utilizing (22) is implemented as follows. For given fuzzy sample $\widetilde{z} = \widetilde{z}(\widetilde{\mathbb{x}}, \widetilde{\mathbb{y}})$ we compute the observed test statistic

$$t_0 = T(\widetilde{\mathbf{x}}, \widetilde{\mathbf{y}}) = n \cdot \mathcal{V}^2_{D^\lambda_\theta}(\widetilde{\mathbf{x}}, \widetilde{\mathbf{y}}). \tag{23}$$

Now let $\widetilde{\mathbf{y}}^*$ denote a permutation of the sample $\widetilde{\mathbf{y}}$. More formally, if π_{ν_Y} is a permutation of ν_Y then $\widetilde{y}^*_i = \widetilde{y}_{\pi_{\nu_Y}(i)}$, for $i = 1, \ldots, n$. Hence, each permutation corresponds to some relabeling of the second row in the data matrix (21) without any change in the first row, so it looks now as $\widetilde{\mathbf{z}}^* = \widetilde{\mathbf{z}}(\widetilde{\mathbf{x}}, \widetilde{\mathbf{y}}^*)$.

Given such permutation we determine the corresponding value of the test statistic (22), i.e.

$$T(\widetilde{\mathbf{x}}, \widetilde{\mathbf{y}}^*) = n \cdot \mathcal{V}^2_{D^\lambda_\theta}(\widetilde{\mathbf{x}}, \widetilde{\mathbf{y}}^*). \tag{24}$$

We repeat the whole procedure, i.e. we draw a permutation and compute a value of the test statistic (24), B times (a typical number of repetitions is $B = 999$). Then the approximate p-value of our test is given by

$$\text{p-value} \approx \frac{1 + \sum_{b=1}^{B} \mathbb{1}\left(T(\widetilde{\mathbf{x}}, \widetilde{\mathbf{y}}^*_b) \geqslant t_0\right)}{B + 1}, \tag{25}$$

where t_0 stands for the test statistic value (23) obtained for the original fuzzy sample. Finally, we reject the null hypothesis H_0 on independence at significance level δ if p-value $\leqslant \delta$.

It is worth noting that the proposed test, as other permutation tests, requires extremly limited assumptions. Actually, the only requirement is *exchangeability* (i.e., under the null hypothesis we can exchange the labels on the observations without affecting the results). Obviously, in the considered case this assumption is satisfied.

6   Empirical Study

To illustrate the behavior of the proposed test we conducted a simulation study. We generated trapezoidal observations using the notation convention indicating the center of the core c, the half of the core's length s and the spread of the left and right arm l and r, respectively, i.e. for $\widetilde{A} = \mathrm{Tra}(a_1, a_2, a_3, a_4)$ we have

$$c = \frac{1}{2}(a_2 + a_3), \quad s = \frac{1}{2}(a_3 - a_2), \quad l = a_2 - a_1, \quad r = a_4 - a_3. \tag{26}$$

Therefore, a fuzzy sample $(\widetilde{\mathbf{x}}, \widetilde{\mathbf{y}}) = (\widetilde{x}_1, \widetilde{y}_1), \ldots, (\widetilde{x}_n, \widetilde{y}_n)$ is obtained by simulating four independent real-valued random variables for each $\widetilde{x}_i = \langle c_{Xi}, s_{Xi}, l_{Xi}, r_{Xi} \rangle$ and four random variables for each $\widetilde{y}_i = \langle c_{Yj}, s_{Yj}, l_{Yj}, r_{Yj} \rangle$, respectively, with the last three random variables in each quartet being nonnegative. In particular, we generated random fuzzy numbers using the following real-valued random variables: c_{Xi}, c_{Yj} from the standard normal distribution and $s_{Xi}, s_{Yj}, l_{Xi}, l_{Yj}, r_{Xi}, r_{Yj}$ from the uniform distribution.

Figure 1 shows a histogram illustrating the null distribution of the test statistic (22) and obtained for a fuzzy sample of size $n = 10$ generated by independent random variables c_X and c_Y from the standard normal distribution $\mathrm{N}(0, 1)$ and s_X, s_Y, l_X, l_Y and r_X, r_Y from the uniform distribution $\mathrm{U}(0.0.5)$. In this case

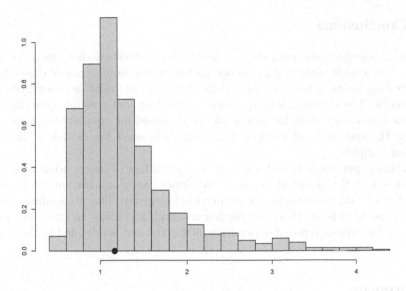

Fig. 1. Empirical null distribution of the permutation test with a black dot indicating the value of the test statistic.

we have obtained $t_0 = 1.1679$, marked on the graph by the black dot. The corresponding p-value = 0.484 leads to the following decision: do not reject H_0 of independence.

Next we examined the proposed test with respect to its size. 1000 repetitions of the test performed on independent variables were generated at the significance level 0.05. In each test $K = 1000$ permutations were drawn and the empirical percentages of rejections under H_0 were determined. The results obtained for a few sample sizes n and gathered in Table 1 show that the size of the proposed test is quite stable.

Table 1. Empirical size of the test for various sample sizes.

n	Empirical size
10	0.051
20	0.048
50	0.058

Finally, we examined the test behavior under a few alternatives related to various dependence types. In particular, we considered situations when

$$\widetilde{Y} = \widetilde{X} + \varepsilon, \qquad \widetilde{Y} = \widetilde{X}^2 + \varepsilon, \qquad \widetilde{Y} = \widetilde{X} \cdot \varepsilon,$$

where $\varepsilon \sim N(0, \sigma)$ for some small σ. The resulting p-values fluctuated within the limits 0.011–0.064, which means that our test discovers different dependencies.

7 Conclusions

Testing independence is not easy in general. Popular tests do not always perform well across a wide variety of monotone and non-monotone types of dependence. This testing problem becomes more difficult when the available observations are not precise. The situation is made worse by the fact that a simple generalization of some well-known tests for independence designed for crisp data is impossible due to the specificity of random fuzzy numbers used for modeling imprecise random samples.

In this paper we proposed a new permutation test of independence for fuzzy data based on the so-called distance covariance. Some simulations to illustrate its behavior and to examine its properties are given. The preliminary results seem to be promising. However, further research including an extensive power study under various types of dependence is desired and is intended in the nearest future.

References

1. Bakirov, N.K., Rizzo, M.L., Székely, G.J.: A multivariate nonparametric test of independence. J. Multivar. Anal. **97**, 1742–1756 (2006)
2. Ban, A.I., Coroianu, L., Grzegorzewski, P.: Fuzzy Numbers: Approximations. Ranking and Applications. Polish Academy of Sciences, Warsaw (2015)
3. Blanco-Fernández, A., et al.: A distance-based statistic analysis of fuzzy number-valued data (with Rejoinder). Int. J. Approximate Reason. **55**, 1487–1501, 1601–1605 (2014)
4. Chukhrova, N., Johannssen, A.: Fuzzy regression analysis: systematic review and bibliography. Appl. Soft Comput. **84**, 105708 (2019)
5. Gibbons, J.D., Chakraborti, S.: Nonparametric Statistical Inference. Marcel Dekker, Inc. (2003)
6. Gil, M.A., Lubiano, M.A., Montenegro, M., López, M.T.: Least squares fitting of an affine function and strength of association for interval-valued data. Metrika **56**, 97–111 (2002)
7. González-Rodríguez, G., Montenegro, M., Colubi, A., Gil, M.A.: Bootstrap techniques and fuzzy random variables: synergy in hypothesis testing with fuzzy data. Fuzzy Sets Syst. **157**, 2608–2613 (2006)
8. Grzegorzewski, P.: Two-sample dispersion problem for fuzzy data. In: Lesot, M.-J., et al. (eds.) IPMU 2020. CCIS, vol. 1239, pp. 82–96. Springer, Cham (2020). https://doi.org/10.1007/978-3-030-50153-2_7
9. Grzegorzewski, P.: Permutation k-sample goodness-of-fit test for fuzzy data, In: Proceedings of the 2020 IEEE International Conference on Fuzzy Systems (FUZZ-IEEE), pp. 1–8 (2020)
10. Grzegorzewski, P., Gadomska, O.: Nearest neighbor tests for fuzzy data, In: Proceedings of the 2021 IEEE International Conference on Fuzzy Systems (FUZZ-IEEE), pp. 1–6 (2021)
11. Hryniewicz, O.: Selection of variables for systems analysis, application of a fuzzy statistical test for independence. In: Information Processing and Management of Uncertainty in Knowledge-Based Systems (IPMU 2014), Proceedings, Part III, pp. 2197–22043. Springer (2004). https://doi.org/10.1007/978-3-319-08855-6

12. Hryniewicz, O.: Goodman-Kruskal γ measure of dependence for fuzzy ordered categorical data. Comput. Stat. Data Anal. **51**, 323–334 (2006)
13. Hukuhara, M.: Integration des applications measurables dont la valeur est un compact convexe. Funkcialaj Ekvacioj **10**, 205–223 (1967)
14. Lubiano, M.A., Gil, M.A., López-Díaz, M., López-García, M.T.: The λ-mean squared dispersion associated with a fuzzy random variable. Fuzzy Sets Syst. **111**, 307–317 (2000)
15. Lubiano, M.A., Montenegro, M., Sinova, B., de la Rosa de Sáa, S., Gil, M.A.: Hypothesis testing for means in connection with fuzzy rating scale-based data: algorithms and applications. Eur. J. Oper. Res. **251** 918–929 (2016)
16. Lyons, R.: Distance covariance in metric spaces. Ann. Probab. **41**, 3284–3305 (2013)
17. Montenegro, M., Colubi, A., Casals, M.R., Gil, M.A.: Asymptotic and Bootstrap techniques for testing the expected value of a fuzzy random variable. Metrika **59**, 31–49 (2004)
18. Puri, M.L., Ralescu, D.A.: Fuzzy random variables. J. Math. Anal. Appl. **114**, 409–422 (1986)
19. Ramos-Guajardo, A.B., Colubi, A., González-Rodríguez, G., Gil, M.A.: One-sample tests for a generalized Fréchet variance of a fuzzy random variable. Metrika **71**, 185–202 (2010)
20. Székely, G.J., Rizzo, M.L., Bakirov, N.K.: Measuring and testing dependence by correlation of distances. Ann. Stat. **35**, 2769–2794 (2007)
21. Székely, G.J., Rizzo, M.L., Bakirov, N.K.: The distance correlation t-test of independence in high dimension. J. Multivar. Anal. **117**, 193–213 (2013)
22. Taheri, S.M., Hesamian, G.: Goodman-Kruskal measure of association for fuzzy-categorized variables. Kybernetika **47**, 110–122 (2011)
23. Taheri, S.M., Hesamian, G., Viertl, R.: Contingency tables with fuzzy information. Commun. Stat. Theory Meth. **45**, 5906–5917 (2016)
24. Trutschnig, W., González-Rodríguez, G., Colubi, A., Gil, M.A.: A new family of metrics for compact, convex (fuzzy) sets based on a generalized concept of mid and spread. Inf. Sci. **179**, 3964–3972 (2009)

Learning from Categorical Data Subject to Non-random Misclassification and Non-response Under Prior Quasi-Near-Ignorance Using an Imprecise Dirichlet Model

Aziz Omar[1,2,4]($\boxtimes$) (iD), Timo von Oertzen[1,3], and Thomas Augustin[2]

[1] Department of Psychology, University of the German Federal Armed Forces
in Munich, Neubiberg, Germany
timo.vonoertzen@unibw.de
[2] Department of Statistics, LMU Munich, Munich, Germany
{aziz.omar,augustin}@stat.uni-muenchen.de
[3] Center for Lifespan Psychology, Max Planck Institute for Human Development,
Berlin, Germany
[4] Department of Mathematics, Insurance and Applied Statistics, Helwan University,
Cairo, Egypt

Abstract. The problem of learning from categorical data is commonly known in many fields. Besides the principal complexity arising from non-correspondence between latent and manifest categories, other common obstacles are non-random non-response and lack of prior knowledge about probabilities of the latent categories. In this paper, we study a situation involving the intersection of these obstacles under the mild condition that one confirmed observation exists in each category; a setting that is highly expected to generally hold in most practices. Under this setting, we derive a quasi-near-ignorance prior for the chances of the latent categories employing the imprecise Dirichlet model. Then, we use this prior to obtain the corresponding posterior predictive probability. We prove that a non-trivial probability interval for the chance of the next observation is obtainable despite the complex nature of the problem.

Keywords: Categorical data · Latent variables · Misclassification · Non-random non-response · Quasi-Near-Ignorance · Imprecise Dirichlet model

1 Introduction

In situations involving a categorical (multinomial) phenomenon, Bayesian framework provides an efficient approach to estimate probabilities of different possible outcomes and to learn about new observations. One major obstacle in applying

D. Ciucci et al. (Eds.): IPMU 2022, CCIS 1602, pp. 532–544, 2022.
https://doi.org/10.1007/978-3-031-08974-9_43

the Bayesian framework is absence of precise knowledge about prior probabilities of possible outcomes. To overcome this problem, different vague priors have been introduced. Priors such as Haldane's improper prior, Jeffreys' prior and Bayes-Laplace's uniform prior have been studied and compared (See, e.g., [3, Sect. 5.6.2], and the comprehensive review by [8] that also discuses selection rules used to choose between these priors).

Non-informative priors are usually introduced based on specific inference principles they satisfy, such as the symmetry principle, the representation invariance principle, the embedding principle, the likelihood principle and the coherence principle (See [7, Sect. 2.3 and 2.4] for details about principles of inference). However, as indicated by [4], with the exception of Haldane's prior, none of these priors satisfies all mentioned inference principles.

As an alternative to using a single prior to convey prior ignorance in multinomial experiments, [19] introduced the imprecise Dirichlet model (IDM) and demonstrated how it satisfies all common inference principles. The IDM expresses a state of prior near-ignorance that results in interval-valued estimates for the classical single-valued estimates resulting from the application of a classical single non-informative prior. Ever since, the IDM has been employed in numerous contexts (See, e.g., [4] for a first overview and e.g., [1,13,17] for recent developments in the field of classification). Furthermore, it is one of the prominent models used under the general framework of "Imprecise Probabilities" (See, e.g., [2,5] and [16]).

Another problem that is frequently encountered while drawing inferences from categorical data is misclassification where a manifest variable masks the true latent one. [15] studied the employment of the IDM to learn about the next observation from potentially misclassified categorical data. Their results showed that under prior near-ignorance, trivial learning outcomes are inevitable, i.e., the resulting interval-valued estimate for the chance of every outcome extends on the whole range from 0 to 1. This result states that when used to model prior near-ignorance in situations that involve misclassification, the IDM produces trivial estimates.

In this paper, we describe a situation characterized by a simple condition that pertaining the categorical variable and describe a Dirichlet prior density that corresponds to it. We show that under this condition, IDM is able to learn from potentially misclassified and incomplete categorical data.

The rest of this paper is organized as follows. In Sect. 2 we discuss previous work on learning based on potentially misclassified categorical data. In Sect. 3 we introduce a Dirichlet prior that refers to a particular setting commonly found in practice. In Sect. 4 we describe our situation involving potentially misclassified and incomplete categorical data and show how the prior derived in Sect. 3 leads to obtaining non-trivial estimates for the chances of the next observation. In Sect. 5 we demonstrate our theoretical results through numeric examples. Finally, Sect. 6 concludes with the discussion and further possible extensions.

2 Bayesian Learning from Potentially Misclassified Categorical Data Under Prior Near-Ignorance

In this section, we introduce the problem of learning from potentially misclassified categorical data in absence of concrete prior information and summarize the main results of previous literature in this regard. We first start with recalling learning from correctly classified categorical data (See, e.g., [3, Sect. 5.2.1]).

2.1 Learning from Correctly Classified Categorical Data

Let a categorical (multinomial) random variable X have $K \geq 2$ possible categories, and let population units belong to category j, $j = 1, 2, \cdots, K$ with probability θ_j, $0 \leq \theta_j$, $\sum_{j=1}^{K} \theta_j = 1$. To learn about $\boldsymbol{\theta} = (\theta_1, \theta_2, \cdots, \theta_K)$, a sample of size n is drawn and a dataset $\boldsymbol{x} = (x_1, x_2, \cdots, x_K)$, $\sum_{j=1}^{K} x_j = n$ is obtained, where x_j, $0 \leq x_j$, is the observed count in category j, $j = 1, 2, \cdots, K$. If the observation process is believed to be perfect, $\boldsymbol{x}$ contains the true number of sample units in each category.

Utilizing the classical Bayesian inference, $\boldsymbol{\theta}$ is assumed a-priori to have the Dirichlet prior density

$$p(\boldsymbol{\theta}) = dir(s, \boldsymbol{t})(\boldsymbol{\theta}) := \frac{\Gamma(s)}{\prod_{j=1}^{K} \Gamma(st_j)} \prod_{j=1}^{K} \theta_j^{st_j - 1}, \tag{1}$$

where $\boldsymbol{t} = (t_1, t_2, \cdots, t_K)$, $0 < t_j < 1$, $\sum_{j=1}^{K} t_j = 1$ and $s > 0$.

Updating the prior in (1) using observed counts in $\boldsymbol{x}$ leads to the conjugate Dirichlet posterior density

$$p(\boldsymbol{\theta}|\boldsymbol{x}) = dir(s + n, \boldsymbol{x} + \boldsymbol{t})(\boldsymbol{\theta}|\boldsymbol{x}) := \frac{\Gamma(n + s)}{\prod_{j=1}^{K} \Gamma(x_j + st_j)} \prod_{j=1}^{K} \theta_j^{x_j + st_j - 1}. \tag{2}$$

The posterior predictive probability $P(X = j|\boldsymbol{x})$ of the next observation given observed data is then

$$P(X = j|\boldsymbol{x}) := \int_{\Theta} \theta_j p(\boldsymbol{\theta}|\boldsymbol{x}) d\theta_j = \frac{x_j + st_j}{n + s}, \qquad j = 1, 2, \cdots, K. \tag{3}$$

2.2 The Imprecise Dirichlet Model (IDM) as a Means to Express Prior Near-Ignorance

As opposite of classical non-informative priors, where a single prior density is employed, the *Imprecise Dirichlet Model* (IDM) introduced by [19] uses a set $\mathcal{M}_s$ of Dirichlet prior densities to represent prior near-ignorance about $\boldsymbol{\theta}$. The densities in $\mathcal{M}_s$ have a common "strength" parameter s that represents the degree of prior near-ignorance about $\boldsymbol{\theta}$.

According to Walley's general coherence theory (see, [18, Ch. 6]), the usual Bayesian approach is applied to the elements of $\mathcal{M}_s$ element-by-element. Thus, every Dirichlet prior density in $\mathcal{M}_s$ is updated using sample counts from $\boldsymbol{x}$ into another Dirichlet posterior density in a set $\mathcal{M}_{s+n}$ of posterior densities whose elements share the strength parameter $s+n$ representing the degree of posterior near-ignorance.

Consequently, inference about certain events and parameters is attainable in the form of interval-valued estimates whose minimum and maximum values are found by optimizing their single-valued counterparts with respect to $\boldsymbol{t}$. Hence the interval-valued estimate for posterior predictive probability $P(X = j|\boldsymbol{x})$ is received as

$$\left(\underline{P}(X = j|\boldsymbol{x}), \overline{P}(X = j|\boldsymbol{x})\right) := \left(\lim_{t_j \to 0} P(X = j|\boldsymbol{x}), \lim_{t_j \to 1} P(X = j|\boldsymbol{x})\right)$$

$$= \left(\frac{x_j}{n+s}, \frac{x_j + s}{n+s}\right), \qquad j = 1, 2, \cdots, K.$$

2.3 Learning from Potentially Misclassified Categorical Data Under Prior Near-Ignorance Using the IDM

Let the correspondence between the latent variable X and its manifest version O be defined through the emission matrix Λ_1 that has the following format

$$\Lambda_1 := \begin{pmatrix} \pi_{11} & \cdots & \pi_{1K} \\ \vdots & \ddots & \vdots \\ \pi_{K1} & \cdots & \pi_{KK} \end{pmatrix}, \tag{4}$$

where $\pi_{ij} := P(O = i|X = j)$, $0 \le \pi_{ij}$, $\sum_{i=1}^{K} \pi_{ij} = 1$ is the probability of a sample unit that belongs to category j, $j = 1, 2, \cdots, K$ of X to be observed as having the category i, $i = 1, 2, \cdots, K$ of O.

In the practically uncommon case when Λ_1 is diagonal, the observation process is believed to be perfect with no misclassification, such that an observed dataset $\boldsymbol{o} = (o_1, o_2, \cdots, o_K)$, $0 \le o_i$, $i = 1, 2, \cdots, K$, $\sum_{i=1}^{K} o_i = n$ corresponds to the true dataset $\boldsymbol{x}$. This correspondence can still be maintained even when Λ_1 is not diagonal, as long as each column in Λ_1 has only one non-zero element. If, however, this is not the case, such that one column of Λ_1 has more than one non-zero element, $\boldsymbol{o}$ is no longer in correspondence with a unique true dataset $\boldsymbol{x}$. Rather, it can correspond to any dataset there is in $\boldsymbol{\chi}^n$, the set of all possible true datasets of size n that could have produced $\boldsymbol{o}$.

[15] discussed learning about the chances of the next sample unit based on the observed dataset $\boldsymbol{o}$ under prior near-ignorance using the IDM.

Theorem 1. *In a multinomial sampling, if the observation process is described by an emission matrix Λ_1 whose j-th column has more than one non-zero element and the IDM is used to express prior near-ignorance about $\boldsymbol{\theta}$, the posterior predictive probability*

$$P(X = j|\boldsymbol{o}) := \int_{\Theta} \theta_j\, p(\boldsymbol{\theta}|\boldsymbol{o}) d\theta_j, \quad j = 1, 2, \cdots, K,$$

where $p(\boldsymbol{\theta}|\boldsymbol{o})$ is the posterior density of $\boldsymbol{\theta}$, has trivial boundaries such that

$$\underline{P}(X = j|\boldsymbol{o}) = 0 \ \ and \ \ \overline{P}(X = j|\boldsymbol{o}) = 1, \ \ \ j = 1, 2, \cdots, K.$$

Proof. See [15].   □

Theorem 1 shows that the IDM can not be used to learn from categorical data subject to misclassification in absence of additional information. The seriousness of this problem stems from commonness of the conditions producing it, which are easily met in many practical situations, as well as, the severity of its direct consequence that the state of prior near-ignorance cannot be expressed using the IDM and hence, other models are needed. In the next section, we propose a modification to the conditions producing the problem that enables learning about the chances of the next sample unit using the IDM.

3 Prior Quasi-Near-Ignorance

In this section, we introduce a subtle condition, under which a Dirichlet prior density can be derived such that the IDM is able to learn from misclassified categorical data. Assume we know a priori that, in each observed count o_j, $j = 1, 2, \cdots, K$ in $\boldsymbol{o}$, there is at least one correctly classified sample unit. Let n_{ij} be the number of sample units with $X = j$ and $O = i$, $i, j = 1, 2, \cdots, K$. The minimal form of the proposed condition requires that

$$n_{jj} = 1, \ \ \ \ j = 1, 2, \cdots, K. \tag{5}$$

Condition (5) resembles Laplace's Rule of Succession (See, e.g. [20]) where a pseudo-count of one unit is added to each possible category. In our condition, however, we regard the sample as a union of two independent sub-samples; one contains a correctly classified unit in each of the K possible categories of X and another one contains $n - K$ potentially misclassified units. In Theorem 2, we derive an IDM prior density that corresponds to condition (5).

Theorem 2. *In an observation process characterized by condition* (5), $p_q(\boldsymbol{\theta})$, *the adjusted prior density of $\boldsymbol{\theta}$ follows the Dirichlet density $dir(s + K, \boldsymbol{t})(\boldsymbol{\theta})$.*

Proof. Let $p(\boldsymbol{o}^{(n)}|\boldsymbol{\theta})$ be the likelihood of the observed dataset under condition (5). It follows then

$$p(\boldsymbol{o}^{(n)}|\boldsymbol{\theta}) \propto p(\boldsymbol{o}^{(n-K)}|\boldsymbol{\theta}) \cdot p(\boldsymbol{o}^{(K)}|\boldsymbol{\theta}), \tag{6}$$

where $p(\boldsymbol{o}^{(K)}|\boldsymbol{\theta})$ is the likelihood of the K correctly classified sample units and $p(\boldsymbol{o}^{(n-K)}|\boldsymbol{\theta})$ is the likelihood of the remaining $n - K$ sample units that might be misclassified and/ or missing. Following the usual Bayesian approach utilizing the prior in (1), $p_q(\boldsymbol{\theta}|o)$, the posterior density of $\boldsymbol{\theta}$ under condition (5), is received as follows

$$p_q(\boldsymbol{\theta}|o) \propto p(o^{(n)}|\boldsymbol{\theta}) \cdot p(\boldsymbol{\theta})$$

$$\propto p(o^{(n-K)}|\boldsymbol{\theta}) \cdot p(o^{(K)}|\boldsymbol{\theta}) \cdot p(\boldsymbol{\theta})$$

$$\propto p(o^{(n-K)}|\boldsymbol{\theta}) \cdot \prod_{j=1}^{K} \theta_j \cdot \prod_{j=1}^{K} \theta_j^{st_j-1}$$

$$\propto p(o^{(n-K)}|\boldsymbol{\theta}) \cdot p_q(\boldsymbol{\theta}),$$

where

$$p_q(\boldsymbol{\theta}) = dir(s+K, t)(\boldsymbol{\theta}) := \frac{\Gamma(s+K)}{\prod_{j=1}^{K} \Gamma(st_j+1)} \prod_{j=1}^{K} \theta_j^{st_j}. \tag{7}$$

$\square$

The prior in (7) represents a state that is a step further from prior near-ignorance represented through the IDM in the direction of prior informativeness. We call this new state *quasi-near-ignorance* to express the presence of some prior information. In what follows, we show how learning under quasi-near-ignorance is possible based on the adjusted prior in (7).

4   Learning from Potentially Misclassified and Incomplete Categorical Data Under Prior Quasi-Near-Ignorance

In this section, we consider a general situation that involves potentially misclassified and incomplete categorical data and show that learning is possible under the state of prior quasi-near-ignorance using the IDM.

4.1   Misclassified and Incomplete Categorical Data

Let the observation process permit sample units to be correctly classified, misclassified or unclassified (missing). Then, the random variable O representing the observed category of the sample units will have an additional category $K+1$ with corresponding count o_{K+1} that represents the number of unclassified units due to missingness. Accordingly, the emission matrix that describes the observation process can be represented as

$$\Lambda_2 := \begin{pmatrix} \pi_{11} & \cdots & \pi_{1K} \\ \vdots & \ddots & \vdots \\ \pi_{K1} & \cdots & \pi_{KK} \\ \pi_{(K+1)1} & \cdots & \pi_{(K+1)K} \end{pmatrix}, \tag{8}$$

where $\pi_{ij} := P(O=i|X=j)$, $0 \le \pi_{ij}$, $\sum_{i=1}^{K+1} \pi_{ij} = 1$, $i = 1, 2, \cdots, K+1$, $j = 1, 2, \cdots, K$ is the probability of a sample unit that belongs to category j of X

to be observed as having the category i of O. The matrix Λ_2 can be seen as an augmented form of the matrix Λ_1 that represents the potentiality for sample units to be unclassified.

4.2   Learning from Potentially Misclassified and Incomplete Categorical Data Under Prior Quasi-Near-Ignorance Using the IDM

Under condition (5), Theorem 3 shows that learning from misclassified and incomplete categorical data is possible using the IDM.

Theorem 3. *In a multinomial sampling, if the observation process is described by the matrix Λ_2 whose j-th column has more than one non-zero element and the IDM is used to express prior quasi-near-ignorance about $\boldsymbol{\theta}$ under condition (5), the posterior predictive probability*

$$P_q(X = j|\boldsymbol{o}) := \int_{\Theta} \theta_j \, p_q(\boldsymbol{\theta}|\boldsymbol{o}) d\theta_j, \quad j = 1, 2, \cdots, K,$$

has non-trivial boundaries such that

$$0 < \underline{P_q}(X = j|\boldsymbol{o}) < \overline{P_q}(X = j|\boldsymbol{o}) < 1, \quad j = 1, 2, \cdots, K. \tag{9}$$

Proof. Under condition (5), the likelihood of the $n - K$ potentially misclassified units in $\boldsymbol{o}$ can be expressed as

$$p(\boldsymbol{o}^{(n-K)}|\boldsymbol{\theta}) = \frac{(n - K)!}{(o_1 - 1)! \cdots (o_K - 1)! o_{K+1}!}$$

$$\overbrace{(\theta_1 \pi_{11} + \cdots + \theta_K \pi_{1K})^{o_1 - 1}}^{o_1 - 1 \text{ potentially misclassified cases observed in category 1}}$$

$$\times \cdots$$

$$\times \overbrace{(\theta_1 \pi_{K1} + \cdots + \theta_K \pi_{KK})^{o_K - 1}}^{o_K - 1 \text{ potentially misclassified cases observed in category K}}$$

$$\times \underbrace{(\theta_1 \pi_{(K+1)1} + \cdots + \theta_K \pi_{(K+1)K})^{o_{K+1}}}_{o_{K+1} \text{ missing cases}}.$$

Let $\tilde{n} := n - K$, $\tilde{o}_i := o_i - 1$, $i = 1, 2, \cdots, K$ and $\tilde{o}_{K+1} := o_{K+1}$. Then, we get

$$
p(o^{(n-K)}|\boldsymbol{\theta}) = \binom{\tilde{n}}{\tilde{o}_1, \cdots, \tilde{o}_{K+1}} \prod_{i=1}^{K+1} \left\{ \left(\sum_{j=1}^{K} \theta_j \pi_{ij} \right)^{\tilde{o}_i} \right\}
$$

$$
= \binom{\tilde{n}}{\tilde{o}_1, \cdots, \tilde{o}_{K+1}} \prod_{i=1}^{K+1} \left\{ \sum_{\substack{b_1 + \cdots + b_K = \tilde{o}_i \\ b_1, \cdots, b_K \geq 0}} \binom{\tilde{o}_i}{b_1, \cdots, b_K} \prod_{j=1}^{K} (\theta_j \pi_{ij})^{b_j} \right\}
$$

$$
= \binom{\tilde{n}}{\tilde{o}_1, \cdots, \tilde{o}_{K+1}} \sum_{\substack{b_1 + \cdots + b_K = \tilde{n} \\ b_1, \cdots, b_K \geq 0}} \prod_{i=1}^{K+1} \left\{ \binom{\tilde{o}_i}{b_1, \cdots, b_K} \prod_{j=1}^{K} \theta_j^{b_j} \pi_{ij}^{b_j} \right\}. \tag{10}
$$

Let $p_q(\boldsymbol{\theta}|o)$ be the posterior density of $\boldsymbol{\theta}$ under condition (5). It follows from Theorem 2 that

$$
p_q(\boldsymbol{\theta}|o) := \frac{p(o^{(n)}|\boldsymbol{\theta})p(\boldsymbol{\theta})}{\int_{\Theta} p(o^{(n)}|\boldsymbol{\theta})p(\boldsymbol{\theta})d\boldsymbol{\theta}} = \frac{p(o^{(n-K)}|\boldsymbol{\theta})p_q(\boldsymbol{\theta})}{\int_{\Theta} p(o^{(n-K)}|\boldsymbol{\theta})p_q(\boldsymbol{\theta})d\boldsymbol{\theta}}. \tag{11}
$$

Combining (7) and (10), we get

$$
p(o^{(n-K)}|\boldsymbol{\theta})p_q(\boldsymbol{\theta}) \propto \sum_{\substack{b_1 + \cdots + b_K = \tilde{n} \\ b_1, \cdots, b_K \geq 0}} \prod_{i=1}^{K+1} \left\{ \binom{\tilde{o}_i}{b_1, \cdots, b_K} \prod_{j=1}^{K} \theta_j^{b_j + st_j} \pi_{ij}^{b_j} \right\}.
$$

Now, since

$$
dir(s + \tilde{n} + K, \boldsymbol{b} + \boldsymbol{t})(\boldsymbol{\theta}) := \frac{\Gamma(s + \tilde{n} + K)}{\prod_{j=1}^{K} \Gamma(b_j + st_j + 1)} \prod_{j=1}^{K} \theta_j^{b_j + st_j},
$$

where $\boldsymbol{b} = (b_1 + 1, \cdots, b_K + 1)$, $\sum_{j=1}^{K} b_j = \tilde{n}$, then,

$$
\prod_{j=1}^{K} \theta_j^{b_j + st_j} = \frac{dir(s + \tilde{n} + K, \boldsymbol{b} + \boldsymbol{t})(\boldsymbol{\theta})}{\Gamma(s + \tilde{n} + K)} \prod_{j=1}^{K} \Gamma(b_j + st_j + 1).
$$

Hence,

$$
p(o^{(n-K)}|\boldsymbol{\theta})p_q(\boldsymbol{\theta}) \propto \sum_{\substack{b_1 + \cdots + b_K = \tilde{n} \\ b_1, \cdots, b_K \geq 0}} dir(s + \tilde{n} + K, \boldsymbol{b} + \boldsymbol{t})(\boldsymbol{\theta})
$$

$$
\prod_{i=1}^{K+1} \left\{ \binom{\tilde{o}_i}{b_1, \cdots, b_K} \prod_{j=1}^{K} \pi_{ij}^{b_j} \Gamma(b_j + st_j + 1) \right\}. \tag{12}
$$

Consequently, we get

$$\int_\Theta p(o^{(n-K)}|\boldsymbol{\theta})p_q(\boldsymbol{\theta})d\boldsymbol{\theta} \propto \sum_{\substack{b_1+\cdots+b_K=\tilde{n} \\ b_1,\cdots,b_K\geq 0}} \prod_{i=1}^{K+1}\left\{\binom{\tilde{o}_i}{b_1,\cdots,b_K} \prod_{j=1}^K \pi_{ij}^{b_j}\Gamma(b_j+st_j+1)\right\}. \quad (13)$$

From (12) and (13), we get

$$p_q(\boldsymbol{\theta}|\boldsymbol{o}) \propto \left(\sum_{\substack{b_1+\cdots+b_K=\tilde{n} \\ b_1,\cdots,b_K\geq 0}} dir(s+\tilde{n}+K,\boldsymbol{b}+\boldsymbol{t})(\boldsymbol{\theta}) \right.$$

$$\left. \prod_{i=1}^{K+1}\left\{\binom{\tilde{o}_i}{b_1,\cdots,b_K}\prod_{j=1}^K \pi_{ij}^{b_j}\Gamma(b_j+st_j+1)\right\}\right) \Bigg/$$

$$\left(\sum_{\substack{b_1+\cdots+b_K=\tilde{n} \\ b_1,\cdots,b_K\geq 0}} \prod_{i=1}^{K+1}\left\{\binom{\tilde{o}_i}{b_1,\cdots,b_K}\prod_{j=1}^K \pi_{ij}^{b_j}\Gamma(b_j+st_j+1)\right\}\right).$$

The posterior predictive probability $P_q(X=j|\boldsymbol{o})$, $j=1,2,\cdots,K$, that satisfies condition (5) can be expressed as

$$P_q(X=j|\boldsymbol{o}) := \int_\Theta \theta_j p_q(\boldsymbol{\theta}|\boldsymbol{o})d\theta_j$$

$$= \left(\sum_{\substack{b_1+\cdots+b_K=\tilde{n} \\ b_1,\cdots,b_K\geq 0}} \frac{b_j+st_j+1}{s+\tilde{n}+K} \right.$$

$$\left. \prod_{i=1}^{K+1}\left\{\binom{\tilde{o}_i}{b_1,\cdots,b_K}\prod_{j=1}^K \pi_{ij}^{b_j}\Gamma(b_j+st_j+1)\right\}\right) \Bigg/$$

$$\left(\sum_{\substack{b_1+\cdots+b_K=\tilde{n} \\ b_1,\cdots,b_K\geq 0}} \prod_{i=1}^{K+1}\left\{\binom{\tilde{o}_i}{b_1,\cdots,b_K}\prod_{j=1}^K \pi_{ij}^{b_j}\Gamma(b_j+st_j+1)\right\}\right).$$

$$= \frac{st_j+1}{s+\tilde{n}+K} + \frac{1}{s+\tilde{n}+K}\cdot$$

$$\frac{\displaystyle\sum_{\substack{b_1+\cdots+b_K=\tilde{n} \\ b_1,\cdots,b_K\geq 0}} \prod_{i=1}^{K+1}\left\{\binom{\tilde{o}_i}{b_1,\cdots,b_K}\prod_{j=1}^K b_j\pi_{ij}^{b_j}\Gamma(b_j+st_j+1)\right\}}{\displaystyle\sum_{\substack{b_1+\cdots+b_K=\tilde{n} \\ b_1,\cdots,b_K\geq 0}} \prod_{i=1}^{K+1}\left\{\binom{\tilde{o}_i}{b_1,\cdots,b_K}\prod_{j=1}^K \pi_{ij}^{b_j}\Gamma(b_j+st_j+1)\right\}}. \quad (14)$$

By positivity of the first summand in (14) and non-negativity of the second one, it is straightforward to verify that $\lim_{t_j \to 0} P_q(X = j|o) > 0$ implying, consequently, that $\lim_{t_j \to 1} P_q(X = j|o) < 1$ since $\sum_{j=1}^{K} P_q(X = j|o) = 1$. $\qquad\square$

The importance of Theorem 3 is that it shows that learning from potentially misclassified and incomplete categorical data is possible using the IDM under quasi-near-ignorance described by condition (5). Since condition (5) is easily met in most practices, the result of Theorem 3 is of a notable importance.

5 Examples

To verify the result of Theorem 3, we provide some examples where the interval-valued estimate for the posterior predictive probability is theoretically known (or easy to calculate) and compare it to the interval-valued estimate for the posterior predictive probability in (14). To make the comparison easier, the examples are presented where $K = 2$, i.e., X is a binary variable with possible categories $j = 1, 2$ and we are interested in obtaining an interval-valued estimate for $P_q(X = 1|o)$. The binary version of the posterior predictive probability $P_q(X = j|o)$ in (14) can be expressed as

$$P_q(X = j|o) = \frac{st_j + 1}{s + n} + \frac{\sum_{h=0}^{o_1-1} \sum_{l=0}^{o_2-1} \sum_{r=0}^{o_3}(h + l + r)f_n}{(s + n)\sum_{h=0}^{o_1-1} \sum_{l=0}^{o_2-1} \sum_{r=0}^{o_3} f_n}, \quad j = 1, 2, \quad (15)$$

where

$$f_n := \binom{o_1 - 1}{h}\binom{o_2 - 1}{l}\binom{o_3}{r}\left(\frac{\pi_{11}}{\pi_{12}}\right)^h \left(\frac{\pi_{21}}{\pi_{22}}\right)^l \left(\frac{\pi_{31}}{\pi_{32}}\right)^r \cdot$$
$$\cdot \Gamma(h + l + r + st_1 + 1)\Gamma(n - h - l - r + st_2 - 1). \quad (16)$$

Moreover, in each example, a sample of size $n \geq 2$ is drawn and the observed dataset $o = (o_1, o_2, o_3)$, $\sum_{i=1}^{3} o_j = n$ is obtained, where o_i, $1 \leq o_i \leq n - 1$, $i = 1, 2$ is the observed count in category i and o_3, $0 \leq o_3 \leq n - 2$ is the number of missing cases. Finally, we set $s = 1$.

Example 1: Correctly Classified Data. Consider the emission matrix

$$\Lambda_{ex1} := \begin{pmatrix} 1 & 0 \\ 0 & 1 \\ 0 & 0 \end{pmatrix}. \quad (17)$$

If the observation process is described by the emission matrix Λ_{ex1}, observed data is assumed to be correctly classified. In this case, the posterior predictive probability $P_q(X = 1|o)$ takes the classical form (3). Hence, its interval-valued estimate is $\left(\frac{o_1}{n+1}, \frac{o_1 + 1}{n+1}\right)$, which is the same result obtained from the interval-valued estimate of the posterior predictive probability (15).

Example 2: Non-informative Data. Consider the emission matrix

$$\Lambda_{ex2} := \begin{pmatrix} 1/3 & 1/3 \\ 1/3 & 1/3 \\ 1/3 & 1/3 \end{pmatrix}. \tag{18}$$

If the observation process is described by the emission matrix Λ_{ex2}, observed data has equal chances to be correctly classified, misclassified or unclassified. Hence, it does not update the prior density of $\boldsymbol{\theta}$. In this case, the posterior predictive probability $P_q(X = 1|\boldsymbol{o})$ takes the classical form (3) based only on the two confirmed cases in each category. Therefore, the equivalent interval-valued estimate is $\left(\dfrac{1}{3}, \dfrac{2}{3}\right)$, which is the same result obtained from the interval-valued estimate of the posterior predictive probability (15).

Example 3: Missing at Random. Consider the emission matrix

$$\Lambda_{ex3} := \begin{pmatrix} 0.5 & 0 \\ 0 & 0.5 \\ 0.5 & 0.5 \end{pmatrix}. \tag{19}$$

If the observation process is described by the emission matrix Λ_{ex3}, observed data has the same chance to be correctly classified or unclassified. In this case, missingness is assumed to occur randomly. Hence, the posterior predictive probability $P_q(X = 1|\boldsymbol{o})$ takes the classical form (3) with a sample size equals to the number of non-missing cases. Therefore, the equivalent interval-valued estimate is $\left(\dfrac{o_1}{n - o_3 + 1}, \dfrac{o_1 + 1}{n - o_3 + 1}\right)$, which is the same result obtained from the interval-valued estimate of the posterior predictive probability (15).

6   Discussion

We considered the problem of learning from potentially misclassified and incomplete categorical data in absence of informative knowledge about the chances of the latent categories, a common situation that arises in many fields of application. We used a treatment developed under the umbrella of the "Imprecise Probability" general framework and employed a simple condition to derive a prior density that expresses a state of quasi-near-ignorance. Afterwards, we deduced the probability of the next observation and showed that its interval-valued estimate is non-trivial. It is worth noting that, although condition (5) is minimal, since it requires only one true case in each category, it ensures non-trivial estimates.

A generalization for the proposed prior is readily feasible when each category has more than one confirmed case by updating the adjusted prior density in (7) to reflect the number of confirmed cases. Consider for example the common situation where one has, on the one hand, a relatively cheap but uncertain diagnostic test with not ideal sensitivity and specificity, and, on the other hand, a rather expensive gold standard testing procedure with high costs per unit. Our result shows, it is worth to invest in applying the gold standard procedure to ensure one true case for every possible outcome, enabling inferences on all the data. Ideally this is done by an adaptive procedure that allows immediate stopping when the required cases are found.

In comparison to other approaches to learn from misclassified categorical data (e.g., the imprecise sample size Dirchlet model (ISSDM) introduced by [11] that specifically satisfies the learning principle introduced in [14]; and the continuos updating rule proposed by [6] in which priors that results in trivial prior estimates are excluded from the set of priors), our approach maintains favourable characteristics of the IDM such as satisfying the representation invariance principle, although it produces not similarly easy to calculate estimates.

An important extension to the current work could be relaxation of the assumptions on misclassification and response probabilities. Throughout this paper the elements of Λ_2 are considered fixed and known. The principled position taken here, expressing prior ignorance by imprecise, i.e., interval-valued probabilities, suggests to look also in this context at approaches from the field of partial identification (see, e.g., [9]), where unjustified assumptions are overcome by interval-valued approaches promising credible results (see, e.g., [12] for misclassification and [10] for propagating cautious treatment of missing data in official statistics).

Acknowledgements. The authors are grateful to all three anonymous reviewers for their helpful comments and suggestions.

References

1. Antonucci, A., Corani, G.: The multilabel Naive Credal classifier. Int. J. Approximate Reasoning **83**, 320–336 (2017)
2. Augustin, T., Coolen, F.-P.A., de Cooman, G., Troffaes, M.-C.M. (eds.): Introduction to Imprecise Probabilities. Wiley, Chichester (2014)
3. Bernardo, J.-M., Smith, A.-F.M.: Bayesian Theory. Wiley, New York (2000)
4. Bernard, J.-M.: An introduction to the imprecise Dirichlet model for multinomial data. Int. J. Approximate Reasoning **39**(2–3), 123–150 (2005)
5. Bradley, S.: Imprecise Probabilities. In: Zalta, E.-N. (eds.) The Stanford Encyclopedia of Philosophy (2019). https://plato.stanford.edu/archives/spr2019/entries/imprecise-probabilities. Accessed 24 February 2022
6. Cattaneo, M.E.G.V.: A continuous updating rule for imprecise probabilities. In: Laurent, A., Strauss, O., Bouchon-Meunier, B., Yager, R.R. (eds.) IPMU 2014. CCIS, vol. 444, pp. 426–435. Springer, Cham (2014). https://doi.org/10.1007/978-3-319-08852-5_44
7. Cox, D.-R., Hinkley, D.-V.: Theoretical Statistics. Chapman & Hall, New York (1974)
8. Kass, R.-E., Wasserman, L.: The selection of prior distributions by formal rules. J. Am. Stat. Assoc. **91**(435), 1343–1370 (1996)

9. Manski, C.-F.: Partial Identification of Probability Distributions. Springer, New York (2003). https://doi.org/10.1007/b97478
10. Manski, C.-F.: Credible interval estimates for official statistics with survey nonresponse. J. Economet. **191**(2), 293–301 (2016)
11. Masegosa, A.-R., Moral, S.: Imprecise probability models for learning multinomial distributions from data. Applications to learning credal networks. Int. J. Approximate Reasoning **55**(7), 1548–1569 (2014)
12. Molinari, F.: Partial identification of probability distributions with misclassified data. J. Economet. **144**(1), 81–117 (2008)
13. Moral-Garcia, S., Mantas, C.-J., Castellano, J.-G., Benitez, M.-D., Abellan, J.: Bagging of credal decision trees for imprecise classification. Expert Syst. Appl. **141**, 112944 (2020)
14. Moral, S.: Imprecise probabilities for representing ignorance about a parameter. Int. J. Approximate Reasoning **53**(3), 347–362 (2012)
15. Piatti, A., Zaffalon, M., Trojani, F., Hutter, M.: Limits of learning about a categorical latent variable under prior near-ignorance. Int. J. Approximate Reasoning **50**(4), 597–611 (2009)
16. Ristic, B., Gilliam, C., Byrne, M., Benavoli, A.: A tutorial on uncertainty modeling for machine reasoning. Inf. Fus. **55**, 30–44 (2020)
17. Utkin, L.-V.: An imprecise extension of SVM-based machine learning models. Neurocomputing **331**, 18–32 (2019)
18. Walley, P.: Statistical Reasoning with Imprecise Probabilities. Chapman & Hall, New York (1991)
19. Walley, P.: Inferences from multinomial data: learning about a bag of marbles. J. Roy. Stat. Soc. **58**(1), 3–57 (1996)
20. Zabell, S.-L.: The rule of succession. Erkenntnis **1975–31**(2/3), 283–321 (1989)

Uncertainty, Heterogeneity, Reliability and Explainability in AI

Uncertainty in Predictive Process Monitoring

Pietro Portolani[1,2](✉) [ID], Alessandro Brusaferri[2] [ID], Andrea Ballarino[2] [ID],
and Matteo Matteucci[1] [ID]

[1] Politecnico di Milano, Milan, Italy
`pietro.portolani@polimi.it`
[2] STIIMA, National Research Council, Milan, Italy

Abstract. Predictive process monitoring focuses on forecasting future events in an ongoing sequence of logs coming from a process. Process owners will then use predictions to make decisions or give users information about the advancement of their case. Estimation of uncertainty is a critical uncovered topic in the task of next activity prediction. In this work, we study uncertainty estimation and model calibration in the case of the next activity prediction exploiting deep learning techniques, suggesting a framework able to take advantage of this information in a real scenario. We used an attention-based neural network as a forecasting algorithm and estimated its uncertainty through an ensemble of networks and Monte Carlo dropout. Two public datasets with different complexities and widely adopted in the Process Mining community are used. We investigate the quality of the models' calibration on the considered datasets and how prediction accuracy is related to uncertainty. We also analyzed test cases in which the model is overconfident and showed that they could be related to uncommon data or sequences not contained in the training set.

Keywords: Uncertainty · Predictive process monitoring · Process mining

1 Introduction

We are all witnessing the increasing pervasiveness of digital technologies in our lives and societies. Cheaper and more powerful hardware with modern software solutions are bringing governments [1] and enterprises to invest in digital transformation to increase productivity and sales, innovate in value creation or better compete in the market [2]. Digital transformation is intrinsically linked to data logging, increasing the amount of data available for analysis.

Process Mining focuses on extracting valuable information from logs produced by an ongoing process, improving its understanding and optimizing its performance [3]. Predictive process monitoring is a branch of Process Mining, providing constant process monitoring through forecasting what will happen in

© Springer Nature Switzerland AG 2022
D. Ciucci et al. (Eds.): IPMU 2022, CCIS 1602, pp. 547–559, 2022.
https://doi.org/10.1007/978-3-031-08974-9_44

every process instance in progress. Usually, predicted variables are the following activity, completion time, and remaining time until a process instance is solved. These quantities help inform the user about his case, identify likely compliance violations or allow better planning.

When a model is part of a decision making pipeline [4,5], it should provide a calibrated confidence measure alongside its forecast to make explicit the uncertainty behind its prediction. Uncertainty quantification is critical in real scenarios because it allows us to evaluate the model's trustworthiness and act accordingly. This work explores uncertainty estimation in the next activity prediction, an uncovered topic, proposing a predictive process monitoring framework that exploits this novel information. We analyze the relation between accuracy and uncertainty, model calibration, and model faults with overconfident output. The paper is organized as follows: Sect. 2 describes other works on uncertainty in predictive process monitoring, and Sect. 3 defines the problem we want to address and the framework to exploit the uncertainty information. Section 4 illustrates the methods composing our approach and, Sect. 5, the datasets used. Section 6 analyses the results obtained while, in Sect. 7, conclusions and future works are reported.

2   Related Works

Next activity prediction in predictive process monitoring has been a rising topic in the last years, mainly due to the spreading of deep learning algorithms, which outperformed existing methods. For instance, authors in [6] and [7] apply two different types of neural networks to reach state-of-the-art performance in this task.

Uncertainty, on the contrary, has not been explored deeply in this field, to the best knowledge of the authors at the moment of writing, with only two works dealing with the topic. In [8], the authors addressed the next activity prediction problem using a bayesian network to directly learn the topology of the process for the event log, obtaining a model also capable of making predictions. The uncertainty is handled indirectly in the network's learning process and embedded in its structure. However, the work does not use deep learning and is not focused on quantifying the prediction uncertainty or analysis but instead on retrieving a transparent and explainable predictive process model. Authors in [9] used a deep learning technique, but they focused on predicting the remaining case time, another main task of predictive process monitoring. They used concrete dropout, a type of Monte Carlo dropout, to estimate uncertainty with neural networks. Despite a similar way to induce stochasticity in the prediction, they focused on a continuous output type while we investigated a categorical one. Therefore, our work differs in the uncertainty estimation methods and analysis of the results.

This work expands the categorical output of neural networks with explicit information about the prediction uncertainty, analyzing its relation with the accuracy and assessing the model calibration. We propose a way to effectively exploit the additional information in a real scenario by observing the probability assigned to the actual predictions plotted against the total uncertainty. We also analyze peculiar model failures in which the model is overconfident, relating them to unusual cases in the training datasets. The methods used are not model-dependent, and it is possible to use them on top of any neural network architecture, such as the ones used in [6,7], with minor or no modifications.

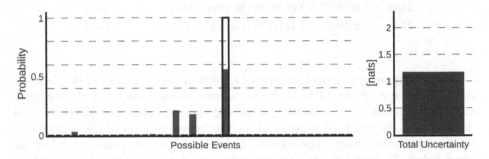

Fig. 1. Example of output from the ensemble of models on the BPIC 2012 dataset. On the left, the probability distribution on possible events (blue) and the actual target (black empty bar). On the right, the indication of the total uncertainty of the prediction. (Color figure online)

3   Problem Statement and Framework

As already mentioned, predictive process monitoring deals with traces contained in an event log. Below we enunciate the definition of the main entities involved in the task.

Definition 1.1 (Event): An *event* is a partial function that maps a subset of all possible process attributes to their values. Attributes are, for example, the timestamp when the event happened or the activity carried on in that specific event.

Definition 1.2 (Process Instance): Let *id* be an identifier associated with every event, such that it is unique for all those related to the work to be done to fulfil a specific request. A *process instance* is the set of all events with the same *id*. A *process instance* can also be called a *case*.

Definition 1.3 (Trace & Event Log): A *trace* is the temporally ordered sequence of a process instance's events. A collection of traces is called an *event log*.

Definition 1.4 (Variant): A *variant* is a unique sequence of activities inside an event log.

Definition 2 (Next Activity Prediction): Given a trace of length n and a subset of length i, the task of *next activity prediction* is to forecast the *"activity"* attribute of the event that is going to happen next, i.e., from $\langle e_0, ..., e_i \rangle$ predict $a_{e_{i+1}}$ with $i < n$.

Table 1. Example of trace

Case ID	Timestamp	Activity	Resource
2748	2010/01/13 12:26:04	Assign seriousness	Value 2
2748	2010/01/19 09:26:05	Take in charge ticket	Value 2
2748	2010/01/19 09:28:50	Resolve ticket	Value 2
2748	2010/02/13 12:00:28	Closed	Value 5

Table 1 reports a practical example of a trace in Helpdesk, one of the datasets used. Every row of the table is the log of one event, composed of attributes such as the timestamp, the activity done and the resource involved. The actual dataset contains other attributes such as the type and level of the service and the resource's workgroup. The column "Case ID" identifies events related to the same ticket. To derive the variant, we only consider the sequence of activities: "Assign seriousness" - "Take in charge ticket" - "Resolve ticket" - "Closed". An event log can contain multiple traces following the same variant depending on how ordinary is that particular process execution. The task of the following activity prediction would be, for example, given the first three events, to predict the activity "Closed".

We propose a framework to exploit the additional uncertainty information at the forecast time. When the model outputs a prediction, the system raises a warning if the uncertainty is higher than a certain level, showing the top-k forecasted possibilities with their associated probabilities to the human decision-maker. As discussed in Sect. 6, we derive the idea from observing the point distribution in Fig. 4. The uncertainty threshold and the number of proposed possible events are user-adjustable parameters, but we found that, for both datasets in our test, top-three accuracy reaches an accuracy of more than 98%. In Fig. 1, we report a possible representation of the model output with the probability distribution over possible outputs and the uncertainty related to the prediction. The model puts the probability mainly on three activities, and consequently, the total uncertainty is relatively high, considering that most test predictions have an uncertainty below 0.5 nats, as reported in Fig. 3.

4   Methods

This section reports all the methods used, divided according to the topics covered in this work.

4.1   Uncertainty Estimation

Two leading techniques have been developed in the machine learning community to estimate a model's categorical output uncertainty. One method directly approximates the uncertainty, evaluating the entropy related to the predicted distribution. The other one empirically verifies if the probabilities associated with the predictions reflect the ground truth likelihood, also called confidence calibration. A good level of confidence calibration is essential in this work due to the uncertainty dependence on the probability distribution, expressed in Eq. 4. Since we only make sure that the network's output sums to one through the *softmax* function, there is no theoretical assurance that it is to all effects a probability distribution over the possible events derived from the training set. Therefore, a good model calibration ensures that the prediction confidence reflects the output likelihood. The following paragraphs describe both ways of dealing with uncertainty.

Confidence Calibration. Confidence calibration indirectly assesses the uncertainty in the predictions of a model, checking if the predicted probability estimates are representative of the actual output likelihood [10]. Having information that correctly reflects the likelihood of the predicted class label supports model trust and consequent adoption in a real-world application and a correct uncertainty estimation, as mentioned before.

The model is perfectly calibrated if the forecasting belief matches its accuracy. In order to evaluate this property, we used reliability diagrams and Expected Calibration Error (ECE).

Reliability Diagrams visually represent model calibration [11]. These diagrams plot the expected accuracy against confidence prediction. If the model is perfectly calibrated, the diagram will show the identity function, and any deviation from it is considered a miscalibration. Accuracy estimates are retrieved by grouping predictions in interval bins based on confidence and then computing the accuracy for each one.

Expected Calibration Error is popularly used as a calibration metric [12]. Although reliability diagrams are handy for a visual inspection, a scalar metric is needed to quantify calibration and compare different models. *ECE* evaluates the average distance from the perfect calibration, also called the calibration gap, weighted by the percentage of samples included in each bin. It is defined as

$$ECE = \sum_{m=1}^{|B_m|} \frac{|B_m|}{N} |acc(B_m) - conf(B_m)| \tag{1}$$

where N is the total number of samples and B_m is the set of all predictions whose confidence falls in a specific interval. Bin accuracy is defined as

$$acc(B_m) = \frac{1}{|B_m|} \sum_{i \in B_m} \mathbf{1}_{\hat{y}_i = y} \tag{2}$$

where $|B_m|$ is the number of elements in the set, $\hat{y}_i$ is the model output with the maximum probability and y is the target. Bin confidence is defined as

$$conf(B_m) = \frac{1}{|B_m|} \sum_{i \in B_m} \hat{p}_i \tag{3}$$

where $\hat{p}_i$ is the maximum probability given to sample i.

Prediction Entropy. Directly estimating uncertainty in predictions requires stochasticity in the forecasting mechanism. We injected variability in the prediction using two different and widely adopted methods: Monte Carlo dropout [13] and an ensemble of neural networks [14].

The first exploits the dropout mechanism, intended initially as a regularization method. A defined percentage of neurons are discarded during training, avoiding weights co-adaptation and forcing the network to generalize better. In Monte Carlo dropout, this technique is preserved at inference time, injecting stochasticity in the output and obtaining at every pass a different distribution. The final prediction is obtained by averaging the sampled distributions.

The latter, instead, makes use of a models ensemble to have different outputs. The central intuition is that independently trained models will converge to different solutions. This phenomenon is beneficial, particularly in unknown data regions, leading to high output variability when a dataset lacks knowledge or contains noise. As in the previous case, the average of the models' output gives the final prediction. We use both methods to compare them and check if there are significant differences in performance.

In order to estimate and analyze uncertainty, we used the quantities reported below.

Total Uncertainty is estimated, following [15], with

$$u_t(\boldsymbol{x}) := -\sum_{y \in Y} \left(\frac{1}{N_m} \sum_{i=1}^{N_m} p(y|h_i, \boldsymbol{x}) \right) log \left(\frac{1}{N_m} \sum_{i=1}^{N_m} p(y|h_i, \boldsymbol{x}) \right) \tag{4}$$

where y is the output, h_i the parameters of the i-th model and $\boldsymbol{x}$ the input sequence. It corresponds to the final prediction's entropy averaged across N_m models.

Accuracy-Uncertainty plot represents the model's accuracy related to its total uncertainty. Accuracy is divided into bins based on the total uncertainty and plotted against it. Bin accuracy is defined similarly to Eq. 2, but we split the bins based on uncertainty. Bin uncertainty is computed as

$$u_t(B_m) = \frac{1}{|B_m|} \sum_{i \in B_m} \hat{u}_{t_i} \tag{5}$$

where $\hat{u}_{t_i}$ is the total uncertainty of a single prediction.

Weighted Correlation Coefficient between bin accuracy and total bin uncertainty is used to numerically verify the existence of a linear relation linking the quantities. We used the data percentage in each bin to weight each contribution to the correlation coefficient, giving more importance to results obtained on more significant data subsets.

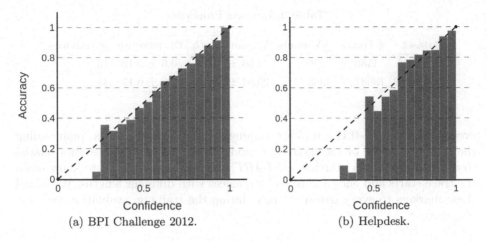

Fig. 2. Reliability diagrams for the ensemble of models on BPIC 2012 and Helpdesk datasets. The diagonal represents perfect calibration.

4.2 Model and Training

The model used in our study is an attention-based neural network, as introduced in [16], composed of just the decoder part of the original Transformer [17]. The model has been adapted to the predictive process monitoring task by [7], outperforming on average accuracy several prior techniques. Still, it is worth remarking that our primary aim is not to improve state-of-the-art performances but to investigate prediction uncertainty, and the methods used can be applied on top of any machine learning algorithm.

Reflecting the different complexity of the two event logs, we used a model with a higher number of parameters in the case of BPI 2012 (4 layers, 4 heads, 1024 units/layer, 0.1 dropout) and a smaller one for Helpdesk (2 layers, 2 heads, 512 units/layer, 0.1 dropout). Section 5 gives more information about the datasets used.

Regarding models' input features, two variable types are present in the data: categorical and continuous. We chose the embedding dimension for the first category based on the unique activity number in the dataset. The latter category has been normalized across all datasets. In the source code are reported all the embedding dimensions used[1]. Temporal features used are the relative time between two events, part of the day and the day of the week in which the event took place, and we extracted them from the event timestamp.

To evaluate the total uncertainty, we chose an ensemble of five models as in [14] or five samples in the case of Monte Carlo dropout. The models have been implemented and trained using the TensorFlow library [18] with 100 epochs and early stopping patience of 10. We used the Adam [19] optimizer in conjunction with a categorical cross-entropy loss function and a batch size of 32. We split the

---

[1] Code available at https://github.com/piepor/uncertain-ppm.git.

Table 2. Datasets Properties

Dataset	#Traces	#Variants	Variant length	DL measure	#Activities
Helpdesk	4580	226	4.66 ± 1.18	0.5 ± 0.16	16
BPI 2012	13087	4366	20.04 ± 19.94	0.39 ± 0.17	38

event logs into 0.7 −0.15 −0.15 for training, validation and test set, maintaining the temporal order of the data. The model needs three special events related to as many particular activities: $\langle START \rangle$, $\langle END \rangle$, and $\langle PAD \rangle$ to know when the trace starts and ends and to pad sequences with different lengths. We added these markers to every datasets' trace during the training, evaluation and test phases.

5 Data

The Process Mining community has established different real-life event logs to test model performance. We chose two public datasets, presented in [20] and [21], which other authors in several papers used to assess their methods against the others. The first event log is referred to a ticketing management process of an Italian software company help desk, from now on called Helpdesk. The second one is related to a loan application process, made available by a German financial institution for the 2012 Business Process Intelligence Challenge, called BPI 2012 in the rest of the manuscript.

Both datasets logs contain the timestamp, the activity executed and the resource taking care of it. Depending on each specific process peculiarity, the event log can report other features; for instance, the Helpdesk also offers the event's seriousness level, the workgroup assigned, the product code, the service level and type provided, and the responsible section. BPI 2012, instead, reports just the amount of money requested in every case. We covered a wide range of complexity, following the categorization in [6], taking Helpdesk as a "simple" event log and BPI 2012 as a "complex" one.

Table 2 confirms the higher complexity of the BPI 2012 event log that has a higher overall number of traces and variants, average length of cases and diversity. To measure the diversity, we used the Damerau-Levenshtein (DL) metric that, given two variants, penalizes every operation needed to recover one sequence starting from the other. The lower the DL, the higher the diversity because more operations are required to make them identical. Moreover, BPI 2012 has more than twice as much possible activities.

Table 3. Models' Calibration Metric and Correlation Coefficient between accuracy and total uncertainty

Dataset	Model Type	ECE	Correlation
Helpdesk	Single	2.24%	–
Helpdesk	MC-dropout	2.07%	−0.9584
Helpdesk	Ensemble	2.17%	−0.9143
BPI 2012	Single	1.04%	–
BPI 2012	MC-dropout	0.98%	−0.9827
BPI 2012	Ensemble	0.48%	−0.9791

Table 4. Model's top-k accuracy

Dataset	Top 1	Top 2	Top 3	Top 4	Top 5	Top 6	Top 7	Top 8
Helpdesk	0.8799	0.9686	0.9926	0.9964	0.999	0.9996	1.000	1.000
BPI 2012	0.8416	0.9264	0.9879	0.9961	0.9975	0.9983	0.9989	0.9993

6   Results

This section reports the results divided according to the topics addressed in this work, models calibration and direct uncertainty estimation, and an analysis of models' overconfident failure cases.

6.1   Calibration

Figure 2 reports only the reliability diagrams of the ensemble of models, but the Monte Carlo dropout case results are similar. Considering these two particular datasets, the models are generally not overconfident. The average calibration, expressed by the ECE column in Table 3, is acceptable compared, for instance, to the ones reported in [10]. The Helpdesk model has the worst behaviour, probably due to fewer data available. On the other hand, BPI 2012 has a good correspondence between confidence and accuracy. In both cases, the single model's calibration is the worst due to the output averaging, which smooths the single predicted distribution and helps to spread the probability mass better. The ensemble of models performs slightly better on the complex dataset at the cost of training more models.

Good calibration is desirable in a real case scenario, and the user's trust in the model should not be a problem in these particular cases. Moreover, it supports the correctness of the uncertainty estimation based on the predicted probability distribution. Future works could extend the same evaluation on other event logs.

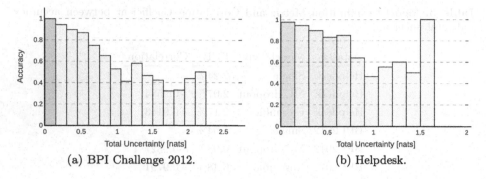

(a) BPI Challenge 2012.   (b) Helpdesk.

Fig. 3. Bin accuracy against total uncertainty. The colour opacity is proportional to the data percentage in the considered bin.

6.2   Total Uncertainty

The main research question of this work regards the relationship between a model's prediction uncertainty and its accuracy to warn the user to take counteraction because the expected outcome may be wrong. Figure 3 and Table 3 report visually and quantitatively the answer. In all the settings, the correlation coefficient is below -0.9, indicating a strong negative relation between accuracy and uncertainty. Figure 3 plots, for the test set, accuracy against uncertainty with the bin opacity being proportional to the data contained in each interval. In both cases, low uncertainty bins hold the vast majority of data and the accuracy seems to stop decreasing for high uncertainty values, but the data scarcity prevents conclusions. Further studies on other datasets are required to verify this behaviour's consistency.

The red and blue dots in Fig. 4 represent the probability assigned to the actual target against the prediction's total uncertainty for the complete test set. Instead, the dotted lines represent the maximum entropy for a multi-class distribution with one class fixed and the remaining mass probability distributed equally onto the others classes, ranging from the binomial to the vocabulary length. The peculiar distribution followed by the points reflects these particular functions, remaining in both cases below the maximum entropy of a subset of the vocabulary considered. For example, in the case of BPI 2012, the model spreads its output on a maximum of thirteen classes.

This property is valuable considering top-k accuracy, reported in Table 4. Both models have a top-three accuracy of more than 98%. The system can suggest multiple options to the user when the uncertainty of the prediction is high to help take countermeasures. Model's top-k accuracy on other event logs and its relation with total uncertainty is left for future works.

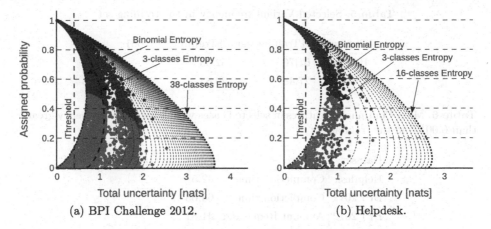

(a) BPI Challenge 2012.   (b) Helpdesk.

Fig. 4. Probability assigned by the ensemble of models to the actual event against total uncertainty (right/blue and wrong/red predictions). Dotted curved lines represent the multi-class entropy while the vertical one is the threshold considered in the analysis of wrong overconfident predictions. (Color figure online)

6.3 Overconfidence

The bottom left part of the diagrams reported in Fig. 4 is particularly interesting since it is characterized by a low uncertainty and low probability assigned to the actual activity. Therefore, the model is quite sure about its prediction, but it is wrong. Considering the peculiar characteristics of the input data, we hypothesize that this overconfidence zone is due to out-of-training-distribution (OOD) examples. The event log does not record why something happened but just the event and some attributes. Imagine two sequences very similar until a specific step in which they diverge: if the model has seen many examples of the first one in the training phase and has to forecast the first event after the divergence of the second one, it will wrongly predict with low uncertainty the following activity of the sequence present in the training set.

In order to verify this OOD hypothesis, we chose a threshold of 0.4 nats on the total uncertainty and analyzed cases and variants containing wrong predictions with an uncertainty below that threshold. Table 5 and Table 6 report the results. For the Helpdesk and BPI 2012 test sets, respectively, 38.24% and 72.52% of selected variants are not present in the training set and could be rightfully inferred as uncommon in a real scenario. 52.38% and 20.83% of the variants included in the training event log are more frequent than 0.1%, leaving 32.34% for Helpdesk and 5.73% for BPI 2012 of the overall variants selected not explained by our hypothesis.

Filtering out variants not present in the event log used for training, we have analyzed the distribution of completion times and the amount requested in Table 5. Most cases' completion times are in the first or last quartile compared to other instances of the same variant. Proportions are not the same for

Table 5. Selected Variant frequency in the training set

Dataset	%Training set	%Training > 0.001
Helpdesk	61.76	52.38
BPI 2012	27.48	20.83

Table 6. Main Continuous Features of selected cases with a variant's frequency greater than 0.001% in the training set

Dataset	Feature	%first/last quartile
Helpdesk	Completion time	62.5
BPI 2012	Completion time	63.63
BPI 2012	Amount Requested	31.81

the amount requested in BPI 2012, around 30% of the considered case. Roughly 12% of Helpdesk's anomalies and 2% of BPI 2012's are left unexplained. In the first case, though, the model could have been influenced by the low number of overall training data available.

In conclusion, most wrong overconfident cases are out of the training dataset, contained with low frequency, or the corresponding features are in the distribution's tail. However, our analysis considers only the overall characteristics of the case in which the model is incorrect: we left to future works studying the relation of the single wrong prediction with the sequence before it and a deeper analysis of the selected cases' features.

7   Conclusions and Future Works

This work focused on the uncertainty in the next activity prediction, a subject not investigated in the predictive process monitoring field, evaluating the model's calibration and estimating the uncertainty through Monte Carlo dropout and an ensemble of models. We showed that the models on the tested datasets are already quite well-calibrated and that the accuracy negatively correlates with the total uncertainty in the output. The results of models calibration sustain the correctness of the uncertainty estimation and models' trustworthiness. We did not notice remarkable differences between Monte Carlo dropout and ensemble of models.

We proposed a framework to exploit the additional uncertainty information, raising a warning and proposing other possible events whenever the uncertainty exceeds a defined limit. We also analyzed cases in which the model is overconfident in his prediction and showed that most of them are either in variants with low frequency or in the tails of feature distributions.

Future works will test other publicly available datasets and explore aleatoric and epistemic uncertainty evaluation. In particular, the latter could identify a concept drift in an event log. Other areas for future studies are a deeper investigation of the remaining wrong and overconfident model predictions analyzing the sequence of preceding events and the cases' features.

References

1. https://www.mise.gov.it/index.php/it/incentivi/impresa/digital-transformation
2. Matt, C., et al.: Digital transformation strategies. Bus. Inf. Syst. Eng. **5**, 339–343 (2015)
3. Van der Aalst, W., et al.: Process mining manifesto. In: BPM Workshops (2011)
4. Bojarski, M., et al.: End to End Learning for Self-Driving Cars. arXiv:1604.07316 (2016)
5. Esteva, A., et al.: Dermatologist-level classification of skin cancer with deep neural networks. Nature **542**(7639), 115–118 (2017)
6. Camargo, M., Dumas, M., González-Rojas, O.: Learning accurate LSTM models of business processes. In: Hildebrandt, T., van Dongen, B.F., Röglinger, M., Mendling, J. (eds.) BPM 2019. LNCS, vol. 11675, pp. 286–302. Springer, Cham (2019). https://doi.org/10.1007/978-3-030-26619-6_19
7. Zaharah, A., et al.: Process Transformer: Predictive Business Process Monitoring with Transformer Network. arXiv:2104.00721 (2021)
8. Prasidis, I., et al.: Handling uncertainty in predictive business process monitoring with Bayesian networks. In: International Conference on Information, Intelligence, Systems & Applications (IISA) (2021)
9. Weytjens, H., De Weerdt, J.: Learning uncertainty with artificial neural networks for improved remaining time prediction of business processes. In: Polyvyanyy, A., Wynn, M.T., Van Looy, A., Reichert, M. (eds.) BPM 2021. LNCS, vol. 12875, pp. 141–157. Springer, Cham (2021). https://doi.org/10.1007/978-3-030-85469-0_11
10. Guo, C., et al.: On calibration of modern neural networks. In: ICML (2017)
11. DeGroot, M.H., Fienberg, S.E.: The comparison and evaluation of forecaster. J. Royal Stat. Soc. Ser. D (The Statistician), **32**(1–2), 12–22 (1983)
12. Bhatt, U., et al.: Uncertainty as a form of transparency: measuring, communicating, and using uncertainty. In: Proceedings of the 2021 AAAI/ACM Conference on AI, Ethics, and Society (2021)
13. Gal, Y., Zoubin, G.: Dropout as a Bayesian approximation: representing model uncertainty in deep learning. In: ICML (2016)
14. Lakshminarayanan, B., et al.: Simple and scalable predictive uncertainty estimation using deep ensembles. In: Advances in NIPS (2017)
15. Hüllermeier, E., Waegeman, W.: Aleatoric and epistemic uncertainty in machine learning: an introduction to concepts and methods. Mach. Learn. **110**, 457–506 (2021)
16. Radford, A., et al.: Improving language understanding by generative pre-training. https://s3-us-west-2.amazonaws.com/openai-assets/research-covers/languageunsupervised/language understanding paper.pdf (2018)
17. Vaswani, A., et al.: Attention is all you need. In: Advances in NIPS (2017)
18. Abadi, M., et al.: TensorFlow: a system for large-scale machine learning. Whitepaper (2015)
19. Kingma, D.P., Ba, J.: Adam: a method for stochastic optimization. In: International Conference on Learning Representations (2015)
20. Verenich, I.: Helpdesk. Mendeley Data (2016)
21. van Dongen, B.: BPI Challenge 2012. 4TU.ResearchData. Dataset

Set-Based Counterfactuals in Partial Classification

Gabriele Gianini[1]([✉])(iD), Jianyi Lin[2](iD), Corrado Mio[3](iD),
and Ernesto Damiani[1,3](iD)

[1] Università degli Studi di Milano, Milan 20133, Italy
gabriele.gianini@unimi.it
[2] Università Cattolica del Sacro Cuore, Milan 20123, Italy
jianyi.lin@unicatt.it
[3] EBTIC/Khalifa University of Science and Technology, 127788 Abu-Dhabi, UAE
{corrado.mio,ernesto.damiani}@ku.ac.ae

Abstract. Given a class label y assigned by a classifier to a point x in feature space, the counterfactual generation task, in its simplest form, consists of finding the minimal edit that moves the feature vector to a new point x', which the classifier maps to a pre-specified target class $y' \neq y$. Counterfactuals provide a local explanation to a classifier model, by answering the questions "Why did the model choose y instead of y': what changes to x would make the difference?". An important aspect in classification is ambiguity: typically, the description of an instance is compatible with more than one class. When ambiguity is too high, a suitably designed classifier can map an instance x to a class set Y of alternatives, rather than to a single class, so as to reduce the likelihood of wrong decisions. In this context, known as set-based classification, one can discuss set-based counterfactuals. In this work, we extend the counterfactual generation problem – normally expressed as a constrained optimization problem – to *set-based counterfactuals*. Using non-singleton counterfactuals, rather than singletons, makes the problem richer under several aspects, related to the fact that non-singleton sets allow for a wider spectrum of relationships among them: (1) the specification of the target set-based class Y' is more varied (2) the target solution x' that ought to be mapped to Y' is not granted to exist, and, in that case, (3) since one might end up with the availability of a number of feasible alternatives to Y', one has to include the degree of partial fulfillment of the solution into the loss function of the optimization problem.

Keywords: Set-based classification · Counterfactual explanations

1 Introduction

The interpretability of the decisions of Artificial Intelligence Algorithms is a desirable requirement in many domains. It is fundamental in those where trust in the outcomes is a critical element, due to the high impact of the decisions.

D. Ciucci et al. (Eds.): IPMU 2022, CCIS 1602, pp. 560–571, 2022.
https://doi.org/10.1007/978-3-031-08974-9_45

The need for interpretability has become stronger after the research in Machine Learning has achieved impressive prediction performance through algorithms, such as those in the Deep Learning area, which are very hard to interpret, and are often metaphorically referred to as "black boxes", or "opaque boxes".

1.1 Explainable Artificial Intelligence

Opening the Black-Box. In recent years the research in eXplainable Artificial Intelligence (XAI) has developed methods of several kinds to answer questions, encompassed by the different forms that interpretability can take [1,6,12–14,23]. Among them are methods for the *design of transparent boxes* (model endowed with *simulatability, decomposability, algorithmic transparency*), and methods for the *post-hoc explanation of black boxes*. Among the latter are: (1) the explanations based on transparent surrogate models (the surrogate is learned not from the training data, but from the opaque model, used as an *oracle*), (2) outcome explanations methods (aimed at understanding the reason for the decision on a given example) and model inspection methods (e.g. sensitivity analysis, feature importance analysis such as those based on the Shapley Value analysis [11,18,24]). Among the models that are transparent by design and that can either be learned directly from the data or be used as surrogate models for post-hoc explanation are: Decision Rules, Decision Trees, Bayesian Models including Bayesian Belief Networks, Linear Models, and Case Based Reasoning.

Case Based Reasoning, and the Counterfactual Generation Task. An approach that, by construction, can be used to explain the outcomes of a decision is Case Based Reasoning (CBR) (one of the most well-known representatives of CBR is the k-Nearest Neighbor algorithm). CBR methods are especially useful in domains that have not been fully understood or modeled. In CBR the explanations for the result of the classification of an instance follow an approach very common among humans, which consists of providing pertinent examples: cases – similar to the one at hand – that support the predicted class, thus providing an answer to the question "Why class P?".

Contrastive Explanations. In recent years the attention of the XAI community has extended to the discussion of *contrastive explanations*, i.e. to explanations that answer the question "Why class P instead of class Q?": in that context P is called the *fact* and Q is called the *foil*. Within CBR a contrastive explanation can be provided by finding nearby examples belonging to class Q and showing what are the differences.

Counterfactual Explanations. A concept connected to contrastive explanations is the one of *counterfactuals*: they answer the question "What are the (possibly small) changes that can be made, starting from the features of the instance under exam, so that the decision is no longer P, but Q?". Finding the candidate edits that take the feature vector beyond the decision boundary, subject to a number of minimality constraints, defines the XAI task of *counterfactuals generation* [25]. The task implicitly assumes that there is a user who can exploit this form of explanation.

Variants of the Task. The task appears in several variants, mainly depending on the constraints (feasibility, actionability, data manifold proximity, etc., see Sect. 2). Another important distinction [7] is whether the foil is a pre-specified class (*targeted counterfactuals*) or any class different form the fact is acceptable (*untargeted counterfactuals*). Other variants depend on whether more than one counterfactual, rather than a single one, are to be returned for a given input data-point (in the latter case one speaks of *multiple counterfactuals generation*).

Related Tasks. The task of counterfactuals generation is a special case of the Machine Learning task of *inverse classification* (given a classifier model and a class, find an input example that the model assigns to that class) and is related to the concept of adversarial attack (a process where the attacker tries to find a perturbation of a given input that leads to a target classification by the ML model, *such that the perturbation goes undetected by a reference observer*, e.g. a human observer or another classifier; this represents an additional constraint to the counterfactual generation task): in this case the user exploiting the counter-factual explanation is the attacker.

1.2   Uncertain Labels and Partial (Set-Based) Classification

In Machine Learning the single-label classification task consists in learning from the data a model able to assign a single label to any previously unseen instance. The result of the learning phase is an explicit or implicit model that partitions the input space into regions, each assigned to a single class: in the subsequent prediction phase, any instance falling into that region is assigned the corresponding class label.

Obviously, except for a few pedagogic cases, the learned model, and with it the learned partition, is always affected by uncertainty. If the uncertainty is modeled within a probabilistic framework, each point in the input space is associated with a probability distribution over the classes, estimated from the training sample, and the output of a single label classifier is normally obtained by choosing the class with the highest probability.

Classification with Rejection. However, in presence of high uncertainty and ambiguity (e.g. due to very small training samples) sharp boundaries between classes can lead to a high misclassification rate. A way for addressing this problem is the use of *rejection*, which consists of abstaining from making a decision when the uncertainty is above a given threshold.

Partial Classification. Another approach is *partial*, i.e. *set-based, classification*. In that case, the outcome of the processing of an instance is a set Y of labels: *only one* of them corresponds to the true state of affairs and we do not know which one in the set Y is the true one.

This set-valued output has a meaning different from the one in multi-label classification, where *all* the labels apply to the instance – in other words in partial classification the output is an *epistemic* set, in multi-label classification is an *ontic* set [8]).

Partial classification is more informative than rejection: the fact that uncertainty is reduced from the universe set Ω encompassing all the singleton classes to a set $Y \subset \Omega$ lowers the likelihood of wrong decisions (in this frame rejection is equivalent to $Y = \Omega$). Obviously, set-based classification contains as a special case the one where Y is a singleton: to distinguish the singleton output from the non-singleton set output, hereafter, where needed, we will use the expressions *singleton class*, and *non-singleton class*. Furthermore, when a set-valued output is a subset of another set-valued output we will say that the former is more precise than the latter.

1.3   Set-Based Counterfactuals

The possibility of set-based classification opens the way to set-based counterfactuals: if one accepts that the output class of a classifier for an input point x can be a set Y, one can also look for a point x', in the feature space X, that, through a classifier mapping c, leads to another set-valued output Y' with desirable properties.

The Task. The task of generating a set-based counterfactual explanation consists of finding an edit to the input point x, which takes the feature vector to another point x', that the classifier c maps to a possibly set-based class Y'. This task encompasses as special cases those in which either Y, or Y', or both, are singletons.

Motivations. Notice that if, on the one end, the imprecision of the fact-class Y, issued by a classifier, can be due to the imperfect information available to the model about the input instance (all those models try to optimize the precision of the output), on the other hand, the imprecision of the foil-class Y' comes as an explicit requirement by the user:

a) either she wants to obtain an instance x' endowed with an ambiguous classification (the ambiguity is *a must*) – this attitude can be motivated by security/privacy reasons (e.g. dictated by location privacy concerns [2–5]);

b) or she wants an x' with a classification falling within Y' and is ready to accept even an ambiguous classification, but would not object to accepting any one of the subsets of Y' (the ambiguity is *a may*) – this attitude can be motivated by the need of moving away for Y and the tolerance for different possible targets.

In setting *(b)* the goal is achieved even if one finds an x' such that its class $c(x') \subset Y'$ (proper subset), provided that x' is the closest point to x satisfying the constraints: the formulation is supported by the tolerance of imprecise results, but, in presence of two classifications satisfying the constraint, the more precise one will be preferred over the less precise one.

Contribution and Structure of the Paper. In this work, we extend the formalization of the counterfactual generation problem to set-based counterfactuals.

The reminder of the paper is structured as follows: in Sect. 2 we recall the definition of the task for singleton class fact and singleton class foil, then in Sect. 3 we formalize the generalized set-based counterfactual problem; a short discussion concludes the paper in Sect. 4.

2   Singleton-class Counterfactual Explanations

Given a classifier (previously defined or learned) that issues only singletons labels, and an instance, mapped by the classifier into a class, one needs to find one or more candidate edits (possibly satisfying distinct additional requirements) that move the instance feature vector across the classifier decision boundary, i.e. into the region of another class.

2.1   Counterfactual Generation Task as an Optimization Problem

The generation of counterfactual explanations is normally formalized as an optimization problem. In a singleton-fact and singleton-foil class setting, the standard approach is the following [25].

2.2   Constrained Form

Let us denote by X the input space and by Ω the set of the classes. Given an instance represented by the feature vector $x \in X$ in the input space, whose class, according to the classifier c, is $c(x) = y \in \Omega$, we need to find a counterfactual $x' \in X$ mapped by the classifier onto the desired label $c(x') \in \Omega$ such that the distance $d(x, x')$ (according to some distance definition, e.g. L^1 or L^2) is minimal.

In *targeted* counterfactual generation we specify a class $c(x') = y'$, thus

$$\underset{x'}{\arg \min} \, d(x, x') \text{ subject to } c(x') = y' \tag{1}$$

In *untargeted* counterfactual generation we impose that the new class is different from the class of x, i.e. $c(x') \neq y$

$$\underset{x'}{\arg \min} \, d(x, x') \text{ subject to } c(x') \neq y \tag{2}$$

Validity Constraint. A counterfactual which belongs to the desired class (the class y', in the targeted case, or a class in the complement of y in the un-targeted case) is a *valid* counterfactual.

Least Effort Constraint. The minimality constraint is sometimes termed *least effort* constraint. As distance d one typically chooses an L_p norm with a suitable standardization: e.g. the standardized Euclidean distance (the Mahalanobis distance for diagonal covariance matrix). Wachter et al. [26] propose to use the L_1 norm (Manhattan distance) weighted by the inverse Median Absolute Deviation (MAD), i.e. given a dataset of N points $x_i \in D \subset X$, with $i = 1, \ldots, N$, where X is an m dimensional feature space, so that a generic point is $x \equiv (x^{(1)}, \ldots, x^{(m)})$, and denoting by mdn the median operator

$$d(x, x') = \sum_{k=1}^{m} \frac{|x^{(k)} - x'^{(k)}|}{MAD_k} \quad \text{with} \quad MAD_k = mdn_{x_i \in D} \left(\left| x_i^{(k)} - mdn_{x_j \in D}(x_j^{(k)}) \right| \right)$$

2.3 Unconstrained Form

The above problems (1) and (2) can be converted into an unconstrained form by inserting a suitable loss into the minimization expression: a typical way consists of adding a strong penalty for the disagreement from y' in the targeted case, and for the agreement with y in the un-targeted case.

For the sake of illustration, in a class space where $(c(x') - y')^2$ makes sense, Wachter et al. [26] use

$$\arg\min_{x'} \left[\max_{\lambda} \left[d(x, x') + \lambda \left(c(x') - y' \right)^2 \right] \right]$$

Maximization over λ is done by solving for x' iteratively, then increasing λ until a sufficiently close solution is found [26].

Further Optional Requirements. To be most effective a counterfactual ideally should change a number of features as small as possible (this requirement is called counterfactual *sparsity*): to this purpose one can introduce another purposely structured distance term $g(x' - x)$ in the objective function; moreover, the requirement that the change remains close to the data manifold (*data manifold closeness*) can be expressed by an appropriate term $h(x', X)$. Further requirements can be added to the problem, for instance *feasibility*, or *actionability* or *meaningfulness* of the edit: this amounts to restricting the choice of x' to a subset $\chi \subset X$, i.e. changing the first operator into $\arg\min_{x' \in \chi}(\ldots)$.

Considerations about the computational complexity of the search for the optimal solution can force the relaxation of the optimality constrains.

The discussion of the precise computational methods for attaining an optimal or a nearly optimal solution falls out of the scope of the present paper.

Counterfactual Generation Algorithms. Algorithms for counterfactual generation apply different optimization methods, depending on the degree of model access. One typically identifies three levels of access to the model: (1) full access to model internals (for instance for a decision tree, the node conditions), (2) access to gradients (applies to differentiable models only, see for instance [10,15,16,20–22,26]) and (3) black-box (access only to the prediction function, in this case, one can use generic optimizers such as Genetic Algorithms, or the Growing Spheres algorithm [17]). Another element that influences the choice of the algorithm is the level of access to the data: one can distinguish between access to training data only, access to test data only, access to both or to none.

3 Towards Set-Based Counterfactuals

In a setting where partial classification is allowed, one can reformulate the optimization problem taking into account the fact that the counterfactual can be set-valued. The various consequences of this assumption are determined by the richer relationships possible between and among non-singleton sets, which are not necessarily disjoint.

In singleton-class classification the possible outputs of the classifier are naturally disjoint and allow only for three cases: an output coincides with the *fact* class y, or it coincides with the counterfactual class, or *foil*, y' or (if more than two singleton classes are available) it coincides with a third class, which is neither y nor y'. Coincidence with a singleton implies the complete exclusion of the others.

In partial classification, on the other hand, the output $c(x')$ of the classifier can coincide, be a superset, be a subset or have a partial overlap with the *fact-class* Y or the *foil-class* Y'. These aspects are relevant both (1) for the precise formulation of the goal of the counterfactual generation and (2) for the assessment of the degree of achievement of that goal. In the singleton-fact singleton-foil setting a solution always exists (if the class y' is represented among the members of the feature space), whereas in the partial classification setting a point x' that is classified ambiguously as a member of a specific non-singleton set Y' is not necessarily available: thus one has to consider also partial satisfaction of the constraints.

Hereafter we formulate the problem of counterfactual generation in set-based classification, as a constrained minimization problem, then we delimit the space of the possible goal constraints, finally, we address the problem of establishing a utility function for partial solutions, a step useful for changing the formulation of the problem into a non-constrained form, taking into account also the partial attainment of the goals.

The Set-Based Counterfactual Generation Problem. The task of counterfactual generation in set-based classification can be described, in rather general terms as follows.

– Given
  • an instance represented by the feature vector $x \in X$ of the input space, mapped by the classifier onto $c(x) = Y \in 2^{\Omega}$,

- and a target set Y'
- one has to find an $x' \in X$,
  - whose image $c(x')$ fulfills a ternary constraint $\mathcal{R}\,(c(x'), Y, Y')$ with the given pair of sets Y and Y', (*validity* requirement)
  - such that the distance $d(x, x')$ is minimal, for a given definition of distance (*least effort* requirement).

In other words, one has to find an x' such that

$$x' = \arg\min_{x'} d(x, x') \text{ subject to } \mathcal{R}\,(c(x'), Y, Y') \tag{3}$$

Goal Constraint Specification In general the constraint $\mathcal{R}$ can involve the three sets $c(x')$, Y' and Y. However, typical cases, it can relax to a binary constraint involving only either Y or Y'. Ternary constraints can be decomposed into the logical *and* of two binary constraints $\mathcal{R}_1(c(x'), Y')$ and $\mathcal{R}_2(c(x'), Y)$ (the fact Y is given, so its relationship with Y' is specified by the latter, which can be disjoint, partially overlapping, etc.).

Notable cases of binary constraints involving Y' are the following:

1. coincidence of $c(x')$ with Y', i.e. $\mathcal{R}$ is $c(x') = Y'$ (*targeted case*)
2. non-overlapping of $c(x')$ to Y, i.e. $\mathcal{R}$ is $c(x') \subseteq \overline{Y}$ (*un-targeted case*)
3. $\mathcal{R}$ expressed by $c(x') \subseteq Y'$ (user satisfied by any *subset* of the target Y')
4. $\mathcal{R}$ expressed by $c(x') \supseteq Y'$ (user satisfied by any *superset* of the target Y')

Notice that the possible requirements are not bound to those sets $c(x')$ that exclude the original fact Y: a user might want to find a nearby x' with a more precise, or with a less precise, classification than the one of x. Binary relations with such characteristics are the following:

5. $\mathcal{R}$ expressed by $c(x') \subset Y$ (the user is satisfied by any subset of the fact Y, i.e. she want to get a nearby more precise fact, provided that it is compatible with Y);
6. $\mathcal{R}$ expressed by $c(x') \supset Y$ (the user is satisfied by any superset of the current fact Y, she wants to obtain a more imprecise fact, provided that it is compatible with Y, this use case can be exploited, for instance, in location obfuscation).

An example of a ternary relation condition that can be decomposed into two binary relation conditions is the following:

7. $\mathcal{R}$ composed by $\mathcal{R}_1(c(x'), Y')$ and $\mathcal{R}_2(c(x'), Y)$ with $\mathcal{R}_1$ given by $c(x') \subset Y'$ and $\mathcal{R}_2$ given by $c(x') \supset Y$ (the user wants to make the classification output more ambiguous, encompassing Y' without excluding the original fact Y).

The possible binary constraints have the form "$c(x') < operator >< set >$", where $< set > \in \{Y, (\neg Y), Y', (\neg Y')\}$ and $< operator > \in \{\subset, \subseteq, =, \supseteq, \supset\}$. As to ternary constrains, they can be obtained by combining binary constraints in pairs: it is true that a subset of the possible constraint pairs is inconsistent for some choices of Y' (for instance condition (7.) cannot be satisfied if Y and Y' are disjoint), however it is reasonable to assume that the user formulates consistent requirements.

Degree of Achievement of the Goals. Another important difference with respect to the singleton-fact singleton-foil setting, is that in that simple case, the constraint (either $c(x') = y'$, in the targeted case, or $c(x') \neq y$, in the un-targeted case) is either fully satisfied or not satisfied. On the contrary, in set-based classification, the constraint can be fulfilled also just partially.

An Illustrative Example. For the sake of illustration, consider the case in which the constraint $\mathcal{R}$ formulated by the user is $c(x') = Y'$. It may happen that the search does not find at any distance a point x' such that $c(x')$ fulfills the constraint, but finds two points located at the same distance from x: point x_* such that $c(x_*) = Y_* \subset Y'$, and point x^* such that $Y' \subset Y^* = c(x^*)$: in such an occurrence, there is no a priori reason why users should prefer the more precise alternative over the less precise. The more precise alternative can be supported by a specific user's avoidance of excessive imprecision; the less precise alternative, could be preferred due to the user's avoidance of a precise classification (e.g. in a privacy related scenario).

A concrete example is the following. A point x has been classified as $Y = \{a\}$. The user is concerned about privacy and asks for a point x' such that $c(x') = Y' \supset Y$, say $c(x') = Y' = \{a, b, c\}$. In the feature space there is no point characterized by an ambiguity among precisely these three classes, but the search finds x_* such that $c(x_*) = Y_* = \{a, b\} \subset Y'$ and x^* such that $c(x^*) = Y^* = \{a, b, c, d\} \supset Y'$. Since the original motivation was attaining a given level of privacy, the user chooses x^*.

A Preference Structure, Conditional to the Constraint $\mathcal{R}$. From the considerations above it follows that, to completely specify the optimization problem, one should also include a preference structure (e.g. a partial order) over the partial fulfillments of a constraint. A reasonable assumption is that all the pairs of sets can be compared with one another in terms of preference with respect to the satisfaction of the constraint, and that ties are allowed, thus the structure, for a given constraint $\mathcal{R}$ should be represented by a partial order. Each constraint in principle could be endowed with a preference structure.

Within a very general approach, one could ask the user to provide a preference on the Boolean lattice 2^{Ω}, however this can be impractical to specify explicitly.

An efficient way for specifying such a structure [19] and for guiding the search for the optimal solution consists in providing an utility function $u(Z|\mathcal{R})$, where $Z \in 2^{\Omega}$ is a candidate set and $\mathcal{R}$ represents the constraint. We can adopt without loss of generality the convention that this utility is expressed by a real number from 0 to 1.

When the constraint $\mathcal{R}$ is expressed under the form of an equality, i.e. $c(x') = Y'$, it is reasonable to set $u(Z|\mathcal{R}) = 1$ if $Z = Y'$, to set $u(Z|\mathcal{R}) = 0$ if $Z \cap Y' = \emptyset$ and to adopt intermediate values for every other $Z \in 2^{\Omega}$.

As a representative example, for subsets and supersets of Y' the utility can be defined by two decreasing functions of the distance $D(Z, Y')$ along the chain of nested sets (i.e. $\emptyset \subset \cdots \subset Y' \subset \cdots \subset \Omega$), measured by the difference of cardinalities: an upper chain decreasing function (whose value decreases as the set cardinality increases) and a lower chain decreasing function (whose value decreases as the set cardinality decreases). A precision avoiding user will adopt a lower chain decreasing function of the distance, whose values start from a lower value, w.r.t. the start values of the upper chain function.

It is clear that the possibilities for the definition of such utility function are many and that they are linked to the application context of the user.

Unconstrained Formulation of the Optimization Problem. We have now the elements for defining the optimization problem in an unconstrained form. We have the term $d(x, x')$ that has to be minimized, and the term $u(Z|\mathcal{R})$ that has to be maximized; we can also introduce a weight w which expresses the relative importance of $d(x, x')$ with respect to $u(Z|\mathcal{R})$.

Thus the problem can be expressed as

$$\min_{x' \in X} [\, w\, d(x, x') - u\left(c(x')|\mathcal{R}\right) \,].$$

4   Conclusions and Future Works

In the classification task the main cause of prediction error is represented by the data that are uncertain to classify. Allowing to express the uncertainty in the output of the classifier can support a more accurate inspection of those data in a later time, and prompt for the elicitation of further knowledge by domain experts. This is especially useful when, as for instance, in the context of health care diagnosis, a delayed decision costs less than an error. A way for expressing that uncertainty consists in letting the classifier provide a set of alternative classifications, instead of a single one (the outcomes of the classifier bear the semantics of the epistemic sets, one of the alternative is the true one, we are uncertain about which one). This is the rationale behind partial, i.e. set-based, classification.

Often is useful to try to understand the results of a classification. Whereas factual explanations focus of why an instance x' has been classified in a class, counterfactual explanations focus on what minimal changes to the input would have brought the classifier to opt for a specific alternative class. Finding such a point x' subject to a number of constraints defines the task of counterfactual generation.

Set-based outcomes, in partial classification, open the possibility to set-based factual and counterfactual generation.

In this work, we discussed the extension of the counterfactual generation problem to *set-based counterfactuals*, in the context of partial, i.e. set-based, classification. While two singleton-classes can simply either coincide or be disjoint, non-singleton classes allow for a wider spectrum of possible relationships. As a consequence: the specification of the goals of counterfactual generation is more varied, and the solution is not granted to exist, hence one has to include the degree of partial fulfillment of the constraints into the loss function of the optimization problem. In this work we discussed various possibilities for the formalization of those concepts.

We formulated the set-based counterfactual generation task and provided a non-constrained form for the corresponding optimization problem.

As to the expression of the uncertainty of classifier results, beyond partial-set based classification, there are several methods. Yuan et al. [27] propose to go further and represent the uncertain output of a classifier using belief functions, within the fame of the Dempster-Shafer Theory, which includes Probability Theory and Possibility Theory as special cases (for a relevant review see for instance [9]). We plan to discuss the counterfactual generation problem in the context of evidential output classifiers.

We plan also to extend the formalization to imprecise input data, where also x and x' are specified up to some degree of epistemic uncertainty, e.g. in the form of evidential input data.

References

1. Adadi, A., Berrada, M.: Peeking inside the black-box: a survey on explainable artificial intelligence (XAI). IEEE Access 6, 52138–52160 (2018)
2. Anisetti, M., Ardagna, C.A., Bellandi, V., Damiani, E., Reale, S.: Map-based location and tracking in multipath outdoor mobile networks. IEEE Trans. Wireless Commun. 10(3), 814–824 (2011)
3. Ardagna, C.A., Cremonini, M., Gianini, G.: Landscape-aware location-privacy protection in location-based services. J. Syst. Architect. 55(4), 243–254 (2009)
4. Ardagna, C.A., Cremonini, M., Vimercati, S.D.C., Samarati, P.: Privacy-enhanced location-based access control. In: Gertz, M., Jajodia, S. (eds.) Handbook of Database Security, pp. 531–552. Springer, US, Boston, MA (2008). https://doi.org/10.1007/978-0-387-48533-1_22
5. Ardagna, C.A., Livraga, G., Samarati, P.: Protecting privacy of user information in continuous location-based services, pp. 162–169 (2012)
6. Arrieta, A.B., et al.: Explainable artificial intelligence (XAI): concepts, taxonomies, opportunities and challenges toward responsible AI. Inf. Fusion, 58, 82–115 (2020)
7. Carlini, N., Wagner, D.: Towards evaluating the robustness of neural networks. In: 2017 IEEE Symposium on Security and Privacy (SP), pp. 39–57. IEEE (2017)
8. Couso, I., Dubois, D.: Statistical reasoning with set-valued information: ontic vs. epistemic views. Int. J. Approximate Reasoning, 55(7), 1502–1518 (2014)
9. Denœux, T., Dubois, D., Prade, H.: Representations of uncertainty in AI: beyond probability and possibility. In: Marquis, P., Papini, O., Prade, H. (eds.) A Guided Tour of Artificial Intelligence Research, pp. 119–150. Springer, Cham (2020). https://doi.org/10.1007/978-3-030-06164-7_4
10. Dhurandhar, A., et al.: Explanations based on the missing: towards contrastive explanations with pertinent negatives. Adv. Neural Inf. Process. Syst. 31 (2018)
11. Gianini, G., Fossi, L.G., Mio, C., Caelen, O., Brunie, L., Damiani, E.: Managing a pool of rules for credit card fraud detection by a game theory based approach. Future Generation Comput. Syst. 102, 549–561 (2020)
12. Guidotti, R., Monreale, A., Pedreschi, D., Giannotti, F.: Principles of explainable artificial intelligence. In: Sayed-Mouchaweh, M. (ed.) Explainable AI Within the Digital Transformation and Cyber Physical Systems, pp. 9–31. Springer, Cham (2021). https://doi.org/10.1007/978-3-030-76409-8_2

13. Gunning, D., Aha, D.: Darpa's explainable artificial intelligence (XAI) program. AI Mag. **40**(2), 44–58 (2019)
14. Gunning, D., Stefik, M., Choi, J., Miller, T., Stumpf, S., Yang, G.Z.: XAI-explainable artificial intelligence. Sci. Robot. **4**(37), 7120 (2019)
15. Joshi, S., Koyejo, O., Vijitbenjaronk, W., Kim, B., Ghosh, J.: Towards realistic individual recourse and actionable explanations in black-box decision making systems. arXiv preprint arXiv:1907.09615 (2019)
16. Karimi, A.-H., Von Kügelgen, J., Schölkopf, B., Valera, I.: Algorithmic recourse under imperfect causal knowledge: a probabilistic approach. Adv. Neural. Inf. Process. Syst. **33**, 265–277 (2020)
17. Laugel, T., Lesot, M.-J., Marsala, C., Renard, X., Detyniecki, M.: Comparison-based inverse classification for interpretability in machine learning. In: Medina, J., Ojeda-Aciego, M., Verdegay, J.L., Pelta, D.A., Cabrera, I.P., Bouchon-Meunier, B., Yager, R.R. (eds.) IPMU 2018. CCIS, vol. 853, pp. 100–111. Springer, Cham (2018). https://doi.org/10.1007/978-3-319-91473-2_9
18. Lundberg, S.M., Lee, S.I.: A unified approach to interpreting model predictions. Adv. Neural Inf. Process. Syst. **30** (2017)
19. Ma, L., Denoeux, T.: Partial classification in the belief function framework. Knowl.-Based Syst. **214**, 106742 (2021)
20. Mothilal, R.K., Sharma, A., Tan, C.: Explaining machine learning classifiers through diverse counterfactual explanations. In: Proceedings of the 2020 Conference on Fairness, Accountability, and Transparency, pp. 607–617 (2020)
21. Pawelczyk, M., Broelemann, K., Kasneci, G.: Learning model-agnostic counterfactual explanations for tabular data. In: Proceedings of The Web Conference, vol. 2020, pp. 3126–3132 (2020)
22. Ramakrishnan, G., Lee, Y.C., Albarghouthi, A.: Synthesizing action sequences for modifying model decisions. In: Proceedings of the AAAI Conference on Artificial Intelligence, vol. 34, pp. 5462–5469 (2020)
23. Rudin, C., Chen, C., Chen, Z., Huang, H., Semenova, L., Zhong, C.: Interpretable machine learning: fundamental principles and 10 grand challenges. Stat. Surv. **16**, 1–85 (2022)
24. Stier, J., Gianini, G., Granitzer, M., Ziegler, K.: Analysing neural network topologies: a game theoretic approach. Procedia Comput. Sci. **126**, 234–243 (2018)
25. Verma, S., Dickerson, J., Hines, K.: Counterfactual explanations for machine learning: a review. arXiv preprint arXiv:2010.10596 (2020)
26. Wachter, S., Mittelstadt, B., Russell, C.: Counterfactual explanations without opening the black box: automated decisions and the GDPR. Harv. JL Tech. **31**, 841 (2017)
27. Yuan, B., Yue, X., Lv, Y., Denoeux, T.: Evidential deep neural networks for uncertain data classification. In: Li, G., Shen, H.T., Yuan, Y., Wang, X., Liu, H., Zhao, X. (eds.) KSEM 2020. LNCS (LNAI), vol. 12275, pp. 427–437. Springer, Cham (2020). https://doi.org/10.1007/978-3-030-55393-7_38

Logic Operators and Sibling Aggregators for Z-grades

Guy De Tré[1]([✉]) [iD], Milan Peelman[1] [iD], and Jozo Dujmović[2] [iD]

[1] Department of Telecommunications and Information Processing, Ghent University, St.-Pietersnieuwstraat 41, B9000 Ghent, Belgium
{Guy.DeTre,Milan.Peelman}@UGent.be
[2] Department of Computer Science, San Francisco State University, San Francisco, CA 94132, USA
jozo@sfsu.edu

Abstract. Trust in data is a crucial aspect of criterion-based flexible query answering and decision making. Inspired by Zadeh's concept Z-number, we introduce the concept of a Z-grade and focus on some elementary aspects of aggregating Z-grades. A Z-grade, z, has two components, $z = (s, c)$. The first component, s, is a satisfaction grade that can for example be used to express to what extent a given data element satisfies a given criterion. The second component, c, is a confidence grade that expresses how confident we can be about s. For example, in case we have less trust in the data element, this could result in a lower confidence in the outcome s of the criterion evaluation. Logical processing and aggregation of satisfaction grades are important aspects of criterion handling. When applied to Z-grades, the computation of the resulting confidence grade depends on the computation of the resulting satisfaction grade. For that purpose, novel logic operators and so-called sibling aggregators for Z-grades are proposed and studied in the paper.

Keywords: Confidence · Veracity · Fuzzy logic · Uncertain computing · Aggregation · Explainability

1 Introduction

In 2011, Zadeh [25] introduced the notion of a Z-number in order to adequately cope with the issue of reliability of (fuzzy) information. A Z-number Z is defined by an ordered pair of fuzzy numbers, i.e. $Z = (A, B)$. The first fuzzy number A acts as a fuzzy restriction $R(X)$ on the values a linguistic variable X can take, written as $X\ IS\ A$ [23]. Here, A is a fuzzy subset of the domain of X playing the role of a possibility distribution, i.e. $R(X) : X\ IS\ A \rightarrow Poss(X = u) = \mu_A(u)$ where μ_A is the membership function of A and u is a generic domain value of X. The second fuzzy number B reflects the certainty, confidence, trust, or strength of belief in $X\ IS\ A$. Here, B is a fuzzy subset of the unit interval $[0, 1]$ modelling a fuzzy restriction on the probability of $X\ IS\ A$, i.e., $Prob(X\ IS\ A)\ IS\ B$.

© Springer Nature Switzerland AG 2022
D. Ciucci et al. (Eds.): IPMU 2022, CCIS 1602, pp. 572–583, 2022.
https://doi.org/10.1007/978-3-031-08974-9_46

Trust in data sources is a pervasive phenomenon in information management and processing [6], which is commonly known as the data veracity problem [4,17]. As veracity propagates from the data sources, through the data processing steps, to the computational outputs it is of utmost importance to properly model and handle it. This is a challenge that is hard to meet. Z-numbers offer a general mathematical framework to model confidence in data, independent on whether these data are uncertain or not. However, computation with Z-numbers is generally known to be quite complex and calculation-intensive [7,18]. Simple forms of Z-numbers, where both components are presented by single numbers have already been studied in, among others, [2,16].

In this work we contribute to the study of the impact of veracity on criteria evaluation and aggregation. Criteria handling is an important component in, among others, criterion-based flexible query answering and decision support systems. In conventional flexible criterion evaluation based on crisply described data, criterion satisfaction is often modelled by so-called satisfaction grades. Here, each satisfaction grade is a number in the unit interval $I = [0,1]$ [13]. In this work, we explore what we call Z-grades, which can be used to model criterion satisfaction when having to handle veracity. A Z-grade $z = (s,c)$ is a simple form of a Z-number, consisting of a satisfaction grade s and its associated confidence grade c, expressing the strength of trust in s. It is important to denote that both components of a Z-grade are interpreted as grades and hence should be processed using a logic model. The scientific contribution of this paper is the study of novel graded logic operators and aggregators for Z-grades that are specifically designed to support reasoning in query handling and decision support. Because such operators and aggregators have to act on both components of a Z-grade and the computation of a resulting confidence grade strongly depends on the computation of a resulting satisfaction grade, we name these operators sibling operators.

Working with Z-grades is one way to handle veracity in criterion processing. Other approaches exist, for example based on intervals [15], type-2 fuzzy sets [5], or R-sets [19]. These mathematical tools can be used to model uncertainty on a satisfaction grade by specifying a kind of upper and lower bound for it. This uncertainty can express veracity. In [22], Yager studied the problem of determining criterion satisfaction in cases where the value under evaluation is uncertain and proposed two approaches: one based on containment and one based on possibility. In [12], skyline querying is extended to cope with data quality. A skyline query aims to filter database records by keeping only those that are not worse than any other. In the study it is assumed that each attribute of a database record can have an associated (known) quality level. Skyline query handling then copes with these quality levels when checking for record dominance in answer set construction. Our preference for using a Z-grade based graded logic approach is motivated by the use of two different components to explicitly distinguish between the modelling of the restrictions on a data value and the modelling of the confidence in these restrictions. This is of pivotal importance in view of (better) interpretable and explainable criterion evaluation and handling, which in its turn is relevant and important in many computational intelligence and explainable artificial intelligence applications [3, 9, 10].

The paper is organized as follows. In Sect. 2 we introduce the notion of a Z-grade and discuss its semantics. Some logic operators for negation, conjunction and disjunction of Z-grades are presented in Sect. 3. In Sect. 4 some sibling aggregators for Z-grades are proposed and discussed. Finally, in Sect. 5 we deal with the conclusions of our work and give some proposals for future research.

2   Z-grades

As we have mentioned in Sect. 1, driven by his ambition to model different manifestations of uncertainty in a general way, Zadeh [24,25] proposed the notion of a Z-number Z as an ordered pair of fuzzy numbers (A, B). The first component A is the representation of a (numerical) value that might be uncertain and is resulting from an observation, computation or other information gathering process to obtain the value of a (linguistic) variable X. Therefore, A is a fuzzy subset of the universe of discourse U of X. The second component B is used to represent the certainty, trust, strength of belief, or reliability one has in A and is therefore a fuzzy subset of the unit interval $I = [0, 1]$. In general Z-numbers lead to complex computations (see, e.g., [1]).

In [7], Dubois and Prade studied some simplified cases of Z-numbers (A, B) where one of the components is crisp and the other is fuzzy. In [18] mixed-discrete Z-numbers are introduced. Driven by observed needs in criteria handling in flexible database querying and multi-criteria decision support [10], we study in this work the simplest form of Z-numbers, which we call Z-grades.

A *Z-grade* z is a Z-number where both components are elements of the unit interval I, and interpreted as grades that will be further processed using gradual logic, i.e.

$$z = (s, c), \text{where } s, c \in I. \tag{1}$$

Each Z-grade is a Z-number because both crisp numbers $s, c \in I$ are special cases of fuzzy numbers, i.e. both can be written as resp. $s = \{(s, 1)\}$ and $c = \{(c, 1)\}$. The first number s of a Z-grade is interpreted as a *satisfaction grade* (or *suitability grade*). The satisfaction grade s is a number form the unit interval with 0 denoting no satisfaction, 1 denoting full satisfaction and all other values denoting partial satisfaction. The second number c reflects the confidence in s and is called a *confidence grade*. As such, c expresses in a simple, gradual way the trust one can have in the associated satisfaction grade. Here, 0 denotes no trust at all, 1 denotes full trust and all other values denote partial trust. For example, when using Z-grades for modelling criterion evaluation results, a Z-grade $(1, 0.4)$ reflects that a given criterion is fully satisfied, but the confidence in the evaluation result is only 0.4, which might for example be caused by the fact that the evaluated data are not fully truthful.

Let $\mathbb{ZG}$ denote the set of all Z-grades, i.e.

$$\mathbb{ZG} = I^2. \tag{2}$$

Z-grades can be totally ordered in a lexicographic way, e.g. giving preference to the satisfaction grades s by

$$(s,c) \leq_s (s',c') \text{ if and only if } s < s' \text{ or } (s = s' \text{ and } c \leq c'), \tag{3}$$

or giving preference to the confidence grades c by

$$(s,c) \leq_c (s',c') \text{ if and only if } c < c' \text{ or } (c = c' \text{ and } s \leq s'). \tag{4}$$

Other orderings are possible.

Our motivation to study Z-grades is both practically and theoretically inspired. From a practical point of view, Z-grades offer great opportunities to model criterion satisfaction in case of evaluations that are based on bad data, which is a pervasive issue when having to deal with veracity in (big) data [4,17]. Indeed, a confidence grade c is a simple way to express the trust we have in the satisfaction grade s, resulting from criterion evaluation or aggregation. From a theoretical point of view it is interesting to study logic operators and aggregators for Z-grades. Moreover, with such operators, it is important to study what kind of impact the computation of the resulting s component should have on the computation of the resulting c component.

As an illustration of the practical applicability of Z-grades consider the Disbiome database [11] that aims to gain new insights on links between microbial composition changes and different disease types. The database collects published microbiota-disease information including the publication (Publication ID), showing a microbial (Organism ID) difference between a sample (Sample ID) of a patient (Disease ID, Host ID) and a control subject (Control ID) using a specific detection method (Method ID). The microbial difference can be presented by absolute quantities between the patient and control or by a ratio. Different detection methods exist. Not all of these can be trusted to the same extend [11]. Currently biomedical experts assess trust in a given detection method by assigning it a percentage that can be translated to a confidence grade. (Fuzzy) queries based on microbial differences should explicitly cope with these confidence grades: each element in the query answer set should be assigned a satisfaction grade (s), expressing how good the element satisfies the query conditions, and a confidence grade (c) expressing how much trust one can have in that element. In this way, e.g., small microbial differences resulting from less trusted detection methods can be meaningfully distinguished from small microbial differences resulting from more trusted detection methods. Using Z-grades $z = (s,c)$ would give users better insight on the interpretation of query results than conventional fuzzy querying would allow.

A further study and implementation to apply Z-grades in the Disbiome database is planned. In the next sections we focus on studying the theoretical aspects of graded logic operators and aggregators for Z-grades.

3   Logic Operators

In this section we propose negation, conjunction and disjunction operators for Z-grades. Special attention goes to studying how the impact of the satisfaction grades of the operands on the computation of the resulting satisfaction grade, influences the computation of the resulting reliability grade.

3.1 Negation Operator

The unary *(sibling) negation operator* $\neg$ for Z-grades is defined by

$$\forall (s, c) \in \mathbb{ZG} : \neg(s, c) = (\sim s, c) \tag{5}$$

where $\sim$ denotes a strong negation, i.e. a fuzzy negation that satisfies

$$\forall s \in [0, 1] : \sim\sim s = s.$$

With this definition it is reflected that the negation of a satisfaction grade has no impact on its associated confidence grade. The double negation property holds for $\neg$, i.e.

$$\neg\neg(s, c) = \neg(\sim s, c) = (\sim\sim s, c) = (s, c).$$

The standard negation defined by $\forall s \in [0, 1] : \sim s = 1 - s$ is an example of a strong fuzzy negation.

3.2 Conjunction Operators

For the conjunction of two Z-grades we use t-norms [14]. A binary *(sibling) conjunction operator* for Z-grades based on a t-norm $\top$ is defined by

$$\forall (s_1, c_1), (s_2, c_2) \in \mathbb{ZG} :$$

$$\top((s_1, c_1), (s_2, c_2)) = (\top(s_1, s_2), a^{conj}((s_1, c_1), (s_2, c_2))) \tag{6}$$

where a^{conj} is a confidence aggregator for conjunction that can be defined by

$$a^{conj}((s_1, c_1), (s_2, c_2)) = \begin{cases} \min(c_1, c_2) & \text{if } s_1 \neq 0 \text{ and } s_2 \neq 0 \\ \max_{i \in \{i | s_i = 0\}} c_i & \text{otherwise.} \end{cases} \tag{7}$$

Hence, the satisfaction grades s_1 and s_2 are aggregated using t-norm $\top$, while an aggregator a^{conj} is used for the confidence grades c_1 and c_2. With the given definition of a^{conj} a rather pessimistic strategy is followed. If none of the satisfaction grades of the operands equals 0 (0 denoting the truth value 'false', which is the absorbing element for conjunction), the resulting confidence grade is considered to be the lowest of the confidence grades of the operands. In these cases, a higher confidence grade in the other operand will not improve the confidence in the result of the conjunction. However, if at least one of the satisfaction grades of the operands equals 0, the t-norm $\top(s_1, s_2)$ will be 0 and a non-zero satisfaction degree in the other operand will not influence the result of the conjunction. Hence, the resulting confidence grade $a^{conj}((s_1, c_1), (s_2, c_2))$ is considered to be the highest confidence grade that occurs in an operand with satisfaction grade 0. A lower confidence grade in the other operand has in that case no impact on the confidence in the result of the conjunction.

Other, more optimistic definitions for the confidence aggregator a^{conj} are possible. These are subject for further research.

3.3 Disjunction Operators

Dually to the definition of conjunction operators, we propose to base the definition of a binary *(sibling) disjunction operator* for Z-grades on a t-conorm $\bot$, i.e.

$$\forall (s_1, c_1), (s_2, c_2) \in \mathbb{ZG} :$$

$$\bot((s_1, c_1), (s_2, c_2)) = (\bot(s_1, s_2), a^{disj}((s_1, c_1), (s_2, c_2))) \quad (8)$$

where a^{disj} is a confidence aggregator for disjunction that can be defined by

$$a^{disj}((s_1, c_1), (s_2, c_2)) = \begin{cases} \min(c_1, c_2) & \text{if } s_1 \neq 1 \text{ and } s_2 \neq 1 \\ \max_{i \in \{i | s_i = 1\}} c_i & \text{otherwise.} \end{cases} \quad (9)$$

The disjunction of the satisfaction grades s_1 and s_2 is computed using t-conorm $\bot$ and an aggregator a^{disj} is used for the confidence grades c_1 and c_2. The definition of a^{disj} is similar to the definition of a^{conj}. If neither s_1, nor s_2 equals 1 (1 denoting 'true', which is the absorbing element for disjunction), the resulting confidence grade is considered to be the minimum of c_1 and c_2 what again reflects a rather pessimistic strategy. A higher confidence grade in the other operand will have no impact on the confidence in the result of the disjunction. In case at least one of the satisfaction grades s_1 and s_2 equals 1, the t-conorm $\bot(s_1, s_2)$ will be 1, independent of an eventual lower satisfaction grade in the other operand. For that reason, the resulting confidence grade $a^{disj}((s_1, c_1), (s_2, c_2))$ is considered to be the highest confidence grade that occurs in an operand with satisfaction grade 1. A lower confidence grade in the other operand has in that case no impact on the confidence in the result of the disjunction.

Other, more optimistic definitions for the confidence aggregator a^{disj} are possible. These are subject for further research.

3.4 De Morgan's Laws

It is generally known that for any t-norm $\top$ and strong negation $\sim$ a corresponding t-conorm $\bot$ can be defined by

$$\forall s_1, s_2 \in [0, 1] : \bot(s_1, s_2) = \sim \top(\sim s_1, \sim s_2). \quad (10)$$

Likewise,

$$\forall s_1, s_2 \in [0, 1] : \top(s_1, s_2) = \sim \bot(\sim s_1, \sim s_2). \quad (11)$$

If the standard negation $\sim (s) = 1 - s$ is used, $\bot$ is called a dual t-conorm to $\top$ and reversely, $\top$ is called a dual t-norm to $\bot$. Equation (10) and Eq. (11) express the fuzzy De Morgan's laws.

Theorem 1. *If $\top$, $\bot$ and $\neg$ are resp. defined as in Eq. (6), Eq. (8) and Eq. (5) and if $\top$ and $\bot$ are dual operators, then the triplet $(\top, \bot, \neg)$ satisfies*

$$\forall z_1, z_2 \in \mathbb{ZG} : \top(z_1, z_2) = \neg\bot(\neg z_1, \neg z_2).$$

Proof. Let $z_1 = (s_1, c_1)$ and $z_2 = (s_2, c_2)$, then

$$\neg\bot(\neg(s_1, c_1), \neg(s_2, c_2)) = \neg\bot((1 - s_1, c_1), (1 - s_2, c_2))$$
$$= \neg(\bot(1 - s_1, 1 - s_2), a^{disj}((1 - s_1, c_1), (1 - s_2, c_2)))$$

where

$$a^{disj}((1 - s_1, c_1), (1 - s_2, c_2))$$
$$= \begin{cases} \min(c_1, c_2) & \text{if } 1 - s_1 \neq 1 \text{ and } 1 - s_2 \neq 1 \\ \max_{i \in \{i | 1 - s_i = 1\}} c_i & \text{otherwise} \end{cases}$$
$$= \begin{cases} \min(c_1, c_2) & \text{if } s_1 \neq 0 \text{ and } s_2 \neq 0 \\ \max_{i \in \{i | s_i = 0\}} c_i & \text{otherwise} \end{cases}$$
$$= a^{conj}((s_1, c_1), (s_2, c_2))$$

So,

$$\neg\bot(\neg(s_1, c_1), \neg(s_2, c_2)) = \neg(\bot(1 - s_1, 1 - s_2), a^{disj}((1 - s_1, c_1), (1 - s_2, c_2)))$$
$$= (\neg\bot(1 - s_1, 1 - s_2), a^{disj}((1 - s_1, c_1), (1 - s_2, c_2)))$$
$$= (\top(s_1, s_2), a^{conj}((s_1, c_1), (s_2, c_2)))$$
$$= \top((s_1, c_1), (s_2, c_2))$$

Q.E.D.

4 Sibling Aggregators

The conjunction and disjunction operators introduced in the previous section are examples of binary aggregators that map $\mathbb{ZG}^2$ onto $\mathbb{ZG}$. In this section we study aggregators h of arity n, $n \geq 2$ that map $\mathbb{ZG}^n$ onto $\mathbb{ZG}$. Such aggregators are, among others, important in advanced querying and multi-criteria decision making [10,20]. In the remainder of this section we study some basic n-ary aggregators h.

4.1 Weighted Mean

Consider a weight vector $\vec{w} = (w_1, \ldots, w_n) \in [0, 1]^n$, such that $\sum_{i=1}^n w_i = 1$, reflecting the relative importance of the n operands. The n-ary *(sibling) weighted mean* for Z-grades is defined by

$$h_{\vec{w}} : \mathbb{ZG}^n \to \mathbb{ZG} : ((s_1, c_1), \ldots, (s_n, c_n)) \mapsto (s, c) \tag{12}$$

where $s = \sum_{i=1}^n w_i \cdot s_i$ and $c = \sum_{i=1}^n w_i \cdot c_i$.

The satisfaction grades s_i, as well as the confidence grades c_i, $i = 1, \ldots, n$ of the operands are each aggregated by the fuzzy weighted mean. The choice for also using weighted mean for aggregating the confidence grades is motivated as follows. The resulting confidence grade c should reflect the trust one can have in the weighted mean of the satisfaction grades s_i, taking into account the trust one has in the satisfaction grades s_i of the operands and considering the importance w_i of each satisfaction grade s_i. The assumption is that each confidence grade c_i should have a similar impact in the computation of the overall confidence grade c as its corresponding 'sibling' satisfaction grade s_i has in the computation of the overall satisfaction grade s. The impact of s_i in the computation of s is determined by w_i, hence the proposal for using w_i to model the impact of c_i in the computation of c.

4.2 Ordered Weighted Average

For a given weight vector $\vec{w} = (w_1, \ldots, w_n) \in [0,1]^n$ with $\sum_{i=1}^n w_i = 1$, the corresponding n-ary ordered weighted averaging (OWA) operator $h_{\vec{w}}^{OWA}$ is defined by

$$h_{\vec{w}}^{OWA} : [0,1]^n \to [0,1] : (s_1, \ldots, s_n) \mapsto \sum_{i=1}^{n} w_i \cdot s_{\rho(i)} \tag{13}$$

where $\rho : \{1, \ldots, n\} \to \{1, \ldots, n\}$ is a permutation on the index set satisfying $s_{\rho(1)} \geq s_{\rho(2)} \geq \cdots \geq s_{\rho(n)}$. So, the largest value of $(s_1, \ldots, s_n)$ is multiplied by the first weight w_1, the second largest to the second weight, and so on [21]. Based on this definition, the n-ary *(sibling) ordered weighted averaging aggregator* for Z-grades is defined by

$$h_{\vec{w}}^{OWA} : \mathbb{ZG}^n \to \mathbb{ZG} : ((s_1, c_1), \ldots, (s_n, c_n)) \mapsto (s, c) \tag{14}$$

where $s = \sum_{i=1}^n w_i \cdot s_{\rho(i)}$ and $c = \sum_{i=1}^n w_i \cdot c_{\rho(i)}$. With this definition each satisfaction grade $s_{\rho(i)}$ is assigned a weight w_i as in the regular OWA aggregator and its corresponding 'sibling' confidence grade $c_{\rho(i)}$ is assigned the same weight w_i. As such, $c_{\rho(i)}$ has the same impact in the computation of the overall confidence grade c as its corresponding 'sibling' satisfaction grade $s_{\rho(i)}$ has in the computation of the overall satisfaction grade s.

4.3 Generalized Conjunction Disjunction Aggregators

As last kind of aggregators we consider Generalized Conjunction Disjunction (GCD) aggregators. A GCD aggregator takes n satisfaction grades s_i, $i = 1, \ldots, n$ as input and is parameterized by (i) a weight vector $\vec{w} = (w_1, \ldots, w_n) \in [0,1]^n$ with $\sum_{i=1}^n w_i = 1$, reflecting the relative importance of the n inputs and (ii) a global andness α, $0 \leq \alpha \leq 1$ determining the logical behaviour of the aggregator. If $\alpha = 1$, then the aggregator behaves like a full conjunction ('and'), whereas for $\alpha = 0$ a full disjunction ('or') is obtained. With the intermediate parameter values, a continuous transition from full conjunction to full disjunction is modelled [8]. Based on α, global orness ω is defined by $\omega = 1 - \alpha$.

A GCD aggregator can be defined in a variety of ways. In a simple special case the GCD aggregator can be implemented as a weighted power mean (WPM), where the exponent $q \in [-\infty, +\infty]$ is a function of the desired andness α:

$$h^{gcd}_{(\vec{w};q)} : [0,1]^n \to [0,1] : (s_1, \ldots, s_n) \mapsto \lim_{p \to q} (w_1 \cdot s_1^p + \cdots + w_n \cdot s_n^p)^{(1/p)}$$

$$= \begin{cases} \min(s_1, \ldots, s_n) & \text{if } q = -\infty, \alpha = 1, \omega = 0; \\ (w_1 \cdot s_1^q + \cdots + w_n \cdot s_n^q)^{(1/q)} & \text{if } -\infty < q < 1, 0 < \omega < 1/2 < \alpha < 1; \\ w_1 \cdot s_1 + \cdots + w_n \cdot s_n & \text{if } q = 1, \alpha = \omega = 1/2; \\ (w_1 \cdot s_1^q + \cdots + w_n \cdot s_n^q)^{(1/q)} & \text{if } 1 < q < +\infty, 0 < \alpha < 1/2 < \omega < 1; \\ \max(s_1, \ldots, s_n) & \text{if } q = +\infty, \alpha = 0, \omega = 1. \end{cases}$$

$$(15)$$

The five cases in Eq. (15) are resp. called (from top to bottom) pure conjunction, partial conjunction, neutrality, partial disjunction and pure disjunction.

For the definition of (sibling) GCD aggregators for Z-grades, it is again assumed that the confidence grades c_i, $i = 1, \ldots, n$ should be aggregated in such a way that each confidence grade has a similar impact in the computation of the aggregated confidence grade, as its corresponding 'sibling' satisfaction grade s_i has in the computation of the aggregated satisfaction grade. However, due to the fact that the logical behaviour of a GCD aggregator to a large extend depends on its andness parameter α, determining the impact of each satisfaction grade s_i, $i = 1, \ldots, n$ in the computation of the aggregated satisfaction grade is not straightforward and should be estimated.

A possible approach is to approximate impact by the 'total range' [10]. For a given GCD aggregator $h^{gcd}_{(\vec{w};q)}$, the impact Δ_i of a satisfaction grade s_i is then determined by

$$\Delta_i = h^{gcd}_{(\vec{w};q)}(s_1, \ldots, s_{i-1}, 1, s_{i+1}, \ldots, s_n)$$
$$- h^{gcd}_{(\vec{w};q)}(s_1, \ldots, s_{i-1}, 0, s_{i+1}, \ldots, s_n).$$

$$(16)$$

So, the 'total range' impact Δ_i of s_i is considered to be the difference between the overall satisfaction grade that is obtained when input s_i is replaced by the maximal satisfaction 1 and the overall satisfaction grade that is obtained when input s_i is replaced by the minimal satisfaction 0.

Using the 'total range' approximation for impact and considering a weight vector $\vec{w} = (w_1, \ldots, w_n) \in [0,1]^n$ with $\sum_{i=1}^n w_i = 1$ and behaviour parameter $q \in [-\infty, +\infty]$, the WPM implementation of an n-ary (sibling) ordered weighted averaging aggregator for Z-grades is defined by

$$h^{gcd}_{(\vec{w};q)} : \mathbb{ZG}^n \to \mathbb{ZG} : ((s_1, c_1), \ldots, (s_n, c_n)) \mapsto (s, c) \qquad (17)$$

where s is obtained by applying Eq. (15), i.e., $s = h^{gcd}_{(\vec{w};q)}(s_1, \ldots, s_n)$ and c is computed from $c_1, \ldots, c_n$ in such a way that each confidence grade c_i, $i = 1, \ldots, n$

has the same impact Δ_i in the computation of c as its corresponding 'sibling' satisfaction grade s_i has in the computation of s, which is obtained by the weighted mean

$$c = \begin{cases} \dfrac{\sum_{i=1}^n \Delta_i \cdot c_i}{\sum_{i=1}^n \Delta_i} & \text{if } \sum_{i=1}^n \Delta_i > 0 \\ 1 & \text{otherwise.} \end{cases} \tag{18}$$

In this and the previous section we introduced and studied some important operators for handling Z-grades. Such operators are required tools for developing applications and definitely require further study. Z-grades are a fundamental contributor to the explainability of evaluation results [9]. Explainability is a human-centric concept. This paper contributes to the infrastructure of the currently relevant and important area of explainability as it is studied in artificial and computational intelligence.

5 Conclusions and Future Work

In this paper we studied the simplest case of Z-numbers, which we call Z-grades. A Z-grade is a Z-number where both components are crisp numbers. The first component is interpreted as a satisfaction grade, whereas following Zadeh's semantics, the second component is considered to be a confidence grade denoting the certainty, trust, or strength of belief that one has with respect to the associated satisfaction grade. Together both components offer more information than conventional satisfaction (or suitability) grades do.

We introduced Z-grades and studied novel logic operators for negation, conjunction and disjunction. Moreover, we proposed novel basic sibling aggregation techniques for weighted mean, ordered weighted average and generalized conjunction disjunction aggregators. The term 'sibling' refers to the underlying assumption that Z-grades should be aggregated in such a way that each confidence grade has a similar impact in the computation of the aggregated confidence grade, as its corresponding sibling satisfaction grade has in the computation of the aggregated satisfaction grade.

When Z-grades are used to express and process criteria evaluation results, the satisfaction grade denotes how good a given (attribute) value satisfies a criterion, whereas the confidence grade expresses to what extent this criterion satisfaction can be trusted. This extra information is relevant in view of veracity handling in (big) data management applications and could also be useful in view of the development of better interpretable artificial intelligence applications.

Future work should be directed towards the further development of ordering functions and sibling aggregation techniques for Z-grades, their mathematical properties, their comparison and refinement, their experimental validation and their practical applicability and support with appropriate software tools.

Funding Information. This research received funding from the Flemish Government under the 'Onderzoeksprogramma Artificiële Intelligentie (AI) Vlaanderen' programme.

References

1. Aliev, R.A., Huseynov, O.H., Zeinalova, L.M.: The arithmetic of continuous Z-numbers. Inf. Sci. **373**(7), 441–460 (2016)
2. Aliev, R.A., Kreinovich, V.: Z-numbers and Type-2 fuzzy sets: a representation result. Intell. Autom. Soft Comput. **24**(1), 205–210 (2018)
3. Barredo Arrieta, A., et al.: Explainable Artificial Intelligence (XAI): concepts, taxonomies, opportunities and challenges toward responsible AI. Inf. Fusion **58**, 82–115 (2020)
4. Berti-Equille, L., Lamine Ba, M.: Veracity of big data: challenges of cross-modal truth discovery. ACM J. Data Inf. Quality, **7**(3), 3 (2016)
5. Biglarbegian, M., Melek, W.W., Mendel, J.M.: On the robustness of Type-1 and Interval Type-2 fuzzy logic systems in modeling. Inf. Sci. **181**(7), 1325–1347 (2011)
6. de Siqueira Braga, D., Niemann, M., Hellingrath, B., Buarque de Lima Neto, F.: Survey on computational trust and reputation models. ACM Comput. Surv. **51**(5), 101 (2018)
7. Dubois, D., Prade, H.: A fresh look at Z-numbers - relationships with belief functions and p-boxes. Fuzzy Inf. Eng. **10**(1), 5–18 (2018)
8. Dujmović, J.: Soft Computing Evaluation Logic: The LSP Decision Method and Its Applications. Wiley, New Jersey, USA (2018)
9. Dujmović, J.: Interpretability and explainability of LSP evaluation criteria. In: Proceeding of the 2020 IEEE International Conference on Fuzzy Systems (FUZZ-IEEE), pp. 1–8, Glasgow, UK (2020)
10. De Tré, G., Dujmović, J.J.: Dealing with data veracity in multiple criteria handling: an LSP-based sibling approach. In: Andreasen, T., De Tré, G., Kacprzyk, J., Legind Larsen, H., Bordogna, G., Zadrożny, S. (eds.) FQAS 2021. LNCS (LNAI), vol. 12871, pp. 82–96. Springer, Cham (2021). https://doi.org/10.1007/978-3-030-86967-0_7
11. Janssens, Y., et al.: Disbiome database: linking the microbiome to disease. BMC Microbiol. **18**(1), 1–6 (2018)
12. Jaudoin, H., Pivert, O., Smits, G., Thion, V.: Data-quality-aware skyline queries. In: Andreasen, T., Christiansen, H., Cubero, J.-C., Raś, Z.W. (eds.) ISMIS 2014. LNCS (LNAI), vol. 8502, pp. 530–535. Springer, Cham (2014). https://doi.org/10.1007/978-3-319-08326-1_56
13. Kacprzyk, J., Zadrozny, S., De Tré, G.: Fuzziness in database management systems: half a century of developments and future prospects. Fuzzy Sets Syst. **218**, 300–307 (2015)
14. Klement, E.P., Mesiar, R., Pap, E.: Triangular Norms. Kluwer, Dordrecht, The Netherlands (2000)
15. Kreinovich, V., Ouncharoen, R.: Fuzzy (and interval) techniques in the age of big data: an overview with applications to environmental science, geosciences, engineering, and medicine. Int. J. Uncertainty, Fuzziness Knowl.-Based Syst. **23**(suppl. 1), 75–89 (2015)
16. Kreinovich, V., Kosheleva, O., Zakharevich, M.: Z-numbers: how they describe student confidence and how they can explain (and improve) Laplacian and Schroedinger eigenmap dimension reduction in data analysis. In: Lesot, M.-J., Marsala, C. (eds.) Fuzzy Approaches for Soft Computing and Approximate Reasoning: Theories and Applications. SFSC, vol. 394, pp. 285–297. Springer, Cham (2021). https://doi.org/10.1007/978-3-030-54341-9_24

17. Lukoianova, T., Rubin, V.L.: Veracity roadmap: is big data objective, truthful and credible? Adv. Classification Res. Online **24**(1), 4–15 (2014)
18. Massanet, S., Riera, J.V., Torrens, J.: A new approach to Zadeh's Z-numbers: mixed-discrete Z-numbers. Inf. Fusion **53**, 35–42 (2020)
19. Seiti, H., Hafezalkotob, A., Martinez, L.: R-sets, comprehensive fuzzy sets risk modeling for risk-based information fusion and decision-making. IEEE Trans. Fuzzy Syst. **29**(2), 385–399 (2021)
20. Vaníček, J., Vrana, I., Aly, S.: Fuzzy aggregation and averaging for group decision making: a generalization and survey. Knowl.-Based Syst. **22**(1), 79–84 (2009)
21. Yager, R.R.: On ordered weighted averaging aggregation operators in multi-criteria decision making. IEEE Trans. Syst. Man Cybern. **18**, 183–190 (1988)
22. Yager, R.R.: Validating criteria with imprecise data in the case of trapezoidal representations. Soft. Comput. **15**, 601–612 (2011)
23. Zadeh, L.A.: Calculus of fuzzy restrictions. In: Zadeh, L.A., Fu, K.S., Tanaka, K., Shimura, M. (eds.) Fuzzy sets and Their Applications to Cognitive and Decision Processes, pp. 1–39. Academic Press, New York (1975)
24. Zadeh, L.A.: From imprecise to granular probabilities. Fuzzy Sets Syst. **154**(3), 370–374 (2005)
25. Zadeh, L.A.: A note on Z-numbers. Inf. Sci. **8**(3), 2923–2932 (2011)

Canonical Extensions of Conditional Probabilities and Compound Conditionals

Tommaso Flaminio[1], Angelo Gilio[2], Lluis Godo[1],
and Giuseppe Sanfilippo[3]($\boxtimes$)

[1] IIIA - CSIC, 08193 Bellaterra, Spain
{tommaso,godo}@iiia.csic.es
[2] Department of SBAI, University of Rome "La Sapienza", Rome, Italy
angelo.gilio@sbai.uniroma1.it
[3] Department of Mathematics and Computer Science, University of Palermo,
Palermo, Italy
giuseppe.sanfilippo@unipa.it

Abstract. In this paper we show that the probability of conjunctions and disjunctions of conditionals in a recently introduced framework of Boolean algebras of conditionals are in full agreement with the corresponding operations of conditionals as defined in the approach developed by two of the authors to conditionals as three-valued objects, with betting-based semantics, and specified as suitable random quantities. We do this by first proving that the canonical extension of a full conditional probability on a finite algebra of events to the corresponding algebra of conditionals is compatible with taking subalgebras of events.

Keywords: Boolean algebras of conditionals · Conditional probability · Conjunction and disjunction of conditionals

1 Introduction

Conditionals play a key role in different areas of logic, probabilistic reasoning and knowledge representation in AI, and they have been studied from many points of view, see, e.g., [1,4,5,7,11,20,21,26–28]. In particular, a three-valued calculus of conditional objects has been given in [8], where a simple semantics for the preferential entailment studied in [10,23,24] has been provided. Further extensions, from the artificial intelligence perspective, have been given, for instance, in [2,3,22].

In a recent paper [9], an algebraic setting for measure-free conditionals has been put forward. More precisely, given a finite Boolean algebra $\mathbf{A} = (\mathbb{A}, \wedge, \vee, \bar{\ }, \bot, \top)$ of events, the authors build another (much bigger but still finite) Boolean algebra $\mathcal{C}(\mathbf{A})$ where *basic conditionals*, i.e. objects of the form $(A|B)$ for

---

A. Gilio—Retired.

© Springer Nature Switzerland AG 2022
D. Ciucci et al. (Eds.): IPMU 2022, CCIS 1602, pp. 584–597, 2022.
https://doi.org/10.1007/978-3-031-08974-9_47

$A \in \mathbb{A}$ and $B \in \mathbb{A}' = \mathbb{A}\backslash\{\bot\}$, can be freely combined with the usual Boolean operations, yielding compound conditional objects, while they are required to satisfy a set of natural properties. Moreover, the set of atoms of $\mathcal{C}(\mathbf{A})$ are fully identified and it is shown they are in a one-to-one correspondence with sequences of pairwise different atoms of $\mathbf{A}$ of maximal length. Finally, it is also shown that any positive probability P on the set of events from $\mathbf{A}$ can be *canonically* extended to a probability μ_P on the algebra of conditionals $\mathcal{C}(\mathbf{A})$ in such a way that the probability $\mu_P(a|b)$ of a basic conditional coincides with the conditional probability $P(a|b) = P(a \wedge b)/P(b)$. This is done by suitably defining the probability of each atom of $\mathcal{C}(\mathbf{A})$ as a certain product of conditional probabilities.

However, we remark that in [9] explicit definitions of conjunction and disjunction of conditionals are not explicitly given. Rather, any compound conditional comes determined by the disjunction of those atoms in $\mathcal{C}(\mathbf{A})$ that lie below it. Similarly, the probability of any compound conditional is computed as the sum of the probabilities of the atoms below the conditional. But no operational and systematic procedure to do these computations avoiding a combinatorial explosion is provided in [9].

In this paper, after this introduction and some preliminaries in Sect. 2, we will first show that the positivity assumption for the probability on $\mathbf{A}$, needed for the canonical extension to the algebra of conditionals $\mathcal{C}(\mathbf{A})$, can be lifted by starting from a conditional probability (in the axiomatic sense) on $\mathbb{A} \times \mathbb{A}'$. This is done in Sect. 3. Then in Sect. 4 we show that, if $\mathbf{B}$ is a subalgebra of events of $\mathbf{A}$ and P a conditional probability on $\mathbb{A} \times \mathbb{A}'$, then the restriction of the canonical extension μ_P on $\mathcal{C}(\mathbf{A})$ to $\mathcal{C}(\mathbf{B})$ is, in fact, the canonical extension of the restriction of P on $\mathbb{B} \times \mathbb{B}'$. This will allow us to prove in Sect. 5 that the probability of the conjunction coincides with McGee and Kaufmann's expressions obtained within the approach developed by two of the authors to conditionals as three-valued objects, with betting-based semantics, and specified as suitable random quantities. We also obtain the probability of the disjunction and the probability sum rule, in agreement with the approach given in [14]. We conclude in Sect. 6 with some remarks and prospects for future work.

2   Preliminaries

In this section we recall basic notions and results from [9] where, for any Boolean algebra of events $\mathbf{A} = (\mathbb{A}, \wedge, \vee, \bar{\ }, \bot, \top)$, a Boolean algebra of conditionals, denoted $\mathcal{C}(\mathbf{A})$, is built. We will also denote a conjunction $A \wedge B$ simply by AB. Intuitively, a Boolean algebra of conditionals over $\mathbf{A}$ allows *basic conditionals*, i.e. objects of the form $(A|B)$ for $A \in \mathbb{A}$ and $B \in \mathbb{A}' = \mathbb{A}\backslash\{\bot\}$, to be freely combined with the usual Boolean operations up to certain extent.

In mathematical terms, the formal construction of the algebra of conditionals $\mathcal{C}(\mathbf{A})$ is done as follows. One first considers the free Boolean algebra $\mathbf{Free}(\mathbb{A}|\mathbb{A}') = (Free(\mathbb{A}|\mathbb{A}'), \sqcap, \sqcup, \bar{\ }, \bot, \top)$ generated by the set $\mathbb{A}|\mathbb{A}' = \{(A|B) : A \in \mathbb{A}, B \in \mathbb{A}'\}$. Then, one considers the smallest congruence relation $\equiv_{\mathfrak{C}}$ on $\mathbf{Free}(\mathbb{A}|\mathbb{A}')$ satisfying the following natural properties:

(C1) $(B|B) \equiv_c \top$, for all $B \in \mathbb{A}'$;
(C2) $(A_1|B) \sqcap (A_2|B) \equiv_c (A_1 A_2|B)$, for all $A_1, A_2 \in \mathbb{A}$, $B \in \mathbb{A}'$;
(C3) $(\overline{A|B}) \equiv_c (\bar{A}|B)$, for all $A \in \mathbb{A}$, $B \in \mathbb{A}'$;
(C4) $(AB|B) \equiv_c (A|B)$, for all $A \in \mathbb{A}$, $B \in \mathbb{A}'$;
(C5) $(A|B) \sqcap (B|C) \equiv_c (A|C)$, for all $A \in \mathbb{A}$, $B, C \in \mathbb{A}'$ such that $A \leqslant B \leqslant C$.

Finally, the algebra $\mathcal{C}(\mathbf{A})$ is defined as follows.

Definition 1. *For every Boolean algebra* $\mathbf{A}$, *the* Boolean algebra of conditionals *of* $\mathbf{A}$ *is the quotient structure* $\mathcal{C}(\mathbf{A}) = \mathbf{Free}(\mathbb{A}|\mathbb{A})/_{\equiv_c}$.

Since $\mathcal{C}(\mathbf{A})$ is a *quotient* of $\mathbf{Free}(\mathbb{A}|\mathbb{A})$, elements of $\mathcal{C}(\mathbf{A})$ are equivalence classes, but without danger of confusion, one can henceforth identify classes $[t]_{\equiv_c}$ with one of its representative elements, in particular, by t itself. Conditionals of the form $(A|\top)$ will also be simply denoted as A.

A basic observation is that if $\mathbf{A}$ is finite, $\mathcal{C}(\mathbf{A})$ is finite as well, and hence atomic. Indeed, if $\mathbf{A}$ is a Boolean algebra with n atoms $at(\mathbf{A}) = \{\alpha_1, \ldots, \alpha_n\}$, i.e. $|at(\mathbf{A})| = n$, it is shown in [9] that the atoms of $\mathcal{C}(\mathbf{A})$ are in one-to-one correspondence with sequences $\bar{\alpha} = \langle \alpha_{i_1}, \ldots, \alpha_{i_{n-1}} \rangle$ of $n-1$ pairwise different atoms of $\mathbf{A}$, each of these sequences giving rise to an atom $\omega_{\bar{\alpha}}$ of $\mathcal{C}(\mathbf{A})$ defined as the following conjunction of $n-1$ basic conditionals:

$$\omega_{\bar{\alpha}} = (\alpha_{i_1}|\top) \sqcap (\alpha_{i_2}|\bar{\alpha}_{i_1}) \sqcap \cdots \sqcap (\alpha_{i_{n-1}}|\bar{\alpha}_{i_1} \cdots \bar{\alpha}_{i_{n-2}}), \tag{1}$$

It is then clear that $|at(\mathcal{C}(\mathbf{A}))| = n!$.

Next we will recall some properties holding in $\mathcal{C}(\mathbf{A})$ that will be useful for next sections. For each subvector $(i_1, \ldots, i_k)$ of $(1, \ldots, n)$ we set

$$\omega_{i_1 \cdots i_k} = \alpha_{i_1} \sqcap (\alpha_{i_2}|\bar{\alpha}_{i_1}) \sqcap \cdots \sqcap (\alpha_{i_k}|\bar{\alpha}_{i_1} \cdots \bar{\alpha}_{i_{k-1}}), \tag{2}$$

that is, $\omega_{i_1 \cdots i_k}$ denotes an initial conjunction of k components of the atom $\omega_{i_1 \cdots i_{n-1}}$. Indeed, as $(\alpha_{i_n}|\bar{\alpha}_{i_1} \cdots \bar{\alpha}_{i_{n-1}}) = (\alpha_{i_n}|\alpha_{i_n}) = \top$, for each permutation $(i_1, \ldots, i_n)$ of $(1, \ldots, n)$, we obtain the following atom of $\mathcal{C}(\mathbf{A})$:

$$\omega_{i_1 \cdots i_n} = \omega_{i_1 \cdots i_{n-1}} = \alpha_{i_1} \sqcap (\alpha_{i_2}|\bar{\alpha}_{i_1}) \sqcap \cdots \sqcap (\alpha_{i_{n-1}}|\bar{\alpha}_{i_1} \cdots \bar{\alpha}_{i_{n-2}}).$$

We hence recall that, from [9, Proposition 4.3], for each k, the conjunctions $\omega_{i_1 \cdots i_k}$'s constitute a partition of the algebra $\mathcal{C}(\mathbf{A})$. In particular this implies that $\bigsqcup_{(i_1, \ldots, i_k) \in \Pi_{\{j_1, \ldots, j_k\}}} \omega_{i_1 \cdots i_k} = \top$, where $\Pi_{\{j_1, \ldots, j_k\}}$ is the set of all permutations $(i_1, \ldots, i_k)$ of the set $\{j_1, \ldots, j_k\}$.

Now, consider a *positive* probability on the algebra of plain events $P : \mathbf{A} \to [0, 1]$. It is shown in [9] that P can be extended to a probability $\mu_P : \mathcal{C}(\mathbf{A}) \to [0, 1]$ on the Boolean algebra of conditionals $\mathcal{C}(\mathbf{A})$, called *canonical extension*, such that $\mu_P(``(A|B)")$, the probability of a basic conditional $(A|B)$, coincides with the conditional probability of $A$ given $B$, i.e. $\mu_P(``(A|B)") = P(A|B) = P(A \wedge B)/P(B)$. In particular, $\mu_P(``(A|\top)") = P(A|\top) = P(A)$ for any $A \in \mathbb{A}$. Actually, the probability μ_P is first defined on the atoms of $\mathcal{C}(\mathbf{A})$ as follows: for any atom $\omega_{i_1 \cdots i_{n-1}} = \alpha_{i_1} \sqcap (\alpha_{i_2} | \bar{\alpha}_{i_1}) \sqcap \cdots \sqcap (\alpha_{i_{n-1}} | \bar{\alpha}_{i_1} \cdots \bar{\alpha}_{i_{n-2}})$, its probability is defined as the following product of conditional probabilities:

$$\mu_P(\omega_{i_1 \cdots i_{n-1}}) = P(\alpha_{i_1}) \cdot P(\alpha_{i_2} | \bar{\alpha}_{i_1}) \cdots P(\alpha_{i_{n-1}} | \bar{\alpha}_{i_1} \cdots \bar{\alpha}_{i_{n-2}}).$$

Then μ_P is extended to the whole algebra $\mathcal{C}(\mathbf{A})$ of conditionals by additivity. Moreover, it is shown in [9] that for any k, the following factorization holds:

$$\begin{aligned} \mu_P(\omega_{i_1 \cdots i_k}) &= \sum_{(i_{k+1}, \ldots, i_n) \in \Pi_{\{1, \ldots, n\} \setminus \{i_1, \ldots, i_k\}}} \mu_P(\omega_{i_1 \cdots i_{n-1}}) \\ &= P(\alpha_{i_1}) \cdot P(\alpha_{i_2} | \bar{\alpha}_{i_1}) \cdots P(\alpha_{i_k} | \bar{\alpha}_{i_1} \cdots \bar{\alpha}_{i_{k-1}}). \end{aligned} \tag{3}$$

We finally notice that, as observed above, since for each k the conjunctions $\omega_{i_1 \cdots i_k}$'s constitute a partition of $\mathcal{C}(\mathbf{A})$, the sum of the probabilities over all of them is 1, that is:

$$1 = \sum_i P(\alpha_i) = \sum_i \mu_P(\omega_i) = \sum_{i \neq j} \mu_P(\omega_{ij}) = \cdots = \sum_{(i_1, \ldots, i_n) \in \Pi_{\{1, \ldots, n\}}} \mu_P(\omega_{i_1 \cdots i_{n-1}}).$$

3   Canonical Extension of a Conditional Probability

In the definition above of the canonical extension μ_P on $\mathcal{C}(\mathbf{A})$, a crucial assumption is that P is positive, i.e. that $P(\alpha) > 0$ for every $\alpha \in \mathrm{at}(\mathbf{A})$, otherwise $\mu_P(\omega)$ can be undefined for some $\omega \in \mathrm{at}(\mathcal{C}(\mathbf{A}))$ (it would be of the form $0/0$). A way to overcome this problem is, instead of starting with a positive (unconditional) probability on $\mathbf{A}$, to start with a *conditional probability* on $\mathbb{A} \times \mathbb{A}'$ in the axiomatic sense, that is to say, a binary map $P : \mathbb{A} \times \mathbb{A}' \to [0, 1]$, where $\mathbb{A}' = \mathbb{A} \setminus \{\bot\}$, such that

(CP1) For all $B \in \mathbb{A}'$, $P(\cdot | B) : \mathbb{A} \to [0, 1]$ is a finitely additive probability on $\mathbf{A}$;

(CP2) For all $A \in \mathbb{A}$ and $B \in \mathbb{A}'$, $P(A|B) = P(A \wedge B|B)$;

(CP3) For all $A \in \mathbb{A}$, $B, C \in \mathbb{A}'$, if $A \leqslant B \leqslant C$, then $P(A|C) = P(A|B) \cdot P(B|C)$.

As usual, we will also denote $P(A|\top)$ by $P(A)$, for every $A \in \mathbb{A}$.

Remark 1. Notice that, differently from the approach in [9], we do not assume positivity of the (conditional) probability P. Then, the function P may be such that $P(A|B) = 0$ and/or $P(B) = 0$ for some $A \in \mathbb{A}$ and $B \in \mathbb{A}'$. Moreover, we recall that, requesting $P : \mathbb{A} \times \mathbb{A}' \to [0, 1]$ to satisfy the above three postulates, assures that P is a coherent conditional probability assessment in the sense of de Finetti to all the conditional objects $(A|B)$, with $A, B \in \mathbb{A}$ and $B \neq \bot$. In fact, a conditional probability assessment on an arbitrary family of (basic) conditional events $P(A_1|B_1) = x_1, \ldots, P(A_n|B_n) = x_n$, is coherent iff it can be extended to a full conditional probability (in the above sense) on $\mathbb{A} \times \mathbb{A}'$ (see, e.g., [6]). In this paper we directly start (not from a coherent probability assessment on an arbitrary family of conditional events, but) with a full conditional probability P defined on $\mathbb{A} \times \mathbb{A}'$.

Then, given a conditional probability $P : \mathbb{A} \times \mathbb{A}' \to [0,1]$, we can proceed as in the previous section and first define a mapping μ_P on $\mathrm{at}(\mathcal{C}(\mathbf{A}))$ as follows: for any atom $\omega = (\alpha_1|\top) \sqcap (\alpha_2|\bar{\alpha}_1) \sqcap \ldots \sqcap (\alpha_{n-1}|\bar{\alpha}_1 \cdots \bar{\alpha}_{n-2})$,

$$\mu_P(\omega) = P(\alpha_1|\top) \cdot P(\alpha_2|\bar{\alpha}_1) \cdot \ldots \cdot P(\alpha_{n-1}|\bar{\alpha}_1 \cdots \bar{\alpha}_{n-2}). \tag{4}$$

Of course, $\mu_P(\omega) = 0$, if $P(\alpha_1|\top) = 0$, or $P(\alpha_2|\bar{\alpha}_1) = 0$, ..., or $P(\alpha_{n-1}|\bar{\alpha}_1 \cdots \bar{\alpha}_{n-2}) = 0$. That is, differently from Sect. 2, as the positivity property has been lifted it may be that $\mu_P(\omega) = 0$ for some ω's.

One can check that μ_P so defined is a probability distribution on $\mathrm{at}(\mathcal{C}(\mathbf{A}))$.

Proposition 1. $\sum_{\omega \in \mathrm{at}(\mathcal{C}(\mathbf{A}))} \mu_P(\omega) = 1$.

Proof. Although one could adapt here the proof of [9, Lemma 6.8], we provide below a direct proof. Let $\mathrm{at}(\mathbf{A}) = \{\alpha_1, \ldots, \alpha_n\}$. First of all, for any subset of atoms $\{\beta_1, \ldots, \beta_k\} \subseteq \mathrm{at}(\mathbf{A})$, with $k < n$, by the law of total probabilities,

$$\sum_{\beta \in \mathrm{at}(\mathbf{A}) \setminus \{\beta_1, \ldots, \beta_k\}} P(\beta|\bar{\beta}_1 \cdots \bar{\beta}_k) = 1.$$

For $k = 1$ it is clear that $\sum_{\alpha \in \mathrm{at}(\mathbf{A})} P(\alpha|\top) = 1$, and for $k = 2$, we have $P(\alpha|\top) = \sum_{\beta \neq \alpha} P(\alpha|\top) \cdot P(\beta|\bar{\alpha})$. More generally, for any $k < 1$ we have:

$$P(\beta_1) \cdot \ldots \cdot P(\beta_k|\bar{\beta}_1 \cdots \bar{\beta}_{k-1}) = \sum_{\beta \notin \{\beta_1, \ldots, \beta_k\}} P(\beta_1) \cdot \ldots \cdot P(\beta_k|\bar{\beta}_1 \cdots \bar{\beta}_{k-1}) \cdot P(\beta|\bar{\beta}_1 \cdots \bar{\beta}_k).$$

Then, we can write: $1 = \sum_{\beta_1} P(\beta_1|\top) = \sum_{\beta_1} \sum_{\beta_2 \neq \beta_1} P(\beta_1|\top) \cdot P(\beta_2|\bar{\beta}_1) = \cdots = \sum_{\langle \beta_1, \ldots, \beta_n \rangle \in Seq(\mathbf{A})} P(\beta_1|\top) \cdot P(\beta_2|\bar{\beta}_1) \cdots P(\beta_{n-1}|\bar{\beta}_1 \cdots \bar{\beta}_{n-2}) = \sum_{\bar{\alpha} \in Seq(\mathbf{A})} \mu_P(\omega_{\bar{\alpha}}) = \sum_{\omega \in \mathrm{at}(\mathcal{C}(\mathbf{A}))} \mu_P(\omega)$. $\square$

Then, we can extend μ_P to a probability on the whole algebra $\mathcal{C}(\mathbf{A})$ in the usual way by additivity, as in the previous case: for any $T \in \mathcal{C}(\mathbf{A})$, $\mu_P(T) = \sum_{\omega \leqslant T} \mu_P(\omega)$. We will keep referring to μ_P as the *canonical extension* of P.

To conclude this section, we check that Eq. (3) keeps holding in this more general setting. Indeed, concerning the canonical extension on the conjunctions $\omega_{i_1 \cdots i_k}$'s, we first observe that, as $\omega_{1 \cdots n-2\,n-1} \sqcup \omega_{1 \cdots n-2\,n} = \omega_{1 \cdots n-2}$, from (4) it holds that:

$$\begin{aligned}
\mu_P(\omega_{1 \cdots n-2}) &= \mu_P(\omega_{1 \cdots n-2\,n-1}) + \mu_P(\omega_{1 \cdots n-2\,n}) \\
&= P(\alpha_1)P(\alpha_2|\bar{\alpha}_1) \cdots P(\alpha_{n-2}|\bar{\alpha}_1 \cdots \bar{\alpha}_{n-3})[P(\alpha_{n-1}|(\alpha_{n-1} \vee \alpha_n)) \\
&\quad + P(\alpha_n|(\alpha_{n-1} \vee \alpha_n))] = P(\alpha_1)P(\alpha_2|\bar{\alpha}_1) \cdots P(\alpha_{n-2}|\bar{\alpha}_1 \cdots \bar{\alpha}_{n-3}).
\end{aligned}$$

Likewise $\mu_P(\omega_{i_1 \cdots i_{n-2}}) = P(\alpha_{i_1})P(\alpha_{i_2}|\bar{\alpha}_{i_1}) \cdots P(\alpha_{i_{n-2}}|\bar{\alpha}_{i_1} \cdots \bar{\alpha}_{i_{n-3}})$. Then, by backward iteration, for each $k \leqslant n - 1$, it holds that

$$\mu_P(\omega_{i_1 \cdots i_k}) = P(\alpha_{i_1})P(\alpha_{i_2}|\bar{\alpha}_{i_1}) \cdots P(\alpha_{i_k}|\bar{\alpha}_{i_1} \cdots \bar{\alpha}_{i_{k-1}}). \tag{5}$$

The question of whether μ_P actually extends P, in the sense that, for any basic conditional $(A|B) \in \mathcal{C}(\mathbf{A})$, it holds $\mu_P(``(A|B)") = P(A|B)$ is deferred to Theorem 2 in the next section.

4   The Canonical Extension for Subalgebras

In this section we examine the restriction of the canonical extension μ_P for conditional subalgebras of $C(\mathbf{A})$. Then, we let $\mathbf{A}$ be a finite algebra whose set of atoms is $\{\alpha_1, \alpha_2, \ldots, \alpha_n\}$. Let $i < n$, and for $i = 1, \ldots, n-1$, let

$$\beta_j = \begin{cases} \alpha_j, & \text{if } j < i \\ \alpha_i \vee \alpha_{i+1}, & \text{if } j = i \\ \alpha_{j+1}, & \text{if } j > i+1 \end{cases}$$

and let $\mathbf{B}$ be the subalgebra of $\mathbf{A}$ generated by $\beta_1, \ldots, \beta_{n-1}$, so that $\mathrm{at}(\mathbf{B}) = \{\beta_1, \ldots, \beta_{n-1}\}$. Now let us consider $P : \mathbb{A} \times \mathbb{A}' \to [0,1]$ a conditional probability and $\mu_P : C(\mathbf{A}) \to [0,1]$ its canonical extension to $C(\mathbf{A})$. Further, let $P' : \mathbb{B} \times \mathbb{B}' \to [0,1]$ be the restriction of P to $\mathbb{B} \times \mathbb{B}'$, and let $\mu_{P'} : C(\mathbf{B}) \to [0,1]$ be its canonical extension to $C(\mathbf{B})$. The question is whether $\mu_{P'}$ is the restriction of μ_P to $C(\mathbf{B})$. Next theorem shows this is actually the case.

We set $\omega'_{j_1 \cdots j_{n-2}} = (\beta_{j_1}|\top) \sqcap (\beta_{j_2}|\bar{\beta}_{j_1}) \sqcap \cdots \sqcap (\beta_{j_{n-2}}|\bar{\beta}_{j_1} \cdots \bar{\beta}_{j_{n-3}})$ and we recall that $\mu_{P'}(\omega'_{j_1 \cdots j_{n-2}}) = P(\beta_{j_1}|\top)P(\beta_{j_2}|\bar{\beta}_{j_1}) \cdots P(\beta_{j_{n-2}}|\bar{\beta}_{j_1} \cdots \bar{\beta}_{j_{n-3}})$. In the next result we show that $\mu_P(\omega'_{j_1 \cdots j_{n-2}}) = \mu_{P'}(\omega'_{j_1 \cdots j_{n-2}})$.

Theorem 1. *For each atom $\omega'_{j_1 \cdots j_{n-2}} \in \mathrm{at}(\mathbf{B})$, the following holds:*

$$\mu_P(\omega'_{j_1 \cdots j_{n-2}}) = \mu_P((\beta_{j_1}|\top) \sqcap (\beta_{j_2}|\bar{\beta}_{j_1}) \sqcap \cdots \sqcap (\beta_{j_{n-2}}|\bar{\beta}_{j_1} \cdots \bar{\beta}_{j_{n-3}}))$$
$$= P(\beta_{j_1}|\top)P(\beta_{j_2}|\bar{\beta}_{j_1}) \cdots P(\beta_{j_{n-2}}|\bar{\beta}_{j_1} \cdots \bar{\beta}_{j_{n-3}}) = \mu_{P'}(\omega'_{j_1 \cdots j_{n-2}}).$$

Proof. The proof is omitted due to lack of space.

Remark 2. We observe that, for each conditional subalgebra $C(\mathbf{B})$ of $C(\mathbf{A})$, by a suitable iterated application of Theorem 1, it can be proved that $\mu_P(\omega') = \mu_{P'}(\omega')$, for every $\omega' \in \mathrm{at}(\mathbf{B})$.

As an illustration of Theorem 1, let us consider the following simple example. Let $\mathbf{A}$ be an algebra with four atoms $\{\alpha_1, \alpha_2, \alpha_3, \alpha_4\}$. Now let us consider the partition defined by the elements $\beta_1 = \alpha_1, \beta_2 = \alpha_2$ and $\beta_3 = \alpha_3 \vee \alpha_4$, and let $\mathbf{B}$ be the subalgebra of $\mathbf{A}$ generated these three elements so that so that $\{\beta_1, \beta_2, \beta_3\}$ become the atoms of $\mathbf{B}$. As above, let P be a conditional probability on $\mathbb{A} \times \mathbb{A}'$, and let P' be its restriction to $\mathbb{B} \times \mathbb{B}'$. According to Theorem 1, let us practically show that $\mu_{P'}$ is the restriction of μ_P on $C(\mathbf{B})$. We have to show that, for any pairwise different $i, j \in \{1, 2, 3\}$, the following condition holds:

$$\mu_P((\beta_i|\top) \sqcap (\beta_j|\bar{\beta}_i)) = P(\beta_i) \cdot P(\beta_j|\bar{\beta}_i) = \mu_{P'}((\beta_i|\top) \sqcap (\beta_j|\bar{\beta}_i)).$$

The cases $(\beta_i|\top) \sqcap (\beta_3|\bar{\beta}_i)$ with $i \in \{1, 2\}$ can be easily verified by exploiting (5). Let us consider the case $(\beta_3|\top) \sqcap (\beta_1|\bar{\beta}_3)$, the other case $(\beta_3|\top) \sqcap (\beta_2|\bar{\beta}_3)$ is analogous. We have to compute the probability $\mu_P((\beta_3|\top) \sqcap (\beta_1|\bar{\beta}_3))$. First of all, note that $(\beta_3|\top) \sqcap (\beta_1|\bar{\beta}_3) = (\alpha_3 \vee \alpha_4|\top) \sqcap (\alpha_1|\alpha_1 \vee \alpha_2)$, so we have to compute the probability $\mu_P((\alpha_3 \vee \alpha_4|\top) \sqcap (\alpha_1|\alpha_1 \vee \alpha_2))$, and for that, we have to find the compound conditionals ω of $C(\mathbf{A})$ such that $\omega \leqslant (\alpha_3 \vee \alpha_4|\top) \sqcap (\alpha_1|\alpha_1 \vee \alpha_2)$. It is not difficult to check that $(\alpha_3 \vee \alpha_4 | \top) \sqcap (\alpha_1 | \alpha_1 \vee \alpha_2) = \omega_{31} \sqcup \omega_{341} \sqcup \omega_{41} \sqcup \omega_{431}$. Then, by recalling (5), we have:

$$\mu_P((\beta_3|\top) \sqcap (\beta_1|\bar{\beta}_3)) = P(\omega_{31}) + P(\omega_{341}) + P(\omega_{41}) + P(\omega_{431})$$
$$= P(\alpha_3) \cdot P(\alpha_1|\bar{\alpha}_3) + P(\alpha_3) \cdot P(\alpha_4|\bar{\alpha}_3) \cdot P(\alpha_1|\bar{\alpha}_3\bar{\alpha}_4)$$
$$+ P(\alpha_4) \cdot P(\alpha_1|\bar{\alpha}_4) + P(\alpha_4) \cdot P(\alpha_3|\bar{\alpha}_4) \cdot P(\alpha_1|\bar{\alpha}_3\bar{\alpha}_4)$$
$$= P(\alpha_3) \cdot P(\alpha_1|\bar{\alpha}_3\bar{\alpha}_4) \cdot P(\bar{\alpha}_3\bar{\alpha}_4|\bar{\alpha}_3) + P(\alpha_3) \cdot P(\alpha_4|\bar{\alpha}_3) \cdot P(\alpha_1|\bar{\alpha}_3\bar{\alpha}_4)$$
$$+ P(\alpha_4) \cdot P(\alpha_1|\bar{\alpha}_3\bar{\alpha}_4) \cdot P(\bar{\alpha}_3\bar{\alpha}_4|\bar{\alpha}_4) + P(\alpha_4) \cdot P(\alpha_3|\bar{\alpha}_4) \cdot P(\alpha_1|\bar{\alpha}_3\bar{\alpha}_4)$$
$$= P(\alpha_1|\bar{\alpha}_3\bar{\alpha}_4) \cdot [P(\alpha_3) \cdot (P(\bar{\alpha}_3\bar{\alpha}_4|\bar{\alpha}_3) + P(\alpha_4|\bar{\alpha}_3))$$
$$+ P(\alpha_4) \cdot (P(\bar{\alpha}_3\bar{\alpha}_4|\bar{\alpha}_4) + P(\alpha_3|\bar{\alpha}_4))]$$
$$= P(\alpha_1|\alpha_1 \vee \alpha_2) \cdot [P(\alpha_3) \cdot P(\alpha_1 \vee \alpha_2 \vee \alpha_4|\bar{\alpha}_3) + P(\alpha_4) \cdot P(\alpha_1 \vee \alpha_2 \vee \alpha_3|\bar{\alpha}_4)]$$
$$= P(\alpha_1|\alpha_1 \vee \alpha_2) \cdot (P(\alpha_3) + P(\alpha_4)) = P(\alpha_1|\alpha_1 \vee \alpha_2) \cdot P(\alpha_3 \vee \alpha_4)$$
$$= P(\beta_3) \cdot P(\beta_1|\bar{\beta}_3) = \mu_{P'}((\beta_3|\top) \sqcap (\beta_1|\bar{\beta}_3)).$$

In the next result we give a proof of [9, Theorem 6.13] where P is (not a positive probability on $\mathbf{A}$, but) a conditional probability on $\mathbb{A} \times \mathbb{A}'$.

Theorem 2. *Let P be a conditional probability on $\mathbb{A} \times \mathbb{A}'$ and μ_P its canonical extension to $\mathcal{C}(\mathbf{A})$. Then, for every basic conditional $(A|H) \in \mathcal{C}(\mathbf{A})$, it holds that $\mu_P(A|H) = P(A|H)$.*

Proof. Given any $(A|H) \in \mathcal{C}(\mathbf{A})$, let $\mathbf{B}$ be the subalgebra of $\mathbf{A}$ generated by the partition $\{\beta_1, \beta_2, \beta_3\} = \{AH, \bar{A}H, \bar{H}\}$. Let $P' : \mathbb{B} \times \mathbb{B}' \to [0,1]$ be the restriction of P to $\mathbb{B} \times \mathbb{B}'$, and let $\mu_{P'} : \mathcal{C}(\mathbf{B}) \to [0,1]$ be its canonical extension to $\mathcal{C}(\mathbf{B})$. Of course $P'(A|H) = P(A|H)$. We notice that $A|H = \omega'_{12} \sqcup \omega'_{13} \sqcup \omega'_{31}$, where $\omega'_{12} = \beta_1 \sqcap (\beta_2|\bar{\beta}_1) = AH \sqcap (\bar{A}H|(\bar{A}H \vee \bar{H}))$, $\omega'_{13} = \beta_1 \sqcap (\beta_3|\bar{\beta}_1) = AH \sqcap (\bar{H}|(\bar{A}H \vee \bar{H}))$, and $\omega'_{31} = \beta_3 \sqcap (\beta_1|\bar{\beta}_3) = \bar{H} \sqcap (A|H)$. Then, by Theorem 1, it holds that

$$\mu_P(A|H) = \mu_P(\omega'_{12}) + \mu_P(\omega'_{13}) + \mu_P(\omega'_{31}) = \mu_{P'}(\omega'_{12}) + \mu_{P'}(\omega'_{13}) + \mu_{P'}(\omega'_{31})$$
$$= P(AH)P(\bar{A}H|(\bar{A}H \vee \bar{H})) + P(AH)P(\bar{H}|(\bar{A}H \vee \bar{H})) + P(\bar{H})P(A|H)$$
$$= P(AH) + P(\bar{H})P(A|H) = P(H)P(A|H) + P(\bar{H})P(A|H) = P(A|H). \quad \square$$

We now generalize the above result to a general element of a conditional subalgebra of $\mathcal{C}(\mathbf{A})$.

Theorem 3. *Given a conditional probability P on $\mathbb{A} \times \mathbb{A}'$, let P' be its restriction to $\mathbb{B} \times \mathbb{B}'$, where $\mathbf{B}$ is a subalgebra of $\mathbf{A}$. For each $\mathcal{C} \in \mathcal{C}(\mathbf{B})$ it holds that*

$$\mu_P(\mathcal{C}) = \mu_{P'}(\mathcal{C}).$$

Proof. Indeed, by observing that $\mathcal{C} = \bigsqcup_{\omega' \leqslant \mathcal{C}} \omega'$, by Theorem 1 it follows that $\mu_P(\mathcal{C}) = \sum_{\omega' \leqslant \mathcal{C}} \mu_P(\omega') = \sum_{\omega' \leqslant \mathcal{C}} \mu_{P'}(\omega') = \mu_{P'}(\mathcal{C})$. $\quad \square$

Theorem 3 shows that μ'_P is the restriction of μ_P to $\mathcal{C}(\mathbf{B})$. This result allows a local approach in order to study properties, as done in the next section.

Remark 3. Given three events A, B, C, with $A \leqslant B \leqslant C$, by (CP3) it holds that $P(A|C) = P(A|B)P(B|C)$. Moreover, by recalling (C5), we observe that

$$(A|B) = [(A|B) \sqcap (B|C)] \sqcup [(A|B) \sqcap (\bar{B}|C)] = (A|C) \sqcup [(A|B) \sqcap (\bar{B}|C)]$$

and hence by Theorem 2

$$P(A|B) = P(A|C) + \mu_P[(A|B) \sqcap (\bar{B}|C)]. \tag{6}$$

Then, as $P(A|B) - P(A|C) = P(A|B) - P(A|B)P(B|C)$, it follows that

$$\mu_P[(A|B) \sqcap (\bar{B}|C)] = P(A|B)P(\bar{B}|C). \tag{7}$$

As we can see, (7) shows that the "independence" between $A|B$ and $B|C$, when $A \leqslant B \leqslant C$, still holds between $A|B$ and $\bar{B}|C$. In particular, given any events E and H, by applying (7) with $A = EH, B = H$,and $C = \top$, as $\bar{H}|\top = \bar{H}$, we obtain

$$\mu_P((E|H) \sqcap \bar{H}) = P(E|H)P(\bar{H}). \tag{8}$$

Formula (8) will be generalized in Theorem 4, where $\bar{H}$ is replaced by any K such that $HK = \perp$.

5   Probability of the Conjunction and the Disjunction

In this section we start by showing a basic property for the probability of the conjunction and then, based on the canonical extension, we obtain the probability for the conjunction and the disjunction of two conditional events, which are related with analogous results given in the setting of coherence in [14–18]. In the next result we generalize formula (8).

Theorem 4. *Given an algebra* $\mathbf{A}$ *and any events* $A, H, K \in \mathbb{A}$, *with* $H \neq \perp$ *and* $HK = \perp$, *given a conditional probability* P *on* $\mathbb{A} \times \mathbb{A}'$ *and its canonical extension* μ_P *to* $C(\mathbf{A})$, *it holds*

$$\mu_P[K \sqcap (A|H)] = P(K)P(A|H). \tag{9}$$

Proof. As $HK = \perp$, it holds that $H\bar{K} = H$, $H \vee \bar{K} = \bar{K}$, and $\bar{H}K = K$; then

$$\top = (AH \vee \bar{A}H \vee \bar{H}) \wedge (K \vee \bar{K}) = AH\bar{K} \vee \bar{H}K \vee \bar{A}H\bar{K} \vee \bar{H}\bar{K}.$$

We consider the partition $\{\beta_1, \ldots, \beta_4\}$, where

$$\beta_1 = AH\bar{K} = AH, \ \beta_2 = \bar{H}K = K, \ \beta_3 = \bar{A}H\bar{K} = \bar{A}H, \ \beta_4 = \bar{H}\bar{K},$$

and the associated subalgebra $\mathbf{B}$; moreover we consider the atoms $\omega'_{i_1 i_2 i_3}$'s of $C(\mathbf{B})$. As $\omega'_{213} \sqcup \omega'_{214} = \omega'_{21}$, it holds that

$$K \sqcap (A|H) = \omega'_{213} \sqcup \omega'_{214} \sqcup \omega'_{241} = \omega'_{21} \sqcup \omega'_{241}.$$

Let P' be the restriction of P to $\mathbb{B} \times \mathbb{B}'$ and $\mu_{P'}$ its canonical extension to $C(\mathbf{B})$. As $H\bar{K} = H$ it holds that $P(AH|\bar{K}) = P(A|H\bar{K})P(H|\bar{K}) = P(A|H)P(H|\bar{K})$. Then, from Theorem 2 and from (5) we obtain

$$\begin{aligned}
\mu_P[K \sqcap (A|H)] &= \mu_{P'}[K \sqcap (A|H)] = \mu_{P'}(\omega'_{21}) + \mu_{P'}(\omega'_{241}) \\
&= P(\beta_2)P(\beta_1|\bar{\beta}_2) + P(\beta_2)P(\beta_4|\bar{\beta}_2)P(\beta_1|\bar{\beta}_2\bar{\beta}_4) = P(K)P(AH|\bar{K}) \\
&\quad + P(K)P(\bar{H}|\bar{K})P(A|H) = P(K)[P(AH|\bar{K}) + P(\bar{H}|\bar{K})P(A|H)] \\
&= P(K)[P(A|H)P(H|\bar{K}) + P(A|H)P(\bar{H}|\bar{K})] = P(K)P(A|H).
\end{aligned}$$
$\square$

In the next result we obtain the probability for the conjunction $(A|H) \sqcap (B|K)$.

Theorem 5. *Given an algebra* $\mathbf{A}$ *and a conditional probability* P *on* $\mathbb{A} \times \mathbb{A}'$, *let* μ_P *be the canonical extension to* $C(\mathbf{A})$. *For any conditional events* $A|H, B|K \in C(\mathbf{A})$ *it holds that*

$$\mu_P[(A|H) \sqcap (B|K)]$$
$$= P(AHBK|(H \vee K)) + P(A|H)P(\bar{H}BK|(H \vee K)) + P(B|K)P(\bar{K}AH|(H \vee K)).$$
$$(10)$$

Proof. We consider the partition $\{\beta_1, \ldots, \beta_9\}$, where

$$\beta_1 = AHBK, \; \beta_2 = AH\bar{B}K, \; \beta_3 = AH\bar{K}, \; \beta_4 = \bar{A}HBK,$$
$$\beta_5 = \bar{A}H\bar{B}K, \; \beta_6 = \bar{A}H\bar{K}, \; \beta_7 = \bar{H}BK, \; \beta_8 = \bar{H}\bar{B}K, \; \beta_9 = \bar{H}\bar{K}, \quad (11)$$

and the associated subalgebra $\mathbf{B}$. Moreover, we consider the compound conditionals $\omega'_{i_1 \cdots i_k}$'s of $C(\mathbf{B})$, $1 \leqslant k \leqslant 8$. Let P' be the restriction of P to $\mathbb{B} \times \mathbb{B}'$ and $\mu_{P'}$ its canonical extension to $C(\mathbf{B})$. We recall that from Theorem 3, $\mu_P(\mathcal{C}) = \mu_{P'}(\mathcal{C})$, for every $\mathcal{C} \in C(\mathbf{B})$. By exploiting the distributivity property, we decompose the conjunction $(A|H) \sqcap (B|K)$ as

$$(A|H) \sqcap (B|K) = [(A|H) \sqcap (B|K) \sqcap HK] \sqcup [(A|H) \sqcap (B|K) \sqcap \bar{H}K]$$
$$\sqcup [(A|H) \sqcap (B|K) \sqcap H\bar{K}] \sqcup [(A|H) \sqcap (B|K) \sqcap \bar{H}\bar{K}].$$

For the compound $(A|H) \sqcap (B|K) \sqcap HK$ it holds that

$$(A|H) \sqcap (B|K) \sqcap HK = ((A|H) \sqcap H) \sqcap ((B|K) \sqcap K) = AHBK = \beta_1 = \omega'_1,$$

with $\mu_P((A|H) \sqcap (B|K) \sqcap HK) = \mu_P(AHBK) = P(AHBK)$.
For the compound $(A|H) \sqcap (B|K) \sqcap \bar{H}K$ it holds that

$$(A|H) \sqcap (B|K) \sqcap \bar{H}K = (A|H) \sqcap ((B|K) \sqcap K) \sqcap \bar{H} = (A|H) \sqcap \bar{H}BK,$$

and, by Theorem 4, $\mu_P((A|H) \sqcap (B|K) \sqcap \bar{H}K) = \mu_P((A|H) \sqcap \bar{H}BK) = P(A|H)P(\bar{H}BK)$.
Likewise, for the compound $(A|H) \sqcap (B|K) \sqcap H\bar{K}$ it holds that

$$(A|H) \sqcap (B|K) \sqcap H\bar{K} = (B|K) \sqcap AH\bar{K},$$

and, by Theorem 4, $\mu_P((A|H) \sqcap (B|K) \sqcap H\bar{K}) = P(B|K)P(AH\bar{K})$.
Thus, by observing that $H \vee K = HK \vee \bar{H}K \vee H\bar{K}$, we obtain

$$\mu_P[(A|H) \sqcap (B|K) \sqcap (H \sqcup K)] = P(ABHK) + P(\bar{H}BK)P(A|H)$$
$$+ P(AH\bar{K})P(B|K) = P(H \vee K)[P(ABHK|(H \vee K))$$
$$+ P(\bar{H}BK|(H \vee K))P(A|H) + P(AH\bar{K}|(H \vee K))P(B|K)] = zP(H \vee K),$$
$$(12)$$

where
$$z = P(ABHK|(H \vee K)) + P(\bar{H}BK|(H \vee K))P(A|H) + P(AH\bar{K}|(H \vee K))P(B|K). \quad (13)$$

For the compound $(A|H) \sqcap (B|K) \sqcap \bar{H}\bar{K}$ it can be verified that

$$(A|H) \sqcap (B|K) \sqcap \bar{H}\bar{K} = \omega'_{91} \sqcup \omega'_{971} \sqcup \omega'_{972} \sqcup \omega'_{973} \sqcup \omega'_{9781} \sqcup \omega'_{9782} \sqcup \omega'_{9783}$$
$$\sqcup \omega'_{931} \sqcup \omega'_{934} \sqcup \omega'_{937} \sqcup \omega'_{9361} \sqcup \omega'_{9364} \sqcup \omega'_{9367}.$$
(14)

By Theorem 4, it holds that
$\mu_P(\omega'_{91}) = \mu_P(\bar{H}\bar{K} \sqcap AHBK|(H \vee K)) = P(\bar{H}\bar{K})P(AHBK|(H \vee K))$. Moreover, as

$$\beta_1|(\bar{\beta}_7\bar{\beta}_9) \sqcap \beta_2|(\bar{\beta}_7\bar{\beta}_9) \sqcap \beta_3|(\bar{\beta}_7\bar{\beta}_9) = (\beta_1 \vee \beta_2 \vee \beta_3)|(\bar{\beta}_7\bar{\beta}_9) = AH|(H \vee \bar{H}\bar{B}K),$$

from (5) it holds that

$$\mu_P(\omega'_{971} \sqcup \omega'_{972} \sqcup \omega'_{973}) = \mu_P(\omega'_{971}) + \mu_P(\omega'_{972}) + \mu_P(\omega'_{973}) = \cdots =$$
$$= P(\bar{H}\bar{K})P(\bar{H}BK|(H \vee K))P(AH|(H \vee \bar{H}\bar{B}K)).$$

Likewise, as

$$\beta_1|(\bar{\beta}_7\bar{\beta}_8\bar{\beta}_9) \sqcap \beta_2|(\bar{\beta}_7\bar{\beta}_8\bar{\beta}_9) \sqcap \beta_3|(\bar{\beta}_7\bar{\beta}_8\bar{\beta}_9) = (\beta_1 \vee \beta_2 \vee \beta_3)|(\bar{\beta}_7\bar{\beta}_8\bar{\beta}_9) = A|H,$$

it holds that

$$\mu_P(\omega'_{9781} \sqcup \omega'_{9782} \sqcup \omega'_{9783}) = \mu_P(\omega'_{9781}) + \mu_P(\omega'_{9782}) + \mu_P(\omega'_{9783}) = \cdots$$
$$= P(\bar{H}\bar{K})P(\bar{H}BK|(H \vee K))P(\bar{H}\bar{B}K|(H \vee \bar{H}\bar{B}K))P(A|H).$$

Then, by observing that $P(AH|(H \vee \bar{H}\bar{B}K)) = P(A|H)P(H|(H \vee \bar{H}\bar{B}K))$, it follows that

$$\mu_P(\omega'_{971} \sqcup \omega'_{972} \sqcup \omega'_{973}) + \mu_P(\omega'_{9781} \sqcup \omega'_{9782} \sqcup \omega'_{9783})$$
$$= P(\bar{H}\bar{K})P(\bar{H}BK|(H \vee K))[P(AH|(H \vee \bar{H}\bar{B}K))$$
$$+ P(\bar{H}\bar{B}K|(H \vee \bar{H}\bar{B}K))P(A|H)] = P(\bar{H}\bar{K})P(\bar{H}BK|(H \vee K))P(A|H).$$

Likewise $\mu_P(\omega'_{931} \sqcup \omega'_{934} \sqcup \omega'_{937}) + \mu_P(\omega'_{9361} \sqcup \omega'_{9364} \sqcup \omega'_{9367}) = \cdots = P(\bar{H}\bar{K})P(AH\bar{K}|(H \vee K))P(B|K)$. Thus, by recalling (13) and (14), it follows that

$$\mu_P[(A|H) \sqcap (B|K) \sqcap (\bar{H}\bar{K})] = P(\bar{H}\bar{K})[P(AHBK|(H \vee K))$$
$$+ P(\bar{H}BK|(H \vee K))P(A|H) + P(\bar{K}AH|(H \vee K))P(B|K)] = zP(\bar{H}\bar{K}).$$
(15)

Finally, by also recalling (12), it follows that

$$\mu_P[(A|H) \sqcap (B|K)] = \mu_P[(A|H) \sqcap (B|K) \sqcap (H \vee K)]$$
$$+ \mu_P[(A|H) \sqcap (B|K) \sqcap (\bar{H}\bar{K})] = zP(H \vee K) + zP(\bar{H}\bar{K}) = z =$$
$$P(AHBK|(H \vee K)) + P(A|H)P(\bar{H}BK|(H \vee K)) + P(B|K)P(\bar{K}AH|(H \vee K)).$$

$\square$

As shown in (12) and in (15), $(A|H) \sqcap (B|K)$ is "independent" from $H \vee K$ and from $\bar{H}\bar{K}$. Notice that formula (10) coincides with the prevision of the conjunction $\mathcal{C} = (A|H) \wedge (B|K)$, introduced in the setting of coherence as the following conditional random quantity (see, e.g.,[14,16])

$$\mathcal{C} = [AHBK + P(A|H)(\bar{H}BK|(H \vee K)) + P(B|K)(AH\bar{K}|(H \vee K))]|(H \vee K), \tag{16}$$

where (conditional) events and their indicators are denoted by the same symbol. Moreover, when $P(H \vee K) > 0$, formula (10) becomes

$$\mu_P[(A|H) \sqcap (B|K)] = \frac{P(AHBK) + P(A|H)P(\bar{H}BK) + P(B|K)P(\bar{K}AH)}{P(H \vee K)},$$

that is the formula obtained by McGee ([25]) and Kaufmann ([21]). We also note that, when $HK = \bot$ and hence $\bar{H}BK = BK, AH\bar{K} = AH$, from (10) it follows that ([13,29])

$$\mu_P[(A|H) \sqcap (B|K)] = P(A|H)P(BK|(H \vee K)) + P(B|K)P(AH|(H \vee K))$$
$$= P(A|H)P(B|K)P(K|(H \vee K)) + P(A|H)P(B|K)P(H|(H \vee K))$$
$$= P(A|H)P(B|K).$$

In the next result we obtain the probability of the disjunction $(A|H) \sqcup (B|K)$.

Theorem 6. *Given an algebra* $\mathbf{A}$ *and a conditional probability* P *on* $\mathbb{A} \times \mathbb{A}'$, *let* μ_P *be the canonical extension to* $\mathcal{C}(\mathbf{A})$. *For any conditional events* $A|H, B|K \in \mathcal{C}(\mathbf{A})$ *it holds that*

$$\mu_P[(A|H) \sqcup (B|K)]$$
$$= P((AH \vee BK)|(H \vee K)) + P(A|H)P(\bar{H}\,\bar{B}K|(H \vee K)) + P(B|K)P(\bar{A}H\bar{K}|(H \vee K)). \tag{17}$$

Proof. We observe that

$$\mu_P((A|H) \sqcup (B|K)) = \mu_P[(A|H) \sqcap (B|K) \sqcup (\bar{A}|H) \sqcap (B|K) \sqcup (A|H) \sqcap (\bar{B}|K)]$$
$$= \mu_P[(A|H) \sqcap (B|K)] + \mu_P[(\bar{A}|H) \sqcap (B|K)] + \mu_P[(A|H) \sqcap (\bar{B}|K)]. \tag{18}$$

From Theorem 5, besides (10), one has $\mu_P[(\bar{A}|H) \sqcap (B|K)] = P(\bar{A}HBK|$ $(H \vee K)) + P(\bar{A}|H)P(\bar{H}BK|(H \vee K)) + P(B|K)P(\bar{K}\bar{A}H|(H \vee K))$ and $\mu_P[(A|H) \sqcap (\bar{B}|K)] = P(AH\bar{B}K|(H \vee K)) + P(A|H)P(\bar{H}\bar{B}K|(H \vee K)) + P(\bar{B}|K)P(\bar{K}AH|(H \vee K))$. As it can be verified, it holds that $AH \vee BK = AHBK \vee AH\bar{B}K \vee \bar{A}HBK \vee AH\bar{K} \vee \bar{H}BK$; then, by recalling (18), it follows that equation (17) is satisfied.   □

We observe that (17) coincides with the prevision of the disjunction of two conditional events obtained in the framework of conditional random quantities in [14]. We also observe that De Morgan Laws are satisfied in $\mathcal{C}(\mathbf{A})$, there-fore, $\mu_P(\overline{(A|H) \sqcup (B|K)}) = \mu_P((\bar{A}|H) \sqcap (\bar{B}|K))$ and $\mu_P(\overline{(A|H) \sqcap (B|K)}) = \mu_P((\bar{A}|H) \sqcup (\bar{B}|K))$ in agreement with formulas (10) and (17). We remark that

$$\mu_P((A|H) \sqcup (B|K)) = P(A|H) + P(B|K) - \mu_P((A|H) \sqcap (B|K)),$$

which coincides with the prevision sum rule obtained in [13,14]. Finally, an aspect to be deepened concerns the notion of iterated conditional, say $(B|K)|(A|H)$, and its probability. If we define $\mu_P((B|K)|(A|H)) =_{def} \frac{\mu_P(A|H)\sqcap(B|K))}{\mu_P(A|H)}$, then, under the hypothesis $P(A|H) > 0$, it holds that

$$\mu_P((B|K)|(A|H)) = \frac{P(AHBK|(H\vee K))+P(A|H)P(\bar{H}BK|(H\vee K))+P(B|K)P(\bar{K}AH|(H\vee K))}{P(A|H)}, \quad (19)$$

which is the prevision of the iterated conditional $(B|K)|(A|H)$, obtained in the setting of coherence in [13, Section 6]. Under the further assumption $P(H \vee K) > 0$, formula (19) coincides with the result given in [21, Thm 3].

Finally we recall that, by suitably generalizing the notion of conjunction to the case of n conditional events, an application of compound conditionals to the probabilistic nonmonotonic reasoning has been given in [15], where a characterization for the p-entailment of Adams ([1]) has been provided. More precisely, given a p-consistent family of n conditional events $\mathcal{F}$ and a further conditional event $E_{n+1}|H_{n+1}$, the following conditions are equivalent:

(i) $\mathcal{F}$ p-entails $E_{n+1}|H_{n+1}$;
(ii) $(E_1|H_1) \wedge \cdots \wedge (E_n|H_n) \leqslant E_{n+1}|H_{n+1}$;
(iii) $(E_1|H_1) \wedge \cdots \wedge (E_n|H_n) \wedge (E_{n+1}|H_{n+1}) = (E_1|H_1) \wedge \cdots \wedge (E_n|H_n)$.

In addition, by introducing a notion of iterated conditional $(E_{n+1}|H_{n+1}) | [(E_1|H_1) \wedge \cdots \wedge (E_n|H_n)]$, which is a suitable conditional random quantity, in [19] it has been shown that

$$\mathcal{F} \text{ p-entails } E_{n+1}|H_{n+1} \iff (E_{n+1}|H_{n+1}) | [(E_1|H_1) \wedge \cdots \wedge (E_n|H_n)] = 1, \tag{20}$$

that is, $\mathcal{F}$ p-entails $E_{n+1}|H_{n+1}$ if and only if the iterated conditional $(E_{n+1}|H_{n+1}) | [(E_1|H_1) \wedge \cdots \wedge (E_n|H_n)]$ is constant and coincides with 1. The characterization of the p-entailment for several inference rules in non-monotonic reasoning has been studied in [12,17] where, in particular, the well known inference rules And, Cut, CM, and Or related to system P have been considered.

6 Conclusions

In this paper we have advanced in the study of conditionals in the setting of the Boolean algebras of conditionals as proposed in [9]. More precisely, given a finite Boolean algebra of events $\mathbf{A}$, we have first considered the canonical extension μ_P of a conditional probability P on $\mathbb{A} \times \mathbb{A}'$ to the Boolean algebra of conditionals $\mathcal{C}(\mathbf{A})$. Our first main result establishes that the process of canonical extensions commutes, in a sense made precise in Sect. 4, with taking subalgebras of $\mathbf{A}$. This fact then allows us to show that μ_P extends P over basic conditionals, and in turn to get an operational computation of the probability of a conjunction and a disjunction of conditionals, in agreement with previous approaches in the literature, in particular with the one developed by Gilio and Sanfilippo by formalising conditionals as random quantities [14].

As for future work, encouraged by the above obtained results, we plan to deepen into the relationship between the approach based on Boolean algebras of conditionals, together with canonical extensions of conditional probabilities on events, and the approach based on interpreting compound and iterated conditionals as random quantities.

Acknowledgments. The authors thank the anonymous referees for their comments. Flaminio and Godo acknowledge partial support by the MOSAIC project (EU H2020-MSCA-RISE-2020 Project 101007627) and also by the Spanish project PID2019-111544GB-C21 funded by MCIN/AEI/10.13039/501100011033. Sanfilippo acknowledges support by the FFR2020-FFR2021 projects of University of Palermo, Italy.

References

1. Adams, E.W.: The Logic of Conditionals. Reidel, Dordrecht (1975)
2. Beierle, C., Eichhorn, C., Kern-Isberner, G., Kutsch, S.: Properties of skeptical c-inference for conditional knowledge bases and its realization as a constraint satisfaction problem. Ann. Math. Artif. Intell. **83**(3), 247–275 (2018). https://doi.org/10.1007/s10472-017-9571-9
3. Benferhat, S., Dubois, D., Prade, H.: Nonmonotonic reasoning, conditional objects and possibility theory. Artif. Intell. **92**, 259–276 (1997)
4. Calabrese, P.: An algebraic synthesis of the foundations of logic and probability. Inf. Sci. **42**(3), 187–237 (1987). https://doi.org/10.1016/0020-0255(87)90023-5
5. Ciucci, D., Dubois, D.: A map of dependencies among three-valued logics. Inf. Sci. **250**, 162–177 (2013)
6. Coletti, G., Scozzafava, R.: Probabilistic Logic in a Coherent Setting. Kluwer, Dordrecht (2002)
7. Coletti, G., Petturiti, D., Vantaggi, B.: Dutch book rationality conditions for conditional preferences under ambiguity. Ann. Oper. Res. **279**(1-2), 115–150 (2019)
8. Dubois, D., Prade, H.: Conditional objects as nonmonotonic consequence relationships. IEEE Trans. Syst. Man Cybern. **24**(12), 1724–1740 (1994)
9. Flaminio, T., Godo, L., Hosni, H.: Boolean algebras of conditionals, probability and logic. Artif. Intell. **286**, 103347 (2020)
10. Freund, M., Lehmann, D., Morris, P.: Rationality, transitivity, and contraposition. Artif. Intell. **52**(2), 191–203 (1991)
11. Gilio, A.: Probabilistic reasoning under coherence in System P. Ann. Math. Artif. Intell. **34**, 5–34 (2002)
12. Gilio, A., Pfeifer, N., Sanfilippo, G.: Probabilistic entailment and iterated conditionals. In: Elqayam, S., Douven, I., Evans, J.S.B.T., Cruz, N. (eds.) Logic and Uncertainty in the Human Mind: A Tribute to David E. Over, pp. 71–101. Routledge, Oxon (2020). https://doi.org/10.4324/9781315111902-6
13. Gilio, A., Sanfilippo, G.: Conjunction, disjunction and iterated conditioning of conditional events. In: Synergies of Soft Computing and Statistics for Intelligent Data Analysis, AISC, vol. 190, pp. 399–407. Springer, Berlin (2013). https://doi.org/10.1007/978-3-642-33042-1_43
14. Gilio, A., Sanfilippo, G.: Conditional random quantities and compounds of conditionals. Stud. Logica. **102**(4), 709–729 (2014)
15. Gilio, A., Sanfilippo, G.: Generalized logical operations among conditional events. Appl. Intell. **49**(1), 79–102 (2018). https://doi.org/10.1007/s10489-018-1229-8
16. Gilio, A., Sanfilippo, G.: Algebraic aspects and coherence conditions for conjoined and disjoined conditionals. Int. J. Approx. Reason. **126**, 98–123 (2020)

17. Gilio, A., Sanfilippo, G.: On compound and iterated conditionals. Argumenta **6**(2), 241–266 (2021)
18. Gilio, A., Sanfilippo, G.: Compound conditionals, Fréchet-Hoeffding bounds, and Frank t-norms. Int. J. Approx. Reason. **136**, 168–200 (2021)
19. Gilio, A., Sanfilippo, G.: Iterated conditionals and characterization of p-entailment. In: Vejnarová, J., Wilson, N. (eds.) ECSQARU 2021. LNCS (LNAI), vol. 12897, pp. 629–643. Springer, Cham (2021). https://doi.org/10.1007/978-3-030-86772-0_45
20. Hailperin, T.: Sentential Probability Logic. Origins, Development, Current Status, and Technical Applications. Lehigh University Press, Bethlehem (1996)
21. Kaufmann, S.: Conditionals right and left: probabilities for the whole family. J. Philosophical Logic **38**, 1–53 (2009)
22. Kern-Isberner, G.: Conditionals in Nonmonotonic Reasoning and Belief Revision - Considering Conditionals as Agents, vol. 2087. Springer, LNAI (2001)
23. Kraus, S., Lehmann, D., Magidor, M.: Nonmonotonic reasoning, preferential models and cumulative logics. Artif. Intell. **44**, 167–207 (1990)
24. Lehmann, D., Magidor, M.: What does a conditional knowledge base entail? Artif. Intell. **55**(1), 1–60 (1992). https://doi.org/10.1016/0004-3702(92)90041-U
25. McGee, V.: Conditional probabilities and compounds of conditionals. Philosophical Rev. **98**(4), 485–541 (1989)
26. Milne, P.: Bruno de Finetti and the logic of conditional events. British J. Philosophy Sci. **48**(2), 195–232 (1997)
27. Over, D., Cruz, N.: Suppositional theory of conditionals and rationality. In: Knauff, M., Spohn, W. (eds.) Handbook of Rationality, pp. 395–404. MIT press (2021)
28. Pfeifer, N., Tulkki, L.: Conditionals, counterfactuals, and rational reasoning: an experimental study on basic principles. Minds Mach. **27**(1), 119–165 (2017)
29. Sanfilippo, G., Gilio, A., Over, D., Pfeifer, N.: Probabilities of conditionals and previsions of iterated conditionals. Int. J. Approx. Reason. **121**, 150–173 (2020)

Classifier Probability Calibration Through Uncertain Information Revision

Sara Kebir$^{(\boxtimes)}$ (ID) and Karim Tabia (ID)

Univ. Artois, CNRS, CRIL, 62300 Lens, France
{kebir,tabia}@cril.fr

Abstract. There has been a lot of interest in explainable and trustworthy machine learning over the past few years. In some problems, it is not enough to predict correctly the true class label, but also to provide the probability that the prediction is true. This probability makes it possible to have confidence or not in the prediction. In this paper, we propose a new approach to calibrate the probabilities of a machine learning model through a post-processing step. The objective is to exploit some positive results such as that calibration is rather better on a small number of categories or subsets of classes than on a large number of classes. Based on this observation, our calibration approach, based on probabilistic belief revision, calibrates the predicted probabilities on the classes with the probabilities of subsets of classes. Preliminary experimental studies show very promising results, especially on certain datasets such as those of hierarchical classification.

Keywords: Classification · Probability calibration · Belief revision

1 Introduction

In AI, the current decade has seen the increasing use of innovative intelligent systems and applications that rely heavily on machine learning (ML). We are then confronted with new and difficult problems and risks induced, for the most part, by the complexity of the systems, their opacity and the sensitivity of certain critical applications. Among the serious questions we find confidence, trust, interpretability and explainability of reasoning and decisions that can be made automatically. Therefore, many issues are currently challenging the ML community to strengthen the explainability, interpretability and reliability of ML models and AI-based systems more generally.

In classification tasks, the goal is typically to learn a model that predicts the class variable as accurately as possible. Sometimes this is not enough since not only do we need to predict the class with good accuracy, but also provide a probability that the prediction is correct. This probability makes it possible to know the confidence that the model has in its prediction, and this can have consequences on how this prediction is managed in the context of automated decision-making as in safety-critical applications. Some models, like Bayesian

© Springer Nature Switzerland AG 2022
D. Ciucci et al. (Eds.): IPMU 2022, CCIS 1602, pp. 598–611, 2022.
https://doi.org/10.1007/978-3-031-08974-9_48

classifiers, can provide directly posterior probabilities while other classifiers use certain techniques and tricks to provide these probabilities. In practice, many models give poor estimates of predictive probabilities [13], often they overestimate these probabilities as in the case of random forests and even modern deep network-based models [5]. Calibration techniques are often used to better calibrate the probabilities of the model when it makes predictions.

We propose in this paper an original idea never used to calibrate the probabilities of a classifier. We see calibration as an uncertain information update task in the light of new uncertain and more reliable inputs. We place ourselves within the framework of Jeffrey's revision [12], a well-known framework for updating probabilistic beliefs with uncertain inputs. The preliminary results obtained confirm this intuition, in particular on certain datasets where there are class taxonomies, as is the case in several fields.

This paper is organized as follows : The second section summarizes some basic background notions about classification, probability calibration of classifiers and updating uncertain information with new uncertain inputs. Section 3 presents our work's main idea, followed in Sect. 4 by details about our proposed Jeffrey's rule-based probability calibration approach. Section 5, presents the experimental results, and Sect. 6 concludes this paper with some concluding remarks.

2 Basic Background Notions

2.1 Classification

Classification is a predictive task consisting in associating input data instances with symbolic labels. A classification task is defined by two sets of variables: A set of features $X = \{X_1,..,X_n\}$ where $|X|=n$, and a discrete target variable denoted C taking values in its domain D_C.

Definition 1. *(Classifier) A classifier f is a function mapping each input data instance x (vector instantiating each variable in X) to one value from the discrete variable domain D_C.*

Note that classification where each data instance x is associated exactly with one class is called multi-class classification, contrary to multi-label classification where one can associate a subset of classes at the same time to a data instance. In this paper, we deal only with calibrating multi-class classifiers.

2.2 Classifier Probability Calibration

We say that a classifier f is calibrated (or provides calibrated prediction probabilities) if, when it predicts a label $c_i \in C$ with probability $\hat{p}_i$, this prediction will be correct with probability $\hat{p}_i$ (intuitively, the probability $p_f(C = c_i | p = \hat{p}_i)$ is calibrated if on average, the prediction is correct with probability $\hat{p}_i$).

In order to visualize the quality of predicted probabilities, one can display a *reliability diagram* (see example of Fig. 1), a visual representation of prediction calibration. This comes down to plotting expected accuracy as a function of confidence. In such a diagram, confidence estimates are grouped into bins to allow computing the sample accuracy. Hence, a well calibrated model corresponds to the plot of the identity function. Calibration errors correspond to the gap between the estimated probabilities and the accuracy ones.

As for measuring classifiers efficiency, there are different measures for measuring miscalibration. Some commonly used metrics are Expected Calibration Error (ECE), Average Calibration Error (ACE), and Maximum Calibration Error (MCE). In few words, miscalibration metrics assess the errors by binning the samples by their confidence then assess the accuracy in each bin. For instance, ECE is simply the weighted mean gap between the confidence of the classifier and the observed accuracy (on a test set) in each bin. Similarly, the Maximum Calibration Error (MCE) gives simply the maximal gap (see for instance [8] for more details on measuring miscalibration). Negative log likelihood (NLL) can also be used to indirectly measure the model calibration since it penalizes high probability scores assigned to incorrect labels and low probability ones assigned to correct labels [1]. The lower these metrics are, the better is the quality of the calibration.

Various approaches have been proposed to perform recalibration. Platt scaling [11] and isotonic regression [9] are the most widely used in the binary classification setting. The most common way of extending these methods to the multi-class setting is treating the problem as k one-versus-all problems, where k is the number of classes [15]. Below, we recall how the two main post-processing methods perform.

Platt scaling is a parametric calibration method which fits a logistic regression model on the validation set, using the non-probabilistic predictions of the initial classifier z_i as features, to learn scalar parameters $a, b \in \mathbb{R}$ and compute the calibrated probabilities $\hat{p}_i$ as $\hat{p}_i = \alpha(az_i + b)$. The parameters a and b can be optimized by minimizing the Negative Log Likelihood. The Platt scaling can be extended to the multi-class setting by applying a linear transformation $Wz_i + b$ to the logits vector z_i. The parameters a and b will be in a higher dimension, $W \in \mathbb{R}^{k*k}$ and $b \in \mathbb{R}^k$ respectively. The resulting method is called **Matrix scaling** or **Vector scaling** if W is a diagonal matrix. There is also another extension of this method, called **Temperature scaling**, which uses a single scalar parameter $T > 0$ to calibrate the probabilistic predictions as follows $\hat{p}_i = \max_l \alpha_{sm}(z_i/T)^{(l)}$, where $\alpha_{sm}(z_i)^{(l)}$ is the predicted probability of the class $l = 1..k$ [5].

Isotonic regression is a non-parametric calibration method which outputs a piecewise constant function f to transform a probability p_i into a calibrated one by minimizing the mean-squared loss function $\sum_{i=1}^n (f(p_i) - y_i)^2$, where y_i is the true label. The Isotonic regression can be extended to the multi-class setting by performing a calibration for each class separately as a one-versus-all problem followed by a postprocessing to normalize the combined calibrated probabilities.

2.3   Updating Uncertain Information with New Uncertain Inputs: Jeffrey's Rule

Jeffrey's rule [6] extends the classical probabilistic conditioning to the case where the new information is uncertain. It allows to update an initial probability distribution p into a posterior one p' given the uncertainty bearing on a set of mutually exclusive and exhaustive events $\lambda=\{\lambda_1,..,\lambda_n\}$ (namely, λ is a partition of the set of possible states Ω). In this setting, the new input is in the form (λ_i, α_i), $i=1..n$ where α_i denotes the new probability of λ_i. Jeffrey's rule lies on the two following principles:

- **Success principle (P1)**

$$\forall \lambda_i \in \lambda, p'(\lambda_i) = \alpha_i \tag{1}$$

After the update operation, the posterior probability of each event λ_i must be equal to α_i as required in the new inputs. The uncertain inputs are seen as constraints or an effect once the new information is fully accepted.

- **Probability kinematics principle (P2)**

$$\forall \lambda_i \in \lambda, \forall \phi \subseteq \Omega, p(\phi|\lambda_i) = p'(\phi|\lambda_i) \tag{2}$$

This principle aims to ensure a kind of minimal change by ensuring that the posterior distribution p' should not change the conditional probability degrees of any event ϕ given the uncertain events λ_i. Jeffrey's rule assumes that in spite of the disagreement about the events λ_i in the prior distribution p and the posterior one p', the conditional probability of any event $\phi \subseteq \Omega$ given any uncertain event λ_i should remain the same in the original and the revised distributions.

Given a probability distribution p encoding the initial beliefs and new inputs in the form (λ_i, α_i) for $i=1..n$, the updated probability degree of any event $\phi \subseteq \Omega$ is obtained as follows:

$$p'(\phi) = \sum_{\lambda_i} \alpha_i * \frac{p(\phi \cap \lambda_i)}{p(\lambda_i)} \tag{3}$$

The posterior distribution p' obtained using Jeffrey's rule always exists and it is unique [4]. Note that in Jeffrey's rule, the events λ_i should be somewhat possible in the prior distribution (namely, $\forall \lambda_i \in \lambda$, $p(\lambda_i)>0$).

Jeffrey's framework for updating uncertain information with uncertain inputs has been used primarily for reasoning with uncertain information and observations [10]. In classification, it has been used instead to make predictions with uncertain observations (often called soft evidence) in some classifiers [2,3] such as those based on Bayesian networks or recently with some neural networks-based classifiers [14]. To the best of our knowledge, none of these works had used Jeffrey's rule of conditioning for probability calibration purposes.

3 Motivating Example

The starting point of this work is the following observation: on large-scale classification problems (involving a large number of classes), it is often difficult to correctly predict certain classes, especially in the case of unbalanced datasets (often, classifiers favor majority classes to the detriment of underrepresented classes). What is true for the accuracy of the predictions is also true for the confidence probabilities of the model in its predictions. This difficulty may be essentially linked to the nature of the data and the specificities of the classifiers used. For example, if we use random forests by limiting the depth of the trees too much, it will be difficult to discriminate certain classes since the number of tests on the features is strongly limited.

Two simple questions then arise: i) If we group classes into categories (kind of super classes, so as to be better represented and reduce the number of classes), can we improve the quality of the predictions (in terms of accuracy and calibration)? ii) If so, would it be possible to exploit the best performances of the predictions on the categories (subsets of the initial classes) to *rectify* or *calibrate* the predictions and the calibration of the classifier f?

For the first question, the answer is positive for most of the datasets and classifiers, although with different results depending on how the classes $\{c_1, .., c_k\}$ are grouped into categories $\{cat_1, .., cat_j\}$ (with $j < k$). In Fig. 1, one can see the reliability diagrams of an SVM classifier learnt on the well-known DBPedia[1] dataset. Clearly, the SVM classifier built on 70 classes is poorly calibrated (see the gap to the perfect calibration line) compared to the SVM built on 9 categories.

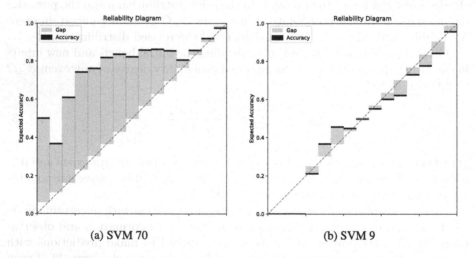

(a) SVM 70   (b) SVM 9

Fig. 1. Reliability diagrams of SVM classifiers on DBPedia

---

[1] This dataset is a cleaned extract of 342,782 wikipedia articles' data providing hierarchical classes (there are 3 levels with 9, 70 and 219 classes respectively), https://www.kaggle.com/danofer/dbpedia-classes.

For the second question, having a more calibrated classifier on the categories provides relevant information on the classes included in each category. How to use these probabilities on the categories and in which case they should be used in order to guarantee an improvement of the probabilities of the initial classifier will be discussed in the following section.

4   Probability Calibration Based on Jeffrey's Rule

Let us denote the classifier to calibrate f and let us denote by f' the one learnt on categories and ensuring better calibrated probabilities on categories. Each category from $\{cat_1, .., cat_j\}$ is a subset of classes from $\{c_1, .., c_k\}$ (note that a given class belongs to only one category). Namely, the categories form a partition of the set of classes.

4.1   Jeffrey's Rule-Based Probability Calibration

We have, on one side, classes and a probability distribution provided by the classifier f, and on the other side, categories (a partition of the classes) and another probability distribution on categories provided by the classifier f'. Moreover, in most cases, the probabilities of the classifiers f' are more calibrated as illustrated in the example of Fig. 1 meaning that the classifier f' is more reliable in terms of calibration. This places our problem somehow in the framework of updating uncertain information with new uncertain information. Note that the prior information is a probability distribution p over $\{c_1, .., c_k\}$ and that the new information is also uncertain and it is in the form of a probability distribution p' over a partition of $\{c_1, .., c_k\}$ and denoted $\{cat_1, .., cat_j\}$.

It fully makes sense to update the distribution p with p' since this latter provides more calibrated probabilities. This comes down to give priority to the new information p' exactly in line with Jeffrey's rule. Hence, the revised probabilities p_c are obtained following Jeffrey's rule as follows : $\forall c_i \in D_C$,

$$p_c(c_i) = p(c_i) * \frac{p'(cat(c_i))}{p(cat(c_i))}, \tag{4}$$

where $cat(c_i)$ denotes the category of class c_i. Note also that $p(cat(c_i))$ is the probability of all classes from category $cat(c_i)$ computed from the prior distribution p. The posterior distribution p_c always exists and it is unique unless the first classifier f associates a zero probability to $cat(c_i)$ (namely, $p(cat(c_i)) = 0$).

4.2   Jeffrey's Rule-Based Probability Calibration in Practice

Up to now, we have briefly presented the main idea to calibrate the probabilities of a classifier f by exploiting the probabilities of another classifier f' on categories in the spirit of Jeffrey's rule. Now, several questions arise regarding the use of our calibration technique in practice :

- *How to group classes into categories ?* In some domains, there are taxonomies and class hierarchies to semantically group classes into categories. This is a first option but it does not necessarily guarantee the best results. The number of categories and the composition of the categories is one of the key points to have well calibrated probabilities on the categories to ensure better results after the calibration. For datasets without class taxonomies, one way would be to cluster the data and obtain the categories corresponding to the clusters using some clustering techniques.
- *In what case there could be an improvement and how much will the improvement be?* The improvement will be all the greater when the initial probabilities are weakly calibrated and when the probabilities on the categories p' are such that they can correct the initial probabilities p. This needs to be formally or empirically characterized to identify the situations where calibration based on the proposed method is advisable. Another idea to improve the results of the calibration is, as we will see in our preliminary experimental study, to start with calibrated probabilities both for the classifier to be calibrated f and for the calibration classifier f'. Indeed, nothing prevents in this case from using existing calibration techniques to pre-calibrate p and p' to finally revise according to our Jeffrey's rule-based Probability Calibration method. We don't even have to use the same classification technique for f and f' in case the latter guarantees a better calibration on the categories.

These questions are not trivial and are beyond the scope of this paper.

Before proceeding to the experimental study section, Fig. 2 below illustrates our Jeffrey's rule-based Probability Calibration process that encompasses both of the above issues, i.e. the part regarding how to group classes into categories and the one on how to further improve the initial calibration using an oracle (for instance, using existing calibration techniques).

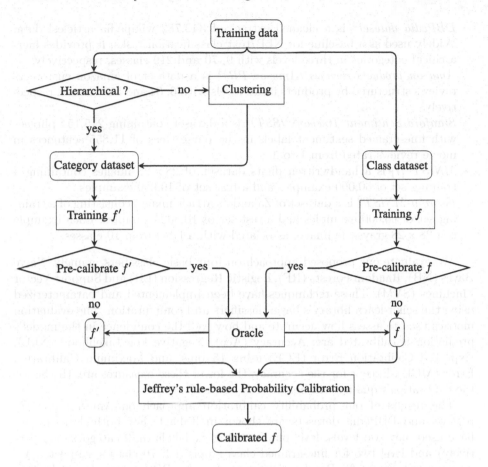

Fig. 2. Jeffrey's rule-based probability calibration

5   Experimental Study

This section presents preliminary results evaluating our approach for classifier probability calibration based on Jeffrey's rule. The experiments are carried out on some well-known datasets where grouping classes into categories is done thanks to existing class taxonomies (for instance, DBPedia and Amazon products reviews datasets) or where it is easy to group manually classes into categories according to the semantics of classes (such as, Stanford Sentiment Treebank dataset) or using a clustering method (as is the case for MNIST and Fashion-MNIST datasets) which are described below:

- *DBPedia dataset*[2] is a cleaned extract of 342,782 wikipedia articles' data. Widely used as a baseline for NLP/text classification tasks, it provides hierarchical categories in three levels with 9, 70 and 219 classes, respectively.
- *Amazon products reviews (Amazon PR)*[3] is a dataset of amazon customers reviews structured by products into 3 levels with 6, 64 and 510 classes, respectively.
- *Stanford Sentiment Treebank (SST)*[4] is a dataset containing 215,154 phrases with fine-grained sentiment labels in the parse trees of 11,855 sentences in movie reviews rated from 1 to 5.
- *MNIST* [7] is a handwritten digits dataset of 28×28 images containing a training set of 60,000 examples and a test set of 10,000 examples.
- *Fashion-MNIST*[5] is a dataset of Zalando's article images consisting of a training set of 60,000 examples and a test set of 10,000 examples. Each example is a 28×28 grayscale image, associated with a label from 10 classes.

We evaluate our proposed approach on four basic classifiers, namely, Naive Bayes (NB), Random Forests (RF), Logistic Regression (LR) and Support Vector Machines (SVM). These techniques have been implemented and parameterized using the scikit-learn library's basic classifiers and configuration. The evaluation metrics used to assess how accurate and how well the confidence in the model's prediction is calibrated are: Accuracy (Acc), Negative Log Likelihood (NLL), Expected Calibration Error (ECE) using 15 bins, and Maximum Calibration Error (MCE). Except for the accuracy, the lower these measures are, the better the calibration's quality will be.

The results of our probability calibration approach on Amazon product reviews and DBPedia datasets are shown in Table 1. For both datasets, we have used only two levels: level one for category labels (6, 9 categories, respectively) and level two for fine-grained class labels (64, 70 classes, respectively). As expected, except for Random Forest on Amazon PR, all tested models show an improvement in confidence quality based on all calibration evaluation metrics, while maintaining, or improving in some cases, the initial accuracy, thus confirming the potential effectiveness of our approach. Indeed, unlike the other cases, one can notice that the Random Forest was already well calibrated at the beginning, with an ECE $\approx 5\%$, which explains the ineffectiveness of the calibration in this situation and confirms that when the initial classifier is already well calibrated, neither the state-of-the-art methods nor ours have much room for improvement, as shown in Table 2.

---

[2] DBPedia dataset, https://www.kaggle.com/danofer/dbpedia-classes.
[3] Amazon PR dataset, https://www.kaggle.com/kashnitsky/hierarchical-text-classification.
[4] SST dataset, https://nlp.stanford.edu/sentiment/treebank.html.
[5] Fashion-MNIST dataset, https://github.com/zalandoresearch/fashion-mnist.

Table 1. Comparison between the performance of the classifiers before (Uncalibrated) and after the use of Jeffrey's rule-based Probability Calibration (Proposed method) on Amazon Products Reviews and DBPedia datasets.

Model		Amazon PR				DBPedia			
		Acc%	NLL	ECE%	MCE%	Acc%	NLL	ECE%	MCE%
NB	Uncalibrated	42.93	2.67	27.63	67.17	71.42	1.22	22.41	34.08
	Proposed method	**53.01**	**2.05**	**26.49**	**53.33**	**71.71**	**1.14**	**20.51**	**32.09**
LR	Uncalibrated	64.00	1.77	23.73	44.90	**92.30**	0.36	10.31	36.25
	Proposed method	**67.44**	**1.42**	**18.25**	**32.79**	91.76	**0.35**	**09.23**	**29.74**
RF	Uncalibrated	67.19	2.50	**05.16**	**10.59**	**90.57**	0.62	26.39	47.24
	Proposed method	**67.39**	**2.40**	14.76	25.32	90.36	**0.60**	**25.66**	**43.03**
SVM	Uncalibrated	62.40	1.65	16.85	40.86	**83.77**	**0.87**	21.16	54.12
	Proposed method	**63.35**	**1.59**	**06.78**	**14.98**	81.01	0.90	**11.05**	**30.08**

Table 2. Comparison between the performance of the classifiers before (Uncalibrated) and after the use of state-of-the-art calibration methods and Jeffrey's rule-based Probability Calibration using the oracle (Proposed method-oracle), on Amazon Products Reviews and DBPedia datasets.

Model		Amazon PR				DBPedia			
		Acc%	NLL	ECE%	MCE%	Acc%	NLL	ECE%	MCE%
NB	Uncalibrated	42.93	2.67	27.63	67.17	71.42	1.22	22.41	34.08
	Isotonic reg	68.19	1.50	13.03	24.46	**88.88**	**0.45**	12.13	**24.8**
	Sigmoid reg	62.32	1.71	22.49	28.37	83.03	0.76	16.62	25.66
	Proposed method-oracle	**70.12**	**1.38**	**9.40**	**20.67**	82.00	0.69	**8.53**	26.41
LR	Uncalibrated	64.00	1.77	23.73	44.90	**92.30**	**0.36**	10.31	36.25
	Isotonic reg	67.32	1.47	11.51	19.33	91.76	0.41	15.99	33.26
	Sigmoid reg	68.05	1.31	13.45	20.50	92.12	0.40	16.35	32.35
	Proposed method-oracle	**70.27**	**1.13**	**9.10**	**18.76**	91.50	**0.36**	**8.38**	**24.76**
RF	Uncalibrated	**67.19**	2.50	**05.16**	10.59	90.57	0.62	26.39	47.24
	Isotonic reg	64.39	5.16	21.14	46.88	**91.99**	0.39	2.71	13.30
	Sigmoid reg	67.02	**1.73**	20.79	41.86	91.47	**0.32**	1.94	11.72
	Proposed method-oracle	66.24	2.63	5.65	**9.12**	90.10	0.49	**1.41**	**8.46**
SVM	Uncalibrated	62.40	1.65	16.85	40.86	83.77	0.87	21.16	54.12
	Isotonic reg	**65.89**	**1.52**	11.03	22.16	**85.49**	**0.64**	18.35	30.88
	Sigmoid reg	24.59	2.65	**6.12**	33.11	33.67	2.55	**8.76**	40.95
	Proposed method-oracle	63.35	1.59	6.78	**14.98**	81.01	0.90	11.06	**30.08**

To further improve the calibration performance, we apply our approach on the resulting models from the oracle which have, in the most cases, a pre-calibrated probabilities p and p'. The pre-calibration is provided through the use of state-of-the-art calibration techniques or another classifier for f', trained on the same training set and showing better calibrated probabilities p' on the categories. The obtained results are shown in Tables 2, 3 and 4.

As expected, the previous results, illustrated in Table 1, improve further with the use of an oracle and outperform state-of-the-art models, namely, Isotonic regression and Sigmoid regression. We can see for instance in Table 2, the Random Forest on DBPedia related ECE being reduced from 26.39 to 1.41%, while ensuring a good accuracy. The same applies to almost all other classifiers on those datasets. In most cases, the oracle result is a pre-calibration of the probabilities p and p' using Isotonic regression since when comparing the results of the two state-of-the-art calibration methods, the latter performs better.

The Stanford Sentiment Treebank (SST) dataset has been processed differently. As it does not have any levels, a manual clustering based on the semantic of its classes, films rating, is used to group them into 2 ([[1,2],[3,4,5]]) and 3 ([[1,2],[3],[4,5]]) clusters. The results illustrated in Table 3 confirm the effectiveness of our calibration approach on this dataset too. In addition to the calibration effect, one may notice that the initial evaluation metrics are much lower compared to the results presented previously.

Table 3. Comparison between the performance of the classifiers before (Uncalibrated) and after the use of state-of-the-art calibration methods and Jeffrey's rule-based Probability Calibration using the oracle (Proposed method-oracle), on SST.

Model		SST			
		Acc%	NLL	ECE%	MCE%
NB	Uncalibrated	36.20	1.49	3.26	5.93
	Isotonic reg	39.77	1.40	3.76	6.53
	Sigmoid reg	**40.14**	1.42	5.79	9.59
	Proposed method-oracle	39.14	**1.38**	**2.80**	**5.67**
LR	Uncalibrated	36.47	1.43	2.37	**7.32**
	Isotonic reg	39.55	1.39	2.95	26.54
	Sigmoid reg	38.46	1.41	2.89	13.33
	Proposed method-oracle	**39.82**	**1.37**	**0.91**	8.12
RF	Uncalibrated	33.94	2.51	20.81	45.87
	Isotonic reg	37.10	1.45	2.83	4.78
	Sigmoid reg	**37.38**	1.47	4.27	**7.20**
	Proposed method-oracle	36.38	**1.44**	**1.48**	21.50
SVM	Uncalibrated	36.83	1.52	12.71	49.87
	Isotonic reg	38.55	1.40	1.71	26.38
	Sigmoid reg	36.61	1.42	**1.43**	**18.75**
	Proposed method-oracle	**39.77**	**1.38**	2.33	32.25

To get the required categories from datasets without class taxonomies, as is the case for MNIST and Fashion-MNIST, and where a semantic categorisation as separating the handwritten digits of MNIST dataset into odd and even numbers proved to be ineffective, a k-means clustering technique is performed to group the classes into 2 and 3 clusters. The results obtained with the application of our proposed calibration approach on this different type of datasets are given in Table 4 and confirm, once again, the effectiveness of our proposed Jeffrey's rule-based Probability Calibration technique.

Table 4. Comparison between the performance of the classifiers before (Uncalibrated) and after the use of state-of-the-art calibration methods and Jeffrey's rule-based Probability Calibration using the oracle (Proposed method-oracle), on MNIST and Fashion-MNIST datasets.

Model		MNIST				Fashion-MNIST			
		Acc%	NLL	ECE%	MCE%	Acc%	NLL	ECE%	MCE%
NB	Uncalibrated	83.57	1.99	13.93	45.26	65.52	5.55	29.52	59.56
	Isotonic reg	84.69	0.50	1.37	7.17	70.17	0.86	3.46	11.76
	Sigmoid reg	83.65	0.73	4.63	38.45	65.68	1.24	8.45	25.90
	Proposed method-oracle	**88.54**	**0.36**	**0.76**	**5.67**	**72.54**	**0.73**	**2.64**	**10.63**
LR	Uncalibrated	92.56	0.27	0.68	**6.98**	84.43	0.44	2.16	7.26
	Isotonic reg	90.95	0.46	17.49	29.74	77.12	0.80	21.18	32.90
	Sigmoid reg	91.33	0.47	20.14	33.76	78.89	0.83	26.42	36.43
	Proposed method-oracle	**94.88**	**0.18**	**0.64**	13.68	**85.43**	**0.41**	**0.91**	**5.90**
RF	Uncalibrated	**97.05**	0.24	13.65	38.27	87.72	**0.41**	7.86	**18.35**
	Isotonic reg	95.62	0.61	0.23	**11.44**	86.16	2.95	0.85	39.85
	Sigmoid reg	97.00	**0.12**	1.75	20.17	**87.73**	0.64	9.53	40.45
	Proposed method-oracle	95.60	0.85	**0.16**	15.44	86.21	2.94	**0.58**	40.38
SVM	Uncalibrated	95.85	0.13	1.55	11.66	**86.87**	**0.37**	0.76	5.49
	Isotonic reg	95.77	0.21	**0.47**	35.37	86.80	0.47	2.72	20.84
	Sigmoid reg	95.78	0.32	2.46	36.38	86.81	0.56	2.12	11.67
	Proposed method-oracle	**97.24**	**0.10**	0.95	**9.96**	86.77	**0.37**	**0.56**	**3.05**

6   Concluding Remarks

In this preliminary work, we have sketched out a novel method to calibrate the probabilities of a classifier through uncertain and more reliable information revision based on Jeffrey's rule of conditioning.

We have noticed during the experimental study that the NLL, ECE and MCE metrics are not sufficient to predict the effect of the probability calibration technique. Using a category model that is either overconfident or underconfident while displaying the same calibration measures in both cases does not lead to the same results after the revision, in other words, trying to calibrate a very overconfident classes model with another overconfident category model is not very useful, the same applies to underconfidence.

One of the most important issues facing the proposed approach is the quality of the category model. Thus, future work may focus on finding a better way of categorisation since the one obtained with the clustering method or the given class taxonomies in hierarchical datasets are not necessarily good.

The computational complexity of our calibration approach depends on the one of calculating the two distributions p and p'. There are two distinct cases. If we place ourselves in the case without using an oracle, we will have to sum the cost of calling the category classifier to get p' and the one of applying our calibration method, which is linear in the number of classes, whereas in the other case we will have to add the cost of the oracle calls.

Even though we have only employed the most basic machine learning classifiers so far, the calibration approach proposed in this paper is quite competitive with state-of-the-art methods and has demonstrated its efficacy. We are aware that a wide range of highly accurate and complex classification models exist, and that highly efficient calibration techniques have already been proposed and applied to them; therefore, our next step is to test our proposed calibration approach on them, as well as to try to expand the range of datasets used to further provide evidence on the effectiveness of our approach.

Acknowledgement. This work has been supported by the Vivah project 'Vers une Intelligence artificielle à VisAge Humain' supported by the ANR.

This work has received support from the ANR CROQUIS (Collecting, Representing, cOmpleting, merging, and Querying heterogeneous and UncertaIn waStewater and stormwater network data) project, grant ANR-21-CE23-0004 of the French research funding agency Agence Nationale de la Recherche (ANR).

References

1. Ashukha, A., Lyzhov, A., Molchanov, D., Vetrov, D.P.: Pitfalls of in-domain uncertainty estimation and ensembling in deep learning. In: 8th International Conference on Learning Representations, ICLR 2020, Addis Ababa, Ethiopia, 26–30 April 2020. OpenReview.net (2020). https://openreview.net/forum?id=BJxI5gHKDr
2. Benferhat, S., Tabia, K.: Inference in possibilistic network classifiers under uncertain observations. Ann. Math. Artif. Intell. **64**(2–3), 269–309 (2012)
3. Benferhat, S., Tabia, K.: Reasoning with uncertain inputs in possibilistic networks. In: Principles of Knowledge Representation and Reasoning: Proceedings of the Fourteenth International Conference, KR 2014, Vienna, Austria, 20–24 July 2014 (2014). http://www.aaai.org/ocs/index.php/KR/KR14/paper/view/7964
4. Chan, H., Darwiche, A.: On the revision of probabilistic beliefs using uncertain evidence. Artif. Intell. **163**(1), 67–90 (2005)
5. Guo, C., Pleiss, G., Sun, Y., Weinberger, K.Q.: On calibration of modern neural networks. In: Precup, D., Teh, Y.W. (eds.) Proceedings of the 34th International Conference on Machine Learning. Proceedings of Machine Learning Research, vol. 70, pp. 1321–1330. PMLR (06–11 Aug 2017). https://proceedings.mlr.press/v70/guo17a.html
6. Jeffrey, R.: The Logic of Decision. McGraw Hill, New York (1965)
7. Lecun, Y., Bottou, L., Bengio, Y., Haffner, P.: Gradient-based learning applied to document recognition. Proc. IEEE **86**(11), 2278–2324 (1998). https://doi.org/10.1109/5.726791
8. Naeini, M.P., Cooper, G.F., Hauskrecht, M.: Obtaining well calibrated probabilities using bayesian binning. In: Proceedings of the Twenty-Ninth AAAI Conference on Artificial Intelligence, pP. 2901–2907. AAAI 2015, AAAI Press (2015)
9. Niculescu-Mizil, A., Caruana, R.: Predicting good probabilities with supervised learning. In: Proceedings of the 22nd International Conference on Machine Learning, pP. 625–632. ICML 2005, Association for Computing Machinery, New York (2005). https://doi.org/10.1145/1102351.1102430

10. Peng, Y., Zhang, S., Pan, R.: Bayesian network reasoning with uncertain evidences. Int. J. Uncertainty Fuzziness Knowl. Based Syst. **18**, 539–564 (2010). https://doi.org/10.1142/S0218488510006696
11. Platt, J.: Probabilistic outputs for support vector machines and comparisons to regularized likelihood methods. Adv. Large Margin Classifiers **10**(3), 61–74 (1999)
12. Shafer, G.: Jeffrey's rule of conditioning. Philos. Sci. **48**(3), 337–362 (1981). http://www.jstor.org/stable/186984
13. de Menezes e Silva Filho, T., Song, H., Perelló-Nieto, M., Santos-Rodríguez, R., Kull, M., Flach, P.A.: Classifier calibration: How to assess and improve predicted class probabilities: a survey. CoRR abs/2112.10327 (2021). https://arxiv.org/abs/2112.10327
14. Yu, E.: Bayesian neural networks with soft evidence. CoRR abs/2010.09570 (2020). https://arxiv.org/abs/2010.09570
15. Zadrozny, B., Elkan, C.: Transforming classifier scores into accurate multiclass probability estimates. In: Proceedings of the Eighth ACM SIGKDD International Conference on Knowledge Discovery and Data Mining. p. 694–699. KDD 2002, Association for Computing Machinery, New York (2002). https://doi.org/10.1145/775047.775151

Evidential Hybrid Re-sampling for Multi-class Imbalanced Data

Fares Grina[1,2(✉)], Zied Elouedi[1], and Eric Lefevre[2]

[1] Institut Supérieur de Gestion de Tunis, LARODEC, Université de Tunis,
Tunis, Tunisia
grina.fares2@gmail.com, zied.elouedi@gmx.fr
[2] Univ. Artois, UR 3926, Laboratoire de Génie Informatique et d'Automatique de
l'Artois (LGI2A), 62400 Béthune, France
eric.lefevre@univ-artois.fr

Abstract. Learning from class-imbalanced datasets has gained substantial attention in the machine learning community, leading to solutions for healthcare, security, banking, etc. Specifically, binary imbalanced problems has received the most interest in the field. Yet, there has been little emphasis given to dealing with multi-class imbalance learning. Data imbalance can significantly worsen the classification performance, especially in the presence of other data difficulties such as uncertainty, i.e., ambiguous samples and noise. In this paper, we present an evidential hybrid re-sampling method for dealing with class imbalance in the multi-class setting. This technique uses the evidence theory to assign a soft label to each object. This evidential modeling provides more information about each object's region, which improves the selection of objects in both undersampling and oversampling. Our approach firstly selects ambiguous majority instances for undersampling, then oversamples minority objects through the generation of synthetic examples in borderline regions to better improve minority class borders. An adjustment has also been integrated in order to avoid excessive oversampling and undersampling. Benchmarking results have shown significant improvement of G-Mean of AUC metrics over other popular re-sampling methods.

Keywords: Resampling · Multi-class imbalance · Evidence theory · Data uncertainty

1 Introduction

Class imbalance is a very common situation in classification problems, specifically in many real-world scenarios such as intrusion detection [4], medical diagnosis [15], and fraud detection [16]. Formally, an imbalanced dataset contains at least one class with much fewer number of examples than the other classes. The underrepresented classes are called minority classes while the others with larger number of instances are referred to as majority classes. In most cases,

© Springer Nature Switzerland AG 2022
D. Ciucci et al. (Eds.): IPMU 2022, CCIS 1602, pp. 612–623, 2022.
https://doi.org/10.1007/978-3-031-08974-9_49

classification models tend to favor majority classes due to their large presence, while incorrectly classifying the instances from the minority classes. This raises a problem since minority classes often have more importance than the majority ones. For instance, in the medical domain, our interest is to detect patients with diseases, which could be a rare pattern (minority).

A lot of studies has been carried out in order to enhance classification performance on binary datasets, in which there is only one minority class and one majority class. Among many strategies, re-sampling is an effective approach for learning from class-imbalanced datasets. It aims at rebalancing the training set at the preprocessing level by adding synthetic minority objects (oversampling), removing majority samples (undersampling), or both (hybrid resampling). As a consequence, it improves the effects. The most traditional resampling techniques are random oversampling (ROS) and random undersampling (RUS). The former randomly selects minority objects and simply replicates them, while the latter randomly removes majority instances. Although it might seem that these random solutions are effective since the class distribution is balanced, they can lead to different issues. For instance, ROS can lead to overfitting, and RUS can potentially remove meaningful samples from the data [11]. Therefore, other methods have been suggested to make re-sampling less naive.

The most popular solution based on this logic is Synthetic Minority Oversampling Technique (SMOTE) [6], which is considered as the baseline for most contributions. Unlike ROS, SMOTE interpolates between minority objects that are close to each other, to create new synthetic minority instances. Unfortunately, SMOTE has some issues regarding overgeneralization and amplification of problems already present in the data, that is, data uncertainty involving class overlapping and noise [23]. In [12], authors suggested BorderlineSMOTE, a SMOTE variant with the goal of identifying borderline minority class objects to generate new samples. In fact, applying oversampling on the borders of the minority class allows the improvement of the visibility of minority objects. Safe-Level-SMOTE [5] is another technique which only selects objects in safe region for oversampling. Clustering has also been introduced in many oversampling methods [8] to smartly select the regions where to create synthetic points. More recently, some interests have been pointed towards generative adversarial networks as the basis for oversampling [24].

Regarding undersampling, contributions focused on intelligent selection of unwanted majority samples to discard, rather than randomly removing them. Traditionally, Editing Nearest Neighbors (ENN) [29] and Tomek Links (TL) [14] are occasionally used for undersampling. Similar to oversampling, clustering-based undersampling was presented in a number of works [27] to improve the selection of majority objects to remove.

The combination of oversampling and undersampling is also a useful strategy to re-balance the dataset. Traditionally, methods combine SMOTE's oversampling with undersampling filtering techniques. In [3], for example, the authors proposed two hybrid sampling approaches: SMOTE-ENN and SMOTE-TOMEKLINKS. Since SMOTE can potentially expand the minority set regions

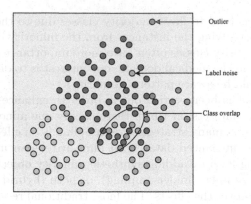

Outlier

Label noise

Class overlap

Fig. 1. Data difficulties in multi-class imbalanced datasets.

by interpolating new synthetic samples into the majority set clusters. This is called over-generalization. The application of ENN [29] to the SMOTE-oversampled training set tends to remove examples from both classes, meaning that every observation misclassified by its nearest neighbors is cleaned from the training data. Similar to SMOTE-ENN, instead of eliminating only the majority examples that form a Tomek Link, examples from both classes are deleted. SMOTE-RSB* [21] is another method that combines SMOTE for oversampling with the Rough Set Theory as a cleaning technique. Nevertheless, learning from multi-class imbalanced datasets has not been as heavily researched. This is due to the complexity of multi-class relationships compared to two-class problems. Rather than directly applying these methods to the multiple classes, many methods focus on decomposing multi-class problems into binary ones. For instance, the One-Versus-One (OVO) approach [13] is a decomposition scheme developed to modify the multi-class problem into multiple binary sub-problems, one for each pair of classes. Each sub-problem is trained on a binary classifier, ignoring the remaining instances not belonging the pair. Similarly to OVO, One-Versus-All (OVA) [22] is another decomposition framework which transforms the multi-class data into multiple binary sub-problems. However, OVA trains a classifier for each class in the training dataset. Other variants of these methods were also suggested [19], mostly to improve the combination of classifiers decisions. Even though binarization is simple and straightforward approach for learning from multiple class problems, it may lead to some regions being ignored and left unlearned. Specifically, when there is high data uncertainty, such as ambiguity created by high overlapping, and noise. These issues cannot be dealt with using decomposition techniques.

To handle the drawbacks of existing re-sampling algorithms, this paper presents an algorithm named Multi-Class Evidential Hybrid re-Sampling (MC-EVHS). It is an extension of the method proposed in [10] to specifically deal with imbalanced datasets with multiple classes and other data difficulties, i.e., overlapping classes, label noise and outliers, as illustrated in Fig. 1. This approach uses evidence theory in order to represent memberships, before combining oversampling and undersampling. Utilizing this theory provides us with more information in order to better choose the locations of newly generated objects and which majority instances to remove. Consequently, we apply an evidential version of SMOTE on the minority classes, and evidential undersampling on majority classes. Since we are dealing with multiple classes, we also propose a mechanism to identify which classes to consider for oversampling or undersampling, in addition to an improved way of controlling the amount of re-sampling that should be performed for each class. This adds an adaptive behavior to our approach.

This paper will be divided as follows. First, evidence theory is recalled in Sect. 2. Section 3 details each step of our idea. Experimental evaluation is discussed in Sect. 4. We finish our paper with a conclusion and an outlook on future work in Sect. 5.

2   Evidence Theory

The theory of evidence [7, 25, 26], also referred to as Dempster-Shafer theory (DST) or belief function theory, is a flexible and well-founded framework for the representation and combination of uncertain knowledge. The frame of discernment defines a finite set of M exclusive possible events, e.g., possible labels for an object to classify, and is denoted as follows:

$$\Omega = \{w_1, w_2, ..., w_M\} \tag{1}$$

A basic belief assignment (bba) denotes the amount of belief stated by a source of evidence, committed to 2^Ω, i.e., all subsets of the frame including the whole frame itself. Precisely, a bba is represented by a mapping function $m : 2^\Omega \to [0, 1]$ such that:

$$\sum_{A \in 2^\Omega} m(A) = 1 \tag{2}$$

Each mass $m(A)$ quantifies the amount of belief allocated to an event A of Ω. A focal element is a subset $A \subseteq \Omega$ where $m(A) \neq 0$.

The *Plausibility* function is another representation of knowledge defined by *Shafer* [25] as follows:

$$Pl(A) = \sum_{B \cap A \neq \emptyset} m(B), \quad \forall \ A \in 2^\Omega \tag{3}$$

$Pl(A)$ represents the total possible support for A and its subsets.

3   Evidential Hybrid Re-sampling for Multi-class Imbalanced Data

MC-EVHS is a re-sampling method which combines oversampling and undersampling to re-balance multi-class datasets. It firstly assigns soft evidential labels to each object in the dataset. The theory of evidence is used here to represent memberships towards each class, and also each meta-class (overlapping between classes). This allows us to represent data uncertainty. The computed soft labels are then used to select unwanted majority samples for undersampling, and pick the right regions for oversampling the minoirty classes.

The definitions of majority and minority classes in multi-class imbalance have been discussed in many works. In this paper, we consider a minority each class that has a number of objects less than the mean s of the number of objects in each class.

For all classes with a number of objects higher than the mean s, the assigned memberships are used to smartly perform undersampling. Our version of undersampling has an adaptive behavior, since the number of removed objects depends on the amount of overlap and noise present in the corresponding majority class. However, each majority class size should not get inferior to the calculated mean s.

Regarding all classes with a number of objects lower than the mean s, we use the calculated evidential memberships in order to perform oversampling in the borders of the minority class. Similarly to undersampling, our version of oversampling adapts to each class and generates synthetic minority instances only in the wanted locations. The only stopping criterion is not exceeding the mean s.

3.1   Computation of Evidential Labels

Our proposed approach proceeds by determining the centers of each class and meta-class (the overlapping region), then creating a *bba* based on the distance between each object and each class center.

The class centers are calculated using the mean value of the training set in the corresponding class. Regarding the overlapping regions represented by meta-classes, the centers are defined by the barycenter of the involved class centers as follows:

$$C_U = \frac{1}{|U|} \sum_{\omega_i \in U} C_i \tag{4}$$

where ω_i are the classes in U, U represents the meta-class, and C_i is the corresponding center.

After the creations of centers, we assign to each example a soft evidential label represented by a *bba* over the frame of discernment $\Omega = \{\omega_1, ..., \omega_M, \omega_0\}$, where the M classes are represented. The proposition ω_0 is included in the frame of discernment to represent the outlier, i.e., assignment of objects that are far from any class in the data. It is important to note that not all meta-classes should be considered as potential focal elements. Indeed, some classes do not overlap, and so no object needs to be assigned to the meta-class involving them. Additionally, it would be more computationally efficient to not calculate the mass value for this type of meta-classes. As enforced in [17], the meta-class center should be closer to the centers of its involved classes than to other incompatible classes' centers.

Let x_s be an object belonging to the training set. The idea is that each class or meta-class center represents a piece of evidence to the evidential membership of x_s. Accordingly, the mass values for each focal element in regard to x_s's memberships should depend on $d(x_s, C)$, that is, the distance between the respective class center C and x_s. The farther the center is, the lower the mass value for the corresponding class. By analogy, the closer x_s is to a class/meta-class center, the more likely it belongs to it. Hence, the initial unnormalized masses should be represented by decreasing distance based functions. We use the Mahalanobis distance [18], in this work, as recommended by [17] in order to deal with anisotropic datasets. Meta-classes U are chosen based on the constraint given above. The unnormalized masses are calculated accordingly:

$$\hat{m}(\{\omega_i\}) = e^{-d(x_s, C_i)} \tag{5}$$

$$\hat{m}(U) = e^{-\gamma \lambda d(x_s, C_U)}, \quad for \ |U| \geq 1 \tag{6}$$

$$\hat{m}(\{\omega_0\}) = e^{-t} \tag{7}$$

where $\lambda = \beta \, 2^\alpha$. A value of $\alpha = 1$ is fixed as recommended to obtain good results on average, and β is a parameter such that $0 < \beta < 1$. It is used to tune the number of objects committed to the overlapping region. The value of γ is equal to the ratio between the maximum distance of x_s to the centers in U and the minimum distance. It is used to measure the degree of distinguishability among classes in U. The smaller γ indicates a poor distinguishability degree between the classes of U for x_s. The outlier class ω_0 is taken into account in order to deal with objects far from all classes, and its mass value is calculated according to an outlier threshold t.

Finally, the unnormalized belief masses are normalized as follows:

$$m(A) = \frac{\hat{m}(A)}{\sum_{B \subseteq \Omega} \hat{m}(B)} \tag{8}$$

3.2   Evidential Adaptive Undersampling

This part consists of downsampling the majority classes. As mentioned above, this is dedicated to the classes whose size is higher than the mean size, corresponding to a majority. The created *bbas* are used here to determine whether an object is necessary for the learning phase or not. The logic behind our idea is to discard the samples which have a high uncertainty, that is, samples which present a relatively higher difficulty to correctly classify. These types of instances involve high ambiguity (class overlapping samples), outliers, and label noise. The evidential membership is used to detect those samples.

Class Overlapping. In our framework, overlapping objects have high masses assigned to meta-class focal elements, i.e., non-singleton propositions. For example, a sample with the maximum mass assigned to $U = \{\omega_1, \omega_2, \omega_3\}$ signifies that it is located in the region intersecting the three classes ω_1, ω_2, and ω_3. This specific instance can be removed in the undersampling phase, in order to reduce the data ambiguity and reduce majority classes' sizes, at the same time.

Since class overlap has not been mathematically well characterized, some control over the number of examples removed should be set up. Consequently, the selected objects for undersampling are sorted in a descending order based on the average mass value attributed to non-singleton elements $\bar{\mu}$. Formally, for a selected object x_i:

$$\bar{\mu}_{x_i} = \frac{\sum_{|A|>1} m(A)}{k}, \qquad A \in 2^\Omega \tag{9}$$

where k represents the number of non-singleton focal elements. In other words, the more ambiguous objects (higher imprecision) are firstly removed until the size of the corresponding majority class reaches the mean s.

Regarding majority objects whose highest mass is not assigned to a non-singleton proposition (meta-class), we can safely say that they are not located in an overlapping region. However, they could be situated far from all classes (outlier), or in a different class (label noise). To further detect those types of samples, the maximum plausibility $Pl_{max} = max_{\omega \in \Omega} Pl(\{\omega\})$ is used.

Outliers. This type of objects are located far from any class in the data. Typically, this could be described as the state of ignorance in our framework. Thus, objects with maximum plausibility assigned to ω_0, i.e., $Pl_{max} = Pl(\{\omega_0\})$, are eliminated from the dataset.

Label Noise. Reasonably, a safe object should have the maximum plausibility assigned to its label. Otherwise, it could be considered as located in another class, which could be described as label noise. Following this logic, each object, with the maximum plausibility affected to another label than its own, is discarded from the dataset.

3.3   Evidential Adaptive Oversampling

In order to strengthen the presence of minority classes in the dataset, an oversampling phase is added to make the borders of each minority class more robust. Our objective, in this phase, is to emphasis the borders of each minority class, much like other oversampling techniques such as BorderlineSMOTE [12]. Another aspect of our approach is avoiding oversampling noisy examples and outliers.

The previously calculated *bbas* are used in the phase to smartly pick the regions where synthetic minority objects should be created. Minority instances are sorted into three probable categories, similar to the cleaning step: overlapping, label noise, or outlier. If an object does not correspond to one of the three categories, it is considered as located in a safe region and is not selected for the creation of new synthetic objects. The same is valid for label noise and outliers. Indeed, selecting noisy objects and outliers to generate new samples could lead to overgeneralization, which is a significant disadvantage of many oversampling techniques [12].

Our evidential approach to oversampling consists of generating synthetic minority data near the borderline objects of the minority class. The idea is to empower the minority class borders in order to avoid the misclassification of difficult objects. Formally speaking, only objects whose highest mass is committed towards an overlapping region are selected for oversampling. This procedure also helps us avoid selecting objects which are committed towards label noise and outlier. Indeed, selecting those objects would amplify the problems already present in the dataset.

As mentioned above, the number of generated examples is also controlled and the size of each minority class should not exceed the mean s. In fact, the objects in the corresponding minority class are sorted in descending order based on Eq. 9. The idea behind this is to give priority towards minority objects with higher uncertainty in order to generate synthetic object in difficult-to-classify locations.

4 Experimental Study

In this section, we will present firstly the setup of the conducted experiments in Subsect. 4.1. Lastly, we will show the results and discuss them in Subsect. 4.2.

4.1 Set-up

Datasets. In this work, seven multi-class imbalanced datasets were selected from the KEEL repository [2]. The details of each dataset are summarized in Table 1, where we describe the number of samples, number of features, number of classes, and the class distributions.

Table 1. Description of the imbalanced datasets selected from the KEEL repository.

Datasets	Class Distrib.	Features	Samples	#Class
Dermatology	112; 61; 72;49; 52; 20	34	366	6
Wine	71; 59; 48	13	178	3
Pageblocks	492; 33; 8; 12; 3	10	548	5
Thyroid	17; 37; 666	21	720	3
Hayes-roth	51; 51; 30	4	132	3
Contraceptive	629; 511; 333	9	1473	3
Yeast	463; 429; 244; 163; 51; 44; 35; 30; 20; 5	8	1484	10

Evaluation Metric. We evelute the performance of the compared techniques multi-class Geometric-Mean measure (G-Mean), which is the standard G-Mean calculated for each class separately by means of One-v-All strategy. The latter strategy was also used to compute the Area Under the ROC Curve (AUC), which is a popular metric to evaluate a model based on how good it separates the classes. For better assessment of the different performances, statistical analysis was run using the Wilcoxon's signed rank tests [28] for the significance level of $\alpha = 0.05$.

Baseline Classifier. As a baseline classifier, we use the decision tree classifier, more specifically CART. For all experiments, the implementation provided in the scikit-learn machine learning python library [20] was used, with the default parameters unchanged.

Compared Methods and Parameters. We aimed at comparing our method against re-sampling approaches specifically proposed for multi-class imbalanced datasets: Mahalanobis Distance Oversampling (MDO) [1], Static-SMOTE (S-SMOTE) [9], the basic version of SMOTE paired with the One-Versus-One strategy (SMOTE-MC), in addition to baseline (BL) with no re-sampling performed.

The considered parameters for our proposed method MC-EVHS are: α was set to 1 as recommended in [17], the tuning parameter t for $m(\{\omega_0\})$ was fixed to 2 to obtain good results in average, and we tested three different values for β in $\{0.3, 0.5, 0.7\}$ and selected the most performing one for each dataset, since the amount of class overlap differs in each case. For the other reference methods, we used the recommended parameters in their respective original papers.

4.2   Results and Discussion

In order to evaluate whether any of the methods perform consistently better than the other, we use a 10-fold stratified cross validation. The G-Mean and AUC results are presented respectively in Table 2 and Table 3. The two tables indicate that our approach MC-EVHS consistently produced the best results, in terms of G-Mean and AUC, when applying to these benchmarking datasets. Our proposal obtains the highest metric in 5 out of 7 datasets for both G-Mean and AUC. We can notice that all performances deteriorates with the increase of noise and overlap present in the dataset. However, our approach performed significantly better in datasets where there are many difficult-to-classify objects, in a consistent manner. It is safe to say that our approach is robust when applied on complex datasets.

The results for Wilcoxon's pairwise test are shown in Table 4. $R+$ represents the sum of ranks in favor of MC-EVHS, $R-$, the sum of ranks in favor of the reference methods, and exact p-values are calculated for each comparison. All pairwise comparisons can be considered as statistically significant with a level of 5% since all p-values are lower than the threshold 0.05. Thus, we can safely say that our method performed significantly better than MDO, Static-SMOTE, and SMOTE-MC.

Table 2. G-Mean results for KEEL datasets using CART.

Datasets	BL	SMOTE-MC	S-SMOTE	MDO	MC-EVHS
Dermatology	0.921	0.937	0.904	0.931	**0.952**
Wine	0.887	**0.936**	0.919	0.887	0.920
Thyroid	0.933	0.919	0.933	0.933	**0.981**
Hayes-roth	0.804	0.788	0.779	0.804	**0.820**
Contraceptive	0.442	0.441	0.448	0.442	**0.460**
Pageblocks	0.492	0.460	**0.553**	0.492	0.542
Yeast	0.599	0.635	0.642	0.599	**0.699**

Table 3. AUC results for KEEL datasets using CART.

Datasets	BL	SMOTE-MC	S-SMOTE	MDO	MC-EVHS
Dermatology	0.917	0.950	0.933	0.947	**0.955**
Wine	0.888	0.938	0.894	0.888	**0.950**
Thyroid	0.983	0.983	**0.989**	0.983	0.985
Hayes-roth	0.796	0.801	0.757	0.796	**0.811**
Contraceptive	0.470	0.464	0.476	0.470	**0.477**
Pageblocks	0.852	0.943	0.938	0.952	**0.954**
Yeast	0.664	0.674	**0.715**	0.664	**0.715**

Table 4. Pairwise comparisons of obtained G-Mean and AUC scores based on Wilcoxon's signed ranks test.

Comparisons	G-Mean			AUC		
	R+	R−	p-value	R+	R−	p-value
MC-EVHS vs BL	28.0	0.0	0.0078125	28.0	0.0	0.0078125
MC-EVHS vs SMOTE-MC	26.0	2.0	0.0234375	28	0.0	0.0078125
MC-EVHS vs S-SMOTE	26.0	2.0	0.0234375	19.0	9.0	0.0373677
MC-EVHS vs MDO	28.0	0.0	0.0078125	28.0	0.0	0.0078125

5 Conclusion

The aim of this paper was to specifically develop an approach for handling multi-class imbalanced datasets. Our method MC-EVHS can exploit the computed evidential memberships to better choose the locations for oversampling minority classes, and improve the selection of objects to eliminate in the undersampling phase. Evidence theory is an ideal tool in this case to express the different data difficulties that could be present in the dataset, i.e., class overlap, label noise and outliers. The conducted experiments confirmed the effectiveness of our proposed method and on the basis of a thorough statistical analysis, we may confirm that MC-EVHS performed significantly better than some popular methods. Further, the combination of undersampling and oversampling paired with evidential memberships, can reduce the multi-class imbalance problem, without excessive use of re-sampling.

As future work, we propose to investigate the wider applicability of MC-EVHS with very large datasets and extreme levels of imbalance.

References

1. Abdi, L., Hashemi, S.: To combat multi-class imbalanced problems by means of over-sampling techniques. IEEE Trans. Knowl. Data Eng. **28**(1), 238–251 (2015)
2. Alcala-Fdez, J., et al.: Keel data-mining software tool: data set repository, integration of algorithms and experimental analysis framework. J. Multiple-Valued Logic Soft Comput. **17**, 255–287 (2010)
3. Batista, G., Prati, R., Monard, M.C.: A study of the behavior of several methods for balancing machine learning training data. SIGKDD Explor. **6**, 20–29 (2004)
4. Bedi, P., Gupta, N., Jindal, V.: I-SiamIDS: an improved Siam-IDS for handling class imbalance in network-based intrusion detection systems. Appl. Intell. **51**(2), 1133–1151 (2020). https://doi.org/10.1007/s10489-020-01886-y
5. Bunkhumpornpat, C., Sinapiromsaran, K., Lursinsap, C.: Safe-level-SMOTE: safe-level-synthetic minority over-sampling TEchnique for handling the class imbalanced problem. In: Theeramunkong, T., Kijsirikul, B., Cercone, N., Ho, T.-B. (eds.) PAKDD 2009. LNCS (LNAI), vol. 5476, pp. 475–482. Springer, Heidelberg (2009). https://doi.org/10.1007/978-3-642-01307-2_43
6. Chawla, N.V., Bowyer, K.W., Hall, L.O., Kegelmeyer, W.P.: SMOTE: synthetic minority over-sampling technique. J. Artif. Intell. Res. **16**, 321–357 (2002)
7. Dempster, A.P.: A generalization of Bayesian inference. J. Roy. Stat. Soc. Ser. B (Methodol.) **30**(2), 205–232 (1968)
8. Douzas, G., Bacao, F., Last, F.: Improving imbalanced learning through a heuristic oversampling method based on k-means and SMOTE. Inf. Sci. **465**, 1–20 (2018)
9. Fernández-Navarro, F., Hervás-Martínez, C., Gutiérrez, P.A.: A dynamic oversampling procedure based on sensitivity for multi-class problems. Pattern Recogn. **44**(8), 1821–1833 (2011)
10. Grina, F., Elouedi, Z., Lefèvre, E.: Uncertainty-aware resampling method for imbalanced classification using evidence theory. In: Vejnarová, J., Wilson, N. (eds.) ECSQARU 2021. LNCS (LNAI), vol. 12897, pp. 342–353. Springer, Cham (2021). https://doi.org/10.1007/978-3-030-86772-0_25
11. Haixiang, G., Yijing, L., Shang, J., Mingyun, G., Yuanyue, H., Bing, G.: Learning from class-imbalanced data: review of methods and applications. Expert Syst. Appl. **73**, 220–239 (2017)
12. Han, H., Wang, W.-Y., Mao, B.-H.: Borderline-SMOTE: a new over-sampling method in imbalanced data sets learning. In: Huang, D.-S., Zhang, X.-P., Huang, G.-B. (eds.) ICIC 2005. LNCS, vol. 3644, pp. 878–887. Springer, Heidelberg (2005). https://doi.org/10.1007/11538059_91
13. Hastie, T., Tibshirani, R.: Classification by pairwise coupling. Adv. Neural. Inf. Process. Syst. **10**, 507–513 (1997)
14. Ivan, T.: Two modifications of CNN. IEEE Trans. Syst. Man Commun. **SMC 6**, 769–772 (1976)

15. Khushi, M., et al.: A comparative performance analysis of data resampling methods on imbalance medical data. IEEE Access **9**, 109960–109975 (2021)

16. Li, Z., Huang, M., Liu, G., Jiang, C.: A hybrid method with dynamic weighted entropy for handling the problem of class imbalance with overlap in credit card fraud detection. Expert Syst. Appl. **175**, 114750 (2021)

17. Liu, Z.g., Pan, Q., Dezert, J., Mercier, G.: Credal classification rule for uncertain data based on belief functions. Pattern Recogn. **47**(7), 2532–2541 (2014)

18. Mahalanobis, P.C.: On the generalized distance in statistics, vol. 2, pp. 49–55. National Institute of Science of India (1936)

19. Murphey, Y.L., Wang, H., Ou, G., Feldkamp, L.A.: OAHO: an effective algorithm for multi-class learning from imbalanced data. In: 2007 International Joint Conference on Neural Networks, pp. 406–411. IEEE (2007)

20. Pedregosa, F., et al.: Scikit-learn: machine learning in Python. J. Mach. Learn. Res. **12**, 2825–2830 (2011)

21. Ramentol, E., Caballero, Y., Bello, R., Herrera, F.: SMOTE-RSB *: a hybrid preprocessing approach based on oversampling and undersampling for high imbalanced data-sets using SMOTE and rough sets theory. Knowl. Inf. Syst. **33**(2), 245–265 (2012)

22. Rifkin, R., Klautau, A.: In defense of one-vs-all classification. J. Mach. Learn. Res. **5**, 101–141 (2004)

23. Sáez, J.A., Luengo, J., Stefanowski, J., Herrera, F.: SMOTE-IPF: addressing the noisy and borderline examples problem in imbalanced classification by a resampling method with filtering. Inf. Sci. **291**(C), 184–203 (2015)

24. Salazar, A., Vergara, L., Safont, G.: Generative adversarial networks and Markov random fields for oversampling very small training sets. Expert Syst. Appl. **163**, 113819 (2021)

25. Shafer, G.: A Mathematical Theory of Evidence, vol. 42. Princeton University Press, Princeton (1976)

26. Smets, P.: The transferable belief model for quantified belief representation. In: Smets, P. (ed.) Quantified Representation of Uncertainty and Imprecision. HDRUMS, vol. 1, pp. 267–301. Springer, Dordrecht (1998). https://doi.org/10.1007/978-94-017-1735-9_9

27. Tsai, C.F., Lin, W.C., Hu, Y.H., Yao, G.T.: Under-sampling class imbalanced datasets by combining clustering analysis and instance selection. Inf. Sci. **477**, 47–54 (2019)

28. Wilcoxon, F.: Individual comparisons by ranking methods. In: Kotz, S., Johnson, N.L. (eds.) Breakthroughs in statistics. PSS, pp. 196–202. Springer, New York (1992). https://doi.org/10.1007/978-1-4612-4380-9_16

29. Wilson, D.L.: Asymptotic properties of nearest neighbor rules using edited data. IEEE Trans. Syst. Man Cybern. **3**, 408–421 (1972)

A Parallel Declarative Framework
for Mining High Utility Itemsets

Amel Hidouri[1,2]([✉]), Said Jabbour[2], Badran Raddaoui[4], Mouna Chebbah[3],
and Boutheina Ben Yaghlane[1]

[1] LARODEC, University of Tunis, Tunis, Tunisia
boutheina.yaghlane@ihec.rnu.tn
[2] CRIL - CNRS UMR 8188, University of Artois, Lens, France
{hidouri,jabbour}@cril.fr
[3] LARODEC, Univ. Manouba, ESEN, Manouba, Tunisia
mouna.chebbah@esen.tn
[4] SAMOVAR, Télécom SudParis, Institut Polytechnique de Paris, Paris, France
badran.raddaoui@telecom-sudparis.eu

Abstract. One of the most active research topics in data mining is pattern discovery involving the well-known task of enumerating interesting patterns from databases. The problem of mining high utility itemsets is to find the set of items with the highest utility values based on a given minimum utility threshold. However, due to the advancement of big data technologies, finding all itemsets is much more harder due to the huge number of patterns and the large required resources. Parallel processing is an effective way to efficiently address the problem of mining patterns from large databases. Based on classical propositional logic, we propose in this paper a parallel method to handle efficiently the problem of discovering high utility itemsets from transaction databases. To do this, a decomposition technique is used to splitting the original problem of mining high utility itemsets into smaller and independent sub-problems that can be handled easily in a parallel manner. Then, empirical evaluations on different real-world datasets show that the proposed method is very efficient while being flexible enough to handle additional user constraints when discovering closed high utility itemsets.

Keywords: Data mining · High utility · Symbolic Artificial Intelligence · Propositional satisfiabilty · Parallel solving

1 Introduction

Pattern extraction is a well-known task in data mining that aims to infer knowledge based on different types of interesting measures. Discovering High Utility Itemsets (HUIM, for short) is one of the fundamental tasks in pattern discovery that generalizes the classical problem of frequent itemsets mining (FIM, for short). In fact, traditional FIM techniques are still insufficient for discovering the most valuable itemsets, e.g., when an itemset with a high profit is regarded as infrequent. Unfortunately, the FIM task is not appropriate to deal with this

© Springer Nature Switzerland AG 2022
D. Ciucci et al. (Eds.): IPMU 2022, CCIS 1602, pp. 624–637, 2022.
https://doi.org/10.1007/978-3-031-08974-9_50

research question because the importance of an itemset is binary and it is only determined by the occurrence of such itemset in the database. To address the previous limitation, the HUIM task was introduced as new paradigm to discover the set of items that appear together in a transaction database and have a high importance to the end-user, i.e., which is expressed by a utility function. So, an itemset is coined a *high utility itemset* (HUI, for short) if its utility value is greater than a user specified threshold.

In the literature, different proposals have been developed to handle HUIM. These approaches mainly differ in terms of data structures and search strategies used to exploring the search space and the database scanning optimization. Nonetheless, the main limitation of this line of research concerns its sequential computational power constraints, which make it more expensive and causes scalability issues, especially when dealing with highly dense and huge databases. Recently, parallel computing with multi-core machine has received a great attention to overcome this limitation and to improve the performance of sequential algorithms. Particularly, the popularity of multi-core architectures allows various data mining problems to be handled in parallel. Parallel computing has been recently used to improve the performance of HUIM algorithms by accelerating the mining of high utility itemsets in transaction databases. Among these proposals, one can cite **pEFIM** [14], an extension of the EFIM algorithm, and **MCHMiner** [17], an extension of iMEFIM (especially exploited for dynamic transaction databases), that used simple static load balancing between different processors. Moreover, the **CLB** and **PLB** algorithms [1], parallel-based extensions of ULB-Miner [4], combine the tree and utility-list structures while applying a parallel processing when mining candidate itemsets. Additionally, other approaches are based on distributed systems to find HUIs from transaction databases. Among these algorithms, we note **P-FHM+** [15] and **PHUI-Growth** [13].

Symbolic Artificial Intelligence has recently been used to solve different data mining problems [3]. It is based on a declarative language to express constraints over patterns. Declarative methods have been then used to model data mining tasks as a constraints network or a propositional formula (Propositional Satisfiability SAT, for short) where the found models correspond exactly to the required patterns. More interestingly, such methods enable users to constraint the desired motifs by adding new constraints without modeling the underlying problem from scratch. Unfortunately, the main challenging point for such approaches is the scalability issue when dealing with large datasets.

Parallel SAT solving has received a lot of attention in the SAT community. Indeed, two kinds of approaches have been proposed. The first category divides the search space using a divide-and-conquer principle that primarily divides the search space using the well-known guiding-path concept [18]. The second one makes use of a Portfolio [7] of DPLL engines, allowing them to compete and collaborate to be the first to solve a given instance. In this paradigm, each solver works on the original formula, and the search space is not divided. In general, the portfolio employs a variety of search engines to increase the likelihood that one of them will solve the problem, as well as various strategies for search space exploration with clause sharing to avoid redundancy. The divide-and-conquer paradigm is clearly the most convenient for our purpose, as our goal is to split the transaction database in order to generate several formulas of reasonable size.

In this paper, we present PSAT-HUIM, a new parallel algorithm for efficiently enumerating high utility itemsets embedded in transaction databases. Based on the divide and conquer paradigm, we present a parallel declarative method, that extends the work of [9], to discover the set of high utility itemsets. We also show through extensive experiments on different real-world datasets the efficiency of our PSAT-HUIM approach.

2   Background

In this section, we present the relevant preliminaries and definitions related to the HUIM and the propositional satisfiability problems, respectively.

2.1   High Utility Itemset Mining Problem

Let Ω represents a universe of unique items (or symbols) that appear in a database. A transaction database $\mathscr{D} = \{T_1, T_2, \ldots, T_m\}$ is a set of m transactions such that each transaction T_i is a set of items (i.e., $T_i \subseteq \Omega$), and T_i has a unique identifier i called its transaction identifier (TID, for short). In what follows, we give some additional definitions related to the high utility mining problem.

Definition 1 (Internal Utility). *For each transaction T_i such that $a \in T_i$, a positive number $w_{int}(a, T_i)$ is called the internal utility of the item a (e.g., purchase quantity).*

Definition 2 (External Utility). *Each item $a \in \Omega$ is associated with a positive number $w_{ext}(a)$, called its external utility (e.g., unit profit).*

Definition 3 (Utility of an item/itemset in a transaction). *Given a transaction database $\mathscr{D}$, the utility of an item a in a transaction $(i, T_i) \in \mathscr{D}$, denoted by $u(a, T_i)$, is $u(a, T_i) = w_{int}(a, T_i) \times w_{ext}(a)$. Then, the itemset's utility X in T_i, written as $u(X, T_i)$ is defined as follows:*

$$u(X, T_i) = \sum_{a \in X} u(a, T_i) \tag{1}$$

Definition 4 (Utility of an itemset in a database). *Let $\mathscr{D}$ be a transaction database. The utility of an itemset X in $\mathscr{D}$, denoted by $u(X, \mathscr{D})$, is defined as:*

$$u(X, \mathscr{D}) = \sum_{(i, T_i) \in \mathscr{D} \ \mid \ X \subseteq T_i} u(X, T_i) \tag{2}$$

Example 1. We assume that Consider the transaction database in Table 1. We assume that this database represents a set of products in a retail store. In addition, the external utility of items, i.e., the price of each item, is given in Table 2. Clearly, the utility of the itemset $\{b, d\}$ is computed as follows: $u(\{b, d\}) = u(\{b, d\}, T_2) + u(\{b, d\}, T_4) = 24.$

Table 1. A sample transaction database

TID	Items				
T_1	$(a, 3)$	$(c, 1)$		$(e, 3)$	
T_2	$(a, 2)$	$(b, 1)$	$(c, 3)$	$(d, 4)$	$(e, 3)$
T_3	$(a, 5)$		$(c, 1)$		$(d, 1)$
T_4	$(b, 4)$		$(d, 3)$	$(e, 1)$	
T_5	$(a, 1)$	$(d, 2)$		$(e, 2)$	

Table 2. The external utilities of the itemsets

Item	Unit profit
a	4
b	2
c	1
d	2
e	3

Definition 5 (High Utility Itemset). *An itemset X is called a high utility itemset (HUI) in a database $\mathscr{D}$ if its utility value is greater than a minimum utility threshold θ, i.e., $u(X) \geq \theta$.*

Definition 6 (Closed high utility itemset). *Let $\mathscr{D}$ be a transaction database. X is called a closed high utility itemset if there exists no high utility itemset X' such that $X \subset X'$, and $\forall (i, T_i) \in \mathscr{D}$, if $X \in T_i$ then $X' \in T_i$.*

Problem Statement. Given a transaction database $\mathscr{D}$ and a user-specified minimum utility threshold θ, the goal of computing (closed) high utility itemsets problem consists of finding the set of all (closed) high utility itemsets in $\mathscr{D}$ with a utility no less than θ, i.e.,
$HUI = \{X : u(X, \mathscr{D}) \mid X \subseteq \Omega, u(X, \mathscr{D}) \geq \theta\}$

Example 2. Let us consider again the transaction database given in Example 1. Given a minimum utility threshold $\theta = 45$, then the set of high utility itemsets with their utility in the database of Table 1 is $\{\{a, c\} : 45, \{a, d\} : 46, \{a, e\} : 59\}$.

In order to prune the search space, existing proposals of HUIM use the so-called *Transaction Weighted Utilization* (TWU, for short), which is an upper bound of the utility measure, together with the property of anti-monotonicity in order to filter out the candidate itemsets that are not high utility. More formally,

Definition 7 (Transaction Utility). *The transaction utility of a transaction T_i in a database $\mathcal{D}$, denoted by $TU(T_i)$, is the sum of the utility of all items in T_i, i.e.,*

$$TU(T_i) = \sum_{a \in T_i} u(a, T_i) \tag{3}$$

Definition 8 (Transaction Weighted Utilization). *The transaction weighted utilization of an itemset X in a transaction database $\mathcal{D}$, denoted by $TWU(X, \mathcal{D})$, is defined as:*

$$TWU(X, \mathcal{D}) = \sum_{(i,T_i) \in \mathcal{D} \ | \ X \subseteq T_i} TU(T_i) \tag{4}$$

2.2 Propositional Logic and SAT Problem

A propositional language $\mathcal{L}$ consists of three main components: a countable set of propositional variables $\mathcal{P}$ for which we use the letters p, q, r, etc. to range over $\mathcal{P}$, a set of logical connectives such as $\neg, \wedge, \vee, \rightarrow$ and the two logical constants $\top$ (*true* or 1) and $\bot$ (*false* or 0). A literal is a propositional variable p or its negation $\neg p$. Propositional formulas of $\mathcal{L}$ will be denoted as Φ, Ψ, etc. We also use $\mathcal{P}(\Phi)$ to denote the set of propositional variables occurring in the propositional formula Φ. A Boolean interpretation Δ of a formula Φ is defined as a function from $\mathcal{P}(\Phi)$ to $(0, 1)$, i.e., $\Delta : \mathcal{P}(\Phi) \rightarrow (0, 1)$. Now, a model of a formula Φ is a Boolean interpretation Δ that satisfies Φ. We use $Mod(\Phi)$ to denote the set of models of Φ. A formula Φ is satisfiable if there exists a model of Φ. Then, Φ is valid or a theorem, if every Boolean interpretation is a model of Φ. As usual, $\models$ refers to the logical inference and $\models_{up}$ the one restricted to unit propagation.

Now, a clause is a disjunction of literals. A formula Φ is in conjunctive normal form (CNF, for short) if Φ can be written as a conjunction of a finite set of clauses. We stress here that any propositional formula can be rewritten as a CNF one by applying the linear Tseitin's encoding [16]. The decision problem verifying the satisfiability of a propositional formula in CNF is called SAT. Given a CNF formula Φ, SAT aims to identify if Φ possesses a model or not. Interestingly, state-of-the-art SAT solvers have been shown to be very useful for handling various real-world problems, including overlapping community detection in graphs [10, 11], data mining [12], etc.

3   SAT Encoding of (Closed) High Utility Itemset Mining

In this section, we review the formulation of closed HUIM problem into propositional satisfiability [8]. The proposed encoding consists in a set of propositional variables to represent both items and transactions of the considered transaction database $\mathscr{D}$. More precisely, each item a (resp. each transaction identifier i), is associated with a propositional variable, denoted by p_a (resp. q_i). Given a Boolean interpretation Δ, the itemset and its cover i.e., the set of transactions in which it appears are simply expressed as the sets $\{a \in \Omega \mid \Delta(p_a) = 1\}$ and $\{i \in \mathbb{N} \mid \Delta(q_i) = 1\}$, respectively. Now, the translation of the HUIM task into propositional satisfiability is obtained by introducing a set of logical constraints as depicted by Fig. 1. The first propositional formula (5) encodes the cover of the candidate itemset. This formula expresses that the itemset appears in the ith transaction, i.e., $q_i = true$. In other words, the candidate itemset is not supported by the ith transaction (i.e., q_i is $false$), when there exists an item a (i.e., p_a is $true$) that does not belong to the transaction ($a \in \Omega \backslash T_i$); when q_i is $false$. This means that at least an item not appearing in the transaction i is set to $true$.

$$\bigwedge_{i=1}^{m}(\neg q_i \leftrightarrow \bigvee_{a \in \Omega \backslash T_i} p_a) \quad (5) \qquad \bigwedge_{a \in \Omega}(p_a \vee \bigvee_{a \notin T_i} q_i) \quad (6) \qquad \sum_{i=1}^{m}\sum_{a \in T_i} u(a, T_i) \times (p_a \wedge q_i) \geqslant \theta \quad (7)$$

$$\sum_{i=1}^{m}\sum_{a \in T_i} u(a, T_i) \times r_{ai} \geqslant \theta \quad (8) \qquad \bigwedge_{i=1}^{m}\bigwedge_{a \in T_i}(r_{ai} \leftrightarrow p_a \wedge q_i) \quad (9)$$

Fig. 1. SAT-based encoding scheme for HUIM.

The constraint over the utility of the itemset X in $\mathscr{D}$ is expressed using the linear inequality (7) requiring at least a threshold θ. Notice that Constraint (7) can be translated into clauses or managed in the solver as proposed in [8]. By the use of additional variables, Constraint (7) can be rewritten as a conjunction of the constraints (8) and (9). Lastly, the propositional formula (6) allows to select the set of closed HUIs in $\mathscr{D}$. It ensures that if the candidate itemset is involved in all transactions containing the item a, then a must belong to the itemset. As shown in [8], the CNF formula (5) $\wedge$ (8) $\wedge$ (9) encodes the HUIM problem, while the formula (5) $\wedge$ (8) $\wedge$ (9) $\wedge$ (6) encodes the closed HUIM problem.

Example 3. The formula encoding the problem of mining closed HUIs of Example 1 with $\theta = 20$ is as follows:

$$\neg q_1 \leftrightarrow (p_b \lor p_d) \qquad \neg q_3 \leftrightarrow (p_b \lor p_e) \qquad \neg q_4 \leftrightarrow (p_a \lor p_c) \qquad \neg q_5 \leftrightarrow (p_b \lor p_c)$$

$$r_{a1} \leftrightarrow p_a \land q_1 \qquad r_{c1} \leftrightarrow p_c \land q_1 \qquad r_{e1} \leftrightarrow p_e \land q_1 \qquad r_{a2} \leftrightarrow p_a \land q_2 \qquad r_{b2} \leftrightarrow p_b \land q_2$$

$$r_{c2} \leftrightarrow p_c \land q_2 \qquad r_{d2} \leftrightarrow p_d \land q_2 \qquad r_{e2} \leftrightarrow p_e \land q_2 \qquad r_{a3} \leftrightarrow p_a \land q_3 \qquad r_{c3} \leftrightarrow p_c \land q_3$$

$$r_{d3} \leftrightarrow p_d \land q_3 \qquad r_{b4} \leftrightarrow p_b \land q_4 \qquad r_{d4} \leftrightarrow p_d \land q_4 \qquad r_{e4} \leftrightarrow p_e \land q_4 \qquad r_{a5} \leftrightarrow p_a \land q_5$$

$$r_{d5} \leftrightarrow p_d \land q_5 \qquad r_{e5} \leftrightarrow p_e \land q_5$$

$$(p_a \lor q_4) \land (p_b \lor q_1 \lor q_3 \lor q_5) \land (p_c \lor q_4 \lor q_5) \land (p_d \lor q_1) \land (p_e \lor q_3)$$

$$12r_{a1} + r_{c1} + 9r_{e1} + 8r_{a2} + 2r_{b2} + 3r_{c2} + 8r_{d2} + 9r_{e2} + 20r_{a3} + r_{c3} + 2r_{d3} + 8r_{b4} + 6r_{d4} + 3r_{e4} +$$

$$4r_{a5} + 4r_{d5} + 6r_{e5} \geq 20$$

4   Parallel High Utility Itemsets Mining Using Propositional Satisfiability

In this section, we present a parallel approach based on propositional logic to computing high utility itemsets from transaction databases. Our proposed algorithm improves the state-of-the-art SAT-based approach presented in [9] by utilizing a multi-core architecture. In fact, two types of solvers have been developed in the literature to solve SAT problems in parallel: divide-and-conquer and portfolios. The first category is of particular interest in this paper, since in pattern mining we deal with an enumeration task. Thus, the divide-and-conquer paradigm aims to split the search space into many sub-spaces, explored in parallel by SAT solvers. Generally, divide-and-conquer based approaches use the well-known guiding-path concept to divide the search space [2]. In fact, a guiding-path consists to restrict the search space to a given region by adding constraints to the original problem. However, the main challenge of parallel processing is the load balancing between processors, which is important to avoid idleness. Our proposed approach relies on an effective static load balancing that initially split the work among processors using a heuristic cost function, while each processor has a direct and equal access to memory, i.e., shared-memory (SMP) architecture. The main advantage of this system is that it is easy to implement. More precisely, in our case, we decompose the enumeration of the whole models of Φ by generating a number of sub-formulas $\{\Phi_1, \ldots, \Phi_n\}$ using the guiding-path concept. Such sub-formulas are then distributed between cores for resolution. The goal is then to avoid encoding the whole transaction database by generating numerous sub-problems with reasonable size that can be solved in a parallel.

Formally, let Φ be a formula and $\Psi_1, \ldots, \Psi_n$ a set of formulas over $\mathscr{P}(\Phi)$. Then, $\Psi_1, \ldots, \Psi_n$ is a guiding-path set if and only if:

$$\top \equiv \Psi_1 \lor \ldots \lor \Psi_n$$

$$\Psi_i \wedge \Psi_j \models \bot, \forall i \neq j$$

Clearly, the satisfiability of Φ is related to the satisfiability of at least $\Phi \wedge \Psi_i$, for $i \in [1..n]$. Moreover, the following result holds:

$$Mod(\Phi) = \bigcup_{i=1}^{n} Mod(\Phi \wedge \Psi_i)$$

In addition, for $i \neq j$, we have $Mod(\Phi \wedge \Psi_i) \cap Mod(\Phi \wedge \Psi_j) = \emptyset$ allowing to enumerate the models of $\Phi \wedge \Psi_i$ independently.

For the high utility itemsets mining task using propositional logic, we define the guiding path set as:

$$\Psi_i = p_{a_i} \wedge \bigwedge_{j<i} \neg p_{a_j}$$

It is easy to check that this set of formulas is a guiding path one. In fact, Ψ_i requires that the literal p_{a_i} is *true*, i.e., the itemset must contain a_i while the literals that represent the items a_j with $j < i$ are *false*. Interestingly, $\Phi(\mathscr{D}, \theta) \wedge \Psi_i$ is equivalent to $\Phi(\mathscr{D}_i, \theta) \wedge \Psi_i$ i.e., the encoding can be restricted to transactions involving a_i where the items a_j, $j < i$ are removed (as depicted in Fig. 2). As performed, the size of the generated sub-formulas $\Phi(\mathscr{D}_i, \theta) \wedge \Psi_i$ can be considerably reduced.

Example 4. Let us consider again the transaction database in Table 1. Figure 2 shows the sub-table obtained by considering the guiding-path set as explained above.

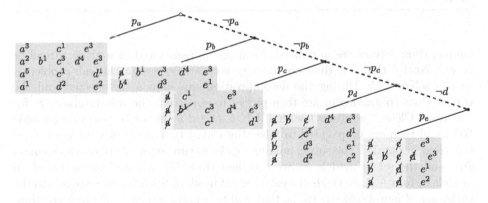

Fig. 2. Item based partitioning tree of the database in Table 1

Now, our parallel SAT-based approach to enumerating all (closed) HUIs from transaction databases is depicted in Algorithm 1. The algorithm takes as inputs a transaction database $\mathscr{D}$, a minimum utility threshold θ, and a number of cores k, and returns all (closed) HUIs embedded in $\mathscr{D}$. First, the SAT-based model

---

Algorithm 1: Parallel SAT for High Utility Itemset Mining (PSAT-HUIM)

Input: $\mathscr{D}$: a transaction database, θ: a minimum utility threshold, k: a number of cores

Output: S: the set of all high-utility itemsets

1 $\Omega = \langle a_1, \dots, a_n \rangle \leftarrow items(\mathscr{D})$;

2 $S \leftarrow \emptyset$;   /* set of models (HUIs) */

3 $\Gamma \leftarrow \emptyset$;

4 **for** i *in* $[0..k-1]$ **do**

5 $\quad$ $initSolver(i)$;

6 $\quad$ $S_i \leftarrow \emptyset$;

7 **end**

8 **for** i *in* $[1..n]$ **do**

9 $\quad$ **if** $\underline{TWU(a_i, \mathscr{D}) \geq \theta}$ **then**

10 $\quad\quad$ $\mathscr{D}_i \leftarrow \{(j, T_j) \in \mathscr{D} \mid a_i \in T_j\}$;

11 $\quad\quad$ **for** $b \in items(\mathscr{D}_i)$ **do**

12 $\quad\quad\quad$ **if** $\underline{TWU(b, \mathscr{D}_i) < \theta}$ **then**

13 $\quad\quad\quad\quad$ $\Gamma \leftarrow \Gamma \wedge \neg p_b$;

14 $\quad\quad\quad$ **end**

15 $\quad\quad$ **end**

16 $\quad\quad$ $\Psi_i \leftarrow p_{a_i} \wedge \bigwedge_{j<i} \neg p_{a_j}$;

17 $\quad\quad$ $S_i \leftarrow dpll\_Enum(\Phi(\mathscr{D}_i, \theta) \wedge \Psi_i \wedge \Gamma, \theta)$;   /* Solved by calling Solver$[i\%k]$ */

18 $\quad\quad$ $S \leftarrow S \cup S_i$;

19 $\quad$ **end**

20 **end**

21 **return** S;

---

enumeration solvers are initialized. Each one is associated to a given thread or core i. Notice that our division strategy consists to assign the sub-problem i to the solver $i\%k$. During the decomposition, an item a_i is selected and the transactions containing a_i are then picked to construct the sub-database $\mathscr{D}_i$ for encoding. Clearly, all items b with $TWU(b, D_i) < \theta$ cannot be part of a (closed) HUI, and then are propagated to *false* (lines 11–14). The set of k solvers is then launched in parallel, the solver number $(i\%k)$ is run successively on the formula Φ_{D_i} and the set of models S_i are returned (line 17). Finally, the union of all models S_i for $i \in [1..k]$ yields the entire set of models. Such set corresponds to the entire set of found (closed) HUIs, that will be finally returned by our algorithm. Let us notice that the guiding path for the HUIM task is generated using an ordering over the variables encoding items. This order must be well chosen for an efficient SAT resolution. The strategy used in this paper consists to sort the items according to their frequency in the initial database. The motivation behind such heuristic is to allow generating small sub-problems by starting with less frequent items.

5   Experimental Results

We conducted an experimental evaluation of our parallel SAT-based formulation of HUIM task using numerous real-world transaction databases. Experiments are carried out on a personal computer equipped with an Intel Core i7 processor and 16 GB of RAM at 2.8 Ghz running macOS 10.13.4. Empirical evaluations were performed on seven real-life datasets commonly used in the HUIM literature. These datasets are available from the Open source Data Mining Library (SPMF) [6]. These benchmarks are *Chess, Foodmart, Mushroom, Retail, Accidents, Kosarak* and *Chainstore*. They have a variety of characteristics and contain data from real-world scenarios. Our algorithm is written in C++, and we used the MiniSAT solver [5], which is slightly modified for the model enumeration problem. Let us note here that the computation time for our approach includes both the encoding and solving time. In Table 3, we report the number of transactions (#Trans), the number of items (#Items), and the average transaction length (AvgTransLen).

Table 3. Datasets characteristics

Instance	#Trans	#Items	AvgTransLen
Chess	3196	75	37
Foodmart	4141	1559	4.42
Mushroom	8124	119	23
Retail	88162	16470	10.3
Accidents	340183	468	33.8
Kosarak	990002	41270	8.1
Chainstore	1112949	46086	7.23

We conduct two types of experiments in order to evaluate the efficiency of our proposed SAT method for computing the set of all HUIs from databases. In the first one, we assess our algorithm's performance on each dataset for various minimum utility thresholds while considering 1, 2 and 4 cores. In the second, we demonstrate the effectiveness of our load balancing strategy among the different cores. Figure 3 displays the running time of our SAT-based algorithm on each dataset.

5.1   Parallel Evaluation

According to Fig. 3, the performance of our SAT-based approach depends on the considered dataset and the minimum utility threshold values as well as the considered cores number. As expected, the parallel based approach allows to significantly reduce the running time of mining the set of HUIs. Clearly, this latter decreases as the number of cores increases. For almost tested datasets, the solving time is divided by two when the number of cores pass from 1 to 2. More specifically, for *Accidents* dataset and $\theta = 17.5 \times 10^6$, the running time with one core exceeds 450 s while this time is approximately equal to 300 seconds with two cores, and for kosarak we move from 90 seconds with one core to 50 seconds with two cores for $\theta = 1.2 \times 10^6$. We can also remark that the gain in terms of solving time is greater when the number of cores increases from 1 to 2 rather than 2 to 4. To summarize, the multi-core architecture outperforms sequential architecture on all tested datasets.

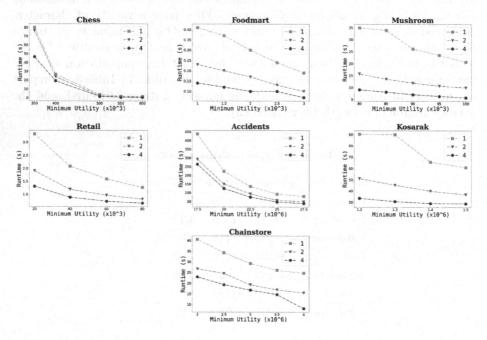

Fig. 3. Performance gain w.r.t. the number of cores

Let us recall here that we use a splitting method where the resulting sub-formulas are shared fairly between cores for a better load balancing. The next sub-section is dedicated to the load balancing analysis.

5.2 Load Balancing

This subsection provides an empirical analysis of the CPU time with different number of cores on all datasets to assess the suitability of our load balancing strategy. For this, we fix the number of cores to 4 for each value of minimum utility threshold. We report in Fig. 4 the average running time over cores as well as the minimum and maximum time to quantify the idleness of some cores. We mention here that the tighter the difference between minimum and maximum running time is, the better the load balancing is. As we can observe the relative

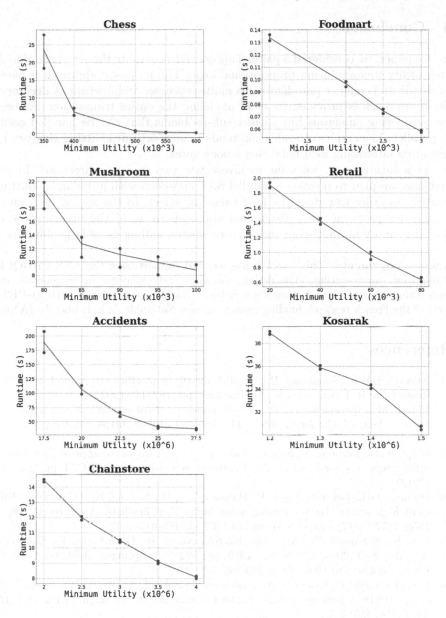

Fig. 4. Load unbalancing between cores

load unbalancing for the datasets *Chainstore*, *Retail* and *Kosarak* is very limited. Thus, all the cores spent almost the same time to solve their assigned sub-problems. Overall, our guiding paths generation principle allows to balance the number of found models between the different threads. These results indicate that our SAT method is both efficient and scalable.

6   Conclusion

In this paper, we considered a parallel approach to compute the set of all (closed) high utility itemsets using propositional logic. The proposed approach is based on divide and conquer paradigm for a multi-processor architecture. A decomposition approach is provided to avoid modeling the entire transaction database by considering numerous but smaller sub-problems that can be handled easily in parallel. The empirical evaluation tends to support our splitting strategy by providing interesting load balancing among cores.

As a future work, we want to investigate two research directions. In the first one, we plan to develop a parallel SAT approach with a dynamic splitting strategy in order to reduce the load balancing effect. In the second, we aim to extend our proposal for a distributed approach to avoid the disadvantage of shared memory in multi-core architecture when handling more large datasets.

Acknowledgements. This research has received support from the ANR CROQUIS (Collecting, Representing, cOmpleting, merging, and Querying heterogeneous and UncertaIn waStewater and stormwater network data) project, grant ANR-21-CE23-0004 of the French research funding agency Agence Nationale de la Recherche (ANR).

References

1. Atmaja, E.H.S., Sonawane, K.: Parallel algorithm to efficiently mine high utility itemset. In: ICT Analysis and Applications, pp. 167–178 (2022)
2. Böhm, M., Speckenmeyer, E.: A fast parallel sat-solver-efficient workload balancing. Ann. Math. Artif. Intell. **17**, 381–400 (1996). https://doi.org/10.1007/BF02127976
3. Coquery, E., Jabbour, S., Sais, L., Salhi, Y., et al.: A sat-based approach for discovering frequent, closed and maximal patterns in a sequence. In: ECAI, pp. 258–263 (2012)
4. Duong, Q.-H., Fournier-Viger, P., Ramampiaro, H., Nørvåg, K., Dam, T.-L.: Efficient high utility itemset mining using buffered utility-lists. Appl. Intell. **48**(7), 1859–1877 (2017). https://doi.org/10.1007/s10489-017-1057-2
5. Eén, N., Sörensson, N.: An extensible SAT-solver. In: Giunchiglia, E., Tacchella, A. (eds.) SAT 2003. LNCS, vol. 2919, pp. 502–518. Springer, Heidelberg (2004). https://doi.org/10.1007/978-3-540-24605-3_37
6. Fournier-Viger, P., Gomariz, A., Gueniche, T., Soltani, A., Wu, C.W., Tseng, V.S., et al.: SPMF: a java open-source pattern mining library. J. Mach. Learn. Res. **15**, 3389–3393 (2014)
7. Hamadi, Y., Jabbour, S., Sais, L.: ManySAT: a parallel sat solver. J. Satisfiability Boolean Model. Comput. **6**, 245–262 (2010)
8. Hidouri, A., Jabbour, S., Raddaoui, B., Yaghlane, B.B.: A sat-based approach for mining high utility itemsets from transaction databases. In: International Conference on Big Data Analytics and Knowledge Discovery, pp. 91–106 (2020)
9. Hidouri, A., Jabbour, S., Raddaoui, B., Yaghlane, B.B.: Mining closed high utility itemsets based on propositional satisfiability. Data Knowl. Eng. **136**, 101927 (2021)
10. Jabbour, S., Mhadhbi, N., Raddaoui, B., Sais, L.: Sat-based models for overlapping community detection in networks. Computing **102**(5), 1275–1299 (2020)

11. Jabbour, S., Mhadhbi, N., Raddaoui, B., Sais, L.: A declarative framework for maximal k-plex enumeration problems. In: AAMAS (2022, to appear)
12. Jabbour, S., Sais, L., Salhi, Y.: Mining top-k motifs with a sat-based framework. Artif. Intell. **244**, 30–47 (2017)
13. Lin, Y.C., Wu, C.W., Tseng, V.S.: Mining high utility itemsets in big data. In: Pacific-Asia Conference on Knowledge Discovery and Data Mining, pp. 649–661 (2015)
14. Nguyen, T.D., Nguyen, L.T., Vo, B.: A parallel algorithm for mining high utility itemsets. In: International Conference on Information Systems Architecture and Technology, pp. 286–295 (2018)
15. Sethi, K.K., Ramesh, D., Edla, D.R.: P-FHM+: parallel high utility itemset mining algorithm for big data processing. Procedia Comput. Sci. **132**, 918–927 (2018)
16. Tseitin, G.S.: On the complexity of derivation in propositional calculus. In: Automation of Reasoning, pp. 466–483 (1983)
17. Vo, B., Nguyen, L.T., Nguyen, T.D., Fournier-Viger, P., Yun, U.: A multi-core approach to efficiently mining high-utility itemsets in dynamic profit databases. IEEE Access **8**, 85890–85899 (2020)
18. Zhang, H., Bonacina, M.P., Hsiang, J.: PSATO: a distributed propositional prover and its application to quasigroup problems. J. Symb. Comput. **21**, 543–560 (1996)

Towards an FCA-Based Approach for Explaining Multi-label Classification

Hakim Radja[1]([⊠]) [iD], Yassine Djouadi[2,3] [iD], and Karim Tabia[3] [iD]

[1] Computer Science Department, Mouloud Mammeri University of Tizi-Ouzou,
BP 17, RP, Tizi-Ouzou, Algeria
hakim.radja@ummto.dz
[2] University of Algiers 1, RIIMA Laboratory, Algiers, Algeria
y.djouadi@univ-alger.dz
[3] Univ. Artois, CNRS, CRIL, 62300 Lens, France
tabia@cril.fr

Abstract. Multi-label classification is a supervised learning task where each data item can be associated with multiple labels simultaneously. Although multi-label classification models seem powerful in terms of prediction accuracy, they have however like mono-label classifiers certain limitations mainly related to their opacity. We propose in this preliminary work a novel approach for explaining multi-label classification models based on formal concept analysis (FCA). The proposed approach makes it possible to answer certain questions that a user may ask such as: *What are the minimum attribute sets allowing the classifier f to make a prediction ?* and *What are the attributes that contribute to a given prediction?*

Keywords: Explainable AI · FCA · Multi-label classification

1 Introduction

Until recently, machine learning (ML) models mainly focused on making accurate predictions. ML is indeed widely used in several areas but regularly comes up against a major issue, its black-box side especially due to the complexity of the models used (e.g. models based on deep learning can have several million parameters). Explainable AI and interpretable ML attempt to address these issues. They aim at equipping ML models with the ability to explain or present their behavior in understandable terms [1]. Most of explainable ML approaches try to assess the influence of attributes in the predictions made by classifiers.

We are interested in this preliminary work in the explanation of the predictions made by multi-label classifiers. To do this, we rely on a powerful mathematical framework, not yet used in explaining multi-label classifiers, that is the one of formal concept analysis (FCA) to answer questions such as:

- *What are the attributes that contributed to the prediction of the class set y predicted by the multi-label classifier f ?*

© Springer Nature Switzerland AG 2022
D. Ciucci et al. (Eds.): IPMU 2022, CCIS 1602, pp. 638–651, 2022.
https://doi.org/10.1007/978-3-031-08974-9_51

- *What is the minimum set allowing the classifier f to make this prediction ?*
- *At what extent does an attribute influence the prediction of a given class belonging to the set y of classes predicted by the classifiers f?*

It should be noted that the majority of existing approaches in explainable ML are interested in the single-label case. This is the first novelty of our work. Moreover, the proposed approach is original insofar as it is based on formal concept analysis which has, to the best of our knowledge, never been used to explain the predictions of a multi-label classifier. Our approach aims to provide some forms of useful symbolic explanations. The other important advantage of our approach is that it is agnostic and can be applied to explain the predictions of any multi-label classifier.

2 Preliminaries

2.1 Multi-label Classification

Multi-label classification is an extension of single-label classification, where classes are not mutually exclusive, and each instance can be assigned to several classes simultaneously. It is encountered in various modern applications such as text categorization [6], scene classification [5], video annotation and bioinformatics [14]. Let $x \in \mathcal{X}$ be a data instance denoted $x = (a_1, a_2, .., a_n)$ where a_i is a binary attribute (feature). Let $\mathcal{C} = \{c_1, c_2, .., c_m\}$ be a finite set of labels. A multi-label classifier f allows to predict for each instance $x \subset \mathcal{X}$ a subset of labels $y \in 2^{|\mathcal{C}|}$.

2.2 Explainable AI

The existing methods in explainable ML mainly focus on how an explanation can be obtained and how the explanation itself can be constructed (for a survey, see [15]). Examples of common methods are: ordering the attributes contributions to a prediction [7], selection, construction and presentation of prototypes [13], summaries with decision trees [17] and decision rules [10]. The two most used methods are **LIME** (Local interpretable model agnostic explanation) [8] and **SHAP** (The Shapley Concept of Value) [4]. LIME is an explanatory technique that explains the predictions of any classifier through learning a locally interpretable model around the prediction. **SHAP** is based on game theory and assesses on average the contribution of each feature to the prediction. In this paper, we rather focus on an alternative and complementary category of explanations that are symbolic and may be very useful to the end-users. More precisely, we focus on some forms of sufficient reason explanations. These latter justify what is enough or necessary to trigger the prediction.

2.3   Formal Concept Analysis

Formal concepts analysis (FCA) [3] consists in learning pairs of subsets (objects, properties) called formal *concepts*, from a binary relation, called *formal context*, between a set of *objects* and a set of *properties*. Let $\mathcal{O}$ and $\mathcal{P}$ be sets of objects and properties respectively, and $\mathcal{R}$ be a binary relation between $\mathcal{O}$ and $\mathcal{P}$ verifying $\mathcal{R} \subseteq \mathcal{O} \times \mathcal{P}$. A pair $(x, a) \in \mathcal{R}$ (also denoted $x\mathcal{R}a$) means that the object $x \in \mathcal{O}$ has the property $a \in \mathcal{P}$. The triplet $\mathcal{K} := (\mathcal{O}, \mathcal{P}, \mathcal{R})$ is called a formal context. Let $\mathcal{K} := (\mathcal{O}, \mathcal{P}, \mathcal{R})$ be a formal context. For all $X \subseteq \mathcal{O}$ and $A \subseteq \mathcal{P}$, the Galois derivation set operator, denoted $(.)^{\Delta}$, is defined as follows:

$$X^{\Delta} = \{p \in \mathcal{P} \mid X \subseteq \mathcal{R}(p), \ A^{\Delta} = \{x \in \mathcal{O} \mid A \subseteq \mathcal{R}(x)\}. \tag{1}$$

Intuitively, X^{Δ} is the set of properties common to all the objects of X and A^{Δ} is the set of objects having all the properties of A.

A formal concept is a pair (X, A) such that $X \subseteq \mathcal{O}$, $A \subseteq \mathcal{P}$, $X^{\Delta} = A$ and $A^{\Delta} = X$. X and A are respectively called *extent* and *intent* of the formal concept (X, A). In this case, we have also $(A^{\Delta})^{\Delta} = A$ and $(X^{\Delta})^{\Delta} = X$.

Example 1. Table 1 provides an example of a formal context $\mathcal{K}$ where the set of items $\mathcal{O} = \{x_1, x_2, x_3, x_4, x_5\}$ and the set of attributes $\mathcal{P} = \{a_1, a_2, a_3, a_4\}$. An example of a formal concept in Table 1 is $\langle X_1, A_1 \rangle = \langle \{x_2, x_3\}, \{a_2, a_3\} \rangle$ where $X_1 = \{x_2, x_3\}$ is the extent of the formal concept and $A_1 = \{a_2, a_3\}$ is its intent.

Table 1. Example of a formal context

$\mathcal{R}$	a_1	a_2	a_3	a_4
x_1	×			×
x_2		×	×	
x_3	×	×	×	
x_4				×
x_5			×	

The set $\mathcal{B}(\mathcal{K})$ of all formal concepts of $\mathcal{K}$ is partially ordered by : $(X_1, A_1) \preceq (X_2, A_2) \Rightarrow X_1 \subseteq X_2 (A_2 \subseteq A_1)$. The subsumption relation $\preceq$ organizes the formal concepts in a concepts lattice (Galois lattice) denoted by $\mathcal{B}(\mathcal{O}, \mathcal{P}, \mathcal{R})$ or $\mathcal{B}(\mathcal{K})$. FCA has several advantages for data analysis. In particular, thanks to the visual representation of the lattice of concepts, it is possible to visually analyze and explore the data and its structure. Another frequent use is the generation of association rules, which also allows data analysis and knowledge extraction. In [16], an overview of classification methods based on formal concepts analysis is provided. In [2], the authors propose a learning classifier system (LCS) based on FCA to generate and exploit multi-label association rules which highlight the different relationships between labels. In [9], a neural network architecture based on concept lattices is used to improve the model explainability.

3   From Multi-label Predictions to FCA Representation

This section briefly presents our FCA-based approach for explaining the predictions of a multi-label classifier f. We first need to associate a *local* or *global* formal context to the classifier.

- *Local explanations*: If one needs to explain locally a prediction $f(x)$, then we need to build a local formal context representing the predictions of the classifier in the neighborhood of data instance x. In case we are given a dataset $\mathcal{D}$, the neighborhood of x, denoted $\mathcal{N}(x)$ is simply obtained by selecting m data instances from $\mathcal{D}$ that are close to x. More precisely, we associate a local formal context composed of data instances $x' \in \mathcal{N}(x)$ and their predictions $f(x')$. Otherwise, one can obtain $\mathcal{N}(x)$ by applying local perturbations to instance x as it is done in most explainability approaches such as LIME [8].
- *Global explanations*: This case corresponds to the use of all the predictions of f over a dataset $\mathcal{D}$ to explain the prediction at hand. Global explanations can also be used in order to explain the global functioning of a classifier in the general case.

Recall that a multi-label classifier f associates with each data instance x described by its feature vector $(a_1, a_2, .., a_n)$ a subset of classes y from $\mathcal{C} = \{c_1, c_2, ..., c_k\}$. Therefore, this relation implies three dimensions: a set of objects $\mathcal{X}$ (namely, data instances), a set of attributes $\mathcal{P}$ and a set of classes $\mathcal{C}$. Thus $\mathcal{R} \subseteq \mathcal{O} \times \mathcal{P} \times \mathcal{C}$. The transformation of multi-label data into a formal context consists in transforming this three-dimensional relation into a two-dimensional one to obtain a formal context with only two dimensions. This is the first contribution of this paper.

3.1   Building a Formal Context for the Classifier Predictions

Let $\mathcal{MLD} = (\mathcal{O}, \mathcal{P}, \mathcal{C}, \mathcal{R})$ be a multi label dataset where $\mathcal{O}$ is a set of instances, $\mathcal{C}$ be the set of classes, $\mathcal{P}$ be the set of features and $\mathcal{R}$ a ternary relationship $\mathcal{R} \subseteq \mathcal{O} \times \mathcal{P} \times \mathcal{C}$. Transforming this multi-label data into a formal context consists in transforming the ternary relation $\mathcal{R} \subseteq \mathcal{O} \times \mathcal{P} \times \mathcal{C}$ into a binary relation $\mathcal{R}' \subseteq \mathcal{O} \times \mathcal{M}$ where $\mathcal{M} = (\mathcal{C} \times \mathcal{P})$ similar to what was done in [12]. Hence, in our approach, a formal context is a triplet $\mathcal{K}' = (\mathcal{O}, \mathcal{M}, \mathcal{R}')$ where $\mathcal{O}$ represents the set of objects, $\mathcal{M} = (\mathcal{C} \times \mathcal{P})$ is a set of pairs (c_k, a_j) with $a_j \in \mathcal{P}$, $c_k \in \mathcal{C}$, and $\mathcal{R}' \subseteq \mathcal{O} \times \mathcal{M}$ a binary relation between the two sets $\mathcal{O}$ and $\mathcal{M}$. In other words, the transformation of a the multi-label data into a context formal $\mathcal{K}'$ is done by flattening (projecting) of the set of classes on all the attributes of the objects, as follows:

- $x_i \in \mathcal{O}$ the i-th instance of the set of objects $\mathcal{O}$, $i = 1, m$
- $a_j \in \mathcal{P}$ the j-th property of the set of properties $\mathcal{P}$, $j = 1, n$
- $c_k \in \mathcal{P}$ the k-th class of the set of classes $\mathcal{C}$, $k = 1, p$
- $f_k(x)$ the prediction for the class c_k for the instance x by the classifier f

$$f_k(x_i) = \begin{cases} 0 & if \quad \mathcal{R}(x_i, a_j, c_k) = 0 \\ 1 & if \quad \mathcal{R}(x_i, a_j, c_k) = 1 \end{cases}$$

Example 2. Table 2 illustrates the use of Algorithm 1, given below, to represent multi-label data in the form of a formal context. The triplet (x_1, a_1a_2, c_1c_3) means that the object x_1 where the attributes a_1 and a_2 are present is predicted in classes c_1 and c_3. In other words, object x_1 is predicted in classes c_1 and c_3 under the conditions a_1 and a_2.

Algorithm 1. Flattening multi-label data into a formal context

---

Require: Multi-label data $\mathcal{D}$
Ensure: Formal context $\mathcal{K}'$
1: **for** $i \leftarrow 1, m$ **do**
2:     **for** $j \leftarrow 1, n$ **do**
3:         **for** $k \leftarrow 1, p$ **do**
4:             **if** $\mathcal{R}(x_i, a_j, c_k) = 1$ **then**
5:                 $\mathcal{R}'(x_i, (a_j \times c_k)) \leftarrow 1$
6:             **else**
7:                 $\mathcal{R}'(x_i, (a_j \times c_k)) \leftarrow 0$
8:             **end if**
9:         **end for**
10:    **end for**
11: **end for**

Table 2. Example of flattening a 3D multi-label data (left side table) into a 2D formal context (right side table).

$\mathcal{O}$	a_1	a_2	a_3	c_1	c_2	c_3	$\mathcal{O}$	$c_1\text{-}a_1$	$c_1\text{-}a_2$	$c_1\text{-}a_3$	$c_2\text{-}a_1$	$c_2\text{-}a_2$	$c_2\text{-}a_3$	$c_3\text{-}a_1$	$c_3\text{-}a_2$	$c_3\text{-}a_3$
	$\mathcal{P}$			$\mathcal{C}$							$\mathcal{M}=\mathcal{C}\times\mathcal{P}$					
x_1	1	1	0	1	0	1	x_1	1	1	0	0	0	0	1	1	0
x_2	1	0	0	1	0	0	x_2	1	0	0	0	0	0	0	0	0
x_3	1	0	1	1	0	0	x_3	1	0	1	0	0	0	0	0	0
x_4	0	0	1	0	1	0	x_4	0	0	0	0	0	1	0	0	0
x_5	1	1	1	1	0	1	x_5	1	1	1	0	0	0	1	1	1
x_6	1	1	0	1	0	1	x_6	1	1	0	0	0	0	1	1	0

Clearly, flattening the 3D formal context ensures that in the new 2D formal context, any formal concept includes a set of objects that are all predicted in the same classes under the same conditions (features values). This allows using existing algorithms and implementations to generate formal concepts directly for explanation purposes. Once the flattened formal context built, one can use it to provide different forms of explanations.

3.2   Computing Explanations as Formal Concepts

In order to compute explanations, let us first adapt the derivation operator to our formal context. Assume we are given a 2D formal context and that our objective is to provide local or global explanations. Let $\mathcal{K}' = (\mathcal{O}, \mathcal{M}, \mathcal{R}')$ a formal context, $\mathcal{O}$ a set of instances, $\mathcal{M} = (\mathcal{C} \times \mathcal{P})$ is a set of pairs (c_i, a_j) with $a_j \in \mathcal{P}$, $c_k \in \mathcal{C}$. The triplet $(x_i, c_k, a_j) \in \mathcal{R}'$ means that the instance $x_i \in \mathcal{O}$ has the class $c_k \in \mathcal{C}$ when the attribute $a_j \in \mathcal{P}$ is present.

For all $X \subseteq \mathcal{O}$ and $Y \subseteq \mathcal{M}$ with $Y = \{(c_k, a_j)/a_j \in \mathcal{P}, c_k \in \mathcal{C}\}$ we define the Galois derivation set operator $(.)^\Delta$, seen in the previous section, as follows:

$$X^\Delta = \{(c_k, a_j) \in \mathcal{M} \mid X \subseteq \mathcal{R}'(c_k, a_j)\} \tag{2}$$

$$= \{(c_k, a_j) \in \mathcal{M} \mid \forall x \in \mathcal{O} x \in X \Rightarrow (x, (c_k, a_j)) \in \mathcal{R}'\}$$

$$Y^\Delta = \{x \in \mathcal{O} \mid Y \subseteq \mathcal{R}'(x)\} \tag{3}$$

$$= \{x \in \mathcal{O} \mid \forall (c, a) \in \mathcal{M}((c, a) \in Y \Rightarrow (x, (c_k, a_j)) \in \mathcal{R}')\}.$$

X^Δ is the set of pairs (c_k, a_j) common to all the objects of X and Y^Δ is the set of objects having all the pairs (c_k, a_j) of Y. The obtained set of all formal concepts is named $\mathcal{L}(K')$

Example 3. Following the new definition of the Galois operator to the formal context of Table 2, we obtain the set $\mathcal{L}(K')$ of the following formal concepts:

- $fc_1 : \langle \{x_1, x_2, x_3, x_4, x_5, x_6\}, \{\} \rangle$
- $fc_2 : \langle \{x_4\}, \{(c_2, a_3)\} \rangle$
- $fc_3 : \langle \{x_1, x_2, x_3, x_5, x_6\}, \{(c_1, a_1)\} \rangle$
- $fc_4 : \langle \{x_1, x_5, x_6\}, \{(c_1, a_1), (c_1, a_2), (c_3, a_1), (c_3, a_2)\} \rangle$
- $fc_5 : \langle \{x_3, x_5\}, \{(c_1, a_1), (c_1, a_3)\} \rangle$
- $fc_6 : \langle \{x_5\}, \{(c_1, a_1), (c_1, a_2), (c_1, a_3), (c_3, a_1), (c_3, a_2), (c_3, a_3)\} \rangle$
- $fc_7 : \langle \{\}, \{(c_1, a_1), (c_1, a_2), (c_1, a_3), (c_2, a_1), (c_2, a_2), (c_2, a_3), (c_3, a_1), (c_3, a_2), (c_3, a_3)\} \rangle$

The formal concept fc_4 can be rewritten as follows $fc_4 : \langle \{x_1, x_5, x_6\}, \{c_1, c_3\}, \{a_1, a_2\} \rangle$ meaning that the classes $\{c_1, c_3\}$ are predicted for the instances $\{x_1, x_5, x_6\}$ when attributes $\{a_1, a_2\}$ are present. In this example, $\{x_1, x_5, x_6\}$, $\{c_1, c_3\}$, $\{a_1, a_2\}$ represent respectively, the extension, the intention and the condition of the formal concept fc_4.

Reducing the Number of Formal Concepts. The number of formal concepts can be very large, then a question arises regarding the relevance of some formal concepts for explanation purposes and whether one can not reduce their number. The particularity of our formal concepts reduction algorithm (Algorithm 2) lies in the fact that it does not only rely on attributes and instances but also on classes, as a third parameter, in the reduction process. For instance, in Example 3, the intention $\{c_1\}$ appears in several formal concepts:

– In fc_3: the class c_1 is predicted for the objects x_1, x_2, x_3, x_4, x_5 and x_6 when the attribute a_1 is present.
– In fc_5: the class c_1 is predicted for the object x_3 and x_5 when the attributes a_1 and a_3 are present.

Clearly the attribute a_1 is sufficient to predict the object x_3 and x_5 in the class c_1 since $\{x_3, x_5\} \subseteq \{x_1, x_2, x_3, x_5, x_6, x_7\}$ and $\{a_1\} \subseteq \{a_1, a_3\}$. Hence, the attribute a_3 is not a necessary condition to predict c_1 for objects $\{x_1, x_5\}$. Then from an explanation point of view, a_3 is not relevant and it suffices to keep only the formal concept fc_3. Based on this observation, Algorithm 2 allows to reduce the number of formal concepts: Let $\mathcal{L}(K') = \{fc_1, fc_2, ..., fc_n\}$ be the set of all formal concepts, $\mathcal{I}nt(fc_i)$ be the intention of the formal concept fc_i, $\mathcal{E}xt(fc_i)$ be the extension of the formal concept fc_i, and $\mathcal{C}ond(fc_i)$ be it condition.

Algorithm 2. Formal concepts reduction

Require: $\mathcal{L}(K')=\{fc_1, fc_2, ..., fc_n\}$
Ensure: $\mathcal{L}'(K')$ ▷ Reduced set of formal concepts
1: **for** $i \leftarrow 1, n-1$ **do**
2:     **for** $j \leftarrow i+1, n$ **do**
3:         **if** $\mathcal{I}nt(fc_i) = \mathcal{I}nt(fc_j)$ **then**
4:             **if** $\mathcal{E}xt(fc_i) \subseteq \mathcal{E}xt(fc_j)$ **then**
5:                 **if** $\mathcal{C}ond(fc_i) \supseteq \mathcal{C}ond(fc_j)$ **then**
6:                     $\mathcal{L}'(K')=\mathcal{L}(K')\setminus fc_i$
7:                 **end if**
8:             **end if**
9:         **end if**
10:     **end for**
11: **end for**

Example 4. Applying Algorithm 2 to the set $\mathcal{L}(K')$ of formal concepts obtained above (Example 3), we obtain the following reduced set of formal concepts $\mathcal{L}'(K') = \{fc_2, fc_3, fc_4\}$.

Up to now, we showed how to build a formal context for explanation purposes and how to reduce the number of the formal concepts. The following section presents explanation generation from the obtained formal concepts.

4   Multi-label Prediction Explanations

Recall that our approach for explaining multi-label classification is agnostic and it allows to provide both *symbolic* and *numerical* of explanations. For lack of space, the presentation is limited to some forms of *symbolic* explanations.

Example 5. In this section, we will illustrate through a real example from the medical field. This example deals with diagnosing some respiratory deceases (labels) given some symptoms of patients (attributes). For the sake of simplicity, we consider only four deceases that are: Asthma (d_1), Bronchiolitis (d_2), COPD (d_3), Covid (d_4). It has been found that the manifestation of these diseases is generally made by the following symptoms: Dry cough (s_1), loose cough (s_2), shortness of breath (s_3), wheezing (s_4), fever(s_5), headaches (s_6), loss of taste (s_7), curvatures (s_8) and Dyspnoea (s_9). Table 3 presents the predictions of a black-box model f.

Table 3. Multi-label data from medical field

$\mathcal{O}$	Symptoms									Diseases			
	s_1	s_2	s_3	s_4	s_5	s_6	s_7	s_8	s_9	d_1	d_2	d_3	d_4
1	1	0	1	1	1	0	0	0	0	1	1	0	0
2	0	1	0	1	1	0	0	1	1	1	0	1	0
3	1	0	1	0	1	1	1	1	0	1	0	0	1
4	1	0	0	0	1	1	1	1	0	0	0	0	1
5	0	1	1	0	0	0	0	1	1	0	0	1	0
6	1	0	1	0	1	1	1	1	0	1	1	0	1
7	0	0	0	0	1	1	1	1	0	0	0	0	1
8	1	0	1	1	1	1	0	1	0	1	1	0	1
9	1	0	1	0	0	0	0	1	0	0	1	0	0
10	1	0	1	0	1	1	1	1	0	0	1	0	1
11	0	1	1	1	1	0	0	0	1	1	1	1	0
12	0	1	0	0	1	0	0	0	1	0	0	1	0
13	1	0	1	0	1	1	1	1	0	0	0	0	1
14	1	0	1	1	0	0	0	0	0	1	0	0	0

Table 3 contains data for 14 patients (instances). Each patient is described by a set of symptoms and the model f diagnosed him with a few diseases (compatible and probable for the observed symptoms). After transforming these multi-label data into a formal context, as shown in the previous section, we obtain the following reduced formal concept set $\mathcal{L}'(K')$:

- fc_1:⟨{11},   {$'Asthma'$, $'Bronchiolitis'$, $'COPD'$},   {$'fever'$, $'Dyspnoa'$, $'Shortness of breath'$, $'Wheezing'$, $'Loose cough'$}⟩
- fc_2:⟨{2, 11}, {$''Asthma'$,$' COPD'$}, {$'Fever'$,$' Loose cough'$,$' Dyspnoa'$,$' Wheezing'$}⟩
- fc_3 :   ⟨{8,6}, {$'Asthma'$, $'Bronchiolitis'$, $'Covid'$}, {$'Fever'$, $'Headache'$, $'Shortness of breath'$, $'Dry cough'$, $'Curvature'$}⟩
- fc_4:⟨{8, 3, 6},{$'Asthma'$, $'Covid'$}, {$'Fever'$, $'Headache'$, $'Shortness of breath'$, $'Dry cough'$, $'Curvature'$}⟩

- fc_5 : $\langle\{8, 10, 6\}, \{'Bronchiolitis', 'Covid'\}, \{'fever', 'Headache', 'Shortnessofbreath', 'Drycough', 'Curvature'\}\rangle$
- fc_6 : $\langle\{8, 1, 11,\}, \{'Asthma','Bronchiolitis''\}, \{'Fever','Shortnessofbreath'\}\rangle$
- fc_7 : $\langle\{2, 11, 12, 5\}, \{'COPD''\}, \{'Loosecough','Dyspnoa'\}\rangle$
- fc_8 : $\langle\{1, 2, 8, 11, 14\}, \{'Asthma''\}, \{'wheezing'\}\rangle$
- fc_9 : $\langle\{1, 2, 3, 6, 8, 11\}, \{'Asthma'\}, \{'Fever'\}\rangle$
- fc_{10} : $\langle\{1, 3, 6, 8, 11, 14\}, \{'Asthma'\}, \{'Shortnessofbreath'\}\rangle$
- fc_{11} : $\langle\{1, 6, 8, 9, 10, 11\}, \{'Bronchiolitis'\}, \{'Shortnessofbreath'\}\rangle$
- fc_{12} : $\langle\{3, 4, 6, 7, 8, 10, 13\}, \{'Covid\}, \{'Fever','Headache','Curvature'\}\rangle$.

Given a data instance x and the prediction $f(x)$, we are first interested in symbolic explanations that are *sufficient reasons*. Broadly speaking, they refer to the subset of features in x that are sufficient to predict $y = f(x)$.

4.1 Sufficient Reason Explanations

In FCA terms, a sufficient reason corresponds to the concept of a *decisive attribute set (Das)*. A Das is the smallest subset of features $a_j \in P$ that allows a model f to predict $y \subseteq C$. This subset is said to be decisive in the sense that this decision remains valid regardless of the values of the other attributes.

Let $x = (a_1, a_2, ..., a_n)$ be the instance to classify and $c_k \in y \subseteq C$ be a class predicted among the labels composing the multi-label prediction $y = f(x)$ where f is the multi label classifier to explain. First, we will define the set $\mathcal{B}(c_k)$ of all the formal concepts having in their intentions the class c_k. Let $\mathcal{L}'(K') = \{fc_1, fc_2, ...\}$ be the set of all formal concepts obtained after reduction. Namely, $\mathcal{B}(c_k) = \{fc_i \in \mathcal{L}'(K')|c_k \in Int(fc_i)\}$. The Das that allows the model f to predict the classes c_k is defined as follows: Let $\mathcal{B}(c_k)$ be the set of all formal concepts having in their intentions the class c_k, $Ext(fc_i) = \{x_i \in \mathcal{O}|fc_i \in \mathcal{B}(c_k)\}$ be the extension of the formal concept whose intention contains the class c_k, $Cond(fc_i) = \{a_i \in P|fc_i \in \mathcal{B}(c_k)\}$ be the condition of the formal concept whose intention contains the class c_k, and $Diff(c_k) = \bigcup_{i=1,n} Ext(fc_i)\backslash\bigcap_{i=1,n} Ext(fc_i)$ is the set difference between the union and the intersection of extensions having the class c_k in intention. The decisive Attribute Set (Das) for the class c_k is given as follows:

$$Das_j(c_k) = \{\bigcup Cond(fc_i)|x_j \in \mathcal{E}xt(fc_i) \text{ and } x_j \in \mathcal{D}iff(c_k)\}$$

The following algorithm summarizes this procedure:

Algorithm 3. Compute $\mathcal{D}as(c_k)$

Require: $\mathcal{B}(c_k)$
Ensure: $\mathcal{D}as(c_k)$
 1: $\mathcal{D}as(c_k) = \emptyset$
 2: **if** $|\mathcal{B}(c_k)|=1$ **then**
 3:     $\mathcal{D}as(c_k) = \mathcal{C}ond(fc_k)$
 4: **else**
 5:     **for** $i \leftarrow 1, n$ **do**
 6:         $\mathcal{D}iff(c_k)=\bigcup \mathcal{E}xt(fc_i)\setminus \bigcap \mathcal{E}xt(fc_i)$
 7:     **end for**
 8: **end if**
 9: **for each** $x \in \mathcal{D}iff(c_k)$ **do**
10:     **for** $i = 1 \leftarrow 1, n$ **do**
11:         **if** $x \in \mathcal{E}xt(fc_i)$ **then**
12:             $\mathcal{D}as_x(c_k)=\bigcup \mathcal{C}ond(fc_i)$
13:         **end if**
14:         **if** $\mathcal{D}as_x(c_k) \notin \mathcal{D}as(c_k)$ **then**
15:             $\mathcal{D}as(c_k) = \mathcal{D}as(c_k) \cup \mathcal{D}as_x(c_k)$ ▷ Add a subset to $\mathcal{D}as(c_k)$)
16:         **end if**
17:     **end for**
18: **end for**

It should be noted that a decisive attribute set ($\mathcal{D}as$), which allows to predict a class c_k, is not necessarily unique and that it is possible to find several decisive attribute sets which allow the same class to be predicted. The set $\mathcal{D}as(c_k)$ is made up of subsets $\mathcal{D}as_x(c_k)$ such that $x \in \mathcal{D}iff(c_k)$.

Example 6. Let us continue our running example. Assume we want to compute the set of all decisive attribute sets that allow f to predict the class $c_k = \{Covid\}$ (label $Covid$ is also denoted (d_4)). First, select the formal concepts having in their intention the class $c_k = \{Covid\}$: $\mathcal{B}(\{Covid(d_4)\}) = \{fc_3, fc_4, fc_5, fc_{12}\}$. Then, we compute the set $\mathcal{D}iff(Covid) = \bigcup \mathcal{E}xt(fc_i) \setminus \bigcap \mathcal{E}xt(fc_i)$ with $fc_i \in \mathcal{B}(\{Covid(d_4)\})$: $\mathcal{D}iff(Covid) = \{3, 4, 7, 10, 13\}$. For each instance $x \in \mathcal{D}iff(Covid)$, we compute the set $\mathcal{D}as_x(Covid)$. For objects $\{10, 3\}$, $\mathcal{D}as_{10}(Covid) = \mathcal{D}as_3(Covid)$ we write:

$\mathcal{D}as_{10,3}(Covid) = \{'Fever','Headache','Shortnessofbreath','Drycough',$ $'Curvature'\}$. For objects $\{4, 13, 7\}$, $\mathcal{D}as_{4,13,7}(Covid) = \{'Fever','Headache',$ $'Curvature'\}$.

Hence the set of all decisive attribute sets allowing the model f to predict the class $y = \{Covid\}$ is: $\mathcal{D}as(Covid) = \{\mathcal{D}as_{4,13,7}(Covid), \mathcal{D}as_{10,3}(Covid)\}$

4.2 Significant Attributes Set

A significant attributes set for a class c_k (denoted $\mathcal{S}sa(c_k)$) is the set of attributes that appear in at least in one decisive attribute set. Namely, $\mathcal{S}sa(c_k)$ is equivalent to the set of attributes appearing in the union of all decisive attributes sets for this class $(\mathcal{D}as(c_k))$. The significant attributes set for a set of classes y is the union of all significant attributes set of the set of classes:

$$\mathcal{S}sa(Y) = \bigcup\nolimits_{c_k \in y} \mathcal{S}sa(c_k)$$

Example 7. The significant attributes set for $y = \{Covid\}$ is the union of all decisive attributes sets of the class *Covid*. Thus,
   $\mathcal{S}sa(Covid)$ = $\{'Fever','Headache','Shortnessofbreath','Drycough',$
$'Curvature'\}$, same for the class $COPD$ $\mathcal{S}sa(COPD) = \{'Wheezing','fever',$
$'Dyspnoa','Loosecough'\}$, and $\mathcal{S}sa(covid, COPD)$ = $\mathcal{S}sa(covid) \cup$
$\mathcal{S}sa(COPD)$ = $\{'Fever','Headache','Shortnessofbreath','Drycough',$
$'Curvature','Wheezing','Dyspnoa','Loosecough'\}$

4.3   Beyond $\mathcal{D}as$ and $\mathcal{S}sa$ Explanations

For lack of space, we have limited our presentation to $\mathcal{D}as$ and $\mathcal{S}sa$ explanations. However, our approach can go further to give other types of explanations. For instance, it can directly provide scoped rules commonly known in explainable AI as *Anchors* [11] (rules are in the form *IF feature1=1 AND feature2=1.. THEN the prediction is y*). The necessary features to trigger a prediction are simply the intersection of $\mathcal{D}as$. Moreover, in addition to symbolic explanations, our FCA-based approach can also provide numerical explanations such as feature importance which can be assessed through the frequency of features in sufficient reasons for instance.

5   Case Study

We present in this section a case study on the well-known Stack Overflow collection of coding questions and answers. It features questions and answers on a wide range of topics in computer programming. Based on the type of tags assigned to questions, the most discussed topics on the site are: Java, PHP, Android, Python, HTML, etc. A question in Stack Overflow contains three segments: *Title, Description* and *Tags* as illustrated in Fig. 2. We consider a prediction

	Title	Tags
0	Flask-SQLAlchemy - When are the tables/databas...	['python', 'mysql']
1	Combining two PHP variables for MySQL query	['php', 'mysql']
2	Counting' the number of records that match a c...	['php', 'mysql']
3	Insert new row in a table and auto id number. ...	['php', 'mysql']
4	Create Multiple MySQL tables using PHP	['php', 'mysql']
...	...	...

Fig. 1. Illustration from stack overflow collection

task consisting in predicting the tags (labels) from the title and description of the question. We selected five hundred questions that were preprocessed into 100 binary features and 5 labels ($C = \{Asp.net.1, java.1, mysql.1, php.1, python.1\}$. In our study, we considered three well-known multi-label techniques that are *Label powerset (LPS)*, *MLkNN* and *RAKEL* classifiers which represent the three multi-label classification methods, namely, the problem transformation method, the problem adaptation method and that combines both methods, respectively (note that in our approach a classifier is considered as a black box). For each of them, we computed the decisive attributes sets (Dfs) as well as the percentage of significant attributes (PF) for each class.

Table 4. Number of decisive feature sets per class (nbDfs) and percent of significant attributes per class (Psf)

$nbDfs$	$Asp.Net.1$	$java.1$	$Mysql.1$	$Php.1$	$Python.1$
$MLKNN$	1	1	78	37	29
LPS	1	1	76	33	26
$RAKEL$	1	1	76	51	40
Psf	$Asp.Net.1$	$java.1$	$Mysql.1$	$Php.1$	$Python.1$
$MLKNN$	1%	1%	53%	29%	18%
LPS	1%	1%	51%	25%	19
$RAKEL$	1%	1%	50%	33%	21%

From these results, we notice that the number of decisive feature sets (nbDfs) as well as the percentage of significant features for each class differs from one classifier to another. This means that each of the Rakel and MLkNN classifiers (considered as black-boxes) rely on different attributes for their predictions concerning a given class or a set of given classes. For example, for the php.1 class the MLkNN classifier used 29% of the features with 37 decisive feature sets, RaKel

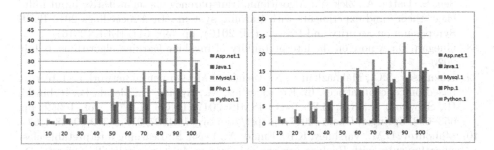

Fig. 2. Average decisive attribute set number per class (left side) and average significant attributes number per class (right side). On the X axis the size (# of instances) of the neighborhood considered (experiments are done of 10,20,...,100 neighbors).

classifier used 33% of the features with 51 decisive feature sets, while the LPS classifier only used 25% of the features with 33 decisive feature sets.

6 Concluding Remarks

In this preliminary work, we proposed an approach based on formal concept analysis to explain multi-label classification, and this by defining the minimal attribute subsets allowing a multi-label classifier to make a given prediction. We have also defined the set of all significant attributes that influence the model predictions concerning a class or a set of classes. As a perspective, we intend to measure the importance of each attribute in the prediction of a set of classes. This work, can also be extended to counterfactual explanations to define, for example, the smallest perturbation (modification) of attribute values that modifies the predictions into a predefined output.

References

1. Berrada, M., Adadi, A.: Peeking inside the black-box: a survey on explainable artificial intelligence (XAI). IEEE Access **6**, 52138–52160 (2018). ieeexplore.ieee.org., 2018
2. Tzima, F., Mitkas, P., Allamanis, M.: Effective rule-based multi-label classification with learning classifier systems. In: 11th International Conference Adaptive and Natural Computing Algorithms. ICANNGA, pp. 466–476 (2013)
3. Ganter, B., Wille, R.: Formal Concept Analysis. Springer, Heidelberg (1999). https://doi.org/10.1007/978-3-642-59830-2
4. Bajorath, J.: Interpretation of machine learning models using shapley values: application to compound potency and multi-target activity predictions. J. Comput.-Aided Mol. Des. **34**, 1013–1026 (2020)
5. Luo, J., Shen, X., Boutell, M.R., Brown, C.M.: Learning multi-label scene classification. Pattern Recogn. **37**, 1757–1771 (2004)
6. Yan, J., Zhang, B., Chen, Z., Chen, W., Yang, Q.: Document transformation for multi-label feature selection in text categorization. In: Seventh IEEE International Conference on Data Mining, pp. 451–456. IEEE (2007)
7. Sen, S., Datta, A., Zick, Y.: Algorithmic transparency via quantitative input influence: theory and experiments with learning systems. In: Proceedings 2016 IEEE Symposium on Security and Privacy (SP 2016), pp. 598–617. IEEE (2016)
8. Guegan, D.: A note on the interpretability of machine learning algorithms, 6 July 2020
9. Kuznetsov, S.O., Makhazhanov, N., Ushakov, M.: On neural network architecture based on concept lattices. In: Kryszkiewicz, M., Appice, A., Ślezak, D., Rybinski, H., Skowron, A., Raś, Z.W. (eds.) ISMIS 2017. LNCS (LNAI), vol. 10352, pp. 653–663. Springer, Cham (2017). https://doi.org/10.1007/978-3-319-60438-1_64
10. Malioutov, D.M., Varshney, K.R., Emad, A., Dash, S.: Learning interpretable classification rules with Boolean compressed sensing. In: Cerquitelli, T., Quercia, D., Pasquale, F. (eds.) Transparent Data Mining for Big and Small Data. SBD, vol. 11, pp. 95–121. Springer, Cham (2017). https://doi.org/10.1007/978-3-319-54024-5_5

11. Guestrin, C., Ribeiro, M.T., Sameer, S.: Anchors: high-precision model-agnostic explanations. In: AAAI Conference on Artificial Intelligence (AAAI) (2018)
12. Emamirad, K., Missaoui, R.: Lattice Miner 2.0: a formal concept analysis tool. In: Supplementary Proceedings of ICFCA, Rennes, France, pp. 91–94 (2017)
13. Dosovitskiy, A., Yosinski, J., Brox, T., Nguyen, A., Clune, J.: Synthesizing the preferred inputs for neurons in neural networks via deep generator networks. In: Advances in Neural Information Processing Systems, 29 (2016)
14. Hua, X.-S., Rui, Y., Tang, J., Mei, T., Qi, G.-J., Zhang, H.-J.: Correlative multi-label video annotation. In: Proceedings of the 15th International Conference on Multimedia - MULTIMEDIA 2007, p. 17. ACM Press (2007)
15. Ruggieri, S., Turini, F., Pedreschi, D., Guidotti, R., Monreale, A., Giannotti, F.: A survey of methods for explaining black box models. arXiv preprint arXiv:1802.01933 (2018)
16. Meddouri, N., Maddouri, M., Trabelsi, M.: New taxonomy of classification methods based on formal concepts analysis. In: What Can FCA Do for Artificial Intelligence, pp. 113–120 (2016)
17. Zhou, Y., Hooker, G.: Interpreting models via single tree approximation. arXiv preprint arXiv:1610.09036 (2016)

Characterizing the Possibilistic Repair for Inconsistent Partially Ordered Assertions

Sihem Belabbes[1]([✉])[iD] and Salem Benferhat[2][iD]

[1] Laboratoire d'Intelligence Artificielle et Sémantique des Données (LIASD),
IUT de Montreuil, Université Paris 8, Saint-Denis, France
belabbes@iut.univ-paris8.fr
[2] Centre de Recherche en Informatique de Lens (CRIL),
Université d'Artois and CNRS, Lens, France
benferhat@cril.fr

Abstract. Ontologies specified in DL-Lite are commonly used to facilitate query answering. Formally, an ontology is a knowledge base composed of a TBox (a set of axioms) and an ABox (a set of assertions). The assertions may be conflicting with respect to the axioms, so the inconsistency in the ABox should be resolved before querying it. This is usually achieved by computing the set of all the conflicts of the ABox. We have recently proposed a method for handling inconsistency in ontologies where the assertions are partially preordered and uncertain. We have defined π-accepted assertions as those assertions that are more certain than at least one assertion of each conflict in the ABox. In DL-Lite ontologies, a conflict is a subset of two assertions, and the set of all the conflicts can be computed in polynomial time. Thus our method is also polynomial in the ABox's size in DL-Lite. We propose here a new equivalent characterization of π-accepted assertions that is also tractable, but without exhibiting the conflicts beforehand. Instead, it is based on a consistency check, such that an assertion is π-accepted if it is consistent with all the assertions that are at least as certain or that are incomparable to it in terms of certainty degrees. This new characterization allows to generalise the method to description logic languages that are more expressive than DL-Lite and where the conflicts may not be computable efficiently.

Keywords: Inconsistency management · Formal ontologies · DL-Lite

1 Introduction

An ontology is a Description Logic [1] knowledge base with two components: a TBox and an ABox. The TBox contains terminological knowledge designed by domain experts and encoded in the form of axioms. The ABox is a dataset composed of ground facts about particular entities. Its elements are called assertions and they are usually obtained from various information sources.

© Springer Nature Switzerland AG 2022
D. Ciucci et al. (Eds.): IPMU 2022, CCIS 1602, pp. 652–666, 2022.
https://doi.org/10.1007/978-3-031-08974-9_52

Answering queries posed over data pieces that are semantically enriched with domain knowledge has the advantage of deriving new facts from the knowledge base. Nonetheless, the drawback is a potential increase in computational complexity, except for the DL-Lite fragments of Description Logics [12] in which query answering is carried out in polynomial time in the ABox's size.

Query answering should be performed over a consistent knowledge base in order to ensure the validity of the derived conclusions. The TBox's axioms can be safely considered as correct and unquestionable. However, the ABox assertions may be prone to errors, incomplete and potentially contradictory when assessed against the axioms. Therefore, the whole knowledge base may be inconsistent and classical Description Logic semantics cannot be used to compute query answers.

Restoring the consistency of the ABox with respect to the TBox can be addressed using the notion of a repair, defined as a maximally-consistent subset of the ABox, and over which query answers can be computed. Since an inconsistent ABox may admit several repairs, a significant body of work has been devoted to designing strategies (a.k.a. inconsistency-tolerant semantics) for choosing which repair(s) should be queried in lieu of the initial ABox [2–4,8–11,14,16,17].

Arguably, one of the most well-known strategies is the "Intersection of ABox Repair" (IAR) semantics [13]. Basically, a query answer is a valid conclusion of the knowledge base, called an IAR-consequence, if it can be derived from a single repair obtained from the intersection of all the repairs of the ABox. Equivalently, an IAR-consequence is an assertion that is not involved in any conflict [5], which is defined as a minimally-inconsistent subset of the ABox.

Another imperfection in the data is uncertainty. Possibility theory has been used as the underlying framework to define a formal setting for standard possibilistic DL-Lite [6]. Each assertion is assigned a real number in the unit interval $]0,1]$ to represent its certainty degree, such that the highest weight where inconsistency is met is called the inconsistency degree of the ABox. The possibilistic repair is defined as a consistent subset of the ABox containing all the assertions that are strictly more certain than the inconsistency degree.

A framework has been recently proposed for the case where the certainty degrees of multi-source data may not be comparable on the same scale [3]. It computes a single repair for the ABox, in the spirit of the IAR semantics. It assumes that the TBox's axioms are fully certain, but that the ABox assertions may be uncertain and are equipped with partially ordered symbolic weights. It proposes a characterization based on the notion of π-accepted assertions, which are the assertions that are more certain than at least one assertion of each conflict. The repair is then the set of all π-accepted assertions. Most notably, it can be computed in polynomial time in the ABox's size in DL-Lite$_{\mathcal{R}}$[1].

This follows directly from the fact that each conflict in DL-Lite involves (at most) two assertions, and that the conflict set can be computed in polynomial time in the ABox's size [11]. However, in more expressive Description Logic

---

[1] The fragment DL-Lite$_{\mathcal{R}}$ is a dialect of DL-Lite that provides the logical underpinnings for the OWL 2 QL profile [15], which is devoted to query answering.

languages, conflicts may involve any number of assertions, and the number of conflicts may be exponential. Hence, the favourable computational properties of this method cannot be guaranteed beyond DL-Lite.

Exhibiting all the conflicts of the ABox is often a prerequisite for computing repairs. In this research, we introduce a new equivalent characterization of π-accepted assertions that is not based on conflicts. It rather performs a consistency check whereby an assertion is π-accepted if it is consistent with all the assertions that are at least as certain or that are incomparable to it in terms of certainty degrees. This way, the method can be generalized to other Description Logic languages, regardless of the computational complexity of computing the conflicts.

This paper is structured as follows. Section 2 presents some preliminaries. Section 3 recalls the method for computing a partially preordered possibilistic repair. Section 4 introduces our new characterization, before concluding.

2 Preliminaries

2.1 Overview of DL-Lite$_\mathcal{R}$

The DL-Lite$_\mathcal{R}$ language is built upon three countably infinite and mutually disjoint sets. These are: a set C_N of *concept names*, a set R_N of *role names* and a set I_N of *individual names*. The syntax is recursively defined as follows:

- $R := P \mid P^-$ is a *basic role*, with $P \in R_N$ and its *inverse* $P^- \in R_N$.
- $E := R \mid \neg R$ denotes a *complex role*.
- $B := A \mid \exists R$, with $A \in C_N$, stands for a *basic concept*.
- $C := B \mid \neg B$ represents a *complex concept*.

In terms of semantics, an interpretation is a tuple $\mathcal{I} = \langle \Delta^\mathcal{I}, \cdot^\mathcal{I} \rangle$, where $\Delta^\mathcal{I} \neq \emptyset$ and $\cdot^\mathcal{I}$ is an interpretation function mapping concept names A to $A^\mathcal{I} \subseteq \Delta^\mathcal{I}$, role names P to $P^\mathcal{I} \subseteq (\Delta^\mathcal{I} \times \Delta^\mathcal{I})$, and individual names a to $a^\mathcal{I} \in \Delta^\mathcal{I}$. We extend the function $\cdot^\mathcal{I}$ to interpret complex concepts and roles of DL-Lite$_\mathcal{R}$ as follows:

$$(P^-)^\mathcal{I} = \{(y, x) \in (\Delta^\mathcal{I} \times \Delta^\mathcal{I}) \mid (x, y) \in P^\mathcal{I}\};$$
$$(\exists R)^\mathcal{I} = \{x \in \Delta^\mathcal{I} \mid \exists y \in \Delta^\mathcal{I} \text{ s.t. } (x, y) \in R^\mathcal{I}\};$$
$$(\neg B)^\mathcal{I} = \Delta^\mathcal{I} \setminus B^\mathcal{I};$$
$$(\neg R)^\mathcal{I} = (\Delta^\mathcal{I} \times \Delta^\mathcal{I}) \setminus R^\mathcal{I}.$$

In this paper, we consider the following vocabulary. Let:

- C_N = {Electric, Thermal, Plugin, Manual, Energy}, representing resp.: electric vehicle, thermal car, rechargeable plug-in car, manual gearbox and energy type.
- R_N = {useFuel}, which links a thermal car to an energy type. Hence, the inverse useFuel$^-$ links an energy type to a thermal car.
- I_N = {v_1, v_2, v_3, p}, where v_i represents a particular vehicle, and "p" stands for the energy type petrol.

An *inclusion axiom* on concepts (resp. on roles) is a statement of the form $B \sqsubseteq C$ (resp. $R \sqsubseteq E$). Concept inclusions with (resp. without) the negation symbol "$\neg$" on the right of the inclusion symbol are called *negative* (resp. *positive*) inclusion axioms. A TBox $\mathcal{T}$ is a finite set of inclusion axioms. An *assertion* is a statement of the form $A(a)$ or $P(a,b)$, where $a, b \in I_N$. An ABox $\mathcal{A}$ is a finite set of assertions. A Knowledge Base (KB) is a pair $\mathcal{K} = \langle \mathcal{T}, \mathcal{A} \rangle$.

An interpretation $\mathcal{I}$ *satisfies* an inclusion axiom $B \sqsubseteq C$ (resp. $R \sqsubseteq E$), denoted by $\mathcal{I} \Vdash B \sqsubseteq C$ (resp. $\mathcal{I} \Vdash R \sqsubseteq E$), if $B^{\mathcal{I}} \subseteq C^{\mathcal{I}}$ (resp. $R^{\mathcal{I}} \subseteq E^{\mathcal{I}}$). Similarly, $\mathcal{I}$ *satisfies* an assertion $A(a)$ (resp. $P(a,b)$), denoted by $\mathcal{I} \Vdash A(a)$ (resp. $\mathcal{I} \Vdash P(a,b)$), if $a^{\mathcal{I}} \in A^{\mathcal{I}}$ (resp. $(a^{\mathcal{I}}, b^{\mathcal{I}}) \in P^{\mathcal{I}}$). An interpretation $\mathcal{I}$ is a *model* of $\mathcal{T}$ (resp. $\mathcal{A}$), denoted by $\mathcal{I} \Vdash \mathcal{T}$ (resp. $\mathcal{I} \Vdash \mathcal{A}$), if $\mathcal{I} \Vdash \alpha$ for every α in $\mathcal{T}$ (resp. in $\mathcal{A}$). We say that $\mathcal{I}$ is a model of a KB $\mathcal{K} = \langle \mathcal{T}, \mathcal{A} \rangle$, if $\mathcal{I} \Vdash \mathcal{T}$ and $\mathcal{I} \Vdash \mathcal{A}$.

A KB $\mathcal{K}$ is *consistent* if it admits at least one model, otherwise it is *inconsistent*. A TBox $\mathcal{T}$ is *incoherent* if there is $A \in C_N$ that is empty in every model of $\mathcal{T}$, otherwise it is *coherent*.

We use the following running example.

Example 1. Let $\mathcal{K} = \langle \mathcal{T}, \mathcal{A} \rangle$ be a DL-Lite$_\mathcal{R}$ KB, where the TBox is given by:

$$\mathcal{T} = \left\{ \begin{array}{ll} \text{1. Electric} \sqsubseteq \neg\text{Manual} & \text{2. Thermal} \sqsubseteq \neg\text{Plugin} \\ \text{3. } \exists\text{useFuel} \sqsubseteq \text{Thermal} & \text{4. } \exists\text{useFuel}^- \sqsubseteq \text{Energy} \end{array} \right\}$$

Axiom 1 indicates that the set of electric vehicles is disjoint from the set of manual transmission vehicles. Axiom 2 indicates that the set of thermal vehicles is disjoint from the set of rechargeable plug-in vehicles. Axiom 3 states that any element using fuel is a thermal vehicle. Axiom 4 specifies that the fuel used by a vehicle is an energy type. Axioms 1 and 2 are negative inclusions on concepts.

Consider the flat ABox (the assertions are equally certain):

$$\mathcal{A} = \left\{ \begin{array}{l} \text{Manual}(v_1), \text{Electric}(v_1), \text{Plugin}(v_1), \text{Thermal}(v_2), \\ \text{Plugin}(v_2), \text{Electric}(v_3), \text{useFuel}(v_2, p), \text{Energy}(p) \end{array} \right\}$$

One can see that $\mathcal{K}$ is inconsistent. For example, the assertions Manual(v_1) and Electric(v_1) violate Axiom 1.

$\square$

2.2 The IAR Semantics

Restoring the consistency of the ABox relies on the notion of ABox repair, inspired from data repair in relational databases to ensure consistent query answering (e.g. see [7]). In the following definitions, we assume $\mathcal{K} = \langle \mathcal{T}, \mathcal{A} \rangle$ is an inconsistent DL-Lite$_\mathcal{R}$ KB.

Definition 1. *A repair, denoted by $\mathcal{R}$, is an inclusion-maximal subset of $\mathcal{A}$ such that $\langle \mathcal{T}, \mathcal{R} \rangle$ is consistent.*

One of the most well-known inconsistency-tolerant semantics is the IAR (Intersection of ABox Repair) semantics [13]. It evaluates queries over a single repair obtained from the intersection of all the repairs of the ABox.

Definition 2. *A query answer is an IAR-consequence of* $\mathcal{K}$ *if it can be derived from the subset:* $\mathsf{IAR}(\mathcal{A}) = \bigcap\{\mathcal{R} \mid \mathcal{R} \text{ is a repair of } \mathcal{A}\}$.

Answers are returned in polynomial time in the ABox's size in DL-Lite$_\mathcal{R}$ [12,13].

Negative inclusion axioms in the TBox allow to specify the disjointness of assertions in the ABox. This is captured by the notion of conflict.

Definition 3. *A conflict, denoted by* $\mathcal{C}$, *is an inclusion-minimal subset of* $\mathcal{A}$ *such that* $\langle \mathcal{T}, \mathcal{C} \rangle$ *is inconsistent.*

We denote the set of all the conflicts in the ABox $\mathcal{A}$ by $\mathsf{Cf}(\mathcal{A})$. We assume that $\mathcal{A}$ does not contain any assertion φ such that $\langle \mathcal{T}, \{\varphi\} \rangle$ is inconsistent. Thus any conflict $\mathcal{C}$ in $\mathsf{Cf}(\mathcal{A})$ is binary, in other words, $|\mathcal{C}| = 2$ [11].

By definition, ABox repairs are conflict-free, so the assertions of the same conflict cannot belong to the same repair. An equivalent characterization for the IAR semantics computes $\mathsf{IAR}(\mathcal{A})$ as the set of free assertions [5,13], i.e., the assertions of $\mathcal{A}$ that are not involved in any conflict.

Example 2. The KB $\mathcal{K}$ contains three conflicts:

- $\{\mathsf{Manual}(v_1), \mathsf{Electric}(v_1)\}$, which contradicts axiom 1.
- $\{\mathsf{Thermal}(v_2), \mathsf{Plugin}(v_2)\}$, which contradicts axiom 2.
- $\{\mathsf{useFuel}(v_2, p), \mathsf{Plugin}(v_2)\}$, which contradicts axioms 2 and 3.

The ABox $\mathcal{A}$ admits the following four repairs:

- $\mathcal{R}_1 = \{\mathsf{Manual}(v_1), \mathsf{Plugin}(v_1), \mathsf{useFuel}(v_2, p), \mathsf{Thermal}(v_2), \mathsf{Electric}(v_3), \mathsf{Energy}(p)\}$.
- $\mathcal{R}_2 = \{\mathsf{Electric}(v_1), \mathsf{Plugin}(v_1), \mathsf{useFuel}(v_2, p), \mathsf{Thermal}(v_2), \mathsf{Electric}(v_3), \mathsf{Energy}(p)\}$.
- $\mathcal{R}_3 = \{\mathsf{Manual}(v_1), \mathsf{Plugin}(v_1), \mathsf{Plugin}(v_2), \mathsf{Electric}(v_3), \mathsf{Energy}(p)\}$.
- $\mathcal{R}_4 = \{\mathsf{Electric}(v_1), \mathsf{Plugin}(v_1), \mathsf{Plugin}(v_2), \mathsf{Electric}(v_3), \mathsf{Energy}(p)\}$.

The intersection of these repairs yields the set:

$$\mathsf{IAR}(\mathcal{A}) = \{\mathsf{Plugin}(v_1), \mathsf{Electric}(v_3), \mathsf{Energy}(p)\}.$$

□

In the rest of the paper, we present the characterization recently proposed in [3]. We then discuss its shortcomings and introduce a characterization that is more efficient computationally.

3 Partially Preordered Possibilistic Repair

Let us recall the method defined in [3] for computing a possibilistic repair for partially preordered DL-Lite$_\mathcal{R}$ ontologies. Consider a partially ordered uncertainty scale $\mathbb{L} = (\mathsf{U}, \rhd)$, defined over:

- a partially ordered set (POS) $\mathsf{U} = \{u_1, \ldots, u_n, \mathbb{1}\}$, and
- a strict partial order $\rhd$ (i.e., an irreflexive and transitive relation).

The element $\mathbb{1}$ represents full certainty, such that: $\forall u_i \in \mathsf{U} \setminus \{\mathbb{1}\}, \mathbb{1} \rhd u_i$.

Intuitively, the elements of a POS denoted by U represent certainty degrees applied to the ABox assertions. When $u_i \not\rhd u_j$ and $u_j \not\rhd u_i$, we say that u_i and u_j are incomparable and we denote it by $u_i \bowtie u_j$.

A partially preordered DL-Lite$_\mathcal{R}$ KB is a triple $\mathcal{K}_\rhd = \langle \mathcal{T}, \mathcal{A}_\rhd, \mathbb{L} \rangle$ with:

$$\mathcal{A}_\rhd = \{(\varphi_i, u_i) \mid \varphi_i \text{ is a DL-Lite}_\mathcal{R} \text{ assertion}, u_i \in \mathsf{U}\},$$

where a single weight u_i is assigned to each assertion φ_i.

Given two assertions $(\varphi_i, u_i), (\varphi_j, u_j) \in \mathcal{A}_\rhd$, we write $\varphi_i \rhd \varphi_j$ to mean $u_i \rhd u_j$ (i.e., φ_i is strictly preferred to φ_j), and write $\varphi_i \bowtie \varphi_j$ to mean $u_i \bowtie u_j$ (i.e., φ_i and φ_j are incomparable). Note that the relation $\rhd$ on U is a strict partial order[2]. However, the ABox $\mathcal{A}_\rhd$ is partially preordered because the same weight can be assigned to more than one assertion.

Computing the partially preordered possibilistic repair proceeds as follows [3]:

(a) Compute the compatible bases of $\mathcal{A}_\rhd$, i.e., consider all the total preorder extensions of $\rhd$ over U.
(b) Compute the possibilistic repair associated with each compatible base.
(c) Intersect all the repairs to obtain a single repair denoted by $\pi(\mathcal{A}_\rhd)$.

Let $\mathcal{WA} = \{(\varphi_i, \alpha_i) \mid (\varphi_i, u_i) \in \mathcal{A}_\rhd, \alpha_i \in]0,1]\}$ be a weighted ABox obtained from $\mathcal{A}_\rhd$ by replacing each symbolic weight $u_i \in \mathsf{U}$ with some real number $\alpha_i \in]0,1]$. Then $\mathcal{WA}$ is compatible with $\mathcal{A}_\rhd$ if it preserves the strict ordering between the assertions. Formally:

$$\forall (\varphi_i, \alpha_i) \in \mathcal{WA}, \forall (\varphi_j, \alpha_j) \in \mathcal{WA}, \text{ if } \varphi_i \rhd \varphi_j \text{ then } \alpha_i > \alpha_j.$$

The set of real numbers that can be assigned to the assertions is infinite, so there are infinitely many compatible bases. However, it suffices to consider a finite number thereof, i.e., only those that express a distinct preference ordering.

The possibilistic repair of a weighted ABox is defined as:

$$\mathcal{R}_\pi(\mathcal{WA}) = \{\varphi \mid (\varphi, \alpha) \in \mathcal{WA}, \alpha > \mathsf{Inc}(\mathcal{WA})\},$$

where $\mathsf{Inc}(\mathcal{WA})$ is the inconsistency degree of $\mathcal{WA}$, i.e., the highest weight attached to an assertion that makes the ABox is inconsistent.

Hence, the partially preordered possibilistic repair of $\mathcal{A}_\rhd$ is defined as:

$$\pi(\mathcal{A}_\rhd) = \bigcap \{\mathcal{R}_\pi(\mathcal{WA}) \mid \mathcal{WA} \text{ is compatible with } \mathcal{A}_\rhd\}.$$

An equivalent characterization [3] of this method produces the same result, without executing the steps (a), (b) and (c) described above. It is based on the notion of π-accepted assertion. Intuitively, this is an assertion that is strictly preferred to at least one assertion of each conflict of $\mathcal{A}_\rhd$. The conflict set $\mathsf{Cf}(\mathcal{A}_\rhd)$ is obtained with a small tweak to Definition 3 to take into account the weights.

---

[2] Namely, $\forall u_i \in \mathsf{U}, \forall u_j \in \mathsf{U}$, if $u_i \rhd u_j$ holds, then $u_j \rhd u_i$ does not hold.

Definition 4. *An assertion* $(\varphi_i, u_i) \in \mathcal{A}_{\triangleright}$ *is* π*-accepted if:*

$$\forall \mathcal{C} \in \mathsf{Cf}(\mathcal{A}_{\triangleright}), \exists (\varphi_j, u_j) \in \mathcal{C}, \varphi_i \neq \varphi_j, \ s.t. \ \varphi_i \triangleright \varphi_j \ (i.e., \ u_i \triangleright u_j).$$

It follows that in DL-Lite$_{\mathcal{R}}$ knowledge bases:

Proposition 1. *The set of all* π*-accepted assertions (without the weights) is equal to* $\pi(\mathcal{A}_{\triangleright})$*. It can be computed in polynomial time in the ABox's size [3].*

Next, we propose a new characterization that is more computationally efficient.

4   Revisiting π-acceptance

4.1   A New Characterization of π-acceptance

In inconsistent knowledge bases that are specified in the lightweight fragment DL-Lite$_{\mathcal{R}}$, each conflict in the ABox is a non-empty subset composed of at most two assertions [11]. In this work, we assume that the ABox does not contain any assertion contradicting itself with respect to the axioms of the TBox. It follows that the conflict set is composed only of pairs of conflicting assertions. Hence, the size of the conflict set (i.e., the number of conflicts) is a square polynomial in the ABox's size, in the worst case.

In order to determine whether some assertion of the ABox is π-accepted, the characterization given in Definition 4 parses all the pairs of assertions in the conflict set, and it does so in linear time in the ABox's size. This means that in the worst case, checking the π-acceptance of an assertion can be achieved in square polynomial time in the ABox's size for DL-Lite$_{\mathcal{R}}$ ontologies. However, despite the fact that a square polynomial time complexity is also polynomial, it may be impractical in applications where the main reasoning task consists in answering queries posed over ontologies with high-dimensional datasets, and especially when the answers need to be computed efficiently.

Furthermore, like most frameworks proposed in the literature for handling inconsistency in Description Logic knowledge bases, the characterization described in Defintion 4 starts from the assumption that the conflict set is computed and available beforehand. This does not constitute an issue in DL-Lite$_{\mathcal{R}}$ ontologies since the time complexity for computing the conflict set is polynomial in the ABox's size [10,11]. However, when dealing with ontologies specified in Description Logic languages that are more expressive than DL-Lite$_{\mathcal{R}}$, there is no assurance that the conflict set can be enumerated in tractable time with respect to the ABox's size.

Two important aspects to take into consideration concern the size of the conflicts (i.e., the number of assertions that constitute each conflict) and the size of the conflict set (i.e., the number of conflicts in the ABox). One advantage of the characterization given in Definition 4 is that it does not place any restrictions on the number of assertions within the conflicts, so long as each conflict contains at least two assertions. Hence, the characterization remains valid in frameworks where the conflicts are not necessarily binary and may be composed of an arbitrary number of assertions. However, as illustrated in Example 4, considering conflicts that may involve any number of assertions entails that the size of the conflict set may be exponential in the ABox's size. This means that the cost of parsing all the elements of the conflict set in order to check the π-acceptance of some assertion can no longer be considered as negligible.

Clearly, the complexity of exhibiting all the conflicts has a direct impact on the computational properties of checking the π-acceptance of an assertion and of determining the set of all π-accepted assertions, i.e., the partially preordered possibilistic repair of the ABox. In order to mitigate these limitations, we propose an equivalent characterization for π-acceptance that does not involve comparisons between an assertion and all the elements of the conflict set. The idea is to rather make use of the consistency checking mechanism that is associated with the Description Logic language in which the ontology is encoded.

The first step is to build, for each assertion (φ_i, u_i) in $\mathcal{A}_\rhd$, the set $\Delta(\varphi_i)$ composed of all the other assertions of $\mathcal{A}_\rhd$ that are either strictly more certain than φ_i or incomparable to φ_i, in terms of the strict partial order $\rhd$. Formally:

$$\Delta(\varphi_i) = \{\varphi_j | (\varphi_j, u_j) \in \mathcal{A}_\rhd \text{ s.t. } u_j \rhd u_i \text{ or } u_j \bowtie u_i\}.$$

Note that we omit the weights associated to the assertions in the set $\Delta(\varphi_i)$ in order to be able to use the standard consistency checking mechanism underlying the ontological language.

Then, the new characterization that we propose in this paper determines the π-acceptance status of any given assertion φ_i by simply checking whether φ_i together with the subset $\Delta(\varphi_i)$ is consistent with respect to the axioms of the TBox. Most importantly, this characterization is equivalent to Definition 4.

Proposition 2. *An assertion (φ_i, u_i) of $\mathcal{A}_\rhd$ is π-accepted in terms of $\rhd$ if and only if $\{\varphi_i\} \cup \Delta(\varphi_i)$ is consistent with respect to $\mathcal{T}$.*

Proof. Consider $(\varphi_i, u_i) \in \mathcal{A}_\rhd$.

(i) Assume that $\{\varphi_i\} \cup \Delta(\varphi_i)$ is consistent w.r.t. $\mathcal{T}$ but that (φ_i, u_i) is not π-accepted. According to Definition 4, this means that there is a conflict $\mathcal{C} \in Cf(\mathcal{A}_\rhd)$ such that for each assertion $(\varphi_j, u_j) \in \mathcal{C}$, $\varphi_i \neq \varphi_j$, we have $\varphi_i \not\rhd \varphi_j$. This means that for each element $(\varphi_j, u_j) \in \mathcal{C}$, we have either $\varphi_j \rhd \varphi_i$ or $\varphi_j \bowtie \varphi_i$ (recall that $\rhd$ is a strict partial order). This means that $\{\varphi_j | (\varphi_j, u_j) \in \mathcal{C}\} \subseteq \Delta(\varphi_i)$. This contradicts the fact that $\{\varphi_i\} \cup \Delta(\varphi_i)$ is consistent.

(ii) Assume that (φ_i, u_i) is π-accepted but that $\{\varphi_i\} \cup \Delta(\varphi_i)$ is inconsistent w.r.t. $\mathcal{T}$. This means that there is a conflict $\mathcal{C}$ such that for each assertion $(\varphi_j, u_j) \in \mathcal{C}$, we have $\varphi_j \in \{\varphi_i\} \cup \Delta(\varphi_i)$. This also means that each element $(\varphi_j, u_j) \in \mathcal{C}$, $\varphi_i \neq \varphi_j$, is such that either $\varphi_j \rhd \varphi_i$ or $\varphi_j \bowtie \varphi_i$. This contradicts the fact that (φ_i, u_i) is π-accepted.

$\square$

An important property of the characterization introduced in Proposition 2 is that it runs in polynomial time in the ABox's size in any Description Logic language where the ABox's consistency can also be checked in polynomial time.

Proposition 3. *Checking whether an assertion* (φ_i, u_i) *of* $\mathcal{A}_\rhd$ *is* π-*accepted can be achieved in polynomial time with respect to the size of* $\mathcal{A}_\rhd$ *in DL-Lite$_\mathcal{R}$.*

Proof. The proof follows directly from the fact that consistency checking is tractable in DL-Lite$_\mathcal{R}$ ontologies. □

We illustrate both characterizations on our running example.

Example 3. Consider the KB $\mathcal{K}_\rhd = \langle \mathcal{T}, \mathcal{A}_\rhd, \mathbb{L} \rangle$ obtained from Example 1 by keeping the same TBox $\mathcal{T}$ and assigning symbolic weights to the assertions as depicted in Fig. 1,(a)-(c). The compatible bases of $\mathcal{A}_\rhd$ are given by Fig. 1,(d)-(f), where the weights in the unit interval are shown on the left side of each sub-figure. The inconsistency degrees are : $\text{Inc}(\mathcal{W}\mathcal{A}_1) = 0.4$, $\text{Inc}(\mathcal{W}\mathcal{A}_2) = 0.4$ and $\text{Inc}(\mathcal{W}\mathcal{A}_3) = 0.2$. The associated possibilistic repairs are by coincidence all the same and they are equal to their intersection. Hence the partially preordered repair corresponds to :

$$\pi(\mathcal{A}_\rhd) = \{\text{Manual}(v_1), \text{Plugin}(v_1), \text{Electric}(v_3), \text{Thermal}(v_2), \text{Energy}(p)\}.$$

First Characterization. It is easy to see that the assertions $(\text{Manual}(v_1), u_4)$, $(\text{Plugin}(v_1), u_4)$, $(\text{Electric}(v_3), u_4)$, $(\text{Thermal}(v_2), u_3)$, $(\text{Energy}(p), u_3)$ are strictly preferred to at least one assertion of each conflict (see Figs. 1(a), 1(b) and 1(c)). Hence, according to Definition 4, these assertions are all π-accepted.

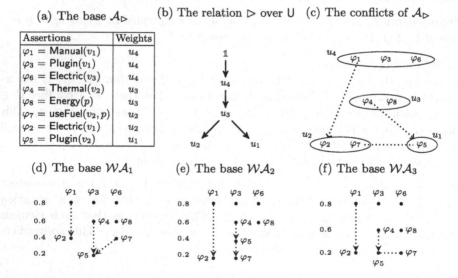

Fig. 1. The base $\mathcal{A}_\rhd$ and its compatible bases $\mathcal{W}\mathcal{A}_1$, $\mathcal{W}\mathcal{A}_2$ and $\mathcal{W}\mathcal{A}_3$. The conflicts are represented with dotted lines. Arrow heads represent strict preference.

Second Characterization. For each assertion in $\mathcal{A}_{\triangleright}$, we determine the corresponding Δ set (see Figs. 1(a), 1(b) and 1(c)).

- $\Delta(\varphi_1) = \Delta(\varphi_3) = \Delta(\varphi_6) = \emptyset$.
  Since $\mathcal{A}_{\triangleright}$ does not contain self-contradictory assertions, each of the three assertions together with the empty set is consistent w.r.t. $\mathcal{T}$. Hence, the assertions (φ_1, u_4), (φ_3, u_4) and (φ_6, u_4) are all π-accepted.
- $\Delta(\varphi_4) = \Delta(\varphi_8) = \{\varphi_1, \varphi_3, \varphi_6\}$.
  Each of the assertions (φ_4, u_3) and (φ_8, u_3) together with the Δ set is consistent w.r.t. $\mathcal{T}$. Hence, both assertions are π-accepted.
- $\Delta(\varphi_2) = \Delta(\varphi_7) = \{\varphi_1, \varphi_3, \varphi_6, \varphi_4, \varphi_8, \varphi_5\}$.
  None of (φ_2, u_2) or (φ_7, u_2) is π-accepted because both assertions are unsatisfiable with the corresponding Δ set.
- $\Delta(\varphi_5) = \{\varphi_1, \varphi_3, \varphi_6, \varphi_4, \varphi_8, \varphi_2, \varphi_7\}$. This Δ set is unsatisfiable, hence the assertion (φ_5, u_1) is not π-accepted.

Hence, the π-accepted assertions using the new characterization are given by the following five assertions: $(\mathsf{Manual}(v_1), u_4)$, $(\mathsf{Plugin}(v_1), u_4)$, $(\mathsf{Electric}(v_3), u_4)$, $(\mathsf{Thermal}(v_2), u_3)$ and $(\mathsf{Energy}(p), u_3)$.
□

This example illustrates that both characterizations return the same π-accepted assertions for $\mathcal{A}_{\triangleright}$, which also correspond to the assertions of the repair $\pi(\mathcal{A}_{\triangleright})$ where the symbolic weights are omitted.

In DL-Lite$_{\mathcal{R}}$ ontologies, $|\mathsf{Cf}(\mathcal{A}_{\triangleright})| = \mathcal{O}(n^2)$, where $n = |\mathcal{A}_{\triangleright}|$. Hence, computing the set of π-accepted assertions using the first characterization requires $\mathcal{O}(n^3)$ steps. Indeed, for each assertion of $\mathcal{A}_{\triangleright}$, the method parses all the conflict pairs in $\mathsf{Cf}(\mathcal{A}_{\triangleright})$ and compares it with both assertions of each pair (in the worst case). In contrast, using the second characterization requires n consistency checks to identify the π-accepted assertions of $\mathcal{A}_{\triangleright}$.

4.2   The Case of Non-binary Conflicts

Our aim in this paper is to compute the set of π-accepted assertions. When dealing with an inconsistent knowledge base, it is desirable to have efficient procedures that allow to:

Task 1 check whether the knowledge base is consistent;
Task 2 compute the set of all the conflicts; and,
Task 3 check whether an assertion is satisfiable with the set of assertions that are either strictly more certain or incomparable.

It is clear that if there exists an efficient procedure that achieves Task 3 in polynomial time, then our method can be extended to Description Logic languages that are richer than DL-Lite$_{\mathcal{R}}$ and in which the conflicts can be of arbitrary size (i.e., not necessarily composed of two assertions like in DL-Lite$_{\mathcal{R}}$). So, checking whether an assertion is π-accepted can be done efficiently.

Note that this is not the case in the original method of calculating π-accepted assertions proposed in [3] for DL-Lite$_\mathcal{R}$ KBs, which is based on Task 2, i.e., it requires the preliminary computation of all the conflicts. However, even if an expressive language has an algorithm for exhibiting all the conflicts in polynomial time in the ABox's size, the time itself may be large. The following simple example illustrates this observation.

Example 4. We are interested in describing the integrity constraints restricting user access to machines (computers) in a large company. We first describe the vocabulary of the language, we then describe the knowledge base.

Suppose that a large company is made up of a number m of departments, simply numbered as $\{d_1, \ldots, d_m\}$. Each department d_i has a number t of machines, denoted by $\{c_{i1}, \ldots, c_{it}\}$ (we assume that all the departments have the same number of machines).

Suppose that we have a set of m role names (one per department) denoted by $\{\text{Access}_1, \ldots, \text{Access}_m\}$. Intuitively, the role $\text{Access}_i(x, c)$ means that in the department d_i, the user x has access to the machine c.

For the sake of simplicity, we are only interested in the permissions granted to a particular employee, for instance the head of the company, denoted by h. Our set of constants is therefore composed of:

$$\{h\} \cup \left(\bigcup_{i=1,\ldots,m} \{c_{i1}, \ldots, c_{it}\} \right).$$

We further assume that the TBox $\mathcal{T}$ is composed of a single negative axiom:

$$\mathcal{T} = \{\exists \text{Access}_1 \sqcap \exists \text{Access}_2 \sqcap \ldots \sqcap \exists \text{Access}_m \sqsubseteq \bot\}.$$

This negative axiom means that there is no user that has access to at least one machine in each department.

The following ABox describes the access permissions granted to the user h, the head of the company. We assume that she has access to all the machines in the company, regardless of the department in which they are located.

$$\begin{aligned}
\mathcal{A} = &\{\text{Access}_1(h, c_{11}), \ldots, \text{Access}_1(h, c_{1t})\} \\
&\cup \{\text{Access}_2(h, c_{21}), \ldots, \text{Access}_2(h, c_{2t})\} \\
&\cup \ldots \\
&\cup \{\text{Access}_m(h, c_{m1}), \ldots, \text{Access}_m(h, c_{mt})\}.
\end{aligned}$$

We assume a partition of the partially preordered ABox $\mathcal{A} = \langle \mathcal{A}_1, \mathcal{A}_2, \ldots, \mathcal{A}_n \rangle$, such that each sub-base $\mathcal{A}_i$ contains the assertions concerning access permissions to the machines of the department d_i:

$$\mathcal{A}_i = \{\text{Access}_i(h, c_{i1}), \ldots, \text{Access}_i(h, c_{it})\}.$$

The preference relation between the sub-bases $\mathcal{A}_i$, $i = 1, \ldots, n$, is defined as:

- for all j, $j = 2, \ldots, n$, we have: $\mathcal{A}_1 \rhd \mathcal{A}_j$, and

– for all k, $k = 2, \ldots, n$ such that $k \neq j$, we have: $\mathcal{A}_j \bowtie \mathcal{A}_k$.

Note that the assertions belonging to same sub-base $\mathcal{A}_i$ are equally certain.

One can easily check that the size of the ABox $\mathcal{A}$ is equal to $t * m$ assertions. Moreover, each m-uple:

$$C = \{\text{Access}_1(h, c_{j_1}), \text{Access}_2(h, c_{j_2}), \ldots, \text{Access}_m(h, c_{j_m})\}$$

obtained by taking exactly one assertion from each role is a conflict. Therefore, the conflict set is:

$$
\begin{aligned}
\mathsf{Cf}(\mathcal{A}) = {} & \{\text{Access}_1(h, c_{11}), \ldots, \text{Access}_1(h, c_{1t})\} \\
& \times \{\text{Access}_2(h, c_{21}), \ldots, \text{Access}_2(h, c_{2t})\} \\
& \times \vdots \\
& \{\text{Access}_m(h, c_{m1}), \ldots, \text{Access}_m(h, c_{mt}\}.
\end{aligned}
$$

where the operator $\times$ denotes the Cartesian product of sets. Hence, the size of the conflict set $\mathsf{Cf}(\mathcal{A})$ is:

$$|\mathsf{Cf}(\mathcal{A})| = \mathcal{O}(t^m).$$

The number of conflicts in the ABox is exponential. This implies that even with reasonable numbers, for instance $m = 10$ and $t = 200$, it is clearly not possible to exhibit all the conflicts in the knowledge base. Hence, the original method for determining π-acceptance [3] is impractical for such a scenario. □

Through Example 4, we argue that even for a Description Logic language that allows to compute the conflict set in polynomial time, it is not sufficient to apply the method for computing π-accepted assertions given in [3]. Indeed, the size of the conflict set also needs to be polynomial in the ABox's size.

Note that if for a given Description Logic language, we have an algorithm that is polynomial in time and space to calculate the conflict set, then checking the consistency is also tractable. However, the converse does not hold. Indeed, assume that there is some language (or at least, special cases of knowledge bases) in which checking consistency is tractable but the number of conflicts in the ABox is not polynomial with respect to the ABox's size. Therefore, the characterization introduced in this paper (Proposition 2) is more efficient than the one introduced in [3] (Definition 4).

In the following example, we apply Proposition 2 in order to determine the π-acceptance of assertions, even in knowledge bases where the size of the conflict set is exponential.

Example 5. We continue Example 4 and illustrate the new characterization with two examples of queries in order to check whether the assertions $\text{Access}_1(h, c_{11})$ and $\text{Access}_3(h, c_{31})$ are π-accepted.

We start with the assertion $\text{Access}_1(h, c_{11})$ and determine $\Delta(\text{Access}_1(h, c_{11}))$. One can determine that:

$$\Delta(\text{Access}_1(h, c_{11})) = \mathcal{A}_1.$$

Indeed, the assertions of $\mathcal{A}_j$ with $j > 1$ are all strictly less certain than any assertion of $\mathcal{A}_1$.

One can easily check that $\Delta(\mathsf{Access}_1(h, c_{11}))$ is consistent. This means that the assertion $\mathsf{Access}_1(h, c_{11})$ is satisfiable with $\Delta(\mathsf{Access}_1(h, c_{11}))$. Therefore, we conclude that $\mathsf{Access}_1(h, c_{11})$ is π-accepted.

Regarding the assertion $\mathsf{Access}_3(h, c_{31})$, one can determine that:

$$\Delta(\mathsf{Access}_3(h, c_{31})) = \mathcal{A}_1 \cup \mathcal{A}_2 \cup \cdots \cup \mathcal{A}_m.$$

Indeed, for any assertion φ of $\mathcal{A}$, either $\mathsf{Access}_3(h, c_{31})$ is incomparable to φ (if $\varphi \in \mathcal{A}_j$ with $j > 1$), or $\mathsf{Access}_3(h, c_{31})$ is strictly less certain than φ (if $\varphi \in \mathcal{A}_1$).

Then, one can also check that each assertion of the ABox $\mathcal{A}$ does not belong to at least one repair of the ABox. Therefore, the assertion $\mathsf{Access}_3(h, c_{31})$ is not satisfiable with the set $\Delta(\mathsf{Access}_3(h, c_{31}))$. Hence, we conclude that $\mathsf{Access}_3(h, c_{31})$ is not π-accepted.

$\square$

5 Conclusion

Handling inconsistency in knowledge bases is an ongoing research topic meeting many applications, such as query answering from ontologies, where the focus is on defining efficient methods and procedures. In this paper, we address this issue by proposing a new characterization for checking (the so-called) π-acceptance in inconsistent and partially preordered ontologies. This characterization is an improvement over the original one. Indeed, the original characterization is based on the conflict set associated with the partially preordered knowledge base, so it assumes that the conflict set is readily available. The new characterization is rather based on a consistency check on a subset of the ABox, and does not require computing the conflict set.

Moreover, the new characterization can be applied to Description Logic languages that are more expressive than DL-Lite$_\mathcal{R}$, and remains efficient for any language where the consistency check can be performed efficiently.

In future work, we plan to investigate methods for producing more productive repairs and that are also tractable. One option is to consider the positive deductive closure but without incurring a computational explosion. We also plan to explore the case where the TBox's axioms may be uncertain and may be ignored or weakened as a means for resolving the inconsistency in the ABox.

Within the research project CROQUIS in collaboration with specialists in hydro-science, we plan to apply our methods of inconsistency handling to knowledge bases representing wastewater and stormwater networks in a large metropolis. The expert knowledge serves to complete the data, which is incomplete, imperfect, fragmented, outdated, multi-source, heterogeneous and uncertain. Data may consist of analog and digital maps of urban networks, geographical data, various types of images, intervention reports, and so on. It may be obtained from public organisations as well as private companies. Hence, it is virtually impossible for such a knowledge base to be consistent, making standard query answering tools inadequate. Moreover, given the sheer volume of data, a reduction in the computational complexity of repairing the inconsistency in the ABox is expected to have a direct positive impact on the experimental performance of our new characterization.

Acknowledgements. This research has received support from the ANR CROQUIS (Collecting, Representing, cOmpleting, merging, and Querying heterogeneous and Uncertain waStewater and stormwater network data) project, grant ANR-21-CE23-0004 of the French research funding agency Agence Nationale de la Recherche (ANR).

References

1. Baader, F., Calvanese, D., Mcguinness, D., Nardi, D., Patel-Schneider, P.: The Description Logic Handbook: Theory, Implementation, and Applications (2007)
2. Baget, J., et al.: A general modifier-based framework for inconsistency-tolerant query answering. In: Principles of Knowledge Representation and Reasoning (KR), Cape Town, South Africa, pp. 513–516 (2016)
3. Belabbes, S., Benferhat, S.: Computing a possibility theory repair for partially preordered inconsistent ontologies. IEEE Trans. Fuzzy Syst. 1–10 (2021). https://doi.org/10.1109/TFUZZ.2021.3107776
4. Belabbes, S., Benferhat, S., Chomicki, J.: Handling inconsistency in partially preordered ontologies: the elect method. J. Logic Comput. **31**(5), 1356–1388 (2021)
5. Benferhat, S., Bouraoui, Z., Tabia, K.: How to select one preferred assertional-based repair from inconsistent and prioritized DL-Lite knowledge bases? In: International Joint Conference on Artificial Intelligence (IJCAI), Buenos Aires, Argentina, pp. 1450–1456 (2015)
6. Benferhat, S., Bouraoui, Z.: Min-based possibilistic DL-Lite. J. Logic Comput. **27**(1), 261–297 (2017)
7. Bertossi, L.: Database Repairing and Consistent Query Answering. Morgan & Claypool Publishers, Synthesis Lectures on Data Management (2011)
8. Bienvenu, M., Bourgaux, C.: Querying and repairing inconsistent prioritized knowledge bases: complexity analysis and links with abstract argumentation. In: Proceedings of the 17th International Conference on Principles of Knowledge Representation and Reasoning, KR, pp. 141–151 (2020). https://doi.org/10.24963/kr.2020/15
9. Bienvenu, M., Bourgaux, C., Goasdoué, F.: Query-driven repairing of inconsistent DL-Lite knowledge bases. In: IJCAI, New York, USA, pp. 957–964 (2016)
10. Bienvenu, M., Bourgaux, C., Goasdoué, F.: Computing and explaining query answers over inconsistent DL-Lite knowledge bases. J. Artif. Intell. Res. **64**, 563–644 (2019)
11. Calvanese, D., Kharlamov, E., Nutt, W., Zheleznyakov, D.: Evolution of DL-Lite knowledge bases. In: International Semantic Web Conference, No. 1, pp. 112–128 (2010)
12. Calvanese, D., De Giacomo, G., Lembo, D., Lenzerini, M., Rosati, R.: Tractable reasoning and efficient query answering in description logics: the DL-Lite family. J. Autom. Reason. **39**(3), 385–429 (2007)

13. Lembo, D., Lenzerini, M., Rosati, R., Ruzzi, M., Savo, D.F.: Inconsistency-tolerant semantics for description logics. In: Hitzler, P., Lukasiewicz, T. (eds.) RR 2010. LNCS, vol. 6333, pp. 103–117. Springer, Heidelberg (2010). https://doi.org/10. 1007/978-3-642-15918-3_9

14. Lukasiewicz, T., Martinez, M.V., Simari, G.I.: Inconsistency handling in datalog+/- ontologies. In: ECAI, Montpellier, France, pp. 558–563 (2012)

15. Motik, B., Grau, B.C., Horrocks, I., Wu, Z., Fokoue, A., Lutz, C.: OWL 2 Web Ontology Language Profiles. W3C Recommendation, 11 December 2012. https:// www.w3.org/TR/owl2-profiles/

16. Trivela, D., Stoilos, G., Vassalos, V.: Querying expressive DL ontologies under the ICAR semantics. In: 31st DL workshop, Tempe, USA (2018)

17. Tsalapati, E., Stoilos, G., Stamou, G., Koletsos, G.: Efficient query answering over expressive inconsistent description logics. In: IJCAI, New York, USA, pp. 1279–1285 (2016)

Coherent Upper Conditional Previsions with Respect to Outer Hausdorff Measures and the Mathematical Representation of the Selective Attention

Serena Doria[✉][ID]

Department of Engineering and Geology, University G.d'Annunzio,
67100 Chieti, Italy
serena.doria@unich.it

Abstract. Coherent upper conditional probabilities defined by Hausdorff outer measure are proposed to represent the unconscious activities of the human brain when information are given. In the model uncertainty measures are defined according to the complexity of the conditioning event that represent the given information. The model is applied to explain mathematically the bias of selective attention described in the so-called "invisible Gorilla" experiment, that is often taken as a characteristic example of the inescapable limitations of human perception. Once people are concentrated on doing a specific action, they do not notice unexpected events (having 0 probability) occurring in the meantime. When applying the model, selective attention is no longer a bias since it is able to explain this function of the human brain mathematically. Moreover different reactions of people to unexpected events can be represented in different metric spaces with metrics which are not bi-Lipschitz. In these metric spaces coherent upper conditional probabilities defined by Hausdorff outer measures are not mutually absolutely continuous and so they do not share the same null events.

Keywords: Coherent upper conditional previsions · Hausdorff outer measures · Bias · Selective attention

1 Introduction

To represent and to explain mathematically some aspects of human brain's activity is one of the aim of AI in the recent years. The model of coherent lower and upper conditional previsions, based on Hausdorff inner and outer measures, ([6–9,11,13,14]) is proposed to represent the preference orderings and the equivalences, respectively assigned by the conscious and unconscious thought in human decision making under uncertainty. In fact for each conditioning event B, a partial strict order , (i.e.an antisymmetric and transitive binary relation) can be defined with respect to a coherent lower conditional prevision $\underline{P}(\cdot|B)$ and an

© Springer Nature Switzerland AG 2022
D. Ciucci et al. (Eds.): IPMU 2022, CCIS 1602, pp. 667–680, 2022.
https://doi.org/10.1007/978-3-031-08974-9_53

equivalence relation, (i.e. a complete reflexive, symmetric and transitive binary relation) can be defined with respect to a coherent upper conditional prevision $\overline{P}(\cdot|B)$. Complexity of partial information is represented by the Hausdorff dimension of the conditioning event. In particular coherent upper conditional probability, which satisfies the symmetric property, can be used to represent the unconscious human brain's activity, called by Matte Blanco [20] symmetrical thought - based upon the principles of symmetry and generalization. While coherent lower conditional probability defined by Hausdorff inner measures represents partial strict preference ordering and so can represent logical conscious/asymmetrical thought - structured on the categories of time and space and ruled by Aristotle's principle of non-contradiction. These two different modes of being are supposed to combine in different human thinking experiences since they yield a bi-logic asset as proposed in [20].

Coherent upper conditional probability defined by Hausdorff outer measures can also explain some biases of human intuition investigated by different experiments in the Prospect Theory [18,19,24]. In this framework experimental methods lead to describe the dual process of the brain activity as regulated by two different ways of thinking namely fast and slow thinking denoted also as System 1 and System 2, respectively. System 1 regulates intuitive, involuntary, unconscious and effortless activities while System 2 is the conscious part of the brain in charge of logical reasoning.

The model has been applied and discussed in Linda's Problem and the conjunction fallacy is resolved ([12,13]).

The model explains mathematically the bias of selective attention described in the so-called invisible gorilla experiment [2] , that is often taken as a characteristic example of the inescapable limitations of human perception. Once people are concentrated on doing a specific action, they do not notice unexpected events (having 0 probability) occurring in the meantime. Unexpected events are represented in the model as sets with Hausdorff dimension less than the Hausdorff dimension of the conditioning event, which represents the specific action which people are concentrated on doing. When applying the model, selective attention is no longer a bias since it is able to explain this function of the human brain mathematically.

2   Coherent Upper and Lower Conditional Previsions Defined by Hausdorff Outer and Inner Measures

Let (Ω, d) be a metric space and let **B** be partition of Ω.

A bounded random variable is a function $X : \Omega \to \Re = (-\infty; +\infty)$ and $L(\Omega)$ is the class of all bounded random variables defined on Ω; for every $B \in \mathbf{B}$ denote by $X|B$ the restriction of X to B and by $\sup(X|B)$ the supremum value that X assumes on B.

Let $L(B)$ be the class of all bounded random variables $X|B$.

Denote by I_A the indicator function of any event $A \in \wp(B)$, i.e. $I_A(\omega) = 1$ if $\omega \in A$ and $I_A(\omega) = 0$ if $\omega \in A^c$.

For every $B \in \mathbf{B}$ coherent upper conditional expectations or previsions $\overline{P}(\cdot|B)$ are functionals defined on $L(B)$ (Walley (1991)).

Definition 1. *Coherent upper conditional previsions are functionals* $\overline{P}(\cdot|B)$ *defined on* $L(B)$, *such that the following axioms of coherence hold for every* X *and* Y *in* $L(B)$ *and every strictly positive constant* λ:

1) $\overline{P}(X|B) \leq \sup(X|B)$;
2) $\overline{P}(\lambda X|B) = \lambda \overline{P}(X|B)$ *(positive homogeneity)*;
3) $\overline{P}(X + Y|B) \leq \overline{P}(X|B) + \overline{P}(Y|B)$ *(subadditivity)*.

Suppose that $\overline{P}(X|B)$ is a coherent upper conditional expectation on $L(B)$. Then its conjugate coherent lower conditional expectation is defined by

$$\underline{P}(X|B) = -\overline{P}(-X|B).$$

Let K be a linear space contained in $L(B)$; if for every X belonging to K we have $P(X|B) = \underline{P}(X|B) = \overline{P}(X|B)$ then $P(X|B)$ is called a coherent linear conditional expectation and it is a linear, positive and positively homogenous functional on K. The unconditional coherent upper expectation $\overline{P} = \overline{P}(\cdot|\Omega)$ is obtained as a particular case when the conditioning event is Ω. Coherent upper conditional probabilities are obtained when only 0–1 valued random variables are considered.

From axioms 1)–3) and by the conjugacy property we have that

$$1 \leq \underline{P}(I_B|B) \leq \overline{P}(I_B|B) \leq 1$$

so that

$$\underline{P}(I_B|B) = \overline{P}(I_B|B) = 1.$$

2.1 Preference Ordering and Indifference Between Random Variables Represented by Coherent Lower and Upper Conditional Previsions

A partial strict order can be represented by the lower conditional prevision $\underline{P}(X|B)$.

Definition 2. *3 We say that* X *is preferable to* Y *given* B *with respect to* $\underline{P}$, *i.e.* $X \succ_* Y$ *given* B *if and only if*

$$\underline{P}((X - Y)|B) > 0$$

In particular we show that
the binary relation $\succ_*$ satisfies the antisymmetric property,

$$X \succ_* Y \Longleftrightarrow \underline{P}((X - Y)|B) > 0 \Longrightarrow$$
$$\underline{P}((Y - X|B) \leq 0 \Longleftrightarrow Y \, not \succ_* X.$$

Two random variables which have previsions equal to zero cannot be compared by the ordering $\succ_*$.

A weak order $\succ^*$ can be defined on $L(B)$ with respect to $\overline{P}$ but it cannot represent a strict preference ordering because it does not satisfied the antisymmetric property.

Definition 3. *We say that $X \succ^* Y$ given B if and only if $\overline{P}((X - Y)|B) > 0$.*

Two complete equivalence relations, which are complete reflexive, symmetric and transitive binary relations on $L(B)$ can be represented by the coherent upper conditional prevision $\overline{P}(X|B)$.

Definition 4. *Two random variables X and $Y \in L(B)$ are equivalent given B with respect to $\overline{P}$ if and only if $\overline{P}(X|B) = \overline{P}(Y|B)$.*

Definition 5. *We say that X and Y are indifferent given B with respect to $\overline{P}$, i.e. $X \approx Y$ in B if and only if*

$$\overline{P}((X - Y)|B) = \overline{P}((Y - X)|B) = 0.$$

If the coherent conditional prevision $P(\cdot|B)$ is linear then

$$P((X - Y)|B) = P((Y - X)|B) = 0 \iff P(X|B) = P(Y|B)$$

and two random variables X and Y are indifferent given B if and only if they are equivalent given B. The notions recalled in this subsection and the results proven in [11] about preference orderings represented by coherent lower previsions and equivalences represented by coherent upper previsions induce to use the former to represent the conscious activity and the latter to describe the unconscious activity of human brain.

2.2   Hausdorff Outer Measures

Outer measures are non-negative, monotone set-functions that are sub-additive so they duplicate basic property of upper probability for sets. Hausdorff outer measures are examples of outer measures defined in a metric space.

Let (Ω, d) be a metric space. The diameter of a non empty set U of Ω is defined as $|U| = sup\left\{d(x, y) : x, y \in U\right\}$ and if a subset A of Ω is such that $A \subset \bigcup_i U_i$ and $0 < |U_i| < \delta$ for each i, the class $\left\{U_i\right\}$ is called a δ-cover of A.

Let s be a non-negative number. For $\delta > 0$ we define $h_{s,\delta}(A) = \inf \sum_{i=1}^{\infty} |U_i|^s$, where the infimum is over all δ-covers $\left\{U_i\right\}$.

The *Hausdorff s-dimensional outer measure* of A ([17,22]) denoted by $h^s(A)$, is defined as

$$h^s(A) = lim_{\delta \to 0} h_{s,\delta}(A).$$

This limit exists, but may be infinite, since $h_{s,\delta}(A)$ increases as δ decreases. The *Hausdorff dimension* of a set A, $dim_H(A)$, is defined as the unique value, such that

$$h^s(A) = \infty \text{ if } 0 \le s < dim_H(A),$$
$$h^s(A) = 0 \text{ if } dim_H(A) < s < \infty.$$

A model of conditioning probability based on Hausdorff outer measures has been introduced [7]. In the model, conditional probability, usually intended as a measure of the probability of an event occurring, given that another event (by assumption, presumption, assertion or evidence) has already occurred, is defined by the Hausdorff measure of order s, or s-dimensional Hausdorff measure, if the conditioning event has Hausdorff dimension equal to s.

Theorem 1. *Let (Ω, d) be a metric space and let $\boldsymbol{B}$ be a partition of Ω. For every $B \in \boldsymbol{B}$ denote by s the Hausdorff dimension of the conditioning event B and by h^s the Hausdorff s-dimensional outer measure. Let m be a 0–1 valued finitely additive, but not countably additive, probability on $\wp(B)$. Thus, for each $B \in \boldsymbol{B}$, the function defined on $\wp(B)$ by*

$$\overline{P}(A|B) = \begin{cases} \frac{h^s(A \cap B)}{h^s(B)} & \text{if } 0 < h^s(B) < +\infty \\ m_B & \text{if } h^s(B) \in \{0, +\infty\} \end{cases}$$

is a coherent upper conditional probability.

If $B \in \boldsymbol{B}$ is a set with positive and finite Hausdorff outer measure in its Hausdorff dimension s the monotone set function μ_B^* defined for every $A \in \wp(B)$ by $\mu_B^*(A) = \frac{h^s(AB)}{h^s(B)}$ is a coherent upper conditional probability, which is submodular, continuous from below and such that its restriction to the σ-field of all μ_B^*-measurable sets is a Borel regular countably additive probability.

Example 1. *Let (Ω, d) be the Euclidean metric space where $\Omega = [0,1]^2$, let P be the Lebesgue measure on Ω and let B be the Sicrpinsky Triangle, which is the attractor of an iterated functions system represented in Fig. 1. It has non-integer Hausdorff dimension less than 2 and so it has P zero probability. According to*

Fig. 1. The Sierpinsky Triangle is an example of conditioning events with non-integer Hausdorff dimension ($s = \frac{log_2}{log_3}$)

the model in Theorem 1 to define the conditional probability $P(\cdot|B)$ we use the Hausdorff measure of order $s = \frac{\log 2}{\log 3}$.

Example 2. Let $\Omega = [0,1]$ and let P be the Lebesgue measure which defines the unconditional probability which represents our knowledge. Let $A = [0, \frac{1}{4}]$ be an event and let $B = \{0, \frac{1}{2}\}$; B is a finite set so it has zero probability with respect to P and so B represents an unexpected event. The Hausdorff dimension of B is 0 and the Hausdorff measure of order 0 is the counting measure and $h^0(B) = 2$. According to the model proposed in Theorem 1 we can calculate the conditional probability $P(A|B)$ by means of the counting measure h^0 so that $P(A|B) = \frac{h^0(A\cap B)}{h^0(B)} = \frac{1}{2}$.

Example 3. Let $\Omega = [0,1]$ and let P be the Lebesgue measure which defines the unconditional probability which represents our knowledge. Let $A = [0, \frac{1}{4}]$ be an event and let B be the set of rational numbers in $[0,1]$; B is an infinite set which has zero probability with respect to P so B represents an unexpected event. The Hausdorff dimension of B is 0 and the Hausdorff measure of order 0 is the counting measure. Since $h^0(B) = +\infty$, according to the model proposed in Theorem 1 we can calculate the conditional probability $P(A|B)$ by a $0-1$ valued finitely, but not countably, additive probability m_B.

In the following theorem, proven in [7], the coherent upper conditional probability defined in Theorem 1 is extended to the class of all bounded random variables and, when the conditioning event B has positive and finite Hausdorff outer measure in its Hausdorff dimension, the coherent upper prevision is defined by the Choquet integral [1].

Theorem 2. Let (Ω, d) be a metric space and let $\boldsymbol{B}$ be a partition of Ω. For every $B \in \boldsymbol{B}$ denote by s the Hausdorff dimension of the conditioning event B and by h^s the Hausdorff s-dimensional outer measure. Let m_B be a 0-1 valued finitely additive, but not countably additive, probability on $\wp(B)$. Then for each $B \in \boldsymbol{B}$ the functional $\overline{P}(X|B)$ defined on $L(B)$ by

$$\overline{P}(X|B) = \begin{cases} \frac{1}{h^s(B)} \int_B X dh^s & \text{if } 0 < h^s(B) < +\infty \\ \int_B X dm_B & \text{if } h^s(B) \in \{0, +\infty\} \end{cases}$$

is a coherent upper conditional prevision.

According to the model the partial knowledge is updated when an event with positive and finite Hausdorff in its Hausdorff dimension represent a new piece of information. In that case if the Hausdorff dimension s of the conditioning event is less than the Hausdorff dimension, t, of Ω the conditioning event is an event with zero probability with respect to the a prior probability h^t and to update the partial knowledge the new Hausdorff outer measure h^s is considered to defined the conditional probability. According to the model the unexpected events are those which really update the knowledge.

3   Absolute Continuity of Coherent Conditional Probability Measures Defined by Hausdorff Measures

In this section probability measures defined on the Borel σ-field of a metric space (Ω, d) by Hausdorff measures as in Theorem 1, are proven to be absolutely continuous with respect to any probability measure defined by Theorem 1 in a metric space (Ω, d') where d' is a bounded metric bi-Lipschitz equivalent to the metric d. It occurs because events which have zero Hausdorff measure in a metric space have also Hausdorff measure equal to zero in a metric space with a bi-Lipschitz equivalent metric.

An example is given to show that probability measures defined by Hausdorff measures in metric spaces that are topological equivalent are not absolutely continuous.

Two different notions of equivalence can be considered for metrics: bi-Lipschitz equivalence and topological equivalence.

Definition 6. *Let (Ω, d) be a metric space; a metric d' on Ω is bi-Lipschitz equivalent to the metric d if there exist two positive real constants α, β such that $\forall x, y \in \Omega$*

$$\alpha d'(x, y) \leq d(x.y) \leq \beta d'(x, y)$$

Definition 7. *Let (Ω, d) be a metric space and let d' be a metric on Ω; d and d' are topological equivalent if they induce the same topology.*

Proposition 1. *Let (Ω, d) be a metric space and let d' be a metric on Ω bi-Lipschitz equivalent to d, then d and d' are topological equivalent.*

The following example shows that the converse is not true.

Example 4. *Let $(\Re^n, d)$ be the Euclidean metric space and let d' be a metric on $\Re^n$ defined $\forall \overline{x}, \overline{y} \in \Re^n$ by*

$$d'(\overline{x}, \overline{y}) = \frac{d(\overline{x}, \overline{y})}{1 + d(\overline{x}, \overline{y})};$$

d' is topological equivalent to the Euclidean metric d but it is not bi-Lipschitz equivalent to d since there not exist two positive real constants constants α, β such that $\alpha d'(x, y) \leq d(x, y) \leq \beta d'(x, y)$

Theorem 3. *Let (Ω, d) be a metric space, let d and d' be two metrics on Ω bi-Lipschitz equivalent and let h^s and h_1^s be the s-dimensional Hausdorff measures defined respectively in the metric space (Ω, d) and (Ω, d'), then there exist two positive real constants α, β such that*

$$\alpha h_1^s(E) \leq h^s(E) \leq \beta h_1^s(E)$$

Proof. The result follows by the definition of Hausdorff outer measures and by the fact that the metrics are bi-Lipschtz equivalent (see Lemma 1.8 of [17]). $\diamond$

Theorem 4. *Let (Ω, d) be a metric space and let d' be a metric on Ω bi-Lipschitz equivalent to d. Then the Hausdorff dimension of any set $A \in \wp(\Omega)$ is invariant in the two metric spaces (Ω, d) and (Ω, d')*

The Hausdorff dimension of any set $A \in \wp(\Omega)$ is not invariant with respect to two topological equivalent metrics which are not bi-Lipschitz equivalent.

Example 5. *Let $\Omega = [0, 1]$ and let d be the Euclidean metric*

$$d(\omega_1, \omega_2) = |\omega_1 - \omega_2|$$

and let d' the discrete distance, that is

$$d'(\omega_1, \omega_2) = \begin{cases} 0 \ if \quad \omega_1 = \omega_2 \\ 1 \quad otherwise. \end{cases}$$

d and d' are not topologically equivalent; in fact all subsets of Ω are open sets in the topology induced by d' since $D_r(x) = \{\omega \in \Omega : d(\omega, x) < r\} = \{x\}$ if $r < 1$ and $D_r(x) = \{\omega \in \Omega : d(\omega, x) < r\} = \Omega$ if $r \geq 1$, while singletons are not open sets in the topology induced by the Euclidean metric.

Example 6. *Let $(\Re^2, d)$ be the Euclidean metric space and let d'' be the metric defined by*

$$d''(x, y) = max \{|x_1 - y_1|; |x_2 - y_2|\}$$

where $x = (x_1, x_2)$ and $y = (y_1, y_2)$.
Then d and d'' are topologically equivalent.

The notion of boundness of a set depends on the metric.

Definition 8. *A metric on Ω is bounded if $diam(\Omega)$ is bounded. A metric space (Ω, d) is bounded if d is bounded.*

Proposition 2. *Let (Ω, d) be a metric space where Ω has positive and finite Hausdorff measures in its Hausdorff dimension then (Ω, d) be a bounded metric space.*

Definition 9. *Let μ and ν be two probabilities measures on the same σ-field $\mathcal{F}$ then ν is absolutely continuous with respect to μ, $\nu << \mu$, if $\mu(A) = 0 \Rightarrow \nu(A) = 0$ for every $A \in \mathcal{F}$.*

Theorem 5. *Let (Ω, d) be a bounded metric space where Ω is a set with positive and finite Hausdorff measure in its Hausdorff dimension, let $\mathbf{B}$ be a partition of Ω and let $\mathcal{B}$ be the Borel σ-field induced by the metric d. Let d' be a bounded metric on Ω bi-Lipschitz equivalent to d. Then for every $B \in \mathbf{B}$ with positive and finite Hausdorff outer measures in its dimensions in both metric spaces, the restrictions μ_B and ν_B on $\mathcal{B}$ of the coherent upper conditional probabilities defined respectively in (Ω, d) and (Ω, d') as in Theorem 1, are countably additive probabilities which are mutually absolutely continuous.*

Proof. Since d' is bi-Lipschitz equivalent, and so topologically equivalent, to d then d and d' induce the same topology and the same Borel σ-field $\mathcal{B}$. Since d' is a bounded metric then by Proposition 2 Ω has positive and finite Hausdorff measure in its Hausdorff dimension also in (Ω, d'). Let s be the Hausdorff dimension of $\mathcal{B}$, let h^s and h_1^s be the s-dimensional Hausdorff measure in the two metric spaces and let μ_B and ν_B be the two probability measures on $\mathcal{B}$ defined by

$$\mu_B(A) = \frac{h^s(A \cap B)}{h^s(B)} \text{ and } \nu_B(A) = \frac{h_1^s(A \cap B)}{h_1^s(B)}.$$

Since d' is bi-Lipschitz equivalent to d by Theorem 2 we have that there exist two positive real constants α and β such that

$$\alpha \nu(A) = \alpha \frac{h_1^s(A)}{h_1^s(\Omega)} \leq \mu(A) = \frac{h^s(A)}{h^s(\Omega)} \leq \beta \frac{h_1^s(A)}{h_1^s(\Omega)} = \beta \nu(A)$$

so that $\nu(A) = 0$ implies $\mu(A) = 0$ and $\mu(A) = 0$ implies $\nu(A) = 0$. $\diamond$

4   The Invisible Gorilla Experiment: A Mathematically Explanation of the Selective Attention

In this section we analyze one of the capacity of the unconscious activity of human brain named selective attention. It consists in the ability to select only some of the numerous information that reach the sense organs when focused on a particular objective. Neglected information is not important to the goal and unexpected events with respect to the goal are not perceived. Selective attention put in evidence the capacity of the unconscious activity of the human brain to manage unexpected events.

In the so called invisible Gorilla experiment [2], participants are asked to watch a short video, in which six people-three in white shirts and three in black shirts-pass basketballs around. While they watch, they must keep a silent count of the number of passes made by the people in white shirts. At some point, a gorilla strolls into the middle of the action, faces the camera and thumps its chest, and then leaves, spending nine seconds on screen. Then, study participants are asked, "But did you see the gorilla?" More than half the time, subjects miss the gorilla entirely. It was as though the gorilla was invisible. More than that, even after the participants are told about the gorilla, they're certain they couldn't have missed it.

This experiment reveals two things: that we are missing a lot of what goes on around us, and that we have no idea that we are missing so much. The experiment can be described in term of coherent upper and lower conditional probabilities defined by Hausdorff outer and inner measures.

Moreover in later versions of the video unexpected events with respect to the goal of counting the passages between the players in white shirts have been added: a girl in a white shirt put on a hat and one in a black shirt goes out. It occurs that if a participant in the experiment knows the video and see Gorilla, he does not notice the girl wearing the hat or the girl who goes out. The experiment therefore confirms that the neglect of unexpected events when focused on a goal is an aspect of the unconscious activity of the brain and on which there is no decision-making capacity.

The experiment can be mathematically described in a metric space (Ω, d) by using the model based on Hausdorff outer measures. Consider the events

E: "You see the Gorilla in the video."
B: "You count the number of passes made by the people in white shirts."

Given a metric space (Ω, d), according to the model of conditional upper conditional probability of Theorem 1, people which do not see the Gorilla asses zero upper probability to the event E, that is it is unexpected given the event B; it means that the event E is represented by a set with Hausdorff dimension less than the Hausdorff dimension s of the event B so that

$$\overline{P}(E|B) = \frac{h^s(E \cap B)}{h^s(B)} = 0$$

According to the model a new unexpected event F with respect to the conditioning event B can be represented by a new set with Hausdorff dimension less than the Hausdorff dimension of B and so a new conditioned event $F|B$ can be considered; by Theorem 1 the coherent upper conditional probability can be extended to a larger domain in a way such that the coherence of the assessment is assured.

Moreover, since

$$0 \le \underline{P}(E|B) \le \overline{P}(E|B) = 0$$

the model describes that the unconscious thought, whose activity is represented by the coherent upper conditional probability, asses zero probability to the event $E \cap B$ and it implies that also, the coherent lower conditional probability of $E \cap B$ is zero.

Since coherent lower conditional probability according to the model represents the activity of the conscious thought, unexpected events for the unconscious thought are unexpected events also for the conscious thought but the converse is not true, that is unexpected events for the conscious thought may be not unexpected events also for the unconscious thought.

The unconscious thought of people who see the Gorilla in the experiment can be represents by a coherent upper probability defined by Theorem 1 in a metric space (Ω, d') where the metric d' is not bi-Lipschitz with respect to d; the conditional upper conditional probabilities are not mutually absolutely continuous and so they do not share the same null events.

4.1   A Mathematically Explanation of the Confirmation Bias

The confirmation bias is the tendency to process information by looking for, or interpreting, information that is consistent with one's existing beliefs. It can be represented by the mathematical model in fact, since $\underline{P}(I_B|B) = \overline{P}(I_B|B) = 1$, we obtain that for any event $A \supseteq B$ by the monotony of the lower conditional probability we have

$$1 \geq \overline{P}(A|B) \geq \underline{P}(A|B) \geq \underline{P}(B|B) = \underline{P}(I_B|B) = 1.$$

It implies that any event contained in the the complementary set of B has zero probability given B.

5   A Comparison Between the Model and Other Conditioning Rules

In probability conditioning is defined by Bayes'formula

$$P(A|B) = \frac{P(A \cap B)}{P(B)}$$

where P is the prior probability and B an event such that $P(B) > 0$. In Theorem 1, if the conditioning event has positive and finite Hausdorff outer measure in its Hausdorff dimension s, conditional upper probability is defined with respect to the Hausdorff outer measure of order s so that a different prior probability is consider for conditioning events with different Hausdorff dimension. If the conditioning event B has Hausdorff outer measure in its Hausdorff dimension equal to zero or infinity then the conditional probability in Theorem 1 is defined by a 0–1 valued finitely, but not countably, additive probability. Also in this case a different probability is considered for each conditioning event B. So the model of posterior probability proposed in Theorem 1 strongly depends on the conditioning event.

Different conditioning rules have been proposed in literature in different framework, namely: Shafer's evidence theory [23], Zadeh's possibility theory [26], Walley's theory [25]. An interesting discussion about different updating rules is proposed in [16]. In Shafer's evidence theory partial knowledge is represented in terms of a basic probability assignment m and belief functions, plausibility functions and their updating are obtained by the Dempster rule defined in terms of focal elements, which are the subsets A of Ω such that $m(A) > 0$. So m-null sets are not involved in the conditioning rule and if the plausibility function of the conditioning event is zero, $Pl(B) = 0$ then the updated plausibility function $Pl(A|B)$ is zero. In possibility theory a possibility measure Π can be defined for all $A \subseteq \Omega$ through a possibility distribution π by

$$\Pi(A) = \sup_{\omega \in A} \pi(\omega)$$

where π is supposed to be normalized, i.e. there exists $\omega \in \Omega, \pi(\omega) = 1$.

The updating formula for a prior possibility distribution π_1 is

$$\pi(\omega|B) = \begin{cases} \frac{\pi_1(\omega)}{\Pi_1(B)} & if \ \omega \in B \ and \ \pi_1(\omega) > 0 \\ 0 & otherwise \end{cases}$$

where $\Pi_1(B) = sup_{\omega \in B} \pi_1(\omega)$ and in particular

$$\Pi(A|B) = \frac{sup_{\omega \in A}\pi_1(\omega)}{\Pi_1(B)} = \frac{\Pi_1(A \cap B)}{\Pi_1(B)}$$

So also in this case if the conditioning event B has possibility measure equal to zero the posterior possibility measure Π is equal to zero.

In Walley's approach coherent upper conditional probabilities are defined for each conditioning event B by the axiom 1–3 of Definition 1 when only indicator functions are considered. This approach allows to consider events with zero probability and it does not define necessary the posterior probability equal to zero if the conditioning event has zero probability. A difference between coherent upper probabilities and the model proposed in Theorem 1 based on Hausdorff outer measures is that in the quoted theorem coherence of the upper conditional probability is proven on the power set; so we can consider any domain and to asses coherent upper conditional probability according to Theorem 1 without checking coherence every time we extend the domain to a larger one. If coherent upper and lower conditional probabilities, which are not defined as in Theorem 1, are considered to represent conscious and unconscious activity of human brain then there is the problem to check coherence every time that a larger domain is considered. Moreover if coherent upper and lower conditional probabilities are defined by Hausdorff outer and inner measures we have that, on the class of all measurable sets, the conscious and unconscious activity of human brain agree and an optimal decision can be interpreted as a decision made by the conscious and unconscious thought. In general for coherent upper and lower conditional probabilities cannot exist a domain, containing subsets A of Ω different from the empty set and Ω, where they coincide; it occurs for the vacuous coherent upper and lower conditional probabilities defined by $\overline{P}(A) = \max I_A$ and $\underline{P}(I_A) = \min I_A$ so that unexpected events cannot be represented and an optimal decision cannot be reached.

6   Conclusions

A new model of Bayesian updating of partial knowledge, based on Hausdorff outer measures, is applied to mathematically represent and explain some human brain's activities which are considered bias of human reasoning such as selective attention and confirmation bias. The model describes these capacities of human brain because it manages unexpected events with respect to the conditioning event. In fact conditional probability, which represents partial knowledge about an event given information represented by the conditioning event B, depends on B since it is defined through the s-dimensional Hausdorff outer measure where s-is the Hausdorff dimension of B. Different Hausdorff outer measures are considered in the assessments of conditional probability when conditioning events have different Hausdorff dimensions. An important aspect is that coherence of the model is proven in Theorem 1 for each conditioning event so that we can extend to the power set the conditional probability $P(\cdot|B)$ for all conditioning events.

References

1. Choquet, G.: Theory of capacity. Ann. Inst. Fourier **5**, 131–295 (1953)
2. Chabris, C.F., Simons, D.J.: Gorillas in our midst: sustained inattentional blindness for dynamic events. Perception **28**, 1059–1074 (1999)
3. de Finetti, B.: Probability Induction and Statistics. Wiley, New York (1970)
4. de Finetti, B.: Theory of Probability. Wiley, London (1974)
5. Denneberg, D.: Non-additive Measure and Integral. Kluwer Academic Publishers (1994)
6. Doria, S.: Probabilistic independence with respect to upper and lower conditional probabilities assigned by Hausdorff outer and inner measures. Int. J. Approx. Reason. **46**, 617–635 (2007)
7. Doria, S.: Characterization of a coherent upper conditional prevision as the Choquet integral with respect to its associated Hausdorff outer measure. Ann. Oper. Res. 33–48 (2012)
8. Doria, S.: Symmetric coherent upper conditional prevision by the Choquet integral with respect to Hausdorff outer measure. Ann. Oper. Res. **229**(1), 377–396 (2014)
9. Doria, S.: On the disintegration property of a coherent upper conditional prevision by the Choquet integral with respect to its associated Hausdorff outer measure. Ann. Oper. Res. **216**(2), 253–269 (2017)
10. Doria, S., Dutta, B., Mesiar, R.: Integral representation of coherent upper conditional prevision with respect to its associated Hausdorff outer measure: a comparison among the Choquet integral, the pan-integral and the concave integral. Int. J. General Syst. **216**(2), 569–592 (2018)
11. Doria, S.: Preference orderings represented by coherent upper and lower previsions. Theory Decis. **87**, 233–259 (2019)
12. Doria, S., Cenci, A.: Modeling decisions in AI: rethinking Linda in terms of coherent lower and upper conditional previsions. In: Torra, V., Narukawa, Y., Nin, J., Agell, N. (eds.) MDAI 2020. LNCS (LNAI), vol. 12256, pp. 41–52. Springer, Cham (2020). https://doi.org/10.1007/978-3-030-57524-3_4
13. Doria, S.: Coherent lower and upper conditional previsions defined by Hausdorff inner and outer measures to represent the role of conscious and unconscious thought in human decision making. Ann. Math. Artif. Intell. **89**(10), 947–964 (2021). https://doi.org/10.1007/s10472-021-09742-6
14. Doria, S.: Disintegration property for coherent upper conditional previsions defined by Hausdorff outer measures for bounded and unbounded random variables. Int. J. General Syst. **50**(3), 262–280 (2021)
15. Dubins, L.E.: Finitely additive conditional probabilities, conglomerability and disintegrations. Ann. Probability **3**, 89–99 (1975)
16. Dubois, D., Prade, H.: Updating with belief functions, ordinal conditional functions and possibility measures. In: Proceedings of the Sixth Annual Conference on Uncertainty in Artificial Intelligence, MIT, Cambridge, MA, USA, 27–29 July (1990)
17. Falconer, K.J.: The Geometry of Fractal Sets. Cambridge University Press, Cambridge (1986)
18. Kahneman, D., Tversky, A.: Prospect theory: an analysis of decision under risk. Econometrica **47**(2), 263–291 (1979)
19. Kahneman, D.: Thinking. Fast and Slow, Farrar, Straus and Giroux (2011)
20. Matte Blanco, I.: The Unconscious as Infinite Sets: An Essay on Bi-Logic. Gerald Duckworth, London (1975)

21. Regazzini, E.: De Finetti's coherence and statistical inference. Ann. Stat. **15**(2), 845–864 (1987)
22. Rogers, C.A.: Hausdorff Measures. Cambridge University Press, Cambridge (1970)
23. Shafer, G.: A Mathematical Theory of Evidence. Princeton University Press, Princeton (2021)
24. Tversky, A., Kahnemann, D.: Extensional versus intuitive reasoning: the conjunction fallacy in probability judgment. Psycological Rev. **90**(4), 293 (1983)
25. Walley, P.: Statistical Reasoning with Imprecise Probabilities. Chapman and Hall, London (1991)
26. Zadeh, L.A.: Fuzzy sets as a basis for a theory of possibility. Fuzzy Sets Syst. **1**(1), 3–28 (1978)

Handling Disagreement in Hate Speech Modelling

Petra Kralj Novak[1,2] ⓘ, Teresa Scantamburlo[3] ⓘ, Andraž Pelicon[2,4] ⓘ,
Matteo Cinelli[5] ⓘ, Igor Mozetič[2] ⓘ, and Fabiana Zollo[3(✉)] ⓘ

[1] Central European University, Vienna, Austria
novakpe@ceu.edu
[2] Jožef Stefan Institute, Ljubljana, Slovenia
{andraz.pelicon,igor.mozetic}@ijs.si
[3] Ca' Foscari University, Venice, Italy
{teresa.scantamburlo,fabiana.zollo}@unive.it
[4] Jožef Stefan International Postgraduate School, Ljubljana, Slovenia
[5] Sapienza University, Rome, Italy
matteo.cinelli@uniroma1.it

Abstract. Hate speech annotation for training machine learning models is an inherently ambiguous and subjective task. In this paper, we adopt a perspectivist approach to data annotation, model training and evaluation for hate speech classification. We first focus on the annotation process and argue that it drastically influences the final data quality. We then present three large hate speech datasets that incorporate annotator disagreement and use them to train and evaluate machine learning models. As the main point, we propose to evaluate machine learning models through the lens of disagreement by applying proper performance measures to evaluate both annotators' agreement and models' quality. We further argue that annotator agreement poses intrinsic limits to the performance achievable by models. When comparing models and annotators, we observed that they achieve consistent levels of agreement across datasets. We reflect upon our results and propose some methodological and ethical considerations that can stimulate the ongoing discussion on hate speech modelling and classification with disagreement.

Keywords: Hate speech · Annotator agreement · Diamond standard evaluation

1 Introduction

Modern research in machine learning (ML) is driven by large datasets annotated by humans via crowdsourcing platforms or spontaneous online interactions [5].

The authors acknowledge financial support from the EU REC Programme (2014–2020) project IMSyPP (grant no. 875263), the Slovenian Research Agency (research core funding no. P2-103), and from the project "IRIS: Global Health Security Academic Research Coalition".

D. Ciucci et al. (Eds.): IPMU 2022, CCIS 1602, pp. 681–695, 2022.
https://doi.org/10.1007/978-3-031-08974-9_54

Most annotation projects assume that a single preferred or even correct annotation exists for each item—the so-called "gold standard". However, this reflects an idealisation of how humans perceive and categorize the world. Virtually, all annotation projects encounter numerous cases in which humans disagree. The reasons behind disagreement can be various. For example, people can disagree because of accidental mistakes or misunderstandings experienced during the annotation process. In other cases, disagreement can originate from the inherent ambiguity of the annotation task or the annotators' subjective beliefs.

When labels represent different (subjective) views, ignoring this diversity creates an arbitrary target for training and evaluating models: If humans cannot agree, why would we expect the correct answer from a machine to be any different [7]? And, if the machine is able to learn an artificial gold standard, would it make it a perfect (infallible) predictor? The acknowledgement of multiple perspectives in the production of ground truth stimulated a reconsideration of the classical gold standard and the growth of a new research field developing alternative approaches. A recent work proposed a data perspectivist approach to ground truthing and suggested a spectrum of possibilities ranging from the traditional gold standard to the so-called "diamond standard", in which multiple labels are kept throughout the whole ML pipeline [3]. It has also been observed that training directly from *soft labels* (i.e., distributions over classes) can achieve higher performance than training from aggregated labels under certain conditions (e.g., large datasets and high quality annotators) [24]. Studies in hate speech classification came to similar conclusions and showed that supervised models informed by different perspectives on the target phenomena outperform a baseline represented by models trained on fully aggregated data [1].

In this paper, we focus on hate and offensive speech detection, which, similarly to other tasks like sentiment analysis, is inherently subjective. Thus, a disagreement between human annotators is not surprising. In sentiment analysis, disagreement ranges between 40–60% for low quality annotations, and between 25–35% even for high quality annotations [13,17]. Until recently, the subjectivity factor has been largely ignored in favor of a gold standard [26,27]. This led to a dramatic overestimation of model performance on human-facing ML tasks [12]. Here we investigate the specifics of hate speech annotation and modelling through the development of three large hate speech datasets and respective ML models. We present the process for data collection and annotation, the training of state-of-the-art ML models and the results achieved during the evaluation step. Our approach is characterized by two elements. First, we embrace disagreement among annotators in all phases of the ML pipeline and use a diamond standard for model training and evaluation. Second, we evaluate annotators' and models' performance through the lens of disagreement by applying the same performance measures to different comparisons (inter-annotator, self-agreement, and annotator vs model). Our experience led us to reflect and discuss a variety of methodological and ethical implications of handling multiple (conflicting) perspectives in hate speech classification. We conclude that disagreement is a genuine and crucial component of hate speech modelling and needs greater consideration within the ML community. A carefully designed annotation procedure supports the study of annotators' disagreement, discerns authentic dissent from spurious differences, and collects additional information that could possibly justify or contextualize the annotators' opinion. Moreover, a greater awareness of disagreement in hate speech datasets can generate more realistic expectations on the performance and limits of the ML models used to make decisions about the toxicity of online contents.

The paper is structured as follows. Section 2 presents the annotation process resulting in three large diamond standard hate speech datasets. Section 3 describes our training and evaluation of neural network-based models from diamond standard data, and reports the results by comparing the models' performance to the annotators' agreement. Finally, in Sect. 4, starting from our own results and experience, we discuss some implications of addressing disagreement in hate speech.

2   Data Selection and Annotation

Annotation campaign design and management drastically influences the quality of the annotated data. In this section, we first introduce the annotation schema used for annotating over 180,000 social media items in three different languages (English, Italian, and Slovenian). Then, we describe our annotation campaign and describe the procedure used to monitor and evaluate the annotation progress.

2.1   Annotation Schema

A simple and intuitive annotation schema facilitates the annotation efforts, and reduces possible errors and misunderstandings. However, since the definition of hate speech is a subtle issue there are other possible categorizations—see [18] for a systematic review. The annotation schema presented in this paper is adapted from the OLID [26] and FRENK [16] schemas, yet it is simpler, while retaining most of their expressiveness. The annotation procedure consists of two steps: first, the type of hate speech is determined, then the target of hate speech, when relevant, is identified. We distinguish between the following four **speech types**:

- **Acceptable**: does not present inappropriate, offensive or violent elements.
- **Inappropriate**: contains terms that are obscene or vulgar; but the text is not directed at any specific target.
- **Offensive**: includes offensive generalizations, contempt, dehumanization, or indirect offensive remarks.
- **Violent**: threatens, indulges, desires or calls for physical violence against a target; it also includes calling for, denying or glorifying war crimes and crimes against humanity.

In the case of offensive or violent speech, the annotation schema also includes a target. There are ten pre-specified targets: Racism, Migrants, Islamophobia,

Antisemitism, Religion (other), Homophobia, Sexism, Ideology, Media, Politics, Individual, and Other. For Italian, an additional "North vs. South" target was included (see Sect. 4.1.). We used the same schema to annotate three datasets: English YouTube, Italian YouTube, and Slovenian Twitter (see Table 1).

Table 1. Description of the datasets used for model training and evaluation. There are data sources, topics covered, timeframe, and the number of annotated items in the training and evaluation sets.

Language	Source	Topic	Period	Training set	Evaluation set
English	**YouTube**	Covid-19	Feb 2020 – May 2020	51,655	10,759
Italian	**YouTube**	Covid-19	Jan 2020 – May 2020	59,870	10,536
Slovenian	**Twitter**	General	Dec 2017 – Oct 2020	50,000	10,000

2.2 Data Selection and Annotation Setup

For each language, we selected two separate sets of data for annotation to be used for training and evaluating machine learning models. To overcome the class-imbalance problem (most hate speech datasets are highly unbalanced [20], see also Table 2), the training data selection was optimized to get hate speech-rich training datasets. This was achieved by selecting the data from large collections based on simple classifiers trained on publicly available hate speech data: we used the FRENK data [16] for Slovenian and English, and a dataset of hate speech against immigrants for Italian [22]. This led to training datasets with about two times more violent hate speech (the minority class) than we would get from a random sample. The evaluation dataset was randomly sampled from a period strictly following the training data time-span.

Table 2. Distribution of hate speech classes across the three application datasets. There is the total size of the collected data, and the classes assigned by the hate speech classification models.

Dataset	No. of tweets/ YT comments	Acceptable	Inappropriate	Offensive	Violent
English YouTube	20,227,765	13,670,748 (67.58%)	226,774 (1.12%)	6,222,405 (30.76%)	107,838 (0.53%)
Italian YouTube	1,273,936	1,047,056 (82.19%)	50,949 (4.00%)	164,600 (12.92%)	11,331 (0.89%)
Slovenian Twitter	12,961,136	9,721,259 (75.00%)	109,348 (0.84%)	3,115,207 (24.03%)	15,322 (0.12%)

Annotators were recruited and selected in Slovenia and Italy. Excellent knowledge of the target language (native speakers of Slovenian and Italian and proficient users of English) as well as an interest in social media and hate speech problems were required. Annotators were provided with written annotations guidelines[1] in their mother tongue. Guidelines included a description of the labels and the instructions on how to select them. They also provided practical information about the annotation interface and contact information to be used in case of doubts or requests. We provided continuous support to the annotators through online meetings and a dedicated group on Facebook.

Based on the number of annotators, we distributed the data according to the following constraints:

- Each social media item should be annotated twice.
- Each annotator gets roughly the same number of items.
- All pairs of annotators have approximately the same overlap (in the number of items) for pair-wise annotator agreement computation.
- For Twitter, each annotator is assigned some items (tweets) twice to compute self-agreement.
- For YouTube: a) Threads (all comments to a video) are kept intact; b) Each annotator is assigned both long and short threads.

Such a careful distribution of work enables continuous monitoring and evaluation of the annotation progress and quality. The annotators were working remotely on their own schedule. Internal deadlines were set to discourage procrastination. We monitored the annotation progress by keeping track of the number of completed annotations and evaluating the self- and inter-annotator agreement measures (see Sect. 3.1). Agreement between (pairs of) annotators

Table 3. The annotator agreement and overall model performance. Two measures are used: Krippendorff's (ordinal) *Alpha* and accuracy (*Acc*). The first column is the self-agreement of individual annotators (available for Twitter data only), and the second column is the aggregated inter-annotator agreement between different annotators. The last two columns are the model evaluation results, on the training and the out-of-sample evaluation sets, respectively. Note that the overall model performance is comparable to the inter-annotator agreement.

Dataset	Agreement				Classification model			
	Self-agreement		Inter-annotator		Training set		Evaluation set	
	Alpha	*Acc*	*Alpha*	*Acc*	*Alpha*	*Acc*	*Alpha*	*Acc*
English YouTube	–	–	0.60	0.78	0.55	0.75	0.60	0.83
Italian YouTube	–	–	0.59	0.78	0.60	0.79	0.58	0.84
Slovenian Twitter	0.79	0.88	0.60	0.79	0.61	0.80	0.57	0.80

---

[1] Hate speech annotation guidelines in English are available as part of IMSyPP D2.1: http://imsypp.ijs.si/wp-content/uploads/IMSyPP-D2.1-Hate-speech-DB-2.pdf, starting from page 16.

(see Table 3) was regularly computed during the process, enabling early detection of poorly-performing annotators, i.e., annotators disagreeing systematically with other annotators, either due to misunderstanding of the task, not following the guidelines or not devoting enough attention.

We used the described schema and protocol for developing three diamond standard datasets, and made them available on the Clarin repository: English YouTube[2], Italian YouTube[3], and Slovenian Twitter[4], summarized in Table 1. In the Slovenian dataset, the tweets are annotated independently, while the English and Italian datasets include contextual information in the form of threads of YouTube comments: Every comment is annotated for hate speech, yet the annotators were also given the context of discussion threads. Furthermore, the YouTube datasets are focused on the COVID-19 pandemic topic.

3   Model Training and Evaluation

We used the three diamond standard datasets to train and evaluate machine learning hate speech models. For each dataset, a state-of-the-art neural model based on a Transformer language model was trained end-to-end [6] to distinguish between the four speech classes. The models were trained directly on the diamond standard data, i.e., the training examples were repeated with several equal or disagreeing labels. For Italian, we used AlBERTo [19], a BERT-based language model pre-trained on a collection of tweets in the Italian language. For English, the base version of English BERT with 12 Transformer blocks [6] was used. For Slovenian, a trilingual CroSloEng-BERT [23], which was jointly pretrained on Slovenian, Croatian and English languages, was used. All three models are available at the IMSyPP project HuggingFace repository[5].

We used the Italian and Slovenian models in two previous analytical studies on hate speech in social media. The Italian model was used in a work investigating relationships between hate speech and misinformation sources on the Italian YouTube [4]. The Slovenian model was used to perform an analysis on the evolution of retweet communities, hate speech and topics on the Slovenian Twitter during 2018–2020 [8–10].

3.1   Evaluation Measures

A distinctive aspect of our approach is to apply the same measures a) to estimate the agreement between the human annotators and b) to estimate the agreement between the results of model classification and the manually annotated data. There are several measures of agreement, and to get robust estimates from different problem perspectives, we apply three well-known measures from the fields

---

[2] English dataset: https://www.clarin.si/repository/xmlui/handle/11356/1454.
[3] Italian dataset: https://www.clarin.si/repository/xmlui/handle/11356/1450.
[4] Slovenian dataset: https://www.clarin.si/repository/xmlui/handle/11356/1398.
[5] IMSyPP HuggingFace model repository: https://huggingface.co/IMSyPP.

of inter-rater agreement and machine learning: Krippendorff's *Alpha*, accuracy (Acc) and F_1 score.

There are several properties of hate speech modelling that require special treatment: i) The four speech types are ordered, from normal to the most hateful, violent speech, and therefore disagreements have very different magnitudes, thus we use ordinal Krippendorff's *Alpha*; ii) The four speech classes are severely imbalanced, a further reason to use Krippendorff's *Alpha*; iii) Since we also need a class-specific measure of (dis)agreement, F_1 is used.

The speech types are modelled by a discrete, ordered 4-valued variable $c \in C$, where $C = \{A, I, O, V\}$, and $A \prec I \prec O \prec V$. The values of c denote acceptable speech (abbreviated A), inappropriate (I), offensive (O) or violent (V) hate speech. The data items that are labelled by speech types are either individual YouTube comments or Twitter posts. The data labeled by different annotators is represented in a reliability data matrix. The data matrix is a n-by-m matrix, where n is the number of items labeled, and m is the number of annotators. An entry in the matrix is a label $c_{iu} \in C$, assigned by the annotator $i \in \{1, \dots, m\}$ to the item $u \in \{1, \dots, n\}$. The data matrix does not have to be full, i.e., some items might not be labelled by all the annotators.

A **coincidence matrix** is constructed from the reliability data matrix. It tabulates all the combined values of c from two different annotators. The coincidence matrix is a k-by-k square matrix, where $k = |C|$, the number of possible values of C, and has the following form:

$$
\begin{array}{c|c|c}
 & c' & \sum \\
\hline
 & \cdot & \cdot \\
c & . \; N(c, c') & . \; N(c) \\
 & \cdot & \cdot \\
\hline
\sum & . \; N(c') & . \; N \\
\end{array}
$$

An entry $N(c, c')$ accounts for all coincidences from all pairs of annotators for all the items, where one annotator has assigned a label c and the other c'. $N(c)$ and $N(c')$ are the totals for each label, and N is the grand total. The coincidences $N(c, c')$ are computed as:

$$
N(c, c') = \sum_u \frac{N_u(c, c')}{m_u - 1} \quad c, c' \in C
$$

where $N_u(c, c')$ is the number of (c, c') pairs for the item u, and m_u is the number of labels assigned to the item u. When computing $N_u(c, c')$, each pair of annotations is considered twice, once as a (c, c') pair, and once as a (c', c) pair. The coincidence matrix is therefore symmetrical around the diagonal, and the diagonal contains all the matching labelling.

We can now define the three evaluation measures that we use to quantify the agreement between the annotators, as well as the agreement between the model and the annotators. Since the annotators might disagree on the labels, there is no "gold standard". The performance of the model can thus only be compared to a (possibly inconsistent) labelling by the annotators.

Krippendorff's *Alpha*[14] is defined as follows:

$$Alpha = 1 - \frac{D_o}{D_e},$$

where D_o is the actual disagreement between the annotators, and D_e is disagreement expected by chance. When annotators agree perfectly, $Alpha = 1$, when there is a baseline agreement as expected by chance, $Alpha = 0$, and when the annotators disagree systematically, $Alpha < 0$. The two disagreement measures, D_o and D_e, are defined as:

$$D_o = \frac{1}{N} \sum_{c,c'} N(c,c') \cdot \delta^2(c,c'), \quad D_e = \frac{1}{N(N-1)} \sum_{c,c'} N(c) \cdot N(c') \cdot \delta^2(c,c').$$

The arguments $N(c,c'), N(c), N(c')$ and N refer to the values in the coincidence matrix, constructed from the labeled data.

$\delta(c,c')$ is a difference function between the values of c and c', and depends on the type of decision variable c (nominal, ordinal, interval, etc.). In our case, c is an ordinal variable, and δ is defined as:

$$\delta(c,c') = \sum_{i=c}^{i=c'} N(i) - \frac{N(c) + N(c')}{2} \quad e.g., \quad c,c',i \in \{1^{st}, 2^{nd}, 3^{rd}, 4^{th}\}.$$

Accuracy (*Acc*) is a common, and the simplest, measure of performance of the model which measures the agreement between the model and the "gold standard". However, it can be also used as a measure of agreement between two annotators. *Acc* is defined in terms of the observed disagreement D_o:

$$Acc = 1 - D_o = \frac{1}{N} \sum_c N(c,c).$$

Accuracy does not account for the (dis)agreement by chance, nor for the ordering of hate speech classes. Furthermore, it can be deceiving in the case of unbalanced class distribution.

F-score (F_1) is an instance of a well-known effectiveness measure in information retrieval [25] and is useful for binary classification. In the case of multi-class problems, it can be used to measure the performance of the model to identify individual classes. $F_1(c)$ is the harmonic mean of precision (*Pre*) and recall (*Rec*) for class c:

$$F_1(c) = 2 * \frac{Pre(c) * Rec(c)}{Pre(c) + Rec(c)}.$$

In the case of a coincidence matrix, which is symmetric, the 'precision' and 'recall' are equal, since false positives and false negatives are both cases of disagreement. $F_1(c)$ thus degenerates into:

$$F_1(c) = \frac{N(c,c)}{N(c)}.$$

In terms of the annotator agreement, $F_1(c)$ is the fraction of equally labelled items out of all the items with label c.

3.2   Annotator Agreement and Model Performance

For the evaluation, we use the same measures to estimate the agreement between the human annotators, and the agreement between the model classification and the manually annotated diamond standard data. Table 3 summarizes the overall annotator agreement and the models' performance in terms of Krippendorff's (ordinal) *Alpha* and accuracy (*Acc*) on all three datasets.

The annotators agree on the hate speech label on nearly 80% of the data points (*Acc* = 0.78–0.79). Our models agree with at least one annotator in over 80% of the cases (*Acc* = 0.80–0.84). Considering the high class imbalance and the ordering of the hate speech classes, a comparison in terms of Krippendorff's (ordinal) *Alpha* is more appropriate: Table 3 shows a very consistent agreement of about 0.6 (*Alpha* = 0.55–0.60) both between the annotators and the models on all three datasets.

The very misleading performance estimates as computed by accuracy are evident from Table 4. We consider two cases of binary classification. In the first case, all three types of speech which are not acceptable (e.g., inappropriate, offensive, or violent) are merged into a single, unacceptable class. In the second case, all types of speech which are not violent (e.g., acceptable, inappropriate, or offensive) are merged into a non-violent class. The performance of such binary classification is then estimated by *Alpha* and *Acc*. The estimates in the first case are comparable to the results in Table 3. In the second case, however, the *Alpha* values drop considerably, while the *Acc* scores rise to almost 100% (*Acc* = 0.97–0.99). This is due to a high imbalance of the non-violent vs. violent items, with a respective ratio of more than 99:1. The *Alpha* score, on the other hand, indicates that the model performance is low, barely above the level of classification by chance (*Alpha* = 0.26–0.39 on the evaluation set).

Class-specific results comparing the model and the annotator agreement in terms of F_1 are available in Table 5. The F_1 scores of the models would in absolute sense not be considered high. Yet they are comparable and in many cases even higher than the F_1 scores between the annotators. The only exception (still consistent in all three datasets) is the relatively low models' performance for the violent class. This is consistent with the binary classification results (Non-violent vs. Violent) in Table 4. We hypothesise, with high degree of confidence, that a poor identification of the violent class is due to the scarcity of training examples.

Table 4. The annotator agreement and model performance for two cases of binary classification: Acceptable (A) vs. Unacceptable class (either I, O, or V), and Violent (V) vs. Non-violent class (either A, I, or O). The performance is measured by the *Alpha* and accuracy (*Acc*) scores. Note the very high and misleading *Acc* scores for the second case, where the class distribution between the Violent and Non-violent classes is highly imbalanced. The *Alpha* scores, on the other hand, are very low, barely above the level of classification by chance.

Agreement		Acceptable vs. Unacceptable		Non-violent vs. Violent	
Dataset	Model	*Alpha*	*Acc*	*Alpha*	*Acc*
Inter-annotator		0.59	0.80	0.54	0.98
English	Train. set	0.55	0.77	0.45	0.98
YouTube	Eval. set	0.60	0.84	0.29	0.99
Inter-annotator		0.60	0.82	0.61	0.98
Italian	Train. set	0.62	0.83	0.51	0.97
YouTube	Eval. set	0.59	0.87	0.39	0.99
Self-agreement		0.79	0.90	0.69	0.99
Inter-annotator		0.60	0.81	0.61	0.99
Slovenian	Train. set	0.62	0.82	0.24	0.99
Twitter	Eval. set	0.57	0.81	0.26	0.99

Table 5. Class-specific annotator agreement and model performance. The classification is done into four hate speech classes (A, I, O, V), and the performance is measured by the F_1 score for each class individually. Note a relatively low model performance for the Violent class ($F_1(V)$).

Agreement		Acceptable	Inappropriate	Offensive	Violent
Dataset	Model	$F_1(A)$	$F_1(I)$	$F_1(O)$	$F_1(V)$
Inter-annotator		0.82	0.32	0.75	0.55
English	Train. set	0.78	0.39	0.74	0.46
YouTube	Eval. set	0.89	0.25	0.69	0.30
Inter-annotator		0.86	0.52	0.63	0.62
Italian	Train. set	0.87	0.53	0.65	0.53
YouTube	Eval. set	0.92	0.59	0.58	0.39
Self-agreement		0.92	0.62	0.85	0.69
Inter-annotator		0.85	0.48	0.71	0.62
Slovenian	Train. set	0.85	0.52	0.73	0.25
Twitter	Eval. set	0.86	0.46	0.69	0.26

4 Discussion

Given the intrinsically subjective nature of judging offensive and violent content, it might be argued that a diamond standard should be preferred in this and other similar contexts through all the phases of the machine learning pipeline. In the following, we discuss methodological and ethical implications of this approach.

4.1 Methodological Implications

Working with diamond standard data influences the data annotation, machine learning training and evaluation. We argue that selecting the data to be annotated, setting up the annotation campaign, monitoring its execution and evaluating the quality of annotations during and after the annotation campaign, are crucial steps that influence the final quality of the annotated data. Yet, the importance of annotation campaigns is often neglected in machine learning pipelines. An important practical dilemma when building diamond standard datasets is still to be investigated: when faced with an intrinsically subjective task (e.g., hate speech detection, sentiment analysis) how should one decide upon how many facets should a diamond have vs. how large should it be? The more diamond faces (i.e., the number of labels per item) ensure better data quality and enable the identification of ambiguous cases. Yet, when limited with the number of labels an annotation campaign can afford, is it better to have more data items labeled (thus a larger dataset with more variety) or more labels to the same items? Is this trade-off the same for the training as well as for the evaluation set?

Our second focus is on model evaluation: we propose a perspectivist view, as we evaluate model performance through the lens of disagreement by applying the same, proper performance measures to evaluate the annotator agreement and the model quality. Standard metrics assume a different meaning in a context where the same object can be assigned to multiple legitimate labels. For example, precision and recall lose the asymmetry that is implicitly assumed between the outcome retrieved from direct observation (also called 'real' outcome) and the prediction provided by the ML models, as we show in Sect. 3.1. In the case of ordered labels (e.g., our speech labels), mutual information, proposed by [24] as a good evaluation measure when learning with disagreement, is not appropriate as it neglects the labels' ordering. Proper performance measures in our case include ordinal Krippendorff's *Alpha*, which accommodates both the ordered nature of the labels (from normal to the most hateful, violent speech, and consequently a varying magnitude of disagreements), and class imbalance (where the Violent class is underrepresented). Furthermore, we use F_1 for the estimation of class-specific disagreement and misclassification, but not macro-F_1. Macro-F_1 is not an appropriate measure to aggregate individual F_1 scores to estimate the overall model performance [11].

In our perspectivist view on model evaluation, model performance is closely tied to the agreement between annotators. This means that annotator agreement poses intrinsic limits to the performance achievable by the ML models. This is implemented by the use of the same measures for all comparisons (e.g., between the annotators and between the annotators and the model). We observed that the level of agreement between our models and the annotators reaches the inter-annotator agreement when applying the overall performance measure (ordinal

Krippendorff's *Alpha*). This indicates that the model is limited by the annotator agreement and can not be drastically improved. However, when considering the class-specific F_1 values, the model reaches the inter-annotator agreement in all classes except for the minority class (i.e., Violent). Without a comparison to the F_1 scores of the annotators, or binary classification Non-violent vs. Violent, this shortcoming of the classification model would not have been detected.

4.2   Ethical Implications

The problem of ground truthing in hate speech modelling has also some ethical and legal implications. Even though the perception and interpretation of offensive and violent speech can vary among people and cultures, it is also true that the lack of respect is a moral violation and can have tangible negative effects on subjects. Some people, for example, can suffer from depression or even physical injuries after being largely exposed to violent and offensive communication [21]. In this regard, many countries impose restrictions to protect individuals from discriminatory and threatening content and digital platforms strive for the limitation of hate speech.

Defining hate speech subsumes important decisions about the ethical and legal boundaries of public debates and bears responsibility for limiting the right of freedom of expression, thereby including or excluding people from democratic participation. Not surprisingly, the introduction of legal boundaries to remove hate speech from the public sphere has raised various criticisms. For example, some consider hate speech bans as a form of paternalism, incompatible with the assumption that humans are responsible and autonomous individuals, while others fear that the power of judging hate speech would put the state in a position to decide what can or cannot be said [2].

The tension between the right to safety and the right to freedom of expression becomes even more controversial when one deals with ML models for hate speech detection and removal. In this context, the decision as to whether accepting or rejecting a potentially harmful content leverages the capacity of ML algorithms to make accurate predictions. However, our results and other studies (e.g. [12]) suggest that measuring hate speech classification in terms of prediction accuracy can be elusive when annotators disagree: a classifier cannot be accurate when the data is inconsistent due to many conflicting views. Deliberating upon items that cannot be classified in a clear-cut way is a questionable practice and requires greater scrutiny among ML developers, managers and policy makers. Achieving a consensus in predictive tasks might not necessarily be an ideal outcome. On the contrary, diversity can improve collective predictions [15]. Moreover, if predictions are accompanied by additional information including the reasons behind the predictions, cultivating a positive disagreement can foster more fruitful judgments.

5   Conclusions and Future Work

In this paper, we adopt a perspectivist approach to data annotation, model training and evaluation of hate speech classification. Our first emphasis is on the annotation process leading to the diamond standard data, as we argue that it influences the final data quality, and thereof the machine learning model quality. As the main point, we propose a perspectivist view on model evaluation, as we evaluate model performance through the lens of disagreement by applying the same, proper performance measures to evaluate the annotator agreement and the model quality. We argue that annotator agreement poses intrinsic limits to the performance achievable by models. By following the same annotation protocol, model training and evaluation, we developed three large scale hate speech datasets and the corresponding machine learning models. All our results are consistent across the three datasets: Trained and reliable annotators disagree in about 20% of the cases, model performance reaches the annotator agreement in the overall evaluation, while for the minority class (Violent) there is still some room for improvement. A broad reflection on the role of disagreement in hate speech detection leads us to consider some methodological and ethical implications that could stimulate the ongoing debate, not limited to hate speech modelling but to subjective classification tasks where disagreement is likely to arise and make a difference.

References

1. Akhtar, S., Basile, V., Patti, V.: Modeling annotator perspective and polarized opinions to improve hate speech detection. In: Proceedings AAAI Conference on Human Computation and Crowdsourcing, vol. 8, pp. 151–154 (2020)
2. Anderson, L., Barnes, M.: Hate speech. In: Zalta, E.N. (ed.) The Stanford Encyclopedia of Philosophy. Metaphysics Research Lab Stanford University (2022)
3. Basile, V., Cabitza, F., Campagner, A., Fell, M.: Toward a perspectivist turn in ground truthing for predictive computing. arXiv:2109.04270 (2021)
4. Cinelli, M., Pelicon, A., Mozetič, I., Quattrociocchi, W., Novak, P.K., Zollo, F.: Dynamics of online hate and misinformation. Sci. Rep. **11**(1), 1–12 (2021). https://doi.org/10.1038/s41598-021-01487-w
5. Cristianini, N., Scantamburlo, T., Ladyman, J.: The social turn of artificial intelligence. AI Soc. 1–8 (2021). https://doi.org/10.1007/s00146-021-01289-8
6. Devlin, J., Chang, M.W., Lee, K., Toutanova, K.: Bert: Pre-training of deep bidirectional transformers for language understanding. arXiv:1810.04805 (2018)
7. Dumitrache, A., Aroyo, L.,Welty, C.: A crowdsourced frame disambiguation corpus with ambiguity. In: Proceedings of NAACL (2019)
8. Evkoski, B., Ljubešić, N., Pelicon, A., Mozetič, I., Kralj Novak, P.: Evolution of topics and hate speech in retweet network communities. Appl. Netw. Sci. **6**(1), 1–20 (2021). https://doi.org/10.1007/s41109-021-00439-7
9. Evkoski, B., Mozetič, I., Ljubešić, N., Novak, P.K.: Community evolution in retweet networks. PLoS One **16**(9), e0256175 (2021). https://doi.org/10.1371/journal.pone.0256175,Non-anonymized version available at arXiv:2105.06214

10. Evkoski, B., Pelicon, A., Mozetič, I., Ljubešić, N., Novak, P.K.: Retweet communities reveal the main sources of hate speech. PLoS ONE 17(3), e0265602 (2022). https://doi.org/10.1371/journal.pone.0265602
11. Flach, P., Kull, M.: Precision-recall-gain curves: PR analysis done right. In: Cortes, C., Lawrence, N.D., Lee, D.D., Sugiyama, M., Garnett, R. (eds.) Advances in Neural Information Processing Systems, pp. 838–846. Curran Associates (2015)
12. Gordon, M.L., Zhou, K., Patel, K., Hashimoto, T., Bernstein, M.S.: The disagreement deconvolution: bringing machine learning performance metrics in line with reality. In: Proceedings CHI Conference on Human Factors in Computing Systems, pp. 1–14 (2021)
13. Kenyon-Dean, K., et al.: Sentiment analysis: It's complicated! In: Proceedings of NAACL, pp. 1886–1895 (2018)
14. Krippendorff, K.: Content Analysis, An Introduction to its Methodology. Sage Publications, 4th edn. (2018)
15. Landemore, H., Page, S.E.: Deliberation and disagreement: problem solving, prediction, and positive dissensus. Politics Philos. Econ. 14(3), 229–254 (2015)
16. Ljubešić, N., Fišer, D., Erjavec, T.: The FRENK datasets of socially unacceptable discourse in Slovene and English (2019), arXiv:1906.02045
17. Mozetič, I., Grčar, M., Smailović, J.: Multilingual Twitter sentiment classification: the role of human annotators. PLoS One 11(5), e0155036 (2016). https://doi.org/10.1371/journal.pone.0155036
18. Poletto, F., Basile, V., Sanguinetti, M., Bosco, C., Patti, V.: Resources and benchmark corpora for hate speech detection: a systematic review. Lang. Res. Eval. 55(2), 477–523 (2020). https://doi.org/10.1007/s10579-020-09502-8
19. Polignano, M., Basile, P., De Gemmis, M., Semeraro, G., Basile, V.: AlBERTo: Italian BERT language understanding model for NLP challenging tasks based on tweets. In: Italian Conference on Computational Linguistics, vol. 2481, pp. 1–6 (2019)
20. Rathpisey, H., Adji, T.B.: Handling imbalance issue in hate speech classification using sampling-based methods. In: IEEE International Conference on Science in Information Technology), pp. 193–198 (2019)
21. Saha, K., Chandrasekharan, E., De Choudhury, M.: Prevalence and psychological effects of hateful speech in online college communities. In: Proceedings 10th ACM Conference on Web Science, pp. 255–264 (2019)
22. Sanguinetti, M., Poletto, F., Bosco, C., Patti, V., Stranisci, M.: An Italian Twitter corpus of hate speech against immigrants. In: Proceedings of 11th International Conference on Language Resources and Evaluation (2018)
23. Sojka, P., Kopeček, I., Pala, K., Horák, A. (eds.): TSD 2020. LNCS (LNAI), vol. 12284. Springer, Cham (2020). https://doi.org/10.1007/978-3-030-58323-1
24. Uma, A.N., Fornaciari, T., Hovy, D., Paun, S., Plank, B., Poesio, M.: Learning from disagreement: a survey. Artif. Intell. Res. 72, 1385–1470 (2021)
25. Van Rijsbergen, C.: Information Retrieval. Butterworth, 2nd edn. (1979)
26. Zampieri, M., Malmasi, S., Nakov, P., Rosenthal, S., Farra, N., Kumar, R.: Predicting the type and target of offensive posts in social media. In: Proceedings of NAACL-HLT, pp. 1415–1420 (2019)
27. Zampieri, M., et al.: SemEval-2020 task 12: Multilingual offensive language identification in social media. arXiv:2006.07235 (2020)

What Is the Cost of Privacy?

Petr Dvořáček(✉) and Petr Hurtik

Centre of Excellence IT4Innovations, Institute for Research and Applications
of Fuzzy Modeling. 30. Dubna 22, University of Ostrava, Ostrava, Czech Republic
{petr.dvoracek,petr.hurtik}@osu.cz
https://gitlab.com/irafm-ai/ipmu-2022-anonymization

Abstract. Grade research has to be replicable, thus the used data need
to be publicly available. Speaking, e.g., about object detection task,
where image data for autonomous driving also contain privacy information such as faces and license plates, the publication of data may be
harmful to captured people. The solution to the moral dilemma is to
anonymize the data. In this study, our aim is to investigate the effect
of various anonymization techniques on the performance of algorithms
that use such data. We discuss anonymization methods that remove
and replace privacy data and select three methods to replace the privacy data: blurring, permutation, and replacing the area with a constant
value. We adopted the Cityscapes dataset from which we extracted areas
containing privacy information and are the manner of the anonymization methods. Our benchmark involves three famous object detectors:
YOLOv3, Mask R-CNN with ResNet-50 backbone, and Mask R-CNN
with Swin-T backbone. The results show that the impact of anonymization methods on the performance is negligible and the impact is similar
for both convolutional-based and transformer-based backbones. Since the
anonymization did not impact the model's performance, we would recommend to anonymize datasets of this kind to prevent potential GDPR
issues.

Keywords: GDPR · Object detection · Dataset anonymization

1 Introduction

Algorithms based on neural networks are often used to handle a wide range
of image processing tasks. To reach good accuracy using neural network, an
appropriate amount of data (images) is needed. Some tasks, such as autonomous
driving, require data collected by a camera in highly occupied areas such as cities
or highways [3,29]. Thus, the collected data contain sensitive information, such
as human faces, licence plate numbers, etc. The ethical research [18] should be
replicable and data publicly available to open the possibility of their reuse and
here raises a question how to handle the sensitive information. According to
valid legals, the data obtained as general ones without focusing on a particular
person/car can be distributed. However, there is a moral problem of providing

© Springer Nature Switzerland AG 2022
D. Ciucci et al. (Eds.): IPMU 2022, CCIS 1602, pp. 696–706, 2022.
https://doi.org/10.1007/978-3-031-08974-9_55

image data [33], where it may be harmful for someone to be publicly captured without consent. In this study, we are aiming at the moral level. Such research is important also regarding GDPR obstacles as there are attempts of making a general position and recommendations [16,27].

The easiest way to solve the moral problem is to anonymize the data, so the information leading to a particular subject such as a face or licence plate number is removed. This makes the data handling easier from the ethics perspective, but it can have a negative influence on the neural network performance. Here, we discuss a scenario where a model is trained on anonymized data and when it is deployed to production. It means that the model realizes the inference on real-time data captured from a camera, thus not anonymized. Our hypothesis is that a certain deidentification method brings bias into the training dataset, so there is a threat of overfitting the bias by the model. In this paper, the influence of different deidentification methods used to anonymize training datasets on the accuracy of the neural network is investigated. In summary, we are answering the question "What is the cost of privacy?".

The contributions of our work are as follows:

- We implemented an algorithm that extracts masks that represent privacy information from the Cityscapes dataset and made them available online. The published extracted masks can be reused in additional experiments.
- We made an overview of approaches anonymizing privacy data and discuss the strengths and weaknesses.
- We established a benchmark releaving the impact of various anonymization techniques on various neural networks.

The main finding of the work are:

- We reveal that the usage of anonymized dataset affects the performance of neural network negligibly compared with original dataset. Moreover, it decreases overfit and has stable behavior considering various neural networks and various anonymization methods.
- We recommend using as an anonymization technique the replacement of privacy data by a constant color. It yields the lowest decrease in neural network performance and is the safest (it is not hackable) of the tested ones.

2   Related Work

CIAGAN [15] focuses on images capturing faces only and on replacing the faces with new, artificially generated identities. The faces are automatically detected using FaceNet [24] and replaced using Conditional GAN [10] with the restriction of producing the same artificial face for the same real face. Munch et al. [17] capture visual data and deal with anonymization of faces with the help of Open-Pose [2] and car plates detected by YOLOv3 object detector [20]. The faces and plates are then blurred. Both studies miss investigation of the impact of the anonymized areas on the performance of a neural network trained on such data. The idea of face anonymization by its detection and replacing by a new one with the help of a GAN is frequently discussed in the literature. The crucial step is the proper design of the GAN: study [21] reports decrease of mAP from 0.568 to 0.523 on the own dataset, while DeepPrivacy [9] decreses in mAP from 0.966 to 0.959 on WiderFace dataset [34].

Schnabel et al. [23] focuse on car plate anonymization for datasets Udacity and BDD 100k [35], where as deidentification are examined the methods of placing solid color, gaussian blur, pixelization, pixel shuffling, and random distortion. They train YOLOv3 [20] object detector and reveal the impact of various anonymizations with the finding that the model trained on pixelized data yields the best performance on non-anonymized validation dataset, even better that training on the original data, which is surprising. We can explain the finding that such anonymization behaves as cutout [4]/random erasing [36] augmentation and therefore reduces overfitting. The mentioned study is restricted to a single problem (car plate) and a single architecture (YOLOv3), so the results cannot be taken as universal.

In contrast to the existing work, we are not focusing on designing one best privacy-preserving method working under restricted conditions, but we investigate the impact of various anonymization methods on the performance of various object detection networks to reveal appropriate couples of anonymization methods and NN architecture. Closest to our aim is the work of Dietlmeier et al. [5] that investigate the effect of annonymization on the network's performance for the following scenarios: trained and tested on original; trained on original, tested on anonymized; trained and tested on anonymized. The scenario described in this study – trained on anonymized, tested on original – is in their work missing.

3   Anonymization Methods

Image anonymization is a process applied to image data with the purpose of hiding privacy data to protect them from misuse. By privacy data, we mean here human faces and vehicle license plates. The image anonymization should satisfy the following:

1. It creates visually safe output, i.e., image data where the original privacy data are hidden for humans.
2. Algorithmically irreversible anonymization procedure is utilized.
3. The performance of algorithms that use the image data is not harmed.

Note that the second point, algorithmically irrevisible anonymization, is a topic of much research. It has been shown that methods that leave a piece of original information, such as blurring or noising, can be broken and the original data can be restored [25,28]. To be complete, we will incorporate even these methods in our study.

3.1   Methods that Remove the Privacy Data

The group consists of blurring, shuffling, pixelization, noising, and replacing by a constant color. An anonymization method can be given by a single algorithm of these or their combination. For a visual comparison, see Fig. 1. The advantage of the algorithms is their simplicity of use and fast processing speed. The disadvantage is the production of salient artificial areas in an image. Such areas may bring in additional bias to the data, harming the generalization of a model that uses the data. For example, we can consider object detection using a neural network; the network may detect a person thanks to the presence of a homogeneous area (created by the anonymization via placing a constant color) and not by the presence of the human itself. The disadvantage of blurring and shuffling methods is that the anonymization can be detected, recognized [22], and using an appropriate algorithm such as deblur [25], re-identification can be obtained.

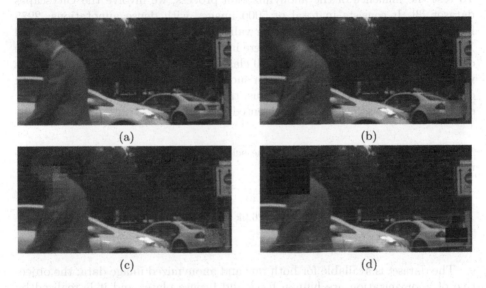

Fig. 1. Original image (a), image anonymized with blur (b), permutation (c), and constant color (d).

3.2   Methods that Replace Privacy Data

The group consists of inpainting, where the inpainting can be given by 'classical' algorithms [31] or by deep neural network-based, namely, by generative neural networks (GANs). The generative networks are trained to produce a new object which is different from the original one but is visually real. GANs achieved remarkable results [14,15], especially in face anonymization. The advantage is that the produced output is nearly safe [11] for the algorithms that use the data. The disadvantage is the high computation cost and the necessity of having additional information, such as the face's keypoints. The current work is focused also on reidentification, i.e., when an image with an artificially replaced face can be restored to the original version using a password [19]. So far, GANs are applied on datasets like WiderFace where the faces are reasonably big. Application to a wide variety of face sizes or even more objects remains an open issue.

4   Experiments

The goal of the experiment is two-fold. Firstly, we want to reveal the impact of various anonymization methods on various neural network performances. Secondly, we will define ideal combinations of them.

4.1   Dataset

To test the influence of the anonymization process, we involve the Cityscapes dataset [3]. It consists in total of 5000 images with dense annotations. 2975 images were used for training, 500 for validation during the training cycle, and 1525 for testing. All original images were in the resolution of 2048×1024 px. The objects in the dataset are split into 30 classes and we used a subset of 8 classes which contain the privacy information such as face or license plate. The chosen classes are person, rider, car, truck, bus, train, bike, and bicycle. The particular distribution of class instances is presented in Table 1.

Table 1. The number of objects for the selected classes in the whole dataset. The numbers are adopted from [7].

Class name	Person	Rider	Car	Truck	Bus	Train	Bike	Bicycle
Nr. of instances	17.9k	1.8k	26.9k	0.5k	0.4k	0.2k	0.7k	3.7k

The dataset is available for both raw and anonymized image data; the objective of anonymization are human faces and license plates and it is realized by blurring. We extracted masks that mark the areas with privacy information as follows. Firstly, we substracted the original and anonymized images, applied thresholding, performed opening/closing mathematical morphology operators to remove false positives and to fill false negatives in the masks. Output of the process is a mask with areas denoting human faces and license plates; those areas are then the aim of anonymization techniques. An example of the output mask is shown in Fig. 2. To make our research transparent, we put the produced masks available online at[1].

For the anonymization realized by pixel permutation, there is a hypothesis that a transformer-based neural network could be able to decode the permutation and thus, reduce the injected bias. The prerequisity is the same permutation

---

[1] https://gitlab.com/irafm-ai/ipmu-2022-anonymization.

scheme. To achieve that, we applied the permutation on 16 × 16px blocks, i.e., the masks were enlarged so the size of each spatial dimension of anonymized area is divisible by 16. Figure 2 shows the original image, the extracted mask, and the mask enlarged to be fully covered by 16 × 16px divisible blocks.

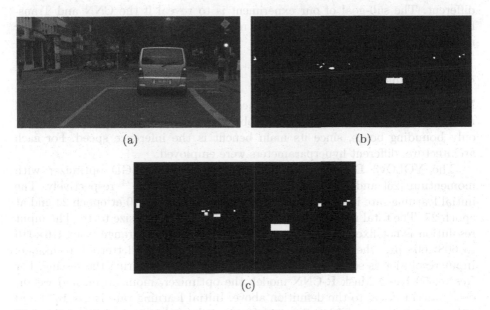

(a)   (b)

(c)

Fig. 2. Original image (a) and extracted mask (b) denoting the privacy data given by the original dataset. Final mask (c) to be fully covered by 16 × 16 tiles.

4.2 Models

We selected a representative of fast detectors, precise detectors with convolutional backbone, and precise detectors with transformer backbone. Namely, YOLOv3 [20] with Darknet53 backbone, Mask R-CNN [7] with ResNet-50 [8] backbone, and Mask R-CNN with Swin [12] backbone. YOLOv3 architecture is the most famous one-stage detector and suitable for tasks in which a compromise between speed and accuracy is needed. Notwithstanding there are YOLOv4 [1,32], YOLOv5, and YOLOx [6] that yield higher accuracy, the major principles valid in v3 are valid also for the newer versions. Mask R-CNN is a two-stage detector concept for the 'instance segmentation' task. The accuracy of two-stage detectors is usually higher compared to one-stage ones, but the inference speed is lower. The ResNet-50 backbone [8] was used for comparison of our results with the original Mask R-CNN paper [7]. Swin [12] is one of the state-of-the-art backbones, and when it was released, it achieved the highest accuracy on the COCO dataset[2].

---

[2] paperswithcode.com/sota/object-detection-on-coco.

The Darknet53 and ResNet-50 backbones fall into the same category of convolutional neural networks (CNN). The Swin backbone is on the other hand Vision Transformer [30] based architecture. Vision Transformer uses so called *self attention* to extract relationships and features from the image instead of determining the values of convolution kernels only, so the primary principle is different. The sub-goal of our experiment is to reveal if the CNN and Transformer backbones react similarly with respect to anonymization techniques or not.

4.3 Training Specification

Mask R-CNN based models were trained to predict both bounding boxes and instance-segmented masks expressing objects. YOLOv3 was trained to predict only bounding boxes, since its main benefit is the inference speed. For each architecture, different hyperparameters were employed.

The YOLOv3 Darknet53 model was trained with SGD optimizer with momentum [26] and weight decay set to 0.9 and 1×10^{-4} respectively. The initial learning rate is 1×10^{-3} and reduced by a factor of 10 at epoch 24 and at epoch 27. The total number of epochs was set to 30, batch size to 16. The input resolution is not fixed during the training, but it is in the range from 416×416 to 608×608 px, where the images were rescaled with the letterbox technique. Input resolution is set to the fixed value of 608×608 px during the testing. For ResNet-50 based Mask R-CNN model, the optimizer, momentum, and weight decay are identical to the definition above. Initial learning rate is 2×10^{-2} and reduced by a factor of 10 at 8th and 10th epoch of the total of 12 epochs. The model is trained on the range of resolutions from 1600×800 to 2048×1024 px. The resolution is fixed during the testing to 2048×1024. Mask R-CNN based on Swin-T backbone was trained with AdamW optimizer [13] with β_1, β_2 set to 0.9, 0.999 and weight decay set to 5×10^{-2}. The input shape was set to range from 1333×480 to 1333×800 px during the training and to the fixed value of 1333×800 px during the testing. The Swin-T based model was trained for 36 epochs, the learning rate was reduced by a factor of 10 at the 30th and 33rd epoch. The batch size was set to the size of 8.

Since the Cityscapes training dataset is relatively small, we employed data augmentation techniques during the training to reduce the overfitting. Namely, the spatial transformations, such as shift, scale, and rotate and intensity transformations such as contrast and brightness change, compression, RGB shift, HUE shift, channel shuffle, and blur. After applying these augmentations, the overfitting of the models decreased, but it was still legible.

4.4 Results

We performed the training procedure as defined above and to reduce the stochasticity of the training, each of the value in Table 2, is computed as the mean of three separate trainings. Based on the results, we can make the following conclusions. The increase of NN's capacity leads to the increase of overfit – in the

Table 2. Influence of various anonymization techniques on train and test accuracy and overfit. Δ expresses the relative decrease of precision with respect to the particular anonymization technique and the original precision.

Architecture	Anonym.	Train		Test		Overfit
		mAP	Δ [%]	mAP	Δ [%]	
YOLOv3 Darknet-53	None	17.5	100.0	15.1	100.0	2.4
	Blur	15.7	89.7	14.1	93.4	1.6
	Permute	16.0	91.4	14.1	93.4	1.9
	Constant	16.5	94.3	14.3	94.7	2.2
Mask R-CNN Resnet-50	None	36.7	100.0	32.5	100.0	4.2
	Blur	34.8	94.8	31.5	96.9	3.3
	Permute	34.4	93.7	31.3	96.3	3.1
	Constant	34.6	94.3	30.8	94.8	3.8
Mask R-CNN Swin-T	None	58.3	100.0	38.5	100.0	19.8
	Blur	55.8	95.7	37.1	96.4	18.7
	Permute	57.0	97.8	37.2	96.6	19.8
	Constant	56.0	96.0	37.4	97.1	18.6

table, it is denoted as the absolute difference between train and test. That can be explained by the insufficient size of the dataset, which is obvious especially for the Swin backbone; it requires a much larger dataset compared to generic convolutional backbones. Regarding anonymization techniques, the differences between them are negligible and the globally most safe one (with the lowest global impact) is blur. Our hypothesis that a transformer-based backbone will be able to decode the permutation and thus, achieve no negative impact was not confirmed.

In summary, the behavior of the convolutional and transformer-based backbone is consistent without major differences. The impact of anonymization itself is insignificant, and to answer the question "What is the cost of privacy?", it is nearly for free and the decrease in precision can be counterbalanced by selecting the backbone with higher learning capacity.

Since one specific dataset was used in the experiment, we could not guarantee generality of our results along the used datasets.

4.5 Discussion and Future Work

The main goal of this work was to reveal the impact of various anonymization techniques on NN's performance. While the decrease is small, there is hypothesis that the injected bias can be suppressed by improving data augmentation. Augmentations used in this work rely on spatial augmentations such as shifting, scaling, and rotating and intensity transformations, namely, distorting contrast and brightness, image compression, color shifts, and blur. In the future work, we can recommend to use such augmentations that reflect the anonymization technique, e.g., cutout [4], random erasing [36], or motion blur.

There is an unknown effect realizing the same anonymization with various settings. In the case of blur, we can control the strength of blurring, so revealing the ideal tradeoff between keeping privacy and reduction of bias is valuable. For the permutation, an ablation study about setting proper blocks' size can be established.

5   Summary

The state-of–of-the–the-art object detection algorithms are based on deep neural networks. To train deep neural networks, a large amount of data (images) must be collected. Personal data could be included in data collected in highly occupied areas such as cities. To avoid GDPR obstacles, the best strategy applicable to storing personal data is simply anonymizing them, so it does not contain any information which leads to a particular person. Here comes the question - how does the anonymization of the image data influence the object detectors trained on them when inferring on the original (not anonymized) data?

We have investigated the influence of different anonymization methods on the accuracy of NN's tested on the original data. Two different detector schemes were used – YOLOv3 and Mask R-CNN. Darknet-53 backbone was used in the YOLOv3 detector, Resnet-50 and Swin-T backbones in the Mask R-CNN detector. In the benchmark, we have shown that the influence of anonymization techniques on test accuracy is negligible and that the effect is stable considering various architectures, backbones, and anonymization methods. As a side effect, it decreases the overfit of the involved neural networks.

Based on the results where only a minor decrease of performance was achieved, releasing anonymized data can become a good practice in the task of object detection. We recommend using as an anonymization technique the replacement of privacy data by a constant color. It yields the lowest decrease in neural network performance and is the safest (it is not hackable) of the tested ones.

Acknowledgement. The work is partially supported by grant SGS17/PřF-MF/2022.

References

1. Bochkovskiy, A., Wang, C.Y., Liao, H.Y.M.: Yolov4: optimal speed and accuracy of object detection. arXiv preprint arXiv:2004.10934 (2020)
2. Cao, Z., Hidalgo, G., Simon, T., Wei, S.E., Sheikh, Y.: Openpose: realtime multi-person 2D pose estimation using part affinity fields. IEEE Trans. Pattern Anal. Mach. Intell. **43**(1), 172–186 (2019)
3. Cordts, M., et al.: The cityscapes dataset for semantic urban scene understanding. In: Proceeding of the IEEE Conference on Computer Vision and Pattern Recognition (CVPR) (2016)

4. DeVries, T., Taylor, G.W.: Improved regularization of convolutional neural networks with cutout. arXiv preprint arXiv:1708.04552 (2017)
5. Dietlmeier, J., Antony, J., McGuinness, K., O'Connor, N.E.: How important are faces for person re-identification? In: 2020 25th International Conference on Pattern Recognition (ICPR), pp. 6912–6919. IEEE (2021)
6. Ge, Z., Liu, S., Wang, F., Li, Z., Sun, J.: Yolox: exceeding yolo series in 2021. arXiv preprint arXiv:2107.08430 (2021)
7. He, K., Gkioxari, G., Dollár, P., Girshick, R.: Mask r-cnn. In: Proceedings of the IEEE International Conference on Computer Vision, pp. 2961–2969 (2017)
8. He, K., Zhang, X., Ren, S., Sun, J.: Deep residual learning for image recognition. In: Proceedings of the IEEE Conference on Computer Vision and Pattern Recognition, pp. 770–778 (2016)
9. Hukkelås, H., Mester, R., Lindseth, F.: DeepPrivacy: a generative adversarial network for face anonymization. In: Bebis, G., et al. (eds.) ISVC 2019. LNCS, vol. 11844, pp. 565–578. Springer, Cham (2019). https://doi.org/10.1007/978-3-030-33720-9_44
10. Isola, P., Zhu, J.Y., Zhou, T., Efros, A.A.: Image-to-image translation with conditional adversarial networks. In: Proceedings of the IEEE Conference on Computer Vision and Pattern Recognition, pp. 1125–1134 (2017)
11. Klomp, S.R., van Rijn, M., Wijnhoven, R.G., Snoek, C.G., de With, P.H.: Safe fakes: evaluating face anonymizers for face detectors. arXiv preprint arXiv:2104.11721 (2021)
12. Liu, Z., et al.: Swin transformer: hierarchical vision transformer using shifted windows. In: International Conference on Computer Vision (ICCV) (2021)
13. Loshchilov, I., Hutter, F.: Decoupled weight decay regularization. arXiv preprint arXiv:1711.05101 (2017)
14. Mandelli, S., Bondi, L., Lameri, S., Lipari, V., Bestagini, P., Tubaro, S.: Inpainting-based camera anonymization. In: 2017 IEEE International Conference on Image Processing (ICIP), pp. 1522–1526. IEEE (2017)
15. Maximov, M., Elezi, I., Leal-Taixé, L.: Ciagan: conditional identity anonymization generative adversarial networks. In: Proceedings of the IEEE/CVF Conference on Computer Vision and Pattern Recognition, pp. 5447–5456 (2020)
16. Milakis, D., Thomopoulos, N., Van Wee, B.: Policy Implications of Autonomous Vehicles. Academic Press (2020)
17. Münch, D., Grosselfinger, A.-K., Krempel, E., Hebel, M., Arens, M.: Data anonymization for data protection on publicly recorded data. In: Tzovaras, D., Giakoumis, D., Vincze, M., Argyros, A. (eds.) ICVS 2019. LNCS, vol. 11754, pp. 245–258. Springer, Cham (2019). https://doi.org/10.1007/978-3-030-34995-0_23
18. Oliver, P.: The Student's Guide to Research Ethics. McGraw-Hill Education (UK) (2010)
19. Pan, Y.L., Chen, J.C., Wu, J.L.: A multi-factor combinations enhanced reversible privacy protection system for facial images. In: 2021 IEEE International Conference on Multimedia and Expo (ICME), pp. 1–6. IEEE (2021)
20. Redmon, J., Farhadi, A.: Yolov3: an incremental improvement. arXiv preprint arXiv:1804.02767 (2018)
21. Ren, Z., Lee, Y.J., Ryoo, M.S.: Learning to anonymize faces for privacy preserving action detection. In: Proceedings of the European conference on computer vision (ECCV), pp. 620–636 (2018)
22. Ruchaud, N., Dugelay, J.L.: Automatic face anonymization in visual data: are we really well protected? Electron. Imaging 2016(15), 1–7 (2016)

23. Schnabel, L., Matzka, S., Stellmacher, M., Pätzold, M., Matthes, E.: Impact of anonymization on vehicle detector performance. In: 2019 Second International Conference on Artificial Intelligence for Industries (AI4I), pp. 30–34. IEEE (2019)

24. Schroff, F., Kalenichenko, D., Philbin, J.: Facenet: a unified embedding for face recognition and clustering. In: Proceedings of the IEEE Conference on Computer Vision and Pattern Recognition, pp. 815–823 (2015)

25. Shen, Z., Lai, W.S., Xu, T., Kautz, J., Yang, M.H.: Exploiting semantics for face image deblurring. Int. J. Comput. Vis. **128**(7), 1829–1846 (2020)

26. Sutskever, I., Martens, J., Dahl, G., Hinton, G.: On the importance of initialization and momentum in deep learning. In: International Conference on Machine Learning, pp. 1139–1147. PMLR (2013)

27. Taeihagh, A., Lim, H.S.M.: Governing autonomous vehicles: emerging responses for safety, liability, privacy, cybersecurity, and industry risks. Transp. Rev. **39**(1), 103–128 (2019)

28. Tu, X., et al.: Joint face image restoration and frontalization for recognition. IEEE Trans. Circ. Syst. Video Technol. (2021)

29. Varma, G., Subramanian, A., Namboodiri, A., Chandraker, M., Jawahar, C.: IDD: a dataset for exploring problems of autonomous navigation in unconstrained environments. In: 2019 IEEE Winter Conference on Applications of Computer Vision (WACV), pp. 1743–1751. IEEE (2019)

30. Vaswani, A., et al.: Attention is all you need. Adv. Neural Inf. Process. Syst. **30** (2017)

31. Vlašánek, P.: F-transform in image inpainting applications. In: 2018 Joint 7th International Conference on Informatics, Electronics & Vision (ICIEV) and 2018 2nd International Conference on Imaging, Vision & Pattern Recognition (icIVPR), pp. 562–566. IEEE (2018)

32. Wang, C.Y., Bochkovskiy, A., Liao, H.Y.M.: Scaled-yolov4: scaling cross stage partial network. In: Proceedings of the IEEE/CVF Conference on Computer Vision and Pattern Recognition, pp. 13029–13038 (2021)

33. Wiles, R., Clark, A., Prosser, J., et al.: Visual research ethics at the crossroads. Sage Vis. Methods **1** (2011)

34. Yang, S., Luo, P., Loy, C.C., Tang, X.: Wider face: a face detection benchmark. In: Proceedings of the IEEE Conference on Computer Vision and Pattern Recognition, pp. 5525–5533 (2016)

35. Yu, F., Xian, W., Chen, Y., Liu, F., Liao, M., Madhavan, V., Darrell, T.: Bdd100k: a diverse driving video database with scalable annotation tooling. **2**(5), 6 (2018). arXiv preprint arXiv:1805.04687

36. Zhong, Z., Zheng, L., Kang, G., Li, S., Yang, Y.: Random erasing data augmentation. In: Proceedings of the AAAI Conference on Artificial Intelligence, vol. 34, pp. 13001–13008 (2020)

Integrating Prior Knowledge in Post-hoc Explanations

Adulam Jeyasothy[1]([✉])(iD), Thibault Laugel[2](iD), Marie-Jeanne Lesot[1](iD),
Christophe Marsala[1](iD), and Marcin Detyniecki[1,2,3](iD)

[1] Sorbonne Université, CNRS, LIP6, Paris 75005, France
{adulam.jeyasothy,marie-jeanne.lesot,christophe.marsala}@lip6.fr
[2] AXA, Paris, France
{thibault.laugel,marcin.detyniecki}@axa.com
[3] Polish Academy of Science, IBS PAN, Warsaw, Poland

Abstract. In the field of explainable artificial intelligence (XAI), post-hoc interpretability methods aim at explaining to a user the predictions of a trained decision model. Integrating prior knowledge into such interpretability methods aims at improving the explanation understandability and allowing for personalised explanations adapted to each user. In this paper, we propose to define a cost function that explicitly integrates prior knowledge into the interpretability objectives: we present a general framework for the optimization problem of post-hoc interpretability methods, and show that user knowledge can thus be integrated to any method by adding a compatibility term in the cost function. We instantiate the proposed formalization in the case of counterfactual explanations and propose a new interpretability method called Knowledge Integration in Counterfactual Explanation (KICE) to optimize it. The paper performs an experimental study on several benchmark data sets to characterize the counterfactual instances generated by KICE, as compared to reference methods.

Keywords: Explainable artificial intelligence · Prior knowledge · Counterfactual explanation · Compatibility

1 Introduction

Machine learning models are now present in many areas of everyday life. The XAI domain [1,9] aims at answering questions raised by these models such as "Why do I get this prediction?" or "What do I need to do to get the prediction I want?". In particular, post-hoc methods focus on generating explanations for the predictions obtained from a trained classifier. They come in different formats such as feature importance (e.g. LIME [16] and SHAP [10]) or counterfactual examples [21] (e.g. Growing Spheres [8] and FACE [15]). They also vary depending on their input data: model agnostic and data agnostic approaches consider that no knowledge is available either on the model, nor on data. This lack of knowledge makes it

© Springer Nature Switzerland AG 2022
D. Ciucci et al. (Eds.): IPMU 2022, CCIS 1602, pp. 707–719, 2022.
https://doi.org/10.1007/978-3-031-08974-9_56

difficult to generate explanations adapted to the context. As a result, they may then not be always understood by the user [17]. To tackle this issue, some studies focus on enriching the considered input and integrating the human in the loop by taking into account expert knowledge [4,11,19], so the explanation is more understandable.

In this context, this paper proposes a general formalization, through the definition of a cost function that integrates such prior knowledge in the search for explanation. It proposes to add to the classical penalty term, that aims at assessing the quality of a candidate explanation, a complementary compatibility term that takes as parameter the considered knowledge. We provide a discussion about this notion of compatibility for which two contradictory semantics can be considered. Indeed, the prior knowledge can be used either to find an explanation that is complementary to the knowledge or an explanation that follows the same language and is in agreement with the knowledge. It can be argued that an explanation in the knowledge language can increase the user's confidence in the explanation, and a complementary explanation can enrich the user's knowledge.

The paper discusses these cases when the considered knowledge provides information regarding features of interest. Then, it focuses on the case of explanation expressed within the knowledge language in the specific framework of counterfactual explanation: it considers that an explanation is compatible with the available knowledge if the features considered by the knowledge and the explanation are similar. In order to generate such a compatible explanation, it proposes a method called KICE which stands for Knowledge Integration in Counterfactual Explanation and presents the conducted illustrative experiments.

To summarize, the paper contribution consists in proposing a general framework for integrating expert knowledge into interpretability methods in Sect. 3, an instantiation of the framework in the context of counterfactual explanations and a new counterfactual explanation method to generate a knowledge-adaptive explanation in Sect. 4.

2   Background

This section provides background information regarding the framework to which the paper refers: it first describes some post-hoc explanation methods; then, it presents some existing approaches to integrate knowledge, after discussing different possible forms the latter can take.

2.1   Post-hoc Explanations

Numerous post-hoc methods which explain the prediction of a trained classifier have been proposed, addressing different problems, see e.g. [6]. A large portion of them relies on generating explanations as a solution to an optimization problem. Among those, we are particularly interested in counterfactuals and surrogates approaches. The former [7,8,12,15,21] aim at answering the question: "What do I have to do to get the prediction I want?". The answer is defined as the closest instance to the studied instance belonging to the desired class. It can be obtained by optimizing, under constraint, a cost function defined as the distance function between the considered instance and the explanation (usually the Euclidean distance [7,8]). On the other hand, surrogate-based explanations [14,16,18] rather answer the question: "Why do I get this prediction?". They rely on an interpretable model, such as a decision tree, a linear regression or classification rules, that approximates the classifier under study to mimic its behavior. In the case of local surrogates we can cite LIME [16] that minimizes the fidelity to the black box classifier.

2.2 Representation of Knowledge

As discussed in the introduction, post-hoc explanations, as well as any explanation building method, can be enriched by available prior knowledge so as to improve their quality and intelligibility. The next section describes existing approaches, we discuss here several forms that can be considered to represent this knowledge; one can cite among others: class prototypes [20], distribution of data [15] or features-based knowledge [19]. We are particularly interested in the latter, which is detailed in this section.

A first type of expert knowledge can be expressed by information on the individual features: the user can for instance indicate the so-called actionable features [19], i.e. the features that can be modified. For instance, they can include the budget in a credit application, as opposed to the age [7]. A second type of expert knowledge on the features can provide information on their interactions through their covariation [3,11] (e.g. "An increase in school level leads to an increase in age") or through a causality graph [4,11]. In this paper, we consider knowledge provided by a set of expected features that can for instance be deduced from a rule-based model.

2.3 Integration of Knowledge

Beyond the question of the form the prior knowledge can take, the question of its exploitation and integration in the explanation must be considered. Existing works differ in the way they answer this question and the problem they address.

Considering the case where knowledge indicates which features are actionable and explanations take the form of counterfactual examples, Ustun et al. [19] restrict the search set to forbid the use of some features or directions. This allows to avoid proposing inapplicable explanation such as "In order to get the credit, you need to decrease your age".

Also considering the case of counterfactual explanations, Mahajan et al. [11] consider knowledge about causal interactions between features. They propose to integrate it in the definition of a new distance measure that quantifies the extent to which the candidate explanation satisfies the causal relationships. Unlike Ustun et al., they do not exclude candidates that do not respect these constraints, but penalize them, in a more flexible approach. This method makes it possible to avoid unrealistic explanations of the form "In order to get the credit, you need to increase your budget and to decrease your salary".

Frye et al. [4] are also interested in the integration of causal relationships, but exploit a different information they induce; they considers the case of explanations in the form of local feature importance vector. He focuses on the asymmetry brought by the causal link orientation and proposes to exploit it by weighting the cost function.

These three approaches integrate different types of expert knowledge into different forms of explanation. They consider that this knowledge is an additional constraint in the explanation generation. This constraint is represented in different forms: reduction of the search space, addition of penalty in the cost function or weight in the cost function. In the following section, we propose a general formalization for all methods integrating knowledge.

3   Proposed General Framework

The considered objective is to explain the prediction of a classifier $f : \mathcal{X} \longrightarrow \mathcal{Y}$ where $\mathcal{X}$ denotes the input space, included in $\mathbb{R}^d$, and $\mathcal{Y}$ the output space. We consider a domain where elements $x \in \mathcal{X}$ are described by means of a set of features (for instance, age, salary, weight,...). In this paper, we consider continuous data, the case of categorical data will be considered in future work. The prior knowledge to be integrated in the generation of the explanation is denoted E. In the following, we consider that an explanation is related to the prediction $f(x)$ for an element $x \in \mathcal{X}$.

3.1   Proposed Optimization Problem

This section describes the proposed general optimization problem to generate explanation integrating user knowledge. Let $\mathcal{E}$ be the set of explanations for a given type (e.g. counterfactual example or surrogate models), we propose the following optimization problem:

$$\underset{e \in \mathcal{E}}{\arg\min} \, agg(penalty_x(e), incompatibility_x(e, E)) \tag{1}$$

where agg, $penalty_x$ and $incompatibility_x$ are three functions described in the following. This optimization problem allows to generate a knowledge-based explanation for any type of explanation and type of knowledge. As discussed below, the choice of the three functions depends on the studied context: the motivations of the user, the type of explanation to be generated or the considered type of knowledge. In our approach, we use a subscript x for functions that are related to the studied instance x.

Penalty Function. As recalled in Sect. 2.1, most existing approaches to generate explanations minimize a cost function. Let $penalty_x : \mathcal{E} \longrightarrow \mathbb{R}$ be such a function: it takes as argument a candidate explanation e, and may depend on the studied instance x in the case of local explainer. It values the quality of the candidate explanation. For instance, it is defined as the distance between the candidate and the considered instance in the case of counterfactual examples and the fidelity of explanation to the black box classifier in the case of surrogate approaches.

Incompatibility Function. To generate an explanation complying with available knowledge, we propose to add a function to measure how compatible the candidate explanation is with the knowledge. This incompatibility function depends on the type of explanation and the type of considered knowledge, as well as the user objective. For instance, an explanation generated for a non-expert customer asking how to get his credit application accepted needs to focus on common knowledge actions such as increasing his salary. On the other hand, in the case where the explanation is destined to a domain expert, the incompatibility can represent the level of fidelity of the explanation with respect to the expert's comprehension of the domain.

As mentioned in the introduction, we propose to distinguish two objectives: to propose an explanation that is complementary to the available knowledge, and to propose an explanation that compete with the knowledge language. We consider an explanation to be in the knowledge language if the features used in the explanation and the knowledge are similar, and an explanation to be complementary in the opposite case (i.e. when there is no redundancy). Below, we discuss these two possibilities in the case of surrogate-based explanations.

Let A_e denote the set of features used in the explanation generated by a surrogate e which is an interpretable model, and $\overline{E}$ the set of features not considered in knowledge E. To build a surrogate-based explanation in the knowledge language, one possibility is to minimize the number of features it takes into account that are not part of E:

$$incompatibility_x(e, E) = Card(A_e \cap \overline{E})$$

On the contrary, where the explanation needs to be complementary to the knowledge, another possibility is to minimize the number of features present in both A_e and E:

$$incompatibility_x(e, E) = Card(A_e \cap E)$$

These two definitions can be both relevant depending on the motivations and needs of the user, as discussed above. Such a case is further discussed in Sect. 4 for the case of counterfactual explanation.

Aggregation Function. The aggregation function combines the penalty and incompatibility functions. There exists an abundant literature on aggregation operators, see e.g. [2,5], which can be divided into three main categories: conjunctive, disjunctive or compromise. A conjunctive operator, for example the min function, returns a high value only if both penalty *and* incompatibility taken as input are high. A disjunctive operator can be represented by the max function, it requires *at least one* value to be high. Finally, trade-off functions such as weighted average result in the compensation of low values by high values.

The choice of this function can be made according to the user preferences, instead of necessarily considering the same function for everyone: one way to personalize the explanation to a user is to let him choose the aggregation he wants to perform.

3.2   An Example of Existing Method Under This Formalism

To show how general the proposed framework is, we propose to express the state-of-the-art approach proposed by Ustun et al. [19] in the proposed formalism highlighting the definition of the three involved functions. In this approach, the user knowledge E is denoted $A(x)$ and defined as the set of modifications that can be applied to instance x, it is integrated to generate an actionable counterfactual explanation. The latter is the solution to an optimization problem of the form:

$$a^* = \underset{a \in A(x)}{\operatorname{argmin}} \, cost\_fct(a, x) \text{ with } f(x + a) \neq f(x)$$

where $cost\_fct$ values the distance between x and $x + a$. The considered explanation expresses the move between the studied instance and the closest instance in the opposite class. This optimization problem can be rewritten as follows: let $\mathcal{E}_x = \{x' \in \mathbb{R}^d \mid f(x') \neq f(x)\}$,

$$e^* = \underset{e \in \mathcal{E}_x}{\operatorname{argmin}} \, cost\_fct(e, x) + \mathbb{1}_{|x-e| \notin A(x)} \times Z$$

where Z is arbitrarily large. Here e^* is the closest instance in the opposite class, and it enables to retrieve a as $a = e^* - x$.

This expression, equivalent to the previous one, makes it possible to identify the *penalty, incompatibility* and *agg* functions. The penalty function equals $cost\_fct$ defined in [19]. The incompatibility function equals $\mathbb{1}_{|x-e| \notin A(x)} \times Z$ and represents the presence or absence of the modified feature in the user knowledge; it takes only two values 0 or Z. An incompatible counterfactual explanation thus has a very high cost function, resulting in only compatible counterfactuals being considered. Finally, aggregation is performed by a weighted sum. However, as the incompatibility function is binary, only compatible counterfactuals are considered, so the resulting counterfactual is both compatible and of good quality.

4   Knowledge Integration in Counterfactual Explanation (KICE)

This section proposes a new method to generate counterfactual explanations taking into account prior knowledge, instantiating for counterfactuals the general framework introduced in the previous section. It then details the algorithm used to solve this problem. In the following, to value a distance we consider the l_2 norm.

The principle of the expected counterfactual example can be illustrated in the following toy scenario; let us consider the case of a patient who risks diabetes if he keeps his daily habits. Without knowledge integration, a plausible explanation is: "To predict a low risk, the patient has to decrease his blood sugar by 0.5 g/L and increase his carbohydrate by 50 g". However, the knowledge that the patient possesses is only reduced to less technical features, such as *sport duration* and *weight*: the proposed counterfactual explanation is thus not understandable to him. We aim to propose a counterfactual explanation more adapted to the user knowledge: "The patient has to run 30 min longer and lose 2 kg".

4.1 Proposed Cost Function: Instantiation of the Framework

We propose in this section an instantiation of the general framework in the counterfactual case and considering prior knowledge in the form of a set of features denoted $E = \{a_i, i = 1, ..., m\}$, considered as a subset of features of the input space $\mathcal{X}$.

Penalty Function. For the penalty function, we propose to use the classical counterfactual function which is the Euclidean square distance:

$$penalty_x(e) = \| x - e \|^2 \tag{2}$$

Incompatibility Function. As specified in Sect. 3.1, the objective is to propose a counterfactual explanation in agreement with the user knowledge, meaning that ideally, the counterfactual modifications must be only performed according to the features appearing in E. However, for some instances, focusing only on a subset features raises the risk of never being able to meet the decision boundary of f, leading to no counterfactual explanation being generated. Therefore, we propose to relax this constraint by penalizing the modifications according to the features from $\overline{E}$ that are not present in the knowledge. This allows, when the boundary cannot be found along the sole features of E, to make sure that a solution can still be found. Thus, we propose the incompatibility function that computes the Euclidean square distance only along absent features:

$$incompatibility_x(e, E) = \| x - e \|_{\overline{E}}^2 = \sum_{i \notin E}(x_i - e_i)^2 \tag{3}$$

Minimizing this incompatibility avoids generating counterfactual explanations that greatly modify the unknown features.

Aggregation Function. The aggregation function combines the penalty function and the incompatibility function, we propose to use a compromise function:

$$agg(u, v) = u + \lambda v \tag{4}$$

with $\lambda \in \mathbb{R}^+$ a hyper-parameter valued by the user. λ defines the weight given to the incompatibility function. In the case of categorical features, the l_2 distance is replaced with an appropriate distance or a dissimilarity measure.

We then obtain the following optimization problem:

Given $\mathcal{E} = \{x' \in \mathbb{R}^d \mid f(x') \neq f(x)\}$ and $\lambda \in \mathbb{R}^+$,

$$e^* = \operatorname*{argmin}_{e \in \mathcal{E}} cost_{x,E}(e) \tag{5}$$

$$\text{with} : cost_{x,E}(e) = \| x - e \|^2 + \lambda \| x - e \|_{\overline{E}}^2 \tag{6}$$

4.2   Algorithm Description

The algorithm we propose to solve optimization problem (5), named KICE for Knowledge Integration in Counterfactual Explanation, uses the principle of iterative generation of instances, inspired by the Growing Spheres algorithm [8], considering the additional term of compatibility to bias the generation. We consider a model and data agnostic case, in which no information about the data distribution nor the decision boundary is available. Therefore, we generate instances around the considered instance x (i.e. for which explanation related to $f(x)$ should be found) in all directions: to find the point that minimizes the cost function, we generate points in increasingly larger spaces until we find one in the opposite class.

The equation $cost_{x,E}(e) = \nu$ defines the equation of an ellipse with center x and radius $\sqrt{\frac{\nu}{1+\lambda}}$ with respect to the features in $\overline{E}$ and $\sqrt{\nu}$ with respect to the features in E. To make the ellipses grow, the value taken by ν is iteratively increased. As we cannot cover all possible values for ν, a hyper-parameter $\epsilon > 0$ is used to generate elliptical layers defined by radius ν and $\nu + \epsilon$ and generate uniformly in these layers.

In order to do so, we use a modified version of the HLG algorithm [13] which consists in generating instances uniformly in the spherical layer $SL(x, a_0, a_1)$ which represents the set of points at a distance greater than a_0 and less than a_1 from x. In the HLG algorithm the rescaling step consists in generating a random variable $U = \{u_i\} \sim \mathcal{U}([0,1])$, and then calculating $a_0 + a_1 U^{\frac{1}{dim(\mathcal{X})}}$. We distinguish the features E and $\overline{E}$, and perform the following operations: $a_0 + a_1 u_i$ for $i \in E$ and $\frac{a_0}{\sqrt{1+\lambda}} + \frac{a_1}{\sqrt{1+\lambda}} u_i$ for $i \in \overline{E}$. The instances are then indeed generated in the desired ellipse.

The proposed KICE algorithm covers the space by generating instances iteratively: in the first step, n instances are generated in the ellipse of center x and radius $\sqrt{\frac{\nu}{1+\lambda}}$ with respect to the features in $\overline{E}$ and $\sqrt{\nu}$ with respect to the features in E. This ellipse is represented in the left part of Fig. 1. If none of these instances belong to the opposite class, we generate instances in the ellipsoidal layer between ν and $\nu + \epsilon$. This layer is represented in blue on Fig. 1.

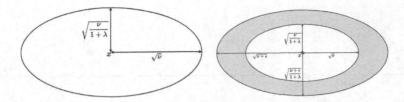

Fig. 1. (left) Generation zone for the initial step, (right) Iteration zone. (Color figure online)

5   Experiments

This section presents the experiments conducted to evaluate the KICE algorithm: the purpose of these experiments is to show that the proposed method finds the expected counterfactual instance, thus it allows to achieve a trade-off between the quality and the compatibility of an explanation.

5.1   Experimental Protocol

Experiments are conducted on three classical benchmark tabular datasets: half-moons, Boston and breast cancer. As a pre-treatment, dataset values are normalized and in the case of the Boston dataset, the regression value is transformed into a binary class by means of a discretization step: the price is "expensive" if it is greater than $21,000, and "cheap" otherwise. Datasets are split into train and test data subsets (80%−20%). Within the considered post-hoc explanation framework, the classifier choice does not matter, we apply SVM with a Gaussian kernel, that obtains good accuracy on the three datasets: half-moons (0.99), Boston (0.98) and breast cancer (0.93). Regarding prior knowledge, we consider the user disposes of less features than the classifier. In order to build somehow realistic knowledge, we train a decision tree with low depth on the train data (more precisely maximal depth is set to half the number of data features), the set of features E is then the ones that occur in this tree. Counterfactual instances e^* are then generated for each instance x of the test data set using KICE, with the value of λ being chosen to observe interesting results, respectively 4, 1 and 6 for the datasets half-moons, Boston and breast cancer.

We compare the proposed method KICE to two competitors. The first one is Growing Spheres [8], which solves the reference counterfactual optimization problem that only minimizes the Euclidean distance and denote e_{ref} the generated point. This corresponds to an extreme case of aggregation of Eq. 5 where the incompatibility term is ignored i.e. $\lambda = 0$. A second competitor is proposed by imposing to strictly comply with the knowledge. Its associated cost function thus minimizes the Euclidean distance according to the knowledge features only, a naive way of integrating knowledge in the explanation.

We denote e_{user} the counterfactual instance that solves the associated problem:

$$e_{user} = \underset{e \in \mathcal{E}}{\operatorname{argmin}} \| x - e \|_E^2$$

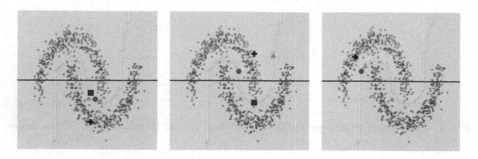

Fig. 2. Examples of obtained results e_{ref}, e_{user} and e^* for three instances x (+: x, ▲: e_{ref}, ■: e_{user}, •: e^*)

We can write this optimization problem in the proposed formulation given in Eq. (6) with λ arbitrarily large:

$$e_{user} = \operatorname*{argmin}_{e\in\mathcal{E}} \| x - e \|^2 + \lambda \| x - e \|_{\overline{E}}^2$$

5.2   Illustrative Examples

First, we illustrate the behavior of the methods with examples in two dimensions denoted X_0 and X_1. Figure 2 shows the counterfactual examples obtained for three different instances of the half-moons dataset. The figure shows the training set, the blue and red regions represent the predicted classes (darker points are the training examples); the decision boundary of the trained SVM classifier is shown in white, it achieves 0.99 accuracy. The considered knowledge system is a rule on a single feature, it is represented by the brown horizontal line : $E = \{X_1\}$.

In the figure, we observe that the counterfactual instance e_{ref} is the closest point belonging to the opposite class. As expected, e_{user} is further away than e_{ref} and only modifies feature X_1 which is the knowledge feature: we observe that e^* is a compromise between e_{ref} and e_{user}. It requires less modifications according to X_0 than e_{ref}, hence it is more compatible. It is also closer to the studied instance than e_{user}. In the graph, on the right, we notice that there is no e_{user}. In this case, there is no counterfactual instance modifying only feature X_1. The proposed method is then useful because it allows to obtain a more compatible counterfactual than e_{ref}.

5.3   Evaluation of the KICE Method

In this section, we further compare the proposed method with its presented competitors and study quantitatively the gain in terms of the proposed cost function offered by KICE.

Table 1. Results with the three considered approaches on the three datasets for metrics: *penalty* defined in Eq. (2), *incompatibility* defined in Eq. (3) and *cost* defined in Eq. (6)

		$Penalty_x(e)$	$Incompatibility_x(e, E)$	$Cost_{x,E}(e)$
Half-moons	e_{ref}	**0.32 ± 0.21**	0.14 ± 0.13	0.86 ± 0.56
	e_{user}	1.48 ± 1.3	**0.0 ± 0.0**	1.48 ± 1.3
	e^*	0.42 ± 0.29	0.08 ± 0.11	**0.73 ± 0.52**
Boston	e_{ref}	**1.48 ± 1.75**	0.7 ± 1.03	3.57 ± 4.72
	e_{user}	2.26 ± 2.71	**0.0 ± 0.0**	2.26 ± 2.71
	e^*	1.72 ± 2.09	0.13 ± 0.19	**2.12 ± 2.54**
Breast cancer	e_{ref}	**8.82 ± 9.22**	7.27 ± 8.25	52.41 ± 58.63
	e_{user}	22.42 ± 24.87	**0.0 ± 0.0**	22.42 ± 24.87
	e^*	10.74 ± 9.85	1.25 ± 1.33	**18.24 ± 16.83**

We apply the three methods described in the previous section on the test set of the different datasets. Among the test set, some instances do not find counterfactual examples with the method that modifies only the features in E (e_{user}). For the datasets half-moons, Boston and breast cancer, this concerns respectively 20%, 0% and 11% of the instances. For the rest of the instances, Table 1 shows the mean and standard deviation values for the penalty, incompatibility and cost functions for the three approaches.

We notice that, as expected, the proposed counterfactual examples have a penalty greater than that of e_{ref} but lower than that of e_{user}. Moreover, the incompatibility function is much smaller than that of e_{ref}. Finally, e^* cost function is the lowest. We notice that the standard deviations are high, which is due the fact that the instances are at different distances from the boundary.

To verify that KICE minimizes the cost function as opposed to the other two on all instances, Fig. 3 shows the value of the cost function obtained by e^* (that is defined as minimizing it) as compared to the value it takes for e_{ref} (left) and e_{user} (right), for each of the test instances of the half-moons dataset. The figures show that, as expected, all points are above the line $y = x$.

On the right hand graph, the points are more scattered but they remain above the line. We notice that the generated counterfactual instances are closer to the cost function of e_{ref} than to e_{user}. Only one feature is consider by the knowledge, it is difficult to get a close counterfactual instance only by modifying this feature. It is then necessary to move away from the studied instance to get 100% compatible one. It is also possible to increase λ so that the points are less scattered.

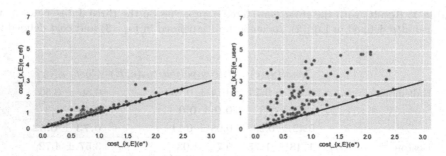

Fig. 3. Cost functions *cost* as defined in Eq. (6) of e_{ref}, e_{user} and e^* for the 80% of the test set for which all three counterfactual instances are defined.

6 Conclusion and Perspectives

In this paper, a general framework is proposed to help defining an optimization problem to integrate prior knowledge in post-hoc model-agnostic explanations. Using this framework, we proposed a new method, KICE, to generate counterfactual explanations in the knowledge language by minimizing the modifications according to the unknown features.

Future works will include a study of the counterfactuals explanations generated for different values of λ. Thus, we will study the λ associated with the expected counterfactual examples. Another direction will focus on exploring and proposing instantiations of the framework for different models and applications, as well as real world experiments including real users.

References

1. Burkart, N., Huber, M.F.: A survey on the explainability of supervised machine learning. J. Artif. Intell. Res. **70**, 245–317 (2021)
2. Calvo, T., Mayor, G., Mesiar, R. (eds.): Aggregation Operators: New Trends and Applications, vol. 97. Springer (2002)
3. Drescher, M., Perera, A.H., Johnson, C.J., Buse, L.J., Drew, C.A., Burgman, M.A.: Toward rigorous use of expert knowledge in ecological research. Ecosphere **4**(7), 1–26 (2013)
4. Frye, C., Rowat, C., Feige, I.: Asymmetric Shapley values: incorporating causal knowledge into model-agnostic explainability. In: Proceeding of Advances in Neural Information Processing Systems, vol. 33 (2020)
5. Grabisch, M., Marichal, J., Mesiar, R., Pap, E.: Aggregation Functions. No. 127 in Encyclopedia of Mathematics and its Applications. Cambridge University Press, Cambridge (2009)
6. Guidotti, R., Monreale, A., Ruggieri, S., Turini, F., Giannotti, F., Pedreschi, D.: A survey of methods for explaining black box models. ACM Comput. Surv. **51**(5), 1–42 (2018)
7. Lash, M.T., Lin, Q., Street, N., Robinson, J.G., Ohlmann, J.: Generalized inverse classification. In: Proceeding of the SIAM International Conference on Data Mining, pp. 162–170 (2017)

8. Laugel, T., Lesot, M.-J., Marsala, C., Renard, X., Detyniecki, M.: Comparison-based inverse classification for interpretability in machine learning. In: Medina, J., Ojeda-Aciego, M., Verdegay, J.L., Pelta, D.A., Cabrera, I.P., Bouchon-Meunier, B., Yager, R.R. (eds.) IPMU 2018. CCIS, vol. 853, pp. 100–111. Springer, Cham (2018). https://doi.org/10.1007/978-3-319-91473-2_9

9. Linardatos, P., Papastefanopoulos, V., Kotsiantis, S.: Explainable AI: a review of machine learning interpretability methods. Entropy **23**(1), 18 (2021)

10. Lundberg, S.M., Lee, S.I.: A unified approach to interpreting model predictions. In: Proceeding of the 31st International Conference on Neural Information Processing Systems, pp. 4768–4777 (2017)

11. Mahajan, D., Tan, C., Sharma, A.: Preserving causal constraints in counterfactual explanations for machine learning classifiers. NeurIPS workshop (2019)

12. Mothilal, R.K., Sharma, A., Tan, C.: Explaining machine learning classifiers through diverse counterfactual explanations. In: Proceeding of the 2020 Conference on Fairness, Accountability, and Transparency. ACM (2020)

13. Muller, M.E.: A note on a method for generating points uniformly on n-dimensional spheres. Commun. ACM **2**(4), 19–20 (1959)

14. Peltola, T.: Local interpretable model-agnostic explanations of Bayesian predictive models via Kullback-Leibler projections. In: Proceeding of the 2nd Workshop on Explainable Artificial Intelligence (XAI 2018) at IJCAI/ECAI 2018 (2018)

15. Poyiadzi, R., Sokol, K., Santos-Rodriguez, R., De Bie, T., Flach, P.: FACE: feasible and actionable counterfactual explanations. In: Proceeding of the AAAI/ACM Conference on AI, Ethics, and Society (2020)

16. Ribeiro, M.T., Singh, S., Guestrin, C.: "Why should I trust you?" explaining the predictions of any classifier. In: Proceeding of the 22nd ACM SIGKDD International Conference on Knowledge Discovery and Data Mining, pp. 1135–1144 (2016)

17. Rudin, C.: Stop explaining black box machine learning models for high stakes decisions and use interpretable models instead. Nat. Mach. Intell. **1**, 206–215 (2019)

18. Sokol, K., Hepburn, A., Santos-Rodriguez, R., Flach, P.: bLIMEy: surrogate prediction explanations beyond LIME. In: Proceeding of the HCML@NeurIPS (2019)

19. Ustun, B., Spangher, A., Liu, Y.: Actionable recourse in linear classification. In: Proceeding of the Conference on Fairness, Accountability, and Transparency, pp. 10–19. Association for Computing Machinery (2019)

20. Van Looveren, A., Klaise, J.: Interpretable counterfactual explanations guided by prototypes. In: Proceeding of European Conference on Machine Learning (2021)

21. Wachter, S., Mittelstadt, B., Russell, C.: Counterfactual explanations without opening the black box: automated decisions and the GDPR. Harvard J. Law Technol. **31**, 841–887 (2018)

PANDA: Human-in-the-Loop Anomaly Detection and Explanation

Grégory Smits[1], Marie-Jeanne Lesot[2(✉)], Véronne Yepmo Tchaghe[1],
and Olivier Pivert[1]

[1] University of Rennes – IRISA, 6074 Lannion, France
{gregory.smits,veronne.yepmo,olivier.pivert}@irisa.fr
[2] Sorbonne Université, CNRS, LIP6, 75005 Paris, France
marie-jeanne.lesot@lip6.fr

Abstract. The paper addresses the tasks of anomaly detection and explanation simultaneously, in the human-in-the-loop paradigm integrating the end-user expertise: it first proposes to exploit two complementary data representations to identify anomalies, namely the description induced by the raw features and the description induced by a user-defined vocabulary. These representations respectively lead to identify so-called data-driven and knowledge-driven anomalies. The paper then proposes to confront these two sets of instances so as to improve the detection step and to dispose of tools towards anomaly explanations. It distinguishes and discusses three cases, underlining how the two description spaces can benefit from one another, in terms of accuracy and interpretability.

Keywords: Outlier detection · Outlier explanation · XAI ·
Human-in-the-loop · Fuzzy vocabulary · Linguistic description

1 Introduction

A common approach to provide users with eXplainable Artificial Intelligence (XAI) tools is to implement the human-in-the-loop paradigm, i.e. to offer the user a crucial role in the mining process itself. This paper considers the case of the anomaly detection task and proposes to take into account user knowledge expressed in the form of a fuzzy vocabulary to describe linguistically the data.

Informally, anomaly or outlier detection aims at identifying, in a data set, instances that are conspicuous and, as put in the commonly accepted definition, "deviate so much from other observations so as to arouse suspicions that they were generated by a different mechanism" [10]. There exist numerous methods to perform this task, as well as multiple surveys and taxonomies, see e.g. [4,15]. However, most of them address the issue as a machine learning task, without taking into account the user who analyses the data. Recently, many methods have been proposed to provide *a posteriori* explanations about the identified outliers within the XAI framework, see e.g. [20] for a survey.

© Springer Nature Switzerland AG 2022
D. Ciucci et al. (Eds.): IPMU 2022, CCIS 1602, pp. 720–732, 2022.
https://doi.org/10.1007/978-3-031-08974-9_57

This paper proposes to take the user into account very early in the outlier detection process, leading to a knowledge-driven method that offers as additional feature an integrated linguistic description of the identified points. It thus opens the way to their interpretation and understanding by the user, i.e. to an outlier explanation method.

In order to do so, the proposed approach called PANDA, that stands for Personalised ANomaly Detection and Analysis, takes as input, in addition to the data set to be processed, a fuzzy vocabulary defined by the user: this vocabulary allows building linguistic descriptions of the data and constitutes precious user knowledge. For instance, the vocabulary defines indistinguishable areas in the data, i.e. values that should be considered as equivalent although they numerically differ: it can lead to distance functions more relevant from the user point of view than the classical Euclidean distance [8].

The PANDA method proposes to exploit such a vocabulary to dispose of a second data representation, complementary to the description induced by the basic data features: it is built as the vector concatenating the membership degrees to all modalities of all features. It thus defines a knowledge-driven representation of the data. In addition to providing a formalization of a subjective interpretation of the data, this vector also provides a normalization (in the unit interval) and a unification of non commensurable values, easing the combination of numerical and categorical attributes within a data mining task.

PANDA then proposes to apply an outlier detection method in these two description spaces. This principle bears similarity with the method proposed in [11] that applies a clustering algorithm in the two data representation spaces: the initial data definition space and the symbolic space induced by the vocabulary. However the aim in [11] is to quantify the adequacy between the vocabulary and the data inner structure. The crucial analysis step of PANDA confronts the two sets of anomalies, identified in the two spaces, so as both to improve the detection step and to dispose of tools towards anomaly explanations: PANDA makes it possible to extend any anomaly detection method with a vocabulary-based personalization of the data and a cross analysis of the outliers detected in the two spaces. The isolation forest method [12] to anomaly detection is used as an illustration in this paper.

The paper is structured as follows: Sect. 2 summarises related works, both on anomaly detection and explanation, Sect. 3 describes the proposed PANDA method, illustrating it with synthetic data and Sect. 4 presents a case study on real data describing car ads. Section 5 concludes the paper.

2   Related Works

This section briefly presents the two tasks to which the proposed PANDA method relates, considering anomaly detection and explanation in turn.

2.1   Anomaly Detection

There exist numerous methods to detect anomalies, i.e. points that deviate from so-called regular phenomena, as well as multiple surveys, see e.g. [4,7,15,20].

Beyond a distinction between supervised and unsupervised approaches, there exists no consensus about a taxonomy or the categories, nor the number of categories, further structuring the domain. It has for instance been proposed to distinguish between approaches based on nearest neighbours, clustering, statistics, subspaces and classifiers [7], or between approaches based on density, distance and models [12], or between approaches based on distance, model and neural networks [20]. To name four, some classical examples of anomaly detection algorithm include LOF (Local Outlying Factor [3] and its numerous variants), Isolation Forests [12], One-class SVM [1] and Auto-Encoder Ensemble [5].

For the implementation of the generic PANDA method described in this paper, the isolation forest (IF) approach [12] is considered. It constitutes an unsupervised ensemble-based method that combines multiple isolation trees. Such an isolation tree recursively draws random features and values to partition the data, until a predefined tree depth is reached or each leaf contains only individual (or indistinguishable) data points. Based on the fact that, by definition, outliers are distant from dense regions, they are likely to be isolated early by the recursively defined node partition. They thus appear in leaves close to the tree root: an isolation score of any data point is defined as the length of the path to the leaf it is assigned to. An isolation forest then combines most often hundreds of such randomly built trees and, for any data point, outputs an anomaly score based on its average isolation score [12].

2.2 Anomaly Explanation

Given a set of identified outliers, a natural question from the user is to ask for the reason why they are considered as such, i.e. what makes them abnormal: this calls for anomaly explanation methods, at the cross-roads of anomaly detection and XAI. According to the recent survey [20], four categories of such methods can be distinguished, depending on the type of provided explanations. The first one, also the most represented one, groups feature importance approaches, that either compute a score for each individual feature, as [14] for instance, or determine relevant subspaces, as e.g. [13]. These approaches can also be distinguished depending on whether they apply locally to single outlier points or globally to sets of outliers, or whether they are detector specific or agnostic, within the so-called outlier aspect mining task [6].

A second category of anomaly explanation methods groups approaches that additionally associate the responsible features with the values they take, as for instance [2]. The latter can for instance be identified by rules expressed as conjunction of predicates, where explanations take a disjunctive normal form. A third category groups approaches based on point comparisons, that underline the difference between an outlying point and regular points, for instance in looking for counterfactual examples [9]. The fourth category focuses on analysing the structure of the data, identifying the relations between subsets, i.e. clusters, of regular points and individual anomalies or sets of anomalies, as e.g. [17]

To the best of our knowledge, none of these methods take into account user knowledge, so as to provide personalised and more understandable explanations.

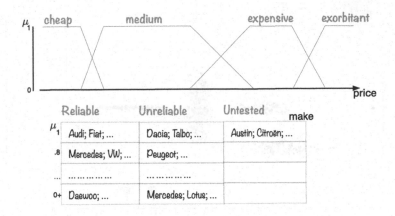

Fig. 1. Examples of fuzzy partitions describing car prices (top) and makes (bottom)

3   Proposed PANDA Approach

After presenting the notations used in the paper, this section describes the two steps of the proposed PANDA approach, respectively corresponding to the machine learning process applied to the considered data set and to the contrastive analysis of the identified anomalies.

3.1   Notations and Illustrative Data Set

$\mathcal{D} = \{x_1, x_2, \ldots, x_n\}$ denotes a set of n data points described by m attributes, A_1 to A_m, with respective domains D_1 to D_m. These attributes can be numerical or categorical.

$\mathcal{V} = \{P_1, \ldots, P_m\}$ denotes a vocabulary defined as a set of linguistic variables: for $i = 1..m$, P_i is a triple $\langle A_i, \{\mu_i\}, \{l_i\}\rangle$ with q_i modalities. The μ_{ij}, $j = 1..q_i$ are the respective membership functions of the modalities defined on universe D_i and the l_{ij} their respective linguistic labels.

Figure 1 depicts two examples of fuzzy partitions: the top part applies to a numerical attribute describing second hand car prices, for which $q = 4$ and with labels 'cheap', 'medium', 'expensive' and 'exorbitant'. The bottom part applies to a categorical attribute describing the car make: it shows a subjective interpretation of their reliability, with $q = 3$ and labels 'reliable', 'unreliable' and 'untested'. The membership functions are defined through their α-cuts: for each term, each row shows the makes whose membership degrees equal the value given on the left of the table.

It is assumed that each P_i defines a strong partition [16], i.e. $\forall y \in D_i$, $\sum_{j=1}^{q_j} \mu_{ij}(y) = 1$. In addition, it is assumed that the partition is such that any value y can partially satisfy up to two modalities only. In the case of features with numerical domains, these two modalities are adjacent.

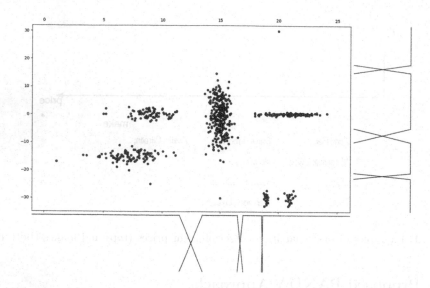

Fig. 2. Considered 2D illustrative data set

Throughout the section, we consider as example the data shown on Fig. 2 with $n = 651$, $m = 2$ with A_1 and A_2 numerical attributes whose respective domains are $D_2 = [0, 25]$ and $D_2 = [-40, 30]$. P_1 contains $q_1 = 4$ modalities whose membership functions are shown below the graph and P_2 contains $q_2 = 4$ modalities as well, whose membership functions are shown on the right side of the graph. The data set contains several dense regions as well as some outliers.

3.2 Data Processing

Data Rewriting with the Fuzzy Vocabulary. Each data point is first rewritten by computing its membership degrees to all modalities of all attributes and concatenating them: $x = \langle x^1, \dots, x^m \rangle$ is represented as the vector of $Q = \sum_{j=1}^{m} q_j$ components:

$$\langle \mu_{v_{11}}(x^1), \dots, \mu_{v_{1q_1}}(x^1), \dots, \mu_{v_{m1}}(x^m), \dots, \mu_{v_{mq_m}}(x^m) \rangle.$$

This vector is sparse, having at most $2m$ non-zero components due to the hypotheses on the partitions described in the previous section.

The whole dataset $\mathcal{D}$ may thus be rewritten according to a vocabulary $\mathcal{V}$ in linear time wrt. $|\mathcal{D}|$ but this process may easily be distributed to handle massive data [19]. The rewritten data $\mathcal{D}^{\mathcal{V}}$ are thus described as vectors of $[0, 1]^Q$.

Double Anomaly Detection. To leverage the expert knowledge about the data embedded in his/her vocabulary, a same anomaly detection method is applied on both $\mathcal{D}$ and $\mathcal{D}^{\mathcal{V}}$. In this paper, the Isolation Forest (IF) [12] method is applied in the two spaces using recommended parameters (100 trees in the forest, anomaly score with threshold 0.5 and a subset minimum size 256).

The resulting sets of identified anomalies are denoted by $\mathcal{A}$ and $\mathcal{A}^{\mathcal{V}}$ respectively. The former, identified in the initial feature space, are interpreted as datadriven anomalies; the latter, identified in the description space induced by the user vocabulary, are interpreted as knowledge-driven anomalies.

3.3   Anomaly Analysis: Cross Comparison of Detected Anomalies

The anomaly analysis step then consists in comparing the two sets $\mathcal{A}$ and $\mathcal{A}^{\mathcal{V}}$, considering their intersection and differences, commented in turn in this section. The goal of this comparison is to help users better understand both the data and the vocabulary. It is shown that it makes it possible to refine the anomaly detection, turning data-driven anomalies into contextual regularities and conversely points looking regular in $\mathcal{D}$ into contextual anomalies. Tools are thus provided towards the explanation of the identified outliers, as discussed below.

Figure 3 shows the result of an IF anomaly detection on $\mathcal{D}$ (top part) and $\mathcal{D}^{\mathcal{V}}$ (bottom part). The blueish zones indicate the anomaly scores for each point of the domain, white zones corresponding to high anomaly scores. Black lines in the top part of Fig. 3 are the separation lines of one isolation tree randomly drawn from the forest.

It can first be observed that, as expected, the general profiles of the anomaly score landscapes differ between the two graphs. In particular, in the rewritten case (bottom part), the regions homogeneous in terms of scores are parallel to the axes: the modalities of the fuzzy variables define indistinguishability zones within which all points have the same representation and are thus treated the same way. As a consequence, the data density is aggregated within each region, with fuzzy boundaries between the Cartesian product of the fuzzy set cores. The anomaly score landscape in the case of the initial representation space obviously follows the observed data density more closely.

Linguistic Description of Anomalies: $\mathcal{A} \cap \mathcal{A}^{\mathcal{V}}$ A first category of anomalies contains the points that are identified as such in both description spaces, i.e. the intersection of the two anomaly sets. These points can be considered as confirmed anomalies, for which in addition a linguistic description is available.

Indeed, a point $x \in \mathcal{A} \cap \mathcal{A}^{\mathcal{V}}$ is a data-driven outlier whose description in the vocabulary-induced space is considered as anomalous as well. Furthermore, this vocabulary-induced description characterises x, as it allows identifying it as an anomaly, and provides a linguistic description.

This case is illustrated with points x_1 with coordinates $(1, -17)$, x_2 $(20, 30)$, x_3 $(9, -25)$ and x_4 $(15, -30)$, on Fig. 3: they are indeed outliers from the data density or separability point of view and from the vocabulary point of view. They illustrate two distinct cases: x_1 and x_2 possess extreme feature values, that make them outliers, whereas x_3 and x_4 possess anomalous features combinations as compared to the other data points. In addition, the vocabulary allows describing linguistically these outliers: x_1 can e.g. be described as *"value on attribute A_1*

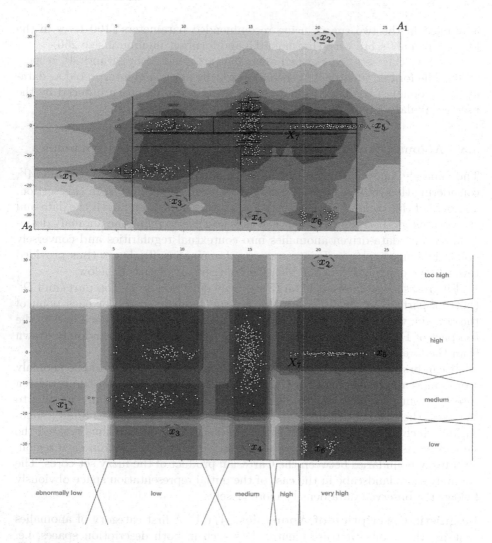

Fig. 3. Profile of the obtained anomaly scores: the lighter the colour, the higher the score. (Top) for $\mathcal{D}$, (bottom) for $\mathcal{D}^{\mathcal{V}}$.

is abnormally low and value on attribute A_2 is medium" and x_3 as *"value on attribute A_1 is low and on attribute A_2 is low".*

The obtained linguistic description can be considered as a step towards an intelligible explanation of the anomalies. However, two caveats require for caution. First, the provided description involves all features, i.e. is of size m. As such it may be the case that it is actually not legible when the number of features is high. Enriching the proposed approach with outlier aspect mining (see Sect. 2.2) is a considered extension to address this issue. Second, the description does not explain the reason why the considered point is an anomaly: for instance this description takes the same form for the two above-mentioned illustrative points, whereas they correspond to different cases. This corresponds to a classic challenge of outlier explanation generation.

Unexpected Anomalies: $\mathcal{A} \backslash \mathcal{A}^\mathcal{V}$ A second category of anomalies contains data points that are identified as outliers in the initial description space, but not in the vocabulary-induced one. This case is illustrated with points x_5 (24, 0) and x_6 (20, –30) on Fig. 3: in $\mathcal{D}^\mathcal{V}$, they are associated with the minimal anomaly score, i.e. they are considered as regular points. Indeed, they are described with terms that make them unanomalous whereas they are isolated from the data density point of view.

Such points can be described as "unexpected anomalies" insofar as the user does not dispose of a vocabulary that allows describing them and thus seems not to expect them. In an interactive information extraction process, it is highly relevant to draw his/her attention to these anomalies, so that the reason why they are not identified as such in the knowledge-driven approach can be explored. Several cases can indeed be distinguished, calling for different treatments.

A first possibility is that the user vocabulary is actually not adequate, i.e. does not correspond to the data distribution and content: it is useful to underline the existence of such specific cases the user may not have envisioned, suggesting to add new linguistic modalities to describe such subspaces specifically. This corresponds to a case of vocabulary data adequacy that is not captured by previous works on this topic, as e.g. [11]. For the considered illustrative data set, x_5 is an example of this case: it may call for splitting the *very high* modality so as to dispose of a term for this specific value of attribute A_1.

On the other hand, a second possibility is that these data point should indeed not be identified as anomalies: the knowledge provided through the vocabulary allow to diagnose issues in the data, suggesting the need to add information so that they are not considered as anomalies. It may for instance be the case that the processed data set is actually incomplete and misses data points, that would e.g. connect the candidate anomaly to a denser data region: the data set may be not representative of the underlying data distribution, about which the user vocabulary provides information. This can again be illustrated by data point x_5, which may be connected to the cluster of regular data observed for lower values of attribute A_1. Similarly, the vocabulary can be considered as suggesting that data point x_6 should not be considered as isolated, further suggesting that there should be no distinction between the two clusters it is inbetween.

Inadequate Vocabulary: $\mathcal{A}^\mathcal{V} \backslash \mathcal{A}$ A third category of anomalies contains, as a reciprocal of the second category, data points that are identified as outliers in the vocabulary-induced description space, but not in the initial one. This case is illustrated with the set of points X_7, around (17, 0) on Fig. 3: they build a minor group of points described with the *high* modality of attribute A_1.

Such points can be characterised as special cases based on the vocabulary, whereas they are not in the raw feature space. They may indicate a type of vocabulary inadequacy, different from the one discussed above: the vocabulary can be interpreted as being too subtle and introducing fine distinctions that are not justified in a data-driven analysis. As illustrated with Fig. 3, such cases e.g. occur when a modality splits a dense data area, here with the distinction between modalities *high* and *very high* of attribute A_1. They may suggest the need for vocabulary revision, in the same manner as the one explored in [18].

On the other hand, the fuzzy vocabulary is a model of the knowledge an expert possesses about a specific applicative context, explaining how subsets of the different attribute domains have to be interpreted [8]. Thus, these points in $\mathcal{A}^\mathcal{V}$ not identified as anomalies in $\mathcal{D}$ could also correspond to contextual false negative. However, the identification of such cases relies on additional contextual knowledge: the user may be interested in detecting the occurrence of such cases and the fine distinction may be required from an expert point of view. The vocabulary then offers the mean to identify them. As a concrete example of such a situation, let us consider the temperature monitoring a combustion engine whose ideal temperature is around 90 °C. Whereas observing operating temperatures in the range [60,91] may not be problematic (it may e.g. be records during the warm-up phase), it may be crucial for the expert to know when the temperature reaches 92 °C. A dedicated vocabulary is a solution to avoid contextual false negatives and false positives.

4   Use Case: Secondhand Car Ads

This section presents preliminary experiments conducted on a real data set describing classified ads about secondhand cars and discusses the results obtained when applying the proposed PANDA method. Identifying anomalies then aims at detecting both possible description errors, e.g. typing errors that make the ads unrealistic, and very specific cars, e.g. vintage cars or rare models.

4.1   Experimental Protocol

The considered real data set contains 49,188 ads about secondhand cars described by six attributes *price, mileage, year, priceNew, make* and *model.* The *priceNew* attribute indicates the price of the car of the considered make and model when sold new. Table 1 gives the labels of the linguistic variables defined for the five first attributes; the associated membership functions, omitted for size constraints, correspond to common sense definitions of the modelled properties.

The proposed PANDA method is applied on this data set $\mathcal{D}$ and its rewritten form $\mathcal{D}^\mathcal{V}$. The Isolation Forest algorithm is run with the hyper-parameter values suggested in [12]: 100 trees are built on random subsamples of the data set each containing 256 data points, the anomaly threshold is set to 0.5.

Table 1. Terms of the vocabulary used to rewrite the car descriptions

Attr.	Linguistic values
Price	almostoffered, veryLow, low, medium,
	expensive, veryExpensive, exorbitant
Mileage	almostNull, veryLow, low, medium, high, veryHigh, huge
Year	vintage, old, acceptable, recent, almostNew
PriceNew	veryLow, low, medium, expensive, veryExpensive, exorbitant
Make	luxury, highClass, mediumClass, lowClass

4.2 Result Analysis

Tables 2 and 3 show the ten instances, respectively in $\mathcal{D}$ and $\mathcal{D}^{\mathcal{V}}$, with the highest anomaly scores. Deviating values and value combinations are shown in bold, based on manual analysis. It can be observed that the first PANDA category, $\mathcal{A} \cap \mathcal{A}^{\mathcal{V}}$, is empty, this section discusses the reason why and comments the two other categories in turn, comparing $\mathcal{A}$ and $\mathcal{A}^{\mathcal{V}}$.

Regarding $\mathcal{A}$ given in Table 2, it can first be observed the ads ranked 1, 4, 5 and 10 can legitimately be considered as anomalies due to their erroneous prices, that take values greater than one million. The analysis of the other ads in this list shows they can be interpreted as anomalies because they correspond to rare luxury sport cars, that despite not being new models are still very expensive even with a medium mileage (see e.g. the third ad).

Observing the results for the rewritten data in Table 3, one can first remark that the integration of expert knowledge using the fuzzy vocabulary leads to a very different list of anomalies. Indeed, due to the fact that luxury makes are now grouped within a dedicated modality, they do not appear anymore as anomalies: *luxury* make having a *very expensive* price despite an *acceptable* year is now a sufficiently frequent conjunction of properties describing a subset of the analyzed ads. As a consequence, the knowledge driven anomaly detection makes it possible to identify other outliers and its combination with the data driven approach to get a better understanding of their respective contents.

Anomalies $\mathcal{A}^{\mathcal{V}}$ can be interpreted as being of two types: typing errors leading to unrealistic values, as a mileage equal to 1 for a *vintage* car (e.g. ad 3), and suspicious combinations of properties. Ads 5, 8 and 9 are examples of the latter: they correspond to cars from a luxury make with an expensive or very expensive price and a very low sale price. Looking more in depth at the ad description reveals that these cars are sold with a broken engine.

Table 2. Top-10 anomalies found in the secondhand cars dataset, $\mathcal{A}$

	Price	Mileage	Year	PriceNew	Make	Model	Score
1	**7,500,000**	112,000	1993	98,754	Mercedes	500 SL A	0.705
2	**110,000**	15,000	**1984**	80,570	Ferrari	BB 512 5	0.696
3	**62,000**	**50,000**	1992	168,174	Ferrari	F 512 4.9i	0.69
4	**42,600,000**	22,000	2010	44,020	Mercedes	Classe C 350 CDI	0.688
5	**17,490,000**	202,000	2005	54,440	Mercedes	Classe CLS 320 CDI	0.682
6	**109,000**	3,800	**2007**	104,719	Porsche	911 3.6i	0.681
7	**93,900**	41,900	**2007**	168,372	Ferrari	F430 Spider V8	0.68
8	**112,000**	21,750	**2009**	136,882	Audi	R8 V10 5.2 FSI 525	0.68
9	**115,000**	22,154	**2009**	136,882	Audi	R8 V10 5.2 FSI 525	0.68
10	**12,500,000**	334,000	2007	32,774	Mercedes	Classe C 220 CDI	0.677

Table 3. Top-10 anomalies found in the rewritten secondhand cars data set, $\mathcal{A}^\mathcal{V}$

	Price	Mileage	Year	PriceNew	Make	Model	Score
1	**450**	**100**	**1988**	8,232	Renault	Super 5 Tiga	0.609
2	**850**	229,000	1983	**8,345**	**Bmw**	315	0.604
3	**25**	**1**	2010	26,798	Audi	A3 Sportback 2.0 TDI	0.602
4	**2,350**	**4,801**	2009	9,639	Dacia	Sandero 1.5 dCi 70	0.599
5	**1,000**	**450,000**	1988	36,550	**Mercedes**	300 TD	0.598
6	**6,999**	**159**	2004	30,387	Jaguar	X	0.597
7	**700**	**10**	**1994**	28,178	**Bmw**	525 TD	0.597
8	**1,000**	**500,000**	1985	36,416	**Bmw**	628 CSi	0.596
9	**2,990**	290,000	1988	**76,441**	**Bmw**	750 iL	0.596
10	**500**	**320**	1991	27,116	**Bmw**	524 TD	0.594

5 Conclusion and Perspectives

Addressing the task of identifying and explaining outliers in a data set, the PANDA approach proposed in this paper makes it possible to integrate user expertise so as to detect and compare both data-driven and knowledge-driven anomalies. Analyses based on an illustrative toy data set and a real data set show how they enrich each other: the PANDA approach provides a personalised outlier detection method, drawing the user attention to different types of specific cases of interest. It thus constitutes a human-in-the-loop outlier detection and offers tools towards outlier explanation.

Future works will aim at including further developments regarding the outlier explanation component, in particular the generation of linguistic description of the identified anomalies, e.g. combining the proposed methodology with outlier aspect mining components. They will also address the question of integrating PANDA within relational data base management systems, as exploratory tool for a user to get a global view on the data content and global structure. Experiments with real data and real users will be conducted to measure the extent to which it contributes to the user understanding and satisfaction when interacting with massive data sets.

References

1. Amer, M., Goldstein, M., Abdennadher, S.: Enhancing one-class support vector machines for unsupervised anomaly detection. In: Proceedings of the ACM SIGKDD Workshop on Outlier Detection and Description, pp. 8–15 (2013)
2. Barbado, A., Corcho, O., Benjamins, R.: Rule extraction in unsupervised anomaly detection for model explainability: application to OneClass SVM. Expert Syst. Appl. **189**, 116100 (2022)
3. Breunig, M.M., Kriegel, H.P., Ng, R.T., Sander, J.: LOF: identifying density-based local outliers. ACM Sigmod Rec. **29**(2), 94–104 (2000)
4. Chandola, V., Banerjee, A., Kumar, V.: Anomaly detection: a survey. ACM Comput. Surv. (CSUR) **41**(3), 1–58 (2009)
5. Chen, J., Sathe, S., Aggarwarl, D., Turaga, D.: Outlier detetion with autoencoder ensembles. In: Proceedings of the SIAM International Conference on Data Mining, pp. 90–98 (2017)
6. Duan, L., Tang, G., Pei, J., Bailey, J., Campbell, A., Tang, C.: Mining outlying aspects on numeric data. Data Min. Knowl. Discov. **29**(5), 1116–1151 (2015). https://doi.org/10.1007/s10618-014-0398-2
7. Goldstein, M., Ushida, S.: A comparative evaluation of unsupervised anomaly detection algorithms for multivariate data. PLoS One 11(4), e0152173 (2016)
8. Guillaume, S., Charnomordic, B., Loisel, P.: Fuzzy partitions: a way to integrate expert knowledge into distance calculations. Inf. Sci. **245**, 76–95 (2013)
9. Haldar, S., Johnand, P.G., Saha, D.: Reliable counterfactual explanations for autoencoder based anomalies. In: Proceedings of the 8th ACM IKDD CODS and 26th COMAD Conference, pp. 83–91. ACM (2021)
10. Hawkins, D.M.: Identification of Outliers, vol. 11. Springer (1980)
11. Lesot, M.J., Smits, G., Pivert, O.: Adequacy of a user-defined vocabulary to the data structure. In: Proceedings of the IEEE International Conference on Fuzzy Systems, pp. 1–8. IEEE (2013)
12. Liu, F.T., Ting, K.M., Zhou, Z.H.: Isolation-based anomaly detection. ACM Trans. Knowl. Discov. Data (TKDD) **6**(1), 3 (2012)
13. Myrtakis, N., Tsamardinos, I., Christophides, V.: Proteus: predictive explanation of anomalies. In: Proceedings of the 37th IEEE International Conference on Data Engineering (ICDE), pp. 1967–1972. IEEE (2021)
14. Pevný, T.: LODA: Lighweight on-line detector of anomaly. Mach. Learn. **102**, 275–304 (2015)
15. Ruff, L., et al.: A unifying review of deep and shallow anomaly detection. Proc. IEEE **109**(5), 756–795 (2021)
16. Ruspini, E.H.: A new approach to clustering. Inf. Control **15**(1), 22–32 (1969)
17. Shukla, A.K., Smits, G., Pivert, O., Lesot, M.J.: Explaining data regularities and anomalies. In: Proceedings of the International Conference on Fuzzy Systems, pp. 1–8. IEEE (2020)
18. Smits, G., Pivert, O., Lesot, M.-J.: A vocabulary revision method based on modality splitting. In: Laurent, A., Strauss, O., Bouchon-Meunier, B., Yager, R.R. (eds.) IPMU 2014. CCIS, vol. 444, pp. 140–149. Springer, Cham (2014). https://doi.org/10.1007/978-3-319-08852-5_15

19. Smits, G., Pivert, O., Yager, R.R., Nerzic, P.: A soft computing approach to big data summarization. Fuzzy Sets Syst. **348**, 4–20 (2018)
20. Tchaghe, V.Y., Smits, G., Pivert, O.: Anomaly explanation: a review. Data Knowl. Eng. **137**, 101946 (2021)

Weak and Cautious Supervised Learning

SSFuzzyART: A Semi-Supervised Fuzzy ART Through Seeding Initialization

Siwar Jendoubi$^{(\boxtimes)}$ (iD) and Aurélien Baelde (iD)

UPSKILLS R&D, 16 Rue Marc Sangnier, 94600 Choisy-Le-Roi, France
{siwar.jendoubi,aurelien.baelde}@upskills.ai
http://www.upskills.com

Abstract. Semi-supervised clustering is a machine learning technique that was introduced to boost clustering performance when labelled data is available. Indeed, labelled data are usually available in real use cases, and can be used to initialize the clustering process to guide it and to make it more efficient. Fuzzy ART is a clustering technique that is proved to be efficient in several real cases, but as an unsupervised algorithm, it cannot use available labelled data. This paper introduces a semi-supervised variant of the Fuzzy ART clustering algorithm (SSFuzzyART). The proposed solution uses the available labelled data to initialize clusters centers. A set of experiments is carried out on some available benchmarks. SSFuzzyART demonstrated better clustering prediction results than its classic counterpart.

Keywords: Semi-supervised learning · Fuzzy ART · Initialization by seeding · Clustering · SSFuzzyART

1 Introduction

Clustering techniques are useful to group similar unlabeled data instances together. It is used to explore unlabeled datasets and to discover similarity relationships between its instances. In some real cases, it is possible to have access to labelled data instances. These instances are, usually, not enough to learn a classifier. In fact, the available labels could represent a subset of possible labels, on only a tiny fraction of the dataset, preventing the training of a learner. However, usual clustering algorithms are not designed to consider this information about labels. In such cases, semi-supervised techniques can use the labelled instances to guide the clustering process.

Fuzzy Adaptive Resonance Theory (Fuzzy ART) is a classic clustering algorithm. It was introduced by Carpenter et al. [4]. Fuzzy ART is an unsupervised learning algorithm that incorporates the min operator from the fuzzy set theory into an ART based neural network. This algorithm can learn stable clusters from continuous input patterns. Besides, it is effective for large scaled data, and it runs fast compared to other clustering algorithms [3,7]. These characteristics

© Springer Nature Switzerland AG 2022
D. Ciucci et al. (Eds.): IPMU 2022, CCIS 1602, pp. 735–747, 2022.
https://doi.org/10.1007/978-3-031-08974-9_58

make Fuzzy ART able to detect clusters with different sizes. Then, small clusters, having few instances, could be detected and separated from larger clusters.

Fuzzy ART is a clustering algorithm that demonstrated good performance in many real cases [3,5,10]. However, in the case where some labelled data instances are available, Fuzzy ART can not consider this information to improve the clustering quality. This paper presents a simple and efficient solution to initialize Fuzzy ART with available labelled data instances. Indeed, the proposed approach is a semi-supervised Fuzzy ART (SSFuzzyART) that considers available labelled data instances to initialize clusters.

The contributions of this paper are the following: 1) the proposition of a semi-supervised version of the Fuzzy ART, SSFuzzyART, clustering algorithm. SSFuzzyART uses a set of available labelled instances to initialize the clusters centers. 2) A set of experiments is made to evaluate the proposed algorithm and to study its strength and weakness.

This paper is organized as follows: Sect. 2 presents an overview of the current state of the art on the semi-supervised clustering and Fuzzy ART. Section 3 details the classic Fuzzy ART algorithm and the proposed SSFuzzyART. Section 4 introduces the considered evaluation measures and discusses the experiment results. Finally, Sect. 5 concludes the paper.

2   Related Works

Adaptive Resonance Theory (ART) and Semi-supervised learning has attracted several researchers' attention. Kim et al. [8] introduced a label propagation ART algorithm. It is a semi-supervised approach. Another semi-supervised Message Passing Adaptive Resonance Theory (MPART), [9], is recently introduced.

Fuzzy Adaptive Resonance Theory algorithm is a classic online clustering algorithm that is known to be sensitive to data order. To remedy this issue, several research papers introduced solutions for this problem. Elnabarawy et al. [6] used evolutionary computation to improve Fuzzy ART. Liew et al. [10] used a genetic algorithm for the same purpose.

Clusters initialization by seeding and semi-supervised clustering are usual machine learning techniques that are always used to make the clustering prediction more efficient. Basu et al. [2] used labelled data to initialize clusters for k-means algorithm. They assume that for each possible cluster in a given problem, they have at least one instance belonging to that cluster. Besides, they initialize each cluster with the mean of instances belonging to that cluster.

According to [1] and [11] that present detailed surveys about the semi-supervised clustering, most available semi-supervised clustering solutions consider the k-means clustering approach. However, to the best of our knowledge, there is no existing work that focus on a semi-supervised Fuzzy ART clustering approach.

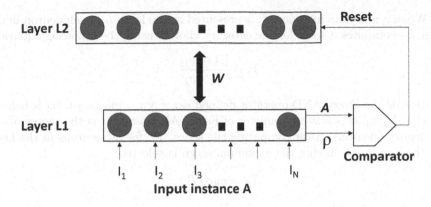

Fig. 1. Fuzzy ART architecture

3   Semi-supervised Fuzzy ART (SSFuzzyART)

In real world problems, some labeled data instances may be available. Fuzzy ART algorithm is not able to consider this available data in its clustering process. To resolve this problem and make Fuzzy ART output more efficient in such cases, this paper introduces the SSFuzzyART algorithm.

This section details first the Fuzzy ART algorithm. Next, it introduces the SSFuzzyART.

3.1   Fuzzy ART Algorithm

Fuzzy ART [4] is a classic clustering algorithm that extends the ART1 algorithm. Indeed, ART1 algorithm categorizes only binary input instances. Then, Fuzzy ART extends this algorithm to categorize both binary and analog instances. For each input instance, the Fuzzy ART network looks for the "nearest" cluster that "resonates" with that instance. Next, it updates the center of the selected cluster with the input instance. The vigilance parameter, $\rho \in [0, 1]$, is an input parameter of Fuzzy ART that determines the similarity of instances in the same cluster. When the vigilance parameter increases, the similarity of instances in the same cluster grows, which leads to a larger number of clusters, [12].

Fuzzy ART network is composed of two layers L_1 and L_2. Figure 1 shows the architecture of Fuzzy ART network. The layer L_1 is the input presentation layer. It has as many neurons as the number of features in the input instance. The layer L_2 is the category representation layer. Then, neurons in the L_2 layer are added while the learning progresses. Indeed, for each new detected category, a new neuron is appended to the L_2 layer. Besides, the layer L_2 maintains one additional neuron in the uncommitted state. The uncommitted neuron is useful to maintain the vigilance criterion (Eq. 3) satisfied when a new category arises [13].

When a new data instance, A, is presented in the layer L_1, each neuron in the layer L_2 computes a score for the category choice, using the following equation:

$$T_j = \frac{|A \wedge w_j|}{\alpha + w_j} \tag{1}$$

where $\wedge$ is the Fuzzy AND operator defined as: $x \wedge y = \min(x, y)$, $|x|$ is defined as $|x| = \sum_i x_i$, $\alpha > 0$ is a parameter of Fuzzy ART, and w_j is the center of the j category (neuron). After computing the score, T_j, for all neurons in the layer L_2, the neuron, J, having the maximum score is selected:

$$T_J = \max_j(T_j) \tag{2}$$

Once a neuron (category) is selected from the layer L_2, Fuzzy ART checks the expectation between the input instance, A, and the selected category. This expectation is compared to the vigilance threshold, ρ, as follows:

$$\rho \leq \frac{|A \wedge w_J|}{A} \tag{3}$$

If this condition is satisfied, then the category neuron center is updated with the instance A using the following equation:

$$w_J = \beta(A \wedge w_J) + (1 - \beta)w_J \tag{4}$$

where $\beta \in [0, 1]$ is the learning rate.

If the condition of the formula 3 is not satisfied, Fuzzy ART re-runs the neuron selection process until finding a neuron that satisfies the expectation condition (Eq. 3), next, it updates that neuron using formula 4. In the case where no committed neuron fits with the expectation condition, then, the data instance represents a new category and the uncommitted neuron is selected and updated to fit the new detected category, then, its weight vector is initialized with the input data instance. Next, a new uncommitted neuron is created. This process is repeated on all available data instances until getting a stable partition in which no instance change its category. To limit the running time, a maximum number of epochs could be defined.

The next section introduces the proposed semi supervised Fuzzy ART algorithm.

3.2   SSFuzzyART Algorithm

The semi-supervised Fuzzy ART (SSFuzzyART) is introduced to consider the case where some labelled data instances are available. These instances may represent all possible labels in a given problem, or just a subset from these labels. Consequently, this data is useful to make Fuzzy ART clustering more efficient and to get better quality results.

The SSFuzzyART has the same steps as the classic Fuzzy ART algorithm, and it differs from it in the algorithm initialization. In fact, Fuzzy ART starts

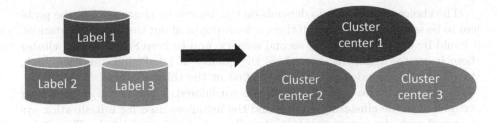

Fig. 2. SSFuzzyART initialization

Table 1. Category weights computation example through applying the Fuzzy min operator on four instances of a given label, the self learning rate $\beta = 1$ in this example

Instance 1	000111
Instance 2	101011
Instance 3	010111
Instance 4	100011
Category weights	**000011**

the clustering process with one neuron in the L_2 layer, which is the commitment neuron. This initialization makes it sensitive to the order of data entry. Indeed, the first coming instances will be used to initialize the clusters and then, will influence the clustering performance. The SSFuzzyART uses the labelled instances to initialize the categories in the L_2 layer. Then, each available category in the labelled data will be used to initialize its corresponding cluster, as shown in Fig. 2.

For a given category in the labelled data, a neuron in the L_2 layer is created to represent that category. Then, available data instances will be used to initialize the created neuron through computing its weights. Consequently, weights are computed through applying the Eq. 4 on the set of data instances for a given category. Table 1 presents a computation example.

Three possible scenarios for the use of the labelled instances after the initialization step are possible:

1. In the first scenario, the labelled data is used to initialize category neurons (layer L_2). Next, it is no longer used in the clustering process.
2. In the second scenario, the labelled data is used to initialize category neurons (layer L_2). Next, their labels are discarded, and initialization instances are clustered with the rest of data regardless of their initial labels.
3. In the third scenario, the labelled data is used to initialize category neurons (layer L_2). Next, initialization instances are clustered with the rest of the data, but their labels are set and cannot change during the clustering process, thus guiding the clustering process.

The choice of the scenario depends on the data to be clustered and the problem to be solved. For example, if there is some doubt about the labelled instances, it could be useful to apply the second scenario and to keep SSFuzzyART cluster these instances again. In case where the quality of labelled instances is high, then it is recommended to consider the first or the third scenario.

In this paper, **the second scenario** is considered. Then, labels are no longer considered in the clustering process and the instances used for initialization are clustered with the rest of the data regardless of their initial labels. This choice allows studying the ability of SSFuzzyART to correctly put the labelled instances in their clusters.

4   Experiments

To evaluate the proposed SSFuzzyART algorithm, a set of experiments is made to compare Fuzzy ART results to the SSFuzzyART results using three benchmark datasets. The considered datasets were selected from UCI machine learning repository[1]. Table 2 presents the considered datasets.

This section details several experiments and the considered evaluation measures.

Table 2. Considered benchmark datasets

Benchmark	Number of categories	Number of instances	Number of features
Iris	3	150	4
Wine	3	178	13
Breast cancer	2	569	30

4.1   Evaluation Measures

Four evaluation measures are considered to evaluate the proposed SSFuzzyART algorithm, and to compare it to the classic Fuzzy ART. The considered measures are defined as follows:

- *Adjusted Rand index*[2]: it is a similarity score between two clustering in the range $[-1, 1]$. It considers all pairs of samples in the prediction and target clustering and checks if pairs are in the same or different clusters. When this score is equal to 1, then the clustering is perfect. When it is near zero, then the clustering is near the random clustering.

---

[1] https://archive.ics.uci.edu/ml/datasets.php.
[2] https://scikit-learn.org/stable/modules/generated/sklearn.metrics.adjusted_rand_score.html.

- *Number of cluster ratio*: the ratio of predicted clusters number by the target clusters number. If this ratio is equal to 1, then the predicted clustering has exactly the same number of clusters as the target clustering. This measure is useful to evaluate Fuzzy ART and SSFuzzyART, as both do not use the number of clusters as input.
- *Cluster purity ratio*: the ratio of pure clusters, where a cluster is considered pure if at least 70% of its elements have the same label.
- *Label detectability ratio*: the ratio of detectable labels, where a label is considered detectable if at least 70% of data instances having that label are in the same cluster.

The considered cluster purity ratio and label detectability ratio have some tolerance on the definition of pure cluster and detectable label, respectively. This tolerance is defined by the fact that a cluster is pure if at least 70% of its elements have the same label, and a label is detectable if at least 70% of data instances having that label are in the same cluster. This tolerance allows these two measures to accept a clustering error of 30%. The choice of the tolerance value is defined according to the clustering problem, then for some problems it could be higher and for other problems it could be lower than 70%. In this paper, the clustering tolerance is fixed to 70%. A good prediction should have high cluster purity ratio and high label detectability ratio. This tolerance is useful to find the best compromise between these two measures. These two measures are influenced by the number of clusters. Consequently, when the number of clusters is high, these clusters are usually pure. However, in that case, the detectability of labels is low. With a high number of clusters, a given label instances could be scattered in several clusters.

The next section presents and discusses experiment results and the impact of the layer L_2 initialization on the final prediction of the SSFuzzyART.

4.2 Results & Discussion

To evaluate SSFuzzyART, a set of experiments on three benchmarks (Table 2) is carried out. The clustering results of SSFuzzyART are compared to Fuzzy ART results according to the four considered evaluation measures. In each experiment, a sample from the dataset is selected to initialize the L_2 layer in SSFuzzyART.

Figure 3, Fig. 4 and Fig. 5 presents the experiment results on respectively Iris, Wine and Breast cancer datasets. In these experiments, the Adjusted Rand score, the number of clusters ratio, the purity ratio and the detectability ratio were measured in function of the vigilance parameter ρ to compare the SSFuzzyART to Fuzzy ART.

Moreover, to study the impact of the number of labels/category used in the initialization sample, a set of experiments was carried out using different samples containing a different number of labels, depending on the selected dataset. The number of selected instances depends also on the number of labels, to ensure that labels are equally present in the sample. For example, the Iris dataset contains three labels. The first sample out of the Iris Dataset contains five instances all having the same label, the second sample contains ten instances belonging to any of two different labels (five instances per label) and the third sample contains fifteen instances from the three available labels (five instances per label). In a given sample, all considered labels are represented with the same number of instances.

According to the results in Fig. 3, Fig. 4 and Fig. 5, the SSFuzzyART leads to better results than Fuzzy ART when the vigilance parameter, ρ, values are low, and they yield similar results when ρ values are higher. Indeed, when the ρ parameter is high, the tolerance of the algorithm to create a new cluster is higher, which increases the number of clusters, then the initialization that is made to guide the clustering process is loosed. However, when the ρ parameter is low, the tolerance to create a new cluster is low too. Then, the number of cluster is also low, and the algorithm will not be allowed to add many clusters, especially when most possible labels are represented with corresponding neurons in the L_2 layer.

Another important conclusion from these experiments is that the available number of labels in the initialization sample has an impact on the output prediction. When only one label is available in the initialization sample, the SSFuzzyART and Fuzzy ART have almost the same output results. However, when the number of available labels in the initialization sample increases, the quality of

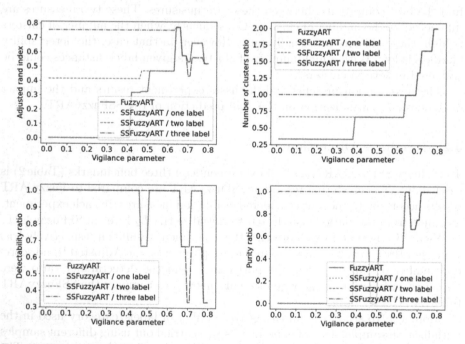

Fig. 3. SSFuzzyART VS Fuzzy ART on Iris dataset. In this experiment, three initialization samples were considered. The first one contains 5 instances having the same label. The second one contains 10 instances having two labels (five instances per label). The third one contains 15 instances having three labels (five instances per label)

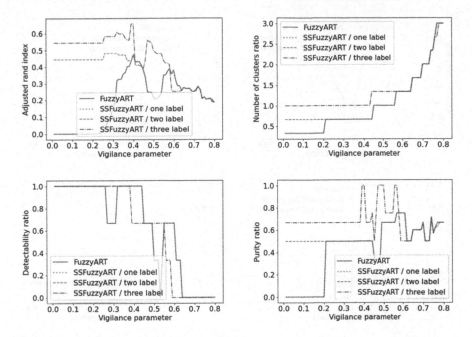

Fig. 4. SSFuzzyART VS Fuzzy ART on Wine dataset. In this experiment, three initialization samples were considered. The first one contains 5 instances having the same label. The second one contains 10 instances having two labels (five instances per label). The third one contains 15 instances having three labels (five instances per label)

the clustering increases too. Besides, according to the results in Fig. 3 and Fig. 4, even with two available labels from three in the labelled data, the SSFuzzyART is able to improve the clustering performance compared to the classic Fuzzy ART results.

A second experiment is made to study the impact of number of instances in the initialization sample if there is only one available label in the data. Nine different samples were considered, and the number of instances in these samples is in the range [1, 9] and all samples are selected from instances that belong to only one label. Figure 6 presents the results of this experiment. Figure 6 shows overlapping curves, i.e., all curves are equals. These results mean that a sample with only one label is not having any impact on the clustering results, even with many instances having that label. This experiment confirms the results of Fig. 3. In fact, having only one label in the initialization sample have no impact on the clustering results of SSFuzzyART. To have some improvement, the number of available labels must be greater than one.

A third experiment is carried out to study the impact of the number of instances used during initialization on the clustering results. Figure 7 shows the Adjusted Rand index with different vigilance threshold values and nine different samples. All these samples contain random instances from the three labels. The number of instances per label is in the range [1, 9]. According to Fig. 7, increasing

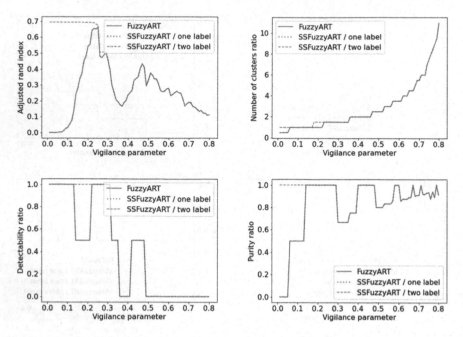

Fig. 5. SSFuzzyART VS Fuzzy ART on Breast cancer dataset. In this experiment, two initialization samples were considered. The first one contains 30 instances having the same label. The second one contains 60 instances having two labels (30 instances per label).

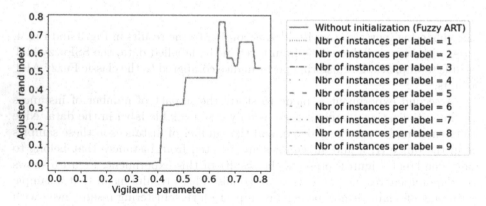

Fig. 6. Impact of the number of considered instances per label in the initialization sample of the SSFuzzyART compared to Fuzzy ART (without initialization). Iris dataset is used, only one label is considered in this experiment.

the number of initialization instances always improves the Adjusted Rand index. Indeed, even with one instance per label, SSFuzzyART has a Rand index that is greater than or equals to the Fuzzy ART Rand index. Furthermore, a number of instances per label that is greater than 2 has always a positive impact on the Rand index, according to these results.

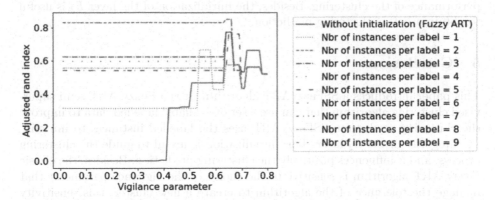

Fig. 7. Impact of the number of considered instances per label in the initialization sample of the SSFuzzyART compared to Fuzzy ART (without initialization). Iris dataset is used, and three labels are considered.

In the previous experiments, the labelled instances that are used to initialize SSFuzzyART are clustered with the rest of the data.

Then, a last experiment is carried out to compare the performance of SSFuzzyART on the labelled data (labelled instances that are used as seeds), the unlabeled data (the rest of the dataset) and the combined dataset. Iris dataset is considered in this experiment. Figure 8 presents the results of this experiment. According to the results in Fig. 8, the highest Adjusted Rand score is obtained on the labelled data. This is an expected result as the labelled data is used

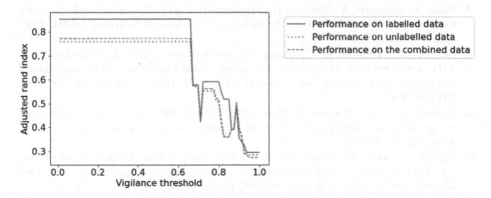

Fig. 8. SSFuzzyART Adjusted Rand index on the labelled data (seeds), the unlabelled data and the combined dataset. Iris dataset is used, and three labels are considered with 7 random instances per label.

to initialize SSFuzzyART second layer neurons. The Adjusted Rand score on the unlabelled data is almost equal to the Adjusted Rand score of the entire dataset. These results confirm that SSFuzzyART has a positive impact on the performance of the clustering. Besides, the initialization of the layer L_2 is useful to obtain a better clustering prediction.

5   Conclusion

This paper introduces a SSFuzzyART algorithm. It is a Fuzzy ART semi supervised clustering algorithm that can use a set of available labelled data to improve the clustering prediction. SSFuzzyART uses the labelled instances to initialize its categories in the L_2 layer. This initialization is useful to guide the clustering process, and it influences positively the clustering prediction. Besides, the classic Fuzzy ART algorithm is sensitive to the choice of the vigilance parameter that manage the tolerance of the algorithm to create a new cluster, this sensitivity has been inherited by SSFuzzyART. Then, it is generally recommended choosing a relatively low ρ to get good performance. The ρ value could depend on the number of categories that are present in the initialization data. Then, if all categories are present, a lower ρ is recommended.

In the future work, it will be interesting to test SSFuzzyART in a real world problem and to study its impact on real data. A second perspective is to compare the three possible scenarios introduced in Sect. 3.2, and to study their impact on real world problems. A Third perspective of this work could be to find a solution to influence the Fuzzy ART clustering process with available labelled data.

References

1. Bair, E.: Semi-supervised clustering methods. Wiley Interdiscipl. Rev. Computat. Statist. **5**(5), 349–361 (2013). https://doi.org/10.1002/wics.1270
2. Basu, S., Banerjee, A., Mooney, R.: Semi-supervised clustering by seeding. In: In: Proceedings of 19th International Conference on Machine Learning (ICML-2002. Citeseer (2002)
3. Bingwen, C., Wenwei, W., Qianqing, Q.: Infrared target detection based on fuzzy ART neural network. In: 2010 Second International Conference on Computational Intelligence and Natural Computing. IEEE (2010). https://doi.org/10.1109/cinc. 2010.5643745
4. Carpenter, G.A., Grossberg, S., Rosen, D.B.: Fuzzy ART: Fast stable learning and categorization of analog patterns by an adaptive resonance system. Neural Netw. **4**(6), 759–771 (1991)
5. Djellali, C., adda, M., Moutacalli, M.T.: A comparative study on fuzzy clustering for cloud computing. taking web service as a case. Procedia Comput. Sci. **184**, 622–627 (2021). https://doi.org/10.1016/j.procs.2021.04.024
6. Elnabarawy, I., Tauritz, D.R., Wunsch, D.C.: Evolutionary computation for the automated design of category functions for fuzzy ART. In: Proceedings of the Genetic and Evolutionary Computation Conference Companion. ACM, July 2017. https://doi.org/10.1145/3067695.3082056

7. Ilhan, S., Duru, N., Adali, E.: Improved fuzzy art method for initializing K-means. Int. J. Comput. Intell. Syst. **3**(3), 274 (2010). https://doi.org/10.2991/ijcis.2010. 3.3.3

8. Kim, T., Hwang, I., Kang, G.C., Choi, W.S., Kim, H., Zhang, B.T.: Label propagation adaptive resonance theory for semi-supervised continuous learning. In: ICASSP 2020–2020 IEEE International Conference on Acoustics, Speech and Signal Processing (ICASSP), pp. 4012–4016. IEEE (2020)

9. Kim, T., Hwang, I., Lee, H., Kim, H., Choi, W.S., Lim, J.J., Zhang, B.T.: Message passing adaptive resonance theory for online active semi-supervised learning. In: International Conference on Machine Learning, pp. 5519–5529. PMLR (2021)

10. Liew, W.S., Loo, C.K., Wermter, S.: Emotion recognition using explainable genetically optimized fuzzy ART ensembles. IEEE Access **9**, 61513–61531 (2021). https://doi.org/10.1109/access.2021.3072120

11. Qin, Y., Ding, S., Wang, L., Wang, Y.: Research progress on semi-supervised clustering. Cogn. Comput. **11**(5), 599–612 (2019). https://doi.org/10.1007/s12559-019-09664-w

12. Sengupta, S., Ghosh, T., Dan, P.K., Chattopadhyay, M.: Hybrid Fuzzy-ART based K-Means Clustering Methodology to Cellular Manufacturing Using Operational Time. arXiv preprint arXiv:1212.5101 (2012)

13. da Silva, L.E.B., Elnabarawy, I., Wunsch, D.C., II.: A survey of adaptive resonance theory neural network models for engineering applications. Neural Netw. **120**, 167–203 (2019)

Informed Weak Supervision for Battery Deterioration Level Labeling

Luciano Sánchez[1](✉) ⓘD, Nahuel Costa[1] ⓘD, David Anseán[2] ⓘD, and Inés Couso[3] ⓘD

[1] Departamento de Informática, Universidad de Oviedo, Gijón, Spain
{luciano,costanahuel}@uniovi.es
[2] Departamento de Ingeniería Eléctrica, Electrónica,
de Comunicaciones y de Sistemas, Gijón, Spain
anseandavid@uniovi.es
[3] Departamento de Estadística e Investigación Operativa y D. M., Gijón, Spain
couso@uniovi.es

Abstract. Learning the deterioration of a battery from charge and discharge data is associated with different non-random uncertainties. A specific methodology is developed, capable of integrating expert knowledge about the problem and of handling the epistemic uncertainty associated with conflicts in the available information. It is shown that the simple concatenation of charge and discharge data in a single training set leads to a biased model. Weak supervision techniques are used to assess the relative importance of subsets of the training data in the empirical loss function.

Keywords: Weak supervision · Battery deterioration · Optimistic losses

1 Introduction

The environmental impact of transportation in vehicles powered by internal combustion engines is high, both in terms of emissions of toxic substances and greenhouse gases. To mitigate the effects of global warming, most of these engines are expected to be replaced by electric motors powered by rechargeable lithium batteries from 2025–2040 [6]. In particular, there is a great interest in the development of non-destructive techniques to analyze the State of Health (SoH) of rechargeable batteries, to anticipate the deterioration of the battery caused during its use. The determination of SoH and lifetime of a battery using machine learning has been extensively studied; see, for example, reference [15], where some of the most relevant review articles are discussed.

In virtually all work applying machine learning to battery condition determination, one of the inputs to the system consists of voltage measurements at the

---

Supported by Ministerio de Economía e Industria de España, grant PID2020-112726RB-I00 and by Principado de Asturias, grant SV-PA-21-AYUD/2021/50994.

D. Ciucci et al. (Eds.): IPMU 2022, CCIS 1602, pp. 748–760, 2022.
https://doi.org/10.1007/978-3-031-08974-9_59

battery terminals. In some cases, voltage measurements are directly the inputs to the model, and in other cases these data are preprocessed to obtain incremental capacity [18] or differential voltage curves on which to base diagnostics [5]. In turn, voltage measurements can be obtained from discharge data, charge data, or both.

It is also common for charging and discharging measurements to have a different importance in the learning task. In fact, in some of the most influential works on machine learning-based diagnostics, battery life is predicted from discharge voltage curves alone [17]. This is because cells have hysteresis and the effects of some types of deterioration do not manifest themselves in the same way in the charge and discharge processes. Although this fact is well known, to the best of our knowledge, the best strategy for combining charge and discharge data in the machine learning problem has not yet been investigated. As will be seen in the following sections, this particular learning problem has different non-random uncertainties associated with it that require the development of a specific methodology capable of integrating expert knowledge about the problem and of handling the epistemic uncertainty associated with conflicts in the available information [13]. We will show that the simple concatenation of charge and discharge data in a single training set can lead to a biased model. The relative importance of the subsets of charge and discharge data depends on the unknown deterioration type, so the use of weak supervision techniques to combine these two sources of information is proposed.

The structure of this paper is as follows: in Sect. 2 we briefly introduce the practical problem that gives rise to the study, indicating the different types of deterioration we want to identify. In Sect. 3 we propose an Informed Learning algorithm [16] that combines the predictions made on different subsamples of the data, resolving conflicts using various criteria. Section 4 performs a comparative study of the results of applying different combinations (optimistic and pessimistic, with and without regularization) of the prediction errors in the learning algorithm, and validates the results on synthetic experimental data (generated by a mechanistic model [9]) and on real batteries. Section 5 presents the conclusions of the study.

2   Battery Deterioration Types

Although most battery health studies focus on capacity loss versus age, the most accurate techniques are based on obtaining the derivative of the voltage with respect to the load (also called incremental capacity curve) and analyzing the evolution of the characteristic points of this curve [8]. In this way, it is not only possible to predict the remaining capacity of the battery, but also to explain the electrochemical processes that have caused the deterioration. This is especially important in the detection of so-called "silent defects", which do not immediately affect the capacity of the battery [7].

In particular, there are equivalent circuit models that can synthesize the incremental capacity curves of a battery from two half-cell models, associated to the positive and negative electrodes of the battery [9]. These equivalent circuit models use a voltage versus load curve for each half-cell, and depend on a set of latent variables (not directly observable) related to the capacity of each half-cell and its lithiation level. The deterioration of a cell is measured by the time evolution of the values of these latent variables.

2.1 Energy and Power Losses

Numerous types of battery degradation mechanisms have been investigated in the literature [11]. The most common aging mode associated to battery degradation is the loss of lithium inventory (LLI), which is related to lithium consumption by parasitic reactions (e.g., electrolyte decomposition, solid electrolyte interface growth, etc.) [14]. The second aging mode is the loss of active material (LAM), which commonly results from particle cracking and loss of electrical contact, among other causes [4]. The LAM can be further classified into four types, depending on the electrode (positive and negative) and the degree of lithiation (predominantly lithiated or delithiated state) [9]. These aging modes (i.e., LLI and LAM) can lead to both capacity and power fade. In batteries with a negative electrode composed of a blended mix of graphite and silicon, such as those used in this study [2], we can distinguish six types of loss of active material: loss of capacity of the delithiated (LAMdPE) or lithiated (LAMlPE) positive electrode and loss of capacity in the graphite and silicon fractions of the negative electrode (LAMlGr, LAMdGr, LAMlSi, LAMdSi). In addition to the deteriorations related to the amount of energy that can be stored in a cell, there are other effects that limit the power that the cell can provide [9]. In the ohmic resistance increase the electrodes offer more resistance to the passage of current, which increases heat losses and reduces charging efficiency. In the faradaic rate degradation, the reaction rate between the electrode and the lithium ions is reduced, which prevents the energy from being extracted from the battery as quickly as necessary.

2.2 Latent Variables of the Deterioration Model

In this study, we analyze the most common battery degradation modes, i.e., LLI and LAM, and consider the following five state variables of the equivalent circuit model:

1. Positive electrode capacity
2. Graphite capacity at the negative electrode
3. Silicon capacity at the negative electrode
4. Initial degree of lithiation of the positive electrode
5. Initial degree of lithiation of the negative electrode

For this simplified problem, we will consider the seven types of degradation considered in this study as influencing the state variables as shown in Table 1.

3 Learning the Deterioration Model

Let us first define a partial problem, that of learning the SoH of a battery from a single charge-discharge cycle. This is a supervised learning task in which the input data is a sequence of charge currents $i_t \in I$, $t = 0, \ldots, N$ and the output data is a sequence of voltage values at the same instants of time, $v_t \in V$, $t = 0, \ldots, N$. The training set $\mathcal{D} \in (I \times V)^N$ is

Table 1. Influence on latent variables of the types of deterioration considered in this study.

Deterioration	Influence on latent variables
LLI	Capacities are maintained, initial electrode lithiation levels are altered
LAMdPE	Positive electrode capacity reduced, initial lithiation levels maintained
LAMdGr	Graphite electrode capacity reduced, initial lithiation levels maintained
LAMdSi	Silicon electrode capacity reduced, initial lithiation levels maintained
LAMlPE	Positive electrode capacity is reduced, initial lithiation levels of the electrodes are altered
LAMlGr	Graphite electrode capacity is reduced, initial lithiation levels of the electrodes are altered
LAMlSi	Silicon electrode capacity is reduced, initial lithiation levels of the electrodes are altered

$$\mathcal{D} := \{(i_0, v_0), \ldots, (i_N, v_N)\}. \tag{1}$$

Let $\mathcal{H}$ be the hypothesis space, let θ be the parameter vector of the mechanistic model, and let $h_\theta \in \mathcal{H}$ be a function of a sequence of currents spanning the values from the beginning of the charge cycle to the instant t:

$$\widehat{v}_t = h_\theta\left((i_0, i_1, \ldots, i_t)\right). \tag{2}$$

It is emphasized that h_θ is not only a function of the current i_t at time t, but of the sequence of current values from the initial instant to the present instant.

The solution to this subproblem is to find the value θ_* that minimizes a risk function, say the mean square error between predictions and measured values:

$$R_{\text{emp}}(h_\theta) := \frac{1}{N} \sum_{t=1}^{N} (\widehat{v}_t - v_t)^2 \tag{3}$$

$$h_{\theta_*} = \arg \min_{h_\theta \in \mathcal{H}} R_{\text{emp}}(h_\theta) \tag{4}$$

Let us now define the problem of learning the deterioration of a battery as obtaining of a list of solutions of different chained partial problems of the above type. Let τ be the lifetime of the battery (suppose, for example, that the subscripts τ are measured as the number of charge-discharge cycles since the battery has been put into use, and that the subscripts t of the above partial problem measure seconds since the start of a charge). Let us then define M learning sets

$$\mathcal{D}_\tau := \{(i_0^\tau, v_0^\tau), \ldots, (i_N^\tau, v_N^\tau)\}, \quad \tau = 1, 2, \ldots, M \tag{5}$$

Note that the parameter sequence $(\theta_*^\tau)_{\tau=1}^M$ obtained by empirical risk minimization for each of the $\mathcal{D}_\tau$ data sets is not a valid solution to this problem. The θ parameter vectors represent latent variables in the battery, and the sequence of $(\theta_*^\tau)_\tau$ estimates may not be physically possible. Given an initial state vector θ_a, let us define a criterion that indicates whether state θ_b is reachable after the application of an arbitrary degradation starting at state θ_a (i.e., that state θ_b corresponds to a SoH worse than θ_a):

$$\text{feasible}(\theta_a, \theta_b) = \begin{cases} 1 & \text{if } \theta_b \text{ is reachable after } \theta_a \\ 0 & \text{if } \theta_b \text{ cannot be reached from } \theta_a. \end{cases} \tag{6}$$

The empirical risk function of the degradation problem is

$$R_{\text{emp}}((h_{\theta^1}, h_{\theta^2}, \ldots, h_{\theta^M})) = \sum_{\tau=1}^M R_{\text{emp}}(h_{\theta^\tau}) \tag{7}$$

and the solution $(\theta_+^\tau)_{\tau=1}^M$ consists of solving the optimization problem

$$(h_{\theta_+^1}, h_{\theta_+^2}, \ldots, h_{\theta_+^M}) = \arg\min_{h_\theta^\tau \in \mathcal{H}} R_{\text{emp}}((h_{\theta^\tau})_{\tau=1}^M) \tag{8}$$

subject to

$$\text{feasible}(\theta_+^a, \theta_+^b) = 1 \quad \text{for all } b > a. \tag{9}$$

3.1   Uncertainty and Inconsistency in the Data

The different types of deterioration influence the observable variables in a non-homogeneous manner. Some deteriorations cause appreciable changes during the phase in which the battery is charged, and other deteriorations have a greater influence in the discharge phase. This behavior introduces an inconsistency into the learning problem.

Consider, for example, an extreme case where the effect of the deterioration is only noticeable in the charging phase. Let us now divide the training set associated with the deteriorated battery into two parts:

$$\mathcal{D}_c := \{(i, v) \in \mathcal{D} : i \geq 0\} \tag{10}$$

$$\mathcal{D}_d := \{(i, v) \in \mathcal{D} : i < 0\} \tag{11}$$

A model fitted to the first part $\mathcal{D}_c$ of the training set would correctly identify capacity losses in the electrodes and changes in the lithium inventory. In contrast, the parameters of a model fitted to the second part $\mathcal{D}_d$ would not detect these deteriorations. A compromise model fitted to the whole set $\mathcal{D}$ would also not be unbiased in most situations.

In real situations, the battery is simultaneously subjected to different types of deterioration. The influences of each deterioration type on each training subset are different. This inconsistency may be resolved by assigning a weight to each of the instances in $\mathcal{D}$ and defining a weighted risk function, but these weights are unknown, since they depend on the also unknown deterioration of the battery.

The solution proposed in this study to reduce the effects of data inconsistency consists of defining a new empirical risk function, in which the risk components corresponding to each of the $\mathcal{D}_c$ and $\mathcal{D}_d$ parts of the training set are combined. For this purpose, we will recall the optimistic and pessimistic criteria defined in superset learning [12]:

$$R_{\text{emp}}^c(h_\theta) := \frac{1}{||\mathcal{D}_c||} \sum_{(i_t, v_t) \in \mathcal{D}_c} (\widehat{v}_t - v_t)^2 \tag{12}$$

$$R_{\text{emp}}^d(h_\theta) := \frac{1}{||\mathcal{D}_d||} \sum_{(i_t, v_t) \in \mathcal{D}_d} (\widehat{v}_t - v_t)^2 \tag{13}$$

$$R_{\text{emp}}^{\text{OPT}}(h_\theta) := \min\{R_{\text{emp}}^c(h_\theta), R_{\text{emp}}^d(h_\theta)\} \tag{14}$$

$$R_{\text{emp}}^{\text{PESS}}(h_\theta) := \max\{R_{\text{emp}}^c(h_\theta), R_{\text{emp}}^d(h_\theta)\} \tag{15}$$

together with their respective versions with L_1 regularization:

$$R_{\text{emp}}^{\text{OPT reg}}(h_\theta) := \min\{R_{\text{emp}}^c(h_\theta), R_{\text{emp}}^d(h_\theta)\} + \lambda||\theta|| \tag{16}$$

$$R_{\text{emp}}^{\text{PESS reg}}(h_\theta) := \max\{R_{\text{emp}}^c(h_\theta), R_{\text{emp}}^d(h_\theta)\} + \lambda||\theta|| \tag{17}$$

Finally, we define a third criterion similar to the min-max regret approach [10] as follows:

$$R_{\text{emp}}^{\text{PESS regret}}(h_\theta) := \max\{R_{\text{emp}}^c(h_\theta) - R_{\text{emp}}^c(h_\theta^{c,*}), R_{\text{emp}}^d(h_\theta) - R_{\text{emp}}^c(h_\theta^{d,*})\} \tag{18}$$

where $h_\theta^{c,*}$ and $h_\theta^{c,*}$ are the empirical minimum risk models on the sets $\mathcal{D}_c$ and $\mathcal{D}_d$, respectively. The version with regularization L_1 of the min-max regret criterion can also be defined as seen above.

3.2 Numerically Efficient Solution Through Informed Learning

Direct solution of the learning problem defined in Eq. (8) and (9) would require a large amount of computational resources. The criterion (9) cannot be expressed by a continuous function of the arguments, and the computing time required to evaluate the h_θ model over a sequence of N current values is high.

Since in this problem the components of the parameter vector have a physical meaning, it is possible to inject knowledge into learning to reduce the search space [1,16]. Let us recall the components of the parameter vector θ:

- θ_1: Positive electrode capacity
- θ_2: Graphite capacity at the negative electrode
- θ_3: Silicon capacity at the negative electrode
- θ_4: Initial degree of lithiation of the positive electrode
- θ_5: Initial degree of lithiation of the negative electrode

The seven types of deterioration introduced in Sect. 2 produce the following changes in the parameter vector; each transformation is characterised by a type and an intensity α:

$$\begin{aligned}
\text{LLI:} \quad & g_1(\theta, \alpha) := (\theta_1, \theta_2, \theta_3, \theta_4, \theta_5 + \alpha) \\
\text{LAMdPE:} \quad & g_2(\theta, \alpha) := (\theta_1 - \alpha, \theta_2, \theta_3, \theta_4, \theta_5) \\
\text{LAMdGr:} \quad & g_3(\theta, \alpha) := (\theta_1, \theta_2 - \alpha, \theta_3, \theta_4, \theta_5) \\
\text{LAMdSi:} \quad & g_4(\theta, \alpha) := (\theta_1, \theta_2, \theta_3 - \alpha, \theta_4, \theta_5) \\
\text{LAMlPE:} \quad & g_5(\theta, \alpha) := (\theta_1 - \alpha, \theta_2, \theta_3, \theta_4, \theta_5 + \alpha) \\
\text{LAMdGr:} \quad & g_6(\theta, \alpha) := (\theta_1, \theta_2 - \alpha, \theta_3, \theta_4, \theta_5 + \alpha) \\
\text{LAMdSi:} \quad & g_7(\theta, \alpha) := (\theta_1, \theta_2, \theta_3 - \alpha, \theta_4, \theta_5 + \alpha)
\end{aligned}$$

Let us call the vector of parameters associated with a non-deteriorated battery θ^0. We will assume that the θ^τ parameter vector of the aged battery consists of the application of a sequence of transformations to the θ^0 vector, where each of the transformations corresponds to one of the seven types of ageing defined above:

$$\theta^\tau = g_{d(D)}(g_{d(D-1)}(\cdots(g_{d(1)}(\theta_0, \alpha_1), \cdots, \alpha_{D-1}), \alpha_D) \tag{19}$$

so that battery ageing between 0 and τ can be encoded as a sequence of pairs (defect type, intensity):

$$S = ((d(1), \alpha_1), (d(2), \alpha_2), \ldots, (d(D), \alpha_D)) \tag{20}$$

In particular, we propose to apply informed learning using functional compositions [3]. The pseudocode of the proposed algorithm for obtaining the evolution of the SoH of a battery in C cycles is as follows:

1. Fit the model h_{θ^0} to a cycle of the undamaged battery.
2. $S \leftarrow []; \ D \leftarrow 1$
3. For τ in $(1, C)$
4.   Repeat
5.   Find the pair $(d(D), \alpha_D)$ that minimizes $R_{\text{emp}}(\theta)$, with

$$\theta = g_{d(D)}(g_{d(D-1)}(\cdots(g_{d(1)}(\theta_0, \alpha_1), \cdots, \alpha_{D-1}), \alpha_D) \text{ and}$$

$$S = [(d(1), \alpha_1), \ldots, (d(D-1), \alpha_{D-1})]$$

6. $S \leftarrow S + (d(D), \alpha_D); \ D \leftarrow D + 1$

7.   Until $R_{\text{emp}}(\theta^{\tau-1}) - R_{\text{emp}}(\theta) < \text{threshold}$
8. $\theta^\tau \leftarrow \theta$
9. End For

The input of the algorithm is a sequence of datasets $\mathcal{D}_1, \ldots, \mathcal{D}_C$ and the "threshold" hyperparameter, which defines the improvement of the risk function below which the intensity of battery degradation is so low that it is considered an error of approximation of the model. The outputs are a sequence of parameter vectors $(\theta^1, \ldots, \theta^C)$ and a sequence of deteriorations $((d(1), \alpha_1), \ldots, (d(C), \alpha_C))$. Step 4 of the algorithm has been solved by successively applying a descent algorithm (L-BFGS-B) for each of the seven possible values of $d(D)$.

4   Numerical Results

This section is divided into two parts. In the first part, the proposed algorithm is applied to Li-Ion batteries simulated using the ALAWA tool [9]. The reason for using simulated batteries is to validate the deterioration learning algorithm in problems with known outcome, to assess the degree of fit of the proposed alternative and to compare it with other uninformed learning techniques. In the second part, the proposed method is applied to a real battery.

4.1   Simulated Batteries

A realistic synthetic 5 Ah battery has been simulated. The positive electrode is NMC (Nickel, Manganese and Cobalt) and the negative electrode consists of a combination of graphite and silicon [2]. The capacity of the positive electrode is 5.55 Ah. The capacities of the graphite and silicon in the negative electrode are 4.5 Ah and 0.5 Ah, respectively. The positive electrode is 90% lithiated at the beginning of the charge and is completely de-lithiated at the end of the charge. The negative electrode of the non-deteriorated battery is fully delithiated at the beginning of the charge and fully lithiated at the end of the charge. The initial values of the parameter vector are therefore

$$\theta^0 = \left(\frac{5}{0.9}, 4.5, 0.5, 5, 0 \right)$$

Starting from the same initial conditions, seven simulations of 2000 cycles were carried out, each with a single type of deterioration. The average of the percentage errors of the electrode capacities and lithium inventory was used as a metric for the model fit:

$$E = 100 \sum_{\tau=1}^{C} \frac{1}{4C} \left(\sum_{k=1}^{3} \frac{|\theta_k^\tau - \widehat{\theta}_k^\tau|}{\theta_k^\tau} + \frac{|\theta_4^\tau + \theta_5^\tau - \widehat{\theta}_4^\tau - \widehat{\theta}_5^\tau|}{\theta_4^\tau + \theta_5^\tau} \right) \tag{21}$$

Table 2 shows the errors obtained by the uninformed algorithms (minimum empirical risk models fitted to each cycle, i.e. the solution of Eq. (8) without applying the constraints defined in equation (9)) and those of the four models proposed in this study. The MSE column contains the results of the empirical risk minimisation over the full charge and discharge cycle. The OPT column is the optimistic/minimin version (Eq. (15)). The PESS column is the pessimistic/minimax version (Eq. (16)). The MMR column is the min-max-regret version (Eq. (17)).

The mean error is better in models with informed learning, and among the versions for which the risk function resolves the inconsistencies in the charge and discharge phases, the version that offers the best results is the minimin/optimistic alternative. This is an expected result, as the optimistic version

Table 2. Comparative results of uninformed and informed learning for the seven defect types. Non-regularised target function.

Deterioration	Non-informed	Informed learning			
		MSE	OPT	PESS	MMR
LLI	0.63	0.42	**0.40**	0.71	0.64
LAMdPE	3.81	2.25	**0.55**	3.28	3.07
LAMdGr	0.99	**0.55**	1.98	1.24	1.23
LAMdSi	1.37	0.66	**0.54**	1.61	1.78
LAMlPE	1.07	0.88	0.88	**0.76**	0.80
LAMlGr	0.81	0.58	0.53	**0.25**	**0.25**
LAMlSi	1.43	0.88	**0.79**	1.48	1.24
Average	1.45	0.89	**0.81**	1.33	1.29
Average PE	2.44	1.56	**0.72**	2.02	1.93
Average NE	1.15	**0.67**	0.96	1.15	1.13

Table 3. Comparative results of uninformed and informed learning for the seven defect types. Objective function with L_1 regularisation.

Deterioration	Non-informed	Regularized informed learning			
		MSE-L_1	OPT-L_1	PESS-L_1	MMR-L_1
LLI	0.63	0.05	**0.02**	0.08	0.06
LAMdPE	3.81	1.64	**0.55**	3.30	2.69
LAMdGr	0.99	0.10	**0.01**	0.71	0.54
LAMdSi	1.37	0.30	**0.22**	1.36	1.33
LAMlPE	1.07	0.18	0.27	0.19	**0.14**
LAMlGr	0.81	0.58	0.39	**0.24**	0.26
LAMlSi	1.43	**0.41**	0.42	1.45	0.46
Average	1.45	0.47	**0.28**	1.05	0.78
Average PE	2.44	0.91	**0.41**	1.74	1.42
Average NE	1.15	0.35	**0.28**	0.93	0.65

Table 4. Deterioration predictions on a real battery. INR18650-35E cell from Samsung-SDI, with standard discharge capacity of 3350 mAh.

Cycle	Complete training set				OPT Weak supervision			
	LLI	LAMPE	LAMGr	LAMSi	LLI	LAMPE	LAMGr	LAMSi
100	0.00	0.00	0.00	0.00	0.00	0.00	0.00	0.00
200	0.00	0.00	0.00	0.00	0.00	0.00	0.00	0.00
300	3.11	1.79	3.09	0.00	1.64	1.03	0.00	0.00
400	4.36	2.27	3.09	0.00	2.92	1.07	4.71	0.00
500	6.09	3.69	6.45	0.00	4.84	1.84	4.72	0.00
600	7.33	4.21	6.45	0.00	6.11	2.33	4.73	0.00
700	8.52	4.92	7.88	0.00	7.34	2.46	7.59	0.00
800	10.07	5.46	7.88	0.00	9.22	3.63	7.66	0.00
900	10.81	5.76	7.88	0.00	9.76	3.69	7.86	0.00
1000	12.25	6.99	7.88	0.00	11.44	4.83	8.22	0.00
1100	13.23	7.50	7.88	0.00	12.71	6.17	12.01	0.00
1200	13.97	7.87	7.88	0.00	13.37	6.69	12.03	0.00
1300	14.79	8.43	7.88	0.00	14.49	7.28	12.03	0.00
1400	15.82	9.41	7.88	0.00	15.60	7.73	12.03	0.00
1500	16.95	9.89	7.88	0.00	17.84	10.40	12.03	0.00
1600	17.69	10.51	7.88	0.00	17.84	10.60	12.16	0.00
1700	17.69	10.51	7.88	0.00	17.84	10.60	12.16	0.00

is robust to coarsening processes [12]. The pessimistic or minimin-regret versions try to minimise the wrong risk in those phases of the cycle where deteriorations do not manifest themselves.

Table 3 shows the errors of the regularised versions, which support the same conclusions drawn from Table 2. In this case, the use of regularisation of type L_1, which leads to a sparse deterioration vector, improves the results because each battery has undergone a single deterioration process in the validation data (i.e., the optimal solution of each problem is a vector in which all components are zero except one; the regularisation penalises solutions with more than one concurrent deterioration and is therefore expected to improve the results of this method for this particular validation set.)

4.2   Actual Batteries

The results obtained in synthetic batteries show that the informed strategy, in its optimistic version, has the best accuracy in estimating the type and intensity of deteriorations. It has also been shown that the addition of a regularization term is beneficial, provided that the assumption that the number of concurrent deterioration types is low is accepted.

In simulated batteries the model prediction error is close to zero. It is worth asking whether the model approximation error in a real battery will cancel out the differences between the techniques studied or whether, on the contrary, the improvement in the accuracy of the weak supervision estimation will be relevant in real problems. To study this case we have used a Samsung-SDI INR18650-35E cell, with a standard discharge capacity of 3350 mAh, which was subjected to a standard set of tests [2]. A multichannel, high-precision-series Arbin LBT was used to perform the tests and record the voltage and current values. The cell was placed at constant 23 °C in a Memmert environmental chamber.

The deterioration predictions produced by the minimum risk models on the full ensemble and with optimistic weak supervision are shown in Table 4. In this table, we have grouped the LAM degradation modes into three categories: LAMPE (loss of capacitance of the positive electrode), LAMGr (loss of capacitance of the graphite component of the negative electrode), and LAMSi (loss of capacitance of the silicon component of the negative electrode). The differences between the lithium inventory estimates are low (–1%) and the same can be said for the differences in the positive electrode capacity, but the active graphite estimates are much different. The estimate with weak supervision detects deterioration of constant magnitude over the entire battery life, whereas the ordinary estimate indicates that no active graphite is lost from cycle 700 onwards, showing that weakly supervised learning techniques can alter the diagnosis that would be obtained by supervised learning in real problems.

5   Concluding Remarks and Future Work

The literature concerning the use of machine learning for nondestructive diagnosis of battery condition is extensive. Numerous algorithms have been published that extract information from current, voltage and temperature measurements under different assumptions. This paper discusses that, in order for the model to converge to the desired solution, it is necessary to weight the instances with a cost that depends on the deterioration that has actually occurred. It is also proposed that, in the absence of knowledge about such weighting, an empirical risk function combining the risks of different subsets of the training data is employed.

Although, as mentioned, there is no background on the use of weak supervision in battery diagnosis, different authors have made an implicit selection of instances and use only discharge data. Other authors go one step further and select voltage and current values within a specific range of states of charge (i.e., they do not take into account the model error when the battery is nearly charged or discharged). In future work we will study the extension of the proposed method to families of subsets of the training data differentiated not only by the sign of the charging current but also by the state of charge. In parallel, we will investigate the extension of the reported procedure to ohmic and faradaic deteriorations, and validate our results with simulations including several concurrent faults, to study in them regularization techniques different from the L_1 criterion analyzed in this contribution.

References

1. Amel, K.R.: From shallow to deep interactions between knowledge representation, reasoning and machine learning. In: Proceedings 13th International Conference Scala Uncertainity Mgmt (SUM 2019), Compiegne, LNCS, pp. 16–18 (2019)
2. Anseán, D., Baure, G., González, M., Cameán, I., García, A., Dubarry, M.: Mechanistic investigation of silicon-graphite/LiNi0.8Mn0.1Co0.1O2 commercial cells for non-intrusive diagnosis and prognosis. J. Power Sour. **459**, 227882 (2020)
3. Bauckhage, C., Ojeda, C., Schücker, J., Sifa, R., Wrobel, S.: Informed machine learning through functional composition. In: LWDA, pp. 33–37 (2018)
4. Birkl, C.R., Roberts, M.R., McTurk, E., Bruce, P.G., Howey, D.A.: Degradation diagnostics for lithium ion cells. J. Power Sour. **341**, 373–386 (2017)
5. Bloom, I., Jansen, A.N., Abraham, D.P., Knuth, J., Jones, S.A., Battaglia, V.S., Henriksen, G.L.: Differential voltage analyses of high-power, lithium-ion cells: 1. technique and application. J. Power Sour. **139**(1–2), 295–303 (2005)
6. Conway, G., Joshi, A., Leach, F., García, A., Senecal, P.K.: A review of current and future powertrain technologies and trends in 2020. Transp. Eng. **5**, 100080 (2021)
7. Dubarry, M., Baure, G., Anseán, D.: Perspective on state-of-health determination in lithium-ion batteries. J. Electrochem. Energy Conv. Storage **17**(4), 044701 (2020)
8. Dubarry, M., Svoboda, V., Hwu, R., Liaw, B.Y.: Incremental capacity analysis and close-to-equilibrium OCV measurements to quantify capacity fade in commercial rechargeable lithium batteries. Electrochem. Solid State Lett. **9**(10), A454 (2006)
9. Dubarry, M., Truchot, C., Liaw, B.Y.: Synthesize battery degradation modes via a diagnostic and prognostic model. J. Power Sour. **219**, 204–216 (2012)
10. Guillaume, R., Dubois, D.: A min-max regret approach to maximum likelihood inference under incomplete data. Int. J. Approximate Reasoning **121**, 135–149 (2020)
11. Han, X., Lu, L., Zheng, Y., Feng, X., Li, Z., Li, J., Ouyang, M.: A review on the key issues of the lithium ion battery degradation among the whole life cycle. ETransportation **1**, 100005 (2019)
12. Hüllermeier, E., Destercke, S., Couso, I.: Learning from Imprecise Data: Adjustments of Optimistic and Pessimistic Variants. In: Ben Amor, N., Quost, B., Theobald, M. (eds.) SUM 2019. LNCS (LNAI), vol. 11940, pp. 266–279. Springer, Cham (2019). https://doi.org/10.1007/978-3-030-35514-2_20
13. Hüllermeier, E., Waegeman, W.: Aleatoric and epistemic uncertainty in machine learning: an introduction to concepts and methods. Mach. Learn. **110**(3), 457–506 (2021)
14. Palacín, M.R.: Understanding ageing in li-ion batteries: a chemical issue. Chem. Soc. Rev. **47**(13), 4924–4933 (2018)
15. Rauf, H., Khalid, M., Arshad, N.: Machine learning in state of health and remaining useful life estimation: Theoretical and technological development in battery degradation modelling. Renew. Sustain. Energy Rev. **156**, 111903 (2022)
16. von Rueden, L., et al.: Informed machine learning - a taxonomy and survey of integrating prior knowledge into learning systems. IEEE Trans. Knowl. Data Eng. (In press)

17. Severson, K.A., et al.: Data-driven prediction of battery cycle life before capacity degradation. Nat. Energy **4**(5), 383–391 (2019)
18. Shim, J., Kostecki, R., Richardson, T., Song, X., Striebel, K.A.: Electrochemical analysis for cycle performance and capacity fading of a lithium-ion battery cycled at elevated temperature. J. Power Sourc. **112**(1), 222–230 (2002)

Rough-set Based Genetic Algorithms for Weakly Supervised Feature Selection

Andrea Campagner$^{(\boxtimes)}$ and Davide Ciucci ⓘ

Dipartimento di Informatica, Sistemistica e Comunicazione,
University of Milano – Bicocca, Viale Sarca 336 – 20126, Milan, Italy
a.campagner@campus.unimib.it

Abstract. In this article, we study the problem of feature selection under weak supervision, focusing in particular on the *fuzzy labels* setting, where the weak supervision is provided in terms of possibility distributions over candidate labels. While traditional Rough Set-based approaches have been applied for tackling this problem, they have high computational complexity and only provide local search heuristic methods. In order to address these issues, we propose a global optimization algorithm, based on genetic algorithms and Rough Set theory, for feature selection under fuzzy labels. Based on a set of experiments, we illustrate the effectiveness of the proposed approach in comparison to state-of-the-art methods.

Keywords: Weak supervision · Feature selection · Fuzzy labels · Genetic algorithms · Rough sets

1 Introduction

Learning from fuzzy labels [9,12] is a weakly supervised learning problem, in which each instance x is associated with a possibility distribution μ over candidate labels, having an epistemic semantics: only one of the labels is the correct one and μ, then, describes the possibility degree of the labels. For example, an image could be tagged with $\{car : 1, bus : 0.8, bicycle : 0.0\}$: the picture then depicts either a *car* or a *bus*, and *car* is deemed more plausible than *bus*.

In the recent years, increasing interest has been devoted to the development of algorithms for the learning from fuzzy labels task [7,9,12,17]. Even though these techniques can be effective on small-scale benchmarks, they can fail to scale to more complex and higher-dimensional problems, as their generalization ability depends (without further assumptions) on the dimensionality of the feature space [3]. While *feature selection* or *data dimensionality* methods could be helpful in mitigating this issue, their development has mostly been ignored.

While Rough Set-theoretic approaches have been applied effectively to address the above mentioned issues for other weakly supervised learning problem [4], their extension to the learning from fuzzy labels case [3] is more difficult,

© Springer Nature Switzerland AG 2022
D. Ciucci et al. (Eds.): IPMU 2022, CCIS 1602, pp. 761–773, 2022.
https://doi.org/10.1007/978-3-031-08974-9_60

due to increased computational complexity costs and the local heuristic nature of the greedy algorithms currently existing in the literature.

To address these limitations, in this article which represents a continuation of our previous work in this line of research [3], we propose a global optimization approach that combines Rough Set-based feature selection with genetic algorithms to solve the feature selection from fuzzy labels problem. In Sect. 2, we provide the necessary background knowledge on possibility theory and Rough Set theory. In Sect. 3, we first introduce the generalization of Rough Set theory to the learning from fuzzy labels setting, as well as the existing methods for performing feature selection in this setting, and then we introduce the proposed genetic algorithm-based approach and discuss its properties; in Sect. 4 we illustrate the effectiveness of the proposed method on a comprehensive set of benchmarks; finally, in Sect. 5, we summarize our results and describe some open problems.

2 Background

In this section, we recall basic notions of rough set theory (RST) and possibility theory, which will be used in the main part of the article.

2.1 Possibility Theory

Possibility theory is a theory of uncertainty which allows for the quantification of degrees of possibility on the basis of a fuzzy set [20]. We recall that a fuzzy set (equivalently, a possibility distribution) F can be seen as a function $F : X \mapsto [0, 1]$, that is, a generalization of the characteristic function representation of classical sets. A possibility measure is a function $pos_F : 2^X \mapsto [0, 1]$ such that

1. $pos_F(\emptyset) = 0$ and $pos_F(X) = 1,$;
2. if $A \cap B = \emptyset$ then $pos_F(A \cup B) = max(pos_F(A), pos_F(B))$.

Thus, every possibility measure is associated with a fuzzy set F, s.t. $pos_F(A) = max_{x \in A} F(x)$: F is the possibility distribution associated with pos_F. We will focus on *normal* possibility distributions, that is on possibility distributions F such that $\exists x \in X, F(x) = 1$. Given $\alpha \in [0, 1]$, the *alpha*-cut of F is defined as $F^\alpha = \{x \in X : F(x) \geq \alpha\}$, while the *strong* α-cut is defined as $F^{\alpha+} = \{x \in X : F(x) > \alpha\}$.

In this article, we will adopt the epistemic interpretation [8] of possibility theory, in which possibility distributions represent the degrees of belief (of an agent) w.r.t. a set of possible alternatives. We refer the reader to [12] for a discussion of epistemic possibility distributions in Machine Learning.

2.2 Rough Set Theory

Rough set theory has been proposed by Pawlak [16] as a framework for representing and managing uncertain data, and has since been widely applied for various problems in the ML domain (see [1] for an overview and survey).

A decision table (DT) is a triple $DT = \langle U, Att, t \rangle$ such that U is a universe of objects and Att is a set of *attributes* employed to represent objects in U. Each attribute $a \in Att$ is a function $a : U \to V_a$, where V_a is the domain of values of a. Moreover, $t \notin Att$ is a distinguished *decision* attribute, which represents the target label (or, decision) associated with each object in the universe.

Given $B \subseteq Att$, we can define the *indiscernibility relation* with respect to B as xI_Bx' iff $\forall a \in B$, $a(x') = a(x)$. Clearly, it is an equivalence relation that partitions the universe U in equivalence classes, also called *granules of information*, $[x]_B$. Then, the *indiscernibility partition* is denoted as $\pi_B = \{[x]_B \mid x \in U\}$.

We say that $B \subseteq Att$ is a *decision reduct* for DT if $\pi_B \leq \pi_t$ (where the order $\leq$ is the refinement order for partitions, that is, π_t is a coarsening of π_B) and there is no $C \subsetneq B$ such that $\pi_C \leq \pi_t$. Then, evidently, a reduct of a decision table represents a set of non-redundant and necessary features: therefore, reduct search can be understood as a process of feature selection. We say that a reduct R is *minimal* if it is among the smallest (with respect to cardinality) reducts. We remark that, given a decision table, the problem of finding minimal reducts is in general NP-HARD.

3 Rough Set-Based Weakly Supervised Feature Selection

In this section we recall the basic definitions regarding the generalization of Rough Set theory to the fuzzy labels setting, and discuss the existing feature selection methods for this setting. Then, we introduce the proposed weakly supervised genetic rough set feature selection method and we discuss its properties.

3.1 Possibilistic Decision Tables and Reducts

In this work, we will refer to the approach for Rough Set-based weakly supervised feature selection proposed in [4]. For other approaches to generalize Rough Set Theory to the case of imprecise data, we refer the reader to [6,15,18].

In the *learning from fuzzy labels* setting, each object $x \in U$ is generally not associated a single annotation $t(x) \in V_t$. Instead, each such object x is associated with a possibility distribution $\pi(x)$, which describes the state of knowledge of the annotating agent (either human or computational): in particular, $\pi(x)_y$ represents the relative plausibility of label y being the true annotation associated with x (as compared to other labels y'). These notions can be modeled within RST by generalizing the definition of a decision table:

Definition 1. *A possibilistic decision table (PDT) is a tuple $P = \langle U, Att, d, t \rangle$, where $d : U \mapsto [0,1]^{|V_t|}$ is a collection of normalized possibility distributions. $t : U \mapsto V_t$ is the true decision attribute, i.e. it is a function s.t. $\langle U, Att, t \rangle$ is a DT and s.t. the weak superset property w.r.t. d holds: $d(x)_{t(x)} > 0$ for all $x \in U$.*

As mentioned in the definition, t is the true decision attribute: for each object x, its true label is $t(x)$. However, t is assumed to be unknown and only the possibility distribution $d(x)$ is available. In regard to this latter, if $|d(x)^{0+}| > 1$ for some $x \in U$, then the correct decision $t(x)$ is not known precisely. Note that, by the weak superset property, the true label $t(x)$ is never considered impossible. Furthermore, if $d(x)_a > d(x)_b$ then a is considered more plausible than b for object x.

A PDT can be associated with a collection of compatible (standard) decision tables, called *instantiations* of the PDT:

Definition 2. *An* instantiation *of a PDT* $P = \langle U, Att, t, d \rangle$ *is a standard decision table* $T = \langle U, Att, t' \rangle$ *such that* $d(x)_{t'(x)} > 0$ *for all* $x \in U$. *The collection of instantiations of* P *is denoted* $\mathcal{I}(P)$. *In particular,* $\langle U, Att, t \rangle \in \mathcal{I}(P)$.

Thus, the collection $\mathcal{I}(P)$ contains all standard decision tables that are compatible (i.e., should not be considered impossible) with the imprecise knowledge descibed by the possibility distribution d. Furthermore, the collection $\mathcal{I}(P)$ inherits a ranking from the definition of the possibilistic decision attribute d:

Definition 3. *Let* $I_1, I_2 \in \mathcal{I}(P)$ *be two instantiations of a PDT* P. *Then we say that* I_1 *is* (conservatively) less possible *than* I_2, *denoted* $I_1 \leq_C I_2$, *if:*

$$min_{x \in U} d(x)_{t'}^{I_1} \leq min_{x \in U} d(x)_{t'}^{I_2} \tag{1}$$

We say that I_1 *is* dominated in possibility *by* I_2, *denoted* $I_1 \leq_D I_2$, *if:*

$$\forall x \in U. \ d(x)_{t'}^{I_1} \leq d(x)_{t'}^{I_2} \tag{2}$$

where, in both definitions $d(x)_{t'}^{I_i}$ *refers to the value of the decision attribute* d *(in* P*) on the label* $t'(x)$ *in the instantiation* I_i.

So as to capture not only the simplicity of the induced model (that is, the size of the reducts), but also the epistemic information encoded by the possibility distribution d, we [3] considered the following definitions of reducts:

Definition 4 ([3]). *For each* $\alpha \in (0, 1]$ *and PDT* P, *let* P^α *be the* α-cut *of* P, *that is* $P^\alpha = \langle U, Att, t, d^\alpha \rangle$, *where* $\forall x \in U, d^\alpha(x) = \{y \in V_t : d(x)_y \geq \alpha\}$. *For each set of attributes* $R \subseteq Att$, *denote by* $\mathcal{I}(R) \subseteq \mathcal{I}(P)$ *the collection of instantiations of* P *for which* R *is a reduct. Then,* $R \subseteq Att$:

- *Is an* α-possibilistic reduct *if it is a reduct for some instantiation of* P^α, *and an* α-MDL reduct *if it is a size-minimal* α-possibilistic reduct;
- *Is a* C-reduct *if it is a possibilistic reduct and* $\nexists R' \subseteq Att$ *s.t. both* $|R'| \leq |R|$ *and* $\exists I_1 \in sup_{\leq_C} \mathcal{I}(R), I_2 \in sup_{\leq_C} \mathcal{I}(R')$. $I_1 <_C I_2$[1];
- *Is a* λ-reduct, *with* $\lambda \in [0, 1]$, *if it is a possibilistic reduct and* $sup_{I \in \mathcal{I}(R)}(1 - \lambda)\mu_{\mathcal{I}(P)}(I) - \lambda \frac{|R|}{|Att|}$ *is maximal among all possibilistic reducts;*
- *Is a* D-reduct *if it is a possibilistic reduct and there is no* $R' \subseteq Att$ *s.t. both* $|R'| \leq |R|$ *and* $\exists I_1 \in sup_{\leq_D} \mathcal{I}(R), I_2 \in sup_{\leq_D} \mathcal{I}(R')$. $I_1 <_D I_2$;

---

[1] Here $sup_{\leq_C} \mathcal{I}(R) = \{I \in \mathcal{I}(R) : \nexists I' \in \mathcal{I}(R) \text{ s.t. } I <_C I'\}$.

Remarkably, the problem of finding C-reducts and λ-reducts can be (poly-nomially) reduced to the problem of finding the α-possibilistic reducts:

Theorem 1 ([3]). *The problem of finding all C-reducts (resp., λ-reducts, for any given value of λ) can be polynomially reduced to the problem of finding all α-MDL reducts (resp., α-possibilistic reducts), for all values of α. In particular, all the problems in the statement are in NP-HARD.*

Even though the reduct search problems in Theorem 1 are unlikely to be computationally feasible [4], a local search greedy algorithm whose runtime is $O(|U|^2|Att|^2)$ has been proposed to find approximated C-reducts or λ-reducts [3]. This latter approach, however, suffers from several limitations. First, it is only a local search algorithm, therefore it does not provide any guarantee about the quality of its results. Second, though polynomial, the complexity of this app-roach is quadratic in both the number of attributes and the number of objects. Consequently, it doesn't scale-well to big data or high-dimensional tasks.

3.2   Genetic Rough Set Selection

The definition of C-reducts, λ-reducts and D-reducts (see Def. 4) is intimately tied to the notion of an instantiation of a PDT. The complexity of finding a reduct for a PDT, therefore, could be understood as stemming from the large size of the search space of all such instantiations. In this section, we show how genetic algorithms can be used to effectively harness the structure of the above mentioned search space, by providing an efficient global search algorithm. In particular, we aim to show (as described also through the results shown in Sect. 4) that a simple global search strategy is sufficient to out-perform the local search methods previously proposed in the literature: for this reason, the approach we propose grounds on basic genetic operators, and does not employ more advanced strategies such as elitism or diversity control.

In the proposed approach, each candidate solution is represented as a pair $\langle I, F \rangle$, where $I \in V_t^{|U|}$ is a vector of decision labels s.t. $\forall x \in U, d(I_x) > 0$, and $F \in \{0, 1\}^{|Att|}$. Intuitively, I represents a candidate instantiation, while F is a corresponding candidate reduct: in particular if $F_a = 1$, then attribute a is included in the candidate reduct. We next define the adopted fitness functions, the mutation and crossover criteria, and the selection algorithm.

In regard to the fitness function, we consider three different functions, in order to take into account the differences among C-reducts, λ-reducts and D-reducts. Namely, the three fitness functions are defined as:

$$Fitness_C(\langle I, F \rangle) = \langle r, p \rangle, \tag{3}$$

$$Fitness_\lambda(\langle I, F \rangle) = (1 - \lambda)p - \lambda \frac{r}{|Att|}, \tag{4}$$

$$Fitness_D(\langle I, F \rangle) = \langle r, s \rangle, \tag{5}$$

where $p = \min_{x \in U} d(x)_{I_x}$, $r = \begin{cases} |F| & F \text{ is a super-reduct for } (U, Att, I) \\ \infty & \text{otherwise} \end{cases}$, and

$s \in [0,1]^{|U|}$ is a vector s.t. $s_x = d(x)_{I_x}$. Note, in particular, that only $Fitness_\lambda$ is single-valued, while the other two fitness functions are multi-valued. Consequently, for these latter two fitness functions we will consider an approach based on multi-objective optimization. In particular, given two candidate solutions $\langle I_1, F_1 \rangle$, $\langle I_2, F_2 \rangle$ we say that:

$$\langle I_1, F_1 \rangle \geq_C^F \langle I_2, F_2 \rangle \text{ iff } r_1 \leq r_2 \wedge p_1 \geq p_2, \tag{6}$$

$$\langle I_1, F_1 \rangle \geq_\lambda^F \langle I_2, F_2 \rangle \text{ iff } Fitness_\lambda(\langle I_1, F_1 \rangle) \geq Fitness_\lambda(\langle I_2, F_2 \rangle), \tag{7}$$

$$\langle I_1, F_1 \rangle \geq_D^F \langle I_2, F_2 \rangle \text{ iff } r_1 \leq r_2 \wedge \forall x \in U, s_x \geq s_x. \tag{8}$$

Given these definitions, selection is performed by non-dominated tournament selection [14], as described in Algorithm 1.

Algorithm 1. The selection algorithm.

---

procedure NON-DOMINATED TOURNAMENT SELECTION (P: population, t: tournament size, c: reduct type)

  $T \leftarrow t$ randomly selected candidate solutions from P
  $a \leftarrow$ randomly selected candidate solution in T
  for all $b \in T$ **do**
    if $b >_c^F a$ **then**
      $a \leftarrow b$
    end if
  end for
  return a ▷ A non-dominated candidate solution
end procedure

In regard to mutation, this is performed separately on the possibility degrees and on the candidate reducts. Specifically, in regard to candidate reducts, features are removed or added randomly according to a Bernoulli distribution with parameter b_{mut}. By contrast, possibility degrees are mutated according to a two step procedure: first, for each instance x, a binary value is randomly sampled from a Bernoulli distribution with parameter b_{mut}; then, if the above mentioned value was equal to 1, a new possibility degree is sampled from the probability distribution $\hat{Pr}_{d(x)}$, given by the possibility-probability transform [10] $\hat{Pr}_{d(x)}(y) = \int_0^{d(x)_y} \frac{d\alpha}{|\{y' \in V_t : d(x)_{y'} \geq \alpha\}|}$. In particular, we decided to adopt this sampling distribution as it is the maximally uncertain distribution among all possible probability distributions Pr compatible with $d(x)$, i.e., satisfying $\forall y \in V_t, Pr(y) \leq d(x)_y$ and $d(x)_y \geq d(x)_{y'} \implies Pr(y) \geq Pr(y')$, and hence has minimum bias [10]. The mutation algorithm is summarized in Algorithm 2.

Finally, single-point crossover is applied to I and F, separately. The complete pseudo-code for the proposed method is reported in Algorithm 3.

Algorithm 2. The mutation algorithm.

---

procedure MUTATION($\langle I, F \rangle$: candidate solution, b_{mut} : mutation probability)
    for all $a \in \{0, ..., |Att| - 1\}$ **do**
        if $Uniform(0, 1) \leq b_{mut}$ **then**
            $F_a \leftarrow 1 - F_a$
        end if
    end for
    for all $x \in \{0, ..., |U| - 1\}$ **do**
        if $Uniform(0, 1) \leq b_{mut}$ **then**
            $I_x \leftarrow$ random label sampled from $\hat{Pr}_{d(x)}$
        end if ·
    end for
    return $\langle I, F \rangle$ ▷ A new candidate solution
end procedure

Algorithm 3. The proposed weakly supervised genetic rough set feature selection algorithm.

---

procedure WEAKLY SUPERVISED GENETIC ROUGH SET SELECTION($\langle U, Att, d \rangle$:
PDT, $Popsize$: population size, b_{mut}, c: reduct type, t: tournament size)
    $Pop \leftarrow Popsize$ randomly initialized candidate solutions
    $Best \leftarrow \emptyset$
    while Not converged and termination criterion not reached **do**
        Compute fitness according to $Fitness_c$
        $Best \leftarrow$ the non-dominated candidate solutions in $Pop \cup Best$
        $NewPop \leftarrow \emptyset$
        for all $i = 1$ to $Popsize$ **do**
            $NewPop.append(Non - DominatedTournamentSelection(Pop, t, c))$
        end for
        $Pop \leftarrow NewPop$
        Apply $Crossover$ on Pop
        Apply $Mutate$ on Pop
    end while
    return $Best$ ▷ The best candidate solutions
end procedure

We next study the convergence and complexity properties of the proposed method. The following result shows that, asymptotically, Algorithm 3 is guaranteed to return all C-reducts (resp., λ-reducts, D-reducts).

Theorem 2. *Let n be the number of iterations for which Algorithm 3 runs before termination. Let P be a PDT. If $n \to \infty$, then almost surely $\exists R \in Best$ s.t. R is a C-reduct. Furthermore, $Best = C(P)$ almost surely, where $C(P)$ is the collection of C-reducts for PDT P. The same result holds for λ-reducts, D-reducts.*

Proof. We prove the result for C-reducts, as the case of λ-reducts and D-reducts is equivalent. A C-reduct corresponds, by definition, to a non-dominated candidate solution according to order $\leq_C^F$. Since at least one C-reduct is guaranteed to exist, $Pr(\exists R \in Best : R$ is a C-reduct$) > 0$. Since as, $n \to \infty$, Algorithm 3 is guaranteed to visit all the non-dominated candidate solutions in the search space (since, at each step, each of these candidate solution has non-zero probability of being added to the population), the result follows.

Thus, in the long run, the proposed method is guaranteed to find all the reducts of the desired class. This property provides an advantage w.r.t. the previously described local search methods that, by contrast, do not provide any such guarantee. Nonetheless, we note two limitations of the previous result: 1) the previous result only holds asymptotically, with no bounds on the expected number of iterations required to achieve convergence; 2) the previous results holds irrespective of the population size, thus, as a degenerate case, also for purely random search. While it is reasonable to expect that larger, or even adaptive, population size could improve speed of convergence, we leave the development of such results as open problem.

The computational complexity of Algorithm 3 can be characterized as follows:

Theorem 3. *Let n be the number of iterations for which Algorithm 3 runs before termination. Then, the complexity of Algorithm 3 is $O(n|U||Att|)$.*

Proof. The mutation and crossover steps have both complexity $O(|U|+|Att|)$ per iteration. The per-iteration complexity of the selection step is $O(|U|)$. The per-iteration complexity required for computing the fitness of the candidate solutions in the population is $O(|U||Att|)$. Therefore the result follows.

The previous theorem ensures that, as long as the number of iterations is $o(|U||Att|)$, the proposed method has better computational complexity than the local search methods described in [3]. We leave as open problem the definition of algorithm to automatically tune the number of iterations, based on the available data, so as to guarantee quick convergence with high probability.

4 Experiments and Results

In this section, we discuss the experiments that we designed to evaluate the proposed method, in comparison with other feature selection methods for the learning from fuzzy labels problem, and present and discuss the obtained results.

4.1 Experimental Design

In order to evaluate the proposed genetic algorithm-based feature selection methods we considered a benchmark suite encompassing 14 different datasets:

- Two fuzzy-labeled datasets, previously described, respectively, in [2] and [5];
- 12 datasets from the UCI repository. The precise labels for these datasets were fuzzified by means of nearest neighbors label smoothing [13]. In particular,

setting the number of neighbors equal to k, for each instance x, the associated possibility distribution is obtained as:

$$d(x)_y = \frac{|\{i \in \{1, \ldots, k\} : N_i \text{ nearest neighbor of } x \wedge t(N_i) = y\}| + 1_{t(x)=y}}{k+1}$$

The full list of datasets, including he number of instances, features and classes, is reported in Table 1.

Table 1. List of datasets

Dataset	Instances	Features	Classes	Dataset	Instances	Features	Classes
Avila	20768	10	10	Myocardial	1700	111	2
Car	864	16	4	Pen	10992	16	10
Crowd	10845	28	6	Sensorless	20000	48	11
Frog family	7195	22	4	Taiwan	6819	94	2
HCV	582	12	4	Wifi	2000	7	4
Iranian	7032	45	2	CTC	617	2500	2
Mushroom	5644	99	6	Kyphosis	120	14	7

We considered the following 3 feature selection algorithms:

- The proposed Rough Set-based genetic algorithms, considering the three fitness functions λ, C, D, denoted respectively as $GRSSL, GRSSC, GRSSD$. We selected, in particular, a budget of $n = 1000$ iterations;
- The greedy Rough Set-based local search algorithms for λ-reducts, C-reducts and D-reducts, denoted respectively as $RSSL, RSSC, RSSD$;
- The DELIN algorithm [19], as a comparison baseline. This latter is a dimensionality reduction algorithm based on linear discriminant analysis, whose runtime is $O(|U||Att|^2)$

Performance was evaluated by means of the following experimental design:

1. For each dataset, split in training set Tr (70%) and test set Ts (30%);
2. Apply the feature selection algorithms on the training set Tr, obtaining a reduct F and the reduced training set Tr_F;
3. Train a kNN classifier on the reduced training set Tr_F;
4. Evaluate the trained kNN classifier on the reduced test set Ts_F.

In regard to performance measures, we measured the balanced accuracy, so as to take into account the label imbalance in the considered datasets. The algorithms were also compared in terms of running time. Differences among algorithms (if any) were analyzed by means of a statistical testing approach. In particular, we applied the Friedman rank test to evaluate whether some global statistically significant difference existed among the considered algorithms, and then applied the post-hoc Nemenyi test for pair-wise comparisons of performance. In both cases, p-values smaller than 0.05 were considered to be significant evidence of performance difference (at a confidence level of 95%).

4.2   Results and Discussion

The results of the experiments are reported in Fig. 1 and 2; both in terms of average ranks and p-values for the post-hoc test. As shown in Fig. 1, the three proposed genetic algorithms reported the best average ranks w.r.t. balanced accuracy, and were significantly more accurate than the Rough Set-based local search algorithm. In particular, the GRSSL algorithm (that is, the genetic rough set selection for λ-reducts) was the feature selection algorithm with best performance, being significantly more accurate than all other algorithms. By contrast, while significantly better than the Rough Set-based local search methods, GRSSC and GRSSD were better than the baseline DELIN only on average. We conjecture that this could be due to the intrinsic complexity of the underlying multi-objective optimization problems: indeed, for example, GRSSD involves the solution of a $|U|+1$-dimensional problem. Future research should thus be devoted at exploring more advanced techniques to address this aspect.

Also in terms of running time, the proposed genetic algorithms compare favorably with the Rough Set-based local search methods, as shown in Fig. 2. We note, though, that DELIN had better run-time than all other considered algorithms, being significantly more efficient than GRSSD, while the difference w.r.t. GRSSL and GRSSC was not significant. This could be explained by two different reasons. First, DELIN only uses matrix operations in its execution, that can be performed very efficiently through numerical linear algebra libraries; by contrast Rough Set-based algorithms also include table manipulation operations, having higher computational costs. Second, the complexity of DELIN is $O(|U||Att|^2)$, while for the genetic algorithms the complexity is $O(n|U||Att|)$, with the iterations' budget set to $n = 1000$ iterations: for all datasets, except CTC, n was much greater than $|Att|$, thus the effective complexity of GRSSL, GRSSC and GRSSD was $o(|U||Att|^2)$. We remark, however that, despite this difference, the proposed

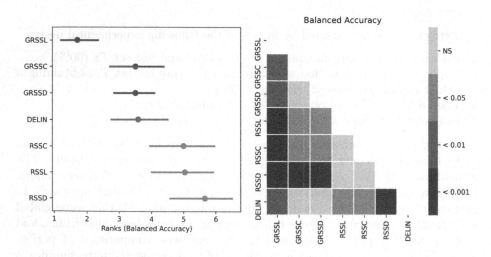

Fig. 1. Average ranks and p-values for balanced accuracy.

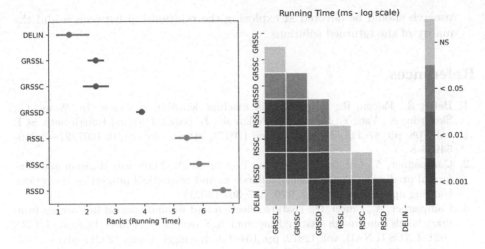

Fig. 2. Average ranks and p-values for running time.

genetic algorithms were significantly more accurate. Nonetheless, future work should be devoted at exploring algorithmic strategies to automatically control the number of iterations n, based on the available data.

5   Conclusion

In this article, we studied the problem of feature selection in the learning from fuzzy label setting, and proposed a method, combining genetic algorithms and Rough Set theory, for efficiently solving this problem. We studied the computational properties of the proposed method and showed its effectiveness in comparison to existing feature selection methods, on a comprehensive set of benchmarks. While this paper provides a promising direction for the application of RST-based feature selection in weakly supervised learning, it naturally leaves many questions open. Specifically, we plan to address the following problems:

- In the proposed method, we did not apply any advanced multi-objective optimization techniques, such as diversity control, elitism or Pareto strength assignment [14]. Future work should evaluate the potential benefits of including such techniques in the proposed method;
- In Theorem 2 we showed that the proposed method is a global search strategy: asymptotically, the best solutions found by the genetic algorithm are exactly the desired reducts. Further research should study non-asymptotic characterizations of the proposed method, in terms of expected time to convergence, or PAC population bounds [11];
- Similarly, as long as the number of iterations n is constant or upper-bounded by the size of the PDT P, the complexity of the proposed algorithm is particularly favourable w.r.t. the standard Rough Set greedy algorithm. Further

research should be devoted at exploring the relationship between n and the quality of the returned solutions.

References

1. Bello, R., Falcon, R.: Rough sets in machine learning: a review. In: Wang, G., Skowron, A., Yao, Y., Ślęzak, D., Polkowski, L. (eds.) Thriving Rough Sets. SCI, vol. 708, pp. 87–118. Springer, Cham (2017). https://doi.org/10.1007/978-3-319-54966-8_5

2. Campagner, A., Cabitza, F., Berjano, P., Ciucci, D.: Three-way decision and conformal prediction: isomorphisms, differences and theoretical properties of cautious learning approaches. Inf. Sci. **579**, 347–367 (2021)

3. Campagner, A., Ciucci, D.: Feature selection and disambiguation in learning from fuzzy labels using rough sets. In: Ramanna, S., Cornelis, C., Ciucci, D. (eds.) IJCRS 2021. LNCS (LNAI), vol. 12872, pp. 164–179. Springer, Cham (2021). https://doi.org/10.1007/978-3-030-87334-9_14

4. Campagner, A., Ciucci, D., Hüllermeier, E.: Rough set-based feature selection for weakly labeled data. Int. J. Approx. Reasoning **136**, 150–167 (2021)

5. Campagner, A., Ciucci, D., Svensson, C.M., Figge, M.T., Cabitza, F.: Ground truthing from multi-rater labeling with three-way decision and possibility theory. Inf. Sci. **545**, 771–790 (2020)

6. Ciucci, D., Forcati, I.: Certainty-based rough sets. In: Polkowski, L., Yao, Y., Artiemjew, P., Ciucci, D., Liu, D., Ślęzak, D., Zielosko, B. (eds.) IJCRS 2017. LNCS (LNAI), vol. 10314, pp. 43–55. Springer, Cham (2017). https://doi.org/10.1007/978-3-319-60840-2_3

7. Côme, E., Oukhellou, L., Denoeux, T., Aknin, P.: Learning from partially supervised data using mixture models and belief functions. Pattern Recogn. **42**(3), 334–348 (2009)

8. Couso, I., Dubois, D., Sánchez, L.: Random Sets and Random Fuzzy Sets as ill-perceived Random Variables. SpringerBriefs in Computational Intelligence (2014)

9. Denœux, T., Zouhal, L.M.: Handling possibilistic labels in pattern classification using evidential reasoning. Fuzzy Sets Syst. **122**(3), 409–424 (2001)

10. Dubois, D., Prade, H., Sandri, S.: On possibility/probability transformations. In: Fuzzy Logic, pp.103–112. Springer (1993)

11. Hernández-Aguirre, A., Buckles, B.P., Martínez-Alcántara, A.: The probably approximately correct (PAC) population size of a genetic algorithm. In: Proceedings of ICTAI 2000, pp. 199–202. IEEE (2000)

12. Hüllermeier, E.: Learning from imprecise and fuzzy observations: data disambiguation through generalized loss minimization. Int. J. Approx. Reasoning **55**(7), 1519–1534 (2014)

13. Lukasik, M., Bhojanapalli, S., Menon, A., Kumar, S.: Does label smoothing mitigate label noise? In: ICML, pp. 6448–6458. PMLR (2020)

14. Luke, S.: Essentials of Metaheuristics. Lulu, 2nd (edn.) (2013)

15. Nakata, M., Sakai, H.: Rule induction based on rough sets from possibilistic data tables. In: Seki, H., Nguyen, C.H., Huynh, V.-N., Inuiguchi, M. (eds.) IUKM 2019. LNCS (LNAI), vol. 11471, pp. 86–97. Springer, Cham (2019). https://doi.org/10.1007/978-3-030-14815-7_8

16. Pawlak, Z.: Rough sets. Int. J. Comput. Inf. Sci. **11**(5), 341–356 (1982)

17. Quost, B., Denoeux, T.: Clustering and classification of fuzzy data using the fuzzy em algorithm. Fuzzy Sets Syst. **286**, 134–156 (2016)
18. Sakai, H., Wu, M., Nakata, M.: Apriori-based rule generation in incomplete information databases and non-deterministic information systems. Fundamenta Informaticae **130**(3), 343–376 (2014)
19. Wu, J.-H., Zhang, M.-L.: Disambiguation enabled linear discriminant analysis for partial label dimensionality reduction. In: Proceedings of the 25th ACM SIGKDD, pp. 416–424 (2019)
20. Zadeh, L.A.: Fuzzy sets as a basis for a theory of possibility. Fuzzy Sets Syst. **1**(1), 3–28 (1978)

Choosing the Decision Hyper-parameter for Some Cautious Classifiers

Abdelhak Imoussaten[✉]

EuroMov Digital Health in Motion, Univ Montpellier, IMT Mines Ales, Ales, France
abdelhak.imoussaten@mines-ales.fr

Abstract. In some sensitive domains where data imperfections are present, standard classification techniques reach their limits. To avoid misclassification that has serious consequences, recent works propose cautious classification algorithms to handle the problem. Despite of the presence of uncertainty, a point prediction classifier is forced to decide which single class to associate to a sample. In such a case, a cautious classifier proposes the appropriate subset of candidate classes that can be assigned to the sample in the presence of imperfect information. On the other hand, cautiousness should not override relevance and a trade-off has to be made between these two criteria. Among the existing cautious classifiers, two classifiers propose to manage this trade-off in the decision step of the classifier algorithm by the mean of a parametrized objective function. The first one is the non-deterministic classifier (ndc) proposed within the framework of probability theory and the second one is eclair (evidential classifier based on imprecise relabelling) proposed within the framework of belief functions. The theoretical aim of the mentioned parameter is to control the size of predictions for both classifiers. This paper proposes to study this parameter in order to select the "best" value in a classification task. First the gain for each prediction candidate is studied related to the values of the hyper-parameter. In the illustration section, we propose a method to choose this hyper-parameter base on the training data and we show the classification results on randomly generated data and we present some comparisons with two other imprecise classifiers on 11 UCI datasets based on five measures of imprecise classification performances used in the state of the art.

Keywords: Cautious classification · Imprecise classification · Belief functions · Supervised machine learning

1 Introduction

In some sensitive applications misclassification can have serious consequences. This is the case in applications having impacts either on people's health or on the environment [6], e.g., in medical diagnosis applications when a classifier is involved to detect early-stage cancer. In such applications cautiousness is necessary when imperfect data are present. This leads some recent works to

D. Ciucci et al. (Eds.): IPMU 2022, CCIS 1602, pp. 774–787, 2022.
https://doi.org/10.1007/978-3-031-08974-9_61

focus on cautious classification. Among the existing cautious classifiers, we focus, in this paper, on those providing a subset of candidate class labels to a new sample to classify and we called them *imprecise classifiers*. Some of them, as the non-deterministic classifier (*ndc*) [3], use the posterior probability when it is known and provide the subset of classes, that minimize/maximize a risk/utility function, as prediction (see Subsect. 2.2 for more details). Other approaches, as the *Naive Credal Classifier* (*ncc*) [12,13] proposed in the framework of imprecise probability, are based on a dominance relation defined on the set of classes using the *credal* set representing the imprecision and uncertainty about the true class label of a sample. Then the subset of the non-dominated classes is considered as the prediction for the sample. The imprecise classifiers proposed within the framework of belief functions utilise the mass function when it is known and a decision procedure. In [7], it is proposed to generalize the utility matrix to the subsets of classes by aggregating the single utilities that are considered as known. The approach in [8] uses the interval dominance approach where the intervals are represented by the values of belief and plausibility functions obtained of each class. In [4,5], the evidential classifier based on imprecise relabelling (*eclair*) uses a generalisation of the gain function proposed in [3] to the case of belief functions framework. An imprecise classifier proposes the appropriate subset of candidate classes that can be assigned to the sample in the presence of imperfect information. But cautiousness should not override relevance and a trade-off has to be made between these two criteria. On one hand, a classifier that predicts always the whole set of the candidate classes is cautious but its predictions are not relevant. On the other hand, a classifier that predicts always a single class for difficult samples is relevant when the prediction is good but it is not cautious. Most of imprecise classifiers cannot control this trade-off except *ndc* and *eclair*. Indeed, the gain function implemented in the decision step of both classifiers *ndc* and *eclair* has an hyper parameter β that is used to control the trade-off between relevance and cautiousness. This hyper parameter is considered as a user-modifiable parameter for the use of these two classifiers and its theoretical aim is to control the size of the predicted subset of classes. The choice of β depends on the level of cautiousness required in the application in which the classifier is going to be used. This paper proposes to study this parameter in the case of the two classifiers and aims to propose a suggestion for the choice of the parameter value in the case of classification task. In the first experiment results, we show, on simulated data, the impact of the selected parameter value on the prediction of the two classifiers when faced to difficult samples, i.e., to which the standard classifiers failed to predict the true class labels. While in the second experiment part, we present some comparison of the *ndc* classifier tuned using our proposition with other imprecise classifiers of the state of the art conduct on 11 UCI data and based on five measures from the state of the art that are usually used to compare imprecise classification performances. The paper is organised as follows. In the second section, the reminders about the decision step in the classifiers *eclair* and *ndc* and the measures of imprecise classification performances are given. The third section presents a study of the expected gain function introduced in the decision step of the two classifiers. Finally, the fourth section presents the experiment results.

2   Reminders and Notations

The imprecise classifiers *eclair* and *ndc* are based on the results of the standard point prediction classifiers to provide respectively the posterior mass function and the posterior probability function for a sample to classify. We focus in this paper on the decision step of those two classifiers that involves these two functions and a gain function that is the F_β score. In this section we give some reminders about the F_β score and it exploitation in the case of imprecise predictions by the two classifiers. Finally, five measures from the state of the art used to evaluate the imprecise classification performances are presented. To simplify notations, we adopt the following notations for the subsets in the rest of the paper: $\theta_i := \{\theta_i\}$, $\theta_{ij} := \{\theta_i, \theta_j\}$.

2.1 F_β Measure

The F_β score used in the decision step of *eclair* and *ndc* to predict a subset of candidate classes is an adaptation of the F_β score introduced in information retrieval and classification to imprecise classification. In the context of binary point prediction for classification, the F_β score is defined as:

$$F_\beta = \frac{(1 + \beta^2)\ \text{recall} \cdot \text{precision}}{(\beta^2 \cdot \text{precision}) + \text{recall}}, \tag{1}$$

where precision $= \frac{true\ positive}{true\ positive + false\ positive}$ and recall $= \frac{true\ positive}{true\ positive + false\ negative}$ are two known performance measures in information retrieval and machine learning.

2.2   The Decision Step in *ndc*

The principle of *ndc* is very simple, a posterior probability is determined using a classification method for point prediction and then a decision rule is applied to determine the imprecise prediction. This subsection presents the decision rule applied in the decision step. The decision step with *ndc* consists in providing for a sample $\mathbf{x}$ a subset of classes as prediction, i.e., precise predictions are given as singletons, by considering as input the posterior probability $p(.|\mathbf{x})$. The predicted subset of classes is the one maximizing the expected gain where the gain associated to each subset of classes is defined using the F_β measure. More precisely, let us consider a set of n class labels $\Theta = \{\theta_1, \ldots, \theta_n\}$. Each subset of candidate classes $A \subseteq \Theta$ is evaluated as the good prediction for $\mathbf{x}$ using the F_β measure as follows:

$$F_\beta(A, \mathbf{x}, \theta) = \frac{(1 + \beta^2) \cdot \mathbb{1}_A(\theta)}{\beta^2 + |A|}. \tag{2}$$

The quantity $F_\beta(A, \mathbf{x}, \theta)$ is interpreted as the gain obtained when predicting the subset of class labels A for the sample $\mathbf{x}$ when its true class label is θ. The Formula in (2) is analogue to the one in (1) where the quantities *precision* and *recall* are redefined as precision$(A) = \frac{\mathbb{1}_A(\theta)}{nb\ of\ classes\ in\ A}$ and recall$(A) = \mathbb{1}_A(\theta)$ but do not have the same meaning. Indeed, in (1) the case of binary classification, the two measures are quantified related to a data test set while in the case of imprecise classification the two measures are quantified related to a subset of classes that is a potential prediction. We can note that when the values of β are close to 0, $F_\beta(A, \mathbf{x}, \theta)$ becomes close to precision(A) thus the size of A is disadvantageous. On the other hand, when β is high, $F_\beta(A, \mathbf{x}, \theta)$ becomes close to recall(A) and in this case the size of A is an advantage. Let us suppose that a posterior probability distribution $p(\cdot|\mathbf{x})$ is known for the sample $\mathbf{x}$, the non-deterministic classifier ndc predicts for $\mathbf{x}$ the subset of candidate classes that maximize the expected gain function $u_\beta((\cdot, \cdot), p(\cdot|\mathbf{x}))$ defined as:

$$u_\beta(A, p(\cdot|\mathbf{x})) = \sum_{i=1}^{n} F_\beta(A, \mathbf{x}, \theta_i) \cdot p(\theta_i|\mathbf{x}). \tag{3}$$

Finally, the predicted subset $\hat{\delta}_{ndc}(\mathbf{x})$ for $\mathbf{x}$ using the classifier ndc is given as:

$$\delta_{ndc}(\mathbf{x}) = arg\max_{A \subseteq \Theta} u_\beta(A, p(\cdot|\mathbf{x})). \tag{4}$$

2.3 The Decision Step in éclair

The decision step with *éclair* consists in providing for a sample $\mathbf{x}$ a subset of classes as prediction, by considering as input the posterior mass function $m(\cdot|\mathbf{x})$. The predicted subset of classes is the one maximizing the expected gain where the gain associated to each subset of classes is defined using a generalisation of the formula (2) [4,5]. The main change is to consider the general case where the available information about the true class of a sample can be partial in the form of a subset $B \subseteq \Theta$. It is the case, for example, when data are coarse [2,9]. This leads to the new gain function defined as follows:

$$F_\beta(A, \mathbf{x}, B) = \frac{(1 + \beta^2) \cdot |A \cup B|}{\beta^2 \cdot |B| + |A|} \tag{5}$$

The quantity $F_\beta(A, \mathbf{x}, B)$ is interpreted as the gain obtained when predicting the subset of class labels A for the sample $\mathbf{x}$ when its true class label is partially known and represented by a subset of classes B. In this case, the precision and recall analogue quantities of ones presented in (1) become: precision$(A) = \frac{|A \cap B|}{|A|}$ and recall$(A) = \frac{|A \cap B|}{nb\ of\ classes\ in\ B}$.

Let us suppose that a posterior mass function $m(\cdot|\mathbf{x})$ is known for the sample $\mathbf{x}$, the *éclair* classifier predicts for $\mathbf{x}$ the subset of candidate classes that maximize the expected gain function $u_\beta((\cdot, \cdot), m(\cdot|\mathbf{x}))$ defined as:

$$u_\beta(A, m(\cdot|\mathbf{x})) = \sum_{B \subseteq \Theta} F_\beta(A, \mathbf{x}, B) \cdot m(B|\mathbf{x}) \tag{6}$$

Finally, the predicted subset $\hat{\delta}_{eclair}(\mathbf{x})$ for $\mathbf{x}$ using the classifier *eclair* is given as:

$$\hat{\delta}_{eclair}(\mathbf{x}) = \arg\max_{A \subseteq \Theta} u_g(A, m(\cdot|\mathbf{x})). \tag{7}$$

2.4 Evaluation Measures for the Imprecise Classifiers

When evaluating an imprecise classifier one ensures that the predicted subset of classes 1) include the "true" class and 2) they are as small as possible depending on the sample data imperfection. Several works have studied this problem and provide some measures to check the two conditions 1) and 2) [1,11,12]. Between the least drastic one that is *imprecise accuracy* which checks if the prediction contains the true class label of the sample and the most drastic one that is *classical accuracy* which checks if the prediction is equal to the true class label of the sample, one can find intermediate measure as *Discounted accuracy* [10] that seems to be an interesting measure as it takes into account the size of the predicted subset. But in order to increase the cautiousness reward to the degree to which the decision maker prefers to fix it depending on his application and the quality of the information obtained for the samples, a family of measure are constructed from *Discounted accuracy* measure that are represented by a function g taking its values in $[0,1]$ and guaranteeing $g(z) \geq z$, i.e., the reward with g is at least the same as the one given by the *discounted accuracy*, $g(0) = 0$ and $g(1) = 1$ (see [14] for more details).

Let us consider a dataset of test samples $dst = (\boldsymbol{x}^l, \theta^l)_{1 \leq l \leq M}$ where $\boldsymbol{x}^l \in \mathcal{X}$ and $\theta^l \in \Theta$ and an imprecise classifier $\hat{\delta}_{ic}$. The five following measures are proposed to evaluate the performance of imprecise classification and applied to the classifier $\hat{\delta}_{ic}$ and the test data dst:

- the *classical accuracy*:

$$accuracy(\hat{\delta}_{ic}, dst) = \frac{1}{M} \sum_{l=1}^{M} \mathbb{1}_{\{\theta^l\}}(\hat{\delta}_{ic}(\boldsymbol{x}^l)).$$

- the *imprecise accuracy* (imprAcc):

$$imprAcc(\hat{\delta}_{ic}, dst) = \frac{1}{M} \sum_{l=1}^{M} \mathbb{1}_{\hat{\delta}_{ic}(\boldsymbol{x}^l)}(\theta^l).$$

- the *discounted accuracy* (discAcc) corresponds to the function $g(z) = z$ [10]:

$$discAcc(\hat{\delta}_{ic}, dst) = \frac{1}{M} \sum_{l=1}^{M} \frac{\mathbb{1}_{\hat{\delta}_{ic}(\boldsymbol{x}^l)}(\theta^l)}{|\hat{\delta}_{ic}(\boldsymbol{x}^l)|},$$

where $|A|$ denotes the size of the subset A. This measure is also denoted u_{50}.

- The u_{65} measure that corresponds to the function $g(z) = -0.6 \cdot z^2 + 1.6 \cdot z$ [14]:

$$u_{65}(\hat{\delta}_{ic}, dst) = -0.6 \cdot [discAcc(\hat{\delta}_{ic}, dst)]^2 + 1.6 \cdot discAcc(\hat{\delta}_{ic}, dst).$$

- The u_{80} measure that corresponds to the function $g(z) = -1.2 \cdot z^2 + 2.2 \cdot z$ [14]:

$$u_{80}(\hat{\delta}_{ic}, dst) = -1.2 \cdot [discAcc(\hat{\delta}_{ic}, dst)]^2 + 2.2 \cdot discAcc(\hat{\delta}_{ic}, dst).$$

3 The Expected Gains Related to β

3.1 The Case of Ndc

Let us consider that the posterior probability distribution of a sample x is known. We denote this distribution by $p(\cdot|x) : \Theta \longrightarrow [0,1]$. We consider the parameter β as a variable and we express the expected gain function in Subsect. 2.2 for a $\beta \in [0, +\infty]$, $A \subseteq \Theta$ and $p(\cdot|x)$ as:

$$u(\beta, A, p(\cdot|x)) = \sum_{i=1}^{n} F_\beta(A, x, \theta_i) \cdot p(\theta_i|x) \qquad (8)$$

In addition, let us consider the situation where the class θ_i is the most likely class of x and some times the class θ_i is confused with the class θ_j, $j \neq i$ due to data imperfection. The Propositions 1 and 2 give some results concerning the predicted subset of classes for x from the three options θ_i, θ_{ij} and θ.

Proposition 1. Let us suppose that $p(\theta_i|x) > p(\theta_i|x)$, $\forall \theta \in \Theta \setminus \theta_i$.
If $p(\theta_j|x) > 0$ then it exists $\beta_1 \geq 0$ such that:

$$\left\{\begin{array}{ll} u(\beta, \theta_{ij}, p(\cdot|x)) \gtrsim u(\beta, \theta_i, p(\cdot|x)) & \text{if } \beta \lesssim \beta_1 \\ u(\beta, \theta_{ij}, p(\cdot|x)) < u(\beta, \theta_i, p(\cdot|x)) & \text{if } \beta > \beta_1. \end{array}\right. \qquad (6)$$

Elsewhere $u(\beta, \Theta, p(\cdot|x)) < u(\beta, \theta_{ij}, p(\cdot|x))$, $\forall \beta \geq 0$.

Proof. We have for all $\beta \geq 0$,

$$u(\beta, \theta_i, p(\cdot|x)) = p(\theta_i|x).$$

and

$$u(\beta, \theta_{ij}, p(\cdot|x)) = \frac{1 + \beta_2^2}{2 + \beta_2^2} \cdot [p(\theta_i|x) + p(\theta_j|x)].$$

On the one hand, the function $u(\cdot, \theta_{ij}, p(\cdot|x))$ increases related to β. Thus $u(\beta, \theta_{ij}, p(\cdot|x)) \gtrsim \frac{1}{2}(p(\theta_i|x) + p(\theta_j|x))$, for all $\beta \geq 0$. On the other hand, $p(\theta_i|x) > p(\theta_j|x)$ then $p(\theta_i|x) > \frac{1}{2}(p(\theta_i|x) + p(\theta_j|x))$. So, $u(\cdot, \theta_{ij}, p(\cdot|x))$ intersects $u(\cdot, \theta_i, p(\cdot|x))$ at $\beta_1 \geq 0$ such that:

$$\frac{1 + \beta_2^2}{2 + \beta_2^2} \cdot [p(\theta_i|x) + p(\theta_j|x)] = p(\theta_i|x).$$

It comes:

$$\beta_1 = \sqrt{\frac{p(\theta_i|x) - p(\theta_j|x)}{p(\theta_j|x)}}.$$

Proposition 2. Let suppose that $p(\theta_i|x) > p(\theta|x)$, $\forall \theta \in \Theta \setminus \theta_i$. If $\mathbb{P}(\theta_{i,j}|x) \in [\frac{2}{3}, 1[$ then it exists $\beta_2 > 0$ such that:

$$\begin{cases} u(\beta,\Theta,p(\cdot|x)) \gtrless u(\beta,\theta_{i,j},p(\cdot|x)) & \text{if } \beta \lessgtr \beta_2 \\ u(\beta,\Theta,p(\cdot|x)) > u(\beta,\theta_{i,j},p(\cdot|x)) & \text{if } \beta > \beta_2. \end{cases} \tag{10}$$

Proof. We have for all $\beta \geq 0$,

$$u(\beta,\Theta,p(\cdot|x)) = \frac{1+\beta^2}{3+\beta^2},$$

and

$$u(\beta,\Theta,p(\cdot|x)) - u(\beta,\theta_{i,j},p(\cdot|x)) = \frac{(1+\beta^2)\cdot(2-3\cdot\mathbb{P}(\theta_{i,j})+(1-\mathbb{P}(\theta_{i,j})\cdot\beta^2))}{(3+\beta^2)\cdot(2+\beta^2)}$$

where $\mathbb{P}(\theta_{i,j}|x) = p(\theta_i|x) + p(\theta_j|x)]$. If $\mathbb{P}(\theta_{i,j}|x) < \frac{2}{3}$, then $u(\beta,\Theta,p(\cdot|x)) = \frac{1+\beta^2}{3+\beta^2} > u(\beta,\theta_{i,j},p(\cdot|x))$, $\forall\beta \geq 0$. Else, if $\mathbb{P}(\theta_{i,j}|x) = 1$, then $u(\beta,\Theta,p(\cdot|x)) = \frac{1+\beta^2}{2+\beta^2} = u(\beta,\theta_{i,j},p(\cdot|x))$, $\forall\beta \geq 0$. Otherwise, let us consider the following value $\beta* \geq 0$ such that:

$$\beta*^2 = \frac{3\,\mathbb{P}(\theta_{i,j}|x) - 2}{1 - \mathbb{P}(\theta_{i,j}|x)}.$$

We can set

$$\beta_2 = \sqrt{\beta*^2}.$$

∎

Example 1. Let us consider the following examples of four samples that obtain the posterior probabilities given in Fig. 1. These distribution express several situation of sharing the masses between the three classes. For the first sample x_1 the mass is uniformly distributed on the classes; for x_2 the mass is totally given to the class θ_1; for x_3 the mass is uniformly distributed to θ_1 and θ_2; and for x_4 the mass distribution is as follows $p(\theta_3|x_4) < p(\theta_1|x_4) < p(\theta_2|x_4)$. As one can see in Fig. 1, for the samples x_1, x_2 and x_3, Θ, θ_1, and $\theta_{1,2}$ are respectively the predictions as they maximize the expected gain regardless the value of β. In the case of x_4, the prediction depends on the value of the parameter β. Indeed, if $\beta < \beta_1 = \sqrt{\frac{p(\theta_1|x_4)}{p(\theta_2|x_4)-p(\theta_1|x_4)}} = 0.5$, i.e., the value of β where the curves of $u(.,\theta_1,x_4)$ and $u(.,\theta_{1,2},x_4)$ intersect, then θ_1 dominates all the other options. When $\beta_2 > \beta > \beta_1$ ($\beta_2 = \sqrt{\frac{3\,\mathbb{P}(\theta_{1,2}|x)-2}{1-\mathbb{P}(\theta_{1,2}|x)}} = 2.65$), then $\theta_{1,2}$ dominates all the other options. When $\beta \geq \beta_2$, it is the turn of Θ to dominate the other options.

3.2 The Case of Eclair

In this subsection, we consider that the posterior mass function of a sample x is known. We denote this mass function by $m(\cdot|x) : 2^\Theta \rightarrow [0,1]$. In this case, the

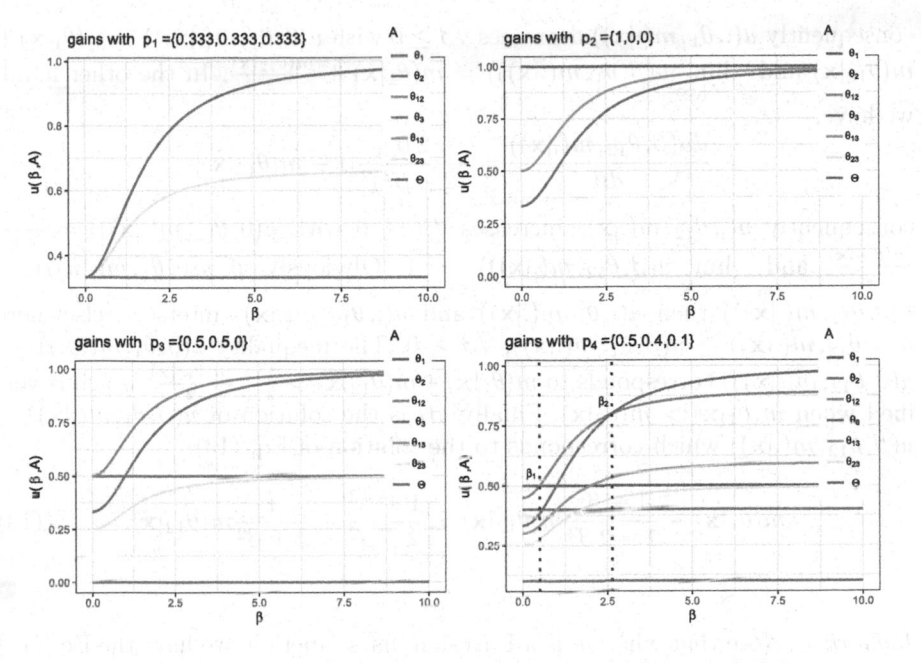

Fig. 1. The expected gain associated to the four posterior probabilities.

expected gain function used as the criterion to choose the subset of classes to associate to **x** is the following:

$$u(\beta, A, m(.|\mathbf{x})) = \sum_{B \subseteq \Theta} F_\beta(A, \mathbf{x}, B) \cdot m(B|\mathbf{x}) \tag{11}$$

The general multi-class case is complicate to treat directly. In this section, we present only the case of two classes. Consequently, the multi-class case can be treated using one-against-one prediction and then infer the final prediction by merging all the one-against-one predictions.

Proposition 3. *Let us consider the case where* $\Theta = \{\theta_1, \theta_2\}$. *If* $m(\theta_1|\boldsymbol{x}) > m(\theta_2|\boldsymbol{x})$, *then it exists* $\beta_3 \geq 0$ *such that:*

$$\begin{cases} u(\beta, \theta_{12}, m(.|\boldsymbol{x})) \leq u(\beta, \theta_1, m(.|\boldsymbol{x})) & \text{if } \beta \leq \beta_3 \\ u(\beta, \theta_{12}, m(.|\boldsymbol{x})) > u(\beta, \theta_1, m(.|\boldsymbol{x})) & \text{if } \beta > \beta_3 \end{cases} \tag{12}$$

Elsewhere, $u(\beta, \theta_{12}, m(.|\boldsymbol{x})) \geq u(\beta, \theta_1, m(.|\boldsymbol{x})), \forall \beta \geq 0.$

Proof. In one hand, we have,

$$\frac{du(\beta, \theta_1, m(.|\mathbf{x}))}{d\beta} = -\frac{2\beta}{(1 + 2\beta^2)^2} m(\theta_{12}|\mathbf{x})$$

consequently $u(., \theta_1, m(.|\mathbf{x}))$ decreases $\forall \beta \gtrsim 0$ with $u(0, \theta_1, m(.|\mathbf{x})) = m(\theta_1|\mathbf{x}) +$
$\frac{m(\theta_{12}|\mathbf{x})}{2}$ and $\lim\limits_{\beta \to +\infty} u(\beta, \theta_1, m(.|\mathbf{x})) = m(\theta_1|\mathbf{x}) + \frac{m(\theta_{12}|\mathbf{x})}{2}$. In the other hand,
we have,

$$\frac{du(\beta, \theta_{12}, m(.|\mathbf{x}))}{d\beta} = \frac{2\beta}{(2 + \beta_2)^2} [1 - m(\theta_{12}|\mathbf{x})]$$

consequently $u(., \theta_{12}, m(.|\mathbf{x}))$ increases $\forall \beta \gtrsim 0$ with $u(0, \theta_{12}, m(.|\mathbf{x})) = \frac{1}{2} + \frac{m(\theta_{12}|\mathbf{x})}{2}$ and $\lim\limits_{\beta \to +\infty} u(\beta, \theta_{12}, m(.|\mathbf{x})) = 1$. Obviously, if $u(0, \theta_1, m(.|\mathbf{x})) >$
$u(0, \theta_{12}, m(.|\mathbf{x}))$ then $u(., \theta_1, m(.|\mathbf{x}))$ and $u(., \theta_{12}, m(.|\mathbf{x}))$ intersect, elsewhere
$u(\beta, \theta_{12}, m(.|\mathbf{x})) \gtrsim u(\beta, \theta_1, m(.|\mathbf{x})), \forall \beta \gtrsim 0$. The inequality $u(0, \theta_1, m(.|\mathbf{x})) >$
$u(0, \theta_{12}, m(.|\mathbf{x}))$ corresponds to $m(\theta_1|\mathbf{x}) + m(\theta_{12}|\mathbf{x}) < \frac{1}{2} + \frac{m(\theta_{12}|\mathbf{x})}{2}$ which is ver-
ified when $m(\theta_1|\mathbf{x}) > m(\theta_2|\mathbf{x})$. Finally, β_3 is the solution of $u(\beta, \theta_1, m(.|\mathbf{x})) =$
$u(\beta, \theta_{12}, m(.|\mathbf{x}))$ which corresponds to the solution of Eq. (13):

$$m(\theta_1|\mathbf{x}) + \frac{1 + \beta_2}{1 + 2\beta_2} m(\theta_{12}|\mathbf{x}) = \frac{1 + \beta_2}{2 + \beta_2} m(\theta_{12}|\mathbf{x}) + \frac{1}{2 + \beta_2}. \tag{13}$$

■

Remark 1. Note that when m is a Bayesian mass function, we have the Eq. (13)
giving β_3 that becomes: $m(\theta_1|\mathbf{x}) = \frac{1 + \beta_2^2}{2 + \beta_3^2}$; which corresponds to

$$\beta_3 = \beta_1 = \sqrt{\frac{m(\theta_1|\mathbf{x}) - m(\theta_2|\mathbf{x})}{m(\theta_2|\mathbf{x})}}.$$

Example 2. To illustrate the different situations, we consider six mass functions
(see Fig. 2). Figure 2 shows that when $m(\theta_1|\mathbf{x}) = m(\theta_2|\mathbf{x})$, e.g. m_1 and m_4,
regardless the mass of θ_{12}, the option θ_{12} obtains the maximal gains for all
$\beta > 0$. In the other cases the higher the mass of ignorance is, the smaller β_3
becomes.

4   Illustration

In this section we present the illustration of the performances of the classifiers
ndc and $eclair$ using generated data and then we present the comparisons of the
ndc classifier tuned using our proposition with other imprecise classifiers on the
UCI data based on the five measures presented in Subsect. 2.4.

4.1   Illustration Using Simulated Data

In this first illustration, we consider a simulated data for three class labels a,
b, and c. For each class label 500 training samples of a bivariate Gaussian distribution are considered, $\mathcal{N}(\mu_a = (0.2, 0.65), \Sigma_a = 0.01 I_2)$ for the class label a,
$\mathcal{N}(\mu_b = (0.5, 0.9), \Sigma_b = 0.01 I_2)$, and for the class label b and $\mathcal{N}(\mu_c = (0.8, 0.6), \Sigma_c =$

$0.01I_2$) for the class label c. In addition, a testing dataset of 50 samples for each label are generated using the same bivariate Gaussian distributions with a Gaussian noise $\mathcal{N}(\mu = (0,0), \Sigma = 0.001I_2)$. First, nine classical classifiers are trained and tested on these data. The standard classifiers considered are the naive Bayes (*nbc*), the k-Nearest Neighbour (*knn*), the evidential k-Nearest Neighbour (*eknn*), the decision tree (*cart*), the random forest (*rfc*), linear discriminant analysis (*lda*), support vector machine (*svm*) and artificial neural networks (*ann*), the logistic classifier (*logistic*). The obtained accuracies are: logistic, ann: 94.67; svm, eknn: 95.33; and knn, nbc, rfc, lda, cart: 96. These classifiers are introduced here to detect the samples that are difficult to predict, i.e., most standard classifiers fail to predict the true class of the those samples.

The idea here for choosing the *ndc* hyper-parameter is to avoid misclassification when the samples are difficult. For the samples that are "certain", i.e., the posterior probability of one of the classes is close to 1, this later class obtain the maximum gain regardless the value given to β (see Subsect. 3.1). Consequently, it is more interesting to set the value of β regarding the less "certain" samples. The proposition of this paper is to consider a active probability distribution p^f where the first component is the mean of the maximal probabilities p^1 obtained for each less "certain" sample of the training data set using leave-one-out technique and the second component is the mean of the second maximal probabilities p^2, and so on. Thus, $p^f = (p^1, p^2, ...)$. To determine the less "certain" sample a threshold is considered and when the maximal probability is lower than this threshold then the sample is considered less certain. In the illustration, this threshold is fixed to

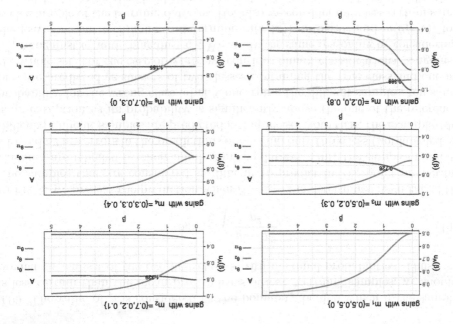

Fig. 2. The gain function for some examples of masses

0.99. The value of β is considered as the boundary behind which if the sample is less certain than the mean probabilities of less "certain" samples, we should predict the subset of the two first classes with maximal probabilities. Thus,

$$\beta_{ndc} = \sqrt{\frac{p^1 - p^2}{p^2}}. \qquad (14)$$

In Fig. 3, we present the prediction when $\beta = 2.571$ is determined as in Eq. (14). The samples that are considered as difficult to predict by the point prediction classifier are labelled by their number in the dataset. Only the samples number 140 and 82 are errors in the predictions of ndc and only three less (not labelled as difficult) difficult samples are predicted as imprecise. Note that, the example 140 is an exception as its probability is significantly above the one of the reference probability for the wrong class label. Concerning the difficult samples, ten samples are predicted as subsets of two classes containing the true class and one as the whole set. For the case of $eclair$, we consider binary classifications "a against b", "a against c" and "b against c". We apply the same reasoning by considering the leave-one-out technique to determine $m(\theta_1|x))$, $m(\theta_2|x)$ and $m(\theta_{12}|x)$ for each example of the learning data set. Her also we consider only less certain samples with the same threshold. From the Subsect. 3.2, to avoid misclassification for difficult samples β should be heigh enough to predict θ_{12} when ignorance is heigh. Let us denote m_{12} the average of $m(\theta_{12}|x))$ obtained for each less certain

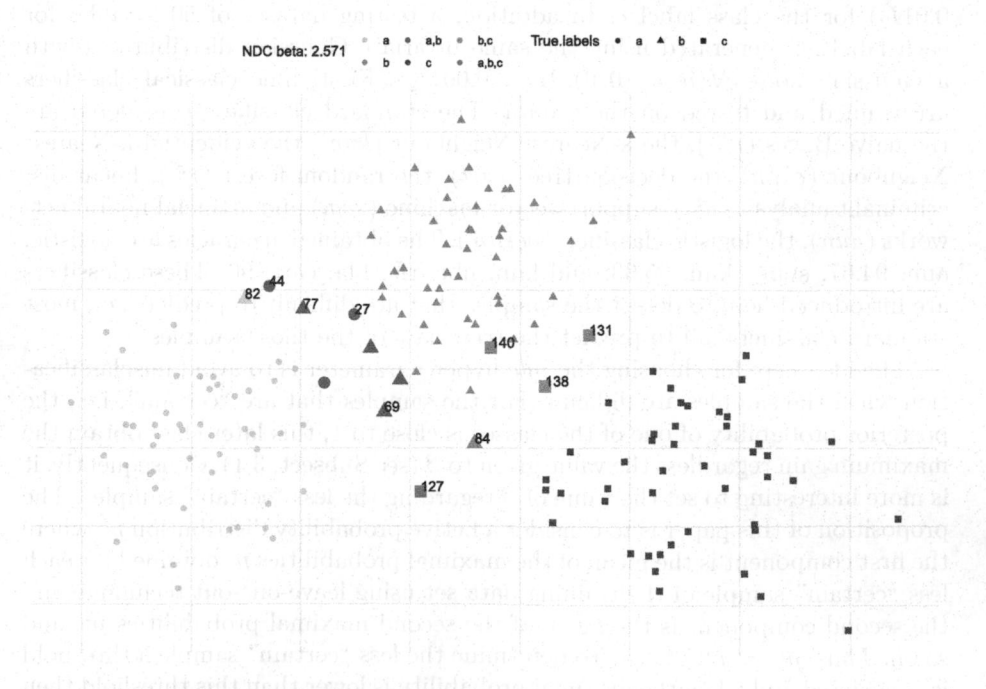

Fig. 3. The predictions obtained with ndc: a large size is given to the point symbols representing predictions that are errors or imprecise.

sample. The proposed value of β is β_{eclair} that is the solution the quadratic Equation (13) with $m(\theta_{12}|x)) = m_{12}$ and $m(\theta_1|x)) = 2\,(1 - m_{12})/3$. In Fig. 4, we can see that, for the case "a against b", two predictions are still errors and four are imprecise. For the case of "a against c", we have only one imprecise prediction. While for the case "b against c", we have four imprecise predictions.

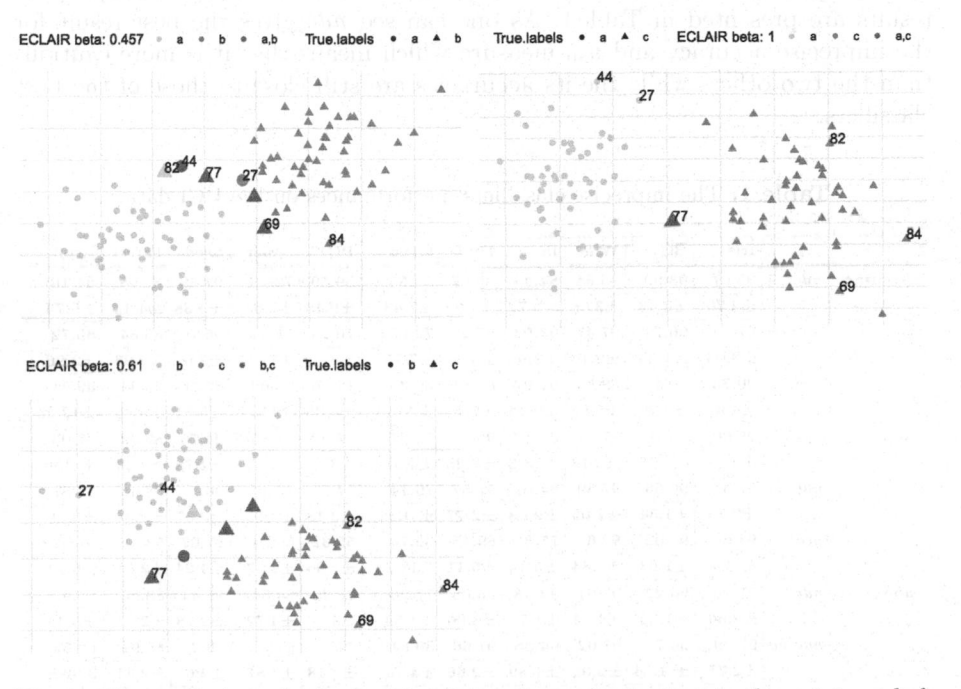

Fig. 4. The predictions obtained with *eclair*: a large size is given to the point symbols representing predictions that are errors or imprecise.

4.2 Illustration Using UCI Data

The second illustration concerns the comparison of the performances of the *ndc* classifier where β is determined as in the Eq. (14) to the *ndc cv*, i.e., classifier tuned using cross-validation, and the naive credal classifier *ncc* using 11 UCI data based on the performances measures presented in the Subsect. 2.4. The experimentation procedure is conduct as follows. Each dataset is split randomly 50 times to obtain a learning set (80%) and a testing set (20%). The parameters are optimized, each time, using the cross-validation technique on the learning dataset. More precisely, for *ndc cv* two hyper-parameters are involved, the point prediction classifier used to obtain the posterior probabilities and the parameter β. For the first parameter, the choice is performed within the nine classifier presented in the Subsect. 4.1 while for the second parameter the choice is performed in the interval $[0, 2]$ with steps of 0.1. Concerning ncc, the choice of parameter s is performed within a set of 20 values $S = \{10^{-30}, 10^{-20}, 10^{-15}, 10^{-10}, 10^{-9}, 10^{-8}, 10^{-7}, 10^{-6}, 10^{-5}, 10^{-4}, 10^{-3}, 10^{-2}, 10^{-1}, 0.2, 0.3, 0.5, 0.6, 1, 1.1, 2\}$. The results are presented in Table 1. As one can see ndc gives the best result for the imprecise accuracy and u_{80} measure which means that it is more cautious than the two others while the its accuracies are still close to those of the best classifiers.

Table 1. The imprecise classifiers' performances on the UCI data.

		Iris	BC	Wine	IS	DBT	Glass	PID	Sonar	Seeds	Forest	Ecoli
Accuracy	ndc	95.07	93.99	95.88	84.49	95.07	55.55	59.59	72.73	92.48	82.08	84.16
		±3.76	±2.43	±3.66	±3.72	±4.5	±7.43	±9.13	±7.63	±3.38	±4.42	±6.74
	$ndc\ cv$	97	96.24	97.47	93.03	97.29	73.65	70.26	75.41	95.76	86.84	85.72
		±3.03	±1.55	±2.14	±2.99	±3.27	±7.75	±6	±7.26	±2.6	±2.96	±3.96
	ncc	90.73	95.47	88.53	61.97	85.71	30.35	15.26	26.63	82.86	25.44	39.41
		±4.92	±2.26	±5.52	±9.14	±7.7	±20.71	±4.06	±11.57	±5.81	±6.75	±12.28
$u50$	ndc	96.60	95.70	97.44	91.17	95.82	72.07	73.92	81.83	94.75	87.33	87.03
		±2.82	±1.47	±2.18	±2.03	±3.28	±4.61	±3.07	±4.77	±2.79	±2.34	± 3.98
	$ndc\ cv$	97.3	96.66	97.59	94.07	97.57	79.14	77.53	82.1	96.1	88.43	87.87
		±2.79	±1.34	±2.05	±2.03	±2.77	± 4.81	±3.18	±4.85	±2.57	±2.34	±3.36
	ncc	93.6	96.02	92.6	75.3	89.45	36.74	55.32	59.34	87.07	53	57.92
		± 3.47	±1.63	±3.84	±5.14	±5.11	±13.24	±1.96	±2.17	±4.24	±2.51	±6.74
$u65$	ndc	97.06	96.22	97.91	93.18	96.05	77.26	78.22	84.56	95.44	88.96	87.9
		±2.66	±1.32	±1.84	±1.7	±3.06	±4.53	±3.17	±4.72	±2.78	±2.01	±3.45
	$ndc\ cv$	97.39	96.79	97.62	94.38	97.66	80.81	79.71	84.1	96.2	88.91	88.52
		±2.77	± 1.34	±2.05	±1.89	±2.66	±4.74	±3.18	±4.87	±2.6	±2.31	3.38±
	ncc	94.51	96.19	94.07	79.3	90.65	39.65	67.34	69.15	88.38	61.83	63.9
		±3.25	±1.51	±3.27	±4.78	±4.75	±9.88	±1.74	±3.79	±4.04	±2.08	± 5.94
$u80$	ndc	97.52	96.73	98.38	95.18	96.27	82.45	82.52	87.29	96.12	90.58	88.77
		±2.58	±1.29	±1.6	±1.56	±2.95	±4.93	±4.6	±5.15	±2.86	±1.96	± 3.22
	$ndc\ cv$	97.48	96.91	97.66	94.7	97.74	82.48	81.89	86.11	96.3	89.39	89.16
		±2.76	±1.36	±2.06	±1.85	±2.57	± 5.17	±3.72	±5.29	±2.64	±2.37	±3.49
	ncc	95.43	96.36	95.54	83.3	91.84	42.56	79.36	78.97	89.7	70.66	69.87
		±3.2	±1.43	±2.77	±5.05	±4.72	±6.99	±1.89	±6.81	±4.05	±2.76	±6.03
imprAcc	ndc	98.13	97.42	99	97.86	96.57	90.95	88.26	90.93	97.05	93.18	89.97
		±2.62	±1.44	±1.53	±1.74	±2.97	±6.31	±7.24	±6.29	±3.1	± 2.31	± 3.5
	$ndc\ cv$	97.6	97.08	97.71	95.11	97.86	84.85	84.79	88.78	96.43	90.04	90.03
		±2.78	±1.44	±2.08	±1.98	±2.5	±6.48	±4.95	±6.33	±2.73	±2.59	±3.79
	ncc	97	96.58	99.18	88.63	93.93	60.85	95.38	92.05	91.76	87.09	81.16
		±3.52	±1.41	±1.79	±6.22	±5.22	± 22.49	±2.54	±11.13	±4.44	±5.38	±8.36

5 Conclusion

In this paper we are interested in the imprecise classification. Especially, we focus on the study of the parameter β involved in the gain function used in the decision step of two imprecise classifiers. More precisely, we studied the predicted subsets depending on this parameter. We proposed a technique to choose the value of this parameter when the classifiers are involved in a classification task. Furthermore, the built classifiers give reasonable good performances related to evaluation measures for imprecise classifier.

References

1. Abellan, J., Masegosa, A.R.: Imprecise classification with credal decision trees. Int. J. Uncertainty Fuzziness Knowl. Based Syst. **20**(05), 763–787 (2012)
2. Couso, I., Sánchez, L.: Machine learning models, epistemic set-valued data and generalized loss functions: an encompassing approach. Inf. Sci. **358**, 129–150 (2016)
3. Coz, J.J.d., Díez, J., Bahamonde, A.: Learning nondeterministic classifiers. J. Mach. Learn. Res. **10**, 2273–2293 (2009)
4. Imoussaten, A., Jacquin, L.: Cautious classification based on belief functions theory and imprecise relabelling. Int. J. Approximate Reasoning **142**, 130–146 (2022)
5. Jacquin, L., Imoussaten, A., Trousset, F., Montmain, J., Perrin, D.: Evidential classification of incomplete data via imprecise relabelling: application to plastic sorting. In: Ben Amor, N., Quost, B., Theobald, M. (eds.) Scalable Uncertainty Management, pp. 122–135. Springer International Publishing, Cham (2019). https://doi.org/10.1007/978-3-030-35514-2_10
6. Jacquin, L., Imoussaten, A., Trousset, F., Perrin, D., Montmain, J.: Control of waste fragment sorting process based on MIR imaging coupled with cautious classification. Resour. Conserv. Recycl. **168**, 105258 (2021)
7. Ma, L., Denœux, T.: Partial classification in the belief function framework. Knowl.-Based Syst. **214**, 106742 (2021)
8. Quost, B., Masson, M.-H., Destercke, S.: Dealing with Atypical Instances in Evidential Decision-Making. In: Davis, J., Tabia, K. (eds.) SUM 2020. LNCS (LNAI), vol. 12322, pp. 217–225. Springer, Cham (2020). https://doi.org/10.1007/978-3-030-58449-8_15
9. Sanchez, L., Couso, I.: A framework for learning fuzzy rule-based models with epistemic set-valued data and generalized loss functions. Int. J. Approximate Reasoning **92**, 321–339 (2018)
10. Tsoumakas, G., Vlahavas, I.: Random k-Labelsets: an ensemble method for multilabel classification. In: Kok, J.N., Koronacki, J., Mantaras, R.L., Matwin, S., Mladenič, D., Skowron, A. (eds.) ECML 2007. LNCS (LNAI), vol. 4701, pp. 406–417. Springer, Heidelberg (2007). https://doi.org/10.1007/978-3-540-74958-5_38
11. Yang, G., Destercke, S., Masson, M.H.: The costs of indeterminacy: how to determine them? IEEE Trans. Cybern. **47**(12), 4316–4327 (2016)
12. Zaffalon, M.: A credal approach to Naive classification. In: ISIPTA. vol. 99, pp. 405–414 (1999)
13. Zaffalon, M.: Statistical inference of the Naive credal classifier. In: ISIPTA. vol. 1, pp. 384–393 (2001)
14. Zaffalon, M., Corani, G., Mauá, D.: Evaluating credal classifiers by utility-discounted predictive accuracy. Int. J. Approximate Reasoning **53**(8), 1282–1301 (2012)

Author Index

inted in the United States
& Taylor Publisher Services